Atomic numbers and atomic weights[a]

Element	Symbol	Number	Weight	Element	Symbol	Number	Weight
Actinium	Ac	89	227.0278	Mercury	Hg	80	200.59
Aluminum	Al	13	26.98154	Molybdenum	Mo	42	95.94
Americium	Am	95	(243)	Neodymium	Nd	60	144.24
Antimony	Sb	51	121.75	Neon	Ne	10	20.179
Argon	Ar	18	39.948	Neptunium	Np	93	237.0482
Arsenic	As	33	74.9216	Nickel	Ni	28	58.70
Astatine	At	85	(210)	Niobium	Nb	41	92.9064
Barium	Ba	56	137.33	Nitrogen	N	7	14.0067
Berkelium	Bk	97	(247)	Nobelium	No	102	(259)
Beryllium	Be	4	9.01218	Osmium	Os	76	190.2
Bismuth	Bi	83	208.9804	Oxygen	O	8	15.9994
Boron	B	5	10.81	Palladium	Pd	46	106.4
Bromine	Br	35	79.904	Phosphorous	P	15	30.97376
Cadmium	Cd	48	112.41	Platinum	Pt	78	195.09
Calcium	Ca	20	40.08	Plutonium	Pu	94	(244)
Californium	Cf	98	(251)	Polonium	Po	84	(209)
Carbon	C	6	12.011	Potassium	K	19	39.0983
Cerium	Ce	58	140.12	Praseodymium	Pr	59	140.9077
Cesium	Cs	55	132.9054	Promethium	Pm	61	(145)
Chlorine	Cl	17	35.453	Protactinium	Pa	91	231.0359
Chromium	Cr	24	51.996	Radium	Ra	88	226.0254
Cobalt	Co	27	58.9332	Radon	Rn	86	(222)
Copper	Cu	29	63.546	Rhenium	Re	75	186.207
Curium	Cm	96	(247)	Rhodium	Rh	45	102.9055
Dysprosium	Dy	66	162.50	Rubidium	Rb	37	85.4678
Einsteinium	Es	99	(254)	Ruthenium	Ru	44	101.07
Erbium	Er	68	167.26	Samarium	Sm	62	150.4
Europium	Eu	63	151.96	Scandium	Sc	21	44.9559
Fermium	Fm	100	(257)	Selenium	Se	34	78.96
Fluorine	F	9	18.99840	Silicon	Si	14	28.0855
Francium	Fr	87	(223)	Silver	Ag	47	107.868
Gadolinium	Gd	64	157.25	Sodium	Na	11	22.98977
Gallium	Ga	31	69.72	Strontium	Sr	38	87.62
Germanium	Ge	32	72.59	Sulfur	S	16	32.06
Gold	Au	79	196.9665	Tantalum	Ta	73	180.9479
Hafnium	Hf	72	178.49	Technetium	Tc	43	(97)
Helium	He	2	4.00260	Tellurium	Te	52	127.60
Holmium	Ho	67	164.9304	Terbium	Tb	65	158.9254
Hydrogen	H	1	1.0079	Thallium	Tl	81	204.37
Indium	In	49	114.82	Thorium	Th	90	232.0381
Iodine	I	53	126.9045	Thulium	Tm	69	168.9342
Iridium	Ir	77	192.22	Tin	Sn	50	118.69
Iron	Fe	26	55.847	Titanium	Ti	22	47.90
Krypton	Kr	36	83.80	Tungsten	W	74	183.85
Lanthanum	La	57	138.9055	Uranium	U	92	238.029
Lawrencium	Lr	103	(260)	Vanadium	V	23	50.9414
Lead	Pb	82	207.2	Xenon	Xe	54	131.30
Lithium	Li	3	6.941	Ytterbium	Yb	70	173.04
Lutetium	Lu	71	174.97	Yttrium	Y	39	88.9059
Magnesium	Mg	12	24.305	Zinc	Zn	30	65.38
Manganese	Mn	25	54.9380	Zirconium	Zr	40	91.22
Mendelevium	Md	101	(258)				

[a] From *Pure Appl. Chem.*, **47**, 75 (1976). A value in parentheses is the mass number of the longest-lived isotope of the element.

WASTEWATER ENGINEERING: TREATMENT DISPOSAL REUSE

McGRAW-HILL SERIES IN
WATER RESOURCES AND ENVIRONMENTAL ENGINEERING

Ven Te Chow, Rolf Eliassen, and Ray K. Linsley
Consulting Editors

WASTEWATER ENGINEERING: TREATMENT DISPOSAL REUSE

SECOND EDITION

METCALF & EDDY, INC.

Revised by
GEORGE TCHOBANOGLOUS
Professor of Civil Engineering
University of California, Davis

McGRAW-HILL BOOK COMPANY

New York   St. Louis   San Francisco   Auckland   Bogotá   Düsseldorf
Johannesburg   London   Madrid   Mexico   Montreal   New Delhi   Panama
Paris   São Paulo   Singapore   Sydney   Tokyo   Toronto

WASTEWATER ENGINEERING
TREATMENT, DISPOSAL, REUSE

34567890KPKP 83210

This book was set in Times Roman. The editors were Frank J. Cerra
and J. W. Maisel; the production supervisor was Dominick Petrellese.
New drawings were done by J & R Services, Inc.
Kingsport Press, Inc., was printer and binder.

Library of Congress Cataloging in Publication Data

Metcalf and Eddy, Boston.
    Wastewater engineering.

    (McGraw-Hill series in water resources and environmental engineering)
    Bibliography: p.
    Includes index.
    1. Sewerage. 2. Sewage disposal.
I. Tchobanoglous, George. II. Title.
TD645.M57 1979   628'.2   78-6173
ISBN 0-07-041677-X

CONTENTS

PREFACE

Following the widespread acceptance and use of *Wastewater Engineering: Collection, Treatment, Disposal*, and the many developments in the field since it was first written, we felt that an update of the book that included use of the metric system was advisable. This second edition has been prepared (1) to keep pace with the new technical development in the field of environmental engineering since the publication of the first edition, (2) to reflect the impact of recent federal legislation dealing with water quality and pollution control, (3) to provide leadership in the wider adoption and use of the metric International System of Units (or SI, for short) in the design and analysis of treatment facilities, and most importantly (4) to make the book more useful for students, teachers, practicing engineers, and other users.

To meet the objectives established for the second edition, it was necessary to revise and rewrite the first edition completely. Because the basic data and information related to the treatment, disposal, and reuse of wastewater and sludge has expanded so dramatically in the past 6 years, the chapters in the first edition that deal with collection and pumping of wastewater have been omitted from this edition. Accordingly, the subtitle of this edition has been changed to *Treatment, Disposal, Reuse.*

The chapters that have been removed are to be issued along with new material as a separate textbook entitled *Collection and Pumping of Wastewater*. The space made available by the omission of these chapters is being taken up by new material and expanded coverage of first-edition material.

Passage of the Federal Water Pollution Control Act Amendments of 1972 (Public Law 92-500) has had a major impact on wastewater engineering. New material presented in this edition reflects the changes brought about by the law. With the establishment for the first time of national goals and objectives, greater emphasis has been placed on reuse of wastewater and land disposal. A new chapter on land-treatment systems addresses the important engineering aspects associated with the treatment and disposal of wastewater and sludge on land.

Most of the world is now using some form of metric units. For this reason and because the United States is using them more and more, this edition uses the SI system along with conversions to U.S. customary units. Inasmuch as both sets of units will be in use for some time to come, we have given complete conversion tables in Appendix A. To increase the usefulness of the text, we have furnished conversions from metric-unit data to U.S. customary units in footnotes to all tables.

To make this second edition more useful as a teaching and reference text, a number of significant changes have been made. To provide the reader with a general introduction to the field of wastewater engineering, a new chapter on wastewater treatment objectives, methods, and design has been provided. The material in this chapter is intended to serve as an introduction to the chapters that follow it. Discussions dealing with fundamentals of process analysis have been gathered together and amplified under the heading "Fundamentals of Process Analysis." And to reflect current knowledge and practice, presentations of the fundamentals of the unit operations and processes used for the treatment of wastewater have all been revised. Sections dealing with the solids flux analysis for the design of secondary settling facilities and the preparation of solids balances for alternative treatment flowsheets are cited as two examples of the many new topics that have been included in this edition.

In addition, more than 60 tables containing a summary of design data and information are included. To illustrate basic concepts and physical applications more clearly, approximately 200 drawings and 90 photographs are furnished. Of this number, about 120 of the drawings and essentially all the photographs are new.

Different example problems have been prepared for this edition, with units carried through all the computational steps to facilitate the reader's understanding of the principles involved. Where appropriate, comments are included at the end of an example problem to elucidate basic concepts and highlight additional applications. To make this edition more useful, it has been reorganized such that the concepts and principles are stated clearly so that the transition from basic principles to design applications follows a more logical sequence. All these factors have contributed to making this book a more complete presentation of wastewater engineering and management by the profession of consulting engineering.

Rolf Eliassen
Chairman of the Board
Metcalf & Eddy, Inc.

George Tchobanoglous
Professor of Civil Engineering
University of California, Davis

ACKNOWLEDGMENTS

Metcalf & Eddy, Inc., is fortunate in having had the services of Dr. George Tchobanoglous as consultant on many important engineering projects over the past decade. But we believe that his greatest contribution to the firm—and, it is hoped, to the profession of wastewater engineering—is this complete revision and expansion of the first edition of our book. His devotion to this task entailed far more than conceptual design and writing. He had responsibility for the entire project, including coordination of the activities, reviews, and contributions of many of the members of the staffs of our Boston and Palo Alto offices.

No undertaking of the magnitude involved in the preparation of this new edition can be accomplished alone. Thus it is with grateful appreciation that we acknowledge the assistance of the following individuals. From the Palo Alto office, Ronald W. Crites served as coordinator for the project, also being responsible for the preparation of Chapter 13 on land-treatment systems. Franklin L. Burton read portions of the manuscript and made technical and editorial comments. Donald J. Schroeder helped secure photographs and checked metric units. Arthur L. Holland coordinated standardization of the figures, most of which were drawn by Diosdado C. Cantimbuhan. In addition to typing the final manuscript, Donald F. Newton made valuable editorial contributions. Marcella S. Tennant served again as general technical editor. Her editing skills are reflected throughout the text, especially in its readability.

Personnel from the Boston office reviewed the manuscript to ensure its consistency with current practice and provided much of the information given in the summary tables. Francis C. Sampson was in charge of the Boston review. David P. Bova served as general coordinator: he reviewed all the chapters and was responsible for providing much of the data used in preparation of tables. Abu M. Z. Alam, Bradley W. Behrman, Stephen L. Bishop, John G. Chalas, Joseph Goss, David S. Graber, Frank M. Gunby, Jr., Winfield A. Peterson, Francis C. Tyler, and Jakobs P. Vittands reviewed various chapters. Randolf A. Johnson helped with the photographs. Allen J. Burdoin, a consultant to Metcalf & Eddy, reviewed and revised portions of the manuscript.

Other individuals who reviewed various sections and contributed to the preparation of this text include Takashi Asano, Max E. Burchett, Jeffrey R. Hauser, Larry J. Karns, Edward D. Schroeder, Sam A. Vigil, and George E. Wilson. Mark R. Matsumoto reviewed all the example problems and checked the metric units. Rosemary Tchobanoglous typed the rough draft and provided moral support. Professor Edward Foree of the University of Kentucky reviewed the manuscript for the publisher and made valuable editorial suggestions.

We also wish to acknowledge the many constructive comments received from teachers, as well as practicing engineers, who used the first edition. Wherever possible, we have incorporated their suggestions in this revised edition.

Finally, we acknowledge our gratitude to Peter J. Gianacakes, president of Metcalf & Eddy, Inc., for his leadership on this edition and his commitment of the resources of the firm to the accomplishment of this important task.

Rolf Eliassen
Chairman of the Board
Metcalf & Eddy, Inc.

WASTEWATER ENGINEERING: TREATMENT DISP SAL REUSE

WASTEWATER ENGINEERING: AN OVERVIEW

Every community produces both liquid and solid wastes. The liquid portion—wastewater—is essentially the water supply of the community after it has been fouled by a variety of uses. From the standpoint of sources of generation, wastewater may be defined as a combination of the liquid or water-carried wastes removed from residences, institutions, and commercial and industrial establishments, together with such groundwater, surface water, and storm water as may be present.

If untreated wastewater is allowed to accumulate, the decomposition of the organic materials it contains can lead to the production of large quantities of malodorous gases. In addition, untreated wastewater usually contains numerous pathogenic or disease-causing microorganisms that dwell in the human intestinal tract or that may be present in certain industrial wastes. It also contains nutrients, which can stimulate the growth of aquatic plants, and it may contain toxic compounds. For these reasons, the immediate and nuisance-free removal of wastewater from its sources of generation, followed by treatment and disposal, is not only desirable but also necessary in an industrialized society. In the United States, it is now mandated by numerous federal and state laws.

Wastewater engineering is that branch of environmental engineering in which the basic principles of science and engineering are applied to the problems of water-pollution control. The ultimate goal—wastewater management—is the protection of the environment in a manner commensurate with economic, social, and political concerns.

To provide an initial perspective of the treatment, disposal, and reuse of wastewater, a brief review of the historical background, current status, and expected new directions in these areas of wastewater engineering is presented in

1

this chapter. Although the subjects of source control, collection, transmission, and pumping will not be covered (see Preface), the role of the engineer in the overall field of wastewater engineering is identified at the end of this chapter.

1-1 WASTEWATER TREATMENT

Wastewater collected from cities and towns must ultimately be returned to receiving waters or to the land. The complex question of which contaminants in wastewater must be removed to protect the environment—and to what extent—must be answered specifically for each case. This requires analyses of local conditions and needs, together with the application of scientific knowledge, engineering judgment based on past experience, and consideration of federal and state requirements and regulations.

Background

Although the collection of storm water and drainage dates from ancient times, the collection of wastewater can be traced only to the early 1800s. The systematic treatment of wastewater followed in the late 1800s and early 1900s. Development of the germ theory in the latter half of the nineteenth century by Koch and Pasteur marked the beginning of a new era in sanitation [6, 7]. Before that time, the relationship of pollution to disease had been only faintly understood, and the science of bacteriology, then in its infancy, had not been applied to the subject of wastewater treatment.

In the United States, the treatment and disposal of wastewater did not receive much attention in the late 1800s because the extent of the nuisance caused by the discharge of untreated wastewater into the relatively large bodies of water (compared to those in Europe) was not severe, and because large areas of land suitable for disposal were available. By the early 1900s, however, nuisance and health conditions brought about an increasing demand for more effective means of wastewater management. The impracticability of procuring sufficient areas for the disposal of untreated wastewater on land, particularly for larger cities, led to the adoption of more intensive methods of treatment.

Current Status

Methods of treatment in which the application of physical forces predominate are known as unit operations. Methods of treatment in which the removal of contaminants is brought about by chemical or biological reactions are known as unit processes. At the present time, unit operations and processes are grouped together to provide what is known as primary, secondary, and tertiary (or advanced) treatment. In primary treatment, physical operations, such as screening and sedimentation, are used to remove the floating and settleable solids found in wastewater. In secondary treatment, biological and chemical processes are used

Table 1-1 Number of municipal wastewater-treatment plants by type of treatment, 1945–1974[a]

Type of treatment	Number of treatment plants reported[b]			
	1945	1957	1968	1974[b]
Minor	60	41	47	79
Primary	2,829	2,730	2,384	2,875
Intermediate	98	100	75	78
Secondary	2,799	4,647	9,951	16,987
Tertiary	[c]	[c]	10	992
National totals	5,786	7,518	12,565[d]	21,011

[a] Adapted from Refs. 4 and 5.

[b] From U.S. Environmental Protection Agency STORET Municipal Waste Facilities Inventory; data elements in inventory current to the period of 1968 to 1974 (some elements were last updated in 1968) as supplied by the U.S. Environmental Protection Agency in September 1974.

[c] Data not given. Assume zero tertiary plants.

[d] Includes 98 plants with unknown treatment.

to remove most of the organic matter. In tertiary treatment, additional combinations of unit operations and processes are used to remove other constituents, such as nitrogen and phosphorus, which are not removed by secondary treatment. Land-treatment processes combine physical, chemical, and biological treatment mechanisms and produce water with quality similar to that from advanced wastewater treatment.

Data on the number and types of municipal wastewater-treatment plants in the United States are reported in Table 1-1. The most significant trend shown is the considerable increase in the number of secondary treatment plants in recent years—from 48 percent in 1945 to 81 percent in 1974 [4, 5].

From an analysis of the data on the sizes of treatment plants [4], it was found that approximately 78 percent of all publicly owned treatment works with treatment needs are smaller than 43.8 L/s (1 Mgal/d); 15 percent are in the range between 43.8 and 219.1 L/s (1 and 5 Mgal/d); and about 7 percent are larger than 219.1 L/s (5 Mgal/d). Correspondingly, approximately 7 percent of the total facility design capacity at publicly owned treatment works with treatment needs is in plants with a design flow of less than 43.8 L/s; 14 percent is in plants with a design flow between 43.8 and 219.1 L/s; and 79 percent is in plants larger than 219.1 L/s (5 Mgal/d).

According to a survey made in 1974 [1], stabilization ponds are the most commonly used means of treatment in facilities with flowrates of 43.8 L/s (1 Mgal/d) or less. As the flowrate increases, however, their use decreases dramatically.

New Directions

With the passage of the Federal Water Pollution Control Act Amendments of 1972 (Public Law 92–500), Congress established a far-reaching program for the control of pollution in U.S. waterways. The implications of this important legislation and the corresponding regulations and guidelines are discussed in Chap. 4. New directions are also evident in various specific areas of wastewater treatment, including (1) treatment operations, processes, and concepts; (2) the changing nature of the wastewater to be treated; (3) the problem of industrial wastes; (4) wastewater treatability studies; (5) environmental and energy concerns; (6) land treatment; and (7) small and individual onsite systems.

Treatment operations, processes, and concepts At the present time, most of the unit operations and processes used for wastewater treatment are undergoing continual and intensive investigation from the standpoint of implementation and application. As a result, many modifications and new operations and processes have been developed and implemented; more need to be made to meet the increasingly stringent requirements for environmental enhancement of water-courses. In addition to the developments taking place with conventional treat-ment methods, alternative treatment systems and technologies are also under study. Land treatment systems and those involving the use of aquatic species are examples. In addition to treatment, aquatic systems can be used for the capture of solar energy and for the utilization of the nutrients in wastewater. The potential for producing usable plant and animal protein is another possible advantage of such systems.

The concept of satellite treatment, though not new, is being revived. It was proposed by Metcalf & Eddy, Inc., for the treatment of wastewater in the Los Angeles area 30 years ago [3], and more recently in the Greater Portland area in Maine as well as in eastern Massachusetts. In practice, small treatment plants would be located throughout a system and would be designed to treat principally domestic wastes. The effluent would be used in local reuse applications or discharged to the environment. The biological solids produced during treatment could be processed for nutrient recovery or would be returned to the sewer for processing at a central location.

New consideration is also being given to the relationship between the design of collection systems and wastewater treatment. As wastewater is transported in collection systems, it undergoes both biological and chemical transformations [9]. To a large extent, the nature of these transformations depends on the design of the collection system. In the future, as the importance of these transformations becomes understood more clearly with respect to wastewater treatment, it is anticipated that the design of wastewater collection systems and treatment facilities will be coordinated to a much greater extent than in the past.

Changing wastewater characteristics The number of organic compounds that have been synthesized since the turn of the century now exceeds half a million, and some 10,000 new compounds are added each year. As a result, many of these compounds are now found in the wastewater from most cities. Although most of them can be treated readily, the number of such compounds that are not amenable to treatment, or that are only slightly amenable to treatment with processes presently used, is increasing. Moreover, in many cases, little or no information is available on the long-term environmental effects caused by their discharge. As these effects become more clearly understood, it is anticipated that more emphasis will be placed on advanced treatment for the removal of specific contaminants or on land treatment.

The problem of industrial wastes The number of industries that now discharge wastes to domestic sewers has increased significantly during the past 20 to 30 years. In view of the toxic effects often caused by the presence of these wastes, the general practice of combining industrial and domestic wastes is now being reevaluated. In the future, many municipalities may either provide separate treatment facilities for these wastes or require that they be treated at the point of origin to render them harmless before allowing their discharge to domestic sewers.

Wastewater treatability studies Because of the problems associated with the transformations that occur as wastewater is transported, the changing characteristics of the wastewater, and the discharge of industrial wastes to domestic sewers, studies of wastewater treatability are increasing. Such studies are especially important where new treatment methods are being proposed. Therefore, the engineer must understand the general approach and methodology involved in (1) assessing the treatability of a wastewater (domestic or industrial), (2) the conduct of laboratory and pilot-plant studies, and (3) the translation of experimental data into design parameters.

Environmental and energy concerns During the past few years, environmental and energy concerns have come to play an increasingly important part in the selection and design of both collection and treatment facilities. Odors are one of the most serious environmental concerns to the public. New techniques for odor measurement are now being used to quantify the development and movement of odors that may emanate from wastewater facilities, and special efforts are being made to design facilities so that the development of odors is minimized.

The need to conserve energy is well documented. Detailed energy analyses are now becoming an important part of any project analysis. More attention is being given to the selection of processes that conserve energy and resources. There is an increasing trend to minimize power usage in the design of wastewater treatment plants by more careful attention to plant siting and by designing facilities to capture solar energy for heating process tanks and buildings. This subject is discussed further in Chap. 4.

Land treatment Although land treatment of wastewater has been practiced for centuries, the processes involved have been recognized in the field of wastewater engineering only in recent years. Land treatment is receiving considerable attention because of the widespread research and development activity that has resulted from the emphasis placed in Public Law 92–500 on water reuse, nutrient recycling, and the use of wastewater for crop production. The processes of land treatment of wastewater involve the use of plants, the soil surface, and the soil matrix. Land treatment is an alternative that will find increasing acceptance from the standpoint of economy as well as engineering efficiency.

Small and individual onsite systems During the past 10 yr, interest in small treatment systems has often been overshadowed by concern over the design, construction, and operation of large regional systems [8]. Small systems were often designed and constructed as small-scale models of large plants. As a consequence, many are operationally energy and resource intensive. Because of economic, environmental, and energy concerns the design, construction, and operation of small systems are coming under careful review. New and innovative designs are being developed, and alternative treatment processes are being used.

Because the ratio of the number of people discharging to wastewater collection systems to the number discharging to individual onsite systems has not changed significantly over the past 20 yr, greater attention is now being focused on the design and operation and maintenance of individual onsite systems. The health and pollutional hazards caused by their use and the limits of their application must be defined and quantified. The organization of local operation and maintenance districts for onsite systems is another recent development.

1-2 EFFLUENT AND SLUDGE DISPOSAL AND REUSE

The ultimate disposal of treated wastewater and the solid and semisolid residuals (sludge) and concentrated contaminants removed by treatment has been and continues to be one of the most difficult and expensive problems in the field of wastewater engineering.

Background

In the past, the disposal of wastewater in most cities was carried out by the easiest method possible, without much regard to unpleasant conditions produced at the place of disposal. Irrigation, practiced in ancient Athens, was probably the first method of wastewater disposal, although dilution was the earliest method adopted by most municipalities. Problems developed when household wastes were admitted to storm sewers because the purification capacity of the watercourse into which they drained was often exceeded. As a result, separate sewers were built and wastewater treatment was instituted. The disposal of sludge became a problem with the application of the more intensive methods of treatment, which resulted in the production of large volumes of sludge.

Current Status and New Directions

The most important recent development relating to the field of wastewater disposal is the establishment by the U.S. Environmental Protection Agency of secondary treatment as the minimum acceptable level of treatment prior to surface water discharge. As a result of the increased amounts of sludge that will be generated and the new requirements to consider land treatment and reuse, effluent and sludge disposal are undergoing considerable study. In addition, as the degree of treatment improves and water shortages become more critical in arid areas, interest in effluent reuse is expected to increase.

Effluent disposal and reuse As reported in Table 1-2, surface water discharge remains today the most common method of wastewater disposal. To protect the aquatic environment, however, the individual states, in conjunction with the federal government, have developed receiving-water standards for the streams, rivers, and estuarine and coastal waters of the United States. It is anticipated that individual states may adopt more stringent requirements in the future. In 1972, Congress mandated the use of effluent-discharge limitations (see Table 4-1), presumably because limitations based on water quality would be difficult if not impossible to enforce.

In a number of localities, treatment plants have been designed and located so that a portion of the treated effluent can be disposed of by land application in conjunction with a variety of reuse applications, such as golf-course irrigation, use as industrial cooling water, and groundwater recharge. This trend is expected to continue to increase in the future.

In many locations where the available supply of fresh water has become inadequate to meet water needs, it is clear that the once-used water collected from towns and cities must be viewed not as a waste to be disposed of but as a resource. This concept is expected to become more widely adopted as other parts of the country experience water shortages. A number of reuse applications have been documented, and many more are currently under study.

Table 1-2 Existing effluent-disposal methods, 1974[a]

Effluent-disposal methods	Number now in use
Surface water outfalls	6858
Ocean outfalls	188
Holding ponds	380
Deep wells	4
Groundwater recharge	105
Other land disposal (not specified)	220
Water-supply recycling	35
Septic-tank fields	366
National totals	8158

[a] Adapted from Ref. 1.

Table 1-3 Sludge-disposal methods[a]

Method	Percent of plants in size category[b]		
	<43.8 L/s	43.8–219.1 L/s	>219.1 L/s
Barged to sea[c]	< 1	1	7
Used for fertilizer	29	32	28
Burned for fuel	< 1	< 1	< 1
Incinerated	1	7	26
Buried in landfill	42	41	23
Unknown	28	19	16

[a] Adapted from Ref. 4.
[b] Based on data contained in STORET Municipal Waste Facilities Inventory, 1973.
[c] Will not be allowed in the future.
Note: L/s × 0.0228 = Mgal/d.

As new reuse applications are developed, the impact on wastewater treatment and sludge disposal will become significant. Improved treatment methods will be needed to provide higher levels of treatment not only for routine wastewater constituents but also for the removal of specific compounds. This, in turn, will lead to the production of larger volumes of sludge that will require disposal. It is estimated that the total volume of municipal wastewater sludge produced in the United States will increase from 4.3 million dry metric tons (4.7 million dry U.S. tons) per year in 1972 to 6.0 million dry metric tons (6.6 million dry U.S. tons) per year in 1985 [2].

Sludge disposal and reuse Data on sludge-disposal methods now in use are reported in Table 1-3. The coincineration and co-pyrolysis of treatment-plant sludges with solid wastes is being considered at large treatment plants. The disposal of sludge with solid wastes in landfills is also increasing, especially where the decomposition gases can be collected for use. Furthermore, the land application of sludge is receiving considerable attention, as a means of disposal, as a means of reclaiming marginal land for productive uses and as a means of utilizing the nutrient content in sludge. Certainly the continuing search for better methods for the processing, disposal, and reuse of sludge will remain high, if not highest, on the list of priorities in the future.

1-3 THE ROLE OF THE ENGINEER

Practicing wastewater engineers are involved in the conception, planning, evaluation, design, construction, and operation and maintenance of the systems that are needed to meet wastewater management objectives. The major elements of

Table 1-4 Major elements of wastewater management systems and associated engineering tasks

Element	Engineering task	See chapter
Source of generation	Estimation of the quantities of wastewater, evaluation of techniques for wastewater reduction, and determination of wastewater characteristics	2, 3
Source control	Design of onsite systems to provide partial treatment of the wastewater before it is discharged to collection systems (principally involves industrial dischargers)	[a]
Collection	Design of sewers used to remove wastewater from the various sources of generation	[a]
Transmission and pumping	Design of large sewers (often called trunk and interceptor sewers) used to transport wastewater to treatment facilities or to other locations for processing	[a]
Treatment (wastewater and sludge)	Selection, analysis, and design of treatment operations and processes to meet specified treatment objectives related to the removal of wastewater contaminants of concern	4–12
Disposal and reuse	Design of facilities used for the disposal and reuse of treated effluent in the aquatic and land environment, and the disposal and reuse of sludge on land	13, 14

[a] Not covered in this text.

wastewater systems and the associated engineering tasks are identified in Table 1-4.

Knowledge of the methods used for the determination of wastewater flowrates and characteristics (Chaps. 2 and 3) is essential to an understanding of all aspects of wastewater engineering. The subjects of source control, collection, and transmission and pumping, although not covered in this text, must also be studied by the engineer if truly integrated wastewater systems are to be designed.

The primary focus of this book (Chaps. 4 to 14) is on the last two elements listed in Table 1-4: (1) treatment and (2) disposal and reuse. These areas of wastewater engineering, like the others, have been and continue to be in a dynamic period of development. Old ideas are being reevaluated, and new concepts are being formulated. To play an active role in the development of this field, the engineer must know the fundamentals on which it is based. The delineation of these fundamentals is the main purpose of this book.

REFERENCES

1. *Cost Estimates for Construction of Publicly Owned Wastewater Treatment Facilities: 1974 Needs Survey*, U.S. Environmental Protection Agency, Final Report to Congress, Washington, D.C., 1975.
2. Farrell, J. B.: Overview of Sludge Handling and Disposal, in *Proceedings of the National Conference on Municipal Sludge Management*, Information Transfer, Inc., Washington, D.C., 1974.
3. Metcalf & Eddy, Inc.: *Sewage Disposal Problem of Los Angeles, California and Adjacent Communities*, Boston, 1944.
4. Metcalf & Eddy, Inc.: *Report to National Commission on Water Quality on Assessment of Technologies and Costs for Publicly Owned Treatment Works*, vols. 1 and 2, prepared under Public Law 92–500, Boston, 1975.
5. *Municipal Waste Facilities in the United States: Statistical Summary, 1968 Inventory*, U.S. Department of the Interior, Federal Water Quality Administration, Publication CWT-6, Washington, D.C., 1970.
6. Sedgwick, W. T.: *Principles of Sanitary Science and the Public Health*, Macmillan, New York, 1903.
7. Stanier, R. Y., J. L. Ingraham, and E. A. Adelberg: *The Microbial World*, 4th ed., Prentice-Hall, Englewood Cliffs, N.J., 1976.
8. Tchobanoglous, G.: Wastewater Treatment for Small Communities, *Public Works*, pt. 1, vol. 105, no. 7, July 1974; pt. 2, vol. 105, no. 8, August 1974.
9. Wood, D. K., and G. Tchobanoglous: Trace Elements in Biological Waste Treatment with Specific Reference to the Activated Sludge Process, *Proceedings of the 29th Industrial Waste Conference*, 1974.

CHAPTER TWO

WASTEWATER FLOWRATES

Determination of the rates of wastewater flows is a fundamental step in the design of wastewater collection, treatment, and disposal facilities. Reliable data on existing and projected flows must be available if these facilities are to be designed properly, and if the associated costs are to be minimized and also shared equitably when the facilities serve more than one community or district.

The purpose of this chapter is to develop a basis for properly assessing wastewater flowrates from a community. The subjects considered include: (1) definition of the various components that make up the wastewater from a community, (2) water used for public supplies and its relationship to wastewater flowrates, (3) wastewater sources and flowrates, (4) analysis of flowrate data, (5) methods of reducing wastewater flows, and (6) methods of measuring wastewater flows.

2-1 COMPONENTS OF WASTEWATER FLOWS

The components that make up the wastewater from a community depend on the type of collection system used and may include:

1. *Domestic (also called sanitary) wastewater.* Wastewater discharged from residences and from commercial, institutional, and similar facilities
2. *Industrial wastewater.* Wastewater in which industrial wastes predominate
3. *Infiltration/inflow (I/I).* Extraneous water that enters the sewer system from the ground through various means, and storm water that is discharged from sources such as roof leaders, foundation drains, and storm sewers
4. *Storm water.* Water resulting from precipitation runoff

Three types of sewer systems are used for the removal of wastewater and storm water: sanitary-sewer systems, storm-sewer systems, and combined-sewer systems. Where separate sewers are used for the collection of wastewater (sanitary sewers) and storm water (storm sewers), wastewater flows in sanitary sewers consist of three major components: (1) domestic wastewater, (2) industrial wastewater, and (3) infiltration/inflow. Where only one sewer system (combined sewer) is used, wastewater flows consist of all four components. In both cases, the percentage of the wastewater components varies with local conditions and the time of the year.

For areas now served with sewers, wastewater flows are commonly determined from existing records or by direct field measurements. For new developments, wastewater flows are derived from an analysis of population data and corresponding projected unit rates of water consumption or from estimates of per capita wastewater flowrates from similar communities. These subjects are considered further in the rest of this chapter.

2-2 WATER USED FOR PUBLIC SUPPLIES

If field measurements are not possible and actual flow data are not available, public-water-supply records can often be used as an aid to estimate wastewater flows. Data on the quantity of water used for public water supplies in the United States and in cities are presented in this section. In addition, typical rates for various users, devices, and industries are given; variations in the rates of usage are described; and the proportion of a municipal water supply that becomes wastewater is discussed.

In considering usage rates of public water supplies, it is important to note the distinction between the amount of water withdrawn or produced (often identified as consumption) and the amount of water actually used. The difference between these two values is the amount of water lost or unaccounted for in the distribution system plus the amount used for various public services, such as fire fighting, street washing, and municipal-park needs.

Water Withdrawals for Public Water Supplies

According to a detailed study by the U.S. Geological Survey, the average quantity of water withdrawn for public water supplies in 1970 was estimated to be about 628 L/capita·d (166 gal/capita·d) [21]. The withdrawals by states are summarized in Table 2-1. The total losses and public uses were estimated to account for about 30 percent of the withdrawals. In a similar study conducted by the U.S. Public Health Service in 1954 [19], the average rate of withdrawal was found to be about 556 L/capita·d (147 gal/capita·d).

It is important to note the wide variations in the withdrawal rates reported in Table 2-1. Because these variations depend on the geographic location, climate, size of the community, degree of industrialization, and other influencing factors, the only way to prepare reliable estimates is to study each area separately.

Table 2-1 Water withdrawals for public supplies by states[a] and by selected municipal systems,[b] 1970

State, city	L/capita·d	gal/capita·d
Alabama:	806	213
Birmingham	576	152
Alaska:	1790	473
Anchorage	769	203
Arizona:	787	208
Phoenix	864	228
Arkansas:	503	133
Little Rock	784	207
California:	685	181
Los Angeles	686	181
San Francisco	1424	376
Colorado:	746	197
Denver	955	252
Connecticut:	541	143
Hartford	564	149
Delaware	700	185
Florida:	617	163
Miami	1208	319
Georgia:	946	250
Atlanta	564	149
Hawaii:	746	197
Honolulu	780	206
Idaho	897	237
Illinois:	772	204
Chicago	871	230
Indiana:	534	141
Indianapolis	508	134
Iowa:	466	123
Des Moines	534	141
Kansas:	587	155
Wichita	508	134
Kentucky:	314	83
Louisville	655	173
Louisiana:	545	144
Shreveport	519	137
Maine:	553	146
Portland	580	153
Maryland:	515	136
Baltimore	648	171
Massachusetts:	530	140
Boston	883	233
Michigan:	636	168
Detroit	671	177
Minnesota:	473	125
St. Paul	515	136

(*continued*)

Table 2-1 (*continued*)

State, city	L/capita·d	gal/capita·d
Mississippi:	507	134
Jackson	432	114
Missouri:	485	128
Kansas City	587	155
Montana:	826	219
Billings	754	199
Nebraska:	636	168
Omaha	742	196
Nevada:	1154	305
Las Vegas	1038	274
New Hampshire	435	128
New Jersey:	526	139
Elizabeth	314	83
New Mexico:	772	204
Albuquerque	746	197
New York:	609	161
New York City	1046	276
Rochester	663	175
North Carolina:	644	170
Greensboro	492	130
North Dakota:	477	126
Fargo	515	136
Ohio:	594	157
Akron	492	130
Oklahoma:	492	130
Tulsa	595	157
Oregon:	712	188
Portland	1129	298
Pennsylvania:	685	181
Pittsburgh	485	128
Rhode Island	462	122
South Carolina:	916	242
Charleston	652	172
South Dakota:	549	145
Sioux Falls	587	155
Tennessee:	488	129
Memphis	549	145
Texas:	587	155
Dallas	610	161
Houston	947	250
Utah:	1113	294
Salt Lake City	523	138
Vermont	553	146
Virginia:	420	111
Richmond	644	170
Washington:	1200	317
Seattle	1091	288
West Virginia:	568	150
Morgantown	549	145

Table 2-1 (*continued*)

State, city	L/capita·d	gal/capita·d
Wisconsin:	587	155
Milwaukee	659	174
Wyoming:	746	197
Cheyenne	841	222
District of Columbia	799	211
Puerto Rico	326	86
United States[c]	628	166

[a] Adapted from Ref. 21.

[b] Per capita consumption of water supplied by municipal systems, based on American Water Works Association data reported in Ref. 22.

[c] Including Puerto Rico.

Note: L × 0.2642 = gal.

Variations in Usage in Cities

A direct comparison of water-supply records from different cities is likely to be misleading. In some cities, large quantities of water used for industrial purposes are obtained from privately owned supplies; in other cities, the industries mainly use the municipal supply. Furthermore, the care taken to reduce the waste of water through leaks in mains, service pipes, and plumbing has a decided effect on the per capita consumption. Metering the individual consumer's supply and billing at established meter rates indirectly prevents waste of water by users and tends to reduce actual water usage. The waste and unaccounted-for water in metered systems ranges from 10 to 20 percent of the total water entering the supply-line system. The corresponding range in unmetered systems is much higher (typically 30 percent). Water consumption in selected U.S. cities is also shown in Table 2-1 [22].

From an analysis of water-supply records and corresponding population data for typical cities in the United States, it appears that, in cities where the population has not stabilized and is increasing, the per capita rate has tended to increase gradually. In cities where the population has become somewhat stabilized, and in some large cities (e.g., Chicago), the per capita rate has even decreased.

Water Use by Various Establishments and Devices

Typical rates of water use for many different establishments and various devices are presented in Tables 2-2 and 2-3. Although these rates vary widely, they are useful in estimating total water use for individual users when no other data are available.

Table 2-2 Typical rates of water use for various establishments[a, b]

User	Range of flow, L/(person or unit)·d
Airport, per passenger	10–20
Assembly hall, per seat	6–10
Automobile service station:	
Per set of pumps	1800–2200
Per vehicle served	40–60
Bowling alley, per alley	600–100
Camp:	
Pioneer type	80–120
Children's, central toilet and bath	160–200
Day, no meals	40–70
Luxury, private bath	300–400
Labor	140–200
Trailer with private toilet and bath, per unit ($2\frac{1}{2}$ persons)[c]	500–600
Country club:	
Resident type	300–600
Transient type, serving meals	60–100
Dwelling unit, residential:	
Apartment house on individual well	300–400
Apartment house on public water supply, unmetered	300–500
Boardinghouse	150–220
Hotel	200–400
Lodging house and tourist home	120–200
Motel	400–600
Private dwelling on individual well or metered supply	200–600
Private dwelling on public water supply, unmetered	400–800
Factory, sanitary wastes, per shift	40–100
Fairground (based on daily attendance)	2–6
Institution:	
Average type	400–600
Hospital	700–1200
Office	40–60
Picnic park, with flush toilets	20–40
Restaurant (including toilet):	
Average	25–40
Kitchen wastes only	10–20
Short order	10–20
Short order, paper service	4–8
Bar and cocktail lounge	8–12
Average type, per seat	120–180
Average type, 24 h, per seat	160–220
Tavern, per seat	60–100
Service area, per counter seat (toll road)	1000–1600
Service area, per table seat (toll road)	600–800
School:	
Day, with cafeteria or lunchroom	40–60
Day, with cafeteria and showers	60–80
Boarding	200–400
Self-service laundry, per machine	1000–3000

Table 2-2 (*continued*)

User	Range of flow, L/(person or unit)·d
Store:	
First 7.5 m (~25 ft) of frontage	1600–2000
Each additional 7.5 m of frontage	1400–1600
Swimming pool and beach, toilet and shower	40–60
Theater:	
Indoor, per seat, two showings per day	10–20
Outdoor, including food stand, per car ($3\frac{1}{3}$ persons)	10–20

[a] Adapted in part from Ref. 18.
[b] It is assumed that water under pressure, flush toilets, and washbasins are provided unless otherwise indicated. These figures are offered as a guide; they should not be used blindly. Add for any continuous flows and industrial usages.
[c] Add 475 L (125 gal) per trailer space for lawn sprinkling, car washing, leakage, etc.

Note: L × 0.2642 = gal.

Table 2-3 Typical rates of water use for various devices[a, b]

Device	Range of flow
Automatic home laundry machine	110–200 L/load
Automatic home-type dishwasher	15–30 L/load
Automatic home-type washing machine	130–200 L/use
Bathtub	90–110 L/use
Continuous-flowing drinking fountain	4–5 L/min
Dishwashing machine, commercial:[b]	
Conveyor type, at 100 kN/m²	15–25 L/min
Stationary rack type, at 100 kN/m²	25–35 L/min
Fire hose, 38 mm, 13 mm nozzle, 20 m head	140–160 L/min
Garbage-disposal unit, home-type	6000–7500 L/wk
Garbage grinder, home-type	4–8 L/person·d
Garden hose, 16 mm, 8 m head	10–12 L/min
Garden hose, 19 mm, 8 m head	16–20 L/min
Lawn sprinkler	6–8 L/min
Lawn sprinkler, 280 m² lawn, 25 mm/wk	6000–7500 L/wk
Shower head, 16 mm, 8 m head	90–110 L/use
Washbasin	4–8 L/use
Water closet, flush valve, 170 kN/m²	90–110 L/min
Water closet, tank	15–25 L/use

[a] Adapted in part from Ref. 18.
[b] Does not include water to fill wash tank.

Note: L × 0.2642 = gal
mm × 0.03937 = in
kN/m² × 0.1450 = lb$_f$/in²
m × 3.2808 = ft
m² × 10.7639 = ft².

Table 2-4 Typical rates of water use for various industries[a]

Industry	Range of flow, m^3/Mg
Cannery:	
Green beans	50–70
Peaches and pears	15–20
Other fruits and vegetables	4–35
Chemical:	
Ammonia	100–130
Carbon dioxide	60–90
Gasoline	7–30
Lactose	600–800
Sulfur	8–10
Food and beverage:	
Beer	10–16
Bread	2–4
Meat packing	15–20[b]
Milk products	10–20
Whisky	60–80
Pulp and paper:	
Pulp	250–800
Paper	120–160
Textile:	
Bleaching	200–300[c]
Dyeing	30–60[c]

[a] Adapted from Ref. 12.
[b] Live weight.
[c] Cotton.

Note: $m^3/Mg \times 239.7 = $ gal/U.S. ton (short).

Industrial water use also varies widely, according to the nature of the manufacturing process. In practical design work, it is therefore desirable to inspect the plant concerned and to make careful estimates of the quantities of both the water used from all sources and the wastes produced. The same is true of use in commercial districts. The typical rates shown in Table 2-4 may be used as an indication of the magnitude of water use to be expected from various industrial operations.

Fluctuations in Water Use

Although it is important to know the average rates of water use, it is even more important to have data on the fluctuations in rates of use. Representative data on the fluctuations in average rates of water use are reported in Table 2-5. The maximum use usually occurs during two seasons: (1) in summer months when water is in demand for street and lawn sprinkling and (2) in winter months when large quantities are allowed to run to prevent freezing of pipes and fixtures.

Table 2-5 Typical fluctuations in water use in community systems[a]

	Percentage of average for year
Daily average in maximum month	120
Daily average in maximum week	140
Maximum consumption in 1 day	180

Hourly variations in water consumption also affect the rate of wastewater flow. In general, the wastewater-discharge curve closely parallels the water-consumption curve, but with a lag of several hours. In some cities, as noted earlier, large quantities of water used by industries and obtained from sources other than the municipal supply are discharged into sewers during the working hours of the day. This tends to increase the peak flow beyond the amount resulting from the normal variation in the draft on the municipal supply.

In the absence of more authoritative information, an additional allowance of 50 percent over the average 24-hour rate of water consumed may be made for the excess of the hourly peak over the average 24-hour rate. The figure will vary with the locality, however, and should be applied only when local conditions warrant it. If this peak hourly consumption is applied to the maximum draft for a single day, or 180 percent of the yearly average, and if it is assumed that the portion of the water supply that finds its way into the sewers averages 380 L/capita·d (100 gal/capita·d), there will be a maximum flow to the sewers in 1 hour from the public and private water supplies of about 1026 L/capita·d ($380 \times 1.8 \times 1.5 = 1026$) (271 gal/capita·d). It should be noted that the extreme peak-flow values recorded in sewers are more often related to storm water infiltration/inflow and periodic industrial discharges.

Proportion of Water Supply Reaching Sewers

Because wastewater consists primarily of used water, the portion of the water supplied that reaches the sewers must be estimated. A considerable portion of the water does not reach the sewers. This includes water used by commercial and manufacturing establishments and power plants and water used for street washing, lawn sprinkling, and extinguishing fires. It also includes water used by consumers whose facilities are not connected with sewers as well as some leakage from water mains and service pipes.

Neglecting infiltration of groundwater, about 60 to 80 percent of the per capita withdrawal (consumption) of water becomes wastewater (the lower percentages are applicable to the semiarid region of the southwestern United States). Often, however, excessive infiltration, roof water, and water used by industries that is obtained from privately owned sources make the quantity of wastewater larger than the quantity of the public water supply. If a community has well-built sewers and if roof water is excluded, the variation from year to year in the ratio of wastewater to water supply is not great, unless there is a substantial change in the industrial uses of water.

2-3 WASTEWATER SOURCES AND FLOWRATES

Data that can be used to estimate average wastewater flows from various domestic and industrial sources and the infiltration/inflow contribution are presented in this section. Variations in the flows that must be established before sewers and treatment facilities are designed are also discussed.

Sources and Rates of Domestic Wastewater Flows

The principal sources of domestic wastewater in a community are the residential and commercial districts. Other important sources include institutional and recreational facilities. For existing districts, flowrate data should be obtained by direct measurement. Methods for areas that are being developed are considered in the following discussion.

Residential districts For small residential districts, wastewater flows are commonly determined on the basis of population density and the average per capita contribution of wastewater. Data on ranges and typical flows are given in Table 2-6. For large residential districts, it is often advisable to develop flowrates on the basis of land-use areas and anticipated population densities. Where possible, these rates should be based on actual flow data from selected typical residential

Table 2-6 Average wastewater flows from residential sources

Source	Unit	Flow, L/unit·d	
		Range	Typical
Apartment	Person	200–340	260
Hotel, resident	Resident	150–220	190
Individual dwelling:			
Average home	Person	190–350	280
Better home	Person	250–400	310
Luxury home	Person	300–550	380
Semimodern home	Person	100–250	200
Summer cottage	Person	100–240	190
Trailer park	Person	120–200	150

Note: L × 0.2642 = gal.

Table 2-7 Population projection methods[a]

Method[b]	Description of method
Graphical	Graphical projections of past population-growth curves are used to estimate future population growth
Decreasing-rate-of-growth	Population is estimated on the basis of the assumption that as the city becomes larger, the rate of growth from year to year becomes smaller
Mathematical or logistic	Population growth is assumed to follow some logical mathematical relationship in which population growth is a function of time
Ratio and correlation	The population-growth rate for a given community is assumed to be related to that of a larger region, such as the county or state
Component	Population is forecast on the basis of a detailed analysis of the components that make up population growth, namely, natural increase and migration. Natural increase represents the increase resulting from the excess of births over deaths
Employment forecast	Population growth is estimated on the basis of various employment forecasts. In actual practice, the relationship between population and the number of jobs is derived by using the techniques of the ratio and correlation method

[a] Additional details on these and other population-projection methods may be found in Refs. 10, 11, and 12.
[b] Methods are arranged in order of increasing complexity.

areas located near the area being considered. In the absence of such data, an estimate of 70 percent of the domestic water-withdrawal rate may be used.

In the past, the preparation of population projections for use in estimating wastewater flowrates was often the responsibility of the sanitary engineer, but today such data are usually available from local, regional, and state planning agencies. If they are not available and must be prepared, Refs. 10, 11, and 12 may be consulted. Population-projection methods that have been used are described in Table 2-7. Saturation population density values that are required for estimating flows from land areas with various use classifications can be obtained from local planning commissions. Even though such values are given or prescribed, they should be checked and assessed in light of possible future changes in land-use patterns.

Commercial districts Commercial wastewater flows are generally expressed in cubic meters per hectare per day (gallons per acre per day) and are based on existing or anticipated future development or comparative data. Unit flows may vary from 42 to more than 1500 m^3/ha·d (4500 to more that 160,000 gal/acre·d). Estimates for certain commercial sources may also be made from the data in Table 2-8.

Table 2-8 Average wastewater flows from commercial sources[a]

Source	Unit	Flow, L/unit·d	
		Range	Typical
Airport	Passenger	8–15	10
Automobile service station	Vehicle served	30–50	40
	Employee	35–60	50
Bar	Customer	5–20	8
	Employee	40–60	50
Hotel	Guest	150–220	190
	Employee	30–50	40
Industrial building			
(excluding industry and cafeteria)	Employee	30–65	55
Laundry (self-service)	Machine	1800–2600	2200
	Wash	180–200	190
Motel	Person	90–150	120
Motel with kitchen	Person	190–220	200
Office	Employee	30–65	55
Restaurant	Meal	8–15	10
Rooming house	Resident	90–190	150
Store, department	Toilet room	1600–2400	2000
	Employee	30–50	40
Shopping center	Parking space	2–8	4
	Employee	30–50	40
	Employee	30–50	40

[a] Adapted in part from Ref. 5.

Note: L × 0.2642 = gal.

Institutional facilities Some typical flows from institutional facilities, which are essentially domestic in nature, are shown in Table 2-9. Again, it is stressed that flows vary with the region, climate, and type of facility. The actual records of institutions are the best sources of flow data for design purposes.

Recreational facilities Flows from many recreational facilities are highly seasonal. Some typical data are presented in Table 2-10.

Sources and Rates of Industrial Wastewater Flows

Industrial wastewater flowrates vary with the type and size of the industry, the supervision of the industry, the degree of water reuse, and the onsite wastewater-treatment methods used, if any. Peak flows that are often encountered may be reduced by the use of detention tanks and equalization basins. A typical design value for estimating the flows from industrial districts that have no wet-process-type industries is about 50 m^3/ha·d (~5,000 gal/acre·d). Alternatively, where the nature of each industry is known, data such as those reported in Table 2-4

Table 2-9 Average wastewater flows from institutional sources[a]

Source	Unit	Flow, L/unit·d Range	Flow, L/unit·d Typical
Hospital, medical	Bed	500–950	650
	Employee	20–60	40
Hospital, mental	Bed	300–550	400
	Employee	20–60	40
Prison	Inmate	300–600	450
	Employee	20–60	40
Rest home	Resident	200–450	350
	Employee	20–60	40
School, day:			
With cafeteria, gym, and showers	Student	60–115	80
With cafeteria, but no gym and no showers	Student	40–80	60
Without cafeteria, gym, and showers	Student	20–65	40
Schools, boarding	Student	200–400	280

[a] Adapted in part from Ref. 5.

Note: L × 0.2642 = gal.

Table 2-10 Wastewater flows from recreational sources

Source	Unit	Flow, L/unit·d Range	Flow, L/unit·d Typical
Apartment, resort	Person	200–280	220
Cabin, resort	Person	130–190	160
Cafeteria	Customer	4–10	6
	Employee	30–50	40
Campground (developed)	Person	80–150	120
Cocktail lounge	Seat	50–100	75
Coffee shop	Customer	15–30	20
	Employee	30–50	40
Country club	Member present	250–500	400
	Employee	40–60	50
Day camp (no meals)	Person	40–60	50
Dining hall	Meal served	15–40	30
Dormitory, bunkhouse	Person	75–175	150
Hotel, resort	Person	150–240	200
Laundramat	Machine	1800–2600	2200
Store, resort	Customer	5–20	10
	Employee	30–50	40
Swimming pool	Customer	20–50	40
	Employee	30–50	40
Theater	Seat	10–15	10
Visitor center	Visitor	15–30	20

Note: L × 0.2642 = gal.

can be used. For industries without internal reuse programs, it can be assumed that about 85 to 95 percent of the water used in the various operations and processes will become wastewater. For large industries with internal water-reuse programs, separate estimates must be made. Average domestic (sanitary) wastewater contributed from industrial activities may vary from 30 to 95 L/capita·d (8 to 25 gal/capita·d).

Infiltration/Inflow

Extraneous flows in sewers have been defined as follows [2]:

Infiltration. The water entering a sewer system, including sewer service connections, from the ground, through such means as, but not limited to, defective pipes, pipe joints, connections, or manhole walls. Infiltration does not include, and is distinguished from, inflow.

Inflow. The water discharged into a sewer system, including service connections, from such sources as, but not limited to, roof leaders, cellar, yard, and area drains, foundation drains, cooling-water discharges, drains from springs and swampy areas, manhole covers, cross connections from storm sewers and combined sewers, catch basins, storm waters, surface runoff, street wash waters, or drainage. Inflow does not include, and is distinguished from, infiltration.

Infiltration/Inflow. The total quantity of water from both infiltration and inflow without distinguishing the source.

Many extensive programs of sewer-system evaluation have been and are being undertaken, because the Federal Water Pollution Control Act Amendments of 1972 require that applicants for treatment works grants must demonstrate that each sewer system that will discharge into the proposed treatment works will not be subject to excessive infiltration/inflow.

Infiltration into sewers One portion of the rainfall in a given area runs quickly into the storm sewers or other drainage channels; another portion evaporates or is absorbed by vegetation; and the remainder percolates into the ground, becoming groundwater. The proportion that thus percolates into the ground depends on the character of the surface and soil formation and on the rate and distribution of the precipitation according to seasons. Any reduction in permeability, such as that due to buildings, pavements, or frost, decreases the opportunity for precipitation to become groundwater and increases the surface runoff correspondingly.

The amount of groundwater flowing from a given area may vary from a negligible amount for a highly impervious district or a district with a dense subsoil, to 25 or 30 percent of the rainfall for a semipervious district with a sandy subsoil permitting rapid passage of water into it. The percolation of water through the ground from rivers or other bodies of water sometimes has considerable effect on the groundwater table, which rises and falls continually.

The presence of high groundwater results in leakage into the sewers and in an increase in the quantity of wastewater and the expense of disposing of it. This leakage from groundwater, or infiltration, may range from 0.0094 to 0.94 $m^3/d \cdot mm \cdot km$ (100 to 10,000 gal/d $\cdot$ in $\cdot$ mi) or more. The number of millimeter-kilometers (inch-miles) in a sewer system is the sum of the products of sewer diameters, in millimeters (inches), times the lengths, in kilometers (miles), of sewers of corresponding diameters. Expressed another way, infiltration may range from 0.2 to 28 $m^3/ha \cdot d$ (20 to 3000 gal/acre $\cdot$ d). During heavy rains, when there may be leakage through manhole covers, or inflow, as well as infiltration, the rate may exceed 470 $m^3/ha \cdot d$ (50,000 gal/acre $\cdot$ d). Infiltration/inflow is a variable part of the wastewater, depending on the quality of the material and workmanship in the sewers and building connections, the character of the maintenance, and the elevation of the groundwater compared with that of the sewers.

The sewers first built in a district usually follow the watercourses in the bottoms of valleys, close to (and occasionally below) the beds of streams. As a result, these old sewers may receive comparatively large quantities of groundwater, while sewers built later at high elevations will receive relatively smaller quantities of groundwater. With an increase in the percentage of area in a district that is paved or built over comes (1) an increase in the percentage of storm water that is conducted rapidly to the storm sewers and watercourses, and (2) a decrease in the percentage of the stormwater that can percolate into the earth and tend to infiltrate the sanitary sewers. A sharp distinction is to be made between maximum and average rates of infiltration into the sewer systems. The maximum rates are necessary to determine required sewer capacities; the average rates are necessary to estimate such factors as annual costs of pumping and treatment of wastewater.

The rate and quantity of infiltration depend on the length of sewers, the area served, the soil and topographic conditions, and to a certain extent, the population density (which affects the number and total length of house connections). Although the elevation of the water table varies with the quantity of rain and snow water percolating into the ground, the leakage through defective joints, porous concrete, and cracks has been large enough, in many cases, to lower the groundwater table to the level of the sewer.

Most of the pipe sewers built during the first half of this century were laid with cement mortar joints or hot poured bituminous compound joints. Manholes were almost always constructed of brick masonry. Deterioration of pipe joints, pipe-to-manhole joints, and the waterproofing of brickwork has resulted in a high potential for infiltration into these old sewers. Modern sewer design calls for the use of high-quality pipe with dense walls, precast manhole sections, and joints sealed with rubber or synthetic gaskets. The use of these improved materials has greatly reduced infiltration into newly constructed sewers, and it is expected that the increase of infiltration rates with time will be much slower than has been the case with the older sewers.

Inflow into sewers For the purpose of analyzing sewer gagings and because of the measuring techniques in use, definition of inflow is usually subdivided into two categories. The first category includes cellar and foundation drainage, cooling-water discharges, and drainage from springs and swampy areas. This type of inflow causes a "steady flow" that cannot be identified separately and so is included in the measured infiltration. The second category consists of those types of inflow which are related directly to storm water runoff and, as a result of rainfall, cause an almost immediate increase in flows in sanitary sewers. Possible sources are roof leaders, yard and areaway drains, manhole covers, cross connections from storm drains and catch basins, and combined sewers.

Infiltration design allowances for sewers When designing for presently unsewered areas or for relief of overtaxed existing sewers, allowance must be made for unavoidable infiltration/inflow as well as for the expected wastewater. Infiltration allowances for existing sewers should be based on flow data, if available, with consideration being given to the expected future leakage. For new sewers, or existing sewers for which no data are available, average rates may be based on data from similar existing sewers, with appropriate modifications to account for differences in materials and construction and for possible differences in expected future conditions.

In the absence of relevant flow data, average infiltration allowances presented in Fig. 2-1 may be used for new sewers or recently constructed sewer systems

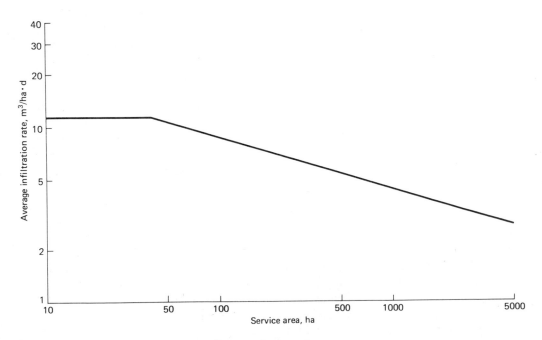

Figure 2-1 Average infiltration-rate allowance for new sewers.

having precast manholes and pipe joints made with gaskets of rubber or rubber-like material. In all cases, the infiltration allowances for design should reflect the expected condition of the sewer system at the end of the period for which it is being designed.

The average rates of flow for use in the design of wastewater-treatment plants and pumping stations may be estimated by adding average domestic and industrial flows and average infiltration allowances. Inflow rates, because of their episodic nature, do not appreciably affect the average design flows.

It is noted that the infiltration design allowance discussed here has little or no relationship to the allowances used for the acceptance of newly constructed sewers. In effect, the acceptance allowance is designed to measure how well the construction job was done, whereas the design allowance is used to account for what may ultimately happen to the sewer, including the construction of building sewers on private property.

Variations in Wastewater Flows

Short-term, seasonal, and industrial variations in wastewater flows are briefly discussed here. The analysis of flowrate data with respect to peak and sustained flows to be expected is discussed in Sec. 2-4. For a more comprehensive discussion of wastewater flow variations, Ref. 3 is recommended.

Short-term variations The variations in wastewater flows observed at treatment plants tend to follow a somewhat diurnal pattern, as shown in Fig. 2-2. Minimum flows occur during the early morning hours when water consumption is lowest and when the base flow consists of leakages, infiltration, and small quantities of sanitary wastewater. The first peak flow generally occurs in the late morning when wastewater from the peak morning water use reaches the treatment plant. A second peak flow generally occurs in the early evening between 7 and 9 P.M., but this varies with the size of the community and the length of the sewers.

When extraneous flows are minimal, the wastewater-discharge curves closely parallel water-consumption curves, but with a lag of several hours. In the absence

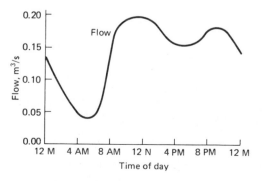

Figure 2-2 Typical hourly variation in domestic wastewater flows.

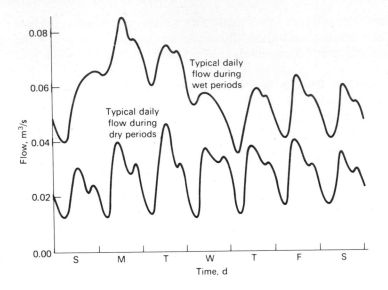

Figure 2-3 Typical daily and weekly variations in domestic wastewater flows. Note: m³/s ×
22.8245 = Mgal/d.

of a typical day when home laundering is done, the variation in weekday flows is
negligible. A plot of typical weekly flows for both wet and dry periods is shown
in Fig. 2-3.

Seasonal variations Seasonal variations in wastewater flows are commonly
observed at resort areas, in small communities with college campuses, and in
communities that have seasonal commercial and industrial activities. The
magnitude of the variations to be expected depends on both the size of the
community and the seasonal activity. An extreme example is the variation at the
city of Modesto, Calif., which occurs because of the substantial amount of
industrial wastes from canneries and other activities related to agriculture.

Infiltration/inflow quantities also vary seasonally. Storm water and groundwater can enter the system through cracks, malformed or broken joints, unauthorized drainage connections, and poorly constructed house connections. The
magnitude of this effect will depend on the type of collection system, whether
separate or combined. (As mentioned earlier, separate sanitary sewers carry only
domestic and industrial wastewater; combined sewers carry storm water in
addition to domestic and industrial wastewater.) The variation of the infiltration/
inflow also depends on location. In the western United States, the rainfall pattern
tends to be cyclical; there is little or no rainfall during the summer. In the
eastern United States, the rainfall pattern tends to be more uniform.

Industrial variations There is no foolproof procedure for predicting industrial
wastewater discharges. While internal process changes may lead to reduced

discharges, plant expansion may lead to increased discharges. Where joint treatment facilities are to be constructed, special attention should be given to industrial discharge projections, whether they are prepared by the industry or jointly with the city's staff or engineering consultant. Industrial discharges are most troublesome in smaller wastewater-treatment plants where there is limited capacity to absorb shock loadings.

2-4 ANALYSIS OF WASTEWATER FLOWRATE DATA

Because the hydraulic design of both collection and treatment facilities is affected by variations in wastewater flows, design values for the expected peak and sustained flows must be developed. Data on peak flows are needed for the design of collection and interceptor sewers; data on sustained flows are needed for the design of treatment facilities.

The best current design practice calls for estimating peaking factors for domestic and industrial wastewater flow and for infiltration and inflow separately; so each category is discussed separately.

Peaking Factors for Wastewater Flows

Ideally, peaking factors (the ratio of peak flow to average flow) would be derived or estimated for each major establishment or for each category of flow in the system. These factors would then be applied to individual average flows, and the resulting peak flows would be combined to obtain maximum expected peak flows. Unfortunately, this degree of refinement is seldom possible; therefore, peaking factors must usually be estimated by more generalized methods.

If flow-measurement records are inadequate to establish peaking factors, the curves given in Fig. 2-4a and b may be used. These curves were developed from analyses of the records of numerous communities throughout the United States. The curve shown in Fig. 2-4a is based on population and should be used only for wastewater that is wholly or predominantly residential in character. The curve in Fig. 2-4b, which is based on average flows, is suited for use where the wastewater contains small amounts of commercial flows and industrial wastes, as well as residential wastewater.

Curves such as those shown in Fig. 2-4 are often used to estimate peak flows from residential areas, as well as mixed flows. When commercial, institutional, or industrial wastewaters make up a significant portion of the average flows (say 25 percent or more of all flows, exclusive of infiltration), peaking factors for the various categories of flow should be estimated separately. If possible, peaking factors for industrial wastewater should be estimated on the basis of average water use, number of shifts worked, and pertinent details of plant operations.

Many state agencies have also set peak design flowrates to be used when no actual measurements are available. Typically, these are about 1500 L/capita·d (400 gal/capita·d) for laterals and 900 L/capita·d (240 gal/capita·d) for trunk

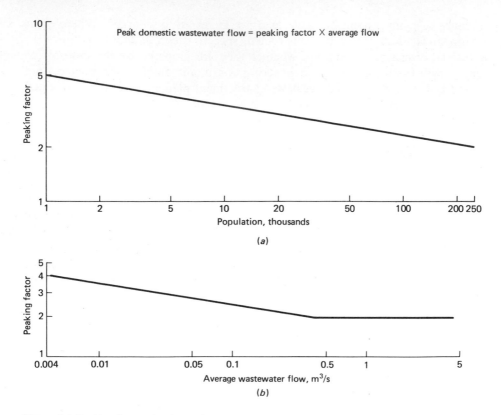

Figure 2-4 Peaking factors for domestic wastewater flows.

sewers (assuming no extraneous flows other than normal infiltration) [7]. One federal agency has set a minimum average design flow of 280 L/capita·d (75 gal/capita·d) to be used unless otherwise justified by sound engineering data. On this basis, laterals are designed for a peaking factor of 4.0, and outfall sewers are then designed for a peaking factor of 2.5 [20].

Peak Infiltration Flows

Peak infiltration allowances for sewer design are often related to the sizes of the areas served by means of curves, such as those shown in Fig. 2-5. In the absence of contradictory measurements, these curves may be considered conservative for most sewer designs. Curve *A* may be used for areas that have old sewers; curve *B* may be used for areas that have either old or new sewers. The choice between curves *A* and *B* for old sewers depends on the present and expected future condition of the sewers, the elevation of the groundwater table, and the method of joint construction. For example, if sewer joints are known or believed to have been formed with cement mortar, and the presence of a high groundwater table is known or expected, curve *A* or higher rates should be used.

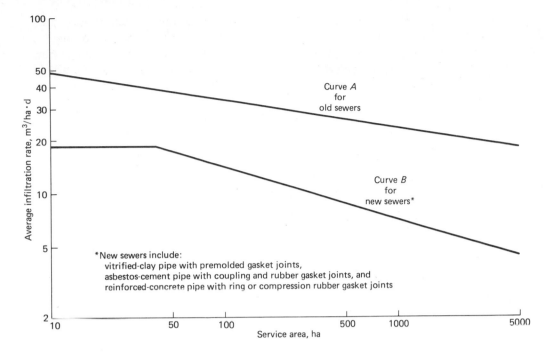

Figure 2-5 Peak infiltration allowances.

In addition to newly designed sewers, the category of "new sewers" includes those recently constructed sewer systems in which precast-concrete manholes were used and those in which the pipe joints were sealed with compression gaskets of rubber or rubberlike materials.

Where curves of average infiltration rates, justified by relevant measurements, are available, the peak infiltration rates for sewer design may be obtained by applying an appropriate peaking factor to the average values read from the curves. Peaking factors for infiltration are properly derived from flow measurements; common values range from 1.5 to 2.0.

Peak Inflow Design Allowance

Separate design allowances for peak inflow rates should be made when designing relief for, or extensions of, existing sewer systems. These allowances should be based on analyses of gagings where possible, with appropriate reductions attributable to corrective measures proposed for the existing system.

Although properly designed and constructed new sewers should be free from rainfall-related inflow, entry points for such inflow may, in time, develop as a result of loose-fitting manhole covers, inadvertent connection of roof leaders, catch basins, and other drains, or other causes. However, conservative infiltration rates would normally be sufficient to allow for such possible occurrences. If it is customary, in the locality to be sewered, for areaway drains or other drains to be connected to the sewers, and there is little prospect of future elimination of this practice, appropriate inflow allowances must be made in the design of new sewers. Also, the use of perforated manhole covers would indicate need for added inflow allowance. It has been estimated that leakage through a manhole cover submerged under 25 mm (1 in) of water varies from 75 to 265 L/min (20 to 70 gal/min), depending on the size and number of openings in the cover [17].

Sustained Flows

Of equal importance to information on the expected peak flows is information on the expected sustained flows, especially in the design of wastewater-treatment facilities. Sustained flows are flows that persist for various time durations (say 2 hours or longer). In wastewater treatment, sustained flows that are both higher and lower than average are of interest. A plot of sustained peak and low flows is shown in Fig. 2-6. The envelope curves were derived by analyzing the records from approximately 46 treatment facilities throughout the country. The length of the records varied from 1 to about 8 years. When developing such curves, the longest possible period of record should be used.

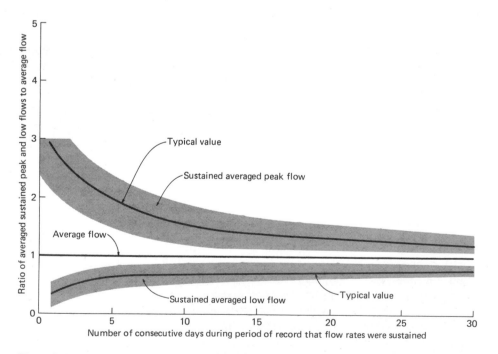

Figure 2-6 Ratio of averaged sustained peak and low flows to average flows for time periods up to 30 d.

The procedure for developing the curves is as follows: First, the average flow is determined for the period of record. Second, the records are searched and the flows for the highest and lowest days are found. These values are divided by the average flow, and the resulting numbers are plotted. Third, the records are searched for the highest and lowest flows occurring in two consecutive days. The two highest and two lowest values are then averaged separately. These averaged values are then divided by the average flow for the period of record. The resulting values are plotted. This process is then repeated until all the ratio values are found for the period of interest (usually 10 to 30 days).

Statistical Analysis

Data on wastewater flows can also be analyzed statistically. The nature of the distribution can best be established by plotting the data on both arithmetic and log probability paper and noting whether they can be fitted with a straight line.

Example 2-1 Estimation of flowrates Estimate the expected average and peak domestic and industrial wastewater flows from the Covell Park development shown in Fig. 2-7.

Data on the expected saturation population densities and wastewater flows for the various types of housing in the Covell Park development were derived from actual records of similar nearby developments and are given in Table 2-11. The commercial (including the shopping area) and industrial wastewater flow rate allowances were estimated to be 35 and 50 m^3/ha·d (3750 and 5000 gal/acre·d), respectively. These estimates were derived from an analysis of the individual types of facilities to be included within the two zones. On the basis of actual flow records of similar activities, the average peaking factors are 1.8 for the commercial flows and 2.1 for the industrial flows.

The school to be built within the Covell Park development is to serve 2000 students at ultimate capacity. The average flow is 75 L/student·d (20 gal/student·d), and the peaking factor for the school is 4.0.

Table 2-11 Saturation population densities and wastewater flows for residential area of the Covell Park development for Example 2-1

Type of development	Saturation population density		Wastewater flows	
	Persons/ha	(Persons/acre)	L/capita·d	(gal/capita·d)
Single-family dwellings	38	(15)	300	(80)
Duplexes	60	(24)	280	(75)
Low-rise apartments	124	(50)	225	(60)

Note:. ha × 2.4711 = acre
L × 0.2642 = gal

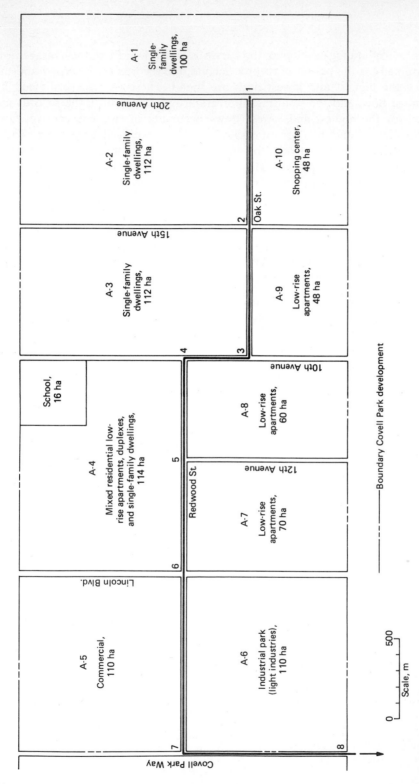

Figure 2-7 Classification of land use for Covell Park development for Example 2-1.

Table 2-12 Determination of average and peak domestic and industrial wastewater flow from the Covell Park development for Example 2-1

Land-use classification (1)	Total area, ha (2)	Population density, persons/ha (3)	Total population (4)	Avg. unit flow Basis (5)	Value (6)	Avg. flow, m³/d (7)	Peaking factor (8)	Peak flow, m³/d (9)
Single-family dwellings	324	38	12,312	L/person/d	300	3,694	2.7[b]	9,974
Mixed residential dwellings	114	74[a]	8,436	L/person/d	268[a]	2,261	2.7[b]	6,105
Low-rise apartments	178	124	22,072	L/person/d	225	4,966	2.7[b]	13,408
School	16		2,000	L/person/d	75	150	4.0	600
Shopping center	48			m³/ha·d	35	1,680	1.8	3,024
Commercial	110			m³/ha·d	35	3,850	1.8	6,930
Industrial park	110			m³/ha·d	50	5,500	2.1	11,550
Total	900					22,101		51,591

[a] Based on unweighted average of data reported in Table 2-11.
[b] From Fig. 2-4a based on a total population of 42,820 (12,312 + 8,436 + 22,072).

Note:

$$\text{ha} \times 2.4711 = \text{acre}$$
$$\text{m}^3/\text{d} \times 2.642 \times 10^{-4} = \text{Mgal/d}$$
$$\text{L} \times 0.2642 = \text{gal}$$
$$\text{m}^3/\text{ha} \cdot \text{d} \times 106.91 = \text{gal/acre} \cdot \text{d}$$

SOLUTION

1 Set up a computation table for estimating domestic and industrial wastewater flows. The required computations are summarized in Table 2-12. Although Table 2-12 is self-explanatory, the following comments are made to clarify some of the specific entries.

   a In cols. 3 and 6, a simple unweighted average of the data in Table 2-11 is used to arrive at the figures used for the population and unit flow for the area of Covell Park with mixed residential dwellings.

   b The peaking factor for the residential areas was obtained from Fig. 2-4a, using the entire residential population. This was done because the flow from the entire area is being determined. Where sewers are to be designed to serve the individual areas, the population of each individual area contributing to a section of sewer would be used to estimate the appropriate peaking factor.

2 Summarize the domestic and industrial wastewater flows.

	Flow, m³/d	
	Average	Peak
Domestic	16,601	40,041
Industrial	5,500	11,550
Total	22,101	51,591

Comment The estimated domestic and industrial wastewater flows determined in this example would be added to the expected infiltration/inflow to determine the size of sewer required to serve the area.

2-5 REDUCTION OF WASTEWATER FLOWS

Because of the importance of conserving both resources and energy, various means for reducing wastewater flows and pollutant loadings from domestic sources are gaining increasing attention. The reduction of flows is considered in this section; the reduction of pollutant loadings is discussed in Chap. 3.

Per capita wastewater flows from conventional domestic devices are given in Table 2-13. The principal devices and systems that are used to reduce domestic wastewater flows are described in Table 2-14. The actual wastewater-flow reduction and the percentage reduction that is possible using these devices and systems as compared with the flows from conventional devices is reported in Table 2-15. The social acceptability and the ease with which these devices can be installed in existing and new systems is discussed in Ref. 13. Another method of achieving flow reductions is to restrict the use of appliances, such as automatic dishwashers and garbage-disposal units, that tend to increase water consumption.

In many communities, the use of one or more of the flow-reduction devices is now specified for all new residential dwellings; in others, the use of garbage grinders has been limited in new housing developments. Further, many individuals

Table 2-13 Per capita wastewater flows from conventional domestic devices [13]

Device	Wastewater flow	
	L/capita·d	%
Bathtub faucet	30.3	12
Clothes-washing machine	34.1	14
Kitchen-sink faucet	26.5	11
Lavatory faucet	11.4	5
Shower head	45.4	19
Toilet	94.6	39
Total	242.3	100

Note: L × 0.2642 = gal.

Table 2-14 Flow-reduction devices and systems[a]

Device/system	Description and/or application
Batch-flush valve	Used extensively in commercial applications. Can be set to deliver between 1.9 L/cycle for urinals and 15 L/cycle for toilets
Brick in toilet tank	A brick or similar device in a toilet tank achieves only a slight reduction in wastewater flow
Dual-cycle tank insert	Insert converts conventional toilet to dual-cycle operation. In new installations, a dual-cycle toilet is more cost-effective than a conventional toilet with a dual-cycle insert
Dual-cycle toilet	Uses 4.75 L/cycle for liquid wastes and 9.5 L/cycle for solid wastes
Faucet aerator	Increases the rinsing power of water by adding air and concentrating flow, thus reducing the amount of wash water used. Comparatively simple and inexpensive to install
Level controller for clothes washer	Matches the amount of water used to the amount of clothes to be washed
Limiting-flow shower head	Restricts and concentrates water passage by means of orifices that limit and divert shower flow for optimum use by the bather
Limiting-flow valve	Restricts flow to a fixed rate that depends on household water-system pressure
Pressure-reducing valve	Maintains home water pressure at a lower level than that of the water-distribution system. Reduces household flows and decreases the probability of leaks and dripping faucets

(*continued*)

Table 2-14 (*continued*)

Device/System	Description and/or application
Recirculating mineral-oil toilet system	Uses mineral oil as a waste-transporting medium and requires no water. Operates in a closed loop in which toilet wastes are collected separately from other household wastes and stored for later pickup by vacuum truck. In the storage tank, wastes are separated from the transporting fluid by gravity. The mineral oil is drawn off by pump, coalesced, and filtered before being recycled to the toilet tank
Reduced-flush device	Toilet-tank insert that either prevents a portion of the tank contents from being dumped during the flush cycle or occupies a portion of the tank volume so that less water is available per cycle
Urinal	Wall-type urinal for home use that requires 5.7 L/cycle
Vacuum-flush toilet system	Uses air as a waste-transporting medium and requires about 1.9 L/cycle
Washwater-recycle system for toilet flushing	Recycles bath and laundry wastewaters for use in toilet flushing

[a] Adapted from Ref. 13.

Note: L × 0.2642 = gal.

who are concerned about conservation have installed such devices as a means of reducing water consumption. It will probably be some time before the actual impact of the use of such devices and methods is known.

2-6 MEASUREMENT OF WASTEWATER FLOWS

The ability to measure wastewater flows is of fundamental importance in the design of all wastewater-management facilities. The two principal categories of methods that are used to measure flowing fluids are direct-discharge methods and velocity-area methods.

Direct-Discharge Methods

Direct-discharge methods are those in which the rate of discharge has been related to one or two easily measured variables. Frequently, where numerous flowrate determinations are to be made, rating curves are developed to simplify the work involved. The principal methods and applications are described in Table 2-16. Sketches of some of the apparatuses are shown in Fig. 2-8. Because Venturi meters and the Palmer-Bowlus flumes are used extensively, they are discussed further in this section. Details on the other methods may be found in the references given in Table 2-16.

Table 2-15 Reductions achieved by flow-reduction devices and systems[a]

Device/system[b]	Wastewater- flow reduction	
	L/capita·d	% of total[c]
Level control for clothes washer	4.5	2
Pressure-reducing valve[d]	60.6	25
Recirculating mineral-oil toilet system	94.6	39
Shower:		
Limiting-flow valve	22.7	9
Limiting-flow shower head	28.4	12
Sink faucet:		
Faucet aerator	1.9	1
Limiting-flow valve	1.9	1
Toilet:		
Single-batch-flush valve	28.4	12
Dual-batch-flush valve	58.7	24
Urinal with batch-flush valve	26.5	11
Toilet and urinal with batch-flush valves	54.9	23
Water-saver toilet	28.4	12
Dual-cycle toilet	66.2	27
Dual-cycle tank insert	37.9	16
Reduced-flush device	37.9	16
Brick in toilet tank	3.8	2
Vacuum-flush toilet system[d]	85.2	35
Washwater-recycle system for toilet flushing	94.6	39

[a] Adapted from Ref. 13.
[b] See Table 2-14 for descriptions and applications of devices and systems.
[c] Percent of total for conventional devices reported in Table 2-13.
[d] Both single and multiple homes.

Note: L × 0.2642 = gal.

Table 2-16 Direct-discharge methods for flow measurement

Method/apparatus	Description/application	References
California pipe	In this method the flowrate is related to the depth of flow from an open end of a partially filled horizontal pipe that is discharging freely to the atmosphere. The discharge pipe should be horizontal and should have a length of at least six pipe diameters. When the pipe is flowing almost full, an air vent should be installed back up the pipe to ensure free circulation of air in the unfilled portion of the discharge pipe	12, 23

(continued)

Table 2-16 (*continued*)

Method/apparatus	Description/application	References
Computation	This method requires field measurements of the depth of flow and slope of the sewer. A value for the coefficient of roughness must also be selected. The method, at best, is an approximation dependent on the steadiness of the flow at the time of observation and the precision with which the coefficient of roughness is assumed for the existing conditions. This method is also based on the assumption that flow is occurring at normal depth. Despite these limitations, this method is used frequently for making wastewater-flow measurements	12
Direct weighing	In this method, which is used to measure small flows, the mass of fluid discharged over a specified time period is weighed and converted to a flowrate using the specific weight of the fluid	
Flow nozzles	Nozzle flowmeters in pipes make use of the Venturi principle but use a nozzle inserted in the pipe instead of the Venturi tube to produce the pressure differential. The form of the nozzle, the method of inserting it in a pipe, and the method of measuring the difference in pressure vary with the manufacturer. Open-flow nozzles attached to the ends of pipes are usually of the Kennison type shown in Fig. 2-8*b*. Because nozzles placed at the ends of pipes are essentially proportional weirs, only a single pressure connection is needed to measure the head	12
Magnetic flowmeters	When an electrical conductor passes through an electromagnetic field, an electromotive force or voltage is induced in the conductor that is proportional to the velocity of the conductor. This statement of Faraday's law serves as the basis of design for electromagnetic flowmeters, as shown in Fig. 2-8*d* In actual operation, the liquid in the pipe (usually water or wastewater) serves as the conductor. The electromagnetic field is generated by placing coils around the pipe. The induced voltage is then measured by electrodes placed on either side of the pipe. If the pipe is a conductor, the electrodes need not penetrate the wall of the pipe. Where the pipe is constructed of nonconductive material, the electrodes must penetrate the pipe wall and, in some cases, protrude into the liquid. Magnetic flowmeters are usually available for pipe sizes varying from 50 to 600 mm in diameter; larger sizes require special order	12

Table 2-16 (*continued*)

Method/apparatus	Description/application	References
Orifice	An orifice is a cylindrical or prismatic opening through which fluid flows. The standard orifice, as generally defined, is one in which the edge of the orifice that determines the jet is such that the jet, upon leaving it, does not again touch the wall of the orifice. Practically, this result is obtained by having the outside of the orifice beveled. The flowrate is determined using Torricelli's theorem	6, 9, 11, 12, 24
Orifice, pipe	A plate with a cylindrical opening in the center is usually inserted into closed pipelines. The flowrate is determined from differential-pressure readings	6, 9, 11, 12, 24
Tracers, chemical and radioactive	In chemical or radioactive gagings, a known concentration of a chemical or radioactive substance is added continuously, at a constant rate, to the stream in which the discharge is to be determined. At a distance downstream sufficient to ensure complete mixing of the tracer and stream, the stream is sampled and the concentration of the chemical or radioactive substances is determined. The flow in the stream can then be determined using a materials-balance equation	12
Venturi flumes	Venturi flumes use the critical-depth principle to measure flows in open channels. The two best-known types are the Parshall and the Palmer-Bowlus flumes. The Parshall flume (see Fig. 2-8c) is usually fixed and is often used to measure flows at treatment plants. The Palmer-Bowlus flume is small and movable and is commonly used to measure wastewater flows in sewers	12, 14, 15, 24
Venturi meters	The Venturi meter (see Fig. 2-8a), which is used to measure flows in closed conduits, consists of three parts: (1) the inlet cone, in which the diameter of the pipe is gradually reduced; (2) the throat or constricted section; and (3) the outlet cone, in which the diameter increases gradually to that of the pipe in which the meter is inserted The throat in standard meter tubes is from one-third to one-half the diameter of the pipe. Its length is but a few inches, sufficient to allow a suitable pressure chamber or piezometer ring to be inserted in the pipe at this point. A piezometer ring is inserted at the upper or large end of the inlet cone, and the determination of the quantity of water flowing is based on the difference in pressures observed or indicated at this point and at the throat of the meter	9, 11, 12, 24

(*continued*)

Table 2-16 (*continued*)

Method/apparatus	Description/application	References
Volumetric measurement	The volume of fluid discharged over a specified time period is measured. Generally, this can be done only with very low flows	
Weir, sharp	A sharp weir is a barrier (usually a metal or plastic plate) over which the fluid to be measured is made to flow. Rectangular, triangular, and trapezoidal weirs are the three types most commonly used. The flowrate is determined by measuring the observed head on the crest of the weir (rectangular and trapezoidal weirs) or over the weir notch (triangular weirs). The head is different in elevation between the top of the crest and the surface of the water in the channel, at a point upstream taken, if possible, just beyond the beginning of the surface curve. The flowrate is determined from a rating curve in which the flowrate is plotted versus the observed head	4, 6, 9, 11, 12, 24,

Note: mm × 0.03937 = in.

Venturi meters A typical Venturi meter used for flow measurement is shown in Fig. 2-8a. The equation used for computing the discharge through a Venturi meter is derived from Bernoulli's equation [12]. For a horizontal meter, the appropriate equation is

$$Q = \frac{A_1 A_2 \sqrt{2g(h_1 - h_2)}}{\sqrt{A_1^2 - A_2^2}}$$

$$= \frac{A_1 A_2 \sqrt{2gH}}{\sqrt{A_1^2 - A_2^2}} \tag{2-1}$$

where A_1 = area at upstream end, m² (ft²)
A_2 = area at throat of meter, m² (ft²)
h_1, h_2 = pressure heads, m (ft)
$H = h_1 - h_2$

Under actual operating conditions and for standard meter tubes, including allowance for friction, Eq. 2-1 reduces to the form

$$Q = C A_2 \sqrt{2gH} \tag{2-2}$$

The coefficient C is made up of two parts, or

$$C = C_1 C_2$$

where $C_1 = A_1 / \sqrt{A_1^2 - A_2^2}$
C_2 = coefficient of friction

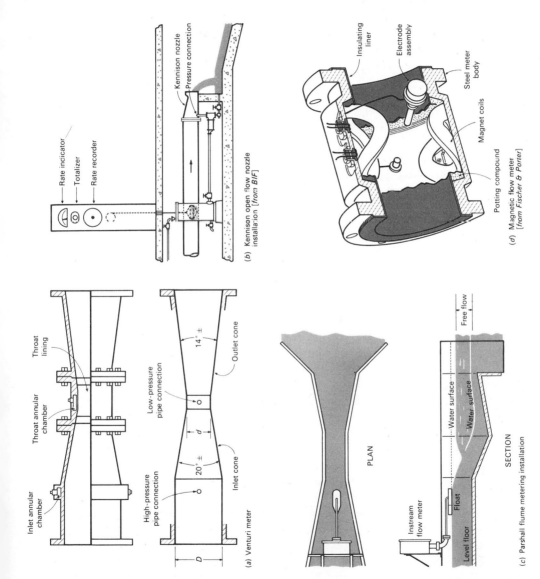

Rate indicator

Totalizer

Rate recorder

Kennison nozzle
Pressure connection

(b) Kennison open flow nozzle
installation [from BIF]

Insulating
liner

Electrode
assembly

Steel meter
body

Magnet coils

Potting compound

(d) Magnetic flow meter
[from Fischer & Porter]

Throat
lining

Throat annular
chamber

Inlet annular
chamber

Low-pressure
pipe connection

Outlet cone

14° ±

d

20° ±

High-pressure
pipe connection

Inlet cone

D

(a) Venturi meter

PLAN

Free flow

Water surface

Water surface

Instream
flow meter

Float

Level floor

SECTION

(c) Parshall flume metering installation

Figure 2-8 Typical direct-discharge flowmeters.

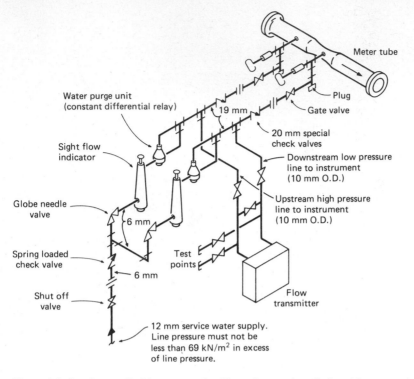

Figure 2-9 Continuous flushing system for Venturi-meter installation. Note: mm × 0.03937 = in; kN/m² × 0.1450 = lb$_f$/in².

For standard meter tubes in which the diameter of the throat is between one-third and one-half that of the pipe, the values of C_1 range between 1.0062 and 1.0328, and the friction coefficient C_2 varies from 0.97 to 0.99. Thus the range of values of C is from 0.98 to 1.02.

Where Venturi meters are used for measuring wastewater, there should be valves at each annular chamber or piezometer ring, so that the pressure openings can be closed. These valves may be so designed that, in closing, a rod is forced through the opening to clean out any matter that may have clogged it. When all these valves have been closed, the plates covering the handholes in the pressure chamber may be removed, and the chamber may be cleaned by flushing with a hose or otherwise. Such flushing at short intervals is usually necessary if Venturi meters for wastewater are to be maintained in good operating condition. A continuous flushing system is shown in Fig. 2-9. A photograph of a Venturi meter is shown in Fig. 2-10.

Palmer-Bowlus flumes The Palmer-Bowlus flume was developed for the measurement of flow in a variety of open channels [14]. The principle of its operation is similar to that of the Parshall flume. The meter is usually placed in the sewer at a manhole, as shown in Fig. 2-11. To function properly, the flume must

Figure 2-10 Venturi-meter installation.

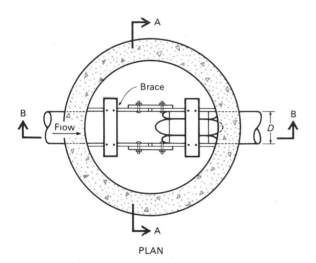

PLAN

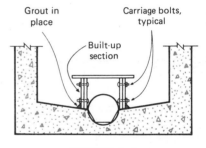

SECTION A–A

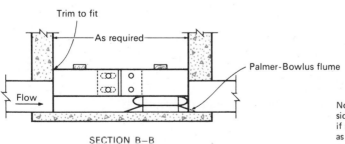

SECTION B–B

Note: Splice boards for channel sides in built-up section only if manhole size prevents placing as a single unit

Figure 2-11 Installation of a Palmer-Bowlus flume in a manhole.

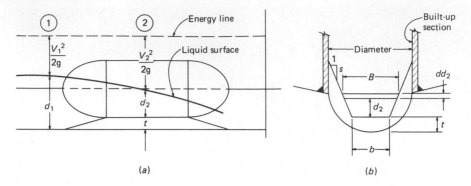

Figure 2-12 Sectional views of a Palmer-Bowlus flume.

act as a hydraulic control in which critical flow is developed. This is usually assured when wastewater is backed up in the sewer above the flume, as a result of its installation, and when discharge from the flume is supercritical.

With critical flow on the flume assured, and with little energy loss, the rate of discharge may be related to the upstream depth. Thus, by measuring the upstream depth, the discharge can be read from a calibration curve which is usually supplied with each unit.

The advantages of the Palmer-Bowlus flume are that it can be installed in existing systems, head loss is insignificant, and it is self-cleansing. Care must be taken to avoid leakage under the flume and conditions in which the flume will be "drowned out." To maintain accuracy of the method, the depth of flow in the upstream should not exceed 0.9 of the pipe diameter, and the point of upstream measurement should be about 0.5 of the pipe diameter from the entrance to the flume. A method of developing a rating curve for Palmer-Bowlus flumes is presented in the following discussion.

Sectional views of a typical flume are shown in Fig. 2-12. Neglecting the friction losses and equating the energies at points 1 and 2 using Bernoulli's theorem, the following equation is obtained:

$$d_1 + \frac{V_1^2}{2g} = d_2 + \frac{V_2^2}{2g} + t \tag{2-3}$$

where d_1 = depth of flow in upstream section, m (ft)
V_1 = flow velocity in upstream section, m/s (ft/s)
g = acceleration due to gravity, 9.81 m/s^2 (32.2 ft/s^2)
d_2 = depth of flow in flume, m
V_2 = flow velocity in flume, m/s (ft/s)
t = depth of flume bottom above channel bottom, m

Solving for d_1 yields

$$d_1 = t + d_2 + \frac{V_2^2}{2g} - \frac{V_1^2}{2g} \tag{2-4}$$

Under free-discharge conditions, the flow will pass through the critical-flow condition in the throat of the flume, in which case d_2 will be equal to d_c.

The specific energy at any section in the throat is given by the following equation:

$$E = d_2 + \frac{V_2^2}{2g} = d_2 + \frac{Q^2}{A_2^2 2g} \tag{2-5}$$

where A = area of cross section through which flow is occurring, m^2 (ft^2)
   Q = discharge, m^3/s (ft^3/s)

Differentiating Eq. 2-5 with respect to depth, d_2 yields

$$\frac{dE}{dd_2} = 1 - \frac{Q^2}{A_2^3 g} \frac{dA_2}{dd_2} \tag{2-6}$$

Because dA is equal to $B \times dd_2$, Eq. 2-6 may be written as follows:

$$\frac{dE}{dd_2} = 1 - \frac{Q^2 B}{A_2^3 g} \tag{2-7}$$

At critical flow, the energy is at a minimum, and dE/dd_2 is equal to zero. This yields

$$\frac{Q_c^2}{g} = \frac{A_c^3}{B_c} \tag{2-8}$$

from which the velocity head is

$$\frac{V_c^2}{2g} = \frac{Q_c^2}{A_c^2 2g} = \left(\frac{A}{2B}\right)_c \tag{2-9}$$

Substituting d_c for d_2 and Eq. 2-9 into Eq. 2-4, the following equation is obtained:

$$d_1 = t + d_c + \left(\frac{A}{2B}\right)_c - \frac{V_1^2}{2g} = t + d_c + \left(\frac{A}{2B}\right)_c - \frac{Q^2}{A_1^2 2g} \tag{2-10}$$

In Eq. 2-10, d_1 depends only on the flowrate because A_1 depends on d_1 and t, d_c, A_c, and B_c are fixed for a given flume and flowrate. A rating curve for the flume can be developed by solving Eq. 2-10 for d_1, for various flowrates. The procedure involved in doing this may be outlined as follows: (1) select a depth of flow in the flume; (2) determine the corresponding values of A and B in the flume for the depth selected in step 1; (3) determine the corresponding rate of flow through the flume using Eq. 2-8; (4) using the rate of flow determined in step 3, solve Eq. 2-10 for d_1 by trial and error. This four-step procedure is repeated for several depths of flow through the flume. The required rating curve is obtained by plotting the computed values of d_1 versus the corresponding flowrates.

To simplify the computations involved in developing the rating curve, either numerical or graphical methods are used. A graphical method for developing a rating curve based on the use of an Arredi diagram [14] is illustrated in Example 2-2. An Arredi diagram is a graphical device for solving Eq. 2-10 without the need for a trial-and-error solution.

Example 2-2 Development of rating curve for Palmer-Bowlus flume Develop a rating curve for a Palmer-Bowlus flume, such as the one shown in Fig. 2-12, using the Arredi-diagram method for an estimated range of flow of 0.025 to 0.15 m³/s. The flume is to be placed in a manhole that connects two 500-mm pipes. Referring to Fig. 2-12, the critical dimensions of the flume are $b = 160$ mm, $t = 40$ mm, and $s = 2$. Determine the flowrate when the depth in the upstream section is 250 mm.

SOLUTION
1 Develop the Arredi diagram for the flow installation. The completed Arredi diagram for this problem is shown in Fig. 2-13; its development is described in the following discussion.
   a Compute values of $Q^2/A^2 2g$ $(= V^2/2g)$ for selected values of Q and A, and plot against A in the upper portion of a piece of millimeter graph paper (see Fig. 2-13). The values selected for Q should cover the anticipated range of flowrates to be measured, and the area values should span a range equal to about twice the area of the sewer.
   b Determine the corresponding cross-sectional areas for selected values of depth in the throat section of the flume and in the upstream section above the flume. The required computations are summarized in Tables 2-17 and 2-18. Note that in the upstream section before the throat, the sides are vertical above a depth of 250 mm, and that zero area in the throat of the flume occurs at depth t, with reference to the bottom of the pipe.
   c Plot separate curves of depth versus area, in the lower portion of the graph for the throat and upstream sections (see Fig. 2-13).
   d Compute values of $A/2B$ $(= V^2/2g)$ for the values of depth in the throat of the flume used in step 2 (see Table 2-17), and plot the computed values against the area in the upper portion of the graph. Note that the intersection of this curve and the velocity head curves plotted in step a represents the solution of Eq. 2-8 for a given flume geometry. Note also that the

---

Table 2-17 Computation table for development of Arredi diagram for throat section of Palmer-Bowlus flume for Example 2-2

d_c mm	$(d_c + t)$, mm	$\left(b + \dfrac{d_c}{2}\right)$, mm	$A = d_c\left(b + \dfrac{d_c}{2}\right)$, m²	$2B = 2b + 4\left(\dfrac{d_c}{2}\right)$, mm	$\dfrac{V^2}{2g} = \dfrac{A}{2B}$, mm
50	90	185	0.0093	420	22
100	140	210	0.0210	520	40
150	190	235	0.0353	620	57
200	240	260	0.052	720	72
250	290	285	0.0713	820	87
300	340	310	0.093	920	101
340	380	330	0.112	1000	112
400	440	360	0.144	1000	127

Note: mm × 0.03937 = in
m² × 10.7639 = ft²

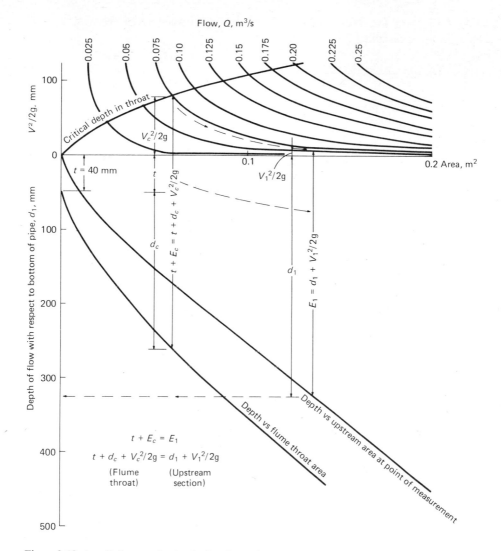

Figure 2-13 Arredi diagram for developing flowrating curve for a Palmer-Bowlus flume for Example 2-2.

Table 2-18 Computation of area in upstream section for Example 2-2

Depth, mm	50	100	150	200	250[a]	450
Area, m²	0.010	0.028	0.050	0.073	0.098	0.198

[a] Rectangular section above this depth.

Note: mm × 0.03937 = in
m² × 10.7639 = ft²

Table 2-19 Data for Palmer-Bowlus flume rating curve derived from Fig. 2-13 for Example 2-2

Flow, m³/s	0.025	0.05	0.075	0.10	0.125	0.150
Depth d, mm	192	270	325	360	390	420

Note: m³/s × 35.3147 = ft³/s
mm × 0.03937 = in

vertical distance from the point of intersection of these curves to the curve of throat area versus depth in the lower portion of the graph represents the total energy in the throat section $(t + d_c + V_c^2/2g)$.

2 Prepare a rating curve for the flume.

    a Starting with the lowest rate of flow to be measured, scale from the graph the value in millimeters for the total energy $(t + d_c + V_c^2/2g)$ for the throat section. Because the total energy in the upstream section (neglecting frictional and other minor losses) must be equal for the same rate of flow, the value of d_1 is found by finding the value of $d_1 + V_1^2/2g$, which is equal to the measured value of $t + d_c + V_c^2/2g$. An example is shown in Fig. 2-13. This procedure is repeated for the range of flowrates to be measured. A summary of the results is presented in Table 2-19.

    b The rating curve is prepared by plotting the depth d_1 versus the corresponding value of flowrate (see Fig. 2-14).

3 Find the flowrate for a depth of flow of 250 mm. From Fig. 2-14, the flowrate is 0.04 m³/s.

Velocity-Area Methods

Using the velocity-area methods, the flowrate is determined by multiplying the velocity of flow (m/s) by the cross-sectional area (m²) through which flow is occurring. The principal methods and apparatuses used for determining velocities are summarized in Table 2-20. Details may be found in the corresponding references given in Table 2-20.

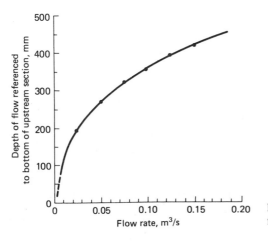

Figure 2-14 Rating curve for Palmer-Bowlus flume shown in Fig. 2-13 for Example 2-2.

Table 2-20 Methods for velocity measurement

Method/apparatus	Description/application	References
Current meters	Current-meter measurements may be used to determine accurately the velocity of flow in large sewers or open channels, provided there is not too much paper or other suspended matter present to clog the meter. Gagings of flow may be made by several methods: the one-, two-, and multiple-point methods, the method of integrating in sections, and the method of integrating in one operation. In the one-point method, the meter is held at 0.6 of the depth, measured from the water surface and in the center of the stream. The result is assumed to indicate the mean velocity of the stream. This is but a rough approximation, suitable only for hasty observations with no pretense of accuracy. In the two-point method, the velocity is observed at 0.2 and 0.8 of the depth, and the average of these two figures is taken to represent the average velocity in the vertical section. The stream can be divided into a number of vertical sections, and the average velocity in each section is approximately determined by this method	9, 11, 12, 24
Electrical methods	Electrical methods to measure the quantity of water flowing in a stream involve the use of equipment such as conductivity cells, hot-wire anemometers, and warm-film anemometers. Although some of these methods have been used in the field, they are not ideally suited for measuring wastewater flows because the floating and suspended material commonly found in wastewater interferes with their operation	11, 12, 24
Float measurements	Float measurements of the flow in sewers are not made routinely, except in rectangular channels or for approximations of the velocity of flow between two manholes. In studies of tidal currents or of wastewater currents in bodies of water into which wastewater may be discharged, floats are used universally. Three types of floats may be used: surface, subsurface, and rod or spar. Only surface velocities can be obtained by the use of surface floats. Owing to the modifying effects of the wind, the results can be considered only as approximations. Subsurface floats consist of relatively large bodies slightly heavier than water, connected by fine wires to surface floats of sufficient size to furnish the necessary flotation, and carrying markers by which their courses may be traced.	11, 12

(continued)

Table 2-20 (*continued*)

Method/apparatus	Description/application	References
	The resistance of the upper float and connecting wire is generally so slight that the combination may be assumed to move with the velocity of the water at the position of the submerged float. Rod floats have been used for measuring flow in open flumes with a high degree of accuracy. They generally consist of metal cylinders so loaded as to float vertically. The velocity of the rod has been found to correspond very closely with the mean velocity of the water in the course followed by the float	
Pitot tubes	The Pitot tube, which has proved so useful in water-pipe gagings, is impractical in extended sewer gagings because the suspended matter in wastewater tends to clog the tube	9, 11, 12, 24
Tracers, chemical and radioactive	Where velocity measurements are to be made, the chemical or radioactive tracers are usually injected into the stream upstream of two control points. The time of passage of the prism of water containing the tracer is noted at these control points, and the velocity is then computed by dividing the distance between the control points by the travel time. When salt (NaCl) is used as the tracer, the time of passage between control points is measured using electrodes connected to an ammeter or recorder. When radioactive tracers are used, the time of passage is noted by radioactive counters attached to the outside of the pipe. The time of passage is the difference between the times when the peak counts were recorded at each counting station	12
Tracers, dye	The use of dyes for measuring the velocity of flow in sewers, particularly in small-pipe sewers, is one of the simplest and most successful methods that has been used. A section of sewer is selected in which the flow is practically steady and uniform, the dye is thrown in at the upper end, and the time of its arrival at the lower end is determined. If a bright-colored dye, such as eosin, is used and a bright plate is suspended horizontally in the sewer at the lower end, the time of appearance and disappearance of the dye at the lower end can be noted with considerable precision, and the mean between these two observed times may be taken as representative of the average time of flow. Other dyes that have been successfully used in tracer studies include fluorescein, congo red, potassium permanganate, rhodamine B, and Pontacyl Brilliant Pink B. Pontacyl Brilliant Pink B is especially useful in the conduct of ocean-outfall dispersion studies	9, 11, 12,

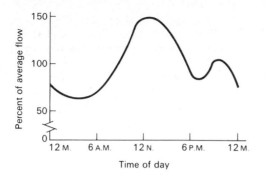

Figure 2-15 Hourly variation of wastewater flow for Prob. 2-1.

DISCUSSION TOPICS AND PROBLEMS

2-1 The flow variation for city A is shown in Fig. 2.15. If the wastewater flows into an aeration tank having a detention time (volume/flow rate) under average flow conditions of 6 h, what would be the average detention time from 8 A.M. to 2 P.M.? From 8 P.M. to 2 A.M.?

2-2 Sewers are to be installed in a recreational camping area that contains a developed campground for 200 persons, lodges and cabins for 100 persons, and resort apartments for 50 persons. Assume that persons staying in lodges use the dining hall for 3 meals per day and that a 50-seat cafeteria with 4 employees and an estimated 100 customers per day has been constructed. Daily attendance at visitor centers is expected to be 50 percent of the campground capacity. Other facilities include a 10-machine laundromat, a 20-seat cocktail lounge, and three gas stations (7.5 $m^3/d \cdot$ station). Determine the average waste flow in cubic meters per day using the design unit flows reported in Table 2-10.

2-3 Estimate the percentage reduction that can be achieved in the domestic flowrate in the Covell Park development considered in Example 2.1 if the developer is required to comply with a new water conservation ordinance that is currently under consideration. The proposed city ordinance will require the use of (1) single-batch flush valve toilets, (2) shower-head-flow limiting devices, and (3) level controls for washing machines in all single-family dwellings, duplexes and low-rise apartments.

2-4 Obtain data for the past 10 years from your local water agency on the amount of water withdrawn per capita for the water-supply system. How does the 1970 value compare with the data given in Table 2.1? If there is a significant variation, explain why it might be expected.

2-5 Using the data from Prob. 2.4, can you discern a trend in water usage? Do you think the existing trend will continue? Identify the pertinent factors that may cause a change in the water-use trend.

2-6 Obtain flow data from your local wastewater treatment facility for a dry period. What is the ratio of the water withdrawn for the water supply for the same period to the measured wastewater flow? How does this value compare with the values reported in the text?

2-7 A 300-mm Venturi meter with a throat diameter of 100 mm is used to measure the flowrate in a 300-mm force main. Assuming the friction coefficient C_2 is equal to 0.982, determine the difference in the pressure head between the entrance and throat when discharging 1000, 5000, and 10,000 m^3/d. Given that the flowrates correspond to low, average, and peak flows, determine if there will be a problem of solids deposition. Assume deposition will occur if the velocity is less than 0.3 m/s.

2-8 Select the throat diameter for a Venturi meter that is to be used in a 250-mm effluent discharge force main from a wastewater treatment plant. The average flow through the plant is 5000 m^3/d and it is desired to have a minimum head differential of 20 cm at low flow. Assume that the flow variation of the treatment plant is as shown in Fig. 2.15, and that the friction coefficient C_2 is 0.982. What is the head differential at maximum flow for the selected Venturi meter?

2-9 Flow measurements are to be made in a manhole using a Palmer-Bowlus flume that connects two 300-mm-diameter pipes. The slope of the upstream sewer is 0.002 and the Manning's n value of

0.013. According to Fig. 2.12, the critical dimensions of the flume are $s = 3$, $b = 75$ mm, and $t = 25$ mm. The lowest expected flow in the pipe is one-tenth of the full capacity; and because of upstream surging, the pipe will occasionally flow full. Develop a rating curve for the flume. Assume that the slope through the manhole is flat.

2-10 The headworks of a small treatment plant are to be updated by the inclusion of a com-munitor (see Sec. 8.2) and a flow metering device. The communitor and flow metering device are to be placed in a rectangular channel that receives the discharge from the trunk sewer. The width of the rectangular channel is 0.5 m and the proposed depth is 1.5 m. The low, average, and peak flowrates are 1500, 4000, and 10,000 m^3/d, respectively.

A Palmer-Bowles flume is to be used for flow metering. If the critical dimensions are $s = 3$, $b = 300$ mm, and $t = 30$ mm, develop a rating curve for the flume. If the approximate headloss through the communitor is 30, 80, and 150 mm at low, average, and peak flows, respectively, determine the expected elevation of the wastewater in front of the communitor at low, average, and peak flowrates. If 0.4 m of freeboard is required at peak flow, is the overall depth of the channel sufficient? If not, what is the required depth?

REFERENCES

1. Borland, S.: New Data on Sewer Infiltration-Exfiltration Ratio, *Public Works*, vol. 87, no. 10, 1956.
2. *Federal Register*, vol. 39, no. 29, sec. 35.905, Feb. 11, 1974.
3. Geyer, J. C., and J. L. Lentz: *An Evaluation of the Problems of Sanitary Sewer System Design*, Final Report, Johns Hopkins University, Baltimore, Md., 1964.
4. Horton, R. E.: *Weir Experiments, Coefficients and Formulas*, U.S. Geological Survey Water Supply and Irrigation Paper 200, 1907.
5. Hubbell, J. W.: Commercial and Institutional Wastewater Loadings, *J. Water Pollut. Control Fed.*, vol. 34, no. 9, 1962.
6. Jaeger, C.: *Engineering Fluid Mechanics*, Blackie, London, 1956.
7. Joint Committee of the American Society of Civil Engineers and the Water Pollution Control Federation: *Design and Construction of Sanitary and Storm Sewers*, ASCE Manual and Report 37, New York, 1969.
8. Kerri, K. D., and J. Brady (eds.): *Operation and Maintenance of Wastewater Collection Systems*, prepared for the U.S. Environmental Protection Agency, Office of Water Program Operations, by Department of Civil Engineering, California State University, Sacramento, 1976.
9. King, H. W., and E. F. Brater: *Handbook of Hydraulics*, 5th ed., McGraw-Hill, New York, 1963.
10. McJunkin, F. E.: Population Forecasting by Sanitary Engineers, *J. Sanit. Eng. Div., ASCE*, vol. 90, no. SA4, 1964.
11. Metcalf, L., and H. P. Eddy: *American Sewerage Practice*, vol. I, 2d ed., McGraw-Hill, New York, 1928.
12. Metcalf & Eddy, Inc.: *Wastewater Engineering: Collection, Treatment, Disposal*, McGraw-Hill, 1972.
13. Metcalf & Eddy, Inc.: *Report to National Commission on Water Quality on Assessment of Technologies and Costs for Publicly Owned Treatment Works*, vol. 2, prepared under Public Law 92-500, Boston, 1975.
14. Palmer, H. K., and F. D. Bowlus: Adaption of Venturi Flumes to Flow Measurements in Conduits, *Trans. ASCE*, vol. 101, p. 1195, 1936.
15. Parshall, R. L.: The Improved Venturi Flume, *Trans. ASCE*, vol. 89, p. 841, 1926.
16. Ramseier, R. E.: Testing New Sewer Pipe Installations, *J. Water Pollut. Control. Fed.*, vol. 44, no. 4, 1972.
17. Rawn, A. W.: What Cost Leaking Manhole? *Waterworks and Sewerage*, vol. 84, no. 12, 1937.
18. Salvato, J. A.: The Design of Small Water Systems, *Public Works*, vol. 91, no. 5, 1960.
19. Select Committee on National Water Resources, United States Senate: *Water Resources Activities in the United States*, Government Printing Office, Washington, D.C., 1960.

20. U.S. Department of Housing and Urban Development: *Minimum Design Standards for Community Sewerage Systems*, FHA 720, Washington, D.C., 1963.
21. U.S. Department of the Interior, Geological Survey: *Estimated Use of Water in the United States in 1970*, C. Richard Murray and E. Bodette Reeves, Geological Survey Circular 676, Washington, D.C., 1972.
22. van der Leeden, F. (ed.): *Water Resources of the World, Selected Statistics*, Water Information Center, Inc., Port Washington, N.Y., 1975.
23. Van Leer, B. R.: The California-Pipe Method of Water Measurement, *Eng. News-Rec.*, Aug. 3, 1922, Aug. 21, 1924.
24. Vennard, J. K., and R. L. Street: *Elementary Fluid Mechanics*, 5th ed., Wiley, New York, 1975.
25. Villemonte, J. R.: Submerged-Weir Discharge Studies, *Eng. News-Rec.*, p. 866, Dec. 26, 1947.

THREE

WASTEWATER CHARACTERISTICS

An understanding of the nature of wastewaters is essential in the design and operation of collection, treatment, and disposal facilities and in the engineering management of environmental quality. To promote this understanding, the information in this chapter is presented in nine sections dealing with (1) the physical, chemical, and biological characteristics of wastewater; (2) wastewater characterization studies; (3) wastewater composition; (4) unit loading factors; (5) variations in concentrations of wastewater constituents; (6) analysis of wastewater loading data; (7) the definition and application of physical characteristics; (8) the definition and application of chemical characteristics; and (9) the definition and application of biological characteristics.

3-1 PHYSICAL, CHEMICAL, AND BIOLOGICAL CHARACTERISTICS OF WASTEWATER

The physical properties and the chemical and biological constituents of wastewater, and their sources, are listed in Table 3-1. The important contaminants of interest in wastewater treatment are listed in Table 3-2.

Secondary treatment standards for wastewater are concerned with the removal of biodegradable organics, suspended solids, and pathogens. Many of the more stringent standards that have been developed recently deal with the removal of nutrients and the improved removal of organics. When wastewater is to be reused,

Table 3-1 Physical, chemical, and biological characteristics of wastewater and their sources

Characteristic	Sources
Physical properties:	
Color	Domestic and industrial wastes, natural decay of organic materials
Odor	Decomposing wastewater, industrial wastes
Solids	Domestic water supply, domestic and industrial wastes, soil erosion, inflow-infiltration
Temperature	Domestic and industrial wastes
Chemical constituents:	
Organic:	
Carbohydrates	Domestic, commercial, and industrial wastes
Fats, oils, and grease	Domestic, commercial, and industrial wastes
Pesticides	Agricultural wastes
Phenols	Industrial wastes
Proteins	Domestic and commercial wastes
Surfactants	Domestic and industrial wastes
Others	Natural decay of organic materials
Inorganic:	
Alkalinity	Domestic wastes, domestic water supply, groundwater infiltration
Chlorides	Domestic water supply, domestic wastes, groundwater infiltration, water softeners
Heavy metals	Industrial wastes
Nitrogen	Domestic and agricultural wastes
pH	Industrial wastes
Phosphorus	Domestic and industrial wastes, natural runoff
Sulfur	Domestic water supply, domestic and industrial wastes
Toxic compounds	Industrial wastes
Gases:	
Hydrogen sulfide	Decomposition of domestic wastes
Methane	Decomposition of domestic wastes
Oxygen	Domestic water supply, surface-water infiltration
Biological constituents:	
Animals	Open watercourses and treatment plants
Plants	Open watercourses and treatment plants
Protista	Domestic wastes, treatment plants
Viruses	Domestic wastes

Table 3-2 Important contaminants of concern in waste-water treatment[a]

Contaminants	Reason for importance
Suspended solids	Suspended solids can lead to the development of sludge deposits and anaerobic conditions when untreated wastewater is discharged in the aquatic environment
Biodegradable organics	Composed principally of proteins, carbohydrates, and fats, biodegradable organics are measured most commonly in terms of BOD (biochemical oxygen demand) and COD (chemical oxygen demand). If discharged untreated to the environment, their biological stabilization can lead to the depletion of natural oxygen resources and to the development of septic conditions
Pathogens	Communicable diseases can be transmitted by the pathogenic organisms in wastewater
Nutrients	Both nitrogen and phosphorus, along with carbon, are essential nutrients for growth. When discharged to the aquatic environment, these nutrients can lead to the growth of undesirable aquatic life. When discharged in excessive amounts on land, they can also lead to the pollution of groundwater
Refractory organics	These organics tend to resist conventional methods of wastewater treatment. Typical examples include surfactants, phenols, and agricultural pesticides
Heavy metals	Heavy metals are usually added to wastewater from commercial and industrial activities and may have to be removed if the wastewater is to be reused
Dissolved inorganic solids	Inorganic constituents such as calcium, sodium, and sulfate are added to the original domestic water supply as a result of water use and may have to be removed if the wastewater is to be reused

[a] Adapted from Ref. 23.

standards normally include requirements for the removal of refractory organics, heavy metals, and in some cases, dissolved inorganic solids. Information on water-quality requirements for various beneficial uses may be found in Refs. 21, 49, and 50.

3-2 WASTEWATER CHARACTERIZATION STUDIES

Wastewater characterization studies are conducted to determine (1) the physical, biological, and chemical characteristics, and the concentrations of constituents in the wastewater, and (2) the best means of reducing the pollutant concentrations.

Procedures for wastewater sampling, methods for sample analysis, and expressions used to present the results are described in this section; methods for flow measurement were described in Chap. 2.

Sampling

The sampling techniques used in a wastewater survey must assure that representative samples are obtained, because the data from the analysis of the samples will ultimately serve as a basis for designing treatment facilities. There are no universal procedures for sampling; sampling programs must be individually tailored to fit each situation. Special procedures are necessary to handle problems when sampling wastes vary considerably in composition. Thus suitable sampling locations must be selected, and the frequency and type of sample to be collected must be determined.

Sampling locations Examination of drawings that show sewers and manholes will help to determine sampling locations where flow conditions encourage a homogeneous mixture. In sewers and in deep, narrow channels, samples should be taken from a point one-third the water depth from the bottom. The collection point in wide channels should be rotated across the channel. The velocity of flow at the sample point should, at all times, be sufficient to prevent deposition of solids. When collecting samples, care should be taken to avoid creating excessive turbulence that may liberate dissolved gases and yield an unrepresentative sample.

Sampling intervals The degree of flowrate variation dictates the time interval for sampling, which must be short enough to provide a true representation of the flow. Even when flowrates vary only slightly, the concentration of waste products may vary widely. Frequent sampling (10- or 15-minute uniform intervals) allows estimation of the average concentration during the sampling period.

Sampling equipment Careful selection of sampling equipment is important if continuous or automatic sampling is appropriate. A simple and inexpensive continuous sampler is shown in Fig. 3-1 [22]. An automatic sampling device is shown in Fig. 3-2. The scope of this chapter does not permit a complete description of the many automatic devices suitable for sampling both domestic and industrial wastewaters. More detailed information may be found in Refs. 2, 11, and 22. A discussion of precautions to be observed in taking samples and using sampling equipment is presented in Ref. 11.

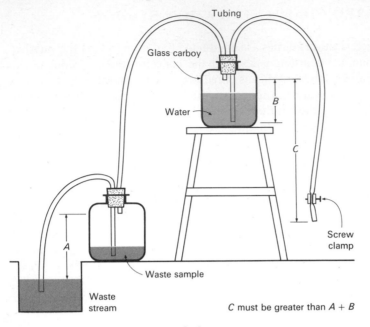

Figure 3-1 Continuous-flow sampler [22].

Sample Preservation

A carefully performed sampling program will be worthless if the physical, chemical, and biological integrity of the samples is not maintained during interim periods between sample collection and sample analysis. Considerable research on the problem of sample preservation has failed to perfect a universal treatment or method or to formulate a set of fixed rules applicable to samples of all types. Prompt analysis is undoubtedly the most positive assurance against error due to sample deterioration. When analytical and testing conditions dictate a lag between collection and analysis, such as when a 24-hour composite sample is collected, provisions must be made for preserving samples. Preservative techniques and maximum holding periods for some selected parameters are shown in Table 3-3 [10]. Recommended methods of sample preservation for the analysis of properties subject to deterioration are also covered in Ref. 10. Probable errors due to deterioration of the sample should be noted in reporting analytical data.

Methods for Sample Analysis

The analyses used to characterize wastewater vary from precise quantitative chemical determinations to the more qualitative biological and physical deter-minations. Many of the parameters are interrelated. For example, temperature, a

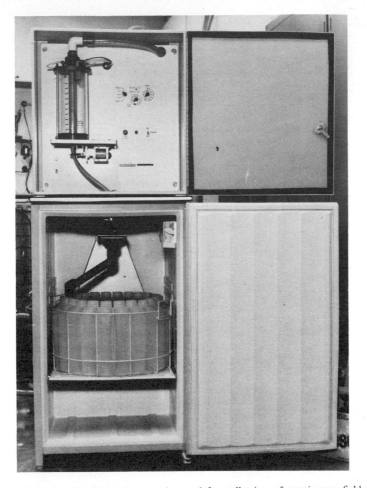

Figure 3-2 Water-quality sampler used for collection of continuous field samples. (*From Manning Environmental Corp.*)

physical parameter. affects both the biological activity in the wastewater and the amounts of gases dissolved in the wastewater.

Details concerning the various analyses may be found in *Standard Methods* [38], the accepted reference that details the conduct of water and wastewater analyses. Simplified techniques of analysis for selected constituents may be found in the U.S. Environmental Protection Agency publication, *Methods of Wastewater Analysis* [10]. Although *Analysis of Water and Sewage* [43] is an older reference, it is still useful. As a general reference, *Chemistry for Environmental Engineering* [32] is recommended. *Aquatic Chemistry* [40], an advanced text, should be consulted for chemical-equilibrium problems, especially in natural waters.

Table 3-3 Preservation of wastewater samples [10]

Parameter	Preservative	Maximum holding period
Acidity-alkalinity	Refrigeration at 4°C	24 h
BOD	Refrigeration at 4°C[a]	6 h
Calcium	None required	
COD	2 mL/L H_2SO_4	7 d
Chloride	None required	
Color	Refrigeration at 4°C	24 h
Cyanide	NaOH to pH 10	24 h
Dissolved oxygen	Determine onsite[b]	No holding
Fluoride	None required	
Hardness	None required	
Metals, total	5 mL/L HNO_3	6 mo
Metals, dissolved	Filtrate: 3 mL/L $1:1$ HNO_3	6 mo
Nitrogen, ammonia	40 mg/L $HgCl_2$, 4°C	7 d
Nitrogen, Kjeldahl	40 mg/L $HgCl_2$, 4°C	Unstable
Nitrogen, nitrate-nitrite	40 mg/L $HgCl_2$, 4°C	7 d
Oil and grease	2 mL/L H_2SO_4, 4°C	24 h
Organic carbon	2 mL/L H_2SO_4 (pH 2)	7 d
pH	None available	
Phenolics	1.0 g $CuSO_4$ + H_3PO_4 to pH 4.0, 4°C	24 h
Phosphorus	40 mg/L $HgCl_2$, 4°C	7 d
Solids	None available	
Specific conductance	None required	
Sulfate	Refrigeration at 4°C	7 d
Sulfide	2 mL/L Zn acetate	7 d
Threshold odor	Refrigeration at 4°C	24 h
Turbidity	None available	

[a] Slow-freezing techniques (to $-25°C$) can be used for preserving samples to be analyzed for organic content.

[b] For some methods of determination, 4 to 8 h preservation can be accomplished with 0.7 mL conc. H_2SO_4 and 20 mg $NaNO_3$. Refer to *Standard Methods* [38] for prescribed applications. (Footnote not in original reference.)

Note: $1.8(°C) + 32 = °F$

$mg/L = g/m^3$

Bacteriology for Sanitary Engineers [17] is an excellent reference on the bacteriology of wastewater and the various testing methods and procedures. For details concerning the biology and microbiology of the various microorganisms encountered in water and wastewater, the following are recommended: *The Ecology of Waste Water Treatment* [12], *The Microbial World* [39], *Microbial Ecology* [3], *Fresh Water Biology* [48], and *A Treatise on Limnology* [15]. Various other specific references are given throughout this chapter.

Expression of Analysis Results

The results of the analysis of samples are expressed in terms of physical and chemical units of measurement. The most common units are reported in Table 3-4. Measurements of chemical parameters are usually expressed in the physical unit of milligrams per liter (mg/L) or grams per cubic meter (g/m^3). For the dilute systems in which one liter weighs approximately one kilogram, such as those encountered in natural waters and wastewater flows, the units of milligrams per liter or grams per cubic meter are interchangeable with parts per million (ppm), which is the mass-to-mass ratio.

Dissolved gases are considered to be chemical constituents and are measured as milligrams per liter or grams per cubic meter. Gases evolved as a by-product

Table 3-4 Units commonly used to express analysis results

Basis	Application	Unit
Physical analyses:		
Density	$\dfrac{\text{Mass of solution}}{\text{Unit volume}}$	kg/m^3
Percent by volume	$\dfrac{\text{Volume of solute} \times 100}{\text{Total volume of solution}}$	% (by vol)
Percent by mass	$\dfrac{\text{Mass of solute} \times 100}{\text{Combined mass of solute} + \text{solvent}}$	% (by mass)
Volume ratio	$\dfrac{\text{Milliliters}}{\text{Liter}}$	mL/L
Mass per unit volume	$\dfrac{\text{Milligrams}}{\text{Liter of solution}}$	mg/L[a]
	$\dfrac{\text{Grams}}{\text{Cubic meter of solution}}$	g/m^3
Mass ratio	$\dfrac{\text{Milligram}}{10^6 \text{ milligram}}$	ppm
Chemical analyses:		
Molality	$\dfrac{\text{Moles of solute}}{1000 \text{ grams of solvent}}$	mol/kg
Molarity	$\dfrac{\text{Moles of solute}}{\text{Liter of solution}}$	mol/L
Normality	$\dfrac{\text{Equivalents of solute}}{\text{Liter of solution}}$	equiv/L
	$\dfrac{\text{Milliequivalents of solute}}{\text{Liter of solution}}$	meq/L

[a] $mg/L = g/m^3$.

of wastewater treatment, such as methane and nitrogen (anaerobic decomposition), are measured in terms of liters (cubic feet). Results of tests, and parameters such as temperature, odor, hydrogen ion, and biological organisms, are expressed in units other than milligrams per liter or grams per cubic meter, as explained in Secs. 3-7, 3-8, and 3-9.

3-3 WASTEWATER COMPOSITION

Composition refers to the actual amounts of physical, chemical, and biological constituents present in wastewater. In this section, data on the constituents found in wastewater and septage are presented. Discussions are also included on the need to characterize wastewater more fully, and on the mineral pickup resulting from water use. Variations in wastewater composition are discussed in Sec. 3-5.

Table 3-5 Typical composition of untreated domestic wastewater

(All values except settleable solids are expressed in mg/L)[a]

Constituent	Concentration		
	Strong	Medium	Weak
Solids, total:	1200	720	350
Dissolved, total	850	500	250
Fixed	525	300	145
Volatile	325	200	105
Suspended, total	350	220	100
Fixed	75	55	20
Volatile	275	165	80
Settleable solids, mL/L	20	10	5
Biochemical oxygen demand, 5-day, 20°C (BOD_5, 20°C)	400	220	110
Total organic carbon (TOC)	290	160	80
Chemical oxygen demand (COD)	1000	500	250
Nitrogen (total as N):	85	40	20
Organic	35	15	8
Free ammonia	50	25	12
Nitrites	0	0	0
Nitrates	0	0	0
Phosphorus (total as P):	15	8	4
Organic	5	3	1
Inorganic	10	5	3
Chlorides[b]	100	50	30
Alkalinity (as $CaCO_3$)[b]	200	100	50
Grease	150	100	50

[a] mg/L = g/m^3.
[b] Values should be increased by amount in domestic water supply.
   Note: 1.8(°C) + 32 = °F.

Constituents in Wastewater and Septage

Typical data on the individual constituents found in domestic wastewater are reported in Table 3-5. Depending on the concentrations of these constituents, wastewater is classified as strong, medium, or weak. Both the constituents and the concentrations vary with the hour of the day, the day of the week, the month of the year, and other local conditions. Therefore, the data in Table 3-5 are intended to serve only as a guide and not as a basis for design.

Septage is the sludge produced in individual onsite wastewater-disposal systems, principally septic tanks and cesspools. The actual quantities and constituents of septage vary widely. The greatest variations are found in communities that do not regulate the collection and disposal of septage [23]. Some data on the constituents found in septage are given in Table 3-6.

Need for Additional Analyses

In general, the constituents reported in Table 3-5 are those that are analyzed more or less routinely. In the past, it was believed that these constituents were sufficient to characterize a wastewater for biological treatment, but as our understanding of the chemistry and microbiology of wastewater treatment has continued to expand, the importance of analyzing additional constituents is becoming more appreciated [53].

These additional constituents that are now analyzed include many of the metals necessary for the growth of microorganisms, such as calcium, cobalt, copper, iron, magnesium, manganese, and zinc. The presence or absence of hydrogen sulfide should be determined to assess whether corrosive conditions may develop and whether any trace metals necessary for the growth of microorganisms are being precipitated [53]. The concentration of sulfate should be determined to assess the suitability of anaerobic waste treatment. The presence of filamentous organisms in the wastewater should also be determined, especially if biological treatment is being considered. Many of the aforementioned constituents were measured in a study made in 1976 for Flagstaff, Ariz. [37], and it was found that

Table 3-6 Characteristics of septage[a]

Constituent	Value	
	Range	Typical
5-day BOD, mg/L	2,000–25,000	8,000
Suspended solids, mg/L	7,000–110,000	30,000
Volatile suspended solids, % of suspended solids	45–80	65
COD, mg/L	5,000–80,000	30,000

[a] Adapted from Ref. 46.

Note: $mg/L = g/m^3$.

the range of values obtained was significant. In fact, if many of the low values obtained for some of the constituents were sustained, problems with effective biological treatment could develop. The need for these analyses is considered further in this chapter and in following chapters that deal with biological treatment.

Mineral Pickup from Water Use

Data on the mineral pickup resulting from water use and the variation of the pickup within a sewerage system are especially important in evaluating the reuse potential of wastewater. Mineral pickup results from domestic use, from the addition of highly mineralized water from private wells and groundwater, and from industrial use. Domestic and industrial water softeners also contribute to the mineral pickup and, in some areas, they may represent the major source of

Table 3-7 Typical mineral pickup from domestic water use[a]

Constituent	Increment range,[b] mg/L
Anions:	
Bicarbonate (HCO_3)	50–100
Carbonate (CO_3)	0–10
Chloride (Cl)	20–50[c]
Nitrate (NO_3)	20–40
Phosphate (PO_4)	20–40
Sulfate (SO_4)	15–30
Cations:	
Calcium (Ca)	15–40[d]
Magnesium (Mg)	15–40[d]
Potassium (K)	7–15
Sodium (Na)	40–70
Other data:	
Aluminum (Al)	0.1–0.2
Boron (B)	0.1–0.4
Fluoride (F)	
Iron (Fe)	0.2–0.4
Manganese (Mn)	0.2–0.4
Silica (SiO_2)	2–10
Total alkalinity	100–150[d]
Total dissolved solids (TDS)	150–400

[a] Adapted in part from Ref. 42.
[b] Reported national average range of mineral pickup by domestic usage. Does not include commercial and industrial additions.
[c] Excluding the addition from domestic water softeners.
[d] Reported as $CaCO_3$.
Note: mg/L = g/m^3.

mineral pickup. Occasionally, water added from private wells and groundwater infiltration will (because of its high quality) serve to dilute the mineral concentration in the wastewater. Typical data on the incremental mineral pickup that can be expected in municipal wastewater resulting from domestic usage are reported in Table 3-7.

3-4 UNIT LOADING FACTORS

When it is impossible to conduct a wastewater characterization study and other data are unavailable, unit per capita loading factors are used to estimate the total waste loadings to be treated.

The total solids in wastewater are derived from the domestic water supply; domestic, commercial, and industrial water use; various nonpoint sources; and groundwater infiltration. Domestic wastewater solids include those derived from toilets, sinks, baths, laundries, garbage grinders, and water softeners. Typical data on the daily per capita quantities of dry solids derived from these and the aforementioned sources are reported in Table 3-8. Assuming that the typical per capita wastewater flow is about 380 L/d (100 gal/d), and using the total

Table 3-8 Estimate of the components of total (dissolved and suspended) solids in wastewater

Component	Dry weight, g/capita·d	
	Range	Typical
Water supply	10–20	15
Domestic wastes:		
Feces (solids, 23%)	30–70	40
Gound food wastes	30–80	45
Sinks, baths, laundries, and other sources of domestic wash waters	60–100	80
Toilet (including paper)	15–25	20
Urine (solids, 3.7%)	40–70	50
Water softeners	a	a
Total for domestic wastewater, excluding contribution from water softeners	185–365	250
Industrial wastes:	150–400	200[b]
Total domestic and industrial wastes	335–765	450
Nonpoint sources	10–40	20[c]
Storm water	20–40	25[c]
Total for domestic, industrial, nonpoint, and storm water	365–845	495

[a] Variable.
[b] Varies with the type and size of facility.
[c] Varies with the season.
   Note: g × 0.0022 = lb.

Table 3-9 Unit waste loading factors

	Value, g/capita·d	
	Range	Typical
Normal domestic wastewater with complete grinding of food wastes:		
BOD$_5$, 20°C	80–120	100
Suspended solids	90–150	120
Total Kjeldahl nitrogen	10–18	15
Total phosphorus	3–6	4
Normal domestic wastewater without grinding of food wastes:[a]		
BOD$_5$, 20°C	60–110	80
Suspended solids	60–115	90
Wastewater from unsewered areas (septic-tank pumpage):		
BOD$_5$, 20°C	6–12	9
Suspended solids	25–50	36

[a] Values for nitrogen and phosphorus are approximately the same as those shown for wastewater with complete grinding.

Note: g × 0.0022 = lb
        1.8(°C) + 32 = °F

solids value reported in Table 3-5 for medium-strength wastewater (720 mg/L), the total solids contribution would be about 274 g/capita·d (0.6 lb/capita·d). Excluding industrial wastes, this value compares well with the data reported in Table 3-8.

From an analysis of data on the composition of wastewater from a number of municipalities, it has been possible to develop unit loading factors for the principal contaminants of concern in wastewater, as reported in Table 3-9. These values must be used with great care, because wastewater constituents vary widely.

If the per capita flow is assumed to be 380 L/d (100 gal/d), the corresponding constituent values with the grinding of food wastes would be as follows: BOD, 263 mg/L; suspended solids, 315 mg/L; total nitrogen, 39 mg/L; and total phosphorus, 11 mg/L. Using the data given in Table 3-5, the wastewater would be classified as being about midway between medium and strong. If the per capita flow were about 500 L/d (130 gal/d), the corresponding wastewater would be classified as medium in strength.

3-5 VARIATIONS IN CONCENTRATIONS OF WASTEWATER CONSTITUENTS

From the standpoint of treatment processes, one of the most serious deficiencies results when the design of a treatment plant is based on average flows and average BOD and suspended-solids loadings, with little or no recognition of peak conditions. In many communities, sustained peak influent flowrates and biochemical oxygen demand (BOD) and suspended-solids loadings can reach two or more times average values. Frequently, peak flowrates and BOD and suspended-solids mass loading rates do not occur at the same time. In such a situation, a design based on the concurrence of peak flowrates and constituent concentrations may result in excess capacity. Analysis of existing records is the best method of arriving at appropriate peak mass loadings. Some of the factors responsible for the variations observed in the BOD and suspended-solids loadings are discussed in this section. Representative data on sustained loading values to be expected are discussed in Sec. 3-6.

The principal factors that are responsible for loading variations are (1) the established habits of community residents, which cause short-term (hourly, daily, and weekly) variations; (2) seasonal conditions, which usually cause longer-term variations; and (3) industrial activities, which cause both long- and short-term variations. These same factors were also discussed in Chap. 2 in connection with flow variations.

Short-Term Variations

Typical data on the hourly variation in domestic wastewater strength are shown in Fig. 3-3. The BOD variation follows the flow variation (same as Fig. 2-2). The peak BOD (organic matter) concentration often occurs in the evening around 9 P.M. Wastewater from combined-sewer systems usually contains more inorganic matter than wastewater from sanitary-sewer systems because of the larger quantities of storm drainage that enter the combined-sewer system. Peak flows and the ratio of peak flow to average flow are higher in combined-sewer systems than in sanitary-sewer systems.

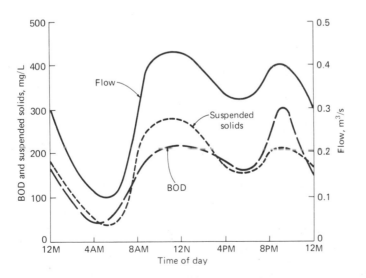

Figure 3-3 Typical hourly variation in flow and strength of domestic wastewater.

Seasonal Variations

For domestic flow only, and neglecting the effects of infiltration, the unit (per capita) loadings and the strength of the wastewater from most seasonal sources, such as resorts, will remain about the same on a daily basis throughout the year even though the total flowrate varies. The total mass of oxygen demand and solids content of the wastewater, however, will increase directly with the population served.

In combined sewers, seasonal variations in BOD and suspended solids are primarily a function of the amount of storm water that enters the system [23]. In the presence of storm water, average concentrations of these constituents will generally be lower than the corresponding concentrations in domestic wastewater. This situation is illustrated in Fig. 3-4, which shows the seasonal BOD variation for the influent to the Calumet Sewage Treatment Works in Chicago [24]. The measured BOD values are below average during the spring and summer months, the period corresponding to the time of the spring thaw and the high summer rainfall.

Although the presence of storm water usually means that the measured concentrations of most constituents will be lower, significantly higher BOD and

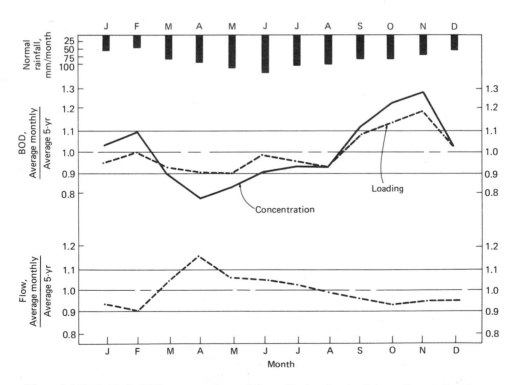

Figure 3-4 Variation in BOD concentration and flow with time for the Calumet Sewage Treatment Works, Chicago [24].

suspended-solids loadings may occur during the early stages of a storm. This is the result of the so-called "first-flush effect," which is most pronounced after a long dry period when material deposited during the dry period is washed away along with any sludge that may have accumulated. When higher initial concentrations are observed, they will seldom be sustained for more than 2 hours. After that, the dilution effect will be observed.

Infiltration/inflow, as explained in Chap. 2, is another source of water flow into a sewer. In most cases, the presence of this extraneous water tends to decrease the concentrations of BOD and suspended solids, but this depends on the characteristics of the water entering the sewer. In some cases, concentrations of some inorganic constituents may actually increase. For example, abnormally high sulfate concentrations have been found in the wastewater of sewers laid in soils where the groundwater contained high levels of sulfates [4].

Industrial Variations

The concentrations of both BOD and suspended solids in industrial wastewater can vary significantly throughout the day. For example, the BOD and suspended-solids concentrations contributed from vegetable-processing facilities during the noon wash-up period may far exceed those contributed during working hours. Problems with high short-term loadings most commonly occur in small treatment plants that do not have adequate reserve capacity to handle these so-called

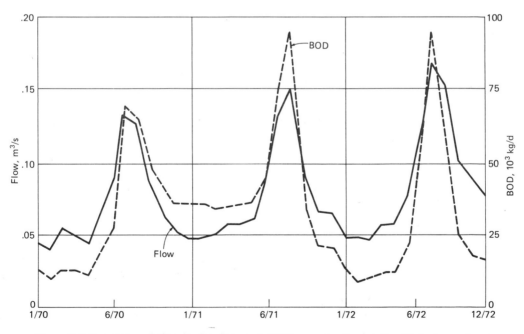

Figure 3-5 Seasonal variation in the flow and BOD mass loading at the Modesto wastewater treatment plant, Modesto, Calif. [6].

"shock loadings." The seasonal impact of industrial wastes is clearly shown in Fig. 3-5, in which both the flow and BOD loading data are presented for a 3-year period for the city of Modesto, Calif. [6]. The variations result from the waste contributions of canneries and other industries related to agriculture.

As noted in Chap. 2, when industrial wastes are to be accommodated in municipal collection and treatment facilities, special attention must be given to developing adequate wastewater characterizations and discharge projections. Further, any proposed future process changes should also be assessed to determine what effects they might have on the wastes to be discharged.

3-6 ANALYSIS OF WASTEWATER LOADING DATA

The analysis of wastewater data involves the determination of simple average or flow-weighted average concentrations, mass loadings, and sustained peak mass loadings. The statistical analysis of the data is the same as discussed in Chap. 2 in connection with the analysis of flow data.

Simple Average

The simple or arithmetic average of a number of individual measurements is given by

$$\bar{x} = \frac{1}{n} \sum_{i=1}^{n} x_i \tag{3-1}$$

where $\bar{x}$ = arithmetic average concentration of the constituent
  n = number of observations
  x_i = average concentration of the constituent during the ith time period

To analyze the BOD and suspended-solids data given in Fig. 3-3, for example, the usual procedure is to divide the day's record into 24 one-hour increments, sum the 24 individual average hourly values, and divide by 24. Although arithmetic averages are still used, they are of little value because the magnitude of the flow at the time of the measurement is not taken into account. If the flowrate remains constant, the use of a simple average is acceptable.

Flow-Weighted Average

To obtain a more representative assessment of constituent concentrations in domestic wastewater, the flow-weighted average is computed using Eq. 3-2:

$$\bar{x}_w = \frac{\sum\limits_{i=1}^{n} x_i q_i}{\sum\limits_{i=1}^{n} q_i} \tag{3-2}$$

where $\bar{x}_w$ = flow-weighted average concentration of the constituent
$\quad n$ = number of observations
$\quad x_i$ = average concentration of the constituent during ith time period
$\quad q_i$ = average flowrate during ith time period

To analyze the data given in Fig. 3-3, for example, the usual procedure is to divide the day's record into 24 one-hour increments and to multiply the corresponding hourly averages of the flowrate and concentration. The 24 values are then summed and divided by the summed values of the 24 individual flowrates.

Mass Loadings

Constituent mass loadings are usually expressed in kilograms per day (pounds per day) and may be computed using Eq. 3-3 when the flowrate is expressed in cubic meters per second or Eq. 3-4 when the flowrate is expressed in liters per day. Note that in the SI system of units, the concentration expressed in milligrams per liter is equivalent to grams per cubic meter.

$$\text{Mass loading, kg/d} = \frac{(\text{concentration, g/m}^3)(\text{flowrate, m}^3/\text{s})(86,400 \text{ s/d})}{1000 \text{ g/kg}} \quad (3\text{-}3)$$

$$\text{Mass loading, kg/d} = \frac{(\text{concentration, mg/L})(\text{flowrate, L/d})}{10^6 \text{ mg/kg}} \quad (3\text{-}4)$$

Sustained Peak Mass Loadings

To design treatment processes to function properly under varying loading conditions, data must be available on the sustained peak mass loadings of constituents that are to be expected. In the past, such information has seldom been available. When the data are not available, the curves shown in Fig. 3-6 can be used. The curves for BOD, suspended solids, total Kjeldahl nitrogen, ammonia, and phosphorus were derived from an analysis of the records of over 50 treatment plants throughout the country. It should be noted that significant variations will be observed from plant to plant, depending on the size of the system, the percentage of combined wastewater, the size and slope of the interceptors, and the types of wastewater contributors.

The procedure used to develop the mass loading curves shown in Fig. 3-6 is the same as that described in Sec. 2-4 for the development of sustained flowrate curves. The daily mass loading rates for the various plants were developed using hourly data and the following expression:

$$\text{Daily mass loading, kg/d} = \sum_{i=1}^{24} \frac{(x_i, \text{ g/m}^3)(q_i, \text{ m}^3/\text{s})(3600 \text{ s/h})}{1000 \text{ g/kg}} \quad (3\text{-}5)$$

The development of a sustained peak mass loading curve is illustrated in Example 3-1. The application of this curve will be discussed in Chaps. 9 and 10.

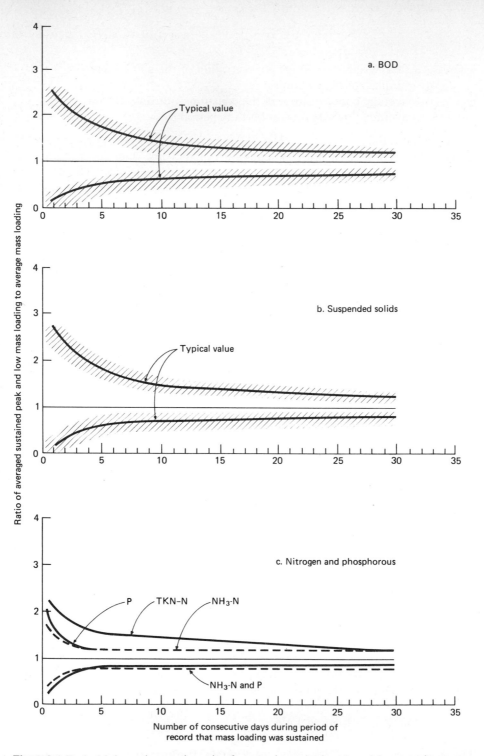

Figure 3-6 Typical information on the ratio of averaged sustained peak and low constituent mass loading rates to average mass loading rate.

Example 3-1 Development of sustained peak mass loading curve for BOD Develop a sustained BOD peak mass loading curve for a treatment plant with a design flowrate of 1 m³/s (22.8 Mgal/d). Assume that the long-term daily average BOD concentration is 200 g/m³.

SOLUTION

1 Compute the daily mass loading value for BOD.

$$\text{Daily BOD mass loading, kg/d} = \frac{(200 \text{ g/m}^3)(1 \text{ m}^3/\text{s})(86,400 \text{ s/d})}{1000 \text{ g/kg}} = 17,280 \text{ kg/d (38,000 lb/d)}$$

2 Set up a computation table for the development of the necessary information for the peak sustained BOD mass loading curve (see Table 3-10).
3 Obtain factors for the sustained peak BOD loading rate from Fig. 3-6a, and determine the sustained mass loading rates for various time periods (see Table 3-10).
4 Develop data for the sustained mass loading curve (see Table 3-10), and prepare a plot of the resulting data (see Fig. 3-7).

Comment The interpretation of the curve plotted in Fig. 3-7 is as follows. If the sustained peak loading period were to last for 12 days, the total amount of BOD that would be received at a treatment facility during the 12-day period would be 280,000 kg. The corresponding amounts for sustained peak periods of 1 and 2 days would be 41,472 and 72,576 kg, respectively.

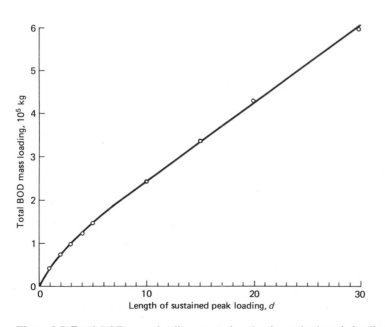

Figure 3-7 Total BOD mass loading versus length of sustained peak loading period in days for Example 3-1.

Table 3-10 Computation table for peak sustained mass loadings for Example 3-1

Length of sustained peak, d (1)	Peaking factor[a] (2)	Peak BOD mass loading, kg/d (3)	Total mass loading, kg[b] (4)
1	2.4	41,472	41,472
2	2.1	36,288	72,576
3	1.9	32,832	98,496
4	1.8	31,104	124,416
5	1.7	29,376	146,880
10	1.4	24,192	241,920
15	1.3	22,464	336,960
20	1.25	21,600	432,000
30	1.15	19,872	596,160
365	1.0	17,280	

[a] From Fig. 3-6a.
[b] Col. 1 × col. 3 = col. 4.
Note: kg × 2.2046 = lb.

3-7 PHYSICAL CHARACTERISTICS: DEFINITION AND APPLICATION

The most important physical characteristic of wastewater is its total solids content, which is composed of floating matter, matter in suspension, colloidal matter, and matter in solution. Other physical characteristics include odor, temperature, and color.

Total Solids

Analytically, the total solids content of a wastewater is defined as all the matter that remains as residue upon evaporation at 103 to 105°C. Matter that has a significant vapor pressure at this temperature is lost during evaporation and is not defined as a solid. Total solids, or residue upon evaporation, can be classified as either suspended solids or filterable solids by passing a known volume of liquid through a filter. The filter is commonly chosen so that the minimum diameter of the suspended solids is about 1 micron (μ). The suspended-solids fraction includes the settleable solids that will settle to the bottom of a cone-shaped container (called an Imhoff cone) in a 60-minute period. Settleable solids are an approximate measure of the quantity of sludge that will be removed by sedimentation.

The filterable-solids fraction consists of colloidal and dissolved solids. The colloidal fraction consists of the particulate matter with an approximate diameter

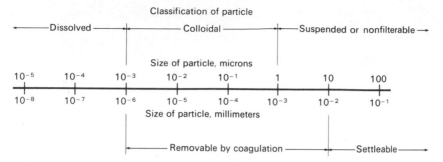

Figure 3-8 Classification and size range of particles found in wastewater. Note: mm × 0.03937 = in.

range of from 1 millimicron (mμ) to 1 μ (see Fig. 3-8). The dissolved solids consist of both organic and inorganic molecules and ions that are present in true solution in water. The colloidal fraction cannot be removed by settling. Generally, biological oxidation or coagulation, followed by sedimentation, is required to remove these particles from suspension.

Each of the categories of solids may be further classified on the basis of their volatility at 600°C. The organic fraction will oxidize and will be driven off as gas at this temperature, and the inorganic fraction remains behind as ash. Thus the terms "volatile suspended solids" and "fixed suspended solids" refer, respectively, to the organic and inorganic (or mineral) content of the suspended solids. At 600°C, the decomposition of inorganic salts is restricted to magnesium carbonate, which decomposes into magnesium oxide and carbon dioxide at 350°C. Calcium carbonate, the major component of the inorganic salts, is stable up to a temperature of 825°C. The volatile-solids analysis is applied most commonly to wastewater sludges to measure their biological stability. The solids content of a medium-strength wastewater may be classified approximately as shown in Fig. 3-9.

Turbidity, a measure of the light-transmitting properties of water, is another test used to indicate the quality of waste discharges and natural waters with respect to colloidal matter. Colloidal matter will scatter or absorb light and thus prevent its transmission.

Odors

Odors in wastewater usually are caused by gases produced by the decomposition of organic matter. Fresh wastewater has a distinctive, somewhat disagreeable odor, which is less objectionable than the odor of septic wastewater. The most characteristic odor of stale or septic wastewater is that of hydrogen sulfide, which is produced by anaerobic microorganisms that reduce sulfates to sulfides. Industrial wastewater may contain either odorous compounds or compounds that produce odors during the process of wastewater treatment.

Odors have been rated as the first concern of the public relative to the

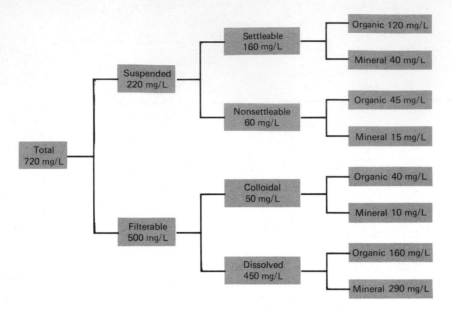

Figure 3-9 Classification of solids found in medium-strength wastewater.

implementation of wastewater-treatment facilities [41]. Within the past few years, the elimination of odors has become a major consideration in the design and operation of wastewater collection, treatment, and disposal facilities, especially with respect to the public acceptance of these facilities. In many areas, projects have been rejected because of the fear of potential odors [52]. In view of the importance of odors in the field of wastewater management, it is appropriate to consider the effects they produce, how they are detected, and their characterization and measurement.

Effects of odors The importance of odors in human terms is related primarily to the psychological stress they produce rather than to the harm they do to the body. Offensive odors can cause poor appetite for food, lowered water consumption, impaired respiration, nausea and vomiting, and mental perturbation [41]. In extreme situations, offensive odors can lead to the deterioration of personal and community pride, interfere with human relations, discourage capital investment, lower socioeconomic status, and deter growth. These problems can result in a decline in market and rental property values, tax revenues, payrolls, and sales [41].

Detection of odors The odorous compounds responsible for producing psychological stress in humans are detected by the olfactory system, but the precise mechanism involved is at present not well understood. Since 1870, about 30 theories have been proposed to explain olfaction. One of the difficulties in

Table 3-11 Major categories of offensive odors [25]

Compound	Typical formula	Odor quality
Amines	CH_3NH_2, $(CH_3)_3N$	Fishy
Ammonia	NH_3	Ammoniacal
Diamines	$NH_2(CH_2)_4NH_2$, $NH_2(CH_2)_5NH_2$	Decayed flesh
Hydrogen sulfide	H_2S	Rotten eggs
Mercaptans	CH_3SH, $CH_3(CH_2)_3SH$	Skunk
Organic sulfides	$(CH_3)_2S$, CH_3SSCH_3	Rotten cabbage
Skatole	$C_8H_5NHCH_3$	Fecal

developing a universal theory has been the inadequate explanation of why compounds with similar structures may have different odors and why compounds with very different structures may have similar odors. Since 1950, most of the major theories have been developed on the premise that the odor of a molecule must be related to the molecule as a whole [41].

Over the years, a number of attempts have been made to classify odors in a systematic fashion. The major categories of offensive odors and the compounds involved are listed in Table 3-11. All these compounds may be found or may develop in domestic wastewater, depending on local conditions.

Odor characterization and measurement It has been suggested that four independent factors are required for the complete characterization of an odor: intensity, character, hedonics, and detectability (see Table 3-12). To date, detectability is the only factor that has been used in the development of statutory regulations for nuisance odors.

Odors can be measured by sensory methods, and specific odorant concentrations can be measured by instrumental methods. It has been shown that, under carefully controlled conditions, the sensory (organoleptic) measurement

Table 3-12 Factors suggested as necessary for the complete characterization of an odor [9]

Factor	Description
Character	Refers to the mental associations made by the subject in sensing the odor. Determination can be quite subjective
Detectability	The number of dilutions required to reduce an odor to its minimum detectable threshold odor concentration (MDTOC)
Hedonics	Relative pleasantness or unpleasantness of the odor sensed by the subject
Intensity	Normally correlated with the concentration of the odor

Table 3-13 Types of errors in the sensory detection of odors[a]

Type of error	Description
Adaptation and cross adaptation	When exposed continuously to a background concentration of an odor, the subject is unable to detect the presence of that odor at low concentrations. When removed from the background odor concentration, the subject's olfactory system will recover quickly. Ultimately, a subject with an adapted olfactory system will be unable to detect the presence of an odor to which his system has adapted
Sample modification	Both the concentration and composition of odorous gases and vapors can be modified in sample-collection containers and in odor-detection devices. To minimize problems associated with sample modification, the period of odor containment should be minimized or eliminated, and minimum contact should be allowed with any reactive surfaces
Subjectivity	When the subject has knowledge of the presence of an odor, random error can be introduced in sensory measurements. Often, knowledge of the odor may be inferred from other sensory signals such as sound, sight, or touch
Synergism	When more than one odorant is present in a sample, it has been observed that it is possible for a subject to exhibit increased sensitivity to a given odor because of the presence of another odor

[a] Adapted from Ref. 52.

of odors by the human olfactory system can provide meaningful and reliable information. Therefore, the sensory method is now used most often to measure the odors emanating from wastewater-treatment facilities.

In the sensory method, human subjects (often a panel of subjects) are exposed to odors that have been diluted with odor-free air, and the number of dilutions required to reduce an odor to its minimum detectable threshold odor concentration (MDTOC) are noted. The detectable odor concentration is reported as the dilutions to the MDTOC. Thus, if 4 volumes of diluted air must be added to 1 unit volume of sampled air to reduce the odorant to its MDTOC, the odor concentration would be reported as 5 dilutions to MDTOC. However, the sensory determination of this minimum threshold concentration can be subject to a number of errors. The principal ones are adaptation and cross adaptation, synergism, subjectivity, and sample modification (see Table 3-13).

The current method of the American Society for Testing and Materials (ASTM) to measure odors in the atmosphere using the dilution procedure (Method D1391-51) has a number of inherent limitations that render it unacceptable for refined threshold odor measurements. Three major limitations are the unnatural breathing situation for the test subject, the lack of adequate control of the test subject's breathing environment during sample evaluation, and the possible sample modification as a consequence of the requirements for

(a)   *(b)*

Figure 3-10 Determination of odors in the field using a direct-reading olfactometer. (*a*) Subject wearing mask (*b*) Assistant adjusting dilutions. (*From EUTEK, Process Development and Engineering.*)

holding discrete air samples. The ASTM test procedure is reportedly being updated to make use of the more reproducible air-dilution technique [52].

To avoid errors in sample modification in containers, a direct-reading olfactometer, as shown in Fig. 3-10, can be used to measure odors at their source without using sampling containers. Descriptions and methods of use of direct-reading olfactometers are available in Ref. 52.

With regard to the instrumental measurement of odors, air-dilution olfactometry provides a reproducible method for measuring threshold odor concentrations. It is often desirable to know the specific compounds responsible for odor. Although gas chromatography has been successfully used for this purpose, it has not been as successfully used in the detection and quantification of odors derived from wastewater collection, treatment, and disposal facilities. The exclusive use of instrumental methods for measuring these odors is ruled out for three reasons: (1) the detectability of odors from wastewater-management facilities often is strongly influenced by other nonodorous compounds (cross adaptation) that may be present, (2) most odors emanating from wastewater-management facilities tend to decay rapidly in storage containers, and (3) the odor molecule must be concentrated before measurement and structural changes can occur.

Temperature

The temperature of wastewater is commonly higher than that of the water supply, because of the addition of warm water from households and industrial activities. As the specific heat of water is much greater than that of air, the observed wastewater temperatures are higher than the local air temperatures during most of the year and are lower only during the hottest summer months. Depending on the geographic location, the mean annual temperature of wastewater varies from about 10 to 21.1°C (50 to 70°F); 15.6°C (60°F) is a representative value.

The temperature of water is a very important parameter because of its effect on the aquatic life, the chemical reactions and reaction rates, and the suitability of the water for beneficial uses. Increased temperature, for example, can cause a change in the species of fish that can exist in the receiving water body. Industrial establishments that use surface water for cooling-water purposes are particularly concerned with the temperature of the intake water.

In addition, oxygen is less soluble in warm water than in cold water. The increase in the rate of biochemical reactions that accompanies an increase in temperature, combined with the decrease in the quantity of oxygen present in surface waters, can often cause serious depletions in dissolved oxygen concentrations in the summer months. When significantly large quantities of heated water are discharged to natural receiving waters, these effects are magnified. It should also be realized that a sudden change in temperature can result in a high rate of mortality of aquatic life. Moreover, abnormally high temperatures can foster the growth of undesirable water plants and wastewater fungus.

Color

Historically, the term "condition" was used along with composition and concentration to describe wastewater. Condition refers to the age of the wastewater which is determined qualitatively by its color and odor. Fresh wastewater is usually gray; however, as organic compounds are broken down by bacteria, the dissolved oxygen in the wastewater is reduced to zero and the color changes to black. In this condition, the wastewater is said to be septic (or stale). Some industrial wastewaters may also add color to domestic wastewater.

3-8 CHEMICAL CHARACTERISTICS: DEFINITION AND APPLICATION

This discussion of chemical characteristics of wastewater is presented in four parts: (1) organic matter; (2) the measurement of organic content; (3) inorganic matter; and (4) gases. The measurement of organic content is discussed separately because of its importance in both the design and operation of wastewater-treatment plants and the management of water quality.

Organic Matter

In a wastewater of medium strength, about 75 percent of the suspended solids and 40 percent of the filterable solids are organic in nature, as shown in Fig. 3-9. These solids are derived from both the animal and plant kingdoms and the activities of man as related to the synthesis of organic compounds.

Organic compounds are normally composed of a combination of carbon, hydrogen, and oxygen, together with nitrogen in some cases. Other important elements, such as sulfur, phosphorus, and iron, may also be present. The principal groups of organic substances found in wastewater are proteins (40 to 60 percent), carbohydrates (25 to 50 percent), and fats and oils (10 percent). Urea, the chief constituent of urine, is another important organic compound contributing to wastewater. Because it decomposes so rapidly, undecomposed urea is seldom found in other than very fresh wastewater.

Along with the proteins, carbohydrates, fats and oils, and urea, wastewater contains small quantities of a large number of different synthetic organic molecules ranging from simple to extremely complex in structure. Typical examples, discussed in this section, include surfactants, phenols, and agricultural pesticides. Further, the number of such compounds is growing yearly as more and more organic molecules are being synthesized. The presence of these substances has, in recent years, complicated wastewater treatment because many of them either cannot be or are very slowly decomposed biologically. This factor also accounts for the renewed interest in the use of chemical precipitation followed by carbon adsorption for the complete treatment of wastewater.

Proteins Proteins are the principal constituents of the animal organism. They occur to a lesser extent in plants. All raw animal and plant foodstuffs contain proteins. The amount present varies from small percentages in watery fruits such as tomatoes and in the fatty tissues of meat to quite high percentages in beans or lean meats. Proteins are complex in chemical structure and unstable, being subject to many forms of decomposition. Some are soluble in water; others are insoluble. The chemistry of the formation of proteins involves the combination or linking together of a large number of amino acids. The molecular weights of proteins are very high, ranging from about 20,000 to 20 million.

All proteins contain carbon, which is common to all organic substances, as well as hydrogen and oxygen. In addition they contain, as their distinguishing characteristic, a fairly high and constant proportion of nitrogen, about 16 percent. In many cases sulfur, phosphorus, and iron are also constituents. Urea and proteins are the chief sources of nitrogen in wastewater. When proteins are present in large quantities, extremely foul odors are apt to be produced by their decomposition.

Carbohydrates Widely distributed in nature, carbohydrates include sugars, starches, cellulose, and wood fiber. All are found in wastewater. Carbohydrates contain carbon, hydrogen, and oxygen. The common carbohydrates contain six or a multiple of six carbon atoms in a molecule, and hydrogen and oxygen in the proportions in which these elements are found in water. Some carbohydrates, notably the sugars, are soluble in water; others, such as the starches, are insoluble. The sugars tend to decompose; the enzymes of certain bacteria and yeasts set up fermentation with the production of alcohol and carbon dioxide.

The starches, on the other hand, are more stable but are converted into sugars by microbial activity as well as by dilute mineral acids. From the standpoint of bulk and resistance to decomposition, cellulose is the most important carbohydrate found in wastewater. The destruction of cellulose in the soil goes on readily, largely as a result of the activity of various fungi, particularly when acid conditions prevail.

Fats, oils, and grease Fats and oils are the third major component of foodstuffs. The term "grease," as commonly used, includes the fats, oils, waxes, and other related constituents found in wastewater. Grease content is determined by extraction of the waste sample with hexane (grease is soluble in hexane). Another group of hexane-soluble substances includes mineral oils, such as kerosene and lubricating and road oils.

Fats and oils are compounds (esters) of alcohol or glycerol (glycerin) with fatty acids. The glycerides of fatty acids that are liquid at ordinary temperatures are called oils, and those that are solids are called fats. They are quite similar, chemically, being composed of carbon, hydrogen, and oxygen in varying proportions.

Fats and oils are contributed to domestic sewage in butter, lard, margarine, and vegetable fats and oils. Fats are also commonly found in meats, in the germinal area of cereals, in seeds, in nuts, and in certain fruits.

Fats are among the more stable of organic compounds and are not easily decomposed by bacteria. Mineral acids attack them, however, resulting in the formation of glycerin and fatty acid. In the presence of alkalies, such as sodium hydroxide, glycerin is liberated, and alkali salts of the fatty acids are formed. These alkali salts are known as soaps, and like the fats, they are stable. Common soaps are made by saponification of fats with sodium hydroxide. They are soluble in water, but in the presence of hardness constituents, the sodium salts are changed to calcium and magnesium salts of the fatty acids, or so-called mineral soaps. These are insoluble and are precipitated.

Kerosene and lubricating and road oils are derived from petroleum and coal tar and contain essentially carbon and hydrogen. These oils sometimes reach the sewers in considerable volume from shops, garages, and streets. For the most part, they float on the sewage, although a portion is carried into the sludge on settling solids. To an even greater extent than fats, oils, and soaps, the mineral oils tend to coat surfaces. The particles interfere with biological action and cause maintenance problems.

As indicated in the foregoing discussion, the grease content of wastewater can cause many problems in both sewers and waste-treatment plants. If grease is not removed before discharge of the waste, it can interfere with the biological life in the surface waters and create unsightly floating matter and films. Limits of 15 to 20 mg/L of grease content and absence of iridescent oil films for wastewaters discharged to natural waters are two examples of standards that have been set by regulating agencies.

Surfactants Surfactants, or surface-active agents, are large organic molecules that are slightly soluble in water and cause foaming in wastewater-treatment plants and in the surface waters into which the waste effluent is discharged. Surfactants tend to collect at the air-water interface. During aeration of wastewater, these compounds collect on the surface of the air bubbles and thus create a very stable foam.

Before 1965, the type of surfactant present in synthetic detergents, called alkyl-benzene-sulfonate (ABS), was especially troublesome because it resisted breakdown by biological means. As a result of legislation in 1965, ABS has been replaced in detergents by linear-alkyl-sulfonate (LAS), which is biodegradable [16]. Since surfactants come primarily from synthetic detergents, the foaming problem has been greatly reduced.

The determination of surfactants is accomplished by measuring the color change in a standard solution of methylene blue dye. Another name for surfactant is methylene blue active substance (MBAS).

Phenols Phenols and other trace organic compounds are also important constituents of water. Phenols cause taste problems in drinking water, particularly when the water is chlorinated. They are produced primarily by industrial operations and find their way to surface waters via wastewater discharges that contain industrial wastes. Phenols can be biologically oxidized at concentrations up to 500 mg/L.

Pesticides and agricultural chemicals Trace organic compounds, such as pesticides, herbicides, and other agricultural chemicals, are toxic to most life forms and therefore can be significant contaminants of surface waters. These chemicals are not common constituents of domestic wastewater but result primarily from surface runoff from agricultural, vacant, and park lands. Concentrations of these chemicals can result in fish kills, in contamination of the flesh of fish that decreases their value as a source of food, and in impairment of water supplies.

The concentration of these trace contaminants is measured by the carbon-chloroform extract method, which consists of separating the contaminants from the water by passing a water sample through an activated-carbon column and then extracting the contaminant from the carbon using chloroform. The chloroform can then be evaporated and the contaminants can be weighed. Pesticides in concentrations of 1 part per billion (ppb) and less can be accurately determined by several methods, including gas chromatography and electron capture or coulometric detectors [32].

Measurement of Organic Content

Over the years, a number of different tests have been developed to determine the organic content of wastewaters. One method, discussed previously, is to measure the volatile-solids fraction of the total solids, but it is subject to many errors and is seldom used [38]. Laboratory methods commonly used today are biochemical oxygen demand (BOD), chemical oxygen demand (COD), and total organic carbon (TOC). Another recently developed test is the total oxygen demand (TOD). Complementing these laboratory tests is the theoretical oxygen demand (ThOD), which is determined from the chemical formula of the organic matter.

Other methods used in the past included (1) total, albuminoid, organic, and ammonia nitrogen, and (2) oxygen consumed. These determinations, with the exception of albuminoid nitrogen and oxygen consumed, are still included in complete wastewater analyses. Their significance, however, has changed. Whereas formerly they were used almost exclusively to indicate organic matter, they are now used to determine the availability of nitrogen to sustain biological activity in industrial waste-treatment processes and to foster undesirable algal growths in receiving waters.

Biochemical oxygen demand The most widely used parameter of organic pollution applied to both wastewater and surface water is the 5-day BOD (BOD_5). This determination involves the measurement of the dissolved oxygen used by microorganisms in the biochemical oxidation of organic matter. Despite the widespread use of the BOD test, it has a number of limitations (which are discussed later in this section). It is hoped that, through the continued efforts of workers in the field, one of the other measures of organic content, or perhaps a new measure, will ultimately be used in its place. Why, then, if the test suffers from serious limitations, is further space devoted to it in this text? The reason is that BOD test results are now used to (1) determine the approximate quantity of oxygen that will be required to biologically stabilize the organic matter present, (2) determine the size of waste-treatment facilities, and (3) measure the efficiency of some treatment processes. Since it is likely that the BOD test will continue to be used for some time, it is important to know as much as possible about the test and its limitations.

To ensure that meaningful results are obtained, the sample must be suitably diluted with a specially prepared dilution water so that adequate nutrients and oxygen will be available during the incubation period. Normally, several dilutions are prepared to cover the complete range of possible values. The ranges of BOD that can be measured with various dilutions based on percentage mixtures and direct pipetting are reported in Table 3-14.

The dilution water is "seeded" with a bacterial culture that has been acclimated, if necessary, to the organic matter present in the water. The seed culture that is used to prepare the dilution water for the BOD test is a mixed culture. Such cultures contain large numbers of saprophytic bacteria and other organisms that oxidize the organic matter. In addition, they contain certain autotrophic bacteria that oxidize noncarbonaceous matter. When the sample contains a large population of microorganisms (untreated wastewater, for example), seeding is not necessary.

The incubation period is usually 5 days at 20°C, but other lengths of time and temperatures can be used. The temperature, however, should be constant

Table 3-14 BOD measurable with various dilutions of samples [32]

By using percent mixtures		By direct pipetting into 300-mL bottles	
% mixture	Range of BOD	mL	Range of BOD
0.01	20,000–70,000	0.02	30,000–105,000
0.02	10,000–35,000	0.05	12,000–42,000
0.05	4,000–14,000	0.10	6,000–21,000
0.1	2,000–7,000	0.20	3,000–10,500
0.2	1,000–3,500	0.50	1,200–4,200
0.5	400–1,400	1.0	600–2,100
1.0	200–700	2.0	300–1,050
2.0	100–350	5.0	120–420
5.0	40–140	10.0	60–210
10.0	20–70	20.0	30–105
20.0	10–35	50.0	12–42
50.0	4–14	100.0	6–21
100.0	0–7	300.0	0–7

throughout the test. After incubation, the dissolved oxygen of the sample is measured and the BOD is calculated using Eq. 3-6a or 3-6b.

For percent mixtures:

$$\text{BOD (mg/L)} = \left[(DO_b - DO_i) \frac{100}{\%} \right] - (DO_b - DO_s) \qquad (3\text{-}6a)$$

For direct pipetting:

$$\text{BOD} = \left[(DO_b - DO_i) \frac{\text{vol of bottle}}{\text{mL of sample}} \right] - (DO_b - DO_s) \qquad (3\text{-}6b)$$

where DO_b, DO_i = dissolved oxygen values found in blank (containing dilution water only) and dilutions of sample, respectively, at end of incubation period

DO_s = dissolved oxygen originally present in undiluted sample

As the value of DO_s approaches DO_b, or when the BOD is over 200 mg/L, the second term in Eqs. 3-6a and 3-6b becomes negligible.

Biochemical oxidation is a slow process and theoretically takes an infinite time to go to completion. Within a 20-day period, the oxidation is about 95 to 99 percent complete, and in the 5-day period used for the BOD test, oxidation is from 60 to 70 percent complete. The 20°C temperature used is an average value for slow-moving streams in temperate climates and is easily duplicated in an incubator. Different results would be obtained at different temperatures because biochemical reaction rates are temperature-dependent.

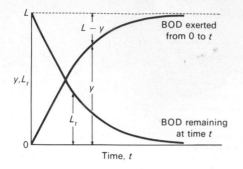

Figure 3-11 Formulation of the first-stage BOD curve.

The kinetics of the BOD reaction are, for practical purposes, formulated in accordance with first-order reaction kinetics and may be expressed as

$$\frac{dL_t}{dt} = -K'L_t \tag{3-7}$$

where L_t is the amount of the first-stage BOD remaining in the water at time t. This equation can be integrated as

$$\ln L_t \Big|_0^{|t} = -K't$$

$$\frac{L_t}{L} = e^{-K't} = 10^{-Kt} \tag{3-8}$$

where L or BOD_L is the BOD remaining at time $t = 0$ (i.e., the total or ultimate first-stage BOD initially present). The relation between K' and K is as follows:

$$K = \frac{K'}{2.303}$$

The amount of BOD remaining at any time t equals

$$L_t = L(10^{-Kt})$$

and y, the amount of BOD that has been exerted at any time t, equals

$$y = L - L_t = L(1 - 10^{-Kt}) \tag{3-9}$$

Note that the 5-day BOD equals

$$y_5 = L - L_5 = L(1 - 10^{-5K})$$

This relationship is shown in Fig. 3-11. Example 3-2 illustrates the use of the BOD equations.

Example 3-2 Calculation of BOD Determine the 1-day BOD and the ultimate first-stage BOD for a wastewater whose 5-day, 20°C BOD is 200 mg/L. The reaction constant $K' = 0.23$ d^{-1}.

SOLUTION

1 Determine ultimate BOD.

$$L_t = Le^{-K't}$$
$$y_5 = L - L_5 = L(1 - e^{-5K'})$$
$$200 = L(1 - e^{-5(0.23)}) = L(1 - 0.316)$$
$$L = 293 \text{ mg/L}$$

2 Determine 1-day BOD.

$$L_1 = Le^{-K't}$$
$$= 293(e^{-0.23(1)}) = 293(0.795) = 233 \text{ mg/L}$$
$$y_1 = L - L_1 = 293 - 233 = 60 \text{ mg/L}$$

For polluted water and wastewater, a typical value of K (base 10, 20°C) is 0.10 d^{-1}. The value of K varies significantly, however, with the type of waste. The range may be from 0.05 to 0.3 d^{-1} or more. For the same ultimate BOD, the oxygen uptake will vary with time and with different K values. The effect of different K values is shown in Fig. 3-12.

As mentioned, the temperature at which the BOD of a wastewater sample is determined is usually 20°C. It is possible, however, to determine the reaction constant K at a temperature other than 20°C. The following approximate equation, which is derived from the van't Hoff Arrhenius relationship (see Eq. 5-9), may be used:

$$K_T = K_{20} \theta^{(T-20°)} \tag{3-10}$$

The value of θ has been found to vary from 1.056 in the temperature range between 20 and 30°C to 1.135 in the temperature range between 4 and 20°C [34]. A value of θ often quoted in the literature is 1.047 [29], but it has been observed that this value does not apply at cold temperatures (e.g., below 20°C) [34].

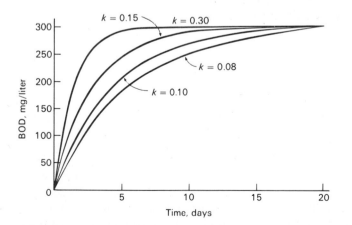

Figure 3-12 Effect of the rate constant K on BOD (for a given L value) [32]. Note: mg/L = g/m^3.

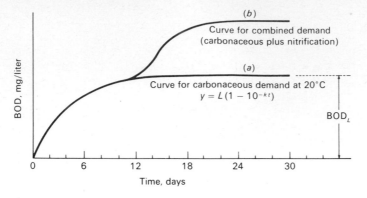

Figure 3-13 The BOD curve [32]. (a) Normal curve for oxidation of organic matter; (b) influence of nitrification. Note: mg/L = g/m³.

Noncarbonaceous matter, such as ammonia, is produced during the hydrolysis of proteins. Some of the autotrophic bacteria are capable of using oxygen to oxidize the ammonia to nitrites and nitrates. The nitrogenous oxygen demand caused by the autotrophic bacteria is called the second-stage BOD. The normal progression of each stage for a domestic wastewater is shown in Fig. 3-13. At 20°C, however, the reproductive rate of the nitrifying bacteria is very slow. It normally takes from 6 to 10 days for them to reach significant numbers and to exert a measurable oxygen demand. The interference caused by their presence can be eliminated by pretreatment of the sample or by the use of inhibitory agents.

Pretreatment procedures include pasteurization [31], chlorination, and acid treatment. Inhibitory agents are usually chemical in nature and include methylene blue, thiourea and allylthiourea, 2-chloro-6-(trichloromethyl)pyridine, and N-Hib, a proprietary product [55]. For a more detailed review of these procedures and a discussion of experimental results obtained using chemical inhibitory agents, Ref. 55 should be consulted.

The value of K is needed if the BOD$_5$ is to be used to obtain L, the ultimate or 20-day BOD. The usual procedure followed when these values are unknown is to determine K and L from a series of BOD measurements. There are several ways of determining K and L from the results of a series of BOD measurements, including the least-squares method [51, 54], the method of moments [26], the daily-difference method [45], the rapid-ratio method [35], and the Thomas method [44]. The least-squares method and the Thomas method are illustrated in the following discussion.

The least-squares method involves fitting a curve through a set of data points, so that the sum of the squares of the residuals (the difference between the observed value and the value of the fitted curve) must be a minimum. Using this method, a variety of different types of curves can be fitted through a set of data points [51]. For example, for a time series of BOD measurements on the same sample, the following equation may be written for each of the various n data points:

$$\frac{dy}{dt}\bigg|_{t=n} = K'(L - y_n) \qquad (3\text{-}11)$$

In this equation both K' and L are unknown. If it is assumed that dy/dt represents the value of the slope of the curve to be fitted through all the data points for a given K' and L value, then because of experimental error, the two sides of Eq. 3-11 will not be equal but will differ by an amount R. Rewriting Eq. 3-11 in terms of R for the general case yields

$$R = K'(L - y) - \frac{dy}{dt} \qquad (3\text{-}12)$$

Simplifying and using the notation y' for dy/dt gives

$$R = K'L - K'y - y' \qquad (3\text{-}13)$$

Substituting a for $K'L$ and $-b$ for K' gives

$$R = a + by - y' \qquad (3\text{-}14)$$

Now, if the sum of the squares of the residuals R is to be a minimum, the following equations must hold:

$$\frac{\partial}{\partial a} \Sigma R^2 = \Sigma 2R \frac{\partial R}{\partial a} = 0$$

$$\frac{\partial}{\partial b} \Sigma R^2 = \Sigma 2R \frac{\partial R}{\partial b} = 0 \qquad (3\text{-}15)$$

If the indicated operations in Eq. 3-15 are carried out using the value of the residual R defined by Eq. 3-14, the following set of equations result:

$$na + b\Sigma y - \Sigma y' = 0 \qquad (3\text{-}16)$$

$$a\Sigma y + b\Sigma y^2 - \Sigma yy' = 0 \qquad (3\text{-}17)$$

where $n =$ number of data points
  $K' = -b$ (base e)
  $L = -a/b$

Application of the least-squares method in the analysis of BOD data is illustrated in Example 3-3.

Example 3-3: Calculation of BOD constants using the least-squares method Compute L and K' using the least-squares method for the following BOD data:

t, d	2	4	6	8	10
y, mg/L	11	18	22	24	26

SOLUTION

1   Set up a computation table and perform the indicated steps.

Time	y	y^2	y'	yy'
2	11	121	4.50	49.5
4	18	324	2.75	49.5
6	22	484	1.50	33.0
8	24	576	1.00	24.0
	$\Sigma 75$	1,505	9.75	156.0

The slope y' is computed as follows:

$$\frac{dy}{dt} = y' = \frac{y_{n+1} - y_{n-1}}{2\Delta t}$$

2   Substituting the values computed in step 1 in Eqs. 3-16 and 3-17 and solving for a and b yields values of 7.5 and -0.271, respectively.

$$4a + 75b - 9.75 = 0$$

$$75a + 1505b - 156.0 = 0$$

3   Determine the values of K' and L.

$$K' = -b = 0.271 \text{ (base } e)$$

$$L = -\frac{a}{b} = \frac{7.5}{0.271} = 27.7 \text{ mg/L}$$

4   Compare these answers to the values obtained using the Thomas method in Example 3-4.

The Thomas method [44], based on the similarity of two series functions, is illustrated here. It is a graphical procedure based on the function

$$\left(\frac{t}{y}\right)^{1/3} = (2.3KL)^{-1/3} + \frac{K^{2/3}}{3.43L^{1/3}} t \qquad (3\text{-}18)$$

where $y =$ BOD that has been exerted in time interval t
$K =$ base 10 reaction-rate constant
$L =$ ultimate BOD

This equation has the form of a straight line,

$$Z = a + bt$$

where $Z = (t/y)^{1/3}$
$a = (2.3KL)^{-1/3}$
$b = K^{2/3}/3.43L^{1/3}$

and Z can then be plotted as a function of t. The slope b and the intercept a

of the line of best fit of the data can then be used to calculate K and L.

$$K = 2.61 \frac{b}{a} \tag{3-19}$$

$$L = \frac{1}{2.3Ka^3} \tag{3-20}$$

To use this method, several observations of y as a function of t are needed. The data observations should be limited to the first 10 days because of nitrogenous interference. The method is illustrated in Example 3-4.

Example 3-4: Calculation of BOD constants using the Thomas method Compute L and K using the Thomas method [44] for the data given in Example 3-3.

SOLUTION
1 Determine the value $(t/y)^{1/3}$ for the data

t, d	2	4	6	8	10
y, mg/L	11	18	22	24	26
$(t/y)^{1/3}$	0.57	0.61	0.65	0.69	0.727

2 Plot the value $(t/y)^{1/3}$ versus t (see Fig. 3-14).
3 From Fig. 3-14 the slope b and intercept a are

$$\text{Slope } b = \frac{0.04}{2} = 0.02$$

$$\text{Intercept } a = 0.53.$$

4 Compute K and L.

$$K = 2.61 \frac{0.02}{0.53} = 0.099 \qquad K' = 0.228$$

$$L = \frac{1}{2.3(0.099)(0.53)^3} = 29.4 \text{ mg/L}$$

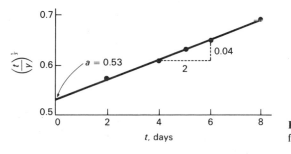

Figure 3-14 Determination of K and L from BOD data by the Thomas method.

The BOD and rate constant K determinations can be made more rapidly in the laboratory by using the Warburg or Gilson respirometer, or with the aid of dissolved-oxygen probes or an electrolysis cell. The Warburg apparatus consists of a constant-temperature water bath, an agitator mechanism, and a set of special flasks equipped with manometers. Each flask has an interior well in which a small quantity of potassium hydroxide solution is placed (see Fig. 3-15). The flask is filled with a measured quantity of waste and biological seed culture and is agitated in the water bath. After the contents are thoroughly mixed, the manometer is connected and readings are taken periodically. Biological respiration within the sample consumes oxygen and produces carbon dioxide. The carbon dioxide is absorbed by the potassium hydroxide solution. Depletion of the dissolved oxygen causes oxygen in the air space over the liquid to enter the solution, thus lowering the pressure in the flask. The quantity of oxygen consumed can then be calculated from the pressure drop as measured with the manometer.

The dissolved-oxygen probe is used in conjunction with the ordinary aqueous BOD technique to reduce the amount of laboratory analysis required. Using an electric strip-chart recorder, the BOD curve can be plotted automatically as it is developed.

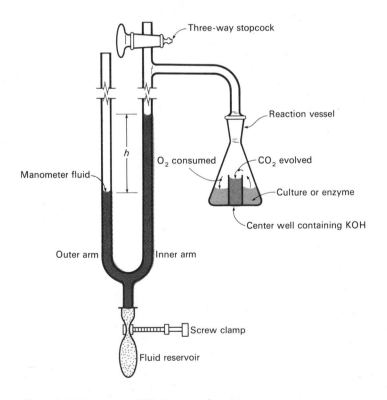

Figure 3-15 Schematic of Warburg respirometer.

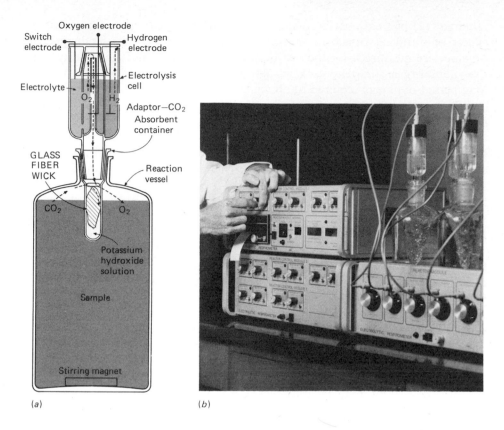

(a)   (b)

Figure 3-16 Electrolysis cell and electrolytic respirometer for BOD determination. (*a*) Schematic of electrolytic cell [57]. (*b*) Commercial electrolytic respirometer apparatus [*A.R.F. Products, Inc.*]

An electrolysis cell (see Fig. 3-16) may also be used to obtain a continuous BOD [7, 56, 57, 58]. In it the oxygen pressure over the sample is maintained constant by continuously replacing the oxygen used by the microorganisms. This is accomplished by producing more oxygen by means of an electrolysis reaction in response to changes in the pressure. The BOD readings are determined by noting the length of time that the oxygen was generated and correlating it to the amount of oxygen produced by the electrolysis reaction. Advantages of the electrolysis cell over the Warburg apparatus are that (1) the use of a large (1-L) sample minimizes the errors of grab sampling and pipetting in dilutions, and (2) the value of the BOD is available directly. A commercially available electrolytic respirometer with multiple electrolysis cells is also shown in Fig. 3-16. A more complete discussion of this method of analysis may be found in Refs. 57 and 58.

Limitations in biochemical oxygen demand test The limitations of the BOD test are as follows: (1) a high concentration of active, acclimated seed bacteria is required; (2) pretreatment is needed when dealing with toxic wastes, and the effects of nitrifying organisms must be reduced; (3) only the biodegradable organics are measured; (4) the test does not have stoichiometric validity after the soluble organic matter present in solution has been used; and (5) an arbitrary, long period of time is required to obtain results.

Perhaps the most serious limitation is that the 5-day period may or may not correspond to the point where the soluble organic matter that is present has been used. This reduces the usefulness of the test results. As a historical note, it was the British Royal Commission of Sewage Disposal that popularized the use of the BOD_5 test measured at 20°C (68°F). The temperature originally used was 18.3°C (65°F). The BOD_5 measured at 18.3°C was chosen because none of the rivers in England has a flow time to the sea of more than 5 days, and the mean summer temperature is 18.3°C [18]. The significance of this limitation of the BOD test is considered further in Chap. 9 and in Ref. 33.

Chemical oxygen demand The COD test is used to measure the content of organic matter of both wastewater and natural waters. The oxygen equivalent of the organic matter that can be oxidized is measured by using a strong chemical oxidizing agent in an acidic medium. Potassium dichromate has been found to be excellent for this purpose. The test must be performed at an elevated temperature. A catalyst (silver sulfate) is required to aid the oxidation of certain classes of organic compounds. Since some inorganic compounds interfere with the test, care must be taken to eliminate them. The principal reaction using dichromate as the oxidizing agent may be represented in a general way by the following unbalanced equation:

$$\text{Organic matter } (C_aH_bO_c) + Cr_2O_7^{--} + H^+ \xrightarrow[\text{heat}]{\text{catalyst}} Cr^{3+} + CO_2 + H_2O \qquad (3\text{-}21)$$

The COD test is also used to measure the organic matter in industrial and municipal wastes that contain compounds that are toxic to biological life. The COD of a waste is, in general, higher than the BOD because more compounds can be chemically oxidized than can be biologically oxidized. For many types of wastes, it is possible to correlate COD with BOD. This can be very useful because the COD can be determined in 3 hours, compared with 5 days for the BOD. Once the correlation has been established, COD measurements can be used to good advantage for treatment-plant control and operation.

Total organic carbon Another means for measuring the organic matter present in water is the TOC test, which is especially applicable to small concentrations of organic matter. The test is performed by injecting a known quantity of sample into a high-temperature furnace. The organic carbon is oxidized to carbon dioxide in the presence of a catalyst. The carbon dioxide that is produced is quantitatively measured by means of an infrared analyzer. Acidification and aeration of the sample prior to analysis eliminates errors due to the presence of inorganic carbon. The test can be performed very rapidly and is becoming more popular. Certain resistant organic compounds may not be oxidized, however, and the measured TOC value will be slightly less than the actual amount present in the sample. Typical TOC values for wastewater were reported in Table 3-5.

Total oxygen demand Another instrumental method that can be used to measure the organic content of wastewater is the TOD test [8]. In this test, organic substances and, to a minor extent, inorganic substances are converted to stable end products in a platinum-catalyzed combustion chamber. The TOD is determined by monitoring the oxygen content present in the nitrogen carrier gas. This test can be carried out rapidly, and the results have been correlated with the COD.

Theoretical oxygen demand Organic matter of animal or vegetable origin in wastewater is generally a combination of carbon, hydrogen, oxygen, and nitrogen. The principal groups of these elements present in wastewater are, as previously noted, carbohydrates, proteins, fats, and products of their decomposition. The biological decomposition of the substances is discussed in Chap. 9. If the chemical formula of the organic matter is known, the ThOD may be computed as illustrated in Example 3-5.

Example 3-5: Calculation of ThOD Determine the ThOD for glycine ($CH_2(NH_2)COOH$) using the following assumptions:
1 In the first step, the carbon is converted to CO_2 and the nitrogen is converted to ammonia.
2 In the second and third steps, the ammonia is oxidized to nitrite and nitrate.
3 The ThOD is the sum of the oxygen required for all three steps.

SOLUTION
1 Carbonaceous demand

$$CH_2(NH_2)COOH + \tfrac{3}{2}O_2 \longrightarrow NH_3 + 2\,CO_2 + H_2O$$

2 Nitrogenous demand

(a) $NH_3 + \tfrac{3}{2}O_2 \xrightarrow{\text{nitrite-forming bacteria}} HNO_2 + H_2O$
(b) $HNO_2 + \tfrac{1}{2}O_2 \xrightarrow{\text{nitrate-forming bacteria}} HNO_3$

3 $ThOD = (\tfrac{3}{2} + \tfrac{4}{2})$ mol O_2/mol glycine $= 112$ g O_2/mol

---

Correlation among measures Establishment of constant relationships among the various measures of organic content depends primarily on the nature of the wastewater and its source. In general, the relationship that exists among these parameters is reported in Table 3-5 and shown graphically in Fig. 3-17. Of all the measures, the most difficult to correlate to the others is the BOD_5 test, because of the problems associated with biological tests (see BOD discussion). For typical untreated domestic wastes, however, the BOD_5/COD ratio varies from 0.4 to 0.8, and the BOD_5/TOC ratio varies from 1.0 to 1.6. Because of the rapidity with which the COD, TOC, and TOD tests can be conducted, it is anticipated that more use will be made of these tests in the future.

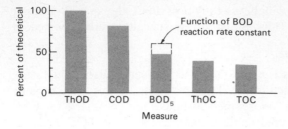

Figure 3-17 Approximate relationship among measures of the organic content of wastewaters.

Inorganic Matter

Several inorganic components of wastewaters and natural waters are important in establishing and controlling water quality. The concentrations of inorganic substances in water are increased both by the geologic formation with which the water comes in contact and by the wastewaters, treated or untreated, that are discharged to it. The natural waters dissolve some of the rocks and minerals with which they come in contact. Wastewaters, with the exception of some industrial wastes, are seldom treated for removal of the inorganic constituents that are added in the use cycle. Concentrations of inorganic constituents also are increased by the natural evaporation process which removes some of the surface water and leaves the inorganic substance in the water. Since concentrations of various inorganic constituents can greatly affect the beneficial uses made of the waters, it is well to examine the nature of some of the constituents, particularly those added to surface water via the use cycle.

pH The hydrogen-ion concentration is an important quality parameter of both natural waters and wastewaters. The concentration range suitable for the existence of most biological life is quite narrow and critical. Wastewater with an adverse concentration of hydrogen ion is difficult to treat by biological means, and if the concentration is not altered before discharge, the wastewater effluent may alter the concentration in the natural waters.

The hydrogen-ion concentration in water is closely connected with the extent to which water molecules dissociate. Water will dissociate into hydrogen and hydroxyl ions as follows:

$$H_2O \rightleftharpoons H^+ + OH^- \tag{3-22}$$

Applying the law of mass action to this equation,

$$\frac{[H^+][OH^-]}{H_2O} = K \tag{3-23}$$

where the brackets indicate concentration of the constituents in moles per liter. Since the concentration of water in a dilute aqueous system is essentially constant, this concentration can be incorporated into the equilibrium constant K to give

$$[H^+][OH^-] = K_w \tag{3-24}$$

K_w is known as the ionization constant, or ion product, of water and is approximately equal to 1×10^{-14} at a temperature of 25°C. Equation 3-24 can be used to calculate the hydroxyl-ion concentration when the hydrogen-ion concentration is known, and vice versa.

The usual means of expressing the hydrogen-ion concentration is as pH, which is defined as the negative logarithm of the hydrogen-ion concentration.

$$pH = -\log_{10}[H^+] \qquad (3\text{-}25)$$

With pOH, which is defined as the negative logarithm of the hydroxyl-ion concentration, it can be seen from Eq. 3-24 that, for water at 25°C,

$$pH + pOH = 14 \qquad (3\text{-}26)$$

The pH of aqueous systems can be conveniently measured with a pH meter. Various indicator solutions that change color at definite pH values are also used. The color of the solution is compared with the color of standard tubes or disks. This method can be used only for relatively clear liquids.

Chlorides Another quality parameter of significance is the chloride concentration. Chlorides in natural water result from the leaching of chloride-containing rocks and soils with which the water comes in contact, and in coastal areas, from saltwater intrusion. In addition, agricultural, industrial, and domestic wastewaters discharged to surface waters are a source of chlorides.

Human excreta, for example, contain about 6 g of chlorides per person per day. In areas where the hardness of water is high, water softeners will also add large quantities of chlorides. Since conventional methods of waste treatment do not remove chloride to any significant extent, higher than usual chloride concentrations can be taken as an indication that the body of water is being used for waste disposal. Infiltration of groundwater into sewers adjacent to saltwater is also a potential source of high chlorides as well as sulfates.

Alkalinity Alkalinity in wastewater results from the presence of the hydroxides, carbonates, and bicarbonates of elements such as calcium, magnesium, sodium, potassium, or of ammonia. Of these, calcium and magnesium bicarbonates are most common. Wastewater is normally alkaline, receiving its alkalinity from the water supply, the groundwater, and the materials added during domestic use. Alkalinity is determined by titrating against a standard acid; the results are expressed in terms of calcium carbonate $CaCO_3$. The concentration of alkalinity in wastewater is important where chemical treatment is to be used (see Chaps. 8 and 12) and where ammonia is to be removed by air stripping (see Chap. 12).

Nitrogen The elements nitrogen and phosphorus are essential to the growth of protista and plants and as such are known as nutrients or biostimulants. Trace quantities of other elements, such as iron, are also needed for biological growth, but nitrogen and phosphorus are, in most cases, the major nutrients of importance.

Since nitrogen is an essential building block in the synthesis of protein, nitrogen data will be required to evaluate the treatability of wastewater by

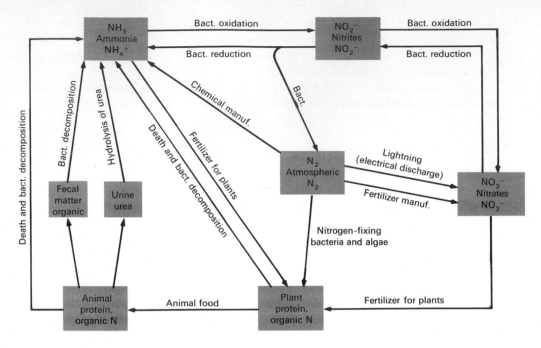

Figure 3-18 Nitrogen cycle [32].

biological processes. Insufficient nitrogen can necessitate the addition of nitrogen to make the waste treatable. Nutrient requirements for biological waste treatment are discussed in Chaps. 9 and 10. Where control of algal growths in the receiving water is necessary to protect beneficial uses, removal or reduction of nitrogen in wastewaters prior to discharge may be desirable (see Chap. 12).

The various forms of nitrogen that are present in nature and the pathways by which the forms are changed are depicted in Fig. 3-18. The nitrogen present in fresh wastewater (see Table 3-5) is primarily combined in proteinaceous matter and urea. Decomposition by bacteria readily changes the form to ammonia. The age of wastewater is indicated by the relative amount of ammonia that is present. In an aerobic environment, bacteria can oxidize the ammonia nitrogen to nitrites and nitrates (see Example 3-5). The predominance of nitrate nitrogen in wastewater indicates that the waste has been stabilized with respect to oxygen demand. Nitrates, however, can be used by algae and other aquatic plants to form plant protein which, in turn, can be used by animals to form animal protein. Death and decomposition of the plant and animal protein by bacteria again yields ammonia. Thus, if nitrogen in the form of nitrates can be reused to make protein by algae and other plants, it may be necessary to remove or to reduce the nitrogen that is present to prevent these growths.

Ammonia nitrogen exists in aqueous solution as either the ammonium ion or ammonia, depending on the pH of the solution, in accordance with the following equilibrium reaction:

$$NH_3 + H_2O \rightleftharpoons NH_4^+ + OH^- \tag{3-27}$$

At pH levels above 7, the equilibrium is displaced to the left; at levels below pH 7, the ammonium ion is predominant. Ammonia is determined by raising the pH, distilling off the ammonia with the steam produced when the sample is boiled, and condensing the steam that absorbs the gaseous ammonia. The measurement is made colorimetrically.

Organic nitrogen is determined by the Kjeldahl method. The aqueous sample is first boiled to drive off the ammonia, and then it is digested. During the digestion the organic nitrogen is converted to ammonia. Total Kjeldahl nitrogen is determined in the same manner as organic nitrogen, except that the ammonia is not driven off before the digestion step. Kjeldahl nitrogen is therefore the total of the organic and ammonia nitrogen.

Nitrite nitrogen is relatively unimportant in wastewater or water-pollution studies because it is unstable and is easily oxidized to the nitrate form. It is an indicator of past pollution in the process of stabilization and seldom exceeds 1 mg/L in wastewater or 0.1 mg/L in surface waters or groundwaters. It is determined by a colorimetric method.

Nitrate nitrogen is the most highly oxidized form of nitrogen found in wastewaters. Where secondary effluent is to be reclaimed for groundwater recharge, the nitrate concentration is important. The U.S. Public Health Service drinking-water standards [47] limit it to 45 mg/L as NO_3^- because of its serious and occasionally fatal effects on infants. Nitrates may vary in concentration from 0 to 20 mg/L as N in wastewater effluents. A typical range is from 15 to 20 mg/L as N. The nitrate concentration is also usually determined by colorimetric methods.

Phosphorus Phosphorus is also essential to the growth of algae and other biological organisms. Because of noxious algal blooms that occur in surface waters, there is presently much interest in controlling the amount of phosphorus compounds that enter surface waters in domestic and industrial waste discharges and natural runoff. Municipal wastewaters, for example, may contain from 4 to 15 mg/L of phosphorus as P (see Table 3-5).

The usual forms of phosphorus that are found in aqueous solutions include the orthophosphate, polyphosphate, and organic phosphate. The orthophosphates, for example, PO_4^{3+}, HPO_4^{--}, $H_2PO_4^-$, and H_3PO_4, are available for biological metabolism without further breakdown. The polyphosphates include those molecules with two or more phosphorus atoms, oxygen atoms, and in some cases, hydrogen atoms combined in a complex molecule. Polyphosphates undergo hydrolysis in aqueous solutions and revert to the orthophosphate forms; however, this hydrolysis is usually quite slow. The organically bound phosphorus is usually of minor importance in most domestic wastes, but it can be an important constituent of industrial wastes and wastewater sludges.

Orthophosphate can be determined by directly adding a substance such as ammonium molybdate which will form a colored complex with the phosphate. The polyphosphates and organic phosphates must be converted to ortho-phosphates before they can be determined in a similar manner.

Sulfur The sulfate ion occurs naturally in most water supplies and is present in wastewater as well. Sulfur is required in the synthesis of proteins and is released in their degradation. Sulfates are chemically reduced to sulfides and to hydrogen sulfide (H_2S) by bacteria under anaerobic conditions, as shown in the following equations:

$$SO_4^{--} + \text{organic matter} \xrightarrow{\text{bacteria}} S^{--} + H_2O + CO_2 \qquad (3\text{-}28)$$

$$S^{--} + 2\,H^+ \longrightarrow H_2S \qquad (3\text{-}29)$$

The H_2S can then be oxidized biologically to sulfuric acid, which is corrosive to sewer pipes.

Sulfates are reduced to sulfides in sludge digesters and may upset the biological process if the sulfide concentration exceeds 200 mg/L. Fortunately, such concentrations are rare. The H_2S gas, which is evolved and mixed with the wastewater gas ($CH_4 + CO_2$), is corrosive to the gas piping, and if burned in gas engines, the products of combustion can damage the engine and severely corrode exhaust-gas heat-recovery equipment, especially if allowed to cool below the dew point.

Toxic Compounds Because of their toxicity, certain cations are of great importance in the treatment and disposal of wastewaters. Copper, lead, silver, chromium, arsenic, and boron are toxic in varying degrees to microorganisms and therefore must be taken into consideration in the design of a biological treatment plant. Many plants have been upset by the introduction of these ions to the extent that the microorganisms were killed and treatment ceased. For instance, in sludge digesters, copper is toxic at a concentration of 100 mg/L, chromium and nickel are toxic at concentrations of 500 mg/L, and sodium is also toxic at high con-centrations [19]. Other toxic cations include potassium and ammonium at 4000 mg/L. The alkalinity present in the digesting sludge will combine with and precipitate the calcium ions before the calcium concentration approaches the toxic level.

Some toxic anions, including cyanides and chromates, are also present in industrial wastes. These are found particularly in metal-plating wastes and should be removed by pretreatment at the site of the factory rather than be mixed with the municipal wastewater. Fluoride is another toxic anion. Organic compounds present in some industrial wastes are also toxic.

Heavy metals Trace quantities of many metals, such as nickel (Ni), manganese (Mn), lead (Pb), chromium (Cr), cadmium (Cd), zinc (Zn), copper (Cu), iron (Fe), and mercury (Hg), are important constituents of most waters. Some of these metals are necessary for growth of biological life, and absence of sufficient quantities of them could limit growth of algae, for example. The presence of any of these metals in excessive quantities will interfere with many beneficial uses of the water because of their toxicity; therefore, it is frequently desirable to measure and control the concentrations of these substances.

Methods for determining the concentrations of these substances vary in complexity according to the interfering substances that may be present [38]. In addition, quantities of many of these metals can be determined at very low concentrations by such instrumental methods as polarography and atomic absorption spectroscopy. For a review of the effects of heavy metals on the environment, the reference work by McKee and Wolf [21] is recommended. Since research in this area is constantly underway, a review of the current literature is also recommended.

Gases

Gases commonly found in untreated wastewater include nitrogen (N_2), oxygen (O_2), carbon dioxide (CO_2), hydrogen sulfide (H_2S), ammonia (NH_3), and methane (CH_4). The first three are common gases of the atmosphere and will be found in all waters exposed to air. The latter three are derived from the decomposition of the organic matter present in wastewater. Although not found in untreated wastewater, other gases with which the sanitary engineer must be familiar include chlorine (Cl_2) and ozone (O_3) (disinfection and odor control), and the oxides of sulfur and nitrogen (combustion processes). The following discussion is limited to those gases that are of interest in untreated wastewater. Under most circumstances, the ammonia in untreated wastewater will be present as the ammonium ion (see "Nitrogen"). Therefore, it will be considered further in Chap. 12 rather than here.

Dissolved oxygen Dissolved oxygen is required for the respiration of aerobic microorganisms as well as all other aerobic life forms. However, oxygen is only slightly soluble in water. The actual quantity of oxygen (other gases too) that can be present in solution is governed by (1) the solubility of the gas, (2) the partial pressure of the gas in the atmosphere, (3) the temperature, and (4) the purity (salinity, suspended solids, etc.) of the water. The interrelationship of these variables is delineated in Chap. 9 and is illustrated in Appendix C, where the effect of temperature and salinity on dissolved oxygen concentration is presented.

Because the rate of biochemical reactions that use oxygen increases with increasing temperature, dissolved oxygen levels tend to be more critical in the summer months. The problem is compounded in summer months because stream flows are usually lower, and thus the total quantity of oxygen available in also lower. The presence of dissolved oxygen in wastewater is desirable because it prevents the formation of noxious odors. The role of oxygen in wastewater treatment is discussed in Chaps. 9 and 10; its importance in water-quality management is discussed in Chap. 14.

Hydrogen sulfide Hydrogen sulfide is formed, as mentioned previously, from the decomposition of organic matter containing sulfur or from the reduction of mineral sulfites and sulfates. It is not formed in the presence of an abundant supply of oxygen. This gas is a colorless, inflammable compound having the characteristic odor of rotten eggs. The blackening of wastewater and sludge usually results from the formation of hydrogen sulfide that has combined with the iron present to form ferrous sulfide (FeS). Although hydrogen sulfide is the most important gas formed from the standpoint of odors, other volatile compounds such as indol, skatol, and mercaptans, which may also be formed during anaerobic decomposition, may cause odors far more offensive than that of hydrogen sulfide.

Methane The principal by-product of the anaerobic decomposition of the organic matter in wastewater is methane gas (see Chap. 11). Methane is a colorless, odorless, combustible hydrocarbon of high fuel value. Normally, large quantities are not encountered in wastewater because even small amounts of oxygen tend to be toxic to the organisms responsible for the production of methane (see Chap. 9). Occasionally, however, as a result of anaerobic decay in accumulated bottom deposits, methane has been produced. Because methane is highly combustible and the explosion hazard is high, manholes and sewer junctions or junction chambers where there is an opportunity for gas to collect should be ventilated with a portable blower during and before the time required for men to work in them on inspection, renewals, or repairs. In treatment plants, notices should be posted about the plant warning of explosion hazards, and plant employees should be instructed in safety measures to be maintained while working in and about the structures where gas may be present.

3-9 BIOLOGICAL CHARACTERISTICS: DEFINITION AND APPLICATION

The sanitary engineer must have considerable knowledge of the biological characteristics of wastewater. He must know (1) the principal groups of micro-organisms found in surface water and wastewaters as well as those responsible for biological treatment, (2) the pathogenic organisms in wastewater, (3) the organisms used as indicators of pollution and their significance, and (4) the methods used to evaluate the toxicity of treated wastewaters. These matters are discussed in this section.

Microorganisms

The principal groups of organisms found in surface water and wastewater are classified as protista, plants, and animals (see Chap. 9). The category protista includes bacteria, fungi, protozoa, and algae. Seed plants, ferns, and mosses and liverworts are classified as plants. Invertebrates and vertebrates are classified as animals [39]. Viruses, which are also found in wastewater, are classified according to the host infected. Because the organisms in the various groups are discussed in detail in the subsequent chapters of this book, the following discussion is meant to serve only as a general introduction to the various groups and their importance in the field of wastewater treatment and water-quality management.

Protista As a class, protista are the most important group of organisms with which the sanitary engineer must be familiar, especially the bacteria, algae, and protozoa.

Because of the extensive and fundamental role played by bacteria in the decomposition and stabilization of organic matter, both in nature and in treatment plants, their characteristics, functions, metabolism, and synthesis must be understood. These subjects are discussed extensively in Chap. 9. Coliform bacteria are also used as an indicator of pollution by human wastes. Their significance and some of the tests used to determine their presence are discussed in a subsequent section.

Algae can be a great nuisance in surface waters because, when conditions are right, they will rapidly reproduce and cover streams, lakes, and reservoirs in large floating colonies called blooms. Algal blooms are usually characteristic of what is called a eutrophic lake, or a lake with a high content of the compounds needed for biological growth. Because effluent from wastewater-treatment plants is usually high in biological nutrients, discharge of the effluent to lakes causes enrichment and increases the rate of eutrophication. The same effects can also occur in streams.

The presence of algae affects the value of water for water supply because they often cause taste and odor problems. Algae can also alter the value of surface waters for the growth of certain kinds of fish and other aquatic life, for recreation, and for other beneficial uses. Determination of the concentration of algae in surface waters involves collecting the sample by one of several possible methods and microscopically counting them. Detailed procedures for algae counts are outlined in *Standard Methods* [38]. Pictures and descriptions of common algae may be found in Refs. 15, 27, 30, 36, and 48.

One of the most important problems facing the sanitary engineering profession in terms of water-quality management is how to treat wastes of various origins so that the effluents do not encourage the growth of algae and other aquatic plants. The solution may involve the removal of carbon, the removal of various forms of nitrogen and phosphorus, and possibly the removal of some of the trace elements, such as iron and cobalt.

Protozoa of importance to sanitary engineers include amoebas, flagellates, and free-swimming and stalked ciliates. These protists feed on bacteria and other microscopic protists and are essential in the operation of biological treatment processes and in the purification of streams because they maintain a natural balance among the different groups of microorganisms. Detailed descriptions of these organisms may be found in Refs. 15, 39, and 48.

Viruses The functioning of viruses with respect to other organisms is explained in Chap. 12. Viruses that are excreted by human beings may become a major hazard to public health. For example, from experimental studies, it has been found that from 10,000 to 100,000 infectious doses of hepatitis virus are emitted from each gram of feces of a patient ill with this disease [1]. It is known that some viruses will live as long as 41 days in water or wastewater at 20°C and for 6 days in a normal river. A number of outbreaks of infectious hepatitis have been attributed to transmission of the virus through water supplies. Much more study is required on the part of biologists and engineers to determine the mechanics of travel and removal of virus in soils, surface waters, and wastewater-treatment plants.

Plants and animals Plants and animals of importance range in size from microscopic rotifers and worms to macroscopic crustaceans. A knowledge of these organisms is helpful in evaluating the condition of streams and lakes, in determining the toxicity of wastewaters discharged to the environment, and in observing the effectiveness of biological life in the secondary treatment processes used to destroy organic wastes. Detailed descriptions of these organisms may be found in Refs. 15, 28, and 48.

Pathogenic Organisms

Pathogenic organisms found in wastewater may be discharged by human beings who are infected with disease or who are carriers of a particular disease. The usual bacterial pathogenic organisms that may be excreted by man cause diseases of the gastrointestinal tract, such as typhoid and paratyphoid fever, dysentery, diarrhea, and cholera. Because these organisms are highly infectious, they are responsible for many thousands of deaths each year in areas with poor sanitation, especially in the tropics [17].

Although bacterial pathogenic organisms are the most numerous, they are by no means the only pathogens in wastewater. Pathogenic organisms that are found in wastewater are reported in Table 3-15. Because the identification of pathogenic organisms in water and wastewater is both extremely time-consuming and difficult, the coliform group of organisms is now used as an indicator of the presence in wastewater of feces and hence pathogenic organisms.

Coliform Organisms

The intestinal tract of man contains countless rod-shaped bacteria known as coliform organisms. Each person discharges from 100 to 400 billion coliform organisms per day, in addition to other kinds of bacteria. Coliforms are harmless to man and are, in fact, useful in destroying organic matter in biological waste-treatment processes.

Because the numbers of pathogenic organisms present in wastes and polluted waters are few and difficult to isolate, the coliform organism, which is more

Table 3-15 Pathogenic organisms commonly found in wastewater [13]

Organism	Disease	Remarks
Ascaris spp., *Enterobius* spp.	Nematode worms	Danger to man from wastewater effluents and dried sludge used as fertilizer
Bacillus anthracis	Anthrax	Found in wastewater. Spores are resistant to treatment
Brucella spp.	Brucellosis. Malta fever in man. Contagious abortion in sheep, goats, and cattle	Normally transmitted by infected milk or by contact. Wastewater is also suspected
Entamoeba histolytica	Dysentery	Spread by contaminated waters and sludge used as fertilizer. Common in hot climates
Leptospira iceterohaemorrhagiae	Leptospirosis (Weil's disease)	Carried by sewer rats
Mycobacterium tuberculosis	Tuberculosis	Isolated from wastewater and polluted streams. Wastewater is a possible mode of transmission. Care must be taken with wastewater and sludge from sanatoriums
Salmonella paratyphi	Paratyphoid fever	Common in wastewater and effluents in times of epidemics
Salmonella typhi	Typhoid fever	Common in wastewater and effluents in times of epidemics
Salmonella spp.	Food poisoning	Common in wastewater and effluents
Schistosoma spp.	Schistosomiasis	Probably killed by efficient wastewater treatment
Shigella spp.	Bacillary dysentery	Polluted waters are main source of infection
Taenia spp.	Tapeworms	Eggs very resistant, present in wastewater sludge and wastewater effluents. Danger to cattle on land irrigated or land manured with sludge
Vibrio cholerae	Cholera	Transmitted by wastewater and polluted waters
Virus	Poliomyelitis, hepatitis	Exact mode of transmission not yet known. Found in effluents from biological wastewater-treatment plants

numerous and more easily tested for, is used as an indicator organism. The presence of coliform organisms is taken as an indication that pathogenic organisms may also be present, and the absence of coliform organisms is taken as an indication that the water is free from disease-producing organisms.

The coliform bacteria include the genera *Escherichia* and *Aerobacter*. The use of coliforms as indicator organisms is complicated by the fact that *Aerobacter* and certain *Escherichia* can grow in soil. Thus, the presence of coliforms does not always mean contamination with human wastes. Apparently, *Escherichia coli* (*E. coli*) are entirely of fecal origin. There is difficulty in determining *E. coli* to the exclusion of the soil coliforms; as a result, the entire coliform group is used as an indicator of fecal pollution. In recent years, tests have been developed that distinguish among total coliforms, fecal coliforms, and fecal streptococci; and all three are being reported in the literature. The use of the ratio of fecal coliforms to fecal streptococci is discussed later in this chapter.

The usual procedure for determining the presence of coliforms consists of the presumptive and the confirmed tests. The presumptive test is based on the ability of the coliform group to ferment lactose broth, producing gas. The confirmed test consists of growing cultures of coliform bacteria on media that suppress the growth of other organisms. The completed test is based on the ability of the cultures grown in the confirmed test to again ferment the lactose broth.

There now are two accepted methods for obtaining the numbers of coliform organisms present in a given volume of water. The most probable number (MPN) technique has been used for a long time and is based on a statistical analysis of the number of positive and negative results obtained when testing multiple portions of equal volume and in portions constituting a geometric series for the presence of coliform. It is emphasized that the MPN is not the absolute concentration of organisms that are present but only a statistical estimate of that concentration. Complete MPN tables are provided in Appendix D, and use of these tables is illustrated in Examples 3-6 and 3-7.

Example 3-6: Calculation of MPN Determine the coliform density (MPN) for a surface water, the bacterial analysis of which yielded the following results for the standard confirmed test.

Size of portion, mL	No. positive	No. negative
10.0	4	1
1.0	4	1
0.1	2	3
0.01	0	5

SOLUTION From Appendix E, eliminating the portion with no positive tubes, as outlined in *Standard Methods* [38], the MPN/100 mL is 47.

Example 3-7: MPN of wastewater overflow A sample of a combined-sewer overflow is tested for coliforms by the MPN method with the following results. Determine the coliform density.

Size of portion, mL	No. positive	No. negative
0.001	5	0
0.0001	5	0
0.00001	5	0
0.000001	3	2

SOLUTION Eliminate the top line, as selection of the top three lines leads to no solution. Using the bottom three lines, by reference to Appendix D, if there were 10, 1, and 0.1 mL of wastewater in the samples, the MPN would have been 920. Since the samples are 100,000 times more dilute, the MPN/100 mL of the original sample is 92,000,000 or 920,000/mL.

The membrane-filter technique can also be used to determine the number of coliform organisms that are present in water. The determination is accomplished by passing a known volume of water sample through a membrane filter that has a very small pore size. The bacteria are retained on the filter because they are larger than the pores. The bacteria are then contacted with an agar that contains nutrients necessary for the growth of the bacteria. After incubation, the coliform colonies can be counted and the concentration in the original water sample determined. The membrane-filter technique has the advantage of being faster than the MPN procedure and of giving a direct count of the number of coliforms. Both methods are subject to limitations, however. Detailed procedures are given for both methods in *Standard Methods* [38].

Ratio of Fecal Coliforms to Fecal Streptococci

It has been observed that the quantities of fecal coliforms and fecal streptococci that are discharged by human beings are significantly different from the quantities discharged by animals. Therefore, it has been suggested that the ratio of the fecal coliform (FC) count to the fecal streptococci (FS) count in a sample can be used to show whether the suspected contamination derives from human or from animal wastes. Typical data on the ratio of FC to FS counts for human beings and various animals are reported in Table 3-16. The FC/FS ratio for domestic animals is less than 1.0, whereas the ratio for human beings is more than 4.0.

If ratios are obtained in the range of 1 to 2, interpretation is uncertain. If the sample is collected near the suspected source of pollution, the most likely interpretation is that the pollution derives equally from human and animal sources. The foregoing interpretations are subject to the following constraints [17]:

1. The sample pH should be between 4 and 9 to exclude any adverse effects of pH on either group of microorganisms.
2. At least two counts should be made on each sample.
3. To minimize errors due to differential death rates, samples should not be

Table 3-16 Estimated per capita contribution of indicator micro-organisms from human beings and some animals [17]

Animal	Average indicator density/g of feces		Average contribution/ capita·24 h		Ratio FC/FS
	Fecal coliform 10^6	Fecal streptococci 10^6	Fecal coliform 10^6	Fecal streptococci 10^6	
Chicken	1.3	3.4	240	620	0.4
Cow	0.23	1.3	5,400	31,000	0.2
Duck	33.0	54.0	11,000	18,000	0.6
Human	13.0	3.0	2,000	450	4.4
Pig	3.3	84.0	8,900	230,000	0.04
Sheep	16.0	38.0	18,000	43,000	0.4
Turkey	0.29	2.8	130	1,300	0.1

Note: g × 0.0022 = lb.

taken farther downstream than 24 hours of flow time from the suspected source of pollution.

4. Only the fecal coliform count obtained at 44°C is to be used to compute the ratio.

Use of the FC/FS ratio can be very helpful in establishing the source of pollution in rainfall-runoff studies and in pollution studies conducted in rural areas, especially where septic tanks are used. In many situations where human pollution is suspected on the basis of coliform test results, the actual pollution may, in fact, be caused by animal discharges. Establishing the source of pollution can be very important, especially where it is proposed or implied that the implementation of conventional wastewater-management facilities will eliminate the measured coliform values.

Bioassay Tests

The results of bioassay tests are used to evaluate the toxicity of wastewaters to the biological life of the receiving waters. The specific objectives of the bioassay test are (1) to determine the concentration of a given waste that will kill 50 percent of the test organisms in a specified time period and (2) to determine the maximum concentration causing no apparent effect on the test organisms in 96 hours. These objectives are achieved by introducing fish or other appropriate test animals into test aquariums (see Fig. 3-19) containing various concentrations of the waste in question and observing their survival with time. Observations are usually made after 24, 48, and 96 hours. General test procedures, the evaluation of test results, and the application of test results are described in the following discussion.

Figure 3-19 Laboratory setup for the conduct of static and flow-through bioassays. (*From J. E. Colt.*)

Test procedures The procedures to be followed in the conduct of routine bioassay tests have been summarized and reported in *Standard Methods* [38]. The routine test is applied widely for detecting and evaluating acute toxicity that is due to substances that are relatively stable and not extremely volatile. This toxicity is not associated with excessive oxygen demand. Death of the test organisms due to a deficiency of dissolved oxygen in polluted water should be distinguished from death due to toxicity. To detect and evaluate the direct lethality of the wastes only, adequate dissolved oxygen must be maintained during the toxicity tests.

When it is suspected that the toxicity of the test solutions declines rapidly during the course of a test, a modification of the routine procedure is recommended. The reduction of the toxicity of the solution may result from the reduction or removal of toxic components. The component reduction may be due to extreme volatility by oxidation, by hydrolysis, by precipitation, by their combination with the metabolic by-products of the test fish, or other reasons.

The validity of the test depends partly on the selection of the proper species and partly on the characteristics of the receiving water where the waste is discharged. For tests relating to estuarine pollution, species such as sticklebacks, killifish, or mosquitofish (*Gambusia*) are suitable for test animals because of their tolerance to a wide variation in salinity, their abundance in many coastal waters, and their size. For freshwater tests, species such as sticklebacks, mosquitofish, minnows, trout, and sunfish have been used successfully.

For details on the size of the test aquariums, temperature, maintenance, selection of test specimens, and other related matters, *Standard Methods* [38] should be consulted.

Evaluation of results The prescribed measure of acute toxicity is the median tolerance limit (TL_m), defined as the concentration of waste (toxicant) in which just 50 percent of the test animals are able to survive for a specified period of exposure. The method of calculating TL_m values is illustrated in Example 3-8. The TL_m values are only estimates of the acute toxicity for the waste under laboratory conditions. The values obtained do not represent the concentrations of the tested waste that may be considered harmless to the diverse aquatic biota. Waste concentrations that are not toxic to selected test organisms in a 96-h period may be very toxic to these same animals or other economically or ecologically important species under conditions of continuous or chronic exposure. A more meaningful approach for evaluating the results of bioassay tests, based on a consideration of threshold toxicity, has been proposed by Chen and Selleck [5].

Example 3-8:. Analysis of bioassay data Using the following hypothetical data, determine the 24- and 48-h TL_m values in percent by volume.

Concentration of waste, % by vol	No. of test animals	No. of test animals surviving	
		After 24 h	After 48 h
40	20	1	0
20	20	8	0
10	20	14	6
5	20	20	13
3	20	20	16

SOLUTION
1 Plot the concentration of waste in percent by volume against test animals surviving in percent.
2 Connect the data points on either side of 50 percent survival for 24 and 48 h.
3 Find the waste concentration for 50 percent survival.
4 The TL_m values, as shown in Fig. 3-20, are 16.0 percent for 24 h and 6.7 percent for 48 h.

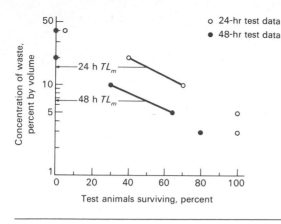

Figure 3-20 Graphical presentation and analysis of bioassay test data for Example 3-8.

Application of results When estimating the permissible waste-dilution ratios on the basis of acute toxicity bioassay test results, some application factor must be used. The application factor is defined as the concentration of the material or waste that is not harmful under long-term or continuous exposure, divided by the 96-h TL_m value for that waste.

In the absence of toxicity data other than the 96-h TL_m, the National Technical Advisory Committee on Water Quality Criteria has recommended application factors varying from 1/10 to 1/100 of the 96-h TL_m, depending on the nature and characteristics of the waste [50]. In some cases, this recommendation is qualified to include a requirement that it be demonstrated that the safe levels prescribed on the basis of the 96-h TL_m and the application factor do not cause decreases in productivity or diversity of the receiving-water biota [50].

DISCUSSION TOPICS AND PROBLEMS

3-1 In a BOD determination, 6 mL of wastewater are mixed with 294 mL of diluting water containing 8.6 mg/L of dissolved oxygen. After a 5-d incubation at 20°C, the dissolved oxygen content of the mixture is 5.4 mg/L. Calculate the BOD of the wastewater. Assume that the initial dissolved oxygen of wastewater is zero.

3-2 The BOD_5 of a waste sample was found to be 40.0 mg/L. The initial oxygen concentration of the BOD dilution water was equal to 9 mg/L, the DO concentration measured after incubation was equal to 2.74 mg/L and the size of sample used was equal to 40 ml. If the volume of the BOD bottle used was equal to 300 ml, estimate the initial DO concentration in the waste sample.

3-3 The 5-day 20°C BOD of a wastewater is 210 mg/L. What will be the ultimate BOD? What will be the 10-day demand? If the bottle had been incubated at 30°C, what would the 5-d BOD have been? $K' = 0.23 d^{-1}$.

3-4 The 5-d BOD at 20°C is equal to 250 mg/L for three different samples, but the 20°C K values are equal to 0.12 d^{-1}, 0.16 d^{-1}, and 0.20 d^{-1}. Determine the ultimate BOD of each sample.

3-5 The BOD value of a wastewater was measured at 2 and 8 d and found to be 125 mg/1 and 225 mg/l, respectively. Determine the 5-d value using the first-order rate model.

3-6 The following BOD results were obtained on a sample of untreated wastewater at 20°C:

t, d	0	1	2	3	4	5
y, mg/L	0	65	109	138	158	172

Compute the reaction constant K and ultimate first-stage BOD using both the least-squares and the Thomas methods.

3-7 Given the following results determined for a wastewater sample at 20°C, determine the ultimate carbonaceous oxygen demand, the ultimate nitrogenous oxygen demand (NOD), the carbonaceous BOD reaction-rate constant (K), and the nitrogenous NOD reaction-rate constant (K_n). Determine K ($\theta = 1.05$) and K_n ($\theta = 1.08$) at 25°C. (Courtesy of Edward Foree.)

Time, d	BOD, mg/L	Time, d	BOD, mg/L
0	0	11	63
1	10	12	69
2	18	13	74
3	23	14	77
4	26	16	82
5	29	18	85
6	31	20	87
7	32	25	89
8	33	30	90
9	46	40	90
10	56		

3-8 Compute the carbonaceous and nitrogenous oxygen demand of a waste represented by the formula $C_9N_2H_6O_2$. (N is converted to NH_3 in the first step.)

3-9 Determine the carbonaceous and nitrogenous oxygen demand in mg/L for a 1-L solution containing 300 mg of acetic acid (CH_3COOH) and 300 mg of glycine ($CH_2(NH_2)COOH$).

3-10 The following data have been obtained from a waste characterization:

$BOD_5 = 400$ mg/L
$K_1 = 0.20$ d^{-1}
$NH_3 = 80$ mg/L

Estimate the total quantity of oxygen in mg/1 that must be furnished to completely stabilize this wastewater. What is the COD and the ThOD for this waste?

3-11 An industrial wastewater is known to contain only stearic acid ($C_{18}H_{36}O_2$), glycine ($C_2H_5O_2N$), and glucose ($C_6H_{12}O_6$). The results of a laboratory analysis are as follows: organic nitrogen = 10 mg/L, organic carbon = 125 mg/L, and COD = 400 mg/1. Determine the concentration of each of the three constituents in mg/L. (*Courtesy of Edward Foree.*)

3-12 The ultimate oxygen demand can be estimated using the following expression: Ultimate oxygen demand (UOD) = 2.67 × organic carbon, mg/L + 4.57 × (organic nitrogen as N + ammonia nitrogen as N), mg/L + 1.14 × nitrite as N, mg/L. Estimate the ultimate oxygen demand for medium-strength wastewater (see Table 3-5).

3-13 The dissolved oxygen of a tidal estuary must be maintained at 4.5 mg/L or more. The average temperature of the water during summer months is 24°C and the chloride concentration is 5000 mg/L. What percent saturation does this represent?

3-14 The variations of flow and BOD with time at a treatment plant are given in Fig. 3-21. Compute both the average and the weighted average BOD.

3-15 If a sample of wastewater from a city with a population of 100,000 and an average flow of 400 L/capita·d contains 5 mL/L settleable solids, how many cubic meters of sludge will be produced per day?

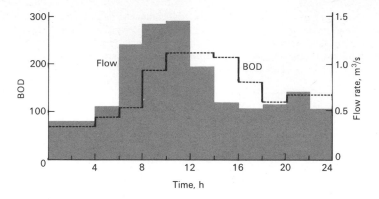

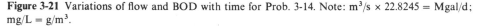

Figure 3-21 Variations of flow and BOD with time for Prob. 3-14. Note: $m^3/s \times 22.8245 = Mgal/d$; $mg/L = g/m^3$.

3-16 An industry currently discharges 20,000 kg of BOD_5 and 15,000 kg of suspended solids per day. If the unit waste loading factors listed in Table 3.9 for domestic wastewater with complete grinding of food waste are used, determine the population equivalent for each parameter in the industrial waste discharge. (Note: Many cities base their charges for the treatment of industrial wastes on the basis of population equivalence.)

3-17 Using the data shown in Fig. 3.5, develop a curve of the ratio of the sustained peak-to-average BOD loadings similar to that shown in Fig. 3.6. Develop the curve for a 24-month period using 1-month time increments.

3-18 The local Air Pollution Control District has threatened to fine and penalize the local wastewater-management agency, your client, because of frequently recurring odor complaints from residents who live downwind of the plant. The plant manager, a full-time employee at the treatment plant, claims that no problem exists. He proves his point by consistently finding less than 5 dilutions to MDTOC at the plant boundary using a Barneyby-Cheny hand-held sniff dilution olfactometer as employed by the local Air Pollution Control District. You, however, live downwind of the plant and have frequently detected odors from it. Why do these differences exist? How would you resolve them objectively?

3-19 You have been asked to review an odor-control system that has apparently failed to control odors from a sludge-dewatering building adequately. The wastewater-management agency, your client, claims the system has failed to perform according to specifications. The engineering contractor who installed the system claims that the specifications were not adequate.

In your investigation you find that the agency employed a reputable odor consultant to develop the odor-control-system specifications. The consultant used the ASTM Panel Method for odor measurement, using evacuated glass cylinders for sample collection. Several measurements were made, and the maximum observed value was doubled to develop the control-system specifications. In this way a 90 percent odor-removal requirement was established to meet the desired final odor-emission limit of 2.8×10^4 odor units per minute (the product of airflow in m^3/min and number of dilutions to MDTOC).

Using a direct-reading olfactometer, you find that the control system removes 99 percent of the odor, and that at a rate of 10^6 odor units per minute the final odor emission is 10^6 odor units per minute. What reasons might explain your findings? How would you resolve the problem?

3-20 You have been asked by a wastewater-management agency to review the adequacy of their odor-control program. What would be your major considerations in making such a review?

3-21 The results of a presumptive coliform test were 4 of 5 tubes of 10-mL portions positive, 3 of 5 tubes of 1-mL portions positive, and 0 of 5 tubes of 0.1-mL portions positive. What is the MPN per 100 mL?

3-22 Four weekly effluent samples have been analyzed for bacterial content using the standard confirmed test. Determine the coliform density, expressed as MPN, for the following test results. Check the answers obtained using the standard MPN tables with the approximate formula proposed by Thomas. (Refer to Ref. 38 for the procedure used to select the appropriate dilutions to be used in determining the MPN value and for the Thomas formula.)

Size of portion, mL	Sample			
	1	2	3	4
100			5/5	5/5
10		4/5	4/5	5/5
1	4/5	5/5	5/5	5/5
0.1	3/5	3/5	3/5	2/5
0.01	1/5	2/5	2/5	3/5

3-23 Discuss the advantages and disadvantages of using the fecal coliform and fecal streptococci tests to indicate bacteriological pollution.

3-24 The following bioassay data were obtained for a treatment effluent. Estimate the 48-h TL_m value.

Concentration of waste, % by volume	No. of test animals	No. of test animals surviving		
		After 24 h	After 60 h	After 96 h
12	20	8	2	0
10	20	10	5	0
8	20	13	8	0
6	20	16	11	0
4	20	20	16	5
2	20	20	20	14

REFERENCES

1. Berg, G.: *Transmission of Viruses by the Water Route*, Wiley, New York, 1965.
2. Black, H. H.: Procedures for Sampling and Measuring Industrial Wastes, *Sewage Ind. Wastes*, vol. 25, p. 45, 1952.
3. Brock, T. D.: *Microbial Ecology*, Prentice-Hall, Englewood Cliffs, N.J., 1966.
4. Burdoin, A. J., and D. F. Seidel: Sulfate In-Digestion, *Calif. Water Pollut. Control Assoc. Bull.*, vol. 4, no. 3, 1968.
5. Chen, C. W., and R. E. Selleck: A Kinetic Model of Fish Toxicity Threshold, *J. Water Pollut. Control Fed.*, vol. 41, no. 8, part 2, 1969.
6. City of Modesto: *Annual Operating Data for Sewage Treatment Plant*, Modesto, Calif., 1970–1972.
7. Clark, J. W.: New Method for Biochemical Oxygen Demand, *N.M. State Univ. Eng. Exp. Sta. Bull.* 11, 1959.
8. Clifford, D.: *Total Oxygen Demand—A New Instrumental Method*, American Chemical Society, Midland, Mich., 1967.
9. Dravnieks, A.: Measuring Industrial Odors, *Chem. Eng. Deskbook Issue*, vol. 81, no. 22, 1974.
10. *FWPCA Methods for Chemical Analysis of Water and Wastes*, U.S. Department of the Interior, Federal Water Pollution Control Administration, 1969.

11. *Handbook for Monitoring Industrial Wastewater*, U.S. Environmental Protection Agency, Technology Transfer, 1973.
12. Hawkes, H. A.: *The Ecology of Waste Water Treatment*, Pergamon, London, 1963.
13. Hawkes, H. A.: *Microbial Aspects of Pollution*, Academic, London, 1971.
14. Huey, N. A.: *Ambient Odor Evaluation*, presented at the 61st Annual Meeting of the Air Pollution Control Association, St. Paul, Minn., 1968.
15. Hutchinson, G. E.: *A Treatise on Limnology*, vol. II, Wiley, New York, 1967.
16. Klein, S. A., and P. H. McGauhey: Degradation of Biologically Soft Detergents by Wastewater Treatment Processes, *J. Water Pollut. Control Fed.*, vol. 37, no. 6, 1965.
17. Mara, D. D.: *Bacteriology for Sanitary Engineers*, Churchill Livingston, Edinburgh, 1974.
18. Mara, D. D.: *Sewage Treatment in Hot Climates*, Wiley-Interscience, London, 1976.
19. McCarty, P. L.: Anaerobic Waste Treatment Fundamentals, *Public Works*, vol. 95, no. 11, 1964.
20. McCord, C. P., and W. M. Witheridge: *Odors: Physiology and Control*, McGraw-Hill, New York, 1949.
21. McKee, J. E., and H. W. Wolf: *Water Quality Criteria*, 2d ed., Report to California State Water Quality Control Board, Publication 3A, 1963.
22. *Manual on Water*, American Society for Testing and Materials, 1969.
23. Metcalf & Eddy, Inc.: *Report to National Commission on Water Quality on Assessment of Technologies and Costs for Publicly Owned Treatment Works*, vol. 2 prepared under Public Law 92-500, Boston, 1975.
24. *Metropolitan Sanitary District of Greater Chicago: Annual Operating Data for Calumet Sewage Treatment Works*, Chicago, 1969–1973.
25. Moncrieff, R. W.: *The Chemical Senses*, 3d ed., Leonard Hill, London, 1967.
26. Moore, E. W., H. A. Thomas, and W. B. Snow: Simplified Method for Analysis of BOD Data, *Sewage Ind. Wastes*, vol. 22, no. 10, 1950.
27. Palmer, E. M.: *Algae in Water Supplies*, U.S. Public Health Service, Pub. 657, Washington, D.C., 1959.
28. Pennak, R. W.: *Fresh-Water Invertebrates of the United States*, Ronald, New York, 1953.
29. Phelps, E. B.: *Stream Sanitation*, Wiley, New York, 1944.
30. Prescott, G. W.: *The Freshwater Algae*, 2d ed., Brown Company, Dubuque, Iowa, 1970.
31. Sawyer, C. N., and L. Bradney: Modernization of the BOD Test for Determining the Efficiency of Sewage Treatment Processes, *Sewage Works J.*, vol. 18, no. 6, 1946.
32. Sawyer, C. N., and P. L. McCarty: *Chemistry for Environmental Engineering*, 3d ed., McGraw-Hill, New York, 1978.
33. Schroeder, E. D.: *Water and Wastewater Treatment*, McGraw-Hill, New York, 1977.
34. Schroepfer, G. J., M. L. Robins, and R. H. Susag: The Research Program on the Mississippi River in the Vicinity of Minneapolis and St. Paul, *Advances in Water Pollution Research*, vol. 1, Pergamon, London, 1964.
35. Sheehy, J. P.: Rapid Methods for Solving Monomolecular Equations, *J. Water Pollut. Control Fed.*, vol. 32, no. 6, 1960.
36. Smith, G. M.: *The Fresh-Water Algae of the United States*, McGraw-Hill, New York, 1950.
37. Speidel, H. K., and G. Tchobanoglous: *Summary Report on Wastewater Characteristics*, prepared for the city of Flagstaff, Flagstaff, Ariz., 1976.
38. *Standard Methods for the Examination of Water and Waste Water*, 14th ed., American Public Health Association, 1975.
39. Stanier, R. Y., J. L. Ingraham, and E. A. Adelberg: *The Microbial World*, 4th ed., Prentice-Hall, Englewood Cliffs, N.J., 1976.
40. Stumm, W., and J. J. Morgan: *Aquatic Chemistry*, Wiley-Interscience, New York, 1970.
41. Sullivan, R. J.: *Preliminary Air Pollution Survey on Odorous Compounds, A Literature Review*, NAPCA Pub. APTD 66-24, 1969.
42. Tchobanoglous, G., and R. Eliassen: The Indirect Cycle of Water Reuse, *Water Wastes Eng.*, vol. 6, no. 2, 1969.
43. Theroux, F. R., E. F. Eldridge, and W. L. Mallmann: *Analysis of Water and Sewage*, McGraw-Hill, New York, 1943.

44. Thomas, H. A., Jr.: Graphical Determination of BOD Curve Constants, *Water & Sewage Works*, vol. 97, p. 123, 1950.
45. Tsivoglou, E. C.: *Oxygen Relationships in Streams*, Robert A. Taft Sanitary Engineering Center, Technical Report W-58-2, 1958.
46. *Upgrading Existing Wastewater Treatment Plants*, U.S. Environmental Protection Agency, Technology Transfer, 1974.
47. U.S. Public Health Service: *Public Health Service Drinking Water Standards*, Washington, D.C., 1962.
48. Ward, H. B., and G. C. Whipple (ed. by W. T. Edmondson): *Fresh Water Biology*, 2d ed., Wiley, New York, 1959.
49. *Water Quality Criteria 1972*, National Academy of Science, Ecological Research Series, U.S. Environmental Protection Agency, Report R3-73-033, Washington, D.C., 1973.
50. *Water Quality Criteria*, National Technical Advisory Committee, Federal Water Pollution Control Administration, Washington, D.C., 1968.
51. Waugh, A. E.: *Elements of Statistical Method*, 2d ed., McGraw-Hill, New York, 1943.
52. Wilson, G.: *Odors, Their Detection and Measurement*, EUTEK, Process Development and Engineering, Sacramento, 1975.
53. Wood, D. K., and G. Tchobanoglous: Trace Elements in Biological Waste Treatment, *J. Water Pollut. Control Assoc.*, vol. 47, no. 7, 1975.
54. Young, H. D.: *Statistical Treatment of Experimental Data*, McGraw-Hill, New York, 1962.
55. Young, J. C.: Chemical Methods for Nitrification Control, *J. Water Pollut. Control Assoc.*, vol. 45, no. 4, 1973.
56. Young, J. C., W. Garner, and J. W. Clark: An Improved Apparatus for Biochemical Oxygen Demand, *Anal. Chem.*, vol. 37, p. 784, 1965.
57. Young, J. C., and E. R. Baumann: The Electrolytic Respirometer—I: Factors Affecting Oxygen Uptake Measurements, *Water Res.*, vol. 10, no. 11, 1976.
58. Young, J. C., and E. R. Baumann: The Electrolytic Respirometer—II: Use in Water Pollution Control Plant Laboratories, *Water Res.*, vol. 10, no. 12, 1976.

CHAPTER FOUR

WASTEWATER-TREATMENT OBJECTIVES, METHODS, AND DESIGN

Since the early 1900s, when the field of environmental engineering was in its infancy in the United States, there has been a steady evolution and development in the methods used for wastewater treatment. Descriptions of the many methods and variations that have been tried to date would fill several large volumes. The approach followed in this text is to identify and discuss basic principles and their application to wastewater treatment.

This introductory chapter is intended to provide perspective and to illustrate how the subject matter to be presented in the following chapters fits into the overall scheme of the design, construction, operation and maintenance, and implementation of wastewater-treatment facilities. The following topics are covered: (1) wastewater-treatment objectives and regulations, (2) methods and concepts of wastewater treatment and disposal, (3) elements of plant analysis and design, and (4) other important considerations.

4-1 WASTEWATER-TREATMENT OBJECTIVES AND REGULATIONS

As noted in Chap. 1, intensive methods of wastewater treatment were first developed in response to the concern for public health and the adverse conditions caused by the discharge of wastewater to the environment. Also important, as cities became larger, was the limited availability of land required for waste-water disposal by irrigation. The purpose of treatment was to accelerate the forces of nature under controlled conditions in treatment facilities of comparatively small size. In general, early treatment objectives were concerned with (1) the removal of suspended and floatable material, (2) the treatment of biodegradable organics, and (3) the elimination of pathogenic organisms. Unfortunately, these objectives were not uniformly met throughout the United States, as is evidenced by the many plants that were discharging partially treated wastewater well into the 1960s.

Since the late 1960s, however, a major effort has been undertaken by both state and federal agencies to achieve more effective and widespread treatment of wastewater. This effort has resulted in part from: (1) increased understanding of the environmental effects caused by the discharge of untreated or partially treated wastewater; (2) developing knowledge of the adverse long-term effects caused by the discharge of some of the specific constituents found in wastewater; (3) the development of national concern for environmental protection; (4) increased scientific knowledge, especially in the areas of chemistry, biochemistry, and microbiology; (5) the need to conserve natural resources and, in many locations, to reuse wastewater; and (6) ever-expanding fields of knowledge of the basic principles and capabilities of the various methods used for wastewater treatment. As a consequence, while the early treatment objectives remain valid today, the required degree of treatment has significantly increased, and additional treatment objectives and goals have been added. The removal of nitrogen, phosphorus, and toxic organic compounds are examples of recent treatment objectives that have been established in selected localities. Other contaminants of concern, identified in Table 3-2, include refractory organics, heavy metals, and dissolved inorganic solids.

A significant event in the field of wastewater management was the passage of the Federal Water Pollution Control Act Amendments of 1972 (Public Law 92-500). Before that date, there were no specific national water-pollution-control goals or objectives. Most regulatory criteria and standards were established by state and local agencies, and they varied significantly from state to state. Public Law 92-500 not only established national goals and objectives ("to restore and maintain the chemical, physical, and biological integrity of the Nation's waters") but also marked a change in water-pollution-control philosophy. No longer was the classification of the receiving stream of ultimate importance, as it had been before. Public Law 92-500 decreed that the quality of the nation's waters is to be improved by the imposition of specific effluent limitations.

In August 1973, the U.S. Environmental Protection Agency published its definition of "secondary treatment" pursuant to Section 304(d) of Public Law 92-500 [1]. This definition, as reported in Table 4-1, includes three major effluent parameters: 5-day BOD, suspended solids, and pH. (A coliform standard, included in the original definition, was deleted on July 26, 1976.) Special interpretations of this definition are permitted for publicly owned treatment works served by combined-sewer systems and for those receiving industrial flows [4]. In addition, to meet the intent of Public Law 92-500, a variety of other guidelines and regulations have been issued.

Table 4-1 Definition of secondary treatment, as promulgated by the U.S. Environmental Protection Agency [1]

Characteristic of discharge	Unit of measurement	Average monthly concentration	Average weekly concentration
BOD_5	mg/L	$30^{a, b}$	45^b
Suspended solids[c]	mg/L	$30^{a, b}$	45^b
Hydrogen-ion concentration	pH units		$6.0–9.0^d$

[a] Or, in no case more than 15 percent of influent value.
[b] Arithmetic mean.
[c] Treatment plants with stabilization ponds and flows $<7,570$ m^3/d (2 Mgal/d) are exempt.
[d] Continuous, only enforced if caused by industrial wastewater or in-plant treatment.

Note: mg/L = g/m^3.

All these guidelines and regulations affect the design of wastewater treatment and disposal facilities and must be adhered to if federal funding is to be secured. Therefore, the practicing engineer must be thoroughly familiar with them and their interpretation. Because they are being revised and changed continuously, however, additional details are not included in this text.

4-2 CLASSIFICATION AND APPLICATION OF WASTEWATER-TREATMENT METHODS

After the treatment objectives have been established for a specific project and the applicable state and federal regulations have been reviewed, the degree of treatment can be determined by comparing the influent-wastewater characteristics to the required effluent-wastewater characteristics. Then, a number of different treatment and disposal or reuse alternatives can be developed and evaluated, and the optimum combination can be selected. It will therefore be helpful at this point to review the classification of the methods used for wastewater treatment (mentioned briefly in Chap. 1) and to consider the application of these methods in achieving treatment objectives.

Classification of Treatment Methods

The contaminants in wastewater are removed by physical, chemical, and biological means. The individual methods usually are classified as physical unit operations, chemical unit processes, and biological unit processes. Although these operations and processes occur in a variety of combinations in treatment systems, it has been found advantageous to study their scientific basis separately because the principles involved do not change.

Physical unit operations Treatment methods in which the application of physical forces predominate are known as physical unit operations. Because most of these methods evolved directly from man's first observations of nature, they were the first to be used for wastewater treatment. Screening, mixing, flocculation, sedimentation, flotation, and filtration are typical unit operations. These methods are considered in detail in Chap. 6, and their application is discussed in Chap. 8.

Chemical unit processes Treatment methods in which the removal or conversion of contaminants is brought about by the addition of chemicals or by other chemical reactions are known as chemical unit processes. Precipitation, gas transfer, adsorption, and disinfection are the most common examples used in wastewater treatment. In chemical precipitation, treatment is accomplished by producing a chemical precipitate that will settle. In most cases, the settled precipitate will contain both the constituents that may have reacted with the added chemicals and the constituents that were swept out of the wastewater as the precipitate settled. Adsorption involves the removal of specific compounds from the wastewater on solid surfaces using the forces of attraction between bodies. Chemical unit processes are considered in detail from a theoretical standpoint in Chap. 7, and their application is also discussed in Chap. 8.

Biological unit processes Treatment methods in which the removal of contaminants is brought about by biological activity are known as biological unit processes. Biological treatment is used primarily to remove the biodegradable organic substances (colloidal or dissolved) in wastewater. Basically, these substances are converted into gases that can escape to the atmosphere and into biological cell tissue that can be removed by settling. Biological treatment is also used to remove the nitrogen in wastewater. With proper environmental control, wastewater can be treated biologically in most cases. Therefore, it is the responsibility of the engineer to ensure that the proper environment is produced and effectively controlled. The factors involved and the treatment kinetics are discussed in Chap. 9, and their application is discussed in Chap. 10.

Application of Treatment Methods

The principal methods now used for the treatment of wastewater and sludge are identified in this section. Detailed descriptions of each method are not presented because the purpose here is only to introduce the many different ways in which treatment can be accomplished [2,5]. The detailed descriptions are presented throughout the remainder of this book.

Wastewater processing It was noted in Chap. 1 that unit operations and processes are grouped together to provide what is known as primary, secondary, and tertiary (or advanced) treatment. The term primary refers to physical unit operations; secondary refers to chemical and biological unit processes; and tertiary refers to combinations of all three. It should be noted that these terms

Table 4-2 Unit operations and processes and treatment systems used to remove the major contaminants found in wastewater[a]

Contaminant	Unit operation, unit process, or treatment system	See chapter
Suspended solids	Sedimentation	6, 8
	Screening and comminution	6, 8
	Filtration variations	6, 8
	Flotation	6, 8
	Chemical-polymer addition	7, 8
	Coagulation/sedimentation	7, 8
	Land treatment systems	13
Biodegradable organics	Activated-sludge variations	9, 10
	Fixed-film: trickling filters	9, 10
	Fixed-film: rotating biological contactors	9, 10
	Lagoon variations	9, 10
	Intermittent sand filtration	6, 10
	Land treatment systems	13
	Physical-chemical systems	7, 8
Pathogens	Chlorination	7, 8
	Hypochlorination	7, 8
	Ozonation	7, 8
	Land treatment systems	13
Nutrients:		
Nitrogen	Suspended-growth nitrification and denitrification variations	12
	Fixed-film nitrification and denitrification variations	12
	Ammonia stripping	12
	Ion exchange	12
	Breakpoint chlorination	7, 12
	Land treatment systems	13
Phosphorus	Metal-salt addition	7, 8
	Lime coagulation/sedimentation	7, 8
	Biological-chemical phosphorus removal	7, 8
	Land treatment systems	13
Refractory organics	Carbon adsorption	12
	Tertiary ozonation	12
	Land treatment systems	13
Heavy metals	Chemical precipitation	7, 8
	Ion exchange	12
	Land treatment systems	13
Dissolved inorganic solids	Ion exchange	12
	Reverse osmosis	12
	Electrodialysis	12

[a] Adapted from Ref. 2.

Table 4-3 Sludge processing and disposal methods[a]

Processing disposal function	Unit operation, unit process, or treatment method	See chapter
Preliminary operations	Sludge pumping and grinding	11
	Sludge blending and storage	11
Thickening	Gravity thickening	6, 11
	Flotation thickening	6, 11
	Centrifugation	11
	Classification	11
Stabilization	Chlorine oxidation	7, 11
	Lime stabilization	11
	Anaerobic digestion	9, 11
	Aerobic digestion	11
	Pure-oxygen aerobic digestion	11
	Heat treatment	11
Disinfection	Disinfection	7, 11
Conditioning	Chemical conditioning	11
	Elutriation	11
Dewatering	Centrifuge	11
	Vacuum filter	11
	Pressure filter	11
	Horizontal-belt filter	11
	Drying bed	11
	Lagoon	11
Drying	Dryer	11
Composting	Composting	11
	Co-composting	11
Thermal reduction	Multiple hearth incineration	11
	Fluidized-bed incineration	11
	Flash combustion	11
	Co-incineration	11
	Co-pyrolysis	11
	Pyrolysis	11
	Wet-air oxidation	11
	Recalcination	11
Ultimate disposal	Landfill	11
	Land application	11, 13
Reuse		14

[a] Adapted in part from Ref. 2.

are arbitrary and in most cases of little value. A more rational approach is first to establish the degree of contaminant removal (treatment) required before the wastewater can be reused or discharged to the environment. The required operations and processes necessary to achieve that required degree of treatment can then be grouped together on the basis of fundamental considerations.

The contaminants of major interest in wastewater and the unit operations and processes or methods applicable to the removal of these contaminants are shown in Table 4-2. Secondary treatment, as defined by the U.S. Environmental Protection Agency, is directed principally toward the removal of biodegradable organics and suspended solids. In general, the methods used to remove these contaminants are well established; they are less costly than those used for the removal of some of the other contaminants of concern.

To further protect the environment in some critical areas, more stringent standards have recently been directed toward the removal of nutrients and toward achieving lower levels of oxygen demand than are now possible with secondary-treatment techniques. When wastewaters are to be reused, standards may include removal requirements for refractory organics, heavy metals, and dissolved inorganic solids. In general, processes used to remove these constituents are not as well documented as secondary-treatment processes, and they are more costly. In most situations, the complexity of the treatment-process flowsheet will depend both on which constituents need to be removed and the required levels of removal.

Sludge processing For the most part, the methods and systems reported in Table 4-2 are used to treat the liquid portion of the wastewater. Of equal if not of more importance in the overall design of treatment facilities are the corresponding unit operations and processes or systems used to process the sludge removed from the liquid portion of the wastewater. The principal methods now in use are reported in Table 4-3. Because the processing and treatment of sludge has become so specialized, Chap. 11 is devoted entirely to this subject.

4-3 ELEMENTS OF PLANT ANALYSIS AND DESIGN

Treatment-plant design is one of the most challenging aspects of environmental engineering. Both theoretical knowledge and practical experience are necessary in the selection and analysis of the process flowsheet. Practical experience is especially important in the design and layout of the physical facilities and appurtenances and in the preparation of plans and specifications. The detailed aspects of process analysis are considered in Chap. 5. The purpose of this section is to describe what is involved in the preparation of treatment-process flowsheets, summary tables of design criteria, solids balances, hydraulic profiles, and plant layouts; how treatment-plant designs are synthesized; and the key steps in their implementation. The important terms are defined as follows:

Flowsheet. A flowsheet is the graphical representation of a particular combination of unit operations and processes used to achieve specific treatment objectives.

Process loading criteria. The process loading (or design) criteria are the key criteria used as the basis for sizing the individual unit operations and processes.

Solids balance. The solids balance is determined by identifying the quantities of solids entering and leaving each unit operation or process.

Hydraulic profile. The hydraulic profile is used to identify the elevation of the free surface of the wastewater as it flows through the various treatment units.

Plant layout. The plant layout is the spatial arrangement of the physical facilities of the treatment plant identified in the flowsheet.

Treatment-Process Flowsheets

Depending on the constituents that must be removed, an almost limitless number of different flowsheets can be developed using the unit operations and processes reported in Tables 4-2 and 4-3. Apart from the analysis of the suitability of the individual treatment methods, as outlined in Table 5-3, the exact flowsheet configuration selected will also depend on factors such as (1) the designer's past experience, (2) company and regulatory agency policies on the application of specific treatment methods, (3) the availability of suppliers of equipment for specific treatment methods, (4) the maximum use that can be made of existing

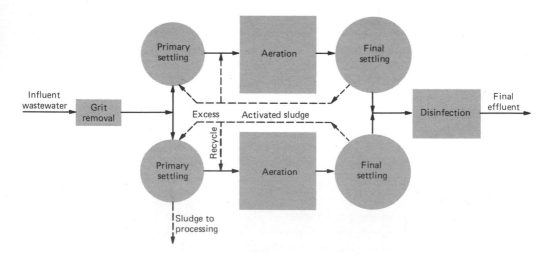

Figure 4-1 Process flowsheet for treatment plant designed to meet secondary-treatment standards of the U.S. Environmental Protection Agency (Westfield, Mass.).

facilities, (5) initial construction costs, and (6) future operation and maintenance costs.

A typical flowsheet for the treatment of wastewater to meet secondary-treatment standards, as defined by the U.S. Environmental Protection Agency (see Table 4-1) is shown in Fig. 4-1. The functions served by the various unit operations and processes are also shown. Alternative flowsheets for processing the sludge generated from these processes are presented in Chap. 11.

Flowsheets for advanced wastewater treatment are more complex than those shown in Fig. 4-1. For these, the reader is referred to Chap. 12.

Process Design Criteria

After one or more preliminary flowsheets have been developed, the next step is to determine the size of the physical facilities needed. The size depends on the process design criteria that are adopted. For example, assume that the hydraulic detention time in the aerated grit chamber shown in Fig. 4-1 is to be 3.5 min at the peak flowrate. For a peak flowrate of 1 m^3/s (22.7 Mgal/d), the corresponding volume would have to be 210 m^3 (7410 ft^3). Often, designers increase the size to account for nonideal inlet and outlet conditions. Similar procedures are followed to determine the size of each unit operation and process.

When the computations have been completed, all the key design criteria should be listed in a summary table. A typical example of such a table (for the process flowsheet shown in Fig. 4-1) is presented in Table E-1 (Appendix E). In this table, data on the basic plant design are shown before the data on unit operations or processes. Typically, these basic data include the population served, the per capita contributions of various waste contaminants, average and peak flowrates, and the average daily loadings of BOD and suspended solids. Because most treatment plants are designed to be effective for some time in the future (10 to 25 years), design criteria are given for the time when the facilities will first be put into operation and for the end of the design period. The latter will be influenced by projections of the population to be served and the economic studies of cost effectiveness for various design periods.

Solids Balance

After the design criteria are established, solids balances should be prepared for each process flowsheet. Ideally, they should be prepared for both the maximum one-day and the average organic loadings to be expected (see Chap. 3). Often, where solids handling is critical, balances for other sustained organic loading periods should also be evaluated. Such information must be available (1) to assess the need for sludge-storage facilities and their capacity, and (2) to determine the proper size of the sludge piping and pumping equipment. The preparation of a solids balance is illustrated in Chap. 11.

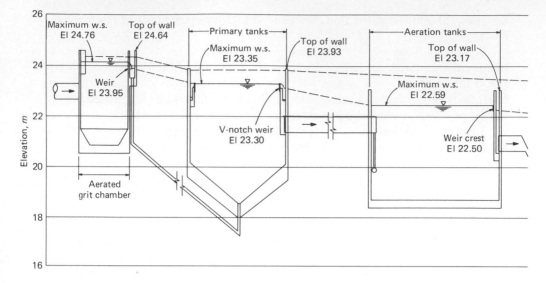

Figure 4-2 Hydraulic profiles for treatment plant shown in Fig. 4-1 (Westfield, Mass.).
Note: ws = water surface.

Hydraulic Profiles

After the flowsheet has been selected and the size of the corresponding physical
facilities and interconnecting piping is determined, hydraulic profiles should be
prepared for both average and peak flowrates. These profiles are prepared for
three reasons: (1) to ensure that the hydraulic gradient is adequate for the
wastewater to flow through the treatment facilities by gravity, (2) to establish
the head requirement for the pumps where pumping will be needed, and (3) to
ensure that the plant facilities will not be flooded or backed up during periods
of peak flow. Profiles for the flowsheet given in Fig. 4-1 are shown in Fig. 4-2.
In preparing a hydraulic profile, distorted vertical and horizontal scales are
commonly used to depict the physical facilities. Specific procedures vary depending
on local conditions. For example, if a downstream discharge condition may be a
control point, some designers prepare the hydraulic profile by working backward
from the control point. Other designers prefer to work from the head end of the
plant. Still others work from the center in each direction, adjusting the elevations
at the end of the computations.

The computations involve the determination of the head loss as the wastewater
flows through each of the physical facilities in the process flowsheet. For example,
consider the flow through mechanically cleaned bar screens. If it is determined
that the maximum head loss just before a cleaning cycle is 75 mm (3 in), the
water-surface elevations shown before and after the bar screens would differ by
75 mm. Typically, the largest head losses occur at control points or at weirs. Head
losses at pipe bends and exits are computed using the expression $h_L = K(V^2/2g)$,
where K is an empirical constant and V is the velocity in the approach piping.

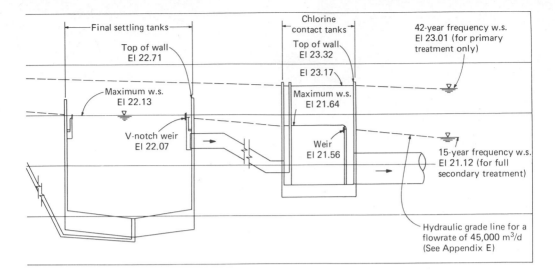

Plant Layout

Plant layout, as defined earlier, refers to the spatial arrangement of the physical facilities required to achieve a given treatment objective. The overall plant layout includes the location of the control and administrative buildings and any other necessary buildings. Several different layouts, using scaled cardboard cutouts of the various treatment facilities, are normally evaluated before a final selection is made.

Among the factors that must be considered when laying out a treatment plant are the following: (1) geometry of the available treatment-plant sites, (2) topography, (3) soil and foundation conditions, (4) location of the influent sewer, (5) location of the point of discharge, (6) transportation access, (7) types of processes involved, (8) effects of the length of process piping on treatment, (9) process performance and efficiency, (10) reliability and economy of operation, (11) aesthetics, (12) environmental control, and (13) an additional area for future plant expansion. The physical layouts of a variety of plants, both small and large, are shown in Figs. 4-3 through 4-6.

4-4 OTHER IMPORTANT CONSIDERATIONS

Because the following important design considerations cannot be dealt with in detail in this book, they are discussed briefly in this section: (1) energy and resource requirements, (2) cost analysis, (3) environmental impact assessment, and (4) the preparation of plans and specifications.

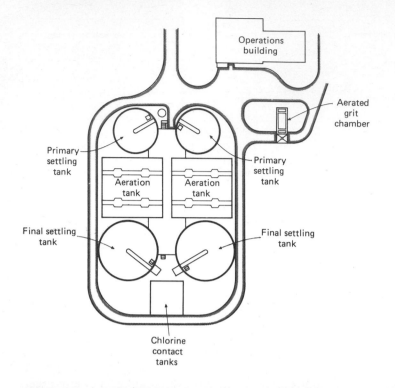

Figure 4-3 Layout and aerial view of treatment plant shown in Fig. 4-1 (Westfield, Mass.).

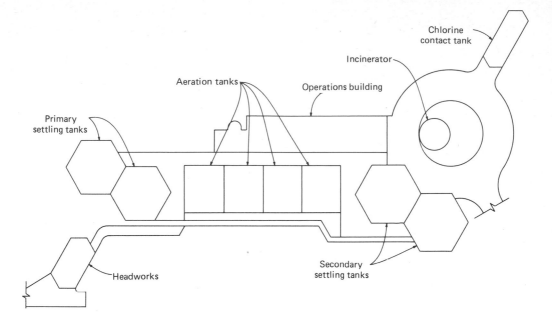

Figure 4-4 Layout and aerial view of small wastewater-treatment plant (Yosemite National Park).

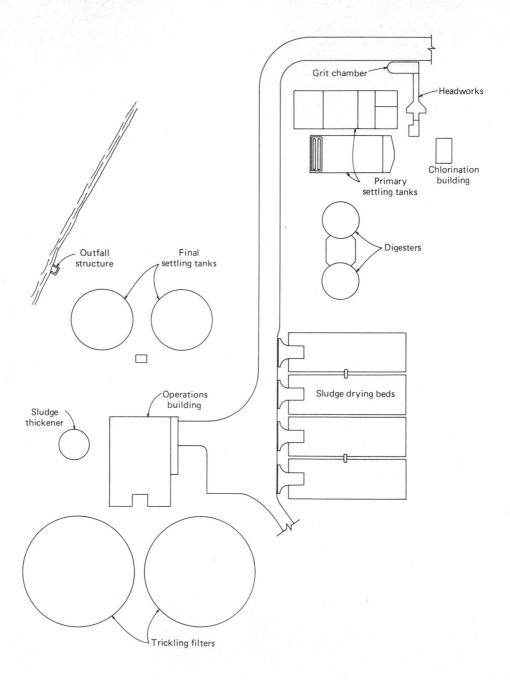

Figure 4-5 Layout (opposite page) and aerial view (above) of small trickling filter wastewater-treatment plant (Greenfield, Mass.).

Figure 4-6 Aerial view of large wastewater-treatment plant (Hartford Ct.).

Energy and Resource Requirements

Concern over the rate of consumption of natural resources and energy has increased during the past few years as shortages have developed and worldwide demands have been assessed. Because the operation of wastewater-management facilities is dependent on energy and resources to some extent, it is important to appraise the requirements realistically.

Energy and resource consumption of treatment plants The operation of facilities accounts for the major component of energy consumption at treatment plants. Because energy consumption of different unit processes and operations varies greatly and because there are innumerable combinations of process flowsheets, data must be available for each prospective treatment operation or process. Significant quantities of energy are also involved in the construction of a treatment plant, in the generation of electricity, and in the manufacture and transportation of construction materials and chemicals used in the treatment of wastewater [3].

Comparison of alternatives The basis of comparison of the alternatives is the consumption of primary energy per unit of time. The consumption of primary energy per unit of plant capacity or per unit of wastewater treated may also be used but can be misleading, especially when several alternatives are compared that involve different flowrates, as would be the case if several sewer-rehabilitation alternatives were being considered.

To illustrate what is involved in making an energy comparison, the two alternative flowsheets shown in Fig. 4-7 will be considered. In alternative 1,

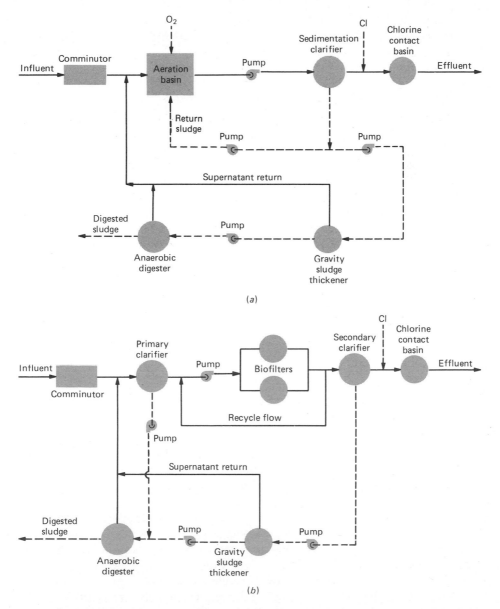

Figure 4-7 Alternative flowsheets used for evaluation of energy requirements as reported in Table 4-4 [3]. (*a*) Activated sludge. (*b*) High rate biofilter.

primary settling is omitted and an activated-sludge process with an earthen lagoon, surface aerators, and recycled sludge is used. In alternative 2, a high-rate biofilter treatment process with rock-media filters and primary sedimentation is used. Preliminary estimates of component sizes were determined using the following design criteria:

1. Average dry-weather flow = 2800 m^3/d (0.75 Mgal/d)
2. Average annual flow = 4800 m^3/d (1.25 Mgal/d)
3. Peak hourly wet-weather flow = 17,000 m^3/d (4.54 Mgal/d)
4. Average dry-weather BOD_5 = 225 mg/L
5. Average dry-weather suspended solids = 250 mg/L

Using these data and the flowsheets shown in Fig. 4-7, the corresponding consumption of primary energy for the two alternatives, as derived in Ref. 3, is shown in Table 4-4. Energy requirements for construction costs and for maintenance labor, parts, and supply costs are reported in addition to the requirements for operations. The basis on which the primary energy consumption (col. 4) is computed is shown in col. 2. The conversion factor (expressed in terms of kilowatthours per dollar) is derived from published data on the energy requirements per dollar of expenditure associated with the construction of such facilities. Additional conversion factors, including those listed in Table 4-4, are presented in Ref. 3.

In reviewing the data presented in Table 4-4, the most striking observation is that the annual energy requirements for the two processes differ by almost a factor of 2. This difference can be significant, especially in areas where power shortages are likely to occur and energy costs are anticipated to increase in the future. It is interesting to note that the energy expended for construction varies from 6 to 13 percent of the total required on an annual basis for the two processes considered. The computed energy requirements are important elements of the environmental-impact assessments discussed at the end of this chapter.

Cost Analysis

Of major significance in the selection and design of alternative wastewater-treatment facilities, especially to the client, is the question of costs—not only initial construction costs but also annual operation and maintenance costs. Although cost estimating is not covered in this text, a few comments about the preparation of cost estimates are in order.

When preparing a cost estimate, the same basis of comparison should be used to evaluate all the alternatives. It is suggested that both capital and annual costs be reported. Regardless of the approach followed, it is imperative that all cost estimates be referenced to some cost index, not only because costs are changing so rapidly, but also to allow effective cost comparisons to be made in the future. Cost data that are not referenced adequately are worthless. The *Engineering News-Record* Construction Cost Index and the Sewer Construction

Table 4-4 Primary energy consumption for alternatives 1 and 2 shown in Fig. 4-7[a]

Item (1)	Energy parameter (2)	Conversion factor (3)	Primary energy consumption (4)	Average annual energy consumption, 10^5 kWh/yr (5)
Alternative 1: activated sludge:				
Construction	$1,050,000[b]	$\frac{900.96[c]}{2100}$ (5.719 kWh/$)	25.76×10^5 kWh	0.86[d]
Operations electricity	2671 kWh/d	1.00	2671 kWh/d	9.75
Operations propane	639.76 kWh/d	1.208	773 kWh/d	2.82
Operations chlorine	42.3 kg/d	5.39 kWh/kg	249 kWh/d	0.83
Maintenance labor parts and supplies	$27,100/yr[b]	$\frac{900.96[c]}{2100}$ (5.08 kWh/$)	0.59×10^5 kWh/yr	0.59
Total				14.83
Alternative 2: high-rate biofilter:				
Construction	$1,244,000[b]	$\frac{900.96[c]}{2100}$ (5.719 kWh/$)	30.5×10^5 kWh	1.02[d]
Operations electricity	723 kWh/d	1.0	723 kWh/d	2.64
Operations propane	534 kWh/d	1.208	644 kWh/d	2.35
Operations chlorine	42.3 kg/d	5.39 kWh/kg	249 kWh/d	0.91
Maintenance labor parts and supplies	$35,900/yr[b]	$\frac{900.96[c]}{2100}$ (5.08 kWh/$)	0.78×10^5 kWh/yr	0.78
Total				7.70

[a] Adapted from Ref. 3.
[b] Estimated using an *Engineering News-Record* Construction Cost Index of 2100.
[c] Value of cost index at time energy factors were developed.
[d] Based on a service life of 30 years.

Note: kg × 2.2046 = lb.

and Sewage Treatment Plant Construction indexes of the U.S. Environmental Protection Agency are the ones most commonly used in the field of wastewater engineering.

Data in engineering reports and in the literature can be adjusted to a common basis for purposes of comparison by using the following relationship:

$$\text{Current cost} = \frac{\text{current value of index}}{\text{value of index at time of report}} \times \text{cost cited in report} \qquad (4\text{-}1)$$

When possible, index values should also be adjusted to reflect current local costs. When using the *Engineering News-Record* Construction Cost Index, if the month of the year that the facilities were built is not given, it is common practice to use the June end-of-the-month index value. To project costs into the future, the following relationship can be used:

$$\text{Future cost} = \frac{\text{projected future value of index}}{\text{current value of index}} \times \text{current cost} \qquad (4\text{-}2)$$

It should be noted, however, that updating or projecting costs for periods of more than 3 to 5 years can result in gross inaccuracies.

Environmental-Impact Assessment

As a result of the National Environmental Policy Act (NEPA) of 1969, Public Law 91-190, an environmental-impact assessment is required for any activity that "significantly affects the quality of the human environment" and that is supported by federal grants, subsidies, loans, permits, or licenses. To comply with this requirement, the U.S. Environmental Protection Agency has directed that all applications for federal assistance grants for wastewater-treatment facilities contain an assessment of the impact of the proposed project on the natural environment. It is important that each new wastewater-management project have a well-conceived, competently prepared, and thoroughly justified environmental-impact report to inform the public and all responsible agencies of government—from local to federal. The report can become the basis of public hearings on the project in an effort to seek the support of the public on which the project will have an impact [6].

The preparation of environmental-impact reports has almost become a new interdisciplinary profession. Every conceivable effect of a project on the environment must be taken into account because the environmental-impact statement becomes a legal document on the basis of which the proposed project may have to be defended in court. Many environmental consulting firms with experts in fields such as environmental engineering, ecology, land-use planning, aquatic and terrestrial biology, soil science, economics, and sociology—to name just a few of the specialties needed—have been created to serve planning agencies responsible for the preparation of these documents. To ensure objectivity, many communities have employed consultants in the various fields to prepare the necessary design and environmental-impact reports.

Plans and Specifications

The final step in the design of treatment-plant facilities is the preparation of construction plans and specifications. This task is usually carried out by the staffs of consulting engineering firms. Some large cities and regional agencies have their own design staffs. The coordinated effort of specialists from many disciplines is involved. These will include engineers specialized in various fields (civil, sanitary, chemical, mechanical, electrical, structural, soils, etc.), architects, draftsmen, and other technical and support personnel. These plans and specifications become the official documents on which contractors base their bids for the construction of the facilities. They are also the documents under which construction administrators hold the contractor responsible for the completion of the project as specified.

DISCUSSION TOPICS AND PROBLEMS

4-1 Prepare a brief summary of the history of wastewater treatment in your community. Identify major events that helped to bring about changes or improvements. If any of the events were related to crisis situations, try to assess whether the same result would have been achieved with proper planning.

4-2 Obtain, or if necessary develop, the flowsheet for you local wastewater-treatment plant. How does it compare to those shown in Fig. 4-1? How does it compare in terms of complexity to those shown in Figs. 12-1 and 12-2?

4-3 At the same time that you are obtaining the flowsheet for your local plant, ask if hydraulic profiles are available. If they are, how do they compare with those shown in Fig. 4-2? If they are not available,

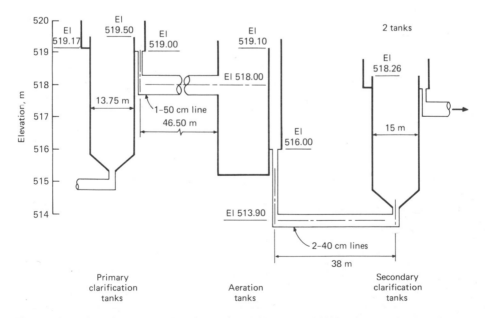

Figure 4-8 Portion of treatment plant for Prob. 4-4. Note: m × 3.2808 = ft; cm × 0.3937 = in.

ask if any unforeseen hydraulic problems have developed. If problems were experienced, could they have been avoided had proper profiles been prepared?

4-4 Develop the hydraulic profile for average and peak flow conditions for the portion of the wastewater-treatment plant shown in Fig. 4-8. The pertinent data and information are as follows:

$$Q_{avg} = 7500 \text{ m}^3/\text{d plus 100 percent sludge recycle}$$

$$Q_{peak} = 15,000 \text{ m}^3/\text{d plus 50 percent sludge recycle}$$

$$\text{Primary clarifier diameter} = 13.75 \text{ m}$$

$$\text{No. of clarifiers} = 2$$

$$\text{Secondary clarifier diameter} = 15 \text{ m}$$

$$\text{No. of clarifiers} = 2$$

$$\text{Width of aeration-tank effluent weir} = 2 \text{ m}$$

4-5 What consideration was given to energy conservation in the design of your local wastewater-treatment facilities? Is anything being done currently to bring about a reduction in the use of energy? Walk around your local wastewater-treatment facilities and try to assess where solar-heating panels could have been used in the construction of the facilities. Make a rough estimate of the energy savings that might have been possible if they had been used. (Although the present cost of solar facilities may exceed the value of the energy, the balance will probably be shifted soon in view of the rising cost of oil and the current research effort on solar energy.)

4-6 Using the current unit cost for energy ($/kW) for your region, determine the annual costs of energy for the two treatment processes given in Table 4.4. Do you feel that the difference between the two processes is significant? Justify your answer.

4-7 The construction cost for a small wastewater treatment plant was estimated to be $5,000,000 in 1978. If the construction of the plant is to be delayed until 1983, estimate the cost in 1983 for this same plant. Use end-of-year ENR values in making your projection.

4-8 Determine the year when your local wastewater treatment plant was constructed or expanded and its construction costs. What would the cost be to construct or expand the plant today? What has been the average rate of inflation from the time your plant was constructed to the present?

4-9 If an environmental impact report (EIR) was prepared for the construction of your local wastewater treatment plant, were the major environmental findings identified in the report borne out after the plant was constructed and put into operation?

4-10 If an EIR was not prepared for the construction of your local wastewater treatment plant, would the preparation of one, in your judgment, have caused any significant differences in the implementation of the treatment facilities?

4-11 If an EIR was prepared for the construction of your local wastewater treatment plant, obtain a copy and review it specifically with respect to the recommendations made concerning odor. Are they consistent with the reality of the situation?

4-12 If an EIR was prepared for the construction of your local wastewater treatment plant, obtain a copy and review it specifically with respect to the recommendations concerning energy utilization. What has actually happened with respect to both the cost and utilization of energy?

REFERENCES

1. *Code of Federal Regulations*, Title 40, Part 35, Appendix A, Sept. 10, 1973.
2. Metcalf & Eddy, Inc.: *Report to the National Commission on Water Quality on Assessment of Technologies and Costs for Publicly Owned Treatment Works*, vols. 1 and 2, prepared under Public Law 92-500, Boston, 1975.

3. Mills, R. A., and G. Tchobanoglous: Energy Consumption in Wastewater Treatment, in W. S. Jewell (ed.), *Proceedings of the Seventh National Agricultural Waste Management Conference*, Ann Arbor Science Publishers, Ann Arbor, Mich., 1976.
4. *Pretreatment of Pollutants Introduced into Publicly Owned Treatment Works*, U.S. Environmental Protection Agency.
5. Tchobanoglous, G.: Wastewater Treatment for Small Communities, *Public Works*, part I, vol. 105, no. 7, July 1974; part II, vol. 105, no. 8, August 1974.
6. Tchobanoglous, G., H. Theisen, and R. Eliassen: *Solid Wastes: Engineering Principles and Management Issues*, McGraw-Hill, New York, 1977.

CHAPTER FIVE

FUNDAMENTALS OF PROCESS ANALYSIS

In Chap. 4, it was noted that the treatment of wastewater involves a number of chemical and biological reactions and conversions. The rate at which these reactions and conversions occur affects the size of the treatment facilities that must be provided. From practical observations, it is known that the type of container (i.e., the reactor) in which these reactions and conversions take place also affects the degree of their completion. Thus both the rate of the reactions and conversions and the type of reactor are important in the selection of treatment processes. Environmental and other physical constraints must also be considered.

The purpose of this chapter is to discuss (1) the nature of basic reactions and reaction rates, (2) the application of materials balances in the analysis of both unit operations and processes, (3) the various types and hydraulic characteristics of reactors that are used to accomplish the various reactions and conversions, (4) the important factors involved in process analysis, (5) the relationship between reaction kinetics and the selection of reactors, and (6) some important practical aspects of process design.

The information in this chapter is intended to serve as an introduction to the subject of process analysis and to the information that will be presented in subsequent chapters. By dealing with the basic concepts first, it will be possible to apply them (without repeating the details) in the remaining chapters. Additional information on process analysis may be found in Refs. 1, 2, 3, 6, 7, and 10 and in the various references cited within this chapter.

5-1 REACTIONS AND REACTION KINETICS

From the standpoint of process selection and design, the controlling stoichiometry and the rate of the reaction are of principal concern. The number of moles of a substance entering into a reaction and the number of moles of the substances produced are defined by the stoichiometry of a reaction. The rate at which a substance disappears or is formed in any given stoichiometric reaction is defined as the rate of reaction. These and other related topics are discussed in this section.

Types of Reactions

The two principal types of reactions that occur in wastewater treatment are classified as homogeneous and heterogeneous (nonhomogeneous).

Homogeneous reactions In homogeneous reactions, the reactants are distributed uniformly throughout the fluid so that the potential for reaction at any point within the fluid is the same. Homogeneous reactions may be either irreversible or reversible. Examples of irreversible reactions are

$$A \rightarrow B$$

$$A + A \rightarrow P$$

$$aA + bB \rightarrow P$$

Examples of reversible reactions are

$$A \rightleftharpoons B$$

$$A + B \rightleftharpoons C + D$$

Heterogeneous reactions Heterogeneous reactions occur between one or more constituents that can be identified with specific sites, such as those on an ion-exchange resin. Reactions that require the presence of a solid-phase catalyst are also classified as heterogeneous. These reactions are more difficult to study because a number of interrelated steps may be involved. The typical sequence of these steps, as quoted from Ref. 7, is as follows:

1. Transport of reactants from the bulk fluid to the fluid-solid interface (external surface of catalyst particle)
2. Intraparticle transport of reactants into the catalyst particle (if it is porous)
3. Adsorption of reactants at interior sites of the catalyst particle
4. Chemical reaction of adsorbed reactants to adsorbed products (surface reaction)
5. Desorption of adsorbed products
6. Transport of products from the interior sites to the outer surface of the catalyst particle
7. Transport of products from the fluid-solid interface into the bulk-fluid stream

Rate of Reaction

The rate of reaction is the term used to describe the change (decrease or increase) in the number of moles of a reactive substance per unit time per unit volume (for homogeneous reactions), or per unit surface area or mass (for heterogeneous reactions) [2].

For homogeneous reactions, the rate of reaction r is given by

$$r = \frac{1}{V}\frac{dN}{dt} = \frac{\text{mol}}{(\text{liquid volume})(\text{time})} \tag{5-1}$$

If N is replaced by the term VC, where V is the volume and C is the concentration, Eq. 5-1 becomes

$$r = \frac{1}{V}\frac{d(VC)}{dt} = \frac{1}{V}\frac{V\,dC + C\,dV}{dt} \tag{5-2}$$

If the volume remains constant (isothermal conditions), Eq. 5-2 reduces to

$$r = \pm\frac{dC}{dt} \tag{5-3}$$

where the plus sign indicates an increase or accumulation of the substance, and the minus sign indicates a decrease of the substance.

For heterogeneous reactions where S is the surface area, the corresponding expression is

$$r = \frac{1}{S}\frac{dN}{dt} = \frac{\text{mol of product}}{(\text{area})(\text{time})} \tag{5-4}$$

For reactions involving two or more reactants with unequal stoichiometric coefficients, the rate expressed in terms of one reactant will not be the same as the rate for the other reactants. For example, for the reaction

$$aA + bB \longrightarrow cC + dD$$

the concentration changes for the various reactants are given by

$$-\frac{1}{a}\frac{d[A]}{dt} = -\frac{1}{b}\frac{d[B]}{dt} = \frac{1}{c}\frac{d[C]}{dt} = \frac{1}{d}\frac{d[D]}{dt} \tag{5-5}$$

Thus, for reactions in which the stoichiometric coefficients are not equal, the rate of reaction is given by

$$r = \pm\frac{1}{c_i}\frac{d[C_i]}{dt} \tag{5-6}$$

where the coefficient term $(1/c_i)$ is negative for reactants and positive for products.

The rate at which a reaction proceeds is an important consideration in wastewater treatment. For example, treatment processes may be designed on the basis of the rate at which the reaction proceeds rather than the equilibrium position of the reaction, because the reaction usually takes too long to go to completion. In this case, quantities of chemicals in excess of the stoichiometric, or exactly reacting amount, may be used to accomplish the treatment step in a reasonable period of time. This is true, for example, with disinfection and with the biological processes for BOD removal and digestion.

Specific Reaction Rate

From the law of mass action, it can be shown that the rate of reaction for a given reaction is proportional to the remaining concentration of the reactants. Thus, for a reaction involving a single component A, the rate of reaction is given by

$$r = \pm kC_A \qquad (5\text{-}7)$$

where k is a constant of proportionality formally defined as the specific reaction rate (also known as the reaction-rate constant, velocity constant, and the rate coefficient). The specific reaction rate has the units of the specific reaction and concentration. For Eq. 5-7, the units of the specific reaction-rate constant are

$$k = \frac{r}{C} = \frac{1}{V}\frac{dN}{dt}\frac{1}{C} = \frac{\text{mol}}{L \cdot s(\text{mol}/L)} = s^{-1} \qquad (5\text{-}8)$$

In application, the rate of reaction r takes into account the effects of concentration, and the specific reaction-rate constant takes into account the effects of all the other variables that may affect the reaction. Of the many variables in any given situation, temperature is usually the most important.

Effects of Temperature on Specific Rate Constants

The temperature dependence of the specific reaction-rate constants is important because of the need to use constants that are determined at one temperature for systems of another temperature. For example, the reaction-rate constant determined for the BOD reaction at 20°C must often be used for systems at temperatures other than 20°C. The temperature dependence of the rate constant is given by the van't Hoff–Arrhenius equation

$$\frac{d(\ln k)}{dT} = \frac{E}{RT^2} \qquad (5\text{-}9)$$

where T = temperature, °K
R = ideal gas constant, 8.314 J/mol/°K (1.987 cal/°K · mol)
E = a constant characteristic of reaction called activation energy

Integration of Eq. 5-9 between the limits T_1 and T_2 gives

$$\ln \frac{k_2}{k_1} = \frac{E(T_2 - T_1)}{RT_1 T_2} \qquad (5\text{-}10)$$

With k_1 known for a given temperature and with E known, k_2 can be calculated from this equation. The activation energy E can be calculated by determining the k at two different temperatures and by using Eq. 5-10. Common values of E for waste-treatment processes are in the range of 8400 to 84,000 J/mol (2000 to 20,000 cal/mol).

Because most wastewater-treatment operations and processes are carried out at or near the ambient temperature, the quantity E/RT_1T_2 may be assumed to be a constant for all practical purposes. If the value of the quantity is designated by C, then Eq. 5-10 can be rewritten as

$$\ln \frac{k_2}{k_1} = C(T_2 - T_1) \tag{5-11}$$

$$\frac{k_2}{k_1} = e^{C(T_2 - T_1)} \tag{5-12}$$

Replacing e^C in Eq. 5-12 with a temperature coefficient θ yields

$$\frac{k_2}{k_1} = \theta^{(T_2 - T_1)} \tag{5-13}$$

which is commonly used in the sanitary engineering field to adjust the value of the operative rate constant to reflect the effect of temperature. An alternative form of the temperature-correction equation may be obtained by expanding Eq. 5-12 as a series and dropping all but the first two terms. It should be noted, however, that although the value of θ is assumed to be constant, it will often vary considerably with temperature. Therefore, caution must be used in selecting appropriate values for θ for different temperature ranges. Typical values for various operations and processes for different temperature ranges are given, where available, in the sections in which the individual topics are discussed.

Analysis of Rate Equations

The rate at which reactions occur usually is determined by measuring the concentration of either a reactant or product as the reaction proceeds to completion. The measured results are then compared to the corresponding results obtained by using various standard types of rate equations to which the reaction under study will proceed.

A first-order reaction is one in which the rate of completion is observed to be directly proportional to the first power of the concentration of the reactant. For example, in the reaction

$$aA + bB + \cdots \rightarrow pP + qQ + \cdots \tag{5-14}$$

if the rate is experimentally found to be proportional to the first power of the concentration of A, then the reaction is said to be first order with respect to A.

When the mechanism of reaction is not known, the reaction rate for Eq. 5-14 may be approximated with the following expression:

$$r = kC_A^a C_B^b, \ldots, C_p^p = kC_A^n \tag{5-15}$$

where a and b are the reaction orders with respect to reactants A and B, and n is the overall reaction order $(n = a + b + \cdots p)$ [2]. The following discussion of irreversible- and reversible-reaction equations illustrates the relationships that are used to identify reactions of various orders.

Irreversible reactions For reactions in which the rate of product formation is found to be independent of the concentration, the reaction is said to be zero-order and is defined as

$$r_A = \frac{d[A]}{dt} = -k_0 \tag{5-16}$$

which for the initial condition $A = A_0$ integrates to

$$[A_t] = [A_0] - k_0 t \tag{5-17}$$

For a first-order reaction, the rate is defined as

$$r_A = \frac{d[A]}{dt} = -k_1[A] \tag{5-18}$$

which for the initial condition $A = A_0$ integrates to

$$\ln \frac{[A_t]}{[A_0]} = -k_1 t \quad \text{or} \quad [A_t] = [A_0]e^{-kt} \tag{5-19}$$

The rate expression for other irreversible reactions, along with the corresponding integrated forms, is reported in Table 5-1.

Reactions for other orders are defined in a similar manner; however, the order of the reaction is an empirical quantity and has no necessary relationship to any stoichiometric equation for the reaction. The determination of the order of a reaction and the corresponding rate constant is illustrated in Example 5-1.

Reversible reactions For a reversible reaction of the form

$$A \underset{k_1}{\overset{k_1}{\rightleftharpoons}} B \tag{5-20}$$

where the specific reaction-rate constant for the conversion of A to B is k_1 and the conversion of B to A is k_1', the rate of the reaction is given by

$$r_A = \frac{d[A]}{dt} = -k_1[A] + k_1'[B] \tag{5-21}$$

Table 5-1 Rate equations for simple irreversible reactions [7]

Reaction	Order	Rate equation	Integrated forms
$A \rightarrow B$	Zero	$\dfrac{d[A]}{dt} = -k_0$	$[A] = [A]_0 - k_0 t$
			$t_{1/2} = \dfrac{[A]_0}{2k_0}$
$A \rightarrow B$	First	$\dfrac{d[A]}{dt} = -k_1[A]$	$\ln \dfrac{[A]}{[A]_0} = k_1 t$
			$t_{1/2} = \dfrac{1}{k_1} \ln 2$
$A + A \rightarrow P$	Second, type I	$\dfrac{d[A]}{dt} = -k_2[A]^2$	$\dfrac{1}{[A]} - \dfrac{1}{[A]_0} = k_2 t$
			$t_{1/2} = \dfrac{1}{k_2[A]_0}$
$aA + bB \rightarrow P$	Second, type II	$\dfrac{d[A]}{dt} = -k_2[A][B]$	$\ln \dfrac{[A]_0 - [B]}{[B]_0 - b/a[X]} = \ln \dfrac{[A]}{[B]}$
			$= \dfrac{b[A]_0 - a[B]_0}{a} k_2 t + \ln \dfrac{[A]_0}{[B]_0}$
			$t_{1/2} = \dfrac{a}{k_2(b[A]_0 - a[B]_0)}$
			$\times \ln \dfrac{a[B]_0}{2a[B]_0 - b[A]_0}$

The integrated form of this reaction and other commonly encountered reversible reactions may be found in Ref. 7 and in most texts dealing with chemical engineering or physical chemistry.

Example 5-1: Determination of reaction order and rate constant Determine if the following substrate-removal data can be described with a first-order reaction, and if so, determine the specific reaction-rate constant.

Time, h	0	1	2	3	4	6
Substrate concentration, mg/L	50	35.6	25.8	18.5	12.8	7.3

SOLUTION

1 Assuming that $C_0 = 50$ mg/L, compute the value of the rate constant at the various times, using Eq. 5-19.

Time	C/C_0	$\ln C/C_0$	k, h^{-1}
0	1.00	0.000	
1	0.71	−0.342	−0.342
2	0.52	−0.654	−0.327
3	0.37	−0.994	−0.331
4	0.27	−1.310	−0.328
6	0.14	−1.966	−0.328

2 Since the specific reaction-rate constant is approximately the same, it can be concluded that the reaction is first-order with respect to substrate and that the average value of the rate constant is about −0.331 h^{-1}.

3 The rate constant could also be determined by plotting log $(C_0 - x)$ versus t on arithmetic paper, where the quantity $(C_0 - x)$ represents the amount of material remaining at each time period.

5-2 MASS-BALANCE ANALYSIS

The fundamental approach used to delineate the changes that take place when a reaction is occurring in a container (reactor) or in some definable portion of a body of liquid is the mass-balance analysis. The basic aspects involved in such an analysis and some applications are described in this section.

The Mass Balance

Because mass is neither created nor destroyed, a mass balance affords a convenient way of defining what occurs within treatment facilities as a function of time. To illustrate the basic concepts involved, a mass-balance analysis will be performed on the contents of the container shown schematically in Fig. 5-1. First, the system boundary must be established so that all the flows of mass into and out of the system can be identified. In Fig. 5-1, the boundary is shown by a

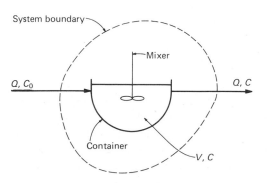

Figure 5-1 Definition sketch for the application of materials mass-balance analysis.

dashed line. The proper selection of the system boundary is extremely important because, in many situations, it will be possible to simplify the mass-balance computations. This subject is considered further in this chapter and in Chaps. 9 and 10.

To apply a mass-balance analysis to the liquid contents of the container shown in Fig. 5-1, it will be assumed that (1) the volumetric flowrate into and out of the container is constant, (2) the liquid within the reactor is not subject to evaporation (isothermal conditions), (3) the liquid within the container is mixed completely, (4) a chemical reaction involving the reactant C is occurring within the reactor, and (5) the rate of change in the concentration of the reactant C that is occurring within the reactor is governed by a first-order reaction ($r_C = -kC$). For the stated assumptions, the materials mass balance can be formulated as follows:

1. General word statement:

$$
\begin{array}{c}
\text{Rate of accumulation} \\
\text{of reactant within} \\
\text{the system boundary}
\end{array}
=
\begin{array}{c}
\text{rate of flow of} \\
\text{reactant into} \\
\text{the system} \\
\text{boundary}
\end{array}
-
\begin{array}{c}
\text{rate of flow of} \\
\text{reactant out of} \\
\text{the system} \\
\text{boundary}
\end{array}
+
\begin{array}{c}
\text{rate of disappearance} \\
\text{(utilization) of} \\
\text{reactant within the} \\
\text{system boundary}
\end{array}
\tag{5-22}
$$

2. Simplified word statement:

$$\text{Accumulation} = \text{inflow} - \text{outflow} + \text{utilization} \tag{5-23}$$

3. Symbolic representation:

$$V \frac{dC}{dt} = QC_0 - QC + V(\text{rate of reaction, } r_c)$$

$$V \frac{dC}{dt} = QC_0 - QC + V(-kC) \tag{5-24}$$

where V = volume of reactor, L^3

$\dfrac{dC}{dt}$ = rate of change of reactant concentration within the container, $ML^{-3}T^{-1}$

Q = volumetric rate of flow into and out of the container, L^3T^{-1}

C_0 = concentration of reactant in influent, ML^{-3}

k = first-order reaction-rate constant, T^{-1}

C = concentration of reactant in reactor and effluent, ML^{-3}

In Eq. 5-22, a positive sign is used for the rate-of-utilization term because the necessary negative sign is part of the rate expression. If the concentration of a reactant is increased instead of utilized or consumed in the reactor, the corresponding mass-balance expression is essentially the same except that the rate-of-utilization term is replaced by a rate-of-generation term. There is no sign change.

Before substituting numerical values in any mass-balance expression, a unit check should always be made to assure that units of the individual quantities are consistent. If the following units are substituted into Eq. 5-24,

$$V = L$$

$$\frac{dC}{dt} = mg/L \cdot s$$

$$Q = L/s$$

$$C_0, C = mg/L$$

$$k = 1/s = s^{-1}$$

the resulting unit check yields

$$V\frac{dC}{dt} = QC_0 - QC + V(-kC)$$

$$L(mg/L/s) = L/s(mg/L) = L/s(mg/L) + L(-1/s)(mg/L)$$

$$mg/s = mg/s - mg/s - mg/s$$

In some situations it may be found that the concentration of a reactant is simultaneously increased through generation and decreased through consumption within the reactor. This is the situation when a portion of the product that is produced is used to sustain the energy requirements of the process. When this occurs, the associated rates usually are not equal. The corresponding mass-balance expressions for this situation are:

1. General word statement:

| Rate of accumulation of reactant within the system boundary | = | rate of flow of reactant into the system boundary | − | rate of flow of reactant out of the system boundary | + | rate of generation of reactant within the system boundary | + | rate of disappearance (utilization) of reactant within the system boundary | (5-25) |

2. Simplified word statement:

$$\text{Accumulation} = \text{inflow} - \text{outflow} + \text{generation} + \text{utilization} \quad (5\text{-}26)$$

3. Symbolic representation:

$$V\frac{dC}{dt} = QC_0 - QC + V(\text{rate of generation}, r_g) + V(\text{rate of utilization}, r_u)$$

$$(5\text{-}27)$$

Mass Balance for Batch Reactor

Before proceeding further, it will be instructive to explore the difference between the rate-of-change term that appears as part of the accumulation term and the rate-of-generation or decay term. In general, these terms are not equal, except in the special case when there is no inflow or outflow from the container or vessel in which the reaction is occurring. Such a container is known as a batch reactor (see Sec. 5-3). In this situation, Q is equal to zero and Eq. 5-24 becomes

$$\frac{dC}{dt} = (\text{rate of utilization } r_u \text{ or generation } r_g) \tag{5-28}$$

The key point to remember is that when flow is not occurring, the concentration per unit volume is changing according to the applicable rate expression. On the other hand, when flow is occurring, the concentration in the reactor is also being modified by the inflow and outflow from the reactor. The form of Eq. 5-28 will be recognized as the form of the equations discussed previously in Sec. 5-1 and in Table 5-1.

Solution Procedures

The analytical procedures that are adopted for the solution of mass-balance equations usually are governed by the mathematical form of the final expression. For example, the general non-steady-state solution for Eq. 5-24 is obtained by first noting that Eq. 5-24 has the form of the standard first-order linear differential equation. The solution procedure for such equations, outlined in Eqs. 5-29 through 5-37, is included in this discussion because these types of equations are encountered frequently in the field of sanitary engineering and in this text. The first step in the solution is to collect terms and rewrite Eq. 5-24 in the form

$$C' + \left(k + \frac{Q}{V}\right)C = \frac{Q}{V}C_0 \tag{5-29}$$

where C' is used to denote the derivative dC/dt. In the next step, both sides of the expression are multiplied by the integrating factor $e^{\beta t}$, where $\beta = (k + Q/V)$.

$$e^{\beta t}(C' + \beta C) = \frac{Q}{V}C_0 e^{\beta t} \tag{5-30}$$

The left side of the expression can now be written as a differential as follows:

$$(Ce^{\beta t})' = \frac{Q}{V}C_0 e^{\beta t} \tag{5-31}$$

The differential sign can be removed by integrating the expression as follows:

$$Ce^{\beta t} = \frac{Q}{V}C_0 \int e^{\beta t}\, dt \tag{5-32}$$

Integration of Eq. 5-32 yields

$$Ce^{\beta t} = \frac{Q}{V} \frac{C_0}{\beta} e^{\beta t} + K \tag{5-33}$$

Dividing by $e^{\beta t}$ results in

$$C = \frac{Q}{V} \frac{C_0}{\beta} + Ke^{-\beta t} \tag{5-34}$$

but when $t = 0$, $C = C_0$, and therefore

$$K = C_0 - \frac{Q}{V} \frac{C_0}{\beta} \tag{5-35}$$

Substituting for K in Eq. 5-33 and simplifying yields the following expression, which is the non-steady-state solution of Eq. 5-24:

$$C = \frac{Q}{V} \frac{C_0}{\beta} (1 - e^{-\beta t}) + C_0 e^{-\beta t} \tag{5-36}$$

When $t \to \infty$, it will be noted that Eq. 5-36 becomes

$$C = \frac{Q}{V} \frac{C_0}{\beta} = \frac{C_0}{1 + k(V/Q)} \tag{5-37}$$

The solution for Eq. 5-24 is one of the very few closed-form solutions that are encountered. Typically, non-steady-state solutions to mass-balance equations must be obtained using numerical methods and digital computers.

Steady-State Simplification

Fortunately, in most applications in the field of wastewater treatment, the solution of mass-balance equations, such as the one given by Eq. 5-24, can be simplified by noting that the long-term (so-called steady-state) concentration is of principal concern. If it is assumed that only the steady-state effluent concentration is desired, then Eq. 5-24 can be simplified by noting that, under steady-state conditions, the rate accumulation is zero ($dC/dt = 0$). Using this fact, Eq. 5-24 can be written as

$$0 = QC_0 - QC - VkC \tag{5-38}$$

When solved for C, Eq. 5-38 yields the following expression, which is the same as Eq. 5-37:

$$C = \frac{C_0}{1 + k(V/Q)}$$

5-3 REACTORS AND THEIR HYDRAULIC CHARACTERISTICS

As mentioned previously, the containers, vessels, or tanks in which chemical and biological reactions are carried out are commonly called reactors. The types of reactors that are available and their hydraulic characteristics are discussed in this section; the selection of reactors is discussed in Sec. 5-6.

Types of Reactors

The principal types of reactors used for the treatment of wastewater are (1) the batch reactor; (2) the plug-flow reactor, also known as a tubular-flow reactor; (3) the continuous-flow stirred-tank reactor, also known as a complete-mix reactor; (4) the arbitrary-flow reactor; (5) the packed-bed reactor; and (6) the fluidized-bed reactor. Descriptions of these reactors are presented in Table 5-2. The classification of the first four reactors is based on their hydraulic characteristics. Homogeneous reactions are usually carried out in such reactors. Heterogeneous reactions are usually carried out in the latter two types of reactors.

In practice, the ideal theoretical plug-flow reactor is approximated by a very long rectangular tank (see Fig. 5-2). In some cases, these tanks exceed 180 m (600 ft) in length. To use space more effectively, such tanks are often folded

Figure 5-2 Empty plug-flow reactor used for the activated-sludge treatment process.

Table 5-2 Principal types of reactors used for the treatment of wastewater

Type of reactor	Identification sketch	Description and/or application
Batch		Flow is neither entering nor leaving the reactor. The liquid contents are mixed completely. For example, the BOD test discussed in Chap. 3 is carried out in a bottle batch reactor
Plug-flow, also known as tubular-flow		Fluid particles pass through the tank and are discharged in the same sequence in which they enter. The particles retain their identity and remain in the tank for a time equal to the theoretical detention time. This type of flow is approximated in long tanks with a high length-to-width ratio in which longitudinal dispersion is minimal or absent
Continuous-flow stirred-tank, also known as complete-mix		Complete mixing occurs when the particles entering the tank are dispersed immediately throughout the tank. The particles leave the tank in proportion to their statistical population. Complete mixing can be accomplished in round or square tanks if the contents of the tank are uniformly and continuously redistributed
Arbitrary-flow		Arbitrary flow is any degree of partial mixing between plug-flow and complete mixing
Packed-bed	Packing medium	Packed-bed reactors are filled with some type of packing medium, such as rock, slag, ceramic, or plastic. With respect to flow, they can be completely filled (anaerobic filter) or intermittently dosed (trickling filter)
Fluidized-bed	Expanded packing medium	The fluidized-bed reactor is similar to the packed-bed reactor in many respects, but the packing medium is expanded by the upward movement of fluid (air or water) through the bed. The porosity of the packing can be varied by controlling the flowrate of the fluid

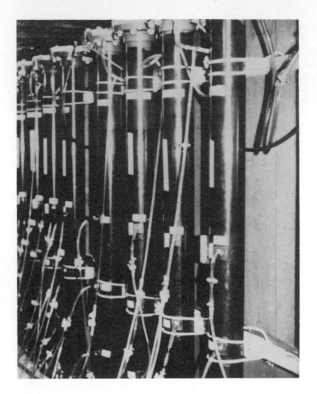

Figure 5-3 Pilot-plant packed-bed reactors used for activated-carbon adsorption.

in a serpentine fashion. Both circular and rectangular tanks have been used to approximate continuous-flow stirred-tank reactors. A typical packed-bed reactor is shown in Fig. 5-3.

Hydraulic Characteristics of Reactors

Plug-flow, continuous-flow stirred-tank, and arbitrary-flow reactors are the types most commonly used in the field of wastewater treatment. The hydraulic characteristics of these reactors are identified in Fig. 5-4, in which dye-tracer response curves are presented for step and impulse (slug-dose) disturbances.

For a plug-flow reactor, if a continuous flow of a dye tracer were injected into such a tank to produce a concentration C_0, the appearance of the tracer in the effluent would occur as shown in Fig. 5-4 (t equals the actual time, and t_0 equals the theoretical detention time V/Q). Under ideal plug-flow conditions, t equals t_0. The effect of an impulse disturbance, such as that caused by a slug injection of tracer, is also shown in Fig. 5-4.

For a continuous-flow stirred-tank reactor, if a continuous flow of conservative

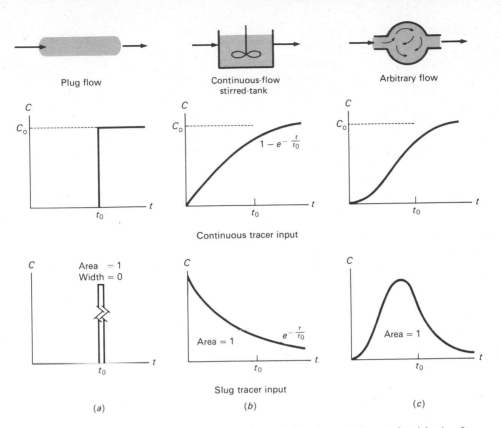

Figure 5-4 Output tracer response curves for step and impulse disturbances for (a) plug-flow, (b) continuous-flow stirred-tank, and (c) arbitrary-flow reactors. (*Adapted from* [3].)

(nonreactive) tracer were injected into the inlet at concentration C_0, the appearance of the tracer at the outlet would occur as shown in Fig. 5-4. The response to a slug input of tracer is also given in Fig. 5-4.

Arbitrary flow represents any degree of partial mixing between plug flow and complete-mix flow. This type of flow is encountered frequently in actual aeration and settling tanks. It is also more difficult to describe mathematically. Therefore, in the mathematical treatment of the chemical and biological unit processes carried out in reactors, ideal models of complete-mix flow or plug flow are usually assumed. The theoretical analysis of the response curves for a continuous-flow stirred-tank reactor and a plug-flow reactor is considered below.

Continuous-flow stirred-tank reactor The effluent concentration as a function of time can be determined from a materials mass balance for the tracer around the reactor as follows:

1. General word statement:

$$
\begin{array}{l}
\text{Rate of} \\
\text{accumulation} \\
\text{of tracer within} \\
\text{the reactor}
\end{array}
=
\begin{array}{l}
\text{rate of flow} \\
\text{of tracer} \\
\text{into the} \\
\text{reactor}
\end{array}
-
\begin{array}{l}
\text{rate of flow} \\
\text{of tracer} \\
\text{out of the} \\
\text{reactor}
\end{array}
\qquad (5\text{-}39)
$$

2. Simplified word statement:

$$\text{Accumulation} = \text{inflow} - \text{outflow} \qquad (5\text{-}40)$$

3. Symbolic representation:

$$V \frac{dC}{dt} = QC_0 - QC \qquad (5\text{-}41)$$

Rewriting and simplifying, Eq. 5-41 yields

$$\frac{dC}{dt} = \frac{Q}{V}(C_0 - C) \qquad (5\text{-}42)$$

which, when integrated between the limits of 0 and C and 0 and t, gives

$$\int_0^C \frac{dC}{C_0 - C} = \frac{Q}{V} \int_0^t dt \qquad (5\text{-}43)$$

which, when solved, yields

$$C = C_0(1 - e^{-t(Q/V)}) = C_0(1 - e^{-t/t_0}) = C_0(1 - e^{-\theta}) \qquad (5\text{-}44)$$

where t_0 equals the theoretical detention time V/Q and θ equals t/t_0.

The corresponding expression for the effluent concentration from a reactor that is being purged of tracer is derived similarly, and is given by

$$C = C_0 e^{-t(Q/V)} = C_0 e^{-t/t_0} = C_0 e^{-\theta} \qquad (5\text{-}45)$$

Plug-flow reactor Using the definition sketch for the plug-flow reactor shown in Fig. 5-5, the effluent tracer concentration as a function of time can be obtained by performing a mass balance as follows:

1. General word statement:

$$
\begin{array}{l}
\text{Rate of} \\
\text{accumulation} \\
\text{of tracer within} \\
\text{the element}
\end{array}
=
\begin{array}{l}
\text{rate of flow} \\
\text{of tracer} \\
\text{into the} \\
\text{element}
\end{array}
-
\begin{array}{l}
\text{rate of flow} \\
\text{of tracer} \\
\text{out of the} \\
\text{element}
\end{array}
\qquad (5\text{-}46)
$$

2. Simplified word statement:

$$\text{Accumulation} = \text{inflow} - \text{outflow} \qquad (5\text{-}47)$$

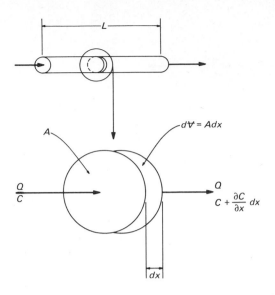

Figure 5-5 Definition sketch for the analysis of a plug-flow reactor.

3. Symbolic representation:

$$dV\, d\left[\frac{C + C + (\partial C/\partial x)\, dx}{2}\right] \Big/ dt = QC - Q\left(C + \frac{\partial C}{\partial x}\, dx\right) \qquad (5\text{-}48)$$

Neglecting second-order terms and following the rules of calculus, the expression in Eq. 5-48 can be simplified to

$$dV\, \frac{\partial C}{\partial t} = -Q\, \frac{\partial C}{\partial x}\, dx \qquad (5\text{-}49)$$

If $A\, dx$ is substituted for dV and AV is substituted for Q, where V is equal to the velocity, the resulting expression is

$$\frac{\partial C}{\partial t} = -V\, \frac{\partial C}{\partial x} \qquad (5\text{-}50)$$

Because both sides of this equation are the same (note that $\partial t = \partial x/V$), except for the minus sign, the only way that the equation can be satisfied is if the change in concentration with distance is zero. Therefore, the influent concentration must equal the effluent concentration.

An alternative approach that is often used in deriving the mass-balance equation for a plug-flow reactor is to consider a volume element ΔV extending over the entire cross section of the reactor with the inlet face of the element located at x and the outlet face at $x + \Delta x$. The mass balance is then written for a time period Δt. Equation 5-50 is then obtained by simplifying and taking the limit of the derivatives.

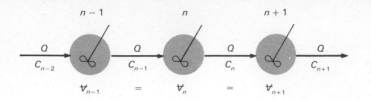

Figure 5-6 Schematic of identical continuous-flow stirred-tank reactors in series.

Continuous-flow stirred-tank reactors in series In Sec. 5-4, which deals with process analysis, it will be shown that, depending on the rate of reaction, the use of a series of continuous-flow stirred-tank reactors may have certain advantages with respect to treatment. It is therefore important to understand the hydraulic characteristics of reactors in series, such as those shown in Fig. 5-6.

Assume that a slug of dye is placed into the first reactor of a series of equally sized reactors so that the resulting concentration of dye in the first reactor is C_0. The total volume of all the reactors is V, and the volume of an individual reactor is V/n. Writing a materials balance for the second reactor results in the following:

1. General word statement:

$$\begin{array}{l}\text{Rate of} \\ \text{accumulation of} \\ \text{tracer within the} \\ \text{second reactor}\end{array} = \begin{array}{l}\text{rate of flow} \\ \text{of tracer} \\ \text{into the} \\ \text{second reactor}\end{array} - \begin{array}{l}\text{rate of flow} \\ \text{of tracer} \\ \text{out of the} \\ \text{second reactor}\end{array} \qquad (5\text{-}51)$$

2. Simplified word statement:

$$\text{Accumulation} = \text{inflow} - \text{outflow} \qquad (5\text{-}52)$$

3. Symbolic representation:

$$\frac{V}{n}\frac{dC_2}{dt} = QC_1 - QC_2 \qquad (5\text{-}53)$$

or

$$\frac{dC_2}{dt} + \frac{nQ}{V}C_2 = \frac{nQ}{V}C_1 \qquad (5\text{-}54)$$

By using Eq. 5-45, the effluent concentration from the first reactor is given by

$$C_1 = C_0 e^{-n(Q/V)t} = C_0 e^{-nt/t_0} = C_0 e^{-n\theta} \qquad (5\text{-}55)$$

Substituting this expression for C_1 in Eq. 5-54 results in

$$\frac{dC_2}{dt} + \frac{nQ}{V}C_2 = \frac{nQ}{V}C_0 e^{-n(Q/V)t} \qquad (5\text{-}56)$$

Eq. 5-56 can be solved using exactly the same procedure as outlined previously

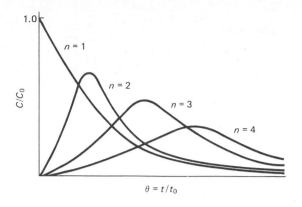

$\theta = t/t_0$

Figure 5-7 Effluent-concentration curves for each of four reactors in series.

for the solution of Eq. 5-24. Carrying through the necessary steps, the resulting solution expressed in terms of θ is

$$C_2 = C_0\, n\theta e^{-n\theta} \tag{5-57}$$

The generalized expression for the effluent concentration from the ith reactor is

$$C_i = \frac{C_0}{(n-1)!}\, (n\theta)^{i-1} e^{-n\theta} \tag{5-58}$$

The effluent-concentration curves that are obtained using Eq. 5-58 for one, two, three, or four reactors in series are shown in Fig. 5-7.

Nonideal plug-flow reactor In most full-scale plug-flow reactors, the flow usually is nonideal because of entrance and exit flow disturbances and axial dispersion. Depending on the magnitude of these effects, the ideal effluent-tracer curves may look like the curves shown in Fig. 5-8.

Because it is difficult to model these effects, the combined nonideal effects often are analytically simulated by replacing the plug-flow reactor with a series of continuous-flow stirred-tank reactors, as shown in Fig. 5-9. In this situation, the hydraulic characteristics of the simulated plug-flow reactor are modeled by plotting the fraction of material remaining in the series of continuous-flow stirred-tank reactors versus the dimensionless detention-time parameter θ. The fraction of tracer remaining in the system F, at any time t, is equal to

$$F = \frac{(V/n)C_1 + (V/n)C_2 + \cdots + (V/n)C_n}{(V/n)C_0}$$

$$F = \frac{C_1 + C_2 + \cdots + C_n}{C_0} \tag{5-59}$$

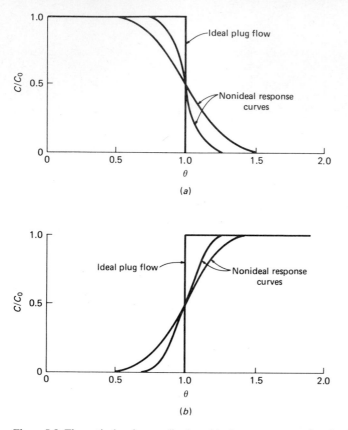

Figure 5-8 Theoretical and generalized nonideal response curves for plug-flow reactor. (*a*) Continuous purging of tracer. (*b*) Continuous input of tracer.

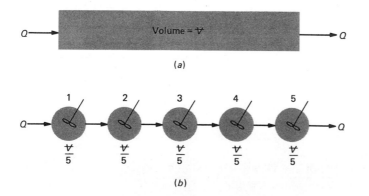

Figure 5-9 Definition sketch for the hydraulic analysis of a plug-flow reactor with dispersion using continuous-flow stirred-tank reactors in series. (*a*) Original plug-flow reactor. (*b*) Substituted reactor composed of continuous-flow stirred-tank reactors in series.

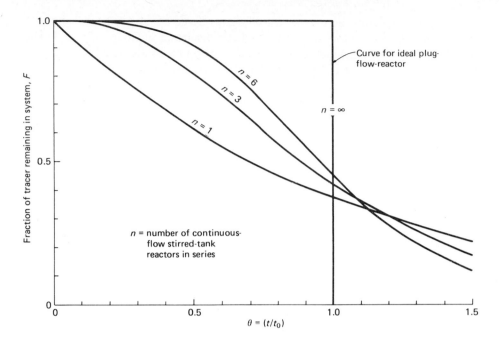

Figure 5-10 Response curves for continuous-flow stirred-tank reactors in series.

Using Eq. 5-59 to obtain the effluent concentration for a series of three continuous-flow stirred-tank reactors, the corresponding expression is

$$F_{3\theta} = \frac{C_0 e^{-3\theta} + C_0(3\theta)e^{-3\theta} + C_0(3\theta)^2 e^{-3\theta}/2}{C_0}$$

$$F_{3\theta} = \left[1 + 3\theta + \frac{(3\theta)^2}{2} \right] e^{-3\theta} \tag{5-60}$$

Curves of the fraction of tracer remaining in the series of continuous-flow stirred-tank reactors made up of one, three, and six continuous-flow stirred-tank reactors in series are shown in Fig. 5-10. For example, using six reactors, about 91 percent of the flow remains in the system of reactors for a time period equal to at least $\theta = 0.5$.

5-4 PROCESS ANALYSIS

Process analysis involves the detailed evaluation of the various factors, and their interrelationships, that must be considered when selecting unit operations and processes and other treatment methods to meet some stated treatment objectives. The purpose of process analysis is to select the most suitable unit operations and processes and the optimum operational criteria.

The most important factors that must be considered are identified in Table 5-3. The first six of these factors have already been discussed in Chaps. 2 and 3. The seventh factor, reaction kinetics and reactor selection, is discussed in Sec. 5-5. The other factors will be discussed throughout the remainder of the book. They are identified here to indicate the diverse nature of the information that must be available to make a proper evaluation of unit operations and processes used for the treatment of wastewater.

Table 5-3 Important factors that must be considered when selecting and evaluating unit operations and processes[a]

Factor	Comment
1. Process applicability	The applicability of a process is evaluated on the basis of past experience, data from full-scale plants, and pilot data from plant studies. If new or unusual conditions are encountered, pilot-plant studies are necessary
2. Applicable flow range	The process should be matched to the expected flow range. For example, stabilization ponds are not suitable for extremely large flows
3. Applicable flow variation	Most unit operations and processes work best with a constant flowrate, although some variation can be tolerated. If the flow variation is too great, flow equalization may be necessary
4. Influent-wastewater characteristics	The characteristics of the influent affect the types of processes to be used (e.g., chemical or biological) and the requirements for their proper operation
5. Inhibiting and unaffected constituents	What constituents are present that may be inhibitory, and under what conditions? What constituents are not affected during treatment?
6. Climatic constraints	Temperature affects the rate of reaction of most chemical and biological processes. Freezing conditions may affect the physical operation of the facilities
7. Reaction kinetics and reactor selection	Reactor sizing is based on the governing reaction kinetics. Data for kinetic expressions usually are derived from experience, the literature, and the results of pilot-plant studies. The effect of reaction kinetics on reactor selection is considered in Sec. 5-5
8. Performance	Performance is most often measured in terms of effluent quality, which must be consistent with the given effluent-discharge requirements
9. Treatment residuals	The types and amounts of solid, liquid, and gaseous residuals produced must be known or estimated. Often, pilot-plant studies are used to identify residuals properly
10. Sludge-handling constraints	Are there any constraints that would make sludge handling expensive or infeasible? In many cases, a treatment method should be selected only after the sludge processing and handling options have been explored

(*continued*)

Table 5-3 (*continued*)

Factor	Comment
11. Environmental constraints	Nutrient requirements must be considered for biological treatment processes. Environmental factors, such as the prevailing winds and wind directions, may restrict the use of certain processes, especially where odors may be produced
12. Chemical requirements	What resources and what amounts must be committed for a long period of time for the successful operation of the unit operation or process?
13. Energy requirements	The energy requirements, as well as probable future energy costs, must be known if cost-effective treatment systems are to be designed
14. Other resource requirements	What, if any, additional resources must be committed to the successful implementation of the proposed treatment system using the unit operation or process in question?
15. Reliability	What is the long-term record of the reliability of the unit operation or process under consideration? Is the operation process easily upset? Can it stand periodic shock loadings? If so, how do such occurrences affect the quality of the effluent?
16. Complexity	How complex is the process to operate under routine conditions and under emergency conditions such as shock loadings? What level of training must the operator have to operate the process?
17. Ancillary processes required	What support processes are required? How do they affect the effluent quality, especially when they become inoperative?
18. Compatibility	Can the unit operation or process be used successfully with existing facilities? Can plant expansion be accomplished easily? Can the type of reactor be modified?

[a] Adapted in part from Refs. 4, 5, and 8.

Selection of Reaction-Rate Expressions or Loading Criteria

On the basis of experience or from data derived from pilot-plant studies, appropriate reaction-rate expressions must be developed to describe the process that is to be designed. Such expressions usually have the form of rate equations, such as those summarized in Table 5-1. The various rate expressions that have been developed for biological waste treatment are considered in Chap. 9, and their application is illustrated in Chap. 10.

If appropriate rate expressions cannot be developed, generalized loading criteria are sometimes used. Such criteria usually have evolved from years of empirical observations. For example, if a process that is loaded at 10 kg/m^3 produces an acceptable effluent and one loaded at 20 kg/m^3 does not, the successful experience tends to be repeated. Unfortunately, records often are not well maintained, and the limits of such loading criteria are seldom defined.

Selection of the Type of Reactor

One of the important considerations in the design of any chemical or biological process is the selection of the type of reactor or reactors to be used in the treatment process. Operational factors that must be considered include (1) the nature of the wastewater to be treated, (2) the reaction kinetics governing the treatment process, (3) process requirements, and (4) local environmental conditions. In practice, initial construction costs and operation and maintenance costs also affect reactor selection. Because the relative importance of these factors varies with each application, each factor should be considered separately when the type of reactor is to be selected. The effects of reaction kinetics on reactor selection are considered in more detail in Sec. 5-5.

5-5 REACTION KINETICS AND REACTOR SELECTION

The purpose of this section is to illustrate the type of analysis involved in considering the effect of reaction kinetics on reactor selection. This will be accomplished using a reactor system composed of (1) a series of identical continuous-flow stirred-tank reactors, and (2) a plug-flow reactor with axial dispersion and arbitrary entrance and exit conditions.

Continuous-Flow Stirred-Tank Reactors in Series with Conversion

Assuming, for the purposes of illustration, that substrate removal is governed by a first-order reaction ($r_C = -kC$), ideal flow in a series of continuous-flow stirred-tank reactors will be considered first. In these reactors, the effluent from one reactor serves as the influent to the next reactor, as shown in Fig. 5-9. A materials balance around the nth reactor yields the following:

1. General word statement:

$$
\begin{matrix}
\text{Rate of} & \text{rate of flow} & \text{rate of flow} & \text{rate of} \\
\text{accumulation of} & \text{of substrate} & \text{of substrate} & \text{substrate} \\
\text{substrate within} = & \text{into the } n\text{th} - & \text{out of the} + & \text{utilization in} \\
\text{the } n\text{th reactor} & \text{reactor} & n\text{th reactor} & \text{the } n\text{th reactor}
\end{matrix}
\tag{5-61}
$$

2. Simplified word statement:

$$\text{Accumulation} = \text{inflow} - \text{outflow} + \text{utilization} \tag{5-62}$$

3. Symbolic representation:

$$\frac{V}{n}\frac{dC_n}{dt} = QC_{n-1} - QC_n + \frac{V}{n}(-kC_n) \tag{5-63}$$

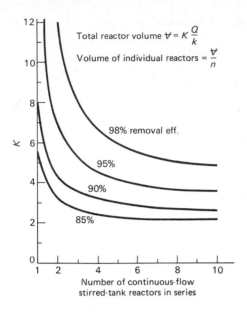

Total reactor volume $V = K \dfrac{Q}{k}$

Volume of individual reactors $= \dfrac{V}{n}$

98% removal eff.

95%

90%

85%

Number of continuous-flow
stirred-tank reactors in series

Figure 5-11 Values of K used to determine the total volume required versus the number of continuous-flow stirred-tank reactors in series for various removal efficiencies for first order removal kinetics $(r_c = -kC)$.

At steady state $(dC_n/dt = 0)$, Eq. 5-63 can be simplified to yield

$$\frac{C_n}{C_{n-1}} = \frac{1}{1 + kV/nQ} \tag{5-64}$$

Applying Eq. 5-64 to n reactors in series results in

$$\frac{C_n}{C_0} = \frac{1}{(1 + kV/nQ)^n} \tag{5-65}$$

where V = volume of all reactors in series
$\quad n$ = number of reactors in series
$\quad C_0$ = influent substrate concentration

The total volume V required for various removal efficiencies can be expressed in terms of the flowrate Q and the specific reaction-rate constant k by rewriting Eq. 5-65 as follows:

$$V = \frac{nQ}{k}\left[\left(\frac{C_0}{C_n}\right)^{1/n} - 1\right] \tag{5-66}$$

The total volume required for various removal efficiencies for first-order kinetics, using 1, 2, 4, 6, 8, or 10 reactors in series (see Fig. 5-9), is reported in Table 5-4. It is also shown graphically in Fig. 5-11. Second-order kinetics are considered in Example 5-2 (following the discussion of plug-flow with conversion).

Table 5-4 Required reactor volumes expressed in terms of Q/k for continuous-flow stirred-tank reactors in series and plug-flow reactors for various removal efficiencies for first-order kinetics[a]

No. of reactors in series	Reactor volume $V = K(Q/k)$			
	85% removal efficiency	90% removal efficiency	95% removal efficiency	98% removal efficiency
1	5.67	9.00	19.00	49.00
2	3.18	4.32	6.96	12.14
4	2.48	3.10	4.48	6.64
6	2.22	2.82	3.90	5.50
8	2.16	2.64	3.60	5.04
10	2.10	2.60	3.50	4.80
Plug flow	1.90	2.30	3.00	3.91

[a] Volume of individual reactors equals value in table divided by the number of reactors in series.

Plug-Flow Reactor with Conversion

In the extreme, as the number of continuous-flow stirred-tank reactors in Fig. 5-9 is increased, the required volume approaches that required for a plug-flow reactor. The required volume can be determined by writing a materials balance around the element shown in Fig. 5-5.

1. General word statement:

$$
\begin{array}{l}
\text{Rate of} \\
\text{accumulation of} \\
\text{substrate within} \\
\text{the } n\text{th reactor}
\end{array}
=
\begin{array}{l}
\text{rate of flow} \\
\text{of substrate} \\
\text{into the } n\text{th} \\
\text{reactor}
\end{array}
-
\begin{array}{l}
\text{rate of flow} \\
\text{of substrate} \\
\text{out of the} \\
n\text{th reactor}
\end{array}
+
\begin{array}{l}
\text{rate of} \\
\text{substrate} \\
\text{utilization in} \\
\text{the } n\text{th reactor}
\end{array}
\tag{5-67}
$$

2. Simplified word statement:

$$
\text{Accumulation} = \text{inflow} - \text{outflow} + \text{utilization}
\tag{5-68}
$$

3. Symbolic representation:

$$
dV \frac{d\left[\dfrac{C + C + (\partial C/\partial x)\, dx}{2}\right]}{dt}
$$

$$
= QC - Q\left(C + \frac{\partial C}{\partial x}\, dx\right) + dV(-k)\left[\frac{C + C + (\partial C/\partial x)\, dx}{2}\right]
\tag{5-69}
$$

Neglecting second-order terms, Eq. 5-69 can be simplified to yield

$$dV \frac{\partial C}{\partial t} = -Q \frac{\partial C}{\partial x} dx - dV k C \tag{5-70}$$

At steady state $(\partial C/\partial t = 0)$. Thus $\tag{5-71}$

$$dV = -\frac{Q}{k} \frac{dC}{C} \tag{5-72}$$

If $A\,dx$ is substituted for dV, Eq. 5-72 can be integrated between the limits of 0 and L and C_0 and C

$$A \int_0^L dx = -\frac{Q}{k} \int_{C_0}^C \frac{dC}{C} \tag{5-73}$$

to obtain

$$AL = V = -\frac{Q}{k} \ln \frac{C}{C_0} \tag{5-74}$$

or

$$\frac{C}{C_0} = e^{-k(V/Q)} \tag{5-75}$$

which is also the same expression that is obtained for a batch reactor. The volume required to achieve the corresponding degrees of treatment reported in Table 5-3 using a plug-flow reactor with a first-order reaction is also reported in Table 5-3. It is noted that in chemical engineering literature, the steady-state expression for a plug-flow reactor is often written as

$$V = Q \int \frac{dC}{r} \tag{5-76}$$

Example 5-2: Comparison of required reactor volumes for second-order kinetics Assuming that second-order kinetics apply $(r_C = -kC^2)$, compare the required volume of a continuous-flow stirred-tank reactor to the volume of a plug-flow reactor to achieve a 90 percent reduction in the concentration $(C_0 = 1$ and $C_e = 0.1)$.

SOLUTION
1 Compute the required volume for a continuous-flow stirred-tank reactor in terms of Q/k.
   a At steady state, a mass balance for a continuous-flow stirred-tank reactor yields (see Eq. 5-63)

$$0 = QC_0 - QC_e - V k C_e^2$$

   b Simplify and substitute the given data.

$$V = \frac{Q}{k} \frac{C_0 - C_e}{C_e^2} = \frac{Q}{k} \frac{1 - 0.1}{(0.1)^2} = 90 \frac{Q}{k}$$

2 Compute the required volume for a plug-flow reactor in terms of Q/k.
   a At steady state, a mass balance for a plug-flow reactor yields (see Eq. 5-70)

$$0 = -Q \frac{dC}{dx} dx + A\,dx(-kC^2)$$

b The integrated form of the steady-state equation is

$$V = \frac{Q}{k} \int_{C_0}^{C_e} \frac{dC}{C^2} = \frac{Q}{k} \frac{1}{C} \Big]_{C_0}^{C_e} = \frac{Q}{k} \left(\frac{1}{C_e} - \frac{1}{C_0} \right)$$

c Substituting the given concentration values yields

$$V = \frac{Q}{k} \left(\frac{1}{0.1} - \frac{1}{1} \right) = \frac{9Q}{k}$$

3 Determine the volume ratio.

$$\frac{V_{\text{CFSTR}}}{V_{\text{PFR}}} = \frac{9Q/k}{9Q/k} = 10$$

Comparison of Continuous-Flow Stirred-Tank and Plug-Flow Reactors

From the previous discussions it can be concluded that, for first-order substrate-removal kinetics, the total volume required for a series of continuous-flow stirred-tank reactors (four or more) is considerably less than that required for a single continuous-flow stirred-tank reactor. Further, the volume differential becomes more pronounced as the removal efficiency increases. The volume differential also increases with increasing reaction order. For example, as shown in Example 5-2, if second-order substrate kinetics governed, the relative ratio of required volumes (single-stage continuous-flow stirred-tank to plug flow) for 90 percent removal would be equal to 10 : 1. On the other hand, if zero-order substrate-removal kinetics were applicable, the ratio of required volumes would be equal to 1. Thus the form of the governing kinetic expression can greatly affect the volume requirements.

Plug-Flow Reactors with Axial Dispersion and Conversion

Consideration of a reactor with axial dispersion (nonideal flow) provides another way of interpreting the data in Table 5-4. It can be assumed that the intermediate values in Table 5-4 represent the required volume for a plug-flow reactor with varying conditions of dispersions. For example, if the contents of a plug-flow reactor were completely dispersed, the result would be equivalent to a continuous-flow stirred-tank reactor. Recognizing that in practice neither a plug-flow reactor nor a continuous-flow stirred-tank reactor functions as assumed, Wehner and Wilhelm [11] derived the following equation from a reactor with axial dispersion, first-order kinetics, and arbitrary entrance and exit conditions:

$$\frac{S}{S_0} = \frac{4a \exp (1/2d)}{(1 + a)^2 \exp (a/2d) - (1 - a)^2 \exp (-a/2d)} \tag{5-77}$$

where S = effluent substrate concentration
S_0 = influent substrate concentration
$a = \sqrt{1 + 4ktd}$
d = dispersion factor = D/uL
D = axial-dispersion coefficient, m^2/h (ft^2/h)
u = fluid velocity, m/h (ft/h)
L = characteristic length, m (ft)
k = first-order reaction constant, 1/h
t = detention time, h

To facilitate the use of Eq. 5-77, Thirumurthi, in connection with his work on stabilization ponds [9], developed Fig. 5-12, in which the term kt is plotted against S/S_0 for dispersion factors varying from zero (0) for an ideal plug-flow reactor to infinity (∞) for a continuous-flow stirred-tank reactor. Dispersion factors for conventional plug-flow activated-sludge reactors are probably within the range from 0 to 0.2. For reactors with mechanical aerators designed to operate as complete-mix systems, values of d are probably in the range from 4.0 to ∞. Most stabilization ponds are somewhere within the range from 0.1 to 2.0.

The relationship of the kt values derived from the dispersion model to the values given in Table 5-4 is as follows. For 90 percent removal efficiency (10 percent remaining) and a dispersion factor of 0.0625, the value of kt is 2.6 (see Fig. 5-12).

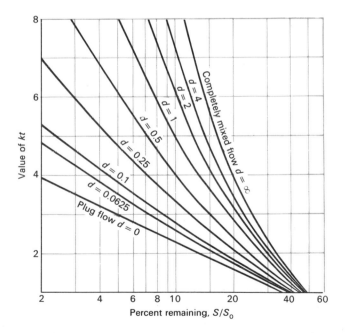

Figure 5-12 Values of kt in the Wehner and Wilhelm equation (Eq. 5-77) versus percent remaining for various dispersion factors [6].

If kt is rewritten as $k(V/Q)$, the volume V is equal to $2.6(Q/k)$. Comparing this value with the data reported in Table 5-4, it is found that the performance of a plug-flow reactor with a dispersion factor of 0.0625 is equivalent to 10 continuous-flow stirred-tank reactors in series. Thus, for a tank with a given degree of dispersion, it is possible to find its equivalent in terms of a series reactor system and to compare the volume required with that for an ideal single-stage continuous-flow stirred-tank or plug-flow reactor. For a further analysis of these and related topics, Refs. 2, 3, and 5 are recommended. The use of Fig. 5-12 is also illustrated in Example 5-3.

Example 5-3: Bacterial die-off in series of stabilization ponds It has been found that the observed die-off coefficient for *E. coli* in biological stabilization ponds can be described adequately with first-order kinetics. Assuming that the value of the specific reaction-rate constant is 1.0 d^{-1}, determine the concentration of *E. coli* in the effluent from a series of three ponds when the initial concentration is 10^6 organisms/mL and the average flowrate is 5000 m³/d (1.32 Mgal/d). The ponds are rectangular and have an average depth of 1.5 m (4.9 ft). The surface areas of the ponds are 1, 2, and 1 ha (2.47, 4.94, and 2.47 acres).

SOLUTION

1 Estimate the dispersion factors from the limited information given. Assume that the dispersion factor is 0.5 for the smaller ponds and about 0.25 for the large pond.
2 Determine the kt values for the ponds.
   a For the smaller ponds:

$$kt = k\frac{V}{Q} = k\frac{A \times d}{Q}$$

$$= 1.0\frac{1}{d}\frac{1.0 \text{ ha} \times 10{,}000 \text{ m}^2/\text{ha} \times 1.5 \text{ m}}{5000 \text{ m}^3/\text{d}}$$

$$= 3.0$$

   b For the larger pond:

$$kt = k\frac{V}{Q} = k\frac{A \times d}{Q}$$

$$= 1.0\frac{1}{d}\frac{2.0 \text{ ha} \times 10{,}000 \text{ m}^2/\text{ha} \times 1.5 \text{ m}}{5000 \text{ m}^3/\text{d}}$$

$$= 6.0$$

3 Determine the corresponding N/N_0 values from Fig. 5-12.
   a For the smaller ponds:

$$N/N_0 = 0.15$$

   b For the large pond:

$$N/N_0 = 0.03$$

4 Estimate the concentration of organisms in the effluent.

$$\frac{N}{N_0} = (0.15)(0.03)(0.15)$$

$$N = 10^6 \text{ organisms/mL } (6.75 \times 10^{-4})$$

$$= 675 \text{ organisms/mL}$$

Other Reactor Flow Regimes and Reactor Combinations

In the previous discussions of plug-flow and continuous-flow stirred-tank reactors, a single-pass straight-through flow pattern has been used for the purpose of analyses. In practice, other flow regimes and reactor combinations are also used.

Some of the more common alternative flow regimes are shown schematically in Fig. 5-13. The flow regime shown in Fig. 5-13a is used to achieve intermediate levels of treatment by blending various amounts of treated and untreated wastewater. The flow regime used in Fig. 5-13b is often adopted to achieve greater process control and will be considered specifically in Chaps. 9 and 10. The flow regime shown in Fig. 5-13c is used to reduce the loading applied to the process. Each of these hydraulic regimes is considered further in the following chapters.

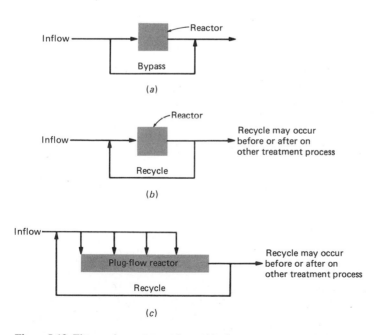

Figure 5-13 Flow regimes commonly used in the treatment of wastewater. (a) Bypass flow (plug-flow or continuous-flow stirred-tank reactor). (b) With recycle flow (plug-flow or continuous-flow stirred-tank reactor). (c) Step input with or without recycle (plug-flow reactor).

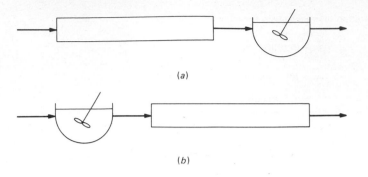

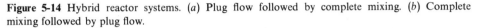

Figure 5-14 Hybrid reactor systems. (*a*) Plug flow followed by complete mixing. (*b*) Complete mixing followed by plug flow.

Among the numerous types of reactor combinations that are possible and that have been used, two combinations using a plug-flow reactor and a continuous-flow stirred-tank reactor are shown in Fig. 5-14. In the arrangement shown in Fig. 5-14*a*, complete mixing takes place later; in the arrangement shown in Fig. 5-14*b*, it occurs first. If no reaction takes place and the reactors are used only to equalize temperature, for example, the result will be identical [2]. If a reaction is occurring, however, the product yields of the two reactor systems can be different (see Prob. 5-14). The use of such hybrid reactor systems will depend on the specific product requirements. Additional details on the analysis of such processes may be found in Refs. 1, 2, 3, and 5.

5-6 PRACTICAL ASPECTS OF REACTOR DESIGN

Although many of the practical aspects of process design are discussed in detail in the chapters that follow, some of the practical aspects of reactor design should be mentioned here. Attention is called to this subject because often it is neglected or not considered properly. As a result, many of the treatment plants that have been built do not perform hydraulically as designed.

One of the more important practical considerations involved in reactor design is how to achieve the ideal conditions postulated in the analysis of their performance. For example, when a continuous-flow stirred-tank reactor is designed, how is the flow to be introduced to satisfy the theoretical requirement of instantaneous and complete dispersion? The answer is that, in practice, there is always some deviation from ideal conditions, and it is the precautions taken to minimize these effects that are important. A typical inlet and outlet arrangement for a continuous-flow stirred-tank reactor is shown in Fig. 5-15.

Another common problem affecting reactor performance is short circuiting. If the incoming wastewater is either colder or warmer than the liquid in the reactor, short circuiting in plug-flow reactors can be caused by density currents (see Fig. 5-16*a*). In warm climates, the performance of large sedimentation tanks can also be affected by density currents. Correction of the short circuiting in plug-flow reactors may require the use of submerged deflection baffles located

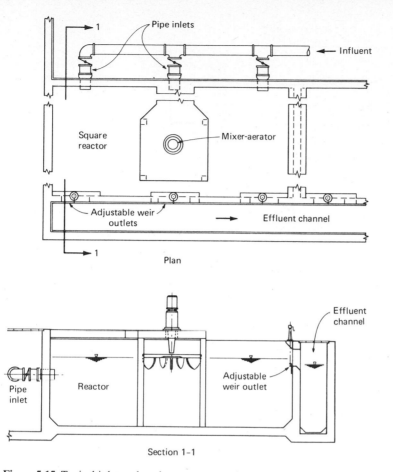

Figure 5-15 Typical inlet and outlet arrangement for continuous-flow stirred-tank reactor.

at either the top or bottom of the bank. Alternatively, some form of energy input may be required to equalize the temperature. Both mechanical mixers and diffused air have been used for this purpose. Inefficient use of the reactor volume and short circuiting caused by inadequate mixing (see Fig. 5-16b) can also be a problem in complete-mix reactors, especially where rectangular tanks are used.

DISCUSSION TOPICS AND PROBLEMS

5-1 The following data were obtained for the reaction $A \rightarrow B + C$. Determine the order of the reaction and the value of the reaction rate constant k.

t, min	0	10	20	40	60
A, mg/L	90	72	57	36	23

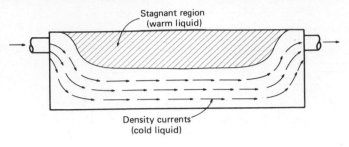

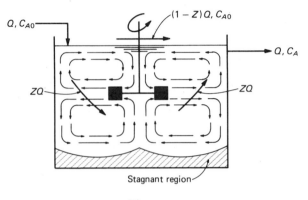

(b)

Figure 5-16 Short circuiting caused by density currents and inadequate mixing. (a) Short circuiting caused by density currents in plug-flow reactors. (b) Short circuiting caused by inadequate mixing in continuous-flow stirred-tank reactor.

5.2 A bimolecular reaction $A + B \rightarrow P$ is 10 percent complete in 10 min. If the initial concentration of A and B is equal to 1.0 mol/L, determine the reaction rate constant and how long it will take for the reaction to be 90 percent complete.

5-3 The following values have been obtained for the rate constant for the reaction $A + B \rightarrow P$.

$$k_{25^\circ C} = 1.5 \times 10^{-2} \text{ L/mol} \cdot \text{min}$$

$$k_{45^\circ C} = 4.5 \times 10^{-2} \text{ L/mol} \cdot \text{min}$$

Using these data, determine the activation energy E and the value of the rate constant at 15°C.

5-4 An aqueous reaction is being studied in a laboratory-sized CFSTR with a volume of 5 L. The stoichiometry of the reaction is $A \rightarrow 2R$, and reactant A is introduced into the reactor at a concentration of 1 mol/L. From the results given in the table, find the rate expression for this reaction. Assume steady-state flow.

Run	Feed rate, cm^3/s	Temperature, °C	Concentration of R in effluent, mol/L
1	2	13	1.8
2	15	13	1.5
3	15	84	1.8

5-5 The rate of reaction for an enzyme-catalyzed substrate in a batch reactor can be described by the following relationship.

$$r_c = -\frac{kC}{K + C}$$

where k = maximum reaction rate, mg/L · min
  C = substrate concentration, mg/L
  K = constant, mg/L

Using this rate expression, derive an equation that can be used to predict the reduction of substrate concentration with time in a batch reactor. If $k = 40$ mg/L · min and $K = 100$ mg/L, determine the time required to decrease the substrate concentration from 1000 mg/L to 100 mg/L.

5-6 A wastewater is to be treated in a continuous-flow stirred-tank reactor. Assuming that the reaction is irreversible and first-order $(r_c = -kC)$ with a reaction rate equal to 0.15 d^{-1}, determine the flowrate that can be treated if the reactor has a volume of 20 m^3 and 98 percent treatment efficiency is required. What volume would be required to treat the flowrate determined above if the required treatment efficiency is 92 percent?

5-7 For first-order removal kinetics, demonstrate that the maximum treatment efficiency in a series of continuous-flow stirred-tank reactors occurs when all the reactors are the same size.

5-8 Determine the number of completely mixed chlorine contact chambers each having a detention time of 30 min that would be required in a series arrangement to reduce the bacterial count of a polluted water sample from 10^6 organism/mL to 14.5 organisms/mL if the first-order removal rate constant is equal to 6.1 h^{-1}. If a plug-flow chlorine contact chamber were used with the same detention time as the series completely mixed chambers, what would the bacterial count be after treatment.

5-9 The concentration of ultimate BOD in a river entering the first of two lakes that are connected in series is equal to 20 mg/L. If the first-order BOD reaction rate constant (K_1) equals 0.35 d^{-1} and complete mixing occurs in each lake, what is the concentration of ultimate BOD at the outlet of each lake? The flow in the river is equal to 4000 m^3/d and the volume of the first and second lake is 20,000 and 12,000 m^3 respectively. Assume steady-state conditions.

5-10 In Prob. 5-9, if the length of the river connection between the two lakes is equal to 3 km and the velocity in the river is equal to 0.4 m/s, determine the concentration of the ultimate BOD in the effluent from the second lake.

5-11 Plot the ratio of required tank volume for a plug-flow reactor to that of a continuous-flow stirred-tank reactor (V_{PFR}/V_{CFSTR}) versus the fraction of the original substrate that is converted for the following reaction rates.

$r = -k$

$r = -kC^{0.5}$

$r = -kC$

$r = -kC^2$

What is the value of the required volume ratio for each of these rates when $C = 0.25$ mg/L and $C_0 = 1.0$ mg/L?

5-12 What explanation can you offer for the tracer curves shown in Fig. 5.17 obtained for the same plug-flow chlorine contact basin?

Figure 5-17 Tracer curves for a plug-flow chlorine contact basin for Prob. 5-12.

5-13 What conclusions can you draw about the performance of the reactors whose hydraulic characteristics are shown in Fig. 5.18?

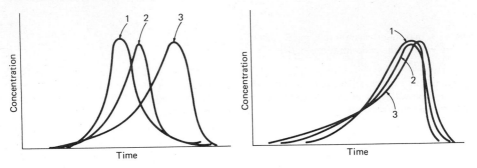

Figure 5-18 Tracer curves for two different plug-flow reactors with the same total volume and geometry for Prob. 5-13.

5.14 If second-order reaction kinetics are applicable ($r_c = -kC^2$), determine the effluent concentration for each of the reactor systems shown in Fig. 5.14. To simplify the computations, assume that the following data apply:

$$k = 1.0 \text{ m}^3/\text{kg} \cdot \text{d}$$
$$Q = 1.0 \text{ m}^3/\text{d}$$
$$V_{PFR} = 1.0 \text{ m}^3$$
$$V_{CFSTR} = 1.0 \text{ m}^3$$
$$C_0 = 1.0 \text{ kg/m}^3$$

Explain your results. What would happen if first- or zero-order kinetics are applicable?

5-15 A portion of the outflow from an ideal plug flow reactor is recycled as shown in the following sketch, where $\alpha \geq 0$. Assume that the rate of conversion can be defined as $r_c = -kC$.

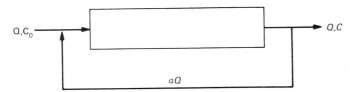

(*a*) Sketch the generalized curve of conversion versus the recycle ratio.
(*b*) Sketch a family of curves showing the effect of the recycle ratio α on the longitudinal concentration gradient.
(*c*) If a continuous-flow stirred tank reactor were substituted for the plug-flow reactor, what effect would the recycle have on conversion?
(*d*) What effect does the recycle ratio α have on the residence time?

REFERENCES

1. Denbigh, K. G., and J. C. R. Turner: *Chemical Reactor Theory*, 2d ed., Cambridge, New York, 1965.
2. Kafarou, V.: *Cybernetic Methods in Chemistry and Chemical Engineering*, MIR Publishers, Moscow, 1976.

3. Levenspiel, O.: *Chemical Reaction Engineering*, 2d ed., Wiley, New York, 1972.
4. Metcalf & Eddy, Inc.: *Report to National Commission on Water Quality on Assessment of Technologies and Costs for Publicly Owned Treatment Works*, vol. 2, prepared under Public Law 92-500, Boston, 1975.
5. Mills, R. A., and G. Tchobanoglous: Energy Consumption in Wastewater Treatment, in W. S. Jewell (ed.), *Proceedings of the Seventh National Agricultural Waste Management Conference*, Ann Arbor Science Publishers, Ann Arbor, Mich., 1976.
6. Schroeder, E. D.: *Water and Wastewater Treatment*, McGraw-Hill, New York, 1977.
7. Smith, J. M.: *Chemical Engineering Kinetics*, 2d ed., McGraw-Hill, New York, 1970.
8. Tchobanoglous, G.: Wastewater Treatment for Small Communities, *Public Works*, parts I, II, vol. 105, no. 7, July 1974; vol. 105, no. 8, August 1974.
9. Thirumurthi, D.: Design Principles of Waste Stabilization Ponds, *J. Sanit. Eng. Div., ASCE*, vol. 95, no. SA2, 1969.
10. Thomas, R. V.: *Systems Analysis and Water Quality Management*, Environmental Sciences Service Corporation, Stamford, Conn., 1972.
11. Wehner, J. F., and R. F. Wilhelm: Boundary Conditions of Flow Reactor, *Chem. Eng. Sci.*, vol. 6, p. 89, 1958.

SIX

PHYSICAL UNIT OPERATIONS

Those operations used for the treatment of wastewater in which change is brought about by means of or through the application of physical forces are known as unit operations. Because they were derived originally from observations of the physical world, they were the first treatment methods to be used. Today, physical unit operations form the basis of most process flowsheets. Those used in a typical flowsheet for wastewater treatment are identified in Fig. 6-1.

The unit operations most commonly used in wastewater treatment include (1) screening, (2) comminution, (3) flow equalization, (4) mixing, (5) flocculation, (6) sedimentation, (7) flotation, and (8) filtration. The principal applications of these operations are summarized in Table 6-1. With the exception of comminution, which is discussed in Chap. 8, a separate section is devoted to each of these operations in this chapter. A discussion of comminution is not included in this chapter because comminutors are complete in themselves as supplied by the manufacturer, and no detailed theoretical analysis is possible. Microscreening, another unit operation that on occasion has been used to remove residual suspended solids, is discussed briefly in Chap. 8. Unit operations associated with the processing of sludge are discussed separately in Chap. 11.

In this chapter, each unit operation to be considered will be described, and the fundamentals involved in the engineering analysis of each one will be discussed. The practical application of these operations in the design of facilities is detailed in Chaps. 8 and 10. This same approach will be used in Chaps. 7 and 9, which deal with the chemical and biological unit processes.

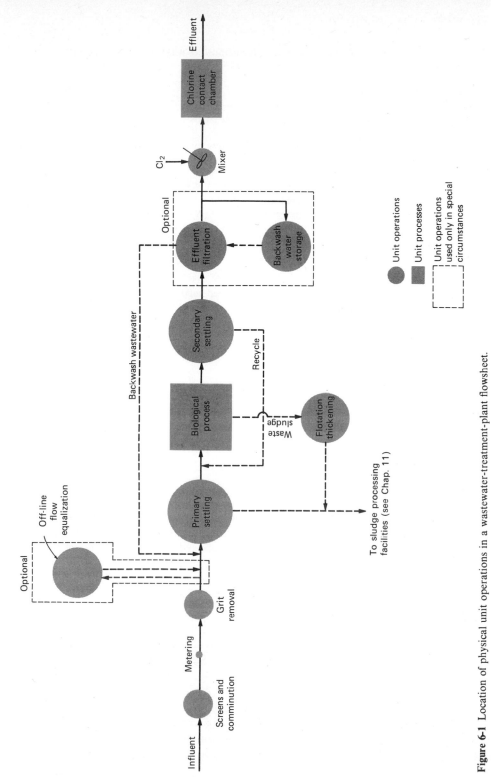

Figure 6-1 Location of physical unit operations in a wastewater-treatment-plant flowsheet.

Table 6-1 Applications of physical unit operations in wastewater treatment

Operation	Application	See Section
Screening	Removal of coarse and settleable solids by interception (surface straining)	6-1
Comminution	Grinding of coarse solids to a more or less uniform size	8-2
Flow equalization	Equalization of flow and mass loadings of BOD and suspended solids	6-2
Mixing	Mixing of chemicals and gases with wastewater, and maintaining solids in suspension	6-3
Flocculation	Promotes the aggregation of small particles into larger particles to enhance their removal by gravity sedimentation	6-4
Sedimentation	Removal of settleable solids and thickening of sludges	6-5
Flotation	Removal of finely divided suspended solids and particles with densities close to that of water. Also thickens biological sludges	6-6
Filtration	Removal of fine residual suspended solids remaining after biological or chemical treatment	6-7
Microscreening	Same as filtration. Also removal of algae from stabilization-pond effluents	8-10

6-1 SCREENING

The first unit operation encountered in wastewater-treatment plants is screening. A screen is a device with openings, generally of uniform size, that is used to retain the coarse solids found in wastewater.

Description

The screening element may consist of parallel bars, rods or wires, grating, wire mesh, or perforated plate, and the openings may be of any shape but generally are circular or rectangular slots. A screen composed of parallel bars or rods is called a rack. Although a rack is a screening device, the term "screen" should be limited to the type with wire cloth or perforated plates. However, the function performed by a rack is called screening, and the materials removed by it are known as screenings or rakings.

According to the method used to clean them, racks and screens are designated as hand-cleaned or mechanically cleaned. Typically, racks have clear openings (spaces between bars) of 25 mm (1 in) or more. Screens have openings of 6 mm (1/4 in) or less.

Racks In wastewater treatment, bar racks are used to protect pumps, valves, pipelines, and other appurtenances from damage or clogging by rags and large objects. Industrial waste plants may or may not need them, depending on the character of the wastes. One type of mechanically cleaned bar rack is shown in

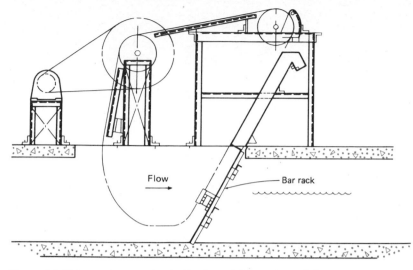

Figure 6-2 Catenary-type mechanically cleaned bar rack used for wastewater treatment. (*From Jeffrey Mfg.*)

Fig. 6-2. The bars run vertically or at a slope varying from 30 to 80° with the horizontal. Large objects are caught on the rack, carried up by traveling rakes, scraped off, and collected.

Screens In the 1920s and earlier, fine screens of the inclined disk or drum type, whose screening media consisted of bronze or copper plates with milled slots, were installed in place of sedimentation tanks for primary treatment. Since the early 1970s, there has been a resurgence of interest in the field of wastewater treatment in the use of screens of all types. The applications range from primary treatment to the removal of the residual suspended solids from biological treatment processes. To a large extent, this renewed interest developed because better screening materials and better screening devices are now available, and research is continuing in this area. The principal types of screens now in use are described in Table 6-2, and typical screening devices are shown in Fig. 6-3. As shown in Fig. 6-3b, where screens operate partially submerged, and rotate continuously or intermittently, the solids are flushed by sprays from the exposed screen surface into a collecting trough each revolution. The traveling screens shown in Fig. 6-3c are used extensively in water-supply systems and, on occasion, for effluent screening. The use of screening devices is considered in Chaps. 8, 10, and 12. Design information can also be found in those chapters.

Analysis

The analysis associated with the use of screens involves the determination of the head loss through them. The approach used for racks differs from that used for fine screens; so they are discussed separately.

Table 6-2 Description of screening devices used in wastewater treatment

Type of screen	Size classification	Screening surface		Application	See Figure
		Size range	Screen material		
Inclined: Fixed	Medium	250–1500 μ	Stainless-steel wedge-wire screen	Primary treatment	6-3a
Rotary	Coarse	0.8–2.4 mm × 50-mm slots	Milled bronze or copper plates	Pretreatment	
	Medium	100–1000 μ	Stainless-steel wire cloth	Primary treatment	
Drum (rotary)	Coarse	0.8–2.4 mm × 50-mm slots	Milled bronze or copper plates, wire screen	Pretreatment	
	Medium	250–1500 μ	Stainless-steel wedge-wire screen	Primary treatment	6-3b
	Fine	15–60 μ	Stainless-steel and polyester screen cloths	Removal of residual secondary suspended solids	
Traveling	Coarse to medium		Stainless steel or other noncorrosive material	Primary treatment	6-3c
Centrifugal	Fine–medium	10–500 μ	Stainless steel, polyester, and various other fabric screen cloths	Primary treatment, secondary treatment with settling tank, and the removal of residual secondary suspended solids	6-3d

Note: mm × 0.03937 = in.

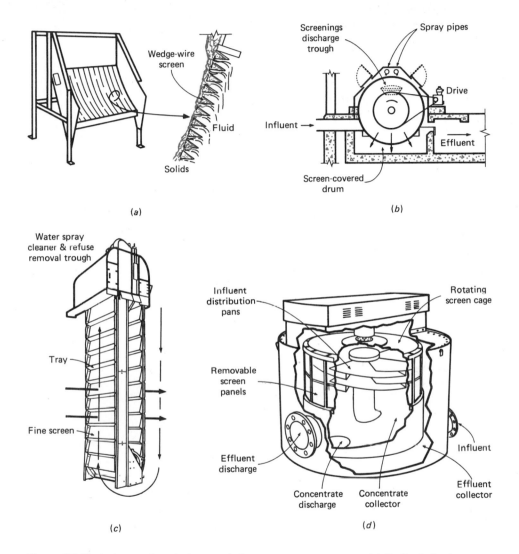

Figure 6-3 Typical screening devices used for wastewater treatment. (*a*) Inclined fixed screen. (*b*) Rotary drum screen. (*c*) Traveling screen. (*From FMC, Link-Belt.*) (*d*) Centrifugal screen. (*From SWECO, Inc.*)

Table 6-3 Kirschmer's values of β [13]

Bar type	β
Sharp-edged rectangular	2.42
Rectangular with semicircular upstream face	1.83
Circular	1.79
Rectangular with semicircular upstream and downstream faces	1.67

Racks Hydraulic losses through bar racks are a function of bar shape and the velocity head of the flow between the bars. Kirschmer [13] has proposed the following equation for head loss:

$$h_L = \beta \left(\frac{w}{b}\right)^{4/3} h_v \sin \theta \tag{6-1}$$

where h_L = head loss, m (ft)
 β = a bar-shape factor (see Table 6-3)
 w = maximum cross-sectional width of bars facing direction of flow, m (ft)
 b = minimum clear spacing of bars, m (ft)
 h_v = velocity head of flow approaching rack, m (ft)
 θ = angle of rack with horizontal

Kirschmer's values of β for several shapes of bars are given in Table 6-3. The head loss calculated from Eq. 6-1 applies only when the bars are clean. Head loss increases with the degree of clogging.

Fine screens The clear-water head loss through fine screens may be obtained from manufacturers' rating tables, or it may be calculated by means of the common orifice formula:

$$h_L = \frac{1}{2g} \left(\frac{Q}{CA}\right)^2 \tag{6-2}$$

where C = coefficient of discharge
 Q = discharge through screen, m^3/s (ft^3/s)
 A = effective submerged open area, m^2 (ft^2)
 g = acceleration due to gravity, m/s^2 (ft/s^2)
 h_L = head loss, m (ft)

Values of C and A depend on screen design factors, such as the size and milling of slots, the wire diameter and weave, and particularly the percent of open area, and must be determined experimentally. A typical value of C for a clean screen is 0.60. The head loss through a clean screen is relatively insignificant. The important determination is the head loss during operation, and this depends on the size and amount of solids in the wastewater, the size of the apertures, and the method and frequency of cleaning.

6-2 FLOW EQUALIZATION

The variations that are observed in the influent-wastewater flowrate and strength at almost all wastewater-treatment facilities were discussed in Chaps. 2 and 3, respectively. Yet in Chap. 5, in the consideration of process fundamentals, it was assumed, for the purpose of simplifying the analysis, that the flowrate was constant. If instead a variable flowrate is substituted for the assumed constant value, it can often be shown that because of the variation involved, there may be a deterioration in performance from the optimum value that can be achieved. Flow equalization is used to overcome the operational problems caused by these variations and to improve the performance of the downstream processes.

Description

Flow equalization simply is the damping of flowrate variations so that a constant or nearly constant flowrate is achieved. This technique can be applied in a number of different situations, depending on the characteristics of the collection system. The principal applications are for the equalization of [17]:

1. Dry-weather flows
2. Wet-weather flows from separate sanitary sewers
3. Combined storm water and sanitary wastewater flows

The application of flow equalization in wastewater treatment is illustrated in the two flowsheets given in Fig. 6-4. In the in-line arrangement (Fig. 6-4a), all of the flow passes through the equalization basin. On the basis of the analyses presented in Chap. 5, it can be shown that this arrangement can be used to achieve a considerable amount of constituent concentration and flowrate damping. In the off-line arrangement (Fig. 6-4b), only the flow above the average daily flowrate is diverted into the equalization basin. Although pumping requirements are minimized in this arrangement, the amount of constituent-concentration damping is considerably reduced.

The principal benefits that are cited as deriving from the application of flow equalization are as follows: (1) wastewater treatability is reportedly enhanced after equalization (this remains to be demonstrated conclusively); (2) biological treatment is enhanced, because shock loadings are eliminated or can be minimized, inhibiting substances can be diluted, and pH can be stabilized; (3) the effluent quality and thickening performance of secondary sedimentation tanks following biological treatment is improved through constant solids loading; (4) effluent-filtration surface-area requirements are reduced, filter performance is improved, and more uniform filter-backwash cycles are possible; and (5) in chemical treatment, damping of mass loadings improves chemical feed control and process reliability [17]. Apart from improving the performance of most treatment

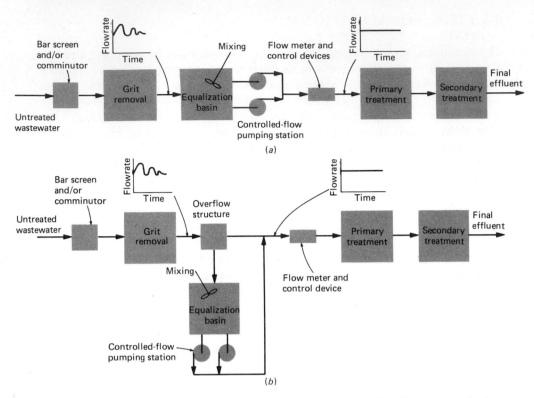

Figure 6-4 Typical wastewater-treatment-plant flowsheets incorporating flowrate equalization. (*Adapted from Ref. 17.*) (*a*) In-line equalization. (*b*) Off-line equalization.

operations and processes, flow equalization is an attractive option for upgrading the performance of overloaded treatment plants because of the relatively low costs involved.

Analysis

The theoretical analysis of flow equalization is concerned with the following questions:

1. Where in the treatment-process flowsheet should the equalization facilities be located?
2. What type of equalization flowsheet should be used—in-line or off-line?
3. What is the required basin volume?

The practical aspects of design (e.g., type of construction, degree of compartmentalization, type of mixing equipment, pumping and control methods, and sludge and scum removal) are discussed in Chap. 8.

Location of equalization facilities The best location for equalization facilities must be determined for each system. Because the optimum location will vary with the type of treatment and the characteristics of the collection system and the wastewater, detailed studies should be performed for several locations throughout the system. Probably the most common location will continue to be at existing and proposed treatment-plant sites. There is also a need to consider location of equalization facilities in the treatment-process flowsheet. In some cases, equalization after primary treatment and before biological treatment may be appropriate. Equalization after primary treatment causes fewer problems with sludge and scum. If flow-equalization systems are to be located ahead of primary settling and biological systems, the design must provide for sufficient mixing to prevent solids deposition and concentration variations, and aeration to prevent odor problems.

In-line or off-line equalization As described earlier and shown in Fig. 6-4, it is possible to achieve considerable damping of constituent mass loadings to the downstream processes with in-line equalization, but only slight damping is achieved with off-line equalization. The analysis of the effect of in-line equalization on the constituent mass loading is illustrated in Example 6-1.

Volume requirements for equalization basin The volume required for flowrate equalization is determined by using an inflow mass diagram in which the cumulative inflow volume is plotted versus the time of day. The average daily flowrate, also plotted on the same diagram, is the straight line drawn from the origin to the end point of the diagram. Diagrams for two typical flowrate patterns are shown in Fig. 6-5.

To determine the required volume, a line parallel to the coordinate axis, defined by the average daily flowrate, is drawn tangent to the mass inflow curve. The required volume is then equal to the vertical distance from the point of tangency to the straight line representing the average flowrate. If the inflow mass curve goes above the line representing the average flowrate (flowrate pattern B), the inflow mass diagram must be bounded with two lines that are parallel to the average flowrate line and tangent to the extremities of the inflow mass diagram. The required volume is then equal to the vertical distance between the two lines. The determination of the required volume for equalization is also illustrated in Example 6-1. This procedure is exactly the same as if the average hourly volume were subtracted from the volume of flow occurring each hour, and the resulting cumulative volumes were plotted. In this case, the low and high points of the curve would be determined using a horizontal line.

The physical interpretation of the diagrams shown in Fig. 6-5 is as follows. At the low point of tangency (flowrate pattern A) the storage basin is empty. Beyond this point, the basin begins to fill because the slope of the inflow mass diagram is greater than that of the average daily flowrate. The basin continues to fill until it becomes full at midnight. For flowrate pattern B, the basin is filled at the upper point of tangency.

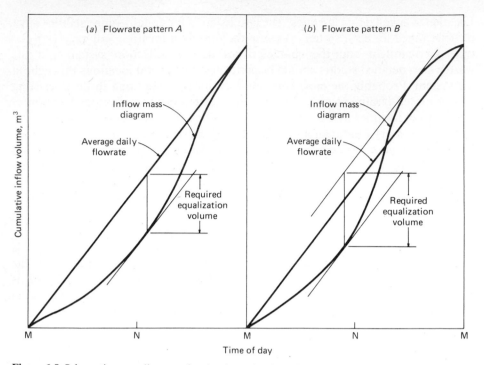

Figure 6-5 Schematic mass diagrams for the determination of the equalization volume required for two typical flowrate patterns.

In practice, the volume of the equalization basin will be made larger than that theoretically determined to account for the following factors [17]:

1. Continuous operation of aeration and mixing equipment will not allow complete drawdown, although special structures can be built.
2. Volume must be provided to accommodate the concentrated plant recycle streams that are expected, if such flows are returned to the equalization basin (a practice that is not recommended).
3. Some contingency should be provided for unforeseen changes in diurnal flow.

Although no fixed value can be given, the additional volume will vary from 10 to 20 percent of the theoretical value.

Example 6-1: Determination of flowrate equalization volume requirements and effects on BOD mass loading For the flowrate and BOD concentration data given in Table 6-4, determine (1) the in-line storage volume required to equalize the flowrate, and (2) the effect of flow equalization on the BOD mass loading rate.

SOLUTION

1 Determine the volume of the basin required for flow equalization.

    a The first step is to develop a cumulative mass curve of the wastewater flowrate expressed in cubic meters. This is accomplished by converting the average flowrate during each hourly

Table 6-4 Flowrate and BOD data for determining effects of flow equalization in Example 6-1[a]

	Given data		Derived data	
Time period	Average flowrate during time period, m^3/s	Average BOD concentration during time period, mg/L	Cumulative volume of flow at end of time period, m^3	BOD mass loading during time period, kg/h
M–1	0.275	150	990	149
1–2	0.220	115	1,782	91
2–3	0.165	75	2,376	45
3–4	0.130	50	2,844	23
4–5	0.105	45	3,222	17
5–6	0.100	60	3,582	22
6–7	0.120	90	4,014	39
7–8	0.205	130	4,752	96
8–9	0.355	175	6,030	223
9–10	0.410	200	7,506	295
10–11	0.425	215	9,036	329
11–N	0.430	220	10,584	341
N–1	0.425	220	12,114	337
1–2	0.405	210	13,572	306
2–3	0.385	200	14,958	277
3–4	0.350	190	16,218	239
4–5	0.325	180	17,388	211
5–6	0.325	170	18,558	199
6–7	0.330	175	19,746	208
7–8	0.365	210	21,060	276
8–9	0.400	280	22,500	403
9–10	0.400	305	23,940	439
10–11	0.380	245	25,308	335
11–M	0.345	180	26,550	224
Average	0.307			213

[a] Data derived from Fig. 3-3.
   Note: $m^3/s \times 35.3147 = ft^3/s$
   $m^3 \times 35.3147 = ft^3$
   $kg \times 2.2046 = lb$
   $mg/L = g/m^3$

period to cubic meters, using the following expression, and then cumulatively summing the hourly values.

$$\text{Volume, } m^3 = (q_i, \, m^3/s)(3600 \text{ s/h})(1.0 \text{ h})$$

For example, for the first three time periods shown in Table 6-4, the corresponding hourly volumes are as follows:
For the time period M–1:

$$V_{M-1} = (0.275 \text{ m}^3/s)(3600 \text{ s/h})(1.0 \text{ h})$$
$$= 990 \text{ m}^3$$

For the time period 1-2:

$$V_{1-2} = (0.220 \text{ m}^3/\text{s})(3600 \text{ s/h})(1.0 \text{ h})$$
$$= 792 \text{ m}^3$$

The cumulative flow, expressed in cubic meters, at the end of each time period is determined as follows:

At the end of the first time period M-1:

$$V_1 = 990 \text{ m}^3$$

At the end of the second time period 1-2:

$$V_2 = 990 + 792 = 1782 \text{ m}^3$$

The cumulative flows for all the hourly time periods are computed in a similar way and are reported in Table 6-4.

b The second step is to prepare a plot of the cumulative flow volumes, as shown in Fig. 6-6. As will be noted, the slope of the line drawn from the origin to the end point of the inflow

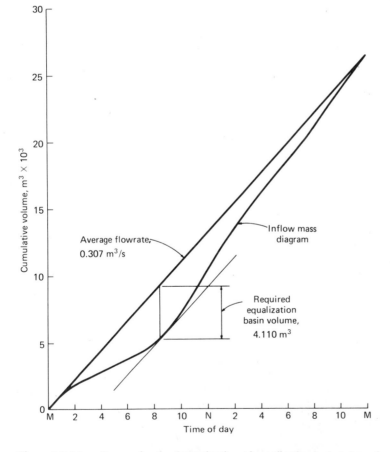

Figure 6-6 Mass diagram for the determination of equalization-tank volume for Example 6-1.

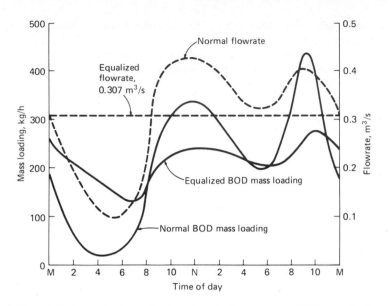

Figure 6-7 Normal and equalized wastewater flowrates and BOD mass loadings for Example 6-1.

mass diagram represents the average flowrate for the day, which in this case is equal to 0.307 m³/s.

c The third step is to determine the required volume. This is done by drawing a line parallel to the average flowrate line tangent to the low point of the inflow mass diagram. The required volume is represented by the vertical distance from the point of tangency to the straight line representing the average flowrate. In this case, the required volume is equal to

$$\text{Volume of equalization basin } V = 4110 \text{ m}^3 \ (145{,}145 \text{ ft}^3)$$

2 Determine the effect of the equalization basin on the BOD mass loading rate. There are a number of ways to do this, but perhaps the simplest way is to perform the necessary computations starting with the time period when the equalization basin is empty. Because the equalization basin is empty at about 8:30 A.M. (see Fig. 6-7), the necessary computations will be performed starting with the 8–9 time period.

a The first step is to compute the liquid volume in the equalization basin at the end of each time period. This is done by subtracting the equalized hourly flowrate expressed as a volume from the inflow flowrate also expressed as a volume. The volume corresponding to the equalized flowrate shown in Fig. 6-6, for a period of 1 h is 1106 m³ (26,550 m³/d) (1.0 h)/(24 h/d). Using this value, the volume in storage is computed using the following expression:

$$V_{sc} = V_{sp} + V_{ic} - V_{oc}$$

where V_{sc} = volume in the equalization basin at the end of current time period
V_{sp} = volume in the equalization basin at the end of previous time period
V_{ic} = volume of inflow during the current time period
V_{oc} = volume of outflow during the current time period

Thus, using the data in Table 6-4, the volume in the equalization basin for the time period 8–9 is as follows:

$$V_{sc} = 0 + 1278 \text{ m}^3 - 1106 \text{ m}^3 = 172 \text{ m}^3$$

Table 6-5 Computation table for determining the equalized BOD mass loading values for Example 6-1

Time period	Volume of flow during time period, m³	Volume in storage at end of time period, m³	Average BOD concentration during time period, mg/L	Equalized BOD concentration during time period, mg/L	Equalized BOD mass loading during time period, kg/h
8–9	1278	172	175	175	193
9–10	1476	542	200	197	218
10–11	1530	966	215	210	232
11–N	1548	1408	220	216	239
N–1	1530	1832	220	218	241
1–2	1458	2184	210	214	237
2–3	1386	2464	200	209	231
3–4	1260	2618	190	203	224
4–5	1170	2680	180	196	217
5–6	1170	2746	170	188	208
6–7	1188	2828	175	184	203
7–8	1314	3036	210	192	212
8–9	1440	3370	280	220	243
9–10	1440	3704	305	245	271
10–11	1368	3966	245	245	271
11–M	1242	4102	180	230	254
M–1	990	3986	150	214	237
1–2	792	3672	115	196	217
2–3	594	3160	75	179	198
3–4	468	2522	50	162	179
4–5	378	1794	45	147	162
5–6	360	1048	60	132	146
6–7	432	374	90	119	132
7–8	738	0	130	126	139
Average					213

Note: m³ × 35.3147 = ft³
kg × 2.2046 = lb
mg/L = g/m³

For the time period 9–10:

$$V_{sc} = 172 \text{ m}^3 + 1476 \text{ m}^3 - 1106 \text{ m}^3 = 542 \text{ m}^3$$

The volume in storage at the end of each time period has been computed in a similar way and is reported in Table 6-5.

b   The second step is to compute the average concentration leaving the storage basin. This is done by using the following expression, which is based on the assumption that the contents of the equalization basin are mixed completely:

$$X_{oc} = \frac{(V_{ic})(X_{ic}) + (V_{sp})(X_{sp})}{V_{ic} + V_{sp}}$$

where X_{oc} = average concentration of BOD in the outflow from the storage basin during the current time period, mg/L

V_{ic} = volume of wastewater inflow during the current period, m^3
X_{ic} = average concentration of BOD in the inflow wastewater volume, mg/L
V_{sp} = volume of wastewater in storage basin at the end of the previous time period, m^3
X_{sp} = concentration of BOD in wastewater in storage basin at the end of the previous time period

Using the data given in Table 6-5, the effluent concentration is computed as follows:
For the time period 8–9:

$$X_{oc} = \frac{(1278 \text{ m}^3)(175 \text{ mg/L}) + (0)(0)}{1278 \text{ m}^3}$$

$$= 175 \text{ mg/L}$$

For the time period 9–10:

$$X_{oc} = \frac{(1476 \text{ m}^3)(200) + (172 \text{ m}^3)(175 \text{ mg/L})}{(1476 + 172) \text{ m}^3}$$

$$= 197 \text{ mg/L}$$

All the concentration values computed in a similar manner are reported in Table 6-5.

c The third step is to compute the hourly mass loading rate using the following expression:

$$\text{Mass loading rate, kg/h} = \frac{(X_{oc}, \text{g/m}^3)(\bar{q}_i, \text{m}^3/\text{s})(3600 \text{ s/h})}{1000 \text{ g/kg}}$$

For example, for the time period 8–9, the mass loading rate is

$$\frac{(175 \text{ g/m}^3)(0.307 \text{ m}^3/\text{s})(3600 \text{ s/h})}{1000 \text{ g/kg}} = 193 \text{ kg/h } (425 \text{ lb/h})$$

All the hourly values are summarized in Table 6-5. The corresponding values without flow equalization are reported in Table 6-4.

d The effect of flow equalization can be shown best graphically by plotting the hourly unequalized and equalized BOD mass loadings as shown in Fig. 6-7. The following table derived from the data presented in Tables 6-4 and 6-5 are also helpful in assessing the benefits derived from flow equalization:

	BOD mass loading	
Ratio	Unequalized	Equalized
$\dfrac{\text{Peak}}{\text{Average}}$	$\dfrac{439}{213} = 2.06$	$\dfrac{271}{213} = 1.27$
$\dfrac{\text{Minimum}}{\text{Average}}$	$\dfrac{17}{213} = 0.08$	$\dfrac{132}{213} = 0.62$
$\dfrac{\text{Peak}}{\text{Minimum}}$	$\dfrac{439}{17} = 25.82$	$\dfrac{271}{132} = 2.05$

Comment Where on-line equalization basins are used, additional damping of the BOD mass loading rate, over that shown in Fig. 6-7, can be obtained by increasing the volume of the basins. Although the flow to a treatment plant was equalized in this example, flow equalization would be used more realistically in locations with high infiltrational inflow or storm water peaks.

6-3 MIXING

Mixing is an important unit operation in many phases of wastewater treatment where one substance must be completely intermingled with another. An example is the mixing of chemicals with wastewater, as shown in Fig. 6-1, where chlorine or hypochlorite is mixed with the effluent from the secondary settling tanks. Chemicals are also mixed with sludge to improve its dewatering characteristics before vacuum filtration. In the digestion tank, mixing is used frequently to assure intimate contact between food and microorganisms. In the biological process tank, air must be mixed with the activated sludge to provide the organisms with the oxygen required. In this case, diffused air is introduced in such a way as to fulfill the mixing requirements, or mechanical turbine aerator-mixers may be used.

Description/Application

Liquid mixing can be carried out in a number of different ways, including (1) hydraulic jumps in open channels, (2) Venturi flumes, (3) pipelines, (4) pumps, and (5) vessels with the aid of mechanical means. In the first four of these ways, mixing is accomplished as a result of turbulence that exists in the flow regime. In the fifth, turbulence is induced through the use of rotating impellers, such as paddles, turbines, and propellers; gases (as in diffused aeration); and air and water jet-lift pumps. A typical mixer used in wastewater-treatment plants is shown in Fig. 6-8.

Paddles generally rotate slowly, as they have a large surface exposed to the liquid. They are used as flocculation devices when coagulants, such as aluminum or ferric sulfate, and coagulant aids, such as polyelectrolytes and lime, are added to wastewater or sludges. The production of a good floc usually requires a detention time of 15 to 30 minutes. On the other hand, a detention time of 2 to 5 minutes is more than adequate for the flash mixing of chemicals in tanks equipped with turbine or propeller mixers.

Turbine mixers are used in sludge-blending tanks (see Fig. 11-7). Large blenders are designed to rotate at moderate speeds (about 25 to 100 r/min). Propeller-type impellers have also been used for high-speed mixing with rotational speeds up to 2000 r/min. They are shaped like ship propellers and generate strong axial currents that rapidly mix chemicals or gases with liquids.

Vortexing or mass swirling of the liquid must be restricted with all types of impellers. Vortexing causes a reduction in the difference between the fluid velocity and the impeller velocity and thereby decreases the effectiveness of mixing. If the mixing vessel is fairly small, vortexing can be prevented by mounting the impellers off-center or at an angle with the vertical, or by having them enter the side of the basin at an angle. The usual method in both circular and rectangular tanks is to install four or more vertical baffles extending approximately one-tenth the diameter out from the wall. These effectively break up the mass rotary motion and promote vertical mixing. Concrete mixing tanks may be made square and the baffles may be omitted.

Figure 6-8 Typical mixer used in wastewater-treatment plants. (*From Lightnin.*)

Analysis

Mixing in sanitary-engineering processes usually occurs in the regime of turbulent flow in which inertial forces predominate. As a general rule, the higher the velocity and the greater the turbulence, the more efficient the mixing. On the basis of inertial and viscous forces, Rushton [18] has developed the following mathematical relationships for power requirements for laminar and turbulent conditions.

Laminar:
$$P = k\mu n^2 D^3 \tag{6-3}$$

Turbulent:
$$P = k\rho n^3 D^5 \tag{6-4}$$

where P = power requirement, W (ft-lb/s)
  k = constant (see Table 6-6)
  μ = dynamic viscosity of fluid, N · s/m² (lb · s/ft²)
  ρ = mass density of fluid, kg/m³ (slug/ft³)
  D = diameter of impeller, m (ft)
  n = revolutions per second, r/s

Table 6-6 Values of k for mixing power requirements [18]

Impeller	Laminar range, Eq. 6-3	Turbulent range, Eq. 6-4
Propeller, square pitch, 3 blades	41.0	0.32
Propeller, pitch of two, 3 blades	43.5	1.00
Turbine, 6 flat blades	71.0	6.30
Turbine, 6 curved blades	70.0	4.80
Fan turbine, 6 blades	70.0	1.65
Turbine, 6 arrowhead blades	71.0	4.00
Flat paddle, 6 blades	36.5	1.70
Shrouded turbine, 2 curved blades	97.5	1.08
Shrouded turbine with stator (no baffles)	172.5	1.12

Values of k, as developed by Rushton, are presented in Table 6-6. For the turbulent range, it is assumed that vortex conditions have been eliminated by four baffles at the tank wall, each 10 percent of the tank diameter, as shown in Fig. 6-9.

Equation 6-3 applies if the Reynolds number is less than 10, and Eq. 6-4 applies if the Reynolds number is greater than 10,000. For intermediate values of

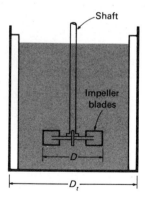

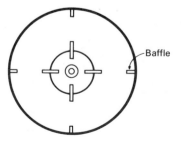

Figure 6-9 Turbine mixer in baffled tank [18].

the Reynolds number, Ref. 14 should be consulted. The Reynolds number is given by

$$N_R = \frac{D^2 n \rho}{\mu} \tag{6-5}$$

where D = diameter of impeller, m (ft)
n = r/s
ρ = mass density of liquid, kg/m^3 $(slug/ft^3)$
μ = dynamic viscosity, $N \cdot s/m^2$ $(lb \cdot s/ft^2)$

The power input per unit volume is a rough measure of mixing effectiveness, based on the reasoning that more input power creates greater turbulence, and greater turbulence leads to better mixing. Mixers are selected on the basis of laboratory or pilot-plant tests or similar data provided by manufacturers. No satisfactory method exists for scaling up from an agitator of one design to a unit of a different design. Geometrical similarity should be preserved, and the power input per unit volume should be kept the same. A small impeller of high speed gives high turbulence but low flow and is best for dispersing gases or small amounts of chemicals in wastewater. A large, slow impeller gives high flow but low turbulence and is best for blending two fluid streams, or for flocculation.

6-4 FLOCCULATION

An essential part of any chemical or chemically aided precipitation system is stirring or agitation to increase the opportunity for particle contact (flocculation), after the chemicals have been added.

Description

Flocculation is promoted by gentle stirring with slow-moving paddles, as shown in Fig. 6-10. The action is sometimes aided by the installation of stationary slats or stator blades, located between the moving blades, that serve to break up

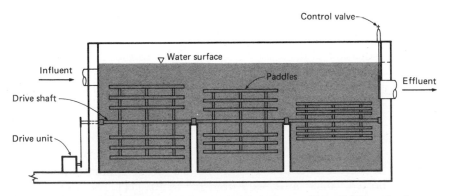

Figure 6-10 Typical flocculator used for wastewater treatment. (*From Ecodyne Corp.*)

the mass rotation of the liquid and promote mixing. Increased particle contact will promote floc growth; however, if the agitation is too vigorous, the shear forces that are set up will break up the floc into smaller particles. Agitation should be carefully controlled so that the floc particles will be of suitable size and will settle readily.

Analysis

Numerous experiments have been performed by equipment manufacturers and plant operators to determine the optimum configuration of paddle size, spacing, and velocity. It has been found that a paddle-tip speed of approximately 0.6 to 0.9 m/s (2 to 3 ft/s) achieves sufficient turbulence without breaking up the floc. Camp and Stein [4] studied the establishment and effect of velocity gradients in coagulation tanks of various types and developed the following equations for use in the design and operation of flocculation systems:

$$F_D = \frac{C_D A \rho v^2}{2} \tag{6-6}$$

$$P = \frac{C_D A \rho v^3}{2} \tag{6-7}$$

$$G = \sqrt{\frac{P}{\mu V}} \tag{6-8}$$

where F_D = drag force N (lb)
C_D = coefficient of drag of flocculator paddles moving perpendicular to fluid
A = area of paddles, m^2 (ft^2)
ρ = mass fluid density, kg/m^3 (slugs/ft^3)
v = relative velocity of paddles in fluid, m/s (ft/s), usually about 0.7 to 0.8 of paddle-tip speed
P = power requirement, W (ft · lb/s)
G = mean velocity gradient, l/s
V = flocculator volume, m^3 (ft^3)
μ = dynamic viscosity, N · s/m^2 (lb · s/ft^2)

In Eq. 6-8, G is a measure of the mean velocity gradient in the fluid. As shown, it depends on the power input, the viscosity of the fluid, and the volume of the basin. Multiplying both sides of Eq. 6-8 by the theoretical detention time $t_d = V/Q$ yields

$$Gt_d = \frac{V}{Q}\sqrt{\frac{P}{\mu V}} = \frac{1}{Q}\sqrt{\frac{PV}{\mu}} \tag{6-9}$$

Typical values for G for a detention time of about 15 to 30 minutes vary from 20 to 75 s^{-1}. Reported values of Gt_d vary from 10^4 to 10^5 [11]. The use of these equations is illustrated in Example 6-2.

Example 6-2: Power requirement and paddle area for a wastewater flocculator Determine the theoretical power requirement and the paddle area required to achieve a G value of 50/s in a tank with a volume of 3000 m³ (10⁵ ft³). Assume that the water temperature is 15°C (60°F), the coefficient of drag for rectangular paddles is 1.8, the paddle-tip velocity v_p is 0.6 m/s (2 ft/s), and the relative velocity of the paddles v is 0.75 v_p.

SOLUTION
1 Determine the theoretical power requirement.

$$P = \mu G^2 V$$

$$= 1.139 \times 10^{-3} \frac{N \cdot s}{m^2} (50/s)^2 (3000 \text{ m}^3)$$

$$= 8543 \text{ W}$$

$$= 8.543 \text{ kW} (6300 \text{ ft} \cdot lb_f/s)$$

2 Determine the required paddle area using Eq. 6-6.

$$A = \frac{2P}{C_D \rho v^3}$$

$$= \frac{2 \times 8543 \text{ kg/m}^2 \cdot s^2}{1.8(999.1 \text{ kg/m}^3)(0.75 \times 0.6 \text{ m/s})^3}$$

$$= 104.3 \text{ m}^2 (1120 \text{ ft}^2)$$

6-5 SEDIMENTATION

Sedimentation is the separation from water, by gravitational settling, of suspended particles that are heavier than water. It is one of the most widely used unit operations in wastewater treatment. The terms sedimentation and settling are used interchangeably. A sedimentation basin may also be referred to as a sedimentation tank, settling basin, or settling tank.

Sedimentation is used for grit removal, particulate-matter removal in the primary settling basin, biological-floc removal in the activated-sludge settling basin, and chemical-floc removal when the chemical coagulation process is used. It is also used for solids concentration in sludge thickeners. In most cases, the primary purpose is to produce a clarified effluent, but it is also necessary to produce sludge with a solids concentration that can be easily handled and treated. In the design of sedimentation basins (see Chap. 8), consideration must be given to production of both a clarified effluent and a concentrated sludge.

Description

On the basis of the concentration and the tendency of particles to interact, four types of settling can occur: discrete particle, flocculant, hindered (also called zone), and compression. These types of settling phenomena are described in Table 6-7. During a sedimentation operation, it is common to have more than one type of

Table 6-7 Types of settling phenomena involved in wastewater treatment

Type of settling phenomenon	Description	Application/occurrence
Discrete particle (type 1)	Refers to the sedimentation of particles in a suspension of low solids concentration. Particles settle as individual entities, and there is no significant interaction with neighboring particles	Removes grit and sand particles from wastewater
Flocculant (type 2)	Refers to a rather dilute suspension of particles that coalesce, or flocculate, during the sedimentation operation. By coalescing, the particles increase in mass and settle at a faster rate	Removes a portion of the suspended solids in untreated wastewater in primary settling facilities, and in upper portions of secondary settling facilities. Also removes chemical floc in settling tanks
Hindered, also called zone (type 3)	Refers to suspensions of intermediate concentration, in which inter-particle forces are sufficient to hinder the settling of neighboring particles. The particles tend to remain in fixed positions with respect to each other, and the mass of particles settles as a unit. A solids-liquid interface develops at the top of the settling mass	Occurs in secondary settling facilities used in conjunction with biological treatment facilities
Compression (type 4)	Refers to settling in which the particles are of such concentration that a structure is formed, and further settling can occur only by compression of the structure. Compression takes place from the weight of the particles, which are constantly being added to the structure by sedimentation from the supernatant liquid	Usually occurs in the lower layers of a deep sludge mass, such as in the bottom of deep secondary settling facilities and in sludge-thickening facilities

settling occurring at a given time, and it is possible to have all four occurring simultaneously.

Because of the fundamental importance of sedimentation in the treatment of wastewater, the analysis of each type of settling will be discussed separately. In addition, after the discussion of flocculant settling, a brief analysis of tube settlers (inclined small-diameter tubes used to improve efficiency of the sedimentation operation) will be presented. Both discrete and flocculant settling can occur in situations where tube settlers are used.

Analysis of Discrete Particle Settling (Type 1)

The settling of discrete, nonflocculating particles can be analyzed by means of the classic laws of sedimentation formed by Newton and Stokes. Newton's law yields the terminal particle velocity by equating the gravitational force of the particle to the frictional resistance, or drag. The gravitational force is given by

$$\text{Gravitational force} = (\rho_s - \rho)g V \tag{6-10}$$

where ρ_s = density of particle
ρ = density of fluid
g = acceleration due to gravity
V = volume of particle

The frictional drag force depends on the particle velocity, fluid density, fluid viscosity, and particle diameter and the drag coefficient C_D (dimensionless) and is defined by Eq. 6-11.

$$\text{Frictional drag force} = \frac{C_D A \rho v^2}{2} \tag{6-11}$$

where C_d = drag coefficient
A = cross-sectional or projected area of particle at right angles to v
v = particle velocity

Equating the gravitational force to the frictional drag force for spherical particles yields Newton's law:

$$V_c = \left[\frac{4}{3} \frac{g(\rho_s - \rho)d}{C_D \rho} \right]^{1/2} \tag{6-12}$$

where V_c = terminal velocity of particle
d = diameter of particle

The drag coefficient takes on different values depending on whether the flow regime surrounding the particle is laminar or turbulent. The drag coefficient is shown in Fig. 6-11 as a function of the Reynolds number. Although particle shape affects the value of the drag coefficient, for spherical particles the curve in Fig. 6-11 is approximated by the following equation (upper limit of $N_R = 10^4$) [11]:

$$C_D = \frac{24}{N_R} + \frac{3}{\sqrt{N_R}} + 0.34 \tag{6-13}$$

For Reynolds numbers less than 0.3, the first term in Eq. 6-13 predominates, and substitution of this drag term into Eq. 6-12 yields Stokes' law:

$$V_c = \frac{g(\rho_s - \rho)d^2}{18 \mu} \tag{6-14}$$

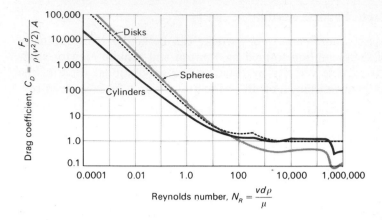

Figure 6-11 Drag coefficients of spheres, disks, and cylinders [14].

For laminar-flow conditions, Stokes found the drag force to be

$$F_d = 3\pi\mu vd \qquad (6\text{-}15)$$

Equating this force to the effective particle weight also yields Eq. 6-14.

In the design of sedimentation basins, the usual procedure is to select a particle with a terminal velocity V_c and to design the basin so that all particles that have a terminal velocity equal to or greater than V_c will be removed. The rate at which clarified water is produced is then

$$Q = AV_c \qquad (6\text{-}16)$$

where A is the surface area of the sedimentation basin. Equation 6-16 yields

$$V_c = \frac{Q}{A} = \text{overflow rate, m}^3/\text{m}^2 \cdot \text{d (gal/ft}^2 \cdot \text{d)}$$

which shows that the overflow rate or surface loading rate, a common basis of design, is equivalent to the settling velocity. Equation 6-16 also indicates that, for type 1 settling, the flow capacity is independent of the depth.

For continuous-flow sedimentation, the length of the basin and the time a unit volume of water is in the basin (detention time) should be such that all particles with the design velocity V_c will settle to the bottom of the tank. The design velocity, detention time, and basin depth are related as follows:

$$V_c = \frac{\text{depth}}{\text{detention time}} \qquad (6\text{-}17)$$

In actual practice, design factors must be included to allow for the effects of inlet and outlet turbulence, short circuiting, sludge storage, and velocity gradients due to the operation of sludge-removal equipment. These factors will be discussed in Chap. 8. The discussion in this chapter refers to ideal settling in which the factors are omitted.

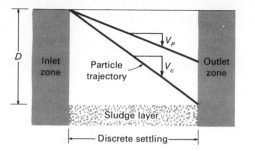

Figure 6-12 Type 1 settling in an ideal settling basin.

Type 1 settling in an ideal settling basin is shown in Fig. 6-12. A full-scale settling basin used in practice is shown in Fig. 8-13. Particles that have a velocity of fall less than V_c will not all be removed during the time provided for settling. Assuming that the particles of various sizes are uniformly distributed over the entire depth of the basin at the inlet, it can be seen from an analysis of the particle trajectory in Fig. 6-12 that particles with a settling velocity less than V_c will be removed in the ratio

$$X_r = \frac{V_p}{V_c} \tag{6-18}$$

where X_r is the fraction of the particles with settling velocity V_p that are removed.

In a typical suspension of particulate matter, a large gradation of particle sizes occurs. To determine the efficiency of removal for a given settling time, it is neccessary to consider the entire range of settling velocities present in the system. This can be accomplished in two ways: (1) by use of sieve analyses and hydrometer tests combined with Eq. 6-12 or (2) by use of a settling column. With either method, a settling-velocity analysis curve can be constructed from the data. Such a curve is shown in Fig. 6-13.

For a given clarification rate Q where

$$Q = V_c A \tag{6-19}$$

only those particles with a velocity greater than V_c will be completely removed.

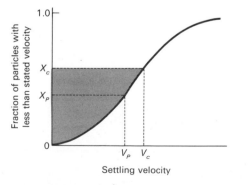

Figure 6-13 Settling-velocity analysis curve for discrete particles.

The remaining particles will be removed in the ratio V_p/V_c. The total fraction of particles removed is given by Eq. 6-20.

$$\text{Fraction removed} = (1 - X_c) + \int_0^{X_c} \frac{V_p}{V_c}\, dx \qquad (6\text{-}20)$$

where $1 - X_c$ = fraction of particles with velocity V_p greater than V_c

$\int_0^{X_c} \dfrac{V_p}{V_c}\, dx$ = fraction of particles removed with V_p less than V_c

The use of Eq. 6-20 is illustrated in Example 6-3.

Example 6-3: Removal of discrete particles (type 1 settling) A particle-size distribution has been obtained from a sieve analysis of sand particles. For each weight fraction, an average settling velocity has been calculated. The data are as follows:

Settling velocity, m/min	3.0	1.5	0.60	0.30	0.22	0.15
Settling velocity, ft/min	10.0	5.0	2.0	1.0	0.75	0.5
Weight fraction remaining	0.55	0.46	0.35	0.21	0.11	0.03

What is the overall removal for an overflow rate of 4000 m³/m² · d (98,160 gal/ft² · d)?

SOLUTION Draw the settling-velocity analysis curve as shown in Fig. 6-14. Compute the settling velocity V_c of the particles that will be completely removed when the rate of clarification is 4000 m³/m² · d.

$$V_c = \frac{4000}{24(60)} = 2.8 \text{ m/min} \ (9.2 \text{ ft/min})$$

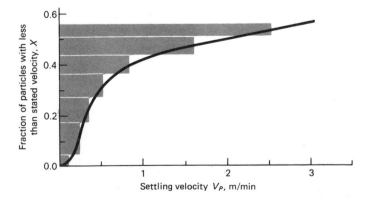

Figure 6-14 Settling-velocity curve for Example 6-3. Note: m × 3.2808 = ft.

From the curve it is found that 0.54 of the particles have a settling velocity less than 2.8 m/min. The graphical integration of the second term in Eq. 6-20 is shown on the curve as a series of rectangles (shaded) and in the following tabulation. (Note that because V_c is constant, it is taken outside the integral or summation sign.)

dx	V_p	$V_p dx$
0.04	0.10	0.004
0.16	0.22	0.035
0.12	0.40	0.048
0.08	0.70	0.056
0.08	1.30	0.104
0.06	2.25	0.135
Σ		0.382

$$\text{Fraction removed} = (1 - X_c) + \frac{1}{V_c} \Sigma V_p dx$$

$$= (1 - 0.54) + \frac{0.382}{2.8} = 0.596$$

Analysis of Flocculant Settling (Type 2)

Particles in relatively dilute solutions sometimes will not act as discrete particles but will coalesce during sedimentation. As coalescence or flocculation occurs, the mass of the particle increases and it settles faster. The extent to which flocculation occurs is dependent on the opportunity for contact, which varies with the overflow rate, the depth of the basin, the velocity gradients in the system, the concentration of particles, and the range of particle sizes. The effects of these variables can be determined only by sedimentation tests.

To determine the settling characteristics of a suspension of flocculant particles, a settling column may be used. Such a column can be of any diameter but should be equal in height to the depth of the proposed tank. Satisfactory results can be obtained with a 150-mm (6-in) -diameter plastic tube about 3 m (10ft) high. Sampling ports should be inserted at 0.6-m (2-ft) intervals. The solution containing the suspended matter should be introduced into the column in such a way that a uniform distribution of particle sizes occurs from top to bottom.

Care should also be taken to ensure that a uniform temperature is maintained throughout the test to eliminate convection currents. Settling should take place under quiescent conditions. At various time intervals, samples are withdrawn from the ports and analyzed for suspended solids. The percent removal is computed for each sample analyzed and is plotted as a number against time and depth, as elevations are plotted on a survey grid. Between the plotted points, curves of equal percent removal are drawn. A settling column and the results of a sedimentation test are shown in Fig. 6-15. The resulting curves are shown, but the plotted numbers representing the individual samples have been omitted from the figure. Determination of removals by using Fig. 6-15 is illustrated in Example 6-4.

Example 6-4: Removal of flocculant suspended solids (type 2 settling) Using the results of the settling test shown in Fig. 6-15, determine the overall removal of solids if the detention time is t_2 and the depth is h_5.

SOLUTION

1 Determine the percent removal.

$$\text{Percent removal} = \frac{\Delta h_1}{h_5} \times \frac{R_1 + R_2}{2} + \frac{\Delta h_2}{h_5} \times \frac{R_2 + R_3}{2}$$

$$+ \frac{\Delta h_3}{h_5} \times \frac{R_3 + R_4}{2} + \frac{\Delta h_4}{h_5} \times \frac{R_4 + R_5}{2}$$

2 For the curves shown in Fig. 6-15 the computations would be

$\dfrac{\Delta h_n}{h_5} \times \dfrac{R_n + R_{n+1}}{2}$ = percent removal	
$0.20 \times \dfrac{100 + 80}{2} =$	18.00
$0.11 \times \dfrac{80 + 70}{2} =$	8.25
$0.15 \times \dfrac{70 + 60}{2} =$	9.75
$0.54 \times \dfrac{60 + 50}{2} =$	29.70
1.00	65.70

yielding a total removal for quiescent settling of 65.7 percent.

To account for the less than optimum conditions encountered in the field, the design settling velocity or overflow rate obtained from column studies often is multiplied by a factor of 0.65 to 0.85, and the detention times are multiplied by a factor of 1.25 to 1.5.

Analysis of Tube Settlers

In the analysis for the settling of discrete particles (type 1) presented earlier in this section, it was shown that the removal efficiency is related directly to the settling velocity and not to the depth of the basin. From this finding, it

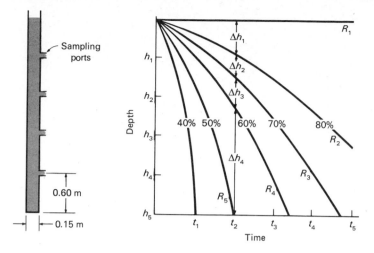

Figure 6-15 Settling column and settling curves for flocculant particles. Note: m × 3.2808 = ft.

can be concluded that sedimentation basins should be constructed as shallow as possible to optimize the removal efficiency. Although this approach is correct theoretically, there are numerous practical considerations that limit the use of extremely shallow basins (see Chap. 8). Tube settlers have been developed as an alternative to shallow basins and are used in conjunction with both existing and specially designed sedimentation basins. Although tube settlers have been used in primary, secondary, and tertiary sedimentation applications, a number of problems have developed with their use. The principal problems are clogging and odors due to biological growths and the buildup of oil and grease.

Tube settlers are shallow settling devices consisting of bundles of small plastic tubes of various geometries. They are used to enhance the settling characteristics of sedimentation basins. Normal practice is to insert modules of tube settlers in sedimentation basins (either rectangular or circular) of sufficient depth. The flow within the basin passes upward through the tube modules and exits from the basin above the tubes. The solids that settle out in the tube move by means of gravity countercurrently downward and out of the tube modules to the basin bottom.

The shape, hydraulic radii, inclination, and length of the tubes vary according to the particular installation. If the tubes are only slightly inclined, the accumulated solids must be flushed out by periodically draining the basin. If the tubes are inclined at sufficiently steep angles, the accumulated solids slide out under the force of gravity. The need for flushing poses a problem with the use of tube settlers where the characteristics of the solids to be removed vary from day to day. From the results of field tests, it appears that angles of inclination steeper than 40° may be required to allow accumulated solids to slide out of the tubes and thus avoid manual flushing.

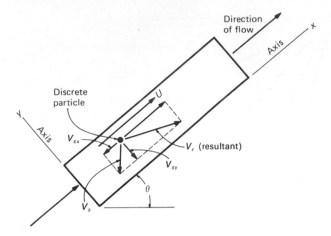

Figure 6-16 Definition sketch for tube settler.

Referring to the definition sketch presented in Fig. 6-16, the analysis of the tube settlers is as follows.

For the inclined-coordinate system, the velocity components for the particle are

$$V_{sx} = U - V_s \sin \theta \tag{6-21}$$

$$V_{sy} = - V_s \cos \theta \tag{6-22}$$

where V_{sx} = velocity component in x direction
U = fluid velocity in x direction
V_s = normal settling velocity of particle
θ = inclination angle for tube with horizontal axis
V_{sy} = settling velocity in y direction

For this system of coordinates, it can be seen that V_{sy} is the critical velocity component, and the analysis for the removal is the same as the analysis presented previously for discrete particles. A more detailed analysis of tube settlers may be found in Ref. 27.

Analysis of Hindered Settling (Type 3)

In systems that contain high concentrations of suspended solids, both hindered or zone settling (type 3) and compression settling (type 4) usually occur in addition to discrete (free) and flocculant settling. The settling phenomenon that occurs when a concentrated suspension, initially of uniform concentration throughout, is placed in a graduated cylinder, is shown in Fig. 6-17.

Because of the high concentration of particles, the liquid tends to move up through the interstices of the contacting particles. As a result, the contacting particles tend to settle as a zone or "blanket," maintaining the same relative

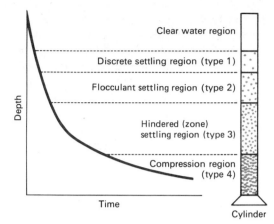

Clear water region

Discrete settling region (type 1)

Flocculant settling region (type 2)

Hindered (zone)
settling region (type 3)

Compression region
(type 4)

Depth

Time

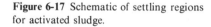

Cylinder

Figure 6-17 Schematic of settling regions for activated sludge.

position with respect to each other. This phenomenon is known as hindered settling. As the particles in this region settle, a relatively clear layer of water is produced above the particles in the settling region.

The scattered, relatively light particles remaining in this region usually settle as discrete or flocculated particles, as discussed previously in this chapter. In most cases, an identifiable interface develops between the more or less clear upper region and the hindered-settling region in Fig. 6-17. The rate of settling in the hindered-settling region is a function of the concentration of solids and their characteristics.

As settling continues, a compressed layer of particles begins to form on the bottom of the cylinder in the compression-settling region. The particles in this region apparently form a structure in which there is close physical contact between the particles. As the compression layer forms, regions containing successively lower concentrations of solids than those in the compression region extend upward in the cylinder. Thus, in actuality the hindered-settling region contains a gradation in solids concentration from that found at the interface of the settling region to that found in the compression-settling region. According to Dick and Ewing [8], the forces of physical interaction between the particles that are especially strong in the compression-settling region lessen progressively with height. They may exist to some extent in the hindered-settling region.

Because of the variability encountered, settling tests are usually required to determine the settling characteristics of suspensions where hindered and compression settling are important considerations. On the basis of data derived from column settling tests, two different design approaches can be used to obtain the required area for the settling/thickening facilities. In the first approach, the data derived from a single (batch) settling test are used. In the second approach, known as the solids flux method, data from a series of settling tests conducted at different solids concentrations are used. Both methods are described in the following discussion.

Area requirement based on single-batch test results For purposes of design, the final overflow rate selected should be based on a consideration of the following factors: (1) the area needed for clarification, (2) the area needed for thickening, and (3) the rate of sludge withdrawal. Column settling tests, as previously described, can be used to determine the area needed for the free-settling region directly. However, because the area required for thickening is usually greater than the area required for settling, the rate of free settling rarely is the controlling factor. In the case of the activated-sludge process where stray, light, fluffy floc particles may be present, it is conceivable that the free or flocculant settling velocity of these particles could control the design.

The area requirement for thickening is determined according to a method developed by Talmadge and Fitch [21]. A column of height H_0 is filled with a suspension of solids of uniform concentration C_0. The position of the interface as time elapses and the suspension settles is given in Fig. 6-18. The rate at which the interface subsides is then equal to the slope of the curve at that point in time. According to the procedure, the area required for thickening is given by Eq. 6-23:

$$A = \frac{Qt_u}{H_0} \qquad (6\text{-}23)$$

where A = area required for sludge thickening, m² (ft²)
$\quad Q$ = flowrate into tank, m³/s (ft³/s)
$\quad H_0$ = initial height of interface in column, m (ft)
$\quad t_u$ = time to reach desired underflow concentrations, s
(*Note*: Any consistent set of units may be used in Eq. 6-23.)

The critical concentration controlling the sludge-handling capability of the tank occurs at a height H_2 where the concentration is C_2. This point is determined by extending the tangents to the hindered-settling and compression regions of the subsidence curve to the point of intersection and bisecting the angle thus formed, as shown in Fig. 6-18. The time t_u can be determined as follows:

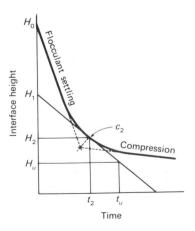

Figure 6-18 Graphical analysis of interface settling curve [21].

1. Construct a horizontal line at the depth H_u that corresponds to the depth at which the solids are at the desired underflow concentration C_u. The value of H_u is determined using the following expression:

$$H_u = \frac{C_0 H_0}{C_u} \qquad (6\text{-}24)$$

2. Construct a tangent to the settling curve at the point indicated by C_2.
3. Construct a vertical line from the point of intersection of the two lines drawn in steps 1 and 2 to the time axis to determine the value of t_u.

With this value of t_u, the area required for thickening is computed using Eq. 6-23. The area required for clarification is then determined. The larger of the two areas is the controlling value. This procedure is illustrated in Example 6-5.

Example 6-5: Calculations for sizing an activated-sludge settling tank In a settling cylinder 0.4 m high, the settling curve shown in Fig. 6-19 was obtained for an activated sludge with an initial solids concentration C_0 of 4000 mg/L. Determine the area to yield a thickened sludge concentration C_u of 24,000 mg/L with an inflow of 400 m^3/d (0.1 Mgal/d). In addition, determine the solids loading in kilograms per day per square meter and the overflow rate in cubic meters per square meter per day.

SOLUTION 1 Determine the area required for thickening using Eq. 6-24.

$$H_u = \frac{C_0 H_0}{C_u}$$

$$= \frac{4000 \text{ mg/L} \times 0.4 \text{ m}}{24,000 \text{ mg/L}} = 0.067 \text{ m}$$

In Fig. 6-19, a horizontal line is constructed at $H_u = 0.067$ m. A tangent is constructed to the settling curve at C_2, the midpoint of the region between hindered and compression settling. Bisecting the angle formed where the two tangents meet determines point C_2. The intersection

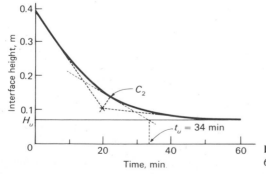

Figure **6-19** Settling curve for Example 6-5. Note: m × 3.281 = ft.

of the tangent at C_2 and the line $H_u = 0.067$ m determines t_u. Thus $t_u = 34$ min, and the required area is

$$A = \frac{Qt_u}{H_0} = \frac{400 \text{ m}^3/\text{d}}{24 \text{ h/d} \times 60 \text{ min/h}} \times \frac{34 \text{ min}}{0.4 \text{ m}}$$

$$= 23.6 \text{ m}^2 \ (254 \text{ ft}^2)$$

2 This area must be adequate for clarification to occur.

  a Determine the subsidence velocity v from the hindered-settling portion of the curve, assuming that the particles present at the interface are those of concern.

$$v = \frac{0.4 \text{ m} - 0.25 \text{ m}}{10 \text{ min} \times (1/60 \text{ min/h})}$$

$$= 0.9 \text{ m/h} \ (2.95 \text{ ft/h})$$

  b Determine the overflow rate. Because the overflow rate is proportional to the liquid volume above the critical sludge zone, it may be computed as follows:

$$Q = 400 \text{ m}^3/\text{d} \times \frac{(0.4 \text{ m} - 0.067 \text{ m})}{0.4 \text{ m}}$$

$$= 333.0 \text{ m}^3/\text{d} \ (0.136 \text{ ft}^3/\text{s})$$

  c Determine the area required for clarification. The required area is obtained by dividing the overflow rate by the settling velocity.

$$A = \frac{Q}{V} = \frac{333.0 \text{ m}^3/\text{d}}{24 \text{ h/d} \times 0.9 \text{ m/h}}$$

$$= 15.4 \text{ m}^2 (166 \text{ ft}^2)$$

3 The controlling requirement is the thickening area of 23.6 m² (254 ft²), because it exceeds the area required for clarification.

4 Determine the solids loading. The solids loading is computed as follows by noting that $g/m^3 = mg/L$:

$$\text{Solids, kg/d} = \frac{400 \text{ m}^3/\text{d} \times 4000 \text{ g/m}^3}{10^3 \text{ g/kg}}$$

$$= 1600 \text{ kg/d}$$

$$\text{Solids loading} = \frac{1600 \text{ kg/d}}{23.6 \text{ m}^2}$$

$$= 67.8 \text{ kg/m}^2 \cdot \text{d} \ (13.9 \text{ lb/ft}^2 \cdot \text{d})$$

5 Determine the hydraulic loading rate.

$$\text{Hydraulic loading rate} = \frac{333 \text{ m}^3/\text{d}}{23.6 \text{ m}^2}$$

$$= 14.1 \text{ m}^3/\text{m}^2 \cdot \text{d} \ (345 \text{ gal/ft}^2 \cdot \text{d})$$

Area requirement based on solids flux analysis An alternative method of arriving at the area required for hindered settling has been delineated by Coe and Clevenger [5], Yoshioka et al. [28], Dick and Young [9], and Dick and Ewing [8].

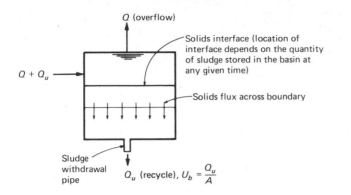

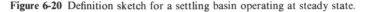

Figure 6-20 Definition sketch for a settling basin operating at steady state.

Data derived from settling tests must be available when applying this method, which is based on an analysis of the mass flux (movement across a boundary) of the solids in the settling basin.

In a settling basin that is operating at steady state, a constant flux of solids is moving downward, as shown in Fig. 6-20. Within the tank, the downward flux of solids is brought about by gravity (hindered) settling and by bulk transport due to the underflow that is being pumped out and recycled. At any point in the tank, the mass flux of solids due to gravity (hindered) settling is

$$SF_g = C_i V_i \times (10^3 \text{ g/kg})^{-1} \qquad (6\text{-}25)$$

where SF_g = solids flux due to gravity, kg/m² · h
$\quad\quad C_i$ = concentration of solids at the point in question, g/m³ (= mg/L)
$\quad\quad V_i$ = settling velocity of the solids at concentration C_i, m/h

The mass flux of solids due to the bulk movement of the suspension is

$$SF_u = C_i U_b(10^3 \text{ g/kg})^{-1} = C_i \frac{Q_u}{A} \times (10^3 \text{ g/kg})^{-1} \qquad (6\text{-}26)$$

where SF_u = solids flux due to underflow, kg/m² ·h
$\quad\quad U_b$ = bulk downward velocity, m/h
$\quad\quad Q_u$ = underflow flowrate, m³/h
$\quad\quad A$ = cross-sectional area, m²

The total mass flux SF_t of solids is the sum of previous components and is given by

$$SF_t = SF_g + SF_u \qquad (6\text{-}27)$$

$$SF_t = (C_i V_i - C_i U_b)(10^3 \text{ g/kg})^{-1} \qquad (6\text{-}28)$$

In this equation, the flux of solids due to gravity (hindered) settling depends on the concentration of solids and the settling characteristics of the solids at that concentration. The procedure used to develop a solids flux curve from column

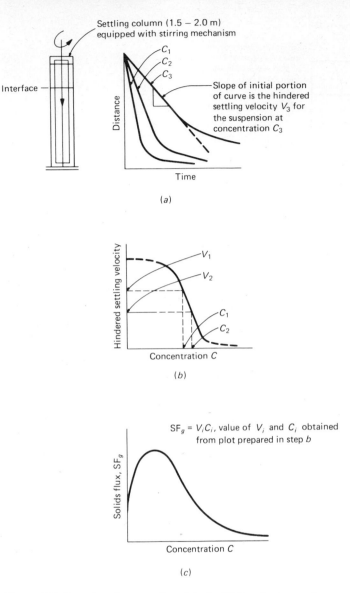

Figure 6-21 Procedure for preparing plot of solid flux due to gravity as a function of solids concentration. (*a*) Hindered settling velocities derived from column settling tests for suspension at different concentrations. (*b*) Plot of hindered settling velocities obtained in step *a* versus corresponding concentration. (*c*) Plot of computed value of solids flux versus corresponding concentration.

settling test data is illustrated in Fig. 6-21. At low concentration (below about 1000 mg/L), the movement of solids due to gravity is small, because the settling velocity of the solids is more or less independent of concentration. If the velocity remains essentially the same as the solids concentration increases, the

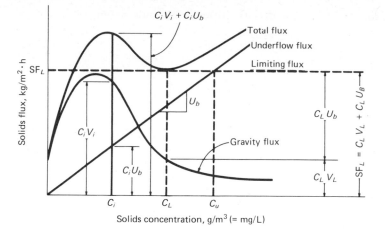

Figure 6-22 Definition sketch for the analysis of settling data using the solids-flux method of analysis.

total flux due to gravity starts to increase as the solids concentration starts to increase. At very high solids concentrations, the hindered-settling velocity approaches zero, and the total solids flux due to gravity again becomes extremely low. Thus it can be concluded that the solids flux due to gravity must pass through a maximum value as the concentration is increased. This is shown schematically in Figs. 6-21c and 6-22.

The solids flux due to bulk transport is a linear function of the concentration with slope equal to U_b, the underflow velocity (Fig. 6-22). The total flux, which is the sum of the gravity and the underflow flux, is also shown in Fig. 6-22. Increasing or decreasing the flow rate of the underflow causes the total-flux curve to shift upward or downward. Because the underflow velocity can be controlled, it is used for process control.

The required cross-sectional area of the thickener is determined as follows: As shown in Fig. 6-22, if a horizontal line is drawn tangent to the low point on the total-flux curve, its intersection with the vertical axis represents the limiting solids flux SF_L that can be processed in the settling basin. The corresponding underflow concentration is obtained by dropping a vertical line to the x axis from the intersection of the horizontal line and the underflow flux line. This can be done because the gravity flux is negligible at the bottom of the settling basin, and the solids are removed by bulk flow. The fact that the gravity flux is negligible at the bottom of the tank can be verified by performing a materials balance around that portion of the settling tank that lies below the depth where the limiting solids flux occurs and comparing the gravity settling velocity of the sludge to the velocity in the sludge withdrawal pipe. If the quantity of solids fed to the settling basin is greater than the limiting solids-flux value defined in Fig. 6-22, the solids will build up in the settling basin and, if adequate storage capacity is not provided, ultimately overflow at the top. Using the limiting solids-flux value, the required area derived from a materials balance is given by

$$A = \frac{(Q + Q_u)C_0}{SF_L} \times (10^3 \text{ g/kg})^{-1} \tag{6-29}$$

$$= \frac{(1 + \alpha)QC_0}{SF_L} \times (10^3 \text{ g/kg})^{-1} \tag{6-30}$$

where A = area, m^2
$(Q + Q_u)$ = total volumetric flowrate to settling basin, m^3/d
C_0 = influent solids concentration, g/m^3 (mg/L)
SF_L = limiting solids flux, kg/m$^2 \cdot$ d
$\alpha = Q_u/Q$

Referring to Fig. 6-22, if a thicker underflow concentration is required, the slope of the underflow flux line must be reduced. This, in turn, will lower the value of the limiting flux and increase the required settling area. In an actual design, the use of several different flowrates for the underflow should be evaluated. Typical values for biological sludges are about 7.1×10^{-5} to 1.4×10^{-4} m/s (150 to 300 gal/ft$^2 \cdot$ d) [25]. The application of this method of analysis is illustrated in Example 6-6.

An alternative graphical method of analysis to that presented in Fig. 6-22 for determining the limiting solids flux is shown in Fig. 6-23. As shown, for a given underflow concentration, the value of the limiting flux on the ordinate is obtained by drawing a line tangent to the flux curve passing through the desired underflow and intersecting the ordinate. The geometric relationship of this method to that given in Fig. 6-22 is shown by the lightly dashed lines in Fig. 6-23. The method detailed in Fig. 6-23 is especially useful where the effect of the use of various underflow concentrations on the size of the treatment facilities (aerator and sedimentation basin) is to be evaluated. The use of this method is illustrated in Example 10-4.

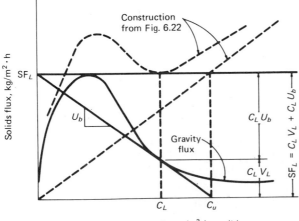

Figure 6-23 Alternative definition sketch for the analysis of settling data using the ϕ solids-flux method of analysis.

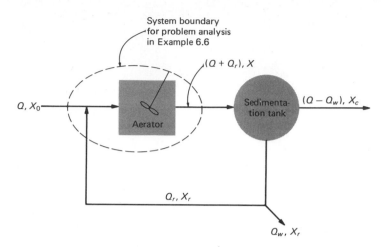

Figure 6-24 Definition sketch for Example 6-6.

Example 6-6: Application of solids flux analysis Given the following settling data for a biological sludge, derived from a pure-oxygen activated-sludge pilot plant, estimate the maximum concentration of the aerator mixed-liquor biological suspended solids that can be maintained if the sedimentation-tank application rate $Q + Q_r$ has been fixed at 25 m³/m² · d (615 gal/ft² · d) and the sludge recycle rate Q_r is equal to 40 percent. The definition sketch for this problem is shown in Fig. 6-24. As shown, settled and thickened biological solids from the sedimentation tank are returned to the aeration tank to maintain the desired level of biological solids in the aerator. Assume that the solids wasting rate Q_w is negligible in this example.

MLSS, X, g/m³ = mg/L[a]	2000	3000	4000	5000	6000	7000	8000	9000	10,000	15,000	20,000	30,000
Initial settling velocity, m/h[a]	4.27	3.51	2.77	2.13	1.28	0.91	0.67	0.49	0.37	0.15	0.07	0.027

[a] Data taken from Ref. 20.

SOLUTION
1   Develop the gravity solids-flux curve from the given data and plot the curve.
   a   Set up a computation table to determine the solids-flux values corresponding to the given solids concentrations.

MLSS, X, g/m³ = mg/L	2000	3000	4000	5000	6000	7000	8000	9000	10,000	15,000	20,000	30,000
Solids flux, kg/m² · h[a]	8.54	10.53	11.08	10.65	7.68	6.37	5.36	4.41	3.70	2.25	1.40	0.81

[a] Data taken from Ref. 20.

$$a \quad \frac{X(\text{g/m}^3)V(\text{m/h})}{10^3 \text{ g/kg}}$$

   b   Plot the solids-flux curve (see Fig. 6-25).

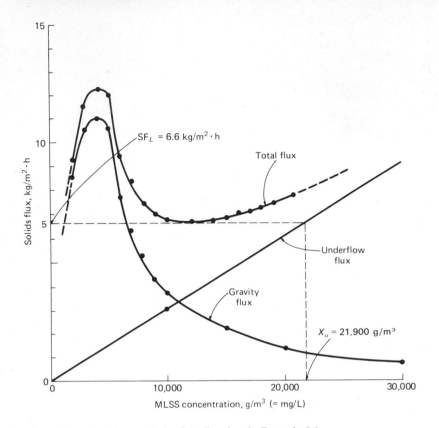

Figure 6-25 Solids-flux analysis of settling data in Example 6-6.

2 Determine the underflow bulk velocity. Referring to Fig. 6-24, the applied loading on the sedimentation facilities equals $(Q + Q_r)$, which per unit area is equal to 25 m³/m²·d. The underflow velocity is therefore equal to

$$U_b = [0.4Q/(Q + 0.4Q)](25 \text{ m}^3/\text{m}^2 \cdot \text{d})$$
$$= 7.14 \text{ m/d}$$
$$= 0.30 \text{ m/h}$$

3 Develop the total-flux curve for the system, and determine the value of the limiting flux and maximum underflow concentration.
  a Plot the underflow-flux curve on Fig. 6-25 using the following relationship:

$$SF_u = X_i U_b \times (10^3 \text{ g/kg})^{-1}$$

   where X_i = MLSS concentration, g/m³
   U_b = bulk underflow velocity, m/h

  b Plot the total-solids-flux curve by summing the values of the gravity and underflow solids flux (see Fig. 6-25).
  c From Fig. 6-25, the limiting solids flux is found to be equal to

$$SF_L = 6.6 \text{ kg/m}^2 \cdot \text{h}$$

d From Fig. 6-25, the maximum underflow solids concentration is equal to 21,900 g/m³

4 Determine the maximum solids concentration that can be maintained in the reactor shown in Fig. 6-24.

    a Write a materials balance for the system within the boundary.

$$QX_0 + Q_r X_t = (Q + Q_r)X$$

    b Assuming that $X_0 = 0$ ($X_0 \ll X_r$) and that $Q_r/Q = 0.4$, solve for the concentration of MLSS in the aerator.

$$0.4Q(21,900 \text{ mg/L}) = (1 + 0.4)QX$$

$$X = 6257 \text{ mg/L}$$

Comment As shown in this analysis, the concentration of the return solids will affect the maximum concentration of solids that can be maintained in the aerator. For this reason, the sedimentation tank should be considered an integral part of the design of an activated-sludge treatment process. This subject is considered in detail in Chap. 10, which deals with the design of biological treatment processes.

Analysis of Compression Settling (Type 4)

The volume required for the sludge in the compression region can also be determined by settling tests. The rate of consolidation in this region has been found to be proportional to the difference in the depth at time t and the depth to which the sludge will settle after a long period of time. This can be represented as Eq. 6-31.

$$H_t - H_\infty = (H_2 - H_\infty)e^{-i(t - t_2)} \qquad (6\text{-}31)$$

where H_t = sludge height at time t
       H_∞ = sludge depth after long period, say, 24 h
       H_2 = sludge height at time t_2
        i = constant for a given suspension

It has been observed that stirring serves to compact sludge in the compression region by breaking up the floc and permitting water to escape. Rakes are often used on sedimentation equipment to manipulate the sludge and thus produce better compaction. Dick and Ewing [8] found that stirring would produce better settling in the hindered-settling region also. With these facts in mind, it is apparent that, when appropriate, stirring should be investigated as an essential part of the settling tests if the proper areas and volumes are to be determined from the tests.

6-6 FLOTATION

Flotation is a unit operation used to separate solid or liquid particles from a liquid phase [10, 16]. Separation is brought about by introducing fine gas (usually air) bubbles into the liquid phase. The bubbles attach to the particulate matter, and the buoyant force of the combined particle and gas bubbles is great enough to cause the particle to rise to the surface. Particles that have a higher density than the liquid can thus be made to rise. The rising of particles with lower density than the liquid can also be facilitated (e.g., oil suspension in water).

In wastewater treatment, flotation is used principally to remove suspended matter and to concentrate biological sludges (see Chap. 11). The principal advantage of flotation over sedimentation is that very small or light particles that settle slowly can be removed more completely and in a shorter time. Once the particles have been floated to the surface, they can be collected by a skimming operation.

Description

The present practice of flotation as applied to municipal wastewater treatment is confined to the use of air as the flotation agent. Air bubbles are added or caused to form in one of the following methods:

1. Injection of air while the liquid is under pressure, followed by release of the pressure (dissolved-air flotation)
2. Aeration at atmospheric pressure (air flotation)
3. Saturation with air at atmospheric pressure, followed by application of a vacuum to the liquid (vacuum flotation)

Further, in all these systems the degree of removal can be enhanced through the use of various chemical additives.

Dissolved-air flotation In dissolved-air flotation systems, air is dissolved in the wastewater under a pressure of several atmospheres, followed by release of the pressure to the atmospheric level (see Fig. 6-26). In small pressure systems, the entire flow may be pressurized by means of a pump to 275 to 350 kPa (40 to 50 lb/in^2 gage) with compressed air added at the pump suction (see Fig. 6-26a). The entire flow is held in a retention tank under pressure for several minutes to allow time for the air to dissolve. It is then admitted through a pressure-reducing valve to the flotation tank where the air comes out of solution in minute bubbles throughout the entire volume of liquid.

In the larger units, a portion of the effluent (15 to 120 percent) is recycled, pressurized, and semisaturated with air (Fig. 6-26b). The recycled flow is mixed with the unpressurized main stream just before admission to the flotation tank, with the result that the air comes out of solution in contact with particulate matter at the entrance to the tank. Pressure types of units have been used mainly for the treatment of industrial wastes and for the concentration of sludges.

Air flotation In air flotation systems, air bubbles are formed by introducing the gas phase directly into the liquid phase through a revolving impeller or through diffusers. Aeration alone for a short period is not particularly effective in bringing about the flotation of solids. The provision of aeration tanks for

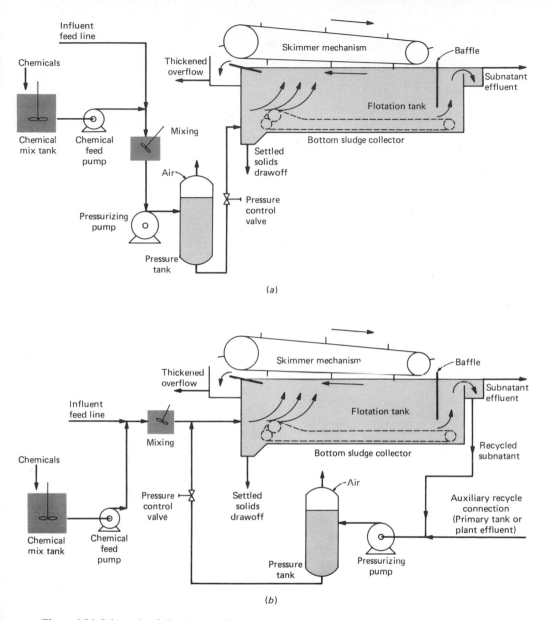

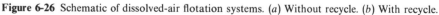

Figure 6-26 Schematic of dissolved-air flotation systems. (*a*) Without recycle. (*b*) With recycle.

flotation of grease and other solids from normal wastewater is usually not warranted, although some success with these units has been experienced on certain scum-forming wastes.

Vacuum flotation Vacuum flotation consists of saturating the wastewater with air either (1) directly in an aeration tank or (2) by permitting air to enter on the section side of a wastewater pump. A partial vacuum is applied, which causes the dissolved air to come out of solution as minute bubbles. The bubbles and the attached solid particles rise to the surface to form a scum blanket, which is removed by a skimming mechanism. Grit and other heavy solids that settle to the bottom are raked to a central sludge sump for removal. If this unit is used for grit removal and if the sludge is to be digested, the grit must be separated from the sludge in a grit classifier before the sludge is pumped to the digesters.

The unit consists of a covered cylindrical tank in which a partial vacuum is maintained. The tank is equipped with scum- and sludge-removal mechanisms. The floating material is continuously swept to the tank periphery, automatically discharged into a scum trough, and removed from the unit to a pump also under partial vacuum. Auxiliary equipment includes an aeration tank for saturating the wastewater with air, a short-period detention tank for removal of large air bubbles, vacuum pumps, and sludge and scum pumps.

Chemical additives Chemicals are commonly used to aid the flotation process. These chemicals, for the most part, function to create a surface or a structure that can easily absorb or entrap air bubbles. Inorganic chemicals, such as the aluminum and ferric salts and activated silica, can be used to bind the particulate matter together and, in so doing, create a structure that can easily entrap air bubbles. Various organic chemicals can be used to change the nature of either the air-liquid interface or the solid-liquid interface, or both. These compounds usually collect on the interface to bring about the desired changes.

Analysis

Because flotation is very dependent on the type of surface of the particulate matter, laboratory and pilot-plant tests must usually be performed to yield the necessary design criteria. Factors that must be considered in the design of flotation units include the concentration of particulate matter, quantity of air used, the particle-rise velocity, and the solids loading rate. In the following analysis, dissolved-air flotation is discussed because it is the method most commonly used. The design of dissolved-air flotation systems as well as other systems is discussed in Chap. 8.

The performance of a dissolved-air flotation system depends primarily on the ratio of the kilograms (pounds) of air to the kilograms of solids required to achieve a given degree of clarification. This ratio will vary with each type of suspension and must be determined experimentally using a laboratory flotation cell. A typical laboratory flotation cell is shown in Fig. 6-27. Procedures for conducting the necessary tests may be found in Ref. 10. Typical air-solids (A/S)

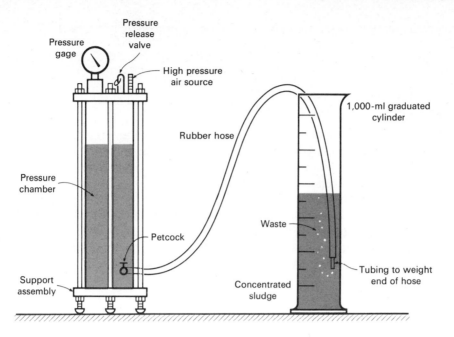

Figure 6-27 Schematic of dissolved-air flotation test apparatus.

ratios encountered in the thickening of sludge in wastewater-treatment plants vary from about 0.005 to 0.060.

The relationship between the A/S ratio and the solubility of air, the operating pressure, and the concentration of sludge solids for a system in which all the flow is pressurized is given in Eq. 6-32 [10].

$$\frac{A}{S} = \frac{1.3s_a(fP - 1)}{S_a} \tag{6-32}$$

where A/S = air-solids ratio, mL (air)/mg(solids)
$\quad\quad s_a$ = air solubility, mL/L

Temp., °C	0	10	20	30
s_a, mL/L	29.2	22.8	18.7	15.7

f = fraction of air dissolved at pressure P, usually 0.8
P = pressure, atm

$$= \frac{p + 101.35}{101.35} \text{ (SI units)}$$

$$= \frac{p + 14.7}{14.7} \text{ (U.S. customary units)}$$

p = gage pressure, kPa (lb/in^2 gage)
S_a = sludge solids, mg/L

The corresponding equation for a system with only pressurized recycle is

$$\frac{A}{S} = \frac{1.3s_a(fP - 1)R}{S_aQ}$$ (6-33)

where R = pressurized recycle, m³/d (Mgal/d)
Q = mixed-liquor flow, m³/d (Mgal/d)

In both the foregoing equations, the numerator represents the weight of air and the denominator the weight of the solids. The factor 1.3 is the weight in milligrams of 1 mL of air and the term (-1) within the brackets accounts for the fact that the system is to be operated at atmospheric conditions. The use of these equations is illustrated in Example 6-7.

The required area of the thickener is determined from a consideration of the rise velocity of the solids, 8 to 160 L/m² · min (0.2 to 4.0 gal/m · ft²), depending on the solids concentration, the degree of thickening to be achieved, and the solids-loading rate (see Table 11-12).

Example 6-7: Flotation thickening of activated-sludge mixed liquor Design a flotation thickener without and with pressurized recycle to thicken the solids in activated-sludge mixed liquor from 0.3 to about 4 percent. Assume that the following conditions apply.
1 Optimum A/S ratio = 0.008 mL/mg
2 Temperature = 20°C, 68°F
3 Air solubility = 18.7 mL/L
4 Recycle-system pressure = 275 kPa (40 lb/in² gage)
5 Fraction of saturation = 0.5
6 Surface-loading rate = 8 L/m² · min (0.2 gal/min · ft²)
7 Sludge flowrate = 400 m³/d (0.1 Mgal/d)

SOLUTION (WITHOUT RECYCLE)
1 Compute the required pressure using Eq. 6-32.

$$\frac{A}{S} = \frac{1.3s_a(fP - 1)}{S_a}$$

$$0.008 = \frac{1.3(18.7 \text{ mL/L})(0.5P - 1)}{3000 \text{ mg/L}}$$

$$0.5P = 0.99 + 1$$

$$P = 3.98 \text{ atm} = \frac{p + 101.35}{101.35}$$

$$p = 302 \text{ kPa (43.8 lb/in}^2 \text{ gage)}$$

2 Determine the required surface area.

$$A = \frac{400 \text{ m}^3/\text{d}(1000 \text{ L/m}^3)}{8 \text{ L/m}^2 \cdot \text{min } (60 \text{ min/h})(24 \text{ h/d})}$$

$$= 34.7 \text{ m}^2 \text{ (374 ft}^2)$$

3 Check the solids-loading rate.

$$kg/m^2 \cdot d = \frac{3000 \text{ mg/L}(400 \text{ m}^3/d)(1000 \text{ L/m}^3)}{34.7 \text{ m}^2 \ (10^6 \text{ mg/kg})}$$

$$= 34.58 \text{ kg/m}^2 \cdot d \ (7.1 \text{ lb/ft}^2 \cdot d)$$

SOLUTION (WITH RECYCLE)
1 Determine pressure in atmospheres.

$$P = \frac{275 + 101.35}{101.35} = 3.71 \text{ atm}$$

2 Determine the required recycle rate.

$$\frac{A}{S} = \frac{1.3 s_a (fP - 1)R}{S_a Q}$$

$$0.008 = \frac{1.3(18.7 \text{ mL/L})[0.5(3.71) - 1]R}{3000 \text{ mg/L } (400 \text{ m}^3/d)}$$

$$R = 461.9 \text{ m}^3/d \ (0.122 \text{ Mgal/d})$$

Alternatively, the recycle flowrate could have been set and the pressure determined. In an actual design, the costs associated with the recycle pumping, pressurizing system, and tank construction can be evaluated to find the most economical combination.
3 Determine the required surface area.

$$A = \frac{461.9 \text{ m}^3/d \ (1000 \text{ L/m}^3)}{8 \text{ L/m}^2 \cdot \text{min } (60 \text{ min/h})(24 \text{ h/d})} = 40.1 \text{ m}^2 \ (432 \text{ ft}^2)$$

6-7 GRANULAR-MEDIUM FILTRATION

Although filtration is one of the principal unit operations used in the treatment of potable water, the filtration of effluents from wastewater-treatment processes is a relatively recent practice. In spite of the short period of its use, however, filtration is already a well-established operation for achieving supplemental removals of suspended solids (including particulate BOD) from wastewater effluents of biological and chemical treatment processes [2]. Filtration has also been used to remove chemically precipitated phosphorus. With the publication of secondary treatment standards by the U.S. Environmental Protection Agency (see Table 4-1), the role of filtration in the removal of effluent suspended solids becomes even more firmly established.

It should be noted that filtration is not the only method available for reducing the concentration of effluent solids. The addition of organic polymers has resulted in effluents containing suspended-solids concentrations of 10 mg/L or less after conventional secondary settling, depending on the characteristics of the wastewater [2]. Microstrainers have also been used, although a number of operational problems have limited their application (see Sec. 8-10). At the present time (1977), effluent filtration using various granular media appears to be one of the most widely adopted method for the removal of residual solids.

The ability to design filters and to predict their performance must be based on (1) an understanding of the variables that control the process and (2) a knowledge of the pertinent mechanism or mechanisms responsible for the removal of particulate matter from a wastewater. The discussion in this section therefore covers the following topics: (1) description of the filtration operation, (2) classifications of filtration systems, (3) filtration-process variables, (4) particle-removal mechanisms, (5) general analysis of filtration operation, (6) analysis of wastewater filtration, and (7) need for pilot-plant studies. The literature dealing with filtration is so voluminous that the information presented in this section can serve only as an introduction to the subject. For additional details, the references in the text should be consulted. The practical implementation of filtration facilities is detailed in Sec. 8-9.

Description of the Filtration Operation

The complete filtration operation essentially involves two phases: filtration and backwashing. Both phases are identified in the definition sketch shown in Fig. 6-28. Filtration is accomplished by passing the wastewater to be filtered through a filter bed composed of granular material with or without the addition of chemicals. Within the granular filter bed, the removal of the suspended solids contained in the wastewater is accomplished by a complex process involving one or more removal mechanisms, such as straining, interception, impaction, sedimentation, and adsorption. The end of the filter run (filtration phase) is reached when the suspended solids in the effluent start to increase (break through) beyond an acceptable level, or when a limiting head loss occurs across the filter bed (see Fig. 6-29). Ideally, both these events should occur at the same time.

Once either of these conditions is reached, the filtration phase is terminated, and the filter must be backwashed to remove the material (suspended solids) that has accumulated within the granular filter bed. Usually, this is done by reversing the flow through the filter (see Fig. 6-28). A sufficient flow of wash water is applied until the granular filtering medium is fluidized (expanded). The material that has accumulated within the bed is then washed away. The wash water moving past the medium also shears away the material attached to the individual grains of the granular medium. In backwashing the filter, care should be taken not to expand the bed to such an extent that the effectiveness of the shearing action of the wash water is reduced. In most wastewater-treatment-plant flowsheets, the wash water containing the suspended solids that are removed from the filter is returned either to the primary settling facilities or to the biological treatment process.

Classifications of Filtration Systems

A number of individual filtration-system designs have been proposed and built. The principal types of granular-medium filters may be classified according to (1) the direction of the flow, (2) types of filter beds, (3) the driving force, and (4) the method of flowrate control.

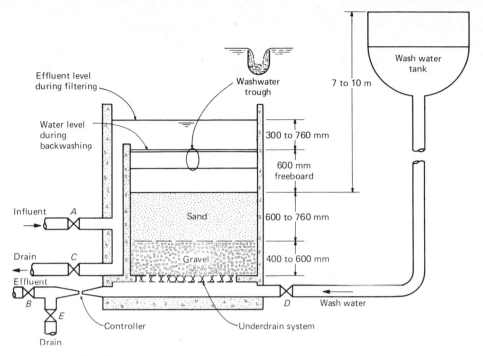

How filter operates

    1. Open valve A. (This allows effluent to flow to filter.)

    2. Open valve B. (This allows effluent to flow through filter.)

    3. During filter operation all other valves are closed.

How filter is backwashed

    1. Close valve A.

    2. Close valve B when water in filter drops down to top of overflow.

    3. Open valves C and D. (This allows water from wash water tank to flow up through the filtering medium, loosening up the sand and washing the accumulated solids from the surface of the sand, out of the filter. Filter backwash water is returned to head end of treatment plant.

How to filter to waste (if used)

    1. Open valves A and E. All other valves closed. Effluent is sometimes filtered to waste for a few minutes after filter has been washed to condition the filter before it is put into service.

Figure 6-28 Definition sketch for operation of downflow, granular-medium, gravity-flow filter.

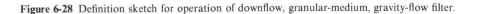

Direction of flow Filters used for the filtration of wastewater effluent may be classified according to the direction of flow as downflow, upflow, or biflow filters [25]. Schematics of these three types of filters are shown in Fig. 6-30. The downflow filter is by far the most common type used. Its operation was described at the beginning of this section.

    In upflow filters, which are proprietary, the flow passes up through the granular medium. The medium stratifies after backwashing, with coarse material on the bottom and fine material on the top. It is claimed that such filters are more efficient because the passage of the liquid to be filtered in the upward

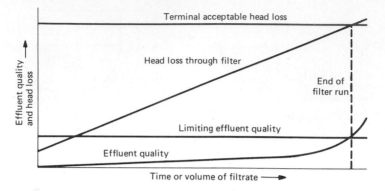

Figure 6-29 Definition sketch for the length of filtration.

direction allows greater penetration of the suspended solids into the bed. During the filtering operation, the granular medium is retained by a metal grid located at the top of the filter. The filtering medium forms an inverted arch with the evenly spaced bars. If the arch structure is disrupted by increasing the flowrate through the filter or by injecting air into the filter, the medium moves up past

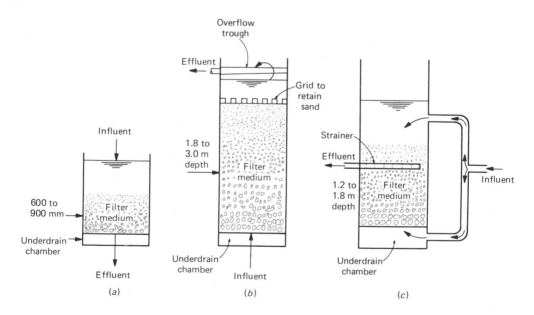

(Shown with single filter medium. Same configuration is used for dual- or tri-medium filter beds. See Fig. 6.31.)

Figure 6-30 Types of filters used for the filtration of treated wastewater, classified by direction of flow (*Adapted from Ref. 25.*) (*a*) Conventional downflow. (*b*) Upflow. (*c*) Biflow.

the retaining bars. This principle is used in backwashing the filter. The same liquid that is being filtered is also used for backwashing, which is another advantage claimed for these filters.

The features of both the downflow and upflow filters are combined in the biflow filter. As shown in Fig. 6-30, the effluent is collected through a strainer placed within the filter bed. Backwashing is accomplished simply by increasing the flowrate to the bottom of the filter. At present, experience with biflow filters in the United States is limited. They are more common in Europe and other parts of the world.

Types of filter beds The principal types of filter beds now used for wastewater filtration may be classified according to the number of filtering media used as single-medium, dual-medium, or tri-medium beds (see Fig. 6-31). In all three types, filtration takes place in the downward direction, and the beds are cleaned by fluidizing in the upward direction, as described previously (see Fig. 6-28). The distribution of grain sizes for each medium after backwashing is from small to

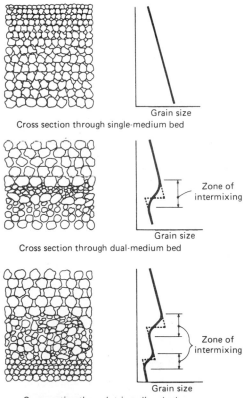

Grain size

Cross section through single-medium bed

Zone of intermixing

Grain size

Cross section through dual-medium bed

Zone of intermixing

Grain size

Cross section through tri-medium bed

Figure 6-31 Schematic diagram of bed stratification after backwash of single-medium, dual-medium, and tri-medium filters. (*Adapted from Ref. 3.*)

large. The degree of intermixing in the dual-medium and tri-medium beds depends on the density and size differences of the various media.

Dual-medium and tri-medium beds were developed to allow the suspended solids in the liquid to be filtered to penetrate farther into the filter bed and thus use more of the solids-storage capacity available within the filter. By comparison, in single-medium beds, most of the removal has been observed to occur in the upper few centimeters of the bed. The penetration of the solids farther into the bed also permits longer filter runs because the buildup of head loss is reduced. Data and design information on the various types of media and the characteristic sizes and depths that have been used are presented in Sec. 8-9.

Filtration driving force Either the force of gravity or an applied pressure force can be used to overcome the frictional resistance to flow offered by the filter bed. Gravity filters of the type shown in Fig. 6-28 are most commonly used for the filtration of treated effluent at large plants. Pressure filters of the type shown in Fig. 6-32 operate in the same manner as gravity filters and are used at smaller plants. The only difference is that, in pressure filters, the filtration operation is carried out in a closed vessel under pressurized conditions achieved by pumping. Pressure filters normally are operated at higher terminal head losses. This generally results in longer filter runs and reduced backwash requirements.

Flow control The rate of flow through a filter may be expressed as follows [25]:

$$\text{Rate of flow} = \frac{\text{driving force}}{\text{filter resistance}} \tag{6-34}$$

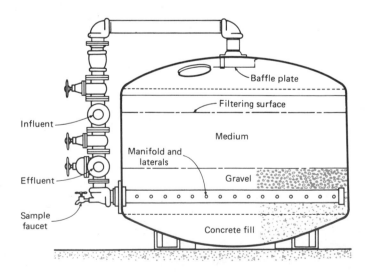

Figure 6-32 Section through pressure filter [27].

In this equation, the driving force represents the pressure drop across the filter. At the start of the filter run, the driving force must overcome only the resistance offered by the clean filter bed and the underdrain system. As solids start to accumulate within the filter, the driving force must overcome the resistance offered by the clogged filter bed and the underdrain system.

The principal methods now used to control the rate of flow through gravity filters may be classified as (1) constant-rate filtration and (2) variable-declining-rate filtration. In constant-rate filtration, the flow through the filter is maintained at a constant rate by means of an effluent-flow control valve that can be operated manually or mechanically. At the beginning of the run, a large portion of the available driving force is dissipated at the valve, which is almost closed. The valve is opened as the head loss builds up within the filter during the run. Because the required control valves are expensive and because they have malfunctioned on a number of occasions, alternative methods of flowrate control have been developed and are coming into wider use.

In variable-declining-rate filtration, the rate of flow through the filter declines and the level of the liquid above the filter bed rises throughout the length of the filter run. When the allowable head loss is reached, the filter is removed from service and backwashed. Additional details on this method of control as well as others may be found in Refs. 25 and 26.

Filtration-Process Variables

The principal variables that must be considered in the design of filters are identified in Table 6-8. In the application of filtration to the removal of residual suspended solids, it has been found that the nature of the particulate matter in the influent to be filtered, the size of the filter material or materials, and the filtration flow-rate are perhaps the most important of the process variables (Table 6-8, items 6, 1, and 4).

Influent characteristics The most important influent characteristics are the suspended-solids concentration, particle size and distribution, and floc strength. Typically, the suspended-solids concentration in the effluent from activated-sludge and trickling-filter plants varies between 6 and 30 mg/L. Because this concentration usually is the principal parameter of concern, turbidity is often used as a practical means of monitoring the filtration process. Within limits, it has been shown that the suspended-solids concentrations found in treated wastewater can be correlated to turbidity measurements [24]

Typical data on the particle size and distribution in the effluent from a pilot-scale activated-sludge plant operated at a mean cell residence time of 10 days are shown in Fig. 6-33. Similar observations have also been made at full-scale plants. As illustrated, the particles fell into two distinct size ranges, small particles varying in areal size (equivalent circular diameter) from 1 to 15 μ and large particles varying in size from 50 to 150 μ. In addition, a few particles larger than about 500 μ are almost always found in settled treated effluent. These

Table 6-8 Principal variables in the design of filters[a]

Variable	Significance

1. Filter-medium characteristics[b]
   a. Grain size
   b. Grain-size distribution
   c. Grain shape, density and
       composition
   d. Medium charge
	Affect particle-removal efficiency and head-loss buildup

2. Filter-bed porosity	Determines the amount of solids that can be stored in the filter
3. Filter-bed depth	Affects head loss, length of run
4. Filtration rate[b]	Used in conjunction with variables 1, 2, 3, and 6 to compute clear-water head loss.
5. Allowable head loss	Design variable

6. Influent wastewater
   characteristics[b]
   a. Suspended-solids
       concentration
   b. Floc or particle size and
       distribution
   c. Floc strength
   d. Floc or particle charge
   e. Fluid properties
	Affect the removal characteristics of a given filter-bed configuration. To a limited extent the listed influent characteristics can be controlled by the designer

[a] Adapted in part from Refs. 23 and 24.
[b] See text for additional discussion of specific variables.

particles are light and amorphous and do not settle readily (see discussion of hindered settling in Sec. 6-6). From the distribution analysis, the mean size for the smaller particles was estimated to be about 3 to 5 μ and that for the larger particles about 80 to 90 μ. The weight fraction of the smaller particles was estimated to be approximately 40 to 60 percent of the total. This percentage will vary, however, depending on the operating conditions of the biological process and the degree of flocculation achieved in the secondary settling facilities.

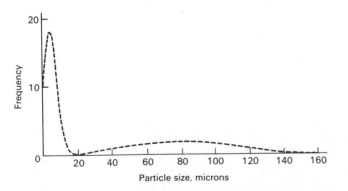

Figure 6-33 Typical particle-size distribution found in settled wastewater effluent [24].

The most significant observation relating to particle size is that the distribution of sizes was found to be bimodal. This is important because it will influence the removal mechanisms that may be operative during the filtration process. For example, it seems reasonable to assume that the removal mechanism for particles 1.0 μ in size would be different from that for particles 80 μ in size or larger. The bimodal particle-size distribution has also been observed in water-treatment plants [12]. This observation has also led to the study of the feasibility of using a two-stage filtration operation for secondary effluents [3].

Floc strength, which will vary not only with the type of process but also with the mode of operation, is also important. For example, the residual floc from the chemical precipitation of biologically processed wastewater may be considerably weaker than the residual biological floc before precipitation. Further, the strength of the biological floc will vary with the mean cell-residence time, increasing with longer mean cell-residence time (see Chap. 9). The increased strength derives in part from the production of extracellular polymers as the mean cell-residence time is lengthened. At extremely long mean cell-residence times (15 days and longer), it has been observed that the floc strength will decrease.

Filter-medium characteristics Grain size is the principal filter-medium characteristic that affects the filtration operation. Grain size affects both the clear-water head loss and the buildup of head loss during the filter run. If too small a filtering medium is selected, much of the driving force will be wasted in overcoming the frictional resistance of the filter bed. On the other hand, if the size of the medium is too large, many of the small particles in the influent will pass directly through the bed (see Fig. 6-33).

Filtration rate The rate of filtration is important because it affects the areal size of the filters that will be required. For a given filter application, the rate of filtration will depend primarily on the strength of the floc and the size of the filtering medium. For example, if the strength of the floc is weak, high filtration rates will tend to shear the floc particles and carry much of the material through the filter. It has been observed that filtration rates in the range of 80 to 320 L/m^2 · min (2 to 8 gal/ft^2 · min) will not affect the quality of the filter effluents because biological floc is strong. This subject is considered further in subsequent discussions.

Particle-Removal Mechanisms

The principal mechanisms that are believed to contribute to the removal of material within a granular-medium filter are identified and described in Table 6-9. The major removal mechanisms (the first five listed in Table 6-9) are illustrated in Fig. 6-34. Straining has been identified as the principal mechanism that is operative in the removal of suspended solids during the filtration of

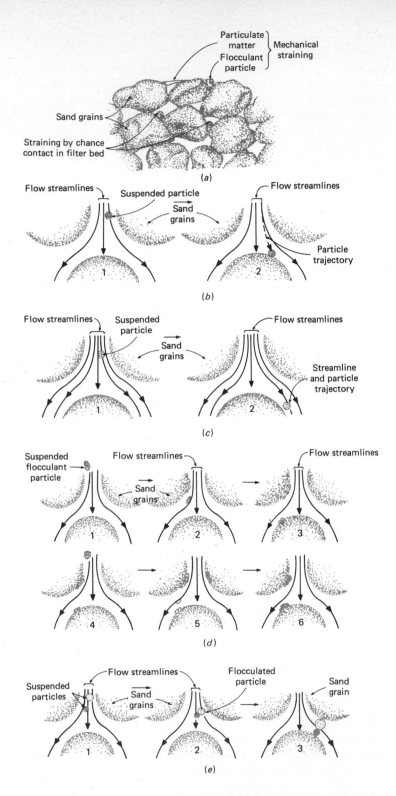

Figure 6-34 Removal of suspended particulate matter within a granular filter. (*a*) By straining. (*b*) By sedimentation or inertial impaction. (*c*) By interception. (*d*) By adhesion. (*e*) By flocculation.

Table 6-9 Mechanisms operative within a granular-medium filter that contribute to the removal of suspended materials[a]

Mechanism	Description
1. Straining[b]	
a. Mechanical	Particles larger than the pore space of the filtering medium are strained out mechanically
b. Chance contact	Particles smaller than the pore space are trapped within the filter by chance contact
2. Sedimentation[b]	Particles settle on the filtering medium within the filter
3. Impaction[b]	Heavy particles will not follow the flow streamlines
4. Interception[b]	Many particles that move along in the streamline are removed when they come in contact with the surface of the filtering medium
5. Adhesion[b]	Flocculant particles become attached to the surface of the filtering medium as they pass by. Because of the force of the flowing water, some material is sheared away before it becomes firmly attached and is pushed deeper into the filter bed. As the bed becomes clogged, the surface shear force increases to a point at which no additional material can be removed. Some material may break through the bottom of the filter, causing the sudden appearance of turbidity in the effluent
6. Chemical adsorption	
a. Bonding	
b. Chemical interaction	
	Once a particle has been brought in contact with the surface of the filtering medium or with other particles, either one of these mechanisms, or both, may be responsible for holding it there
7. Physical adsorption	
a. Electrostatic forces	
b. Electrokinetic forces	
c. van der Waals forces	
8. Flocculation	Large particles overtake smaller particles, join them, and form still larger particles. These particles are then removed by one or more of the above removal mechanisms (1 through 5)
9. Biological growth	Biological growth within the filter will reduce the pore volume and may enhance the removal of particles with any of the above removal mechanisms (1 through 5)

[a] Adapted from Refs. 23 and 24.
[b] Usually identified in the literature as removal mechanisms.

settled secondary effluent from biological treatment processes. This statement is based on data reported in Ref. 22, on the results of subsequent studies conducted by the author of this text, and on other literature references.

Other mechanisms are probably also operative even though their effects are small and, for the most part, masked by the straining action. These other mechanisms include interception, impaction, and adhesion. In fact, it is reasonable to assume that the removal of some of the smaller particles shown in Fig. 6-33 must be accomplished in two steps involving (1) the transport of the particles

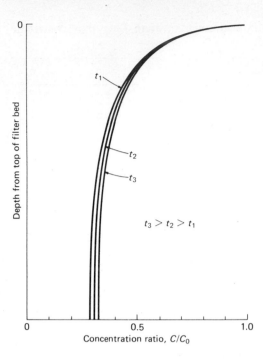

Figure 6-35 Concentration-ratio curves for granular-medium filter where straining is the principal particle-removal mechanism.

Depth from top of filter bed

$t_3 > t_2 > t_1$

0   0.5   1.0

Concentration ratio, C/C_0

to the surface where they will be removed and (2) the removal of the particles by one or more of the operative removal mechanisms. O'Melia and Stumm have identified these two steps as transport and attachment [15].

The removal of suspended material by straining can be identified by noting (1) the variation in the normalized concentration-removal curves through the filter as a function of time and (2) the shape of the head-loss curve for the entire filter or an individual layer within the filter. If straining is the principal removal mechanism, the shape of the normalized removal curve will not vary significantly with time (see Fig. 6-35), and the head-loss curves will be curvilinear. The theoretical basis for these statements is presented in Refs. 22 and 24.

General Analysis of Filtration Operation

In general, the mathematical characterization of the time-space removal of particulate matter within the filter is based on a consideration of the equation of continuity, together with an auxiliary rate equation. This is illustrated in the following discussion, which is adapted from Ref. 24.

Equation of continuity The equation of continuity for the filtration operation may be developed by considering a suspended-solids mass balance for a section of filter of cross-sectional area A, and of thickness dx, measured in the direction of flow. Following the approach delineated in Chap. 5, the mass balance would be as follows:

1. General word statement:

Rate of accumulation of   rate of flow of solids into   rate of flow of solids out
   solids within the volume $=$ the volume element $-$ of the volume element   (6-35)
   element

2. Simplified word statement:

$$\text{Accumulation} = \text{inflow} - \text{outflow}$$

3. Symbolic representation:

$$\left(\frac{\partial q}{\partial t} + \alpha \frac{\partial \overline{C}}{\partial t}\right) dV = Q(C) - Q\left(C + \frac{\partial C}{\partial x} dx\right) \qquad (6\text{-}36)$$

where $\partial q/\partial t =$ change in quantity of solids deposited within the filter with time
             mg/cm$^3 \cdot$ min
           $\alpha =$ average porosity
      $\partial \overline{C}/\partial t =$ change in average concentration of solids in pore space with time,
             mg/cm$^3 \cdot$ min
         $dV =$ differential volume, cm^3
          $Q =$ filtration rate, L/min
          $C =$ concentration of suspended solids, mg/L
     $\partial C/\partial x =$ change in concentration of suspended solids in fluid stream with
             distance, mg/L $\cdot$ cm

Substituting $A\,dx$ for dV and Av for Q where v is the filtration velocity $(L/cm^2 \cdot min)$ and simplifying, Eq. 6-36 yields

$$-v\frac{\partial C}{\partial x} = \frac{\partial q}{\partial t} + \alpha \frac{\partial \overline{C}}{\partial t} \qquad (6\text{-}37)$$

In Eq. 6-37, the first term represents the difference between the mass of suspended solids entering and leaving the section; the second term represents the time rate of change in the mass of suspended solids present on the solid portion of the filter; and the third term represents the time rate of change in the suspended-solids concentration in the fluid portion of the filter volume.

    In a flowing process, the quantity of fluid contained within the bed is usually small compared with the volume of liquid passing through the bed. In this case, the materials balance becomes

$$-v\frac{\partial C}{\partial x} = \frac{\partial q}{\partial t} \qquad (6\text{-}38)$$

This equation is the one most commonly found in the literature dealing with filtration theory.

Rate equation To solve Eq. 6-38, an additional independent equation is required. The most direct approach is to derive a relationship that can be used to describe the change in concentration of suspended matter with distance, such as

$$\frac{\partial C}{\partial x} = \phi(V_1, V_2, V_3, \ldots) \tag{6-39}$$

in which V_1, V_2, and V_3 are the variables governing the removal of suspended matter from solution.

An alternative approach is to develop a complementary equation in which the pertinent process variables are related to the amount of material removed within the filter at various depths. In equation form, this may be written as

$$\frac{\partial q}{\partial t} = \phi(V_1, V_2, V_3, \ldots) \tag{6-40}$$

Analysis of Wastewater Filtration

The following analysis, also adapted from Ref. 24, is based on the assumption that straining is the operative removal mechanism.

Equation of continuity Because the shape of the removal curve within the filter does not vary with time, the equation of continuity (Eq. 6-38) may be written as an ordinary differential equation:

$$-v \frac{dC}{dx} = \frac{dq}{dt} \tag{6-41}$$

Rate equation From the size and distribution of the influent particles (see Fig. 6-33) and the shape of the normalized curves (see Fig. 6-35), it can be concluded that the rate of change of concentration with distance must be proportional to some removal coefficient that is changing with the degree of treatment or removal achieved in the filter. For example, the entire particle-size distribution in the influent is passed through the first layer. The probability of removing particles from the wastestream is p_1. In the second layer, the probability of removing particles is p_2; p_2 is less than p_1, assuming that some of the larger particles will be removed by the first layer. Continuing this argument, it can be reasoned that the rate of removal must always be changing as a function of the degree of treatment. This phenomenon can be expressed mathematically using the following equation:

$$\frac{dC}{dx} = \left[\frac{1}{(1 + ax)^n}\right] r_0 C \tag{6-42}$$

where C = concentration, mg/L
   x = distance, cm
   r_0 = initial removal rate, cm^{-1}
   a, n = constants

In Eq. 6-42, the term within brackets is sometimes called a retardation factor. When the exponent n is equal to zero, the term within brackets is equal to one; under these conditions, Eq. 6-42 represents a logarithmic removal curve. When n equals one, the value of the term within brackets drops off rapidly in the first 12.7 cm (5 in) and then more gradually as a function of distance. Therefore, it appears that the exponent n may be related to the distribution of particle sizes in the influent. For example, when dealing with a uniform filter medium and filtering particles of one size, it would be expected that the value of the exponent n would be equal to zero and that the initial removal could be described as a first-order removal function. It should be noted that this equation was verified only for filtration rates up to 400 L/m^2 · min (10 gal/ft^2 · min) [24].

The value of r_0 is obtained by computing the slope of the removal curve at or near zero depth, since $[1/(1 + ax)^n] \simeq 1$. The constants a and n must be determined using a trial-and-error procedure. The easiest way to do this is to rewrite Eq. 6-42 as follows:

$$\left(\frac{Cr_0}{dC/dx}\right)^{1/n} = 1 + ax \qquad (6\text{-}43)$$

If Eq. 6-43 is plotted functionally, the value of n is equal to the value that results in a straight-line plot. The slope of the line describing the experimental data will be equal to the constant a.

Generalized rate equation On the basis of experimental results derived from this study and data reported in the literature, there appear to be five major factors that affect the time-space removal of the residual suspended matter from a flocculation-sedimentation process within a granular filter for a given temperature. These factors are the size of the filter medium, the rate of filtration, the influent particle size and size distribution, the floc strength, and the amount of material removed within the filter. Therefore, a generalized rate equation must account for the effect of these factors.

Although a number of different formulations are possible, a generalized rate equation in which all five factors are considered can be developed by multiplying Eq. 6-42 by a factor that takes into account the effect of the material accumulated in the filter. The proposed equation is

$$\frac{dC}{dx} = -\frac{1}{(1 + ax)^n} r_0 C \left(1 - \frac{q}{q_u}\right)^m \qquad (6\text{-}44)$$

where q = quantity of suspended solids deposited in the filter
$\quad q_u$ = ultimate quantity of solids that can be deposited in the filter
$\quad m$ = a constant related to floc strength

Initially, when the amount of material removed by the filter is low, $q = 0$; $(1 - q/q_u)^m = 1$, and Eq. 6-44 is equivalent to Eq. 6-42. As the upper layers begin to clog, the term $(1 - q/q_u)^m$ approaches zero, and the rate of change in concentration with distance is equal to zero. At the lower depths, the amount of material removed is essentially zero, and the previous analysis applies.

Head-loss development In the past, the most commonly used approach to determine the head loss in a clogged filter was to compute it with a modified form of the equations used to evaluate the clear-water head loss (see Table 6-10 and Example 6-8). In all cases, the difficulty encountered in using these equations is that the porosity must be estimated for various degrees of clogging. Unfortunately, the complexity of this approach renders most of these formulations useless or, at best, extremely difficult to use.

An alternative approach is to relate the development of head loss to the

Table 6-10 Formulas governing the flow of clear water through a granular medium[a]

Equation	Definition of terms
Carmen-Kozeny: $h = \dfrac{f}{\phi} \dfrac{1-\alpha}{\alpha^3} \dfrac{L}{d} \dfrac{v^2}{g}$ $f = 150 \dfrac{1-\alpha}{N_r} + 1.75$ $N_R = \dfrac{dv\rho}{\mu}$ Fair-Hatch: $h = kvS^2 \dfrac{(1-\alpha)^2}{\alpha^3} \dfrac{L}{d^2} \dfrac{v}{g}$ Rose: $h = \dfrac{1.067}{\phi} C_d \dfrac{1}{\alpha^4} \dfrac{L}{d} \dfrac{v^2}{g}$ $C_d = \dfrac{24}{N_R} + \dfrac{3}{\sqrt{N_R}} + 0.34$ Hazen: $h = \dfrac{1}{C} \dfrac{60}{T + 10} \dfrac{L}{d_{10}^2} v$	C = coefficient of compactness (600 to 1200) C_d = coefficient of drag d = grain diameter, m d_{10} = effective size grain diameter, mm f = friction factor g = gravity constant, 9.807 m/s^2 h = head loss, m k = filtration constant, 5 based on sieve openings, 6 based on size of separation L = depth, m N_R = Reynolds number S = shape factor (varies between 6.0 and 7.7) T = temperature, °F v = filtration velocity, m/s α = porosity μ = viscosity, N · s/m^2 v = kinematic velocity, m^2/s ρ = density, kg/m^3 ϕ = shape factor (usually 1)

[a] Adapted from Ref. 22.

Note: m × 3.2808 = ft
mm × 0.03937 = in
m/s × 3.2808 = ft/s
kg/m^3 × 0.0624 = lb/ft^3

amount of material removed by the filter. The head loss would then be computed using the expression

$$H_t = H_0 + \sum_{i=i}^{n} (h_i)_t \qquad (6\text{-}45)$$

where H_t = total head loss at time t, m
H_0 = total initial clear-water head loss, m
$(h_i)_t$ = head loss in the ith layer of the filter at time t, m

From an evaluation of the incremental head-loss curves for uniform sand and anthracite, the buildup of head loss in an individual layer of the filter was found to be related to the amount of material contained within the layer. The form of the resulting equation for head loss in the ith layer is

$$(h_i)_t = a(q_i)_t^b \qquad (6\text{-}46)$$

where $(q_i)_t$ = amount of material deposited in the ith layer at time t, mg/cm^3
a, b = constants

In this equation, it is assumed that the buildup of head loss is only a function of the amount of material removed. The validity of this assumption has been checked by Baumann and Huang [1]. Representative data for uniform sand and anthracite are presented in Fig. 6-36.

Computation of the clear water head loss through a filter is illustrated in Example 6-8. The determination of the buildup of head loss during the filtration process using the data presented in Fig. 6-36 is illustrated in Example 6-9 at the end of this chapter.

Example 6-8: Computation of clear-water head loss in a granular-medium filter Determine the clear-water head loss in a filter bed composed of 30 cm of uniform anthracite with an average size of 1.6 mm and 30 cm of uniform sand with an average size of 0.5 mm for a filtration rate of 160 L/m^2 · min (4 gal/ft^2 · min). Assume that the operating temperature is 20°C. Use the Rose equation given in Table 6-10 for computing the head loss.

SOLUTION
1 Determine the Reynolds number for the anthracite and sand layers.
    a Anthracite layer

$$N_R = \frac{d v}{\nu}$$

$$d = 1.6 \text{ mm} = 1.6 \times 10^{-3} \text{ m}$$

$$v = 0.160 \text{ m/min} = 2.67 \times 10^{-3} \text{ m/s}$$
(Note that the filtration rate is converted
to an equivalent linear velocity by
converting the volume expressed in liters
to cubic meters.)

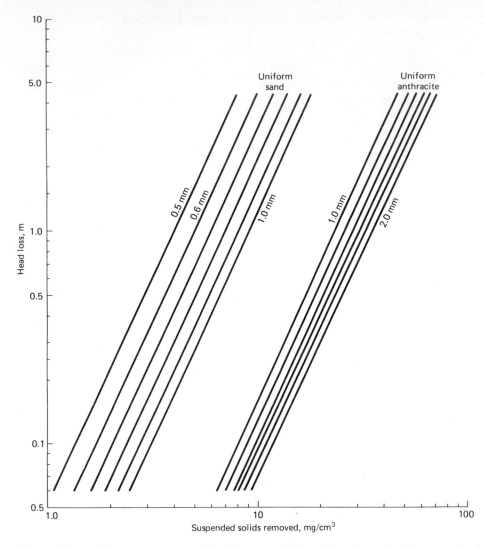

Figure 6-36 Head loss versus suspended solids removed for various sizes of uniform sand and anthracite (*Adapted from Refs. 22, 24.*)

$$v = 1.003 \times 10^{-6} \text{ m}^2/\text{s (see Appendix B)}$$

$$N_R = \frac{1.6 \times 10^{-3} \text{ m} \times 2.67 \times 10^{-3} \text{ m/s}}{1.003 \times 10^{-6} \text{ m}^2/\text{s}}$$

$$= 4.26$$

b Sand layer

$$N_R = \frac{0.5 \times 10^{-3} \text{ m} \times 2.67 \times 10^{-3} \text{ m/s}}{1.003 \times 10^{-6} \text{ m}^2/\text{s}}$$

$$= 1.33$$

2 Determine the coefficient of drag C_D.
   a Anthracite layer

$$C_D = \frac{24}{N_R} + \frac{3}{\sqrt{N_R}} + 0.34$$

$$= \frac{24}{4.26} + \frac{3}{\sqrt{4.26}} + 0.34$$

$$= 7.43$$

   b Sand layer

$$C_D = \frac{24}{1.33} + \frac{3}{\sqrt{1.33}} + 0.34$$

$$= 20.99$$

3 Determine the head loss through the anthracite and sand layers.
   a Anthracite layer

$$h = \frac{1.067}{\phi} C_D \frac{1}{\alpha^4} \frac{L}{d} \frac{v^2}{g}$$

$\phi = 1$ (assumed)
$C_D = 7.43$
$\alpha - 0.4$ (assumed), $\alpha^4 - 0.0256$
$L = 30$ cm $= 0.3$ m
$d = 1.6 \times 10^{-3}$ m
$v = 2.67 \times 10^{-3}$ m/s
$g = 9.81$ m/s²

$$h = \frac{1.067}{1.0} 7.43 \frac{1}{0.0256} \frac{0.3 \text{ m}}{1.6 \times 10^{-3}} \times \frac{(2.67 \times 10^{-3} \text{ m/s})^2}{9.81 \text{ m/s}^2}$$

$$= 0.042 \text{ m}$$

   b Sand layer

$\phi = 1$ (assumed)
$C_D = 20.99$
$\alpha = 0.4$ (assumed), $\alpha^4 = 0.0256$
$L = 30$ cm $= 0.3$ m
$d = 0.5 \times 10^{-3}$ m
$v = 2.67 \times 10^{-3}$ m/s
$g = 9.81$ m/s²

$$h = \frac{1.067}{1.0} 20.99 \frac{1}{0.0256} \frac{0.3 \text{ m}}{0.5 \times 10^{-3}} \times \frac{(2.67 \times 10^{-3} \text{ m/s})^2}{9.81 \text{ m/s}^2}$$

$$= 0.382 \text{ m}$$

4 Determine the total head loss H_T

$H_T =$ head loss through anthracite layer + head loss through sand layer
$H_T = 0.042$ m $+ 0.382$ m
$\quad = 0.424$ m (1.39 ft)

Comment The head-loss computations in this example were simplified by assuming that the anthracite and sand layers were of uniform size. The same computation procedure can be used for stratified filter beds by considering the total head loss to be the sum of the head losses in successive layers.

Need for Pilot-Plant Studies

Although the information presented earlier in this section will help the reader understand the nature of the filtration operation as it is applied to the filtration of treated wastewater, it must be stressed that there is no generalized approach to the design of full-scale filters. The principal reason is the inherent variability in the characteristics of the influent suspended solids to be filtered. For example, changes in the degree of flocculation of the suspended solids in the secondary settling facilities will significantly affect the particle sizes and their distribution in the effluent. This, in turn, will affect the performance of the filter. Further, because the characteristics of the effluent suspended solids will also vary with the organic loading on the process as well as with the time of day, filters must be designed to function under a rather wide range of operating conditions. The best way to ensure that the filter configuration selected for a given application will function properly is to conduct pilot-plant studies.

In most situations, the principal factors to be determined in a pilot testing program aimed at producing a design capable of achieving a given set of effluent criteria are (1) the selection of the optimally sized filter medium or media and their respective depths, (2) the determination of the most appropriate filtration rate and terminal head loss, and (3) the establishment of the expected duration of filter run. The effects of adding filter aids, such as polymers, are also studied in some pilot tests. In these instances, jar tests are useful as a preliminary screening method [2].

From past studies, it has been found that the performance of a pilot filter column with a diameter of at least 150 mm (6 in) simulates the operating results obtained from a full-scale filter. Pilot filters with cross-sectional areas of about 0.1 m² (1.0 ft²) or larger can be used to determine backwashing characteristics. The influent wastewater supply system should be capable of providing any filtration rates between 80 and 400 L/m² · min (2 and 10 gal/ft² · min). It must also be compatible with the expected maximum head loss, typically at least 3 m (10 ft) for gravity filters and 7.5 to 10 m (25 to 32 ft) for pressure filters. All filter columns should be piped together in parallel to ensure that they receive the same wastewater. Ideally, there should be as many pilot filter columns as there are variables to be tested.

In special cases where it is important to refine the design of a filter bed, pilot filter columns equipped with piezometer sampling taps should be used so that those areas within the filter where excessive head losses are occurring can be identified. This information can then be used to modify the design of the filter bed.

Because of the many variables that can be analyzed, care must be taken not to change more than one variable at a time so as not to confound the results in a statistical sense. Testing should be carried out at several intervals, ideally throughout a full year, to assess seasonal variations in the characteristics of the effluent to be filtered. All test results should be summarized and evaluated in several different ways to ensure their proper analysis. Because the specific details of each test program will be different, no generalizations on the best method of analysis can be given. Details on the conduct of pilot testing programs may be found in Refs. 1, 2, 3, 6, 22, 23, and 24. The analysis of some typical pilot-plant data is illustrated in Example 6-9.

Example 6-9: Analysis of filtration data from a pilot plant The normalized suspended-solids-removal-ratio curves shown in Fig. 6-37 were derived from a filtration pilot-plant study conducted at an activated-sludge wastewater-treatment plant. Using these curves and the following data, develop curves that can be used to estimate (1) the head-loss buildup as a function of the length of run and (2) the length of run to a terminal head loss of 3.0 m as a function of the filtration rate.

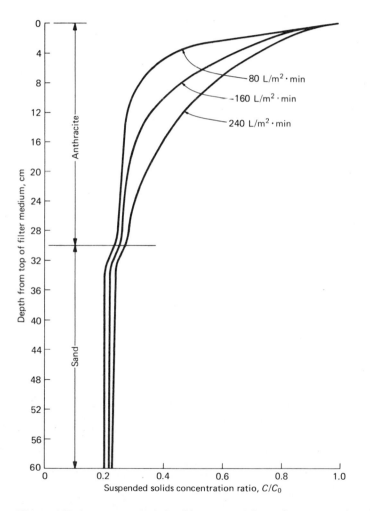

Figure 6-37 Average suspended-solids concentration-ratio curves versus filter depth for various filtration rates for Example 6-9.

Biological treatment process:
1 Mean cell residence time $\theta_c = 10$ d
2 Average suspended-solids concentration in effluent from secondary settling tank = 20 mg/L
3 Particle-size distribution in effluent = similar to that shown in Fig. 6-33
Pilot plant:
1 Type of filter bed = dual-medium
2 Filter media = uniform anthracite and sand
3 Filter-medium characteristics
   a Anthracite, $d = 1.6$ mm, $UC \simeq 1.0$
   b Sand, $d = 0.5$ mm, $UC \simeq 1.0$
4 Filter-bed depth = 60 cm
   a Anthracite = 30 cm
   b Sand = 30 cm
5 Filtration rates = 80, 160, and 240 L/m² · min
6 Temperature = 20°C
7 General observation: average concentration-ratio curves plotted in Fig. 6-37 did not vary significantly with time

Table 6-11 Computation table to determine value of $\Delta C(C_{x-1} - C_x)$ for various filter depths in Example 6-9

Depth, cm (1)	Filtration rate, L/cm² · min								
	0.008			0.016			0.024		
	C_x/C_0 (2)	C_x (3)	ΔC (4)	C_x/C_0 (5)	C_x (6)	ΔC (7)	C_x/C_0	C_x (9)	ΔC (10)
0	1.0	20.0		1.0	20.0		1.0	20.0	
			11.4			7.4			5.6
4	0.43	8.6		0.63	12.6		0.72	14.4	
			2.2			3.2			2.8
8	0.32	6.4		0.47	9.4		0.58	11.6	
			0.8			2.2			2.0
12	0.28	5.6		0.36	7.2		0.48	9.6	
			0.2			1.2			1.6
16	0.27	5.4		0.30	6.0		0.40	8.0	
			0.2			0.4			1.0
20	0.26	5.2		0.28	5.2		0.35	7.0	
			0.2			0.2			0.8
24	0.25	5.0		0.27	5.4		0.31	6.2	
			0.2			0.2			0.6
28	0.24	4.8		0.26	5.2		0.28	5.6	
			0.0			0.0			0.2
30	0.24	4.8		0.26	5.2		0.27	5.4	
			0.5			0.6			0.5
32	0.215	4.3		0.23	4.6		0.245	4.9	
			0.2			0.1			0.1
34	0.205	4.1		0.225	4.5		0.24	4.8	
			0.1			0.1			0.0
36	0.20	4.0		0.22	4.4		0.24	4.8	

SOLUTION

1 To analyze the curves given in Fig. 6-37, rewrite Eq. 6-43 in a form suitable for numerical analysis:

$$-v\frac{\Delta C}{\Delta x} = \frac{\Delta q}{\Delta t}$$

$$-v\frac{C_{x-1} - C_x}{X_{x-1} - X_x} = \frac{q_2 - q_1}{t_2 - t_1}$$

2 Set up a computation table and determine the value of $\Delta C(C_{x-1} - C_x)$ for various depths throughout the filter. The required computations are summarized in Table 6-11. As shown,

Table 6-12 Computation table to determine buildup of suspended solids and head loss for filter runs in Example 6-9

		Run length, h					
		10		15		20	
Depth, cm (1)	ΔC, mg/L (2)	Δg mg/cm³ (3)	Δh, m (4)	Δq, mg/cm³ (5)	Δh, m (6)	Δg, mg/cm³ (7)	Δh, m (8)
0							
	7.4	17.76	0.31	26.64	0.78	35.52	1.30
4							
	3.2	7.68	0.05	11.52	0.13	15.36	0.23
8							
	2.2	5.28	....	7.92	0.06	10.56	0.10
12							
	1.2	2.88	....	4.32	....	5.76	
16							
	0.4	0.96	....	1.44	....	1.92	
20							
	0.2	0.48	....	0.72	....	0.96	
24							
	0.2	0.48	....	0.72	....	0.96	
28							
	0.0	0.00	....	0.00	....	0.00	
30							
	0.6	2.88	0.5	4.32	1.20	5.76	2.20
32							
	0.1	0.48	....	0.72	....	0.96	0.05
34							
	0.1	0.48	....	0.72	....	0.96	0.05
36							
$\sum \Delta h$, m			0.86		2.17		3.93

Note: cm × 0.3937 = in
mg/L = g/m³
m × 3.2808 = ft

values of C/C_0 from Fig. 6-37 corresponding to the depths given in col. 1 are entered in cols. 1, 5, and 8 for each filtration rate. The value of the concentration at each depth is entered in cols. 3, 6, and 9 for each filtration rate. The concentration difference $\Delta C(C_{x-1} - C_x)$ between the depths given in col. 1 is entered in cols. 4, 7, and 10 for each filtration rate.

3 Set up a computation table and determine the buildup of suspended solids and head loss within each layer of the filter for filter runs of various lengths. The necessary computations for a filtration rate of 160 L/m² · min (4 gal/ft² · min) are summarized in Table 6-12. Although the required computations for the other filtration rates are not shown, they are the same. The values of ΔC given in col. 2 are taken from col. 7 of Table 6-11. The values of q shown in cols. 3, 5, and 7 are determined by using the difference equation given in step 1. To illustrate, for the anthracite layer between 4 and 8 cm from the top of the column, the value of Δq after 20 h is as follows:

$$ -v\frac{\Delta C}{\Delta X} = \frac{\Delta q}{\Delta t} $$

where $v = 160$ L/m² · min = 0.016 L/cm² · min
$\Delta C = 3.2$ mg/L
$\Delta X = 4$ cm $- 8$ cm $= -4$ cm
$\Delta t = (20$ h $\times 60$ min/h $- 0) = 1200$ min

$$ \Delta q = -0.016 \text{ L/cm}^2 \cdot \text{min} \frac{3.2 \text{ mg/L}}{-4 \text{ cm}} 1200 \text{ min} $$

$$ = 15.36 \text{ mg/L} $$

The value of incremental head-loss buildup Δh (col. 8) for the anthracite layer between 4 and 8 cm is obtained from Fig. 6-36 by entering with the value of Δq for this layer. The value of the head loss in the sand layers is determined in a similar manner. To simplify the computations, it is assumed that no intermixing occurs between the anthracite and sand. Once all of the Δh head-loss values are entered, the entire column is summed to obtain the total head loss in the filter bed. The total head loss for other time periods and filtration rates is determined in exactly the same manner. Summary data for the other flowrates are as follows:

	Head loss, m	
Time, h	20 L/m² · min	240 L/m² · min
10	0.26	1.45
15	. . . .	3.29
20	1.19	6.33
30	2.78	

4 Plot curves of head loss versus run length for the three flowrates. The required curves, which are plotted using the data given in step 3, are shown in Fig. 6-38.
5 Plot the curve of run length to reach a head loss of 3 m versus filtration rate. The required curve is shown in Fig. 6-39. The data needed to plot this curve are obtained from Fig. 6-38 by finding the time required to reach a head loss of 3 m for each filtration rate.

Comment   The use of unstratified filter beds for the filtration of treated effluents has been studied by Dahab and Young [6]. They found that unstratified filter beds having the same effective size as that used in the top layer of a dual-medium filter were essentially equivalent to dual-medium filters in terms of effluent quality and length of run. The use of unstratified filter beds is considered further in Chap. 8.

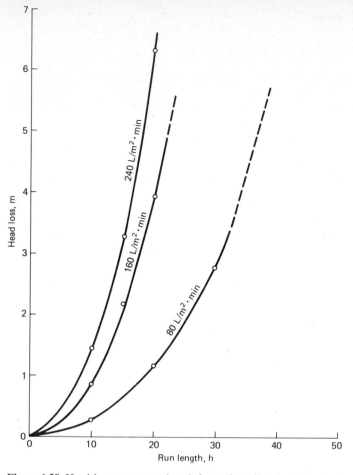

Figure 6-38 Head loss versus run length for various filtration rates for Example 6-9.

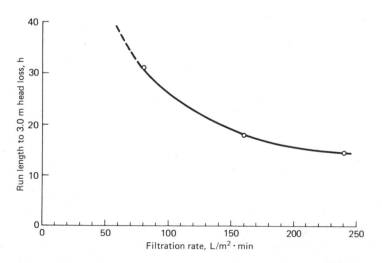

Figure 6-39 Run length versus filtration rate for a head loss of 3.0 m for Example 6-9.

DISCUSSION TOPICS AND PROBLEMS

6-1 A bar rack is inclined at a 50° angle with the horizontal. The circular bars have a diameter of 20 mm and a clear spacing of 25 mm. Determine the head loss when the bars are clean and the velocity approaching the rack is 1.0 m/s. Is this a very realistic computation in terms of what actually happens at a treatment plant?

6-2 Using the data given in Table 6-4 determine (a) the off-line storage volume needed to equalize the flowrate and (b) the effect of flow equalization on the BOD_5 mass loading rate. How does the BOD_5 mass loading-rate curve determined in this problem compare with the curve shown in Fig. 6-7? In your estimation, does the difference in the mass loading rate justify the cost of the larger basin required for in-line storage?

6-3 Using the data given in Table 6-4, determine the in-line volume required to reduce the variation in the BOD_5 mass loading rate between the maximum and minimum from the existing ratio of 25.8 : 1 (439 : 17) to a peak value of 5 : 1.

6-4 The contents of a tank are to be mixed with a turbine impeller that has six flat blades. The diameter of the impeller is 2.0 m, and the impeller is installed 1.2 m above the bottom of the 6.0 m tank. If the temperature is 30°C and the impeller is rotated at 30 r/min, what will be the power consumption? Find the Reynolds number using Eq. 6.5.

6-5 It is desired to flash-mix some chemicals with incoming wastewater that is to be treated. Mixing is to be accomplished using a flat paddle mixer 0.5 m in diameter having six blades. If the temperature of the incoming water is 10°C and the mixing chamber power number is 1.70, determine
(a) The speed of rotation when the Reynolds number is approximately 100,000
(b) Why it is desirable to have a high Reynolds number in most mixing operations
(c) The required mixer motor size, assuming an efficiency factor of 20%
(d) The Froude number and its significance $(F = n^2 D/g)$.

6-6 Assuming that a given flocculation process can be defined by a first order reaction $(r_N = -kN)$, complete the following table assuming the process is occurring in a plug-flow reactor with a detention time of 10 min. What would the value be after 10 min if a batch reactor were used instead, assuming the rate constant is the same?

Time, t	0	5	10
Particles, no/unit volume	10	(?)	3

6-7 If the steady-state effluent from a continuous-flow stirred-tank reactor used as a flocculator contained 3 particles/unit volume, determine the concentration of particles in the effluent 5 min after the process started before steady-state conditions are reached. Assume that the influent contains 10 particles/unit volume, the detention time in the CFSTR is equal to 10 min, and that first order kinetics apply $(r_N = -kN)$.

6-8 An air flocculation system is to be designed. Derive an expression that can be used to estimate the power imparted to the liquid as a function of the air flowrate. If a G value of 60 s^{-1} is to be used, estimate the air flowrate that will be necessary for a 175 m^3 flocculation chamber. Assume the depth of the flocculation basin is to be 3.0 m.

6.-9 Derive Stoke's law by equating Eq. 6-15 to the effective particle mass.

6-10 Determine the settling velocity in meters per second of a sand particle with a specific gravity of 2.6 and a diameter of 1 mm. Assume that the Reynolds number is 175.

6-11 Determine the removal efficiency for a sedimentation basin with a critical velocity V_0 of 2.0 m/h in treating a wastewater containing particles whose settling velocities are distributed as given in the table below. Plot the particle histogram for the influent and effluent wastewater.

Number of particles	2	4	8	12	10	7	4	2	1
Velocity m/h	0.0–0.4	0.4–0.8	0.8–1.2	1.2–1.6	1.6–2.0	2.0–2.4	2.4–2.8	2.8–3.2	3.2–3.6

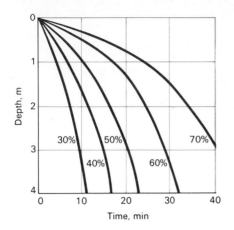

Time, min

Figure 6-40 Settling-tank curves for Prob. 6-13.
Note: m × 3.2808 = ft.

6-12 The rate of flow through an ideal clarifier is 8000 m^3/d, the detention time is 1 h, the depth is 3 m, and the length-to-width ratio for the basin is 3. If a full-length movable horizontal tray is set 1 m below the surface of the water, determine the percent removal of particles having a settling velocity of 1.0 m/h. Could the removal efficiency of the clarifier be improved by moving the tray? If so, where should the tray be located and what would be the maximum removal efficiency? What effect would moving the tray have if the particle settling velocity were equal to 0.3 m/h?

6-13 Using the settling test curves shown in Fig. 6-40, determine the efficiency of a settling tank in removing flocculant particles if the depth is 2 m and the detention time is 20 min.

6-14 For a flocculant suspension, determine the removal efficiency for a basin 3 m deep with an overflow rate V_0 equal to 3.0 m/h using the laboratory settling data presented in the following table.

Time, min	Percent suspended solids removed at indicated depth				
	0.5	1.0	1.5	2.0	2.5
20	61				
30	71	63	55		
40	81	72	63	61	57
50	90	81	73	67	63
60	...	90	80	74	68
70	...	...	86	80	75
80	...	...	...	86	81

6-15 The curve shown in Fig. 6-41 was obtained from a settling test in a 2 m cylinder. The initial solids concentration was 3600 mg/L. Determine the thickener area required for a concentration C_u of 22,500 mg/L with a sludge flow of 1500 m^3/d.

6-16 Given the settling data in the following table from an activated-sludge pilot plant (see Fig. 6-24), determine the percent recycle rate if the sedimentation tank application rate is 20 $m^3/m^2 \cdot d$ and the concentration of the recycled solids is 12,500 mg/L. What will the recycle rate be if the concentration of the recycled solids is 20,000 mg/L?

6-17 Using the equations (developed by Rose), determine the head loss through a 75-cm sand bed. Assume that the sand bed is composed of spherical unsized sand with a diameter of 0.6 mm, the kinematic viscosity is equal to 1.306×10^{-6} m^2/s, the porosity is 0.40, and the filtration velocity is 240 $L/m^2 \cdot min$.

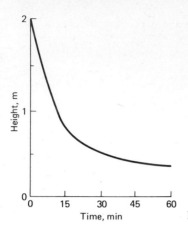

Figure 6-41 Interface settling curve for Prob. 6-15.

6-18 If a 0.3-m layer of anthracite is placed on top of the sand bed in Prob. 6-17, determine the ratio of the head loss through the anthracite to that of the sand. Assume that the grain-size diameter is 2.0 mm and porosity of the anthracite is 0.50.

6-19 For a given filtration operation it has been found that straining is the operative particulate-matter-removal mechanism and that the change in concentration with distance can be approximated with a first-order equation $(dC/dx = -rC)$. If the initial concentration of particulate matter is 10 mg/L, the removal-rate constant is equal to 2.0 cm^{-1}, and the filtration velocity is equal to 100 L/m$^2 \cdot$ min. determine the amount of material arrested within the filter in the layer between 3 and 6 cm over a 1-hr period. Express your answer in mg/cm $\cdot$ m^2. Estimate the head loss in the layer at the end of 1 h.

Height of sludge interface, m for Prob. 6-16

Time, min	Sludge concentration, mg/L					
	1000	2000	3000	5000	10000	15000
0	0	0	0	0	0	0
10	1.17	0.85	0.41	0.17	0.05	0.03
20	1.89	1.67	0.84	0.34	0.10	0.06
30	1.92	1.83	1.25	0.51	0.15	0.09
40	1.93	1.88	1.56	0.68	0.20	0.12
50	1.93	1.89	1.66	0.85	0.26	0.14
60	1.94	1.90	1.72	1.02	0.31	0.17
80	1.95	1.91	1.80	1.27	0.41	0.23
100	1.96	1.92	1.84	1.38	0.50	0.29
120	1.97	1.94	1.88	1.47	0.58	0.34

6-20 The data in the table were obtained from a pilot-plant study on the filtration of settled secondary effluent from an activated-sludge treatment plant. Using these data, estimate the length of run that is possible with and without the addition of polymer if the maximum allowable head loss is 3 m, the filtration rate is 150 L/m$^2 \cdot$ min, and the influent suspended-solids concentration is 15 mg/L. Uniform sand with a diameter of 0.55 mm and a depth of 60 cm was used in the pilot filters.

	Concentration ratio C/C_0	
Depth, cm	With polymer addition	Without polymer addition
0	1.00	1.00
4	0.65	0.70
8	0.26	0.52
12	0.20	0.46
16	0.16	0.42
20	0.15	0.40
24	0.13	0.39
28	0.12	0.38
32	0.12	0.38
36	0.12	0.38
40	0.12	0.38
44	0.12	0.38
48	0.12	0.38
52	0.12	0.38
56	0.12	0.38
60	0.12	0.38

REFERENCES

1. Baumann, E. R., and J. Y. C. Huang: Granular Filters for Tertiary Wastewater Treatment, *J. Water Pollut. Control Fed.* no. 8, 1974.
2. Bishop, S. L., and B. W. Behrman: *Filtration of Wastewater Using Granular Media*, paper presented at the 1976 Thomas R. Camp Lecture Series on Wastewater Treatment and Disposal, Boston Society of Civil Engineers, Boston, 1976.
3. Biskner, C. D., and J. C. Young: Two Stage Filtration of Secondary Effluent, *J. Water Pollut. Control Fed.*, vol. 49, no. 2, 1977.
4. Camp, T. R., and P. C. Stein: Velocity Gradients and Internal Work in Fluid Motion, *J. Boston Soc. Civ. Eng.*, vol. 30, p. 219, 1943.
5. Coe, H. S., and G. H. Clevenger: Determining Thickener Unit Areas, *Trans. AIME*, vol. 55, no. 3, 1916.
6. Dahab, M. F., and J. C. Young: Unstratified-Bed Filtration of Wastewater, *J. Environ. Eng. Div.*, *ASCE*, vol. 103, EE 12714, February 1977.
7. Dick, R. I.: Folklore in the Design of Final Settling Tanks, *J. Water Pollut. Control Fed.*, vol. 48, no. 4, 1976.
8. Dick, R. I., and B. B. Ewing: Evaluation of Activated Sludge Thickening Theories, *J. Sanit. Eng. Div.*, *ASCE*, vol. 93, no. SA-4, 1967.
9. Dick, R. I., and K. W. Young: *Analysis of Thickening Performance of Final Settling Tanks*, presented at the Purdue Industrial Waste Conference, 27th Annual Meeting, Lafayette, Ind., May 24, 1972.
10. Eckenfelder, W. W., Jr: *Industrial Water Pollution Control*, McGraw-Hill, New York, 1966.
11. Fair, G. M., J. C. Geyer, and D. A. Okun: *Water and Wastewater Engineering*, vol. 2, Wiley, New York, 1966.
12. Harris, H. S., W. S. Kaufman, and R. B. Krone: Othokinetic Flocculation in Water Purification, *J. Sanit. Eng. Div.*, *ASCE*, vol. 92, no. SA6, proc. paper 5027, pp. 95–111, December 1966.
13. Jaeger, C.: *Engineering Fluid Mechanics*, Blackie, London, 1956.

14. McCabe, W. L., and J. C. Smith: *Unit Operations of Chemical Engineering*, 3d ed., McGraw-Hill, New York, 1976.
15. O'Melia, C. R., and W. Stumm: Theory of Water Filtration, *J. Am. Water Works Assoc.*, vol. 59, no. 11, November 1967.
16. Perry, J. H.: *Chemical Engineers' Handbook*, 5th ed., McGraw-Hill, New York, 1973.
17. *Process Design Manual for Upgrading Existing Wastewater Treatment Plants*, U.S. Environmental Protection Agency, Technology Transfer, October 1974.
18. Rushton, J. H.: Mixing of Liquids in Chemical Processing, *Ind. Eng. Chem.*, vol. 44, no. 12, 1952.
19. Schroeder, E. D.: *Water and Wastewater Treatment*, McGraw-Hill, New York, 1977.
20. Speece, R. E., and J. F. Malina (eds.): *Applications of Commercial Oxygen to Water and Wastewater Systems*, Center for Research in Water Resources, University of Texas at Austin, Texas, 1973.
21. Talmadge, W. P., and E. B. Fitch: Determining Thickener Unit Areas, *Ind. Eng. Chem.*, vol. 47, no. 1, 1955.
22. Tchobanoglous, G.: *A Study of the Filtration of Treated Sewage Effluent*, Ph.D. thesis, Stanford University, Stanford, Calif., 1968.
23. Tchobanoglous, G.: Filtration Techniques in Tertiary Treatment, *J. Water Pollut. Control Fed.*, vol. 42, no. 4, 1970.
24. Tchobanoglous, G., and R. Eliassen: Filtration of Treated Sewage Effluent, *J. Sanit. Eng. Div.*, *ASCE*, vol. 96, no. SA2, 1970.
25. *Wastewater Filtration—Design Considerations*, U.S. Environmental Protection Agency, Technology Transfer Seminar Publication, 1974.
26. *Wastewater Filtration—Design Considerations*, U.S. Environmental Protection Agency, Technology Transfer Report, 1977.
27. *Water Quality and Treatment: A Handbook of Public Water Supplies*, prepared by the American Water Works Association, Inc., 3d ed., McGraw-Hill, New York, 1971.
28. Yoshioka, N., et al.: Continuous Thickening of Homogeneous Flocculated Slurries, *Kagaku Kogaku*, vol. 26, 1957 (also in *Chem. Eng.*, vol. 21, Tokyo, 1957).

CHAPTER SEVEN

CHEMICAL UNIT PROCESSES

Those processes used for the treatment of wastewater in which change is brought about by means of or through chemical reaction are known as chemical unit processes. In the field of wastewater treatment, chemical unit processes usually are used in conjunction with the physical unit operations discussed in Chap. 6 and the biological unit processes to be discussed in Chap. 9 to meet treatment objectives.

The chemical processes considered in this chapter and their principal applications are reported in Table 7-1. The use of various chemicals to improve the results of other operations and processes is also noted briefly. Here, as in Chap. 6, each unit process will be described and the fundamentals involved in the engineering analysis of each unit process will be discussed. In these discussions, knowledge of the fundamentals of chemistry is assumed. The practical application of these processes, including such matters as facility design and dosage requirements, is considered in Chap. 8. Some of the unit processes, such as precipitation for phosphorus removal, activated-carbon adsorption for the removal of organic compounds, and breakpoint chlorination for nitrogen removal, are also considered in Chap. 12, which deals with advanced wastewater treatment.

In considering the application of the chemical unit processes to be discussed in this chapter, it is important to remember that one of the inherent disadvantages associated with the use of chemical unit processes, as compared with the physical unit operations, is that they are additive processes. In most cases, something is added to the wastewater to achieve the removal of something else. As a result, there is usually a net increase in the dissolved constituents in the wastewater. For example, where chemicals are added to enhance the removal

Table 7-1 Applications of chemical unit processes in wastewater treatment

Process	Application	See section
Chemical precipitation	Removal of phosphorus and enhancement of suspended-solids removal in primary sedimentation facilities used for physical-chemical treatment	7-1
Gas transfer	Addition and removal of gases	7-2
Adsorption	Removal of organics not removed by conventional chemical and biological treatment methods. Also used for dechlorination of wastewater before final discharge of treated effluent	7-3
Disinfection	Selective destruction of disease-causing organisms (can be accomplished in various ways)	7-4
Disinfection with chlorine	Selective destruction of disease-causing organisms. Chlorine is the most commonly used chemical	7-5
Dechlorination	Removal of total combined chlorine residual that exists after chlorination (can be accomplished in various ways)	7-6
Disinfection with ozone	Selective destruction of disease-causing organisms	7-7
Others	Various other chemicals can be used to achieve specific objectives in wastewater treatment	7-8

efficiency of plain sedimentation, the total dissolved-solids concentration of the wastewater is always increased. If the treated wastewater is to be reused, this can be a significant factor. This additive aspect is in contrast to the physical unit operations (Chap. 6) and the biological unit processes (Chap. 9), which may be described as being subtractive, in that material is removed from the wastewater. Another disadvantage of chemical unit processes is that they are all intensive in local operating costs. The costs of some of these chemicals are tied to the costs of energy and can be expected to increase similarly.

7-1 CHEMICAL PRECIPITATION

Chemical precipitation in wastewater treatment involves the addition of chemicals to alter the physical state of dissolved and suspended solids and facilitate their removal by sedimentation. In some cases the alteration is slight, and removal is effected by entrapment within a voluminous precipitate consisting primarily of the coagulant itself. Another result of chemical addition is a net increase in the dissolved constituents in the wastewater.

In the past, chemical precipitation was used to enhance the degree of suspended solids and BOD removal (1) where there were seasonal variations in the concentration of the wastewater (such as in cannery wastewater), (2) where an intermediate degree of treatment was required, and (3) as an aid to the sedimentation process. Additional historical details on the use of chemicals for wastewater treatment may be found in Refs. 6, 20, and 21.

Since about 1970, the need to provide more complete removal of the organic compounds and nutrients (nitrogen and phosphorus) contained in wastewater has brought about renewed interest in chemical precipitation. Chemical processes, in conjunction with various physical operations, have been developed for the complete secondary treatment of untreated wastewater, including the removal of either nitrogen or phosphorus, or both (see Ref. 25). Other chemical processes have also been developed to remove phosphorus by chemical precipitation and are designed to be used in conjunction with biological treatment.

The purpose in this section is to identify and discuss (1) the precipitation reactions that occur when various chemicals are added to improve the performance of wastewater-treatment facilities, (2) the chemical reactions involved in the precipitation of phosphorus from wastewater, and (3) some of the more important theoretical aspects of chemical precipitation. The computations used to determine the quantities of sludge produced as a result of the addition of various chemicals are illustrated in Chap. 8. The removal of phosphorus is considered further in Chap. 12, which deals with advanced wastewater treatment.

Description of Chemical Precipitation for Improving Plant Performance

Over the years a number of different substances have been used as precipitants. The most common ones are listed in Table 7-2. The degree of clarification obtained depends on the quantity of chemicals used and the care with which the process is controlled. It is possible by chemical precipitation to obtain a clear effluent, substantially free from matter in suspension or in the colloidal state. From 80 to 90 percent of the total suspended matter, 40 to 70 percent of the BOD_5, 30 to 60 percent of the COD, and 80 to 90 percent of the bacteria can be removed by chemical precipitation. In comparison, when plain sedimentation is used, only 50 to 70 percent of the total suspended matter and 30 to 40 percent of the organic matter settles out.

The chemicals added to wastewater interact with substances that are either normally present in the wastewater or added for this purpose. The reactions

Table 7-2 Chemicals used in wastewater treatment

Chemical	Formula	Molecular weight
Alum	$Al_2(SO_4)_3 \cdot 18H_2O^a$	666.7
Ferrous sulfate (copperas)	$FeSO_4 \cdot 7H_2O$	278.0
Lime	$Ca(OH)_2$	56 as CaO
Ferric chloride	$FeCl_3$	162.1
Ferric sulfate	$Fe_2(SO_4)_3$	400

^{a} Number of bound water molecules will vary from 13 to 18.

involved with (1) alum, (2) lime, (3) ferrous sulfate (copperas) and lime, (4) ferric chloride, (5) ferric chloride and lime, and (6) ferric sulfate and lime are considered in the following discussion.

Alum When alum is added to wastewater containing calcium and magnesium bicarbonate alkalinity, the reaction that occurs may be illustrated as follows:

$$\overset{666.7}{Al_2(SO_4)_3 \cdot 18H_2O} + \overset{3 \times 100CaCO_3}{3Ca(HCO_3)_2} \rightleftharpoons$$

Aluminum sulfate · · · · · Calcium bicarbonate

$$\overset{3 \times 136}{3CaSO_4} + \overset{2 \times 78}{2Al(OH)_3} + \overset{6 \times 44}{6CO_2} + \overset{18 \times 18}{18H_2O} \quad (7\text{-}1)$$

Calcium sulfate · · · Aluminum hydroxide · · · Carbon dioxide

The numbers above the chemical formulas are the combining molecular weights of the different substances and therefore denote the quantity of each one involved. The insoluble aluminum hydroxide is a gelatinous floc that settles slowly through the wastewater, sweeping out suspended material and producing other changes. The reaction is exactly analogous when magnesium bicarbonate is substituted for the calcium salt.

Because alkalinity in Eq. 7-1 is reported in terms of calcium carbonate ($CaCO_3$), the molecular weight of which is 100, the quantity of alkalinity required to react with 10 mg/L of alum is

$$10.0 \text{ mg/L} \times \frac{3 \times 100 \text{ g/mol}}{666.7 \text{ g/mol}} = 4.5 \text{ mg/L}$$

If less than this amount of alkalinity is available, it must be added. Lime is commonly used for this purpose when necessary, but it is seldom required in the treatment of wastewater.

Lime When lime alone is added as a precipitant, the principles of clarification are explained by the following reactions:

$$\overset{56CaO}{Ca(OH)_2} + \overset{44CO_2}{H_2CO_3} \rightleftharpoons \overset{100}{CaCO_3} + \overset{2 \times 18}{2H_2O} \quad (7\text{-}2)$$

Calcium hydroxide · · · Carbonic acid · · · Calcium carbonate

$$\overset{56CaO}{Ca(OH)_2} + \overset{100CaCO_3}{Ca(HCO_3)_2} \rightleftharpoons \overset{2 \times 100}{2CaCO_3} + \overset{2 \times 18}{2H_2O} \quad (7\text{-}3)$$

Calcium hydroxide · · · Calcium bicarbonate · · · Calcium carbonate

A sufficient quantity of lime must therefore be added to combine with all the free carbonic acid and with the carbonic acid of the bicarbonates (half-bound carbonic acid) to produce calcium carbonate, which acts as the coagulant. Much more lime is generally required when it is used alone than when sulfate of iron is also used (see the following discussion). Where industrial wastes introduce mineral acids or acid salts into the wastewater, these must be neutralized before precipitation can take place.

Ferrous sulfate and lime In most cases, ferrous sulfate cannot be used alone as a precipitant because lime must be added at the same time to form a precipitate. The reaction with ferrous sulfate alone is illustrated in Eq. 7-4.

$$
\underset{\substack{\text{Ferrous}\\\text{sulfate}}}{\overset{278}{FeSO_4 \cdot 7H_2O}} + \underset{\substack{\text{Calcium}\\\text{bicarbonate}}}{\overset{100CaCO_3}{Ca(HCO_3)_2}} \rightleftharpoons \underset{\substack{\text{Ferrous}\\\text{bicarbonate}}}{\overset{178}{Fe(HCO_3)_2}} + \underset{\substack{\text{Calcium}\\\text{sulfate}}}{\overset{136}{CaSO_4}} + \overset{7 \times 18}{7H_2O} \quad (7\text{-}4)
$$

If lime in the form $Ca(OH)_2$ is now added, the reaction that takes place is

$$
\underset{\substack{\text{Ferrous}\\\text{bicarbonate}}}{\overset{178}{Fe(HCO_3)_2}} + \underset{\substack{\text{Calcium}\\\text{hydroxide}}}{\overset{2 \times 56CaO}{2Ca(OH)_2}} \rightleftharpoons \underset{\substack{\text{Ferrous}\\\text{hydroxide}}}{\overset{89.9}{Fe(OH)_2}} + \underset{\substack{\text{Calcium}\\\text{carbonate}}}{\overset{2 \times 100}{2CaCO_3}} + \overset{2 \times 18}{2H_2O} \quad (7\text{-}5)
$$

The ferrous hydroxide is next oxidized to ferric hydroxide, the final form desired, by the oxygen dissolved in the wastewater:

$$
\underset{\substack{\text{Ferrous}\\\text{hydroxide}}}{\overset{4 \times 89.9}{4Fe(OH)_2}} + \underset{\substack{\text{Oxygen}}}{\overset{32}{O_2}} + \overset{2 \times 18}{2H_2O} \rightleftharpoons \underset{\substack{\text{Ferric}\\\text{hydroxide}}}{\overset{4 \times 106.9}{4Fe(OH)_3}} \quad (7\text{-}6)
$$

The insoluble ferric hydroxide is formed as a bulky, gelatinous floc similar to the alum floc. The alkalinity required for a 10 mg/L dosage of ferrous sulfate is

$$
10.0 \text{ mg/L} \times \frac{100 \text{ g/mol}}{278 \text{ g/mol}} = 3.6 \text{ mg/L}
$$

The lime required is

$$
10.0 \text{ mg/L} \times \frac{2 \times 56 \text{ g/mol}}{278 \text{ g/mol}} = 4.0 \text{ mg/L}
$$

The oxygen required is

$$
10.0 \text{ mg/L} \times \frac{32 \text{ g/mol}}{4 \times 278 \text{ g/mol}} = 0.29 \text{ mg/L}
$$

Because the formation of ferric hydroxide is dependent on the presence of dissolved oxygen, the reaction given in Eq. 7-6 cannot be completed with septic wastewater or industrial wastes devoid of oxygen. Ferric sulfate may take the place of ferrous sulfate, and its use often avoids the addition of lime and the requirement of dissolved oxygen.

Ferric chloride The reactions for ferric chloride are

$$\underset{\substack{\text{Ferric}\\\text{chloride}}}{\overset{162.1}{\text{FeCl}_3}} + \underset{\substack{\text{Water}}}{\overset{3\times18}{3\text{H}_2\text{O}}} \rightleftharpoons \underset{\substack{\text{Ferric}\\\text{hydroxide}}}{\overset{106.9}{\text{Fe(OH)}_3}} + 3\text{H}^+ + 3\text{Cl}^- \qquad (7\text{-}7)$$

$$3\text{H}^+ + \underset{\text{Bicarbonate}}{3\text{HCO}_3^-} \rightleftharpoons \underset{\substack{\text{Carbonic}\\\text{acid}}}{3\text{H}_2\text{CO}_3} \qquad (7\text{-}8)$$

Ferric chloride and lime The reactions for ferric chloride and lime are

$$\underset{\substack{\text{Ferric}\\\text{chloride}}}{\overset{2\times162}{2\text{FeCl}_3}} + \underset{\substack{\text{Calcium}\\\text{hydroxide}}}{\overset{3\times56\text{CaO}}{3\text{Ca(OH)}_2}} \rightleftharpoons \underset{\substack{\text{Calcium}\\\text{chloride}}}{\overset{3\times111}{3\text{CaCl}_2}} + \underset{\substack{\text{Ferric}\\\text{hydroxide}}}{\overset{2\times106.9}{2\text{Fe(OH)}_3}} \qquad (7\text{-}9)$$

Ferric sulfate and lime The reactions for ferric sulfate and lime are

$$\underset{\substack{\text{Ferric}\\\text{sulfate}}}{\overset{400}{\text{Fe}_2(\text{SO}_4)_3}} + \underset{\substack{\text{Calcium}\\\text{hydroxide}}}{\overset{3\times56\text{CaO}}{3\text{Ca(OH)}_2}} \rightleftharpoons \underset{\substack{\text{Calcium}\\\text{sulfate}}}{\overset{408}{3\text{CaSO}_4}} + \underset{\substack{\text{Ferric}\\\text{hydroxide}}}{\overset{2\times106.9}{2\text{Fe(OH)}_3}} \qquad (7\text{-}10)$$

Description of Chemical Precipitation for Phosphate Removal and for Physical-Chemical Treatment

Phosphate can be removed by chemical precipitation with various multivalent metal ions. Reactions with calcium [Ca(II)], aluminum [Al(III)], and iron [Fe(III)] ions are presented in Eqs. 7-11, 7-12, and 7-13. (The Roman numerals following the metal refer to the oxidation state.)

Calcium:

$$10\text{Ca}^{++} + 6\text{PO}_4^{3-} + 2\text{OH}^- \rightleftharpoons \underset{\text{Hydroxylapatite}}{\text{Ca}_{10}(\text{PO}_4)_6(\text{OH})_2} \qquad (7\text{-}11)$$

Aluminum:

$$\text{Al}^{3+} + \text{H}_n\text{PO}_4^{3-n} \rightleftharpoons \text{AlPO}_4 + n\text{H}^+ \qquad (7\text{-}12)$$

Iron:

$$\text{Fe}^{3+} + \text{H}_n\text{PO}_4^{3-n} \rightleftharpoons \text{FePO}_4 + n\text{H}^+ \qquad (7\text{-}13)$$

The chemistry of the removal of phosphate with lime is quite different from that of alum or iron. From the equations presented previously, it will be noted that when lime is added to water it reacts with the natural bicarbonate alkalinity to precipitate CaCO_3. Excess calcium ions will then react with the phosphate, as shown in Eq. 7-11, to precipitate hydroxylapatite $\text{Ca}_{10}(\text{PO}_4)_6(\text{OH})_2$. Therefore, the quantity of lime required will, in general, be independent of the amount of phosphate present and will depend primarily on the alkalinity of the wastewater.

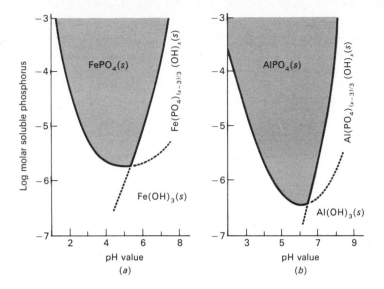

Figure 7-1 Concentration of ferric and aluminum phosphate in equilibrium with soluble phosphorus. (a) Fe(III)-phosphate; (b) Al(III)-phosphate [12].

In the case of alum and iron, 1 mole will precipitate 1 mole of phosphate; however, these reactions are deceptively simple and must be considered in light of the many competing reactions and their associated equilibrium constants, and the effects of alkalinity, pH, trace elements, and ligands found in wastewater. Because of this, Eqs. 7-12 and 7-13 cannot be used to estimate the required chemical dosages directly. Therefore, dosages are generally established on the basis of bench-scale tests and occasionally by full-scale tests, especially if polymers are used. For example, for equimolar initial concentrations of Al(III), Fe(III), and phosphate, the total concentration of soluble phosphate in equilibrium with both insoluble $FePO_4$ and $AlPO_4$ is shown in Fig. 7-1. The solid lines trace the concentration of residual soluble phosphate after precipitation. Pure metal phosphates are precipitated within the shaded area, and mixed complex precipitates are formed outside toward the higher pH values [12].

Because a discussion of the various theoretical details of the precipitation of phosphate with metallic ions is beyond the scope of this text, the reader is advised to consult Stumm and Morgan [34]. Practical details on the removal of phosphate, including process flowsheets and required chemical dosages, are presented in Chap. 12.

Theoretical Aspects of Chemical Precipitation

The theory of chemical precipitation reactions is very complex. The reactions that have been presented explain it only in part, and even they do not necessarily proceed as indicated. They are often incomplete, and numerous side reactions with other substances in wastewater may take place. Therefore, the following discussion is of necessity incomplete, but will serve as an introduction to the nature of the phenomena involved.

Nature of particles in wastewater There are two general types of colloidal solid particle dispersions in liquids. When water is the solvent, these are called the hydrophobic or "water-hating" and the hydrophilic or "water-loving" colloids. These two types are based on the attraction of the particle surface for water. Hydrophobic particles have relatively little attraction for water; hydrophilic particles have a great attraction for water. It should be noted, however, that water can interact to some extent with hydrophobic particles. Some water molecules will generally adsorb on the typical hydrophobic surface, but the reaction between water and hydrophilic colloids occurs to a much greater extent.

Surface charge An important factor in the stability of colloids is the presence of surface charge. It develops in a number of different ways, depending on the chemical composition of the medium (wastewater in this case) and the colloid. Regardless of how it is developed, this stability must be overcome if these particles are to be aggregated (flocculated) into larger particles with enough mass to settle easily.

Surface charge develops most commonly through preferential adsorption, ionization, and isomorphous replacement. For example, oil droplets, gas bubbles, or other chemically inert substances dispersed in water will acquire a negative charge through the preferential adsorption of anions (particularly hydroxyl ions). In the case of substances such as proteins or microorganisms, surface charge is acquired through the ionization of carboxyl and amino groups [32]. This can be represented as $R_{NH_2}^{COO^-}$ at high pH, $R_{NH_3^+}^{COOH}$ at low pH, and $R_{NH_2^+}^{COO^-}$ at the isoelectric point where R represents the bulk of the solid [12]. Charge development through isomorphous replacement occurs in clay and other soil particles, in which ions in the lattice structure are replaced with ions from solution (e.g., the replacement of Si with Al).

When the colloid or particle surface becomes charged, some ions of the opposite charge (known as counter ions) become attached to the surface. They are held there through electrostatic and van der Waals forces strongly enough to overcome thermal agitation. Surrounding this fixed layer of ions is a diffuse layer of ions, which is prevented from forming a compact double layer by thermal agitation. This is illustrated schematically in Fig. 7-2. As shown, the double layer consists of a compact layer (Stern) in which the potential drops from ψ_0 to ψ_s and a diffuse layer in which the potential drops from ψ_s to 0 in the bulk solution.

If a particle such as shown in Fig. 7-2 is placed in an electrolyte solution, and an electric current is passed through the solution, the particle, depending on its surface charge, will be attracted to one or the other of the electrodes, dragging with it a cloud of ions.

The potential at the surface of the cloud (called the surface of shear) is sometimes measured in wastewater-treatment operations. The measured value is

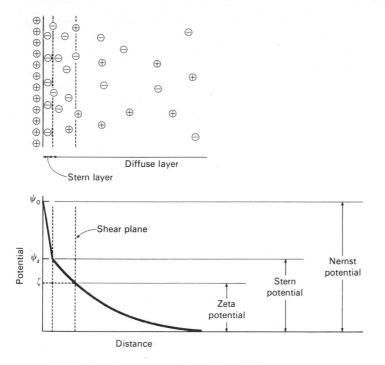

Figure 7-2 Stern model of electrical double layer [32].

often called the zeta potential. Theoretically, however, the zeta potential should correspond to the potential measured at the surface enclosing the fixed layer of ions attached to the particle, as shown in Fig. 7-2. The use of the measured zeta potential value is limited because it will vary with the nature of the solution components, and it therefore is not a repeatable measurement.

Particle aggregation To bring about particle aggregation, steps must be taken to reduce particle charge or to overcome the effect of this charge. This can be accomplished by (1) the addition of potential-determining ions, which will be taken up by or will react with the colloid surface to lessen the surface charge, or the addition of electrolytes, which have the effect of reducing the thickness of the diffuse electric layer and thereby reduce the zeta potential; (2) the addition of long-chained organic molecules (polymers), whose subunits are ionizable and are therefore called polyelectrolytes, that bring about the removal of particles through adsorption and bridging; and (3) the addition of chemicals that form hydrolyzed metal ions.

Addition of potential-determining ions to promote coagulation can be illustrated by the addition of strong acids or bases to reduce the charge of metal oxides or hydroxides to near zero, so that coagulation can occur. Electrolytes can also be added to coagulate colloidal suspensions. Increased concentration of a given electrolyte will cause a decrease in zeta potential and a corresponding decrease in repulsive forces. Similar effects are observed if the electrolyte charge is increased.

Polyelectrolytes may be divided into two categories: natural and synthetic. Important natural polyelectrolytes include polymers of biological origin and those derived from starch products, cellulose derivatives, and alginates. Synthetic polyelectrolytes consist of simple monomers that are polymerized into high-molecular-weight substances. Depending on whether their charge, when placed in water, is negative, positive, or neutral, these polyelectrolytes are classified as anionic, cationic, and nonionic, respectively.

The action of polyelectrolytes may be divided into three general categories. In the first category, polyelectrolytes act as coagulants that lower the charge of the wastewater particles. Since wastewater particles normally are charged negatively, cationic polyelectrolytes are used for this purpose. In this application, the cationic polyelectrolytes are considered to be primary coagulants.

The second mode of action of polyelectrolytes is interparticle bridging. In this case, polymers that are anionic and nonionic (usually anionic to a slight extent when placed in water) become attached at a number of adsorption sites to the surface of the particles found in the settled effluent. A bridge is formed when two or more particles become adsorbed along the length of the polymer. Bridged particles become intertwined with other bridged particles during the flocculation process. The size of the resulting three-dimensional particles grows until they can be removed easily by sedimentation.

The third type of polyelectrolyte action may be classified as a coagulation-bridging phenomenon, which results from using cationic polyelectrolytes of extremely high molecular weight. Besides lowering the charge, these poly-electrolytes also form particle bridges.

Metal salt polymer formation In contrast with the aggregation brought about by the addition of chemicals that act as electrolytes and polymers, aggregation brought about by the addition of alum or ferric sulfate is a more complex process. In the past, it was thought that free Al^{3+} and Fe^{3+} were responsible for the effects observed during particle aggregation; however, it is now known that their hydrolysis products are responsible [35, 36]. Although the effect of these hydrolysis products is only now appreciated, it is interesting to note that their chemistry was first elucidated in the early 1900s by Pfeiffer (1902–1907), Bjerrum (1907–1920), and Werner (1907) [38]. For example, Pfeiffer proposed that the hydrolysis of trivalent metal salts, such as chromium, aluminum, and iron, could be represented as

$$\begin{bmatrix} H_2O & OH_2 \\ H_2O-Me-OH_2 \\ H_2O & OH_2 \end{bmatrix}^{3+} \rightleftharpoons \begin{bmatrix} H_2O & OH_2 \\ H_2O-Me-OH_2 \\ H_2O & OH_2 \end{bmatrix}^{++} + H^+ \quad (7\text{-}14)$$

with the extent of the dissociation depending on the anion associated with the metal and on the physical and chemical characteristics of the solution. Further, it was proposed that, upon the addition of sufficient base, the dissociation can proceed to produce a negative ion [38], such as

$$
\left[
\begin{array}{c}
H_2O \quad\; OH \\
H_2O{-}Me{-}OH \\
HO \quad\; OH
\end{array}
\right]^{-}
$$

Recently, however, it has been observed that the intermediate hydrolysis reactions of Al(III) are much more complex than would be predicted on the basis of a model in which a base is added to the solution. A hypothetical model proposed by Stumm [12] for Al(III) is shown in Eq. 7-15.

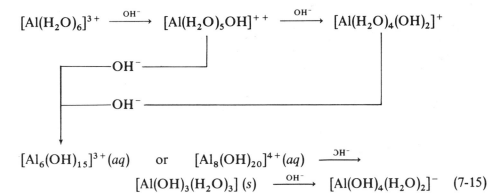

$$[Al(OH)_3(H_2O)_3]\,(s) \xrightarrow{\;OH^-\;} [Al(OH)_4(H_2O)_2]^- \quad (7\text{-}15)$$

Before the reaction proceeds to the point where a negative ion is produced, polymerization as depicted in the following formula will usually take place [38].

$$
2\left[
\begin{array}{c}
H_2O \quad\; OH \\
H_2O{-}Me{-}OH_2 \\
H_2O \quad\; OH_2
\end{array}
\right]^{++}
\rightleftharpoons
\left[
\begin{array}{c}
H \\
O \\
(H_2O)_4\,Me \quad Me(H_2O)_4 \\
O \\
H
\end{array}
\right]^{4+}
+\,2H_2O \quad (7\text{-}16)
$$

The possible combinations of the various hydrolysis products is endless, and their enumeration is not the purpose here. What is important, however, is the realization that one or more of the hydrolysis products may be responsible for the observed action of aluminum or iron. Further, because the hydrolysis reactions follow a stepwise process, the effectiveness of aluminum and iron will vary with time. For example, an alum slurry that has been prepared and stored will behave differently from a freshly prepared solution when it is added to a wastewater. For a more detailed review of the chemistry involved, the excellent articles on this subject by Stumm and Morgan [35] and Stumm and O'Melia [36] are recommended.

7-2 GAS TRANSFER

Gas transfer may be defined as the process by which gas is transferred from one phase to another, usually from the gaseous to the liquid phase. It is a vital part of a number of wastewater-treatment processes. For example, the functioning of aerobic processes, such as activated-sludge biological filtration and aerobic digestion, depends on the availability of sufficient quantities of oxygen. Chlorine, when used as a gas, must be transferred to solution in the water for disinfection purposes. Oxygen is often added to treated effluent after chlorination (postaeration). One process for removing nitrogen compounds consists of converting the nitrogen to ammonia and transferring the ammonia gas from the water to air.

Description

The most common application of gas transfer in the field of wastewater treatment is in the transfer of oxygen in the biological treatment of wastewater. Because of the low solubility of oxygen and the consequent low rate of oxygen transfer, sufficient oxygen to meet the requirements of aerobic waste treatment does not enter water through normal surface air-water interfaces. To transfer the large quantities of oxygen that are needed, additional interfaces must be formed. Either air or oxygen can be introduced into the liquid, or the liquid in the form of droplets can be exposed to the atmosphere. The most commonly used aeration devices are described in Table 7-3 and illustrated in Fig. 7-3. The design and application of many of these devices are considered in Chap. 10 in connection with the design of biological treatment processes.

Oxygen can be supplied by means of air or pure-oxygen bubbles introduced to the water to create additional gas-water interfaces. In wastewater-treatment plants, submerged-bubble aeration is most frequently accomplished by dispersing air bubbles in the liquid at depths up to 10 m ($\sim$30 ft). Depths up to 30 m (100 ft) have been used in some European designs. As summarized in Table 7-3 and shown in Fig. 7-3, aerating devices include porous plates and tubes, perforated pipes, and various configurations of metal and plastic diffusers. Hydraulic shear devices may also be used to create small bubbles by impinging a flow of liquid at an orifice to break up the air bubbles into smaller sizes. Turbine mixers may be used to disperse air bubbles introduced below the center of the turbine.

In the alternative method of introducing large quantities of oxygen into the liquid, surface aerators are used to expose the liquid to the atmosphere. Mechanical aerators generally consist of either low- or high-speed turbines or high-speed floating units operating at the surface of the liquid, partially submerged. They are designed both to mix the liquid in the basin and to expose it to the atmosphere in the form of small liquid droplets.

Table 7-3 Description of commonly used devices for wastewater aeration[a]

Classification	Description	Use or application
Submerged:		
Diffused air		
Fine-bubble system	Bubbles generated with ceramic, vitreous, or resin-bonded porous plates and tubes	All types of activated-sludge processes
Medium-bubble system	Bubbles generated with plastic-wrapped or cloth-covered tubes	All types of activated-sludge processes
Coarse-bubble system	Bubbles generated with orifices, injectors and nozzles, or shear plates	All types of activated-sludge processes
Sparger turbine	Consists of low-speed turbine and compressed-air injection system	All types of activated-sludge processes
Static-tube mixer	Short tubes with internal baffles designed to retain air injected at bottom of tube in contact with liquid	Aerated lagoons and activated-sludge processes
Jet	Compressed air injected into mixed liquor as it is pumped under pressure through jet device	All types of activated-sludge processes
Surface:		
Low-speed turbine aerator	Large-diameter turbine used to expose liquid droplets to the atmosphere	Conventional activated-sludge processes and aerated lagoons
High-speed floating aerator	Small-diameter propeller used to expose liquid droplets to the atmosphere	Aerated lagoons
Rotor-brush aerator	Blades mounted on central shaft are rotated through liquid. Oxygen is induced into the liquid by the splashing action of blades and by exposure of liquid droplets to the atmosphere	Oxidation ditch, channel aeration, and aerated lagoons

[a] Adapted in part from Ref. 15.

Analysis of Gas Transfer

Over the past 50 years a number of mass-transfer theories have been proposed to explain the mechanism of gas transfer. The simplest and the one most commonly used is the two-film theory proposed by Lewis and Whitman in 1924 [19]. The penetration model proposed by Higbie [14] and the surface-renewal model proposed by Danckwerts [7] are more theoretical and take into account more of the physical phenomena involved. The two-film theory remains popular because, in more than 95 percent of the situations encountered, the results obtained are essentially the same as those obtained with the more complex theories. Even in the 5 percent where there is disagreement between the two-film theory and other theories, it is not clear which approach is more correct. For these reasons the two-film theory will be described in the following discussion. Details on the other gas-transfer theories may be found in Refs. 7, 14, 19, and 29.

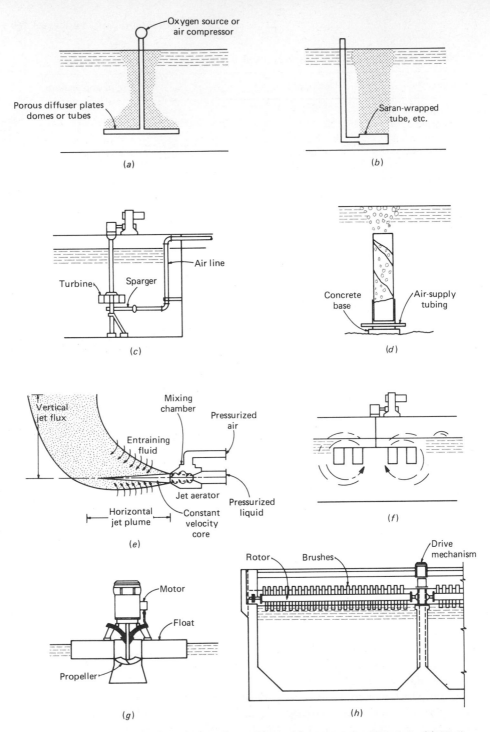

Figure 7-3 Typical devices used for the transfer of oxygen. (*a*) Fine bubble diffused air. (*b*) Medium bubble diffused air. (*c*) Sparger turbine. (*d*) Static tube mixer. (*e*) Jet aerator. (*f*) Low-speed turbine. (*g*) High-speed floating aerator. (*h*) Rotor-brush aerator.

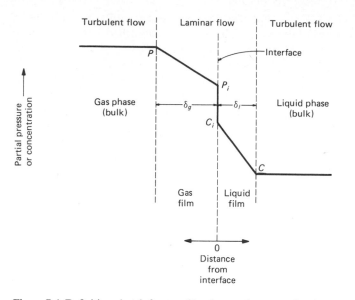

Figure 7-4 Definition sketch for two-film theory of gas transfer. (*Adapted from Ref. 19.*)

The two-film theory The two-film theory is based on a physical model in which two films exist at the gas-liquid interface, as shown in Fig. 7-4. The two films, one liquid and one gas, provide the resistance to the passage of gas molecules between the bulk-liquid and the bulk-gaseous phases. For the transfer of gas molecules from the gas phase to the liquid phase, slightly soluble gases encounter the primary resistance to transfer from the liquid film, and very soluble gases encounter the primary resistance to transfer from the gaseous film. Gases of intermediate solubility encounter significant resistance from both films.

In the systems used in the field of wastewater treatment, the rate of gas transfer generally is proportional to the difference between the existing concentration and the equilibrium concentration of the gas in solution. In equation form, this relationship can be expressed as

$$\frac{dm}{dt} = K_g A (C_s - C) \qquad (7\text{-}17)$$

where dm/dt = rate of mass transfer
$\quad K_g$ = coefficient of diffusion for gas
$\quad A$ = area through which gas is diffusing
$\quad C_s$ = saturation concentration of gas in solution
$\quad C$ = concentration of gas in solution

Noting that $dm/dt = V \, dC/dt$, Eq. 7-17 can be written as

$$\frac{dC}{dt} = K_g \frac{A}{V} (C_s - C) \qquad (7\text{-}18)$$

In practice, the term $K_g(A/V)$ is replaced by a proportionality factor that is related to existing conditions of exposure. This factor is identified in the literature as $K_L a$. If $K_L a$ is used, Eq. 7-18 can be rewritten as

$$\frac{dC}{dt} = K_L a(C_s - C) \tag{7-19}$$

where dC/dt = change in concentration, mg/L · s
$\quad K_L a$ = overall mass-transfer coefficient, s^{-1}
$\quad C_s$ = saturation concentration of gas in solution, mg/L
$\quad C$ = concentration of gas in solution, mg/L

The integrated form of Eq. 7-19 is obtained by integrating between the limits of C_0 and C and 0 and t as follows:

$$\int_{C_0}^{C} \frac{dC}{C_s - C} = K_L a \int_0^t dt \tag{7-20}$$

which, when solved, yields

$$\frac{C_s - C_t}{C_s - C_0} = e^{-K_L at} \tag{7-21}$$

In Eq. 7-21, the terms $(C_s - C_t)$ and $(C_s - C_0)$ represent the final and initial oxygen-saturation deficits. Where a supersaturated solution is to be degassed, the following alternative form of Eq. 7-21 is used:

$$\frac{C_t - C_s}{C_0 - C_s} = e^{-K_L at} \tag{7-22}$$

The derivation of an expression similar to Eq. 7-21 for estimating the amount of oxygen required for the postaeration of treated wastewater is illustrated in Example 7-1.

Example 7-1: Derivation of equation for estimating diffused-air requirements Develop an expression that can be used to estimate the diffused-air requirement for the postaeration of effluent following chlorination. Assume that aeration will be accomplished in a plug-flow reactor. (Note: The following development is based on a derivation given by Graber [13].)

SOLUTION
1 Write an equation for the oxygen-solution rate. The appropriate expression is

$$\frac{dm}{dt} = K_T'(C_s - C)$$

where K_T' = overall mass-transfer coefficient for the given conditions

$$K_T' = K_{20}' \times (1.024)^{T-20}$$

2 Write an expression for the oxygen-transfer efficiency. The efficiency may be defined as

$$E = \frac{(dm/dt)_{20°C, C=0}}{M}$$

where $\qquad\qquad E$ = oxygen transfer efficiency

$(dm/dt)_{20°C, C=0}$ = oxygen-solution rate at 20°C and zero dissolved oxygen

M = mass rate at which oxygen is introduced

3 Develop a differential expression for the mass rate at which oxygen is introduced. The mass rate at which oxygen is introduced is given by

$$M = \frac{1}{E} \left(\frac{dm}{dt}\right)_{20°C, C=0}$$

$$= \frac{1}{E} \left(\frac{dm}{dt}\right)_T \frac{(dm/dt)_{20°C, C=0}}{(dm/dt)_T}$$

Substituting for $(dm/dt)_{20, C=0}$ and $(dm/dt)_T$ yields

$$M = \frac{1}{E} \left(\frac{dm}{dt}\right)_T \frac{(C_s)_{20°C}}{(C_s - C)_T (1.024)^{T-20}}$$

If the above expression is applied to an infinitesimal transverse segment of the tank and $Q\,dC$ is substituted for dm/dt [note that $V(dC/dt) = dm/dt$ and $Q = V/dt$], then the differential form of the above expression can be rewritten as

$$dM = \frac{Q}{E} \frac{(C_s)_{20°C}}{(1.024)^{T-2}} \left(\frac{dC}{C_s - C}\right)_T$$

4 Derive the integrated form of the differential expression that was derived in step 3. The integrated form of the equation can be obtained by integrating the expression from the inlet of the tank where $C = C_i$ to the outlet of the tank where $C = C_o$:

$$\int_0^M dM = \frac{Q}{E} \frac{(C_s)_{20°C}}{(1.024)^{T-20}} \int_{C_i}^{C_o} \frac{dc}{C_s - C}$$

$$M = \frac{Q}{E} \frac{(C_s)_{20°C}}{(1.024)^{T-20}} \left(\ln \frac{C_s - C_i}{C_s - C_o}\right)_T$$

where M = required mass rate of oxygen input, g/s

Q = wastewater flowrate, m³/s

C_s = saturation concentration of oxygen at 20°C, g/m³

5 Rewrite the equation derived in step 4 in a more practical format. This can be done by noting that the density of air is 1.23 kg/m³ and that air contains about 23 percent oxygen by weight. Using these values, the rate of oxygen input, in terms of the equivalent air flowrate expressed in m³/s, is equal to

$$Q_a = 3.53 \times 10^{-3} \frac{Q}{E} \frac{(C_s)_{20°C}}{(1.024)^{T-20}} \left(\ln \frac{C_s - C_i}{C_s - C_o}\right)_T$$

where Q_a = required air flowrate, m³/s

Comment The value of Q_a is usually multiplied by a factor of 1.1 to account for the fact that the saturation value of oxygen in wastewater is about 95 percent of that in distilled water, and to account for the difference in the transfer rates.

Evaluation of transfer coefficient In Eq. 7-19, $K_L a$ includes the effect of the resistance of either or both films and is also a function of the area of liquid-gas interface that exists per unit volume of fluid. For a given amount of air introduced to a liquid system, the available surface through which gas transfer can take place increases with decreasing bubble size. The generalization can then be made that the value of $K_L a$ increases as the bubble size decreases because of the advantageous change in the surface area-volume ratio, but this principle holds only within certain limits.

Efficient gas transfer also depends on the agitation of the wastewater. This turbulence reduces the thickness of the liquid film and lowers resistance to transfer and to dispersion of the dissolved gas once transfer has taken place. Where air bubbles are used, they promote turbulence and circulation of the liquid because of the lifting effect from viscous drag. The value of $K_L a$ also increases with temperature. This effect is most often expressed as

$$(K_L a)_T = (K_L a)_{20} \times (1.024)^{T-20} \qquad (7\text{-}23)$$

where $(K_L a)_T$ = value of coefficient at the given temperature
$(K_L a)_{20}$ = value of coefficient at 20°C
T = temperature, °C

Because of the many variables involved, studies of the transfer coefficient are usually made on water and then corrected for the wastewater. For instance, Sawyer and Lynch [28] studied the effect of various synthetic detergents on $K_L a$ and found values as low as 40 percent of the value in pure water with 15 mg/L of detergents. King [16] reported that, for relatively fresh wastewater, values of 26 to 46 percent of the fresh-water transfer coefficient were observed. In addition, the saturation value of oxygen in domestic wastewater is about 95 percent of that in distilled water.

The equilibrium or saturation concentration of gas dissolved in a liquid is a function of the partial pressure of the gas adjacent to the liquid. This relationship is given by Henry's law

$$P_g = Hx_g \qquad (7\text{-}24)$$

where P_g = partial pressure of gas, atm
H = Henry's law constant
x_g = equilibrium mole fraction of dissolved gas

$$= \frac{\text{mol gas } (n_g)}{\text{mol gas } (n_n) + \text{mol water } (n_w)}$$

Henry's law constant is a function of the type, temperature, and constituents of the liquid. Values of H for various gases are listed in Table 7-4, and the use of Henry's law is illustrated in Example 7-2.

Example 7-2: Saturation concentration of nitrogen in water What is the saturation concentration of nitrogen in water in contact with dry air at 1 atm and 20°C?

SOLUTION

1 Dry air contains about 79 percent nitrogen. Therefore, $p_g = 0.79$.

Table 7-4 Henry's law constants for several gases that are slightly soluble in water [24]

T, °C	Air	CO_2	CO	H_2	H_2S	CH_4	N_2	O_2
				$H \times 10^{-4}$, atm/mol fraction				
0	4.32	0.0728	3.52	5.79	0.0268	2.24	5.29	2.55
10	5.49	0.104	4.42	6.36	0.0367	2.97	6.68	3.27
20	6.64	0.142	5.36	6.83	0.0483	3.76	8.04	4.01
30	7.71	0.186	6.20	7.29	0.0609	4.49	9.24	4.75
40	8.70	0.233	6.96	7.51	0.0745	5.20	10.4	5.35
50	9.46	0.283	7.61	7.65	0.0884	5.77	11.3	5.88
60	10.1	0.341	8.21	7.65	0.103	6.26	12.0	6.29

Note: $1.8(°C) + 32 = °F$.

2 From Table 7-4, $H = 8.04 \times 10^4$, and

$$x_g = \frac{P_g}{H} = \frac{0.79}{8.04 \times 10^4}$$

$$= 9.84 \times 10^{-6}$$

3 One liter of water contains $1000/18 = 55.6$ g mol; thus

$$\frac{n_g}{n_g + n_w} = 9.84 \times 10^{-6}$$

$$n_g = (n_g + 55.6)9.84 \times 10^{-6}$$

Because the quantity $(n_g)9.84 \times 10^{-6}$ is very much less than n_g,

$$n_g \approx (55.6)9.84 \times 10^{-6}$$

$$\approx 5.47 \times 10^{-4} \text{ mol/L nitrogen}$$

4 Determine the saturation concentration of nitrogen.

$$C_s \approx \frac{5.47 \times 10^{-4} \text{ mol}}{L} \left(\frac{28 \text{ g}}{\text{mol}}\right)\left(10^3 \frac{\text{mg}}{\text{g}}\right)$$

$$\approx 15.3 \text{ mg/L}$$

Evaluation of Aerator Performance

For a given volume of water being aerated, aerators are evaluated on the basis of the quantity of oxygen transferred per unit of air introduced to the water for equivalent conditions (temperature and chemical composition of the water, depth at which the air is introduced, etc.). Usual procedures for analyzing efficiencies may involve:

1. Measuring the rate of sulfite oxidation when a sodium sulfite solution is aerated. Sulfite is rapidly oxidized by oxygen, and thus C in Eq. 7-19 is zero for this analysis.

2. Measuring the rate at which oxygen is added to the water by direct analysis for the oxygen. The oxygen content of the water is usually lowered to approximately zero prior to initiating the test. The oxygen level is then measured continuously or at known time intervals after the start of aeration. To determine $K_L a$, the oxygen data are plotted on arithmetic paper according to Eq. 7.19 (see Prob. 7-5).

Alternatively, if the log of Eq. 7-21 is taken, the value of $K_L a$ can be determined by plotting $C_s - C_t$ versus t on log-arithmetic paper. When this form of analysis is used, it should be noted that the value of C_s for oxygen for diffused-air aeration systems will be greater than the value computed using Henry's law. The appropriate value of C_s can be determined by allowing the contents of the test tank to come to equilibrium and then measuring the resulting dissolved-oxygen concentration.

3. Measuring the uptake of oxygen by microorganisms. In the activated-sludge system, oxygen is usually maintained at a level of 1 to 3 mg/L, and the oxygen is used by the microorganisms as rapidly as it is supplied. In equation form,

$$\frac{dC}{dt} = K_L a(C_s - C) - r_m \qquad (7\text{-}25)$$

where r_m is the rate of oxygen used by the microorganisms. Typical values of r_m vary from 2 to 7 g/d per gram of mixed-liquor volatile suspended solids ($MLVSS$). If the oxygen level is maintained at a constant level, dC/dt is zero and

$$r_m = K_L a(C_s - C) \qquad (7\text{-}26)$$

C in this case is constant also. Values of r_m can be determined in a laboratory by means of the Warburg apparatus. In this case, $K_L a$ can easily be determined as follows:

$$K_L a = \frac{r_m}{C_s - C} \qquad (7\text{-}27)$$

7-3 ADSORPTION

Adsorption, in general, is the process of collecting soluble substances that are in solution on a suitable interface. The interface can be between the liquid and a gas, a solid, or another liquid. Although adsorption is used at the air-liquid interface in the flotation process, only the case of adsorption at the liquid-solid interface will be considered in this discussion. In the past, the adsorption process has not been used extensively in wastewater purification, but demands for a better quality of treated wastewater effluent have led to an intensive examination and use of the process of adsorption on activated carbon.

Activated-carbon treatment of wastewater is usually thought of as a polishing process for water that has already received normal biological treatment. The

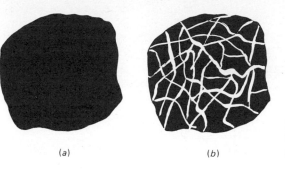

(a)   (b)

Figure 7-5 Sketch of activated carbon before and after activation. (*a*) Before activation. (*b*) After activation.

carbon in this case is used to remove a portion of the remaining dissolved organic matter. Depending on the means of contacting the carbon with the water, the particulate matter that is present may also be removed. Complete treatment with activated carbon is also being studied as a possible substitute for biological treatment of municipal wastewater when site limitations or industrial waste components pose problems for biological processes.

Description

The nature of activated carbon, the use of granular carbon and powdered carbon for wastewater treatment, and carbon regeneration are discussed below.

Activated-carbon production Activated carbon is prepared by first making a char from material such as wood or coal. This is accomplished by heating the material to a red heat in a retort to drive off the hydrocarbons but with an insufficient supply of air to sustain combustion. The char particle is then activated by exposure to an oxidizing gas a high temperature. This gas develops a porous structure in the char and thus creates a large internal surface area (see Fig. 7-5). The surface properties that result are a function of both the initial material used and the exact preparation procedure, so that many variations are possible. The type of base material from which the activated carbon is derived may also affect the pore-size distribution and the regeneration characteristics. Typical pore sizes and pore-size distributions are given in Allen, Joyce, and Kasch [1]. After activation, the carbon can be separated into, or prepared in, different sizes with different adsorption capacity. The two size classifications generally are powdered, which has a diameter of less than 200 mesh, and granular, which has a diameter greater than 0.1 mm.

Treatment with granular carbon A fixed-bed column is often used as a means of contacting wastewater with granular carbon. A typical column used in the treatment of wastewater is shown in Fig. 7-6. The water is applied to the top of the column and withdrawn at the bottom. The carbon is held in place with an underdrain system at the bottom of the column. Provision for backwash and surface wash is usually necessary. Backwashing is necessary to keep excessive

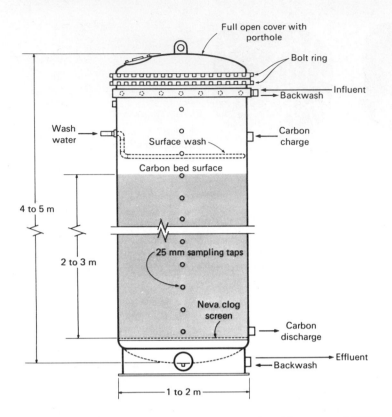

Figure 7-6 Typical activated-carbon adsorption column. Note: mm × 0.03937 = in; m × 3.2808 = ft.

head loss from building up. Fixed-bed columns can be operated singly, in series, or in parallel.

One of the most serious problems encountered with the use of fixed carbon beds is the surface clogging caused by the suspended solids in the wastewater effluent to be treated. The problem of clogging in fixed beds is partially overcome by using a surface wash or air scour, or both. Expanded-bed and moving-bed carbon contactors have also been developed to overcome the problem of surface clogging. In the expanded-bed system, the influent is introduced at the bottom of the column and is allowed to expand, much as a filter bed expands during backwash. In the moving-bed system, spent carbon is displaced continuously with fresh carbon. In such a system, head loss does not build up with time after the operating point has been reached.

Treatment with powdered activated carbon An alternative means of application is that of adding powdered activated carbon. It has been added to the effluent from biological treatment processes, directly to the various biological treatment processes, and in physical-chemical treatment process flowsheets. In the case of biological-treatment-plant effluent, the carbon is added to the effluent in a contacting basin. After a certain amount of time for contact, the carbon is allowed to settle to the bottom of the tank, and the treated water is then removed from the tank. Because carbon is very fine, a coagulant, such as a polyelectrolyte, may be needed to aid the removal of the carbon particles, or filtration through rapid sand filters may be required. The addition of powdered activated carbon directly in the aeration basin of an activated-sludge treatment process has proved to be effective in the removal of some soluble refractory organics.

Carbon regeneration Economical application of carbon depends on an efficient means of regenerating the carbon after its adsorptive capacity has been reached. Granular carbon can be regenerated easily in a furnace by oxidizing the organic matter and thus removing it from the carbon surface. Some of the carbon (about 5 to 10 percent) is also destroyed in this process during transport and must be replaced with new or virgin carbon. The capacity of regenerated carbon is slightly less than that of virgin carbon. A major problem with the use of powdered activated carbon is that the methodology for its regeneration is not well defined. It is anticipated that its use will increase as regeneration problems are solved. The use of powdered activated carbon produced from solid wastes may obviate the need to regenerate the spent carbon. References 1, 25, 39, 40, and 42 may be consulted for a more complete discussion of the use of activated carbon for wastewater treatment.

Analysis of Activated-Carbon Adsorption

The adsorption process can be pictured as one in which molecules leave solution and are held on the solid surface by chemical and physical bonding. The molecules are called the adsorbate and the solid is called the adsorbent. If the bonds that form between the adsorbate and adsorbent are very strong, the process is almost always irreversible, and chemical adsorption or chemisorption is said to have occurred. On the other hand, if the bonds that are formed are very weak, as is characteristic of bonds formed by van der Waals forces, physical adsorption is said to have occurred. The molecules adsorbed by this means are easily removed, or desorbed, by a change in the solution concentration of the adsorbate, and for this reason, the process is said to be reversible. Physical adsorption is the process that occurs most frequently in the removal of wastewater constituents by activated carbon. (The terms "adsorption" and "physical adsorption" are often used interchangeably. The practice will be followed in this discussion.)

The quantity of adsorbate that can be taken up by an adsorbent is a function of both the concentration of adsorbate C and the temperature T. Generally, the amount of material adsorbed is determined as a function of C at a constant temperature, and the resulting function is called an adsorption isotherm. Equations that are often used to describe the experimental isotherm data were developed by Freundlich, by Langmuir, and by Brunauer, Emmet, and Teller (BET isotherm) [29]. The isotherms developed by Freundlich and Langmuir are most commonly used in the application of activated carbon for the treatment of wastewater.

Freundlich isotherm equation Derived from empirical considerations, the Freundlich isotherm is

$$\frac{X}{M} = kC^{1/n} \tag{7-28}$$

where X/M = amount adsorbed per unit weight of adsorbent (carbon)
C = equilibrium concentration of adsorbate in solution after adsorption
k, n = empirical constants

The constants in this equation can be evaluated by plotting X/M versus C on double logarithmic paper.

Langmuir isotherm equation Derived from rational considerations, the Langmuir adsorption isotherm is

$$\frac{X}{M} = \frac{abC}{1 + bC} \tag{7-29}$$

where X/M = amount adsorbed per unit weight of adsorbent (carbon)
a, b = empirical constants
C = equilibrium concentration of adsorbate in solution after adsorption

This equation was developed on the basis of the assumptions that (1) a fixed number of accessible sites are available on the adsorbent surface, all of which have the same energy and that (2) adsorption is reversible. Equilibrium is reached when the rate of adsorption of molecules onto the surface is the same as the rate of desorption of molecules from the surface. The rate at which adsorption proceeds, then, is proportional to the driving force, which is the difference between the amount adsorbed at a particular concentration and the amount that can be adsorbed at that concentration. At the equilibrium concentration, this difference is zero.

Correspondence of experimental data to the Langmuir equation does not mean that the stated assumptions are valid for the particular system being studied, because departures from the assumptions can have a canceling effect. A Langmuir isotherm is shown in Fig. 7-7. The constants in the Langmuir equation can be

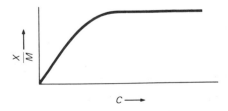

Figure 7-7 Sketch showing generalized form of Langmuir isotherm.

determined by plotting $C/(X/M)$ versus C and making use of Eq. 7-29 rewritten as

$$\frac{C}{X/M} = \frac{1}{ab} + \frac{1}{a}C \tag{7-30}$$

Application of the Langmuir adsorption isotherm is illustrated in Example 7-3.

Example 7-3: Analysis of activated-carbon adsorption data Determine if the carbon adsorption data presented below follow the Langmuir adsorption isotherm, and if they do, determine the constants a and b.

C, mg/L	10	20	30
X/M, g/g	0.133	0.187	0.220

SOLUTION
1 Plot $C/(X/M)$ versus C (see Fig. 7-8).

C, mg/L	10	20	30
$\dfrac{C}{X/M}$	75	107	136

2 Because the data plot as a straight line, it can be concluded that they can be described with the Langmuir adsorption isotherm.
3 Determine the constant a. From Fig. 7-8,

$$\frac{1}{a} = \frac{37}{12} = 3.08$$

Therefore

$$a = \frac{1}{3.08} = 0.325$$

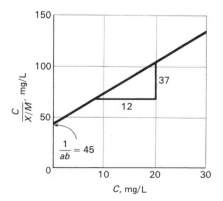

Figure 7-8 Plot of Langmuir adsorption data for Example 7-3. Note: mg/L = g/m³.

4   Determine the constant b. From Fig. 7-8,

$$\frac{1}{ab} = 45$$

Therefore

$$b = \frac{1}{45(0.325)} = 0.0685$$

Rate of adsorption The adsorption process can be divided into three steps: (1) transfer of the adsorbate molecules through the film that surrounds the adsorbent, (2) diffusion through the pores if the adsorbent is porous, and (3) uptake of the adsorbate molecules by the active surface, including formation of the bonds between the adsorbate and the carbon. Step 3 is considered to be very rapid, since equilibrium on nonporous adsorbents can be accomplished in a matter of minutes. Steps 1 and 2 are generally held to be rate-limiting. The thickness of the stagnant aqueous film that surrounds the adsorbent depends on the flow regime maintained in the system. The rate of adsorption then depends on the rate at which the molecules move or diffuse in solution or the rate at which the molecules can reach the available surface by diffusing through the film and the pore.

Adsorption of mixtures In the application of adsorption to wastewater treatment, mixtures of organic compounds are encountered. Weber [40] found that, although there will be a depression of the adsorptive capacity of any individual compound in a solution of many compounds, the total adsorptive capacity of the adsorbent may be larger than the adsorptive capacity with a single compound. The amount of inhibition due to competing adsorbates is related to the size of the molecules being adsorbed, their adsorptive affinities, and their relative concentrations.

Process Analysis

As noted previously, both granular carbon (in downflow and upflow columns) and powdered activated carbon are used for wastewater treatment. The analysis procedures for both types are described briefly in the following discussion.

Granular carbon Where carbon columns are used, the adsorptive capacity of the carbon in the column can be estimated from isotherm data as follows. If the isotherm data presented in the first two columns of Table 7-5 are analyzed by applying the Freundlich isotherm equation, the data in the last two columns are obtained. If these isotherm data are plotted, the resulting isotherm will be as shown in Fig. 7-9. Using this figure, the adsorptive capacity of the carbon can be estimated by extending a vertical line from the point on the horizontal scale corresponding to the initial concentration C_0 and extrapolating the isotherm to intersect this line. The $(X/M)_{C_0}$ value at the point of intersection can be read from the vertical scale. This value $(X/M)_{C_0}$ represents the amount of constituent

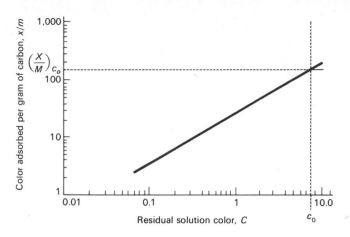

Figure 7-9 Typical decolorization isotherm.

adsorbed per unit weight of carbon when the carbon is at equilibrium with the initial concentration of constituent. This condition should exist in the upper section of a carbon bed during column treatment, and it therefore represents the ultimate capacity of the carbon for the particular waste. From the $(X/M)_{C_0}$ value, the ultimate capacity, in terms of the volume of waste treated per gram of carbon, can be calculated simply by dividing $(X/M)_{C_0}$ by the amount of impurity adsorbed per unit volume.

Because of the breakthrough phenomenon (see Fig. 7-10), the usual practice is either to use two columns in series and rotate them as they become exhausted, or to use multiple columns in parallel so that breakthrough in a single column will not significantly affect the effluent quality. With proper sampling from points within the column, BOD breakthrough can be anticipated. As shown in Fig. 7-10, the capacity of the upper portion of the column is exhausted with the passage of time. The active zone for adsorption is designated by the symbol δ and corresponds to the length of column required to achieve the degree of removal required. The thickness of the active zone varies with the flowrate because

Table 7-5 Analysis of isotherm data

M Weight of carbon, g/100 mL solution	C Residual- solution color	X Color adsorbed	X/M Color adsorbed per unit weight
0	7.70		
0.05	3.67	4.03	80.6
0.1	2.20	5.50	55.0
0.3	0.87	6.83	22.8
1.0	0.25	7.45	7.5

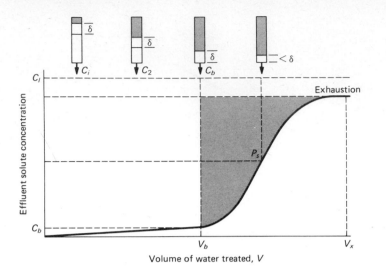

Figure 7-10 Typical breakthrough curve for activated carbon [29].

dispersion, diffusion, and channeling in a granular medium are directly related to the flowrate. The only way to use the capacity at the bottom of the column is to have two columns in series and switch them as they are exhausted, or to use multiple columns in parallel. The optimum flowrate and bed depth, as well as the operating capacity of the carbon, must be established to determine the dimensions and the number of columns necessary for continuous treatment. Because these parameters can be determined only from dynamic column tests, pilot-plant studies are recommended.

Powdered activated carbon For a powdered-carbon application, the isotherm adsorption data can be used in conjunction with a materials-balance analysis to obtain an approximate estimate of the amount of carbon that must be added [29]. Here again, because of the many unknown factors involved, pilot-plant tests to develop the optimum design data are recommended.

7-4 DISINFECTION

Disinfection refers to the selective destruction of disease-causing organisms. All the organisms are not destroyed during the process. This differentiates disinfection from sterilization, which is the destruction of all organisms. In the field of waste-water treatment, the three categories of human enteric organisms of the greatest consequence in producing disease are bacteria, viruses, and amoebic cysts. Diseases caused by waterborne bacteria include typhoid, cholera, paratyphoid, and bacillary dysentery; diseases caused by waterborne viruses include poliomyelitis and infectious hepatitis (see Table 3-16).

At present, the most common method of disinfecting wastewater is by the addition of chlorine. However, some of the adverse effects that may be caused by the addition of chlorine, including the possible formation of carcinogenic compounds, are only now becoming appreciated (see Sec. 7-6), and a variety of other means for achieving the disinfection of wastewater are currently under investigation. For this reason, the detailed discussion of this important subject is presented in four separate sections. The purpose in this section is to introduce the reader to the general concepts involved in the disinfection of microorganisms. The remaining sections deal with disinfection using chlorine (Sec. 7-5), de-chlorination (Sec. 7-6), and ozone (Sec. 7-7).

Description of Disinfection Methods and Means

Disinfection is most commonly accomplished by the use of (1) chemical agents, (2) physical agents, (3) mechanical means, and (4) radiation.

Chemical agents The requirements for an ideal chemical disinfectant are reported in Table 7-6. As shown, an ideal disinfectant would have to possess a wide range of characteristics. Although such a compound may not exist, the requirements set forth in Table 7-6 should be considered in evaluating proposed or recommended disinfectants. To sanitary engineers, it is also important that the disinfectant be safe to handle and apply, and that its strength or concentration in treated waters be measurable.

Chemical agents that have been used as disinfectants include (1) chlorine and its compounds, (2) bromine, (3) iodine, (4) ozone, (5) phenol and phenolic compounds, (6) alcohols, (7) heavy metals and related compounds, (8) dyes, (9) soaps and synthetic detergents, (10) quaternary ammonium compounds, (11) hydrogen peroxide, and (12) various alkalies and acids.

Of these, the most common disinfectants are the oxidizing chemicals, and chlorine is the one most universally used. Bromine and iodine occasionally are used for swimming-pool water but have not been used for treated wastewater. Ozone is a highly effective disinfectant, and its use is increasing even though it leaves no residual (see Sec. 7-7). Highly acid or alkaline water can also be used to destroy pathogenic bacteria, because water with a pH greater than 11 or less than 3 is relatively toxic to most bacteria.

Physical agents Physical disinfectants that can be used are heat and light. Heating water to the boiling point, for example, will destroy the major disease-producing non-spore-forming bacteria. Heat is commonly used in the beverage and dairy industry, but it is not a feasible means of disinfecting large quantities of wastewater because of the high cost. However, pasteurization of sludge is used extensively in Europe.

Sunlight is also a good disinfectant. In particular, ultraviolet radiation can be used. Special lamps that emit ultraviolet rays have been used successfully to sterilize small quantities of water. The efficiency of the process depends on

Table 7-6 Comparison of ideal and actual characteristics of chemical disinfectants[a]

Characteristic	Ideal disinfectant	Chlorine	Sodium hypochlorite	Calcium hypochlorite	Chlorine dioxide	Ozone
Toxicity to microorganisms	Should be highly toxic at high dilutions	High	High	High	High	High
Solubility	Must be soluble in water or cell tissue	Slight	High	High	High	High
Stability	Loss of germicidal action on standing should be low	Stable	Slightly unstable	Relatively stable	Unstable, must be generated as used	Unstable, must be generated as used
Nontoxic to higher forms of life	Should be toxic to microorganisms and nontoxic to man and other animals	Highly toxic to higher life forms	Toxic	Toxic	Toxic	Toxic
Homogeneity	Solution must be uniform in composition	Homogeneous	Homogeneous	Homogeneous	Homogeneous	Homogeneous
Interaction with extraneous material	Should not be absorbed by organic matter other than bacterial cells	Oxidizes organic matter	Active oxidizer	Active oxidizer	High	Oxidizes organic matter
Toxicity at ambient temperatures	Should be effective in ambient temperature range	High	High	High	High	Very high
Penetration	Should have the capacity to penetrate through surfaces	High	High	High	High	High
Noncorrosive and nonstaining	Should not disfigure metals or stain clothing	Highly corrosive	Corrosive	Corrosive	Highly corrosive	Highly corrosive
Deodorizing ability	Should deodorize while disinfecting	High	Moderate	Moderate	High	High
Availability	Should be available in large quantities and reasonably priced	Low cost	Moderately low cost	Moderately low cost	Moderate cost	High cost

[a] Adapted from Refs. 18 and 23.

Table 7-7 Removal or destruction of bacteria by different treatment processes

Process	Percent removal
Coarse screens	0–5
Fine screens	10–20
Grit chambers	10–25
Plain sedimentation	25–75
Chemical precipitation	40–80
Trickling filters	90–95
Activated sludge	90–98
Chlorination of treated wastewater	98–99

the penetration of the rays into water. The contact geometry between the ultraviolet-light source and the water is extremely important, because suspended matter, dissolved organic molecules, and water itself will absorb the radiation, in addition to the microorganisms. It is therefore difficult to use ultraviolet radiation in aqueous systems, especially when particulate matter is present.

Mechanical means Bacteria and other organisms are also removed by mechanical means during wastewater treatment. Typical removal efficiencies for various treatment operations and processes are reported in Table 7-7. The first four operations listed may be considered to be physical. The removals accomplished are a by-product of the primary function of the process.

Radiation The major types of radiation are electromagnetic, acoustic, and particle. Gamma rays are emitted from radioisotopes, such as cobalt 60. Because of their penetration power, gamma rays have been used to disinfect (sterilize) both water and wastewater. A schematic diagram of a high-energy electron-beam device for the irradiation of wastewater or sludge is shown in Fig. 7-11 [8].

Mechanisms of Disinfectants

Four mechanisms that have been proposed to explain the action of disinfectants are (1) damage to the cell wall, (2) alteration of cell permeability, (3) alteration of the colloidal nature of the protoplasm, and (4) inhibition of enzyme activity [23].

Damage or destruction of the cell wall will result in cell lysis and death. Some agents, such as penicillin, inhibit the synthesis of the bacterial cell wall.

Agents such as phenolic compounds and detergents alter the permeability of the cytoplasmic membrane. These substances destroy the selective permeability of the membrane and allow vital nutrients, such as nitrogen and phosphorus, to escape.

Heat, radiation, and highly acid or alkaline agents alter the colloidal nature of the protoplasm. Heat will coagulate the cell protein and acids or bases will denature proteins, producing a lethal effect.

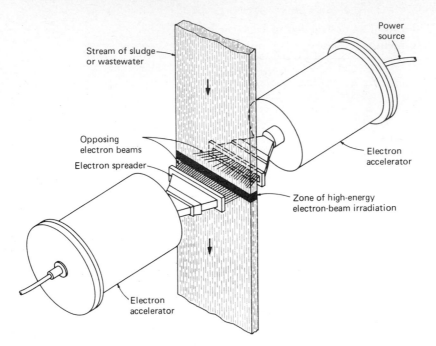

Power source

Stream of sludge or wastewater

Opposing electron beams

Electron spreader

Electron accelerator

Electron accelerator

Zone of high-energy electron-beam irradiation

Figure 7-11 Schematic diagram of high-energy electron-beam device for the irradiation of wastewater or sludge [8].

Another mode of disinfection is the inhibition of enzyme activity. Oxidizing agents, such as chlorine, can alter the chemical arrangement of enzymes and inactivate the enzymes.

Analysis of Factors Influencing the Action of Disinfectants

In applying the disinfection agents or means that have been described, the following factors must be considered: (1) contact time, (2) concentration and type of chemical agent, (3) intensity and nature of physical agent, (4) temperature, (5) number of organisms, (6) types of organisms, and (7) nature of suspending liquid [23].

Contact time Perhaps one of the most important variables in the disinfection process is contact time. In general, as shown in Fig. 7-12, it has been observed that for a given concentration of disinfectant, the longer the contact time, the greater the kill. This observation was first formalized in the literature by Chick [3]. In differential form, Chick's law is

$$\frac{dN}{dt} = -kN_t \qquad (7\text{-}31)$$

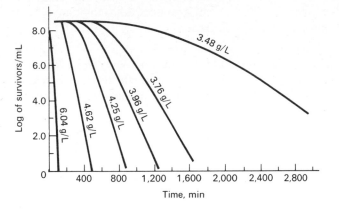

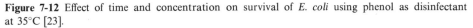

Figure 7-12 Effect of time and concentration on survival of *E. coli* using phenol as disinfectant at 35°C [23].

where N_t = number of organisms at time t

t = time

k = constant, time^{-1}

If N_0 is the number of organisms when t equals 0, Eq. 7-31 can be integrated to

$$\frac{N_t}{N_0} = e^{-kt} \tag{7-32}$$

or

$$\ln \frac{N_t}{N_0} = -kt$$

Departures from this rate law are common. Rates of kill have been found to increase with time in some cases and to decrease with time in other cases. To formulate a valid relationship for the kill of organisms under a variety of conditions, an assumption often made is that

$$\ln \frac{N_t}{N_0} = -kt^m \tag{7-33}$$

where m is a constant. If m is less than 1, the rate of kill decreases with time, and if m is greater than 1, the rate of kill increases with time. The constants in Eq. 7-33 can be obtained by plotting $-\ln (N/N_0)$ versus the contact time t on log-log paper. The straight-line form of the equation is

$$\log \left(-\ln \frac{N_t}{N_0} \right) = \log k + m \log t \tag{7-34}$$

Another formulation that has been used to describe the observed effects of contact time is

$$\frac{N_t}{N_0} = kt^m \tag{7-35}$$

Equation 7-35 results from the analysis of chlorination data that have been found to plot as straight lines on double-log paper.

Concentration and type of chemical agent Depending on the type of chemical agent, it has been observed that, within limits, disinfection effectiveness is related to concentration (see Fig. 7.17). The effect of concentration has been formulated empirically [11]:

$$C^n t_p = \text{constant} \tag{7-36}$$

where C = concentration of disinfectant
n = constant
t_p = time required to effect a constant percentage kill

The constants in Eq. 7-36 can be evaluated by plotting, on log-log paper, the concentration versus the time required to effect a given percentage kill. The slope of the line then corresponds to the value of $-1/n$. In general, if n is greater than 1, contact time is more important than the dosage; if n equals 1, the effects of time and dosage are about the same [11].

Intensity and nature of physical agent As noted earlier, heat and light are physical agents that have been used from time to time in the disinfection of wastewater. It has been found that their effectiveness is a function of intensity. For example, if the decay of organisms can be described with a first-order reaction, such as

$$\frac{dN}{dt} = -kN \tag{7-37}$$

where N = number of organisms
t = time, min
k = reaction velocity of constant, 1/min

then the effect of the intensity of the physical disinfectant is reflected in the constant k through some functional relationship.

Temperature The effect of temperature on rate of kill can be represented by a form of the van't Hoff–Arrhenius relationship. Increasing the temperature results in a more rapid kill. In terms of the time t required to effect a given percentage kill, the relationship is

$$\ln \frac{t_1}{t_2} = \frac{E(T_2 - T_1)}{R T_1 T_2} \tag{7-38}$$

where t_1, t_2 = time for given percentage kill at temperatures T_1 and T_2, °K, respectively
E = activation energy, J/mol (cal/mol)
R = gas constant, 8.314 J/mol · °K (1.99 cal/°K · mol)

Table 7-8 Activation energies for aqueous chlorine and chloramines at normal temperatures [11]

Compound	pH	E, J/mol
Aqueous chlorine	7.0	34,332
	8.5	26,796
	9.8	50,242
	10.7	62,802
Chloramines	7.0	50,242
	8.5	58,615
	9.5	83,736

Note: J × 0.2388 = cal.

Some typical values for the activation energy for various chlorine compounds at different pH values are reported in Table 7-8.

Number of organisms In a dilute system such as wastewater, the concentration of organisms is seldom a major consideration. However, it can be concluded from Eq. 7-36 that the larger the organism concentration, the longer the time required for a given kill. An empirical relationship that has been proposed to describe the effect of organism concentration is [11]

$$C^q N_p = \text{constant} \tag{7-39}$$

where C = concentration of disinfectant
N_p = concentration of organisms reduced given percentage in given time
q = constant related to strength of disinfectant

Types of organisms The effectiveness of various disinfectants will be influenced by the nature and condition of the microorganisms. For example, viable growing bacteria cells are killed easily. In contrast, bacterial spores are extremely resistant, and many of the chemical disinfectants normally used will have little or no effect. Other disinfecting agents, such as heat, may have to be used. This subject is considered further in Sec. 7-5.

Nature of suspending liquid In addition to the foregoing factors, the nature of the suspending liquid must be evaluated carefully. For example, extraneous organic material will react with most oxidizing disinfectants and reduce their effectiveness. Turbidity will reduce the effectiveness of disinfectants by absorption and by protecting entrapped bacteria.

7-5 DISINFECTION WITH CHLORINE

As noted earlier, of all the chemical disinfectants, chlorine is perhaps the one most commonly used throughout the world. The reason is that it satisfies most of the requirements specified in Table 7-6. Because the practical aspects of chlorination are discussed in Chap. 8, the following discussion is limited to a brief description of chlorine chemistry and breakpoint chlorination, and an analysis of the performance of chlorine as a disinfectant and the factors that may influence the effectiveness of the chlorination process.

Description of Chlorine Chemistry

The most common chlorine compounds used in wastewater-treatment plants are chlorine gas (Cl_2), calcium hypochlorite [$Ca(OCl)_2$], sodium hypochlorite (NaOCl), and chlorine dioxide (ClO_2). Calcium and sodium hypochlorite are most often used in very small treatment plants, such as package plants, where simplicity and safety are far more important than cost. However, in 1965 New York City completed a changeover from liquid chlorine in ton containers to sodium hypochlorite, primarily for reasons of safety as influenced by local conditions. In 1967, Chicago, faced with the need to install chlorination facilities to meet upgraded standards for effluent and receiving-water quality, installed sodium hypochlorite storage tanks and feed equipment at its North Side and Calumet treatment plants. The decision to use hypochlorite was based solely on safety considerations. The use of chlorine dioxide for wastewater treatment is not well defined at the present time. However, it has some unusual properties (does not react with ammonia), and its application may increase in the future. Nevertheless, the discussion in this section is based primarily on the use of chlorine in the form of a gas, because it is the most commonly used form.

Reactions in water When chlorine in the form of Cl_2 gas is added to water, two reactions take place: hydrolysis and ionization.

Hydrolysis may be defined as

$$Cl_2 + H_2O \rightleftharpoons HOCl + H^+ + Cl^- \qquad (7\text{-}40)$$

The stability constant for this reaction is

$$K = \frac{[HOCl][H^+][Cl^-]}{[Cl_2]} = 4.5 \times 10^{-4} \text{ at } 25°C \qquad (7\text{-}41)$$

Because of the magnitude of this coefficient, large quantities of chlorine can be dissolved in water.

Ionization may be defined as

$$HOCl \rightleftharpoons H^+ + OCl^- \qquad (7\text{-}42)$$

Table 7-9 Values of the ionization constant of hypochlorous acid at different temperatures [42]

Temperature, °C	0	5	10	15	20	25
$K_i \times 10^8$, mol/L	2.0	2.3	2.6	3.0	3.3	3.7

Note: $1.8(°C) + 32 = °F$.

The ionization constant for this reaction is

$$K_i = \frac{[H^+][OCl^-]}{[HOCl]} = 3.7 \times 10^{-8} \text{ at } 25°C \tag{7-43}$$

The variation in the value of K_i with temperature is reported in Table 7-9.

The quantity of HOCl and OCl⁻ that is present in water is called the free available chlorine. The relative distribution of these two species (see Fig. 7-13) is very important because the killing efficiency of HOCl is about 40 to 80 times that of OCl⁻. The percentage distribution of HOCl at various temperatures can be computed using Eq. 7-44 and the data in Table 7-9.

$$\frac{HOCl}{HOCl + OCl} = \frac{1}{1 + OCl/HOCl} = \frac{1}{1 + K_i/H} \tag{7-44}$$

Free chlorine can also be added to water in the form of hypochlorite salts. The pertinent reactions are as follows:

$$Ca(OCl)_2 + 2H_2O \rightarrow 2HOCl + Ca(OH)_2 \tag{7-45}$$

$$NaOCl + H_2O \rightarrow HOCl + NaOH \tag{7-46}$$

Reactions with ammonia As noted in Chap. 3, untreated wastewater contains nitrogen in the form of ammonia and various combined organic forms. The effluent from most treatment plants also contains significant amounts of nitrogen, usually in the form of ammonia, or nitrate if the plant is designed to achieve nitrification (see Chaps. 9 and 10). Because hypochlorous acid is a very active

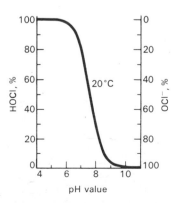

Figure 7-13 Distribution of hypochlorous acid and hypochlorite in water at different pH values.

oxidizing agent, it will react readily with ammonia in the wastewater to form three types of chloramines in the successive reactions:

$$NH_3 + HOCl \rightarrow NH_2Cl \text{ (monochloramine)} + H_2O \tag{7-47}$$

$$NH_2Cl + HOCl \rightarrow NHCl_2 \text{ (dichloramine)} + H_2O \tag{7-48}$$

$$NHCl_2 + HOCl \rightarrow NCl_3 \text{ (nitrogen trichloride)} + H_2O \tag{7-49}$$

These reactions are very dependent on the pH, temperature, and contact time, and on the initial ratio of chlorine to ammonia [42]. The two species that predominate, in most cases, are monochloramine (NH_2Cl) and dichloramine ($NHCl_2$). The chlorine in these compounds is called combined available chlorine. As will be discussed subsequently, these chloramines also serve as disinfectants, although they are extremely slow-reacting.

Breakpoint Reaction

The maintenance of a residual (combined or free) for the purpose of wastewater disinfection is complicated by the fact that free chlorine not only reacts with ammonia, as noted previously, but also is a strong oxidizing agent. The stepwise phenomena that result when chlorine is added to wastewater containing ammonia can be explained by referring to Fig. 7-14.

As chlorine is added, readily oxidizable substances, such as Fe^{++}, Mn^{++}, H_2S, and organic matter, react with the chlorine and reduce most of it to the chloride ion (point A in Fig. 7-14). After meeting this immediate demand, the

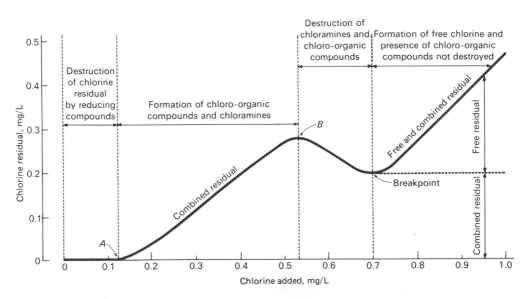

Figure 7-14 Generalized curve obtained during breakpoint chlorination. Note: $mg/L = g/m^3$.

chlorine continues to react with the ammonia to form chloramines between points A and B. For mole ratios of chlorine to ammonia less than 1, monochloramine and dichloramine will be formed. The distribution of these two forms is governed by their rates of formation, which are dependent on the pH and temperature. Between point B and the breakpoint, some chloramines will be converted to nitrogen trichloride (see Eq. 7-49), the remaining chloramines will be oxidized to nitrous oxide (N_2O) and nitrogen (N_2), and the chlorine will be reduced to the chloride ion. With continued addition of chlorine, most of the chloramines will be oxidized at the breakpoint. Theoretically, as determined in Example 7-4, the weight ratio of chlorine to ammonia nitrogen at the breakpoint is 7.6 : 1.

Possible reactions to account for the appearance of the aforementioned gases and the disappearance of chloramines are as follows (see also Eq. 7-49):

$$NH_2Cl + NHCl_2 + HOCl \rightarrow N_2O + 4HCl \tag{7-50}$$

$$4NH_2Cl + 3Cl_2 + H_2O \rightarrow N_2 + N_2O + 10HCl \tag{7-51}$$

$$2NH_2Cl + HOCl \rightarrow N_2 + H_2O + 3HCl \tag{7-52}$$

$$NH_2Cl + NHCl_2 \rightarrow N_2 + 3HCl \tag{7-53}$$

Continued addition of chlorine past the breakpoint, as shown in Fig. 7-15a, will result in a directly proportional increase in the free available chlorine (unreacted hypochlorite).

The main reason for adding enough chlorine to obtain a free chlorine residual is that usually disinfection can then be assured. Occasionally, serious odor problems have developed during breakpoint-chlorination operations because of the formation of nitrogen trichloride and related compounds. The presence of additional compounds during chlorination will react with the alkalinity of the wastewater, and under most circumstances, the pH drop will be slight. The presence of additional compounds that will react with chlorine, such as organic nitrogen, may greatly alter the shape of the breakpoint curve, as shown in Fig. 7-15b. The amount of chlorine that must be added to reach a desired level of residual is called the chlorine demand.

Example 7-4: Breakpoint chlorination Determine the stoichiometric weight ratio of chlorine to ammonia nitrogen at the breakpoint.

SOLUTION

1 Write an overall reaction to describe the breakpoint phenomenon. This can be done using Eqs. 7-47 and 7-52.

$$2NH_3 + 2HOCl \rightarrow 2NH_2Cl + 2H_2O \tag{7-47}$$

$$2NH_2Cl + HOCl \rightarrow N_2 + H_2O + 3HCl \tag{7-52}$$

$$\overline{2NH_3 + 3HOCl \rightarrow N_2 + 3H_2O + 3HCl}$$

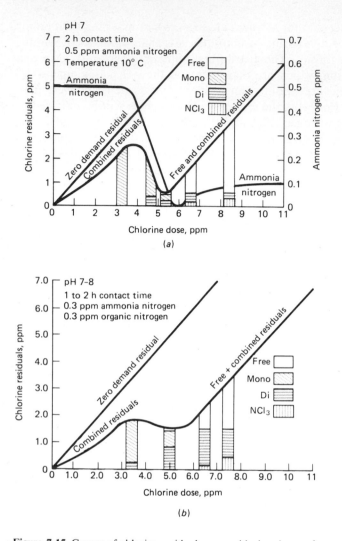

Figure 7-15 Curves of chlorine residual versus chlorine dosage for wastewater [42]. (*a*) For waste-water containing nitrogen in the form of ammonia (NH_3). (*b*) For wastewater containing nitrogen in the form of ammonia and organic nitrogen. Note: $1.8(°C) + 32 = °F$.

2 Determine the molecular weight of the ammonia (NH_3) expressed as N and the hypochlorous acid (HOCl) expressed as Cl_2.

$$\text{Molecular weight of } NH_3 \text{ expressed as N} = \frac{17}{17}(14 \text{ g/mol})$$

$$= 14$$

$$\text{Molecular weight of HOCl expressed as } Cl_2 = \frac{52.45}{52.45}(70.9 \text{ g/mol})$$

$$= 70.9$$

3 Determine the weight ratio of chlorine to ammonia nitrogen.

$$\frac{Cl_2}{NH_3\text{-}N} = \frac{3(70.9)}{2(14)} = \frac{7.6}{1}$$

Comment The ratio computed in step 3 will vary somewhat, depending on the actual reactions involved, which at present are unknown. In practice, the actual ratio has been found to vary from 8 : 1 to 10 : 1.

Acid generation In practice, the hydrochloric acid formed during chlorination (see reaction given in step 1 of Example 7-3) will react with the alkalinity of the wastewater, and under most circumstances, the pH drop will be slight. Stoichiometrically, 14.3 mg/L of alkalinity, expressed as $CaCO_3$, will be required for each 1.0 mg/L of ammonia nitrogen that is oxidized in the breakpoint-chlorination process. In practice, it has been found that about 15 mg/L of alkalinity are actually required because of the hydrolysis of chlorine [26].

Buildup of total dissolved solids In addition to the formation of hydrochloric acid, the chemicals added to achieve the breakpoint reaction will also contribute an increment to the total dissolved solids of the wastewater. In situations where the level of total dissolved solids may be critical with respect to reuse applications, this incremental buildup from breakpoint chlorination should always be checked. The total dissolved-solids contribution for each of several chemicals that may be used in the breakpoint reaction is summarized in Table 7-10. The magnitude of the possible buildup of total dissolved solids is illustrated in Example 12-3, in which the use of breakpoint chlorination is considered for the seasonal control of nitrogen.

Factors That Affect Disinfection Efficiency of Chlorine

The purpose of the following discussion is to explore the important factors that affect the disinfection efficiency of chlorine to the extent that they are now known. These include (1) the germicidal efficiency of chlorine, (2) the germicidal efficiency

Table 7-10 Effects of chemical addition on total dissolved solids in breakpoint chlorination [26]

Chemical addition	Total dissolved solids increase: NH_4^+-N consumed
Breakpoint with chlorine gas	6.2 : 1
Breakpoint with sodium hypochlorite	7.1 : 1
Breakpoint with chlorine gas— neutralization of all acidity with lime (CaO)	12.2 : 1
Breakpoint with chlorine gas— neutralization of all acidity with sodium hydroxide (NaOH)	14.8 : 1

of the various chlorine compounds, (3) the importance of initial mixing, (4) the breakpoint reaction, (5) the contact time, (6) the characteristics of the wastewater, and (7) the characteristics of the microorganisms. To provide a framework in which to view these factors, it will be appropriate to consider first how the effectiveness of the chlorination process is now assessed and how the results are analyzed.

Germicidal efficiency of chlorine When using chlorine for the disinfection of wastewater, the principal parameters that can be measured, apart from environmental variables such as pH and temperature, are the number of organisms and the chlorine residual remaining after a specified period of time. The coliform group of organisms can be determined using the most probable number (MPN) procedure as discussed in Chap. 3. The organisms remaining can also be determined by the plate-count procedure using an agar mixture as the plating medium [37]. Either the standard "pour-plate" method or the "spread-plate" method can be used. The plates should be incubated at 37°C (98.6°F), because this temperature results in the optimum growth of *E. coli*, and the colonies should be counted after a 24-hour incubation period.

The chlorine residual (free and combined) should be measured using the amperometric method, which has proved to be the most consistently reliable method now available. Also, because almost all the commercial analyzers of residual chlorine use it, the adoption of this method will allow the results of independent studies to be compared directly. Numerous tests have shown that when all the physical parameters controlling the chlorination process are held constant, the germicidal efficiency of disinfection, as measured by bacterial survival, depends primarily on the residual bactericidal chlorine present R and the contact time t. It has also been found that by increasing either one of the two variables R or t and simultaneously decreasing the other one, it is possible to achieve approximately the same degree of disinfection. Thus the efficiency of disinfection may be expressed as a function of the product $(R \times t)$. This is in agreement with the discussion presented in Sec. 7-4 and with Collins [5], who in 1971 described the bacterial survival ratio N/N_0 as a function of R and t.

Using a batch reactor whose contents were well stirred, Selleck, Collins, and White [31] found that the reduction of coliform organisms in a chlorinated primary treated effluent can be defined by the following relationship:

$$\frac{N_t}{N_0} = (1 + 0.23\, C_t t)^{-3} \tag{7-54}$$

where N_t = number of coliform organisms at time t
N_0 = number of coliform organisms at time t_0
C_t = total amperometric chlorine residual at time t, mg/L
t = residence time, min

The data from which this relationship was developed are shown in Fig. 7-16. The application of Eq. 7-54 is considered in Probs. 7-11 and 7-12.

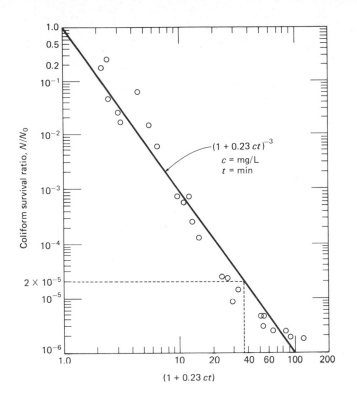

Figure 7-16 Coliform survival in a batch reactor as a function of amperometric chlorine residual and contact time (temperature range 15–18°) [30]. Note: $mg/L = g/m^3$.

Germicidal efficiency of various chlorine compounds A comparison of the germicidal efficiency of hypochlorous acid (HOCl), hypochlorite ion (OCl), and monochloramine (NH_2Cl) is presented in Fig. 7-17. For a given contact time or residual, the germicidal efficiency of hypochlorous acid, in terms of either time or residual, is significantly greater than that of either the hypochlorite ion or monochloramine. It should be noted, however, that given an adequate contact time, monochloramine is nearly as effective as chlorine in achieving disinfection.

Referring to Fig. 7-17, it is clear that hypochlorous acid offers the most positive way of achieving disinfection. For this reason, with proper mixing, the formation of hypochlorous acid following breakpoint is most effective in achieving wastewater chlorination. If sufficient chlorine cannot be added to achieve the breakpoint reaction, great care must be taken to ensure that the proper contact time is maintained. Because of the equilibrium between hypochlorous acid and the hypochloric ion, maintenance of the proper pH is also important if effective disinfection is to be achieved.

Initial mixing Only recently has the importance of initial mixing on the disinfection process been demonstrated conclusively [4]. It was shown that the

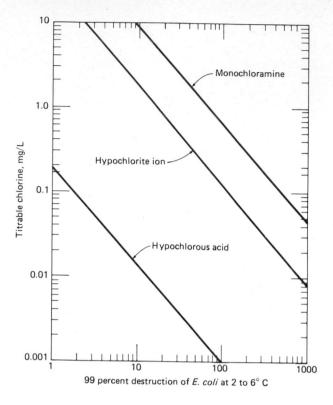

Figure 7-17 Comparison of the germicidal efficiency of hypochlorous acid, hypochlorite ion, and monochloramine for 99 percent destruction of *E. coli* at 2 to 6°C [42]. Note: $1.8(°C) + 32 = °F$.

application of chlorine in a highly turbulent regime $(N_R = 10^4)$ resulted in kills two orders of magnitude greater than when chlorine was added separately to a continuous-flow stirred-tank reactor under similar conditions [4]. The need for proper initial mixing has also been documented on a practical basis in a study of treatment plants discharging to San Francisco Bay [4]. Although the importance of initial mixing is well delineated, the optimum level of turbulence is not known. The design of proper mixing facilities is considered in Chap. 8.

Breakpoint reaction The basic aspects of the breakpoint reaction and its effects on the disinfection process have been discussed previously. The discussion here is concerned with the practice of using chlorinated wastewater for the chlorine injection water (see Chap. 8). The contention is that if nitrogeneous compounds are present in the wastewater, a portion of the chlorine that is added will react with these compounds, and by the time it is injected, it will be in the form of monochloramine or dichloramine. This can be a problem in small installations or where the chlorine solution lines from the chlorinator to the point of injection are quite long. It has been shown, however, that with proper initial mixing bacterial kills are the same whether untreated or treated effluent is used for the injector water supply [42]. From the evidence to date, proper initial mixing appears to be more important in achieving effective disinfection than the form in which the chlorine is injected. Again, it should be remembered that hypochlorous acid (HOCl) and monochloramine (NH_2Cl) are equally effective as disinfecting compounds; only the contact time required is different (see Fig. 7-17).

Contact time Because of the reaction of chlorine with the nitrogenous compounds found in untreated and treated wastewater, and because chlorination beyond the breakpoint to obtain free hypochlorous acid is not economically feasible in many situations, the fundamental importance of contact time in the disinfection of wastewater cannot be overemphasized.

In Sec. 7-4 it was noted that the effect of contact time has at one time or another been described with each of the following relationships:

$$\ln \frac{N_t}{N_0} = -kt \tag{7-32}$$

$$\ln \frac{N_t}{N_0} = -kt^m \tag{7-33}$$

$$\frac{N_t}{N_0} = kt^m \tag{7-35}$$

Of these relationships, Eq. 7-35 appears to provide the best fit for the data obtained from the chlorination of wastewater. The probable reason that Eq. 7-35 applies to wastewater data, as opposed to Eq. 7-32, is that in most cases the chlorine residual is made up of chloramines.

Because of the importance of contact time, either a batch or plug-flow reactor should be used to achieve effective disinfection, and since a batch reactor for chlorination is impractical, plug-flow reactors are used at most treatment plants. The proper design of chlorination basins is considered in Chap. 8.

Characteristics of the wastewater It has often been observed that, for treatment plants of similar design with exactly the same effluent characteristics measured in terms of BOD, COD, and nitrogen, the effectiveness of the chlorination process varies significantly from plant to plant. To investigate the reasons for this observed phenomenon, and to assess the effects of the compounds present on the chlorination process, Sung [37] studied the characteristics of the compounds in untreated and treated wastewater. Among the more important conclusions derived from Sung's study are the following:

1. In the presence of interfering organic compounds, the total chlorine residual cannot be used as a reliable measure for assessing the bactericidal efficiency of chlorine.

2. The degree of interference of the compounds studied depended on their functional groups and their chemical structure.
3. Saturated compounds and carbohydrates exert little or no chlorine demand and do not appear to interfere with the chlorination process.
4. Organic compounds with unsaturated bonds may exert an immediate chlorine demand, depending on their functional groups. In some cases, the resulting compounds may titrate as chlorine residual and yet may possess little or no disinfection potential.
5. Compounds with polycyclic rings containing hydroxyl groups and compounds containing sulfur groups react readily with chlorine to form compounds which have little or no bactericidal potential, but which still titrate as chlorine residual.
6. To achieve low bacterial counts in the presence of interfering organic compounds, additional chlorine and longer contact times will be required.

While these conclusions must be considered preliminary until additional work is performed to substantiate the reported observations, they nevertheless provide insight into the chlorination process. From the results of this work, it is easy to see why the efficiency of chlorination at plants with the same effluent characteristics can be quite different. Clearly, it is not the value of the BOD or COD that is significant, but the nature of the compounds that make up the measured values. Thus the nature of the treatment process used in any plant will also have an effect on the chlorination process.

Characteristics of the microorganisms Another important variable in the chlorination process is the age of the microorganisms. For example, in the study by Sung that has just been described [37], it was found that there was a

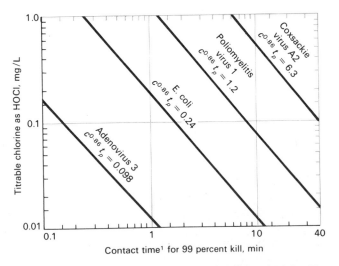

Figure 7-18 Concentration of chlorine as HOCl required for 99 percent kill of *E. coli* and three enteric viruses at 0 to 6°C [2]. Note: mg/L = g/m³.

noticeable difference in the resistance of bacterial cultures to chlorine. For a young bacterial culture (1 d old or less) with a chlorine dosage of 2 mg/L, only 1 min was needed to reach a low bacterial number. When the bacterial culture was 10 d old or more, approximately 30 min was required to achieve a comparable reduction for the same applied chlorine dosage. It is likely that the resistance offered by the polysaccharide sheath, which the microorganisms develop as they age, accounts for this observation. In the activated-sludge treatment process, the operating mean cell residence time, which to some extent is related to the age of the bacterial cells in the system, will thus affect the performance of the chlorination process (see Chap. 9).

In view of the renewed interest in wastewater reclamation, the viricidal efficiency of the chlorination process is of great concern. Unfortunately, definitive data on this subject are not available at present. Some representative data on the effectiveness of chlorine in killing *E. coli* and three enteric viruses are reported in Fig. 7-18. From the evidence available on the viricidal effectiveness of the chlorination process, it appears that chlorination beyond the breakpoint to obtain free chlorine will be required to kill many of the viruses of concern. Where breakpoint chlorination is used, it will be necessary to dechlorinate the treated wastewater before reuse to reduce any residual toxicity that may remain after chlorination. Additional details may be found in Ref. 2.

7-6 DECHLORINATION

Dechlorination is the practice of removing the total combined chlorine residual that exists after chlorination. Effluent requirements for the removal of the total combined chlorine residual, with the objective of reducing the toxic effects of chlorinated effluents on receiving-stream biota (see Ref. 33), are a relatively recent criterion imposed by regulatory agencies in certain parts of the United States (e.g., the San Francisco Bay area). Where effluent requirements are applicable, or where dechlorination is used as a polishing step following the breakpoint chlorination process for the removal of ammonia nitrogen, sulfur dioxide is a leading candidate for dechlorination. Activated carbon has also been used. Both these means are discussed in this section. Other chemicals that have been used are sodium sulfite (Na_2SO_3) and sodium metabisulfite ($Na_2S_2O_5$).

Toxicity of Chlorine Residuals

Chlorination is one of the most commonly used methods for the destruction of pathogenic and other harmful organisms that may endanger human health. As noted in the previous discussion, however, certain organic constituents in wastewater interfere with the chlorination process. Many of these organic compounds may react with the chlorine to form toxic compounds that can have long-term adverse effects on the beneficial uses of the waters to which they are discharged. The controversy concerning the possible formation of carcinogenic by-products is also unresolved (as of 1977).

In 1971, the Michigan Department of Natural Resources [22] conducted studies at four municipal wastewater-treatment plants in Michigan to determine the effects of discharging chlorinated wastewater to the environment and found that chlorine residuals were toxic to both minnows and trout at distances up to 0.8 mile downstream from the effluent-discharge points. Concurrently, Esvelt et al. [9] reported that chlorine increased the toxicity of municipal wastewaters that had undergone either primary activated-sludge treatment or chemical-precipitation treatment. To minimize the effects of these potentially toxic chlorine residuals on the environment, it has been found necessary to dechlorinate wastewater treated with chlorine [33].

Analysis of Dechlorination

Sulfur dioxide Sulfur dioxide gas successively removes free chlorine, monochloramine, dichloramine, nitrogen trichloride, and poly-n-chlor compounds. When sulfur dioxide is added to wastewater, the following reactions occur [25]:

Reactions with chlorine:

$$SO_2 + H_2O \rightarrow HSO_3^- + H^+ \tag{7-55}$$

$$HOCl + HSO_3^- \rightarrow Cl^- + SO_4^= + 2H^+ \tag{7-56}$$

$$SO_2 + HOCl + H_2O \rightarrow Cl^- + SO_4^= + 3H^+ \tag{7-57}$$

Reaction with chloramines:

$$SO_2 + H_2O \rightarrow HSO_3^- + H^+ \tag{7-58}$$

$$NH_2Cl + HSO_3^- + H_2O \rightarrow Cl^- + SO_4^= + NH_4^+ + H^+ \tag{7-59}$$

$$SO_2 + NH_2Cl + 2H_2O \rightarrow Cl^- + SO_4^= + NH_4 + 2H^+ \tag{7-60}$$

For the overall reaction between sulfur dioxide and chlorine (Eq. 7-57), the stoichiometric weight ratio of sulfur dioxide to chlorine is 0.9 : 1. In practice, it has been found that about 1.0 ppm of sulfur dioxide will be required for the dechlorination of 1.0 ppm of chlorine residue (expressed as Cl_2). Because the reactions of sulfur dioxide with chlorine and chloramines are nearly instantaneous, contact time is not usually a factor and contact chambers are not used, but rapid and positive mixing at the point of application is an absolute requirement.

The ratio of free chlorine to the total combined chlorine residual before dechlorination determines whether the dechlorination process is partial or proceeds to completion. A ratio of less than 85 percent normally indicates that significant organic nitrogen is present and interferes with the free residual chlorine process.

In most situations, sulfur dioxide dechlorination is a very reliable unit process in wastewater treatment, provided that the precision of the combined chlorine residual monitoring service is adequate. Excess sulfur dioxide dosages should be avoided, not only because of the chemical wastage, but also because of the oxygen demand exerted by the excess sulfur dioxide. The relatively slow reaction between excess sulfur dioxide and dissolved oxygen is given by the following expression [25]:

$$HSO_3^- + 0.5O_2 \rightarrow SO_4^= + H^+ \qquad (7\text{-}61)$$

The result of this reaction is a reduction in the dissolved oxygen contained in the wastewater, a corresponding increase in the measured BOD and COD, and a possible drop in the pH. All these effects can be eliminated by proper control of the dechlorination system.

Sulfur dioxide dechlorination systems are similar to chlorination systems because sulfur dioxide equipment is interchangeable with chlorination equipment. The components of these systems are discussed in Chap. 8. The key control parameters of this process are (1) proper dosage based on precise (amperometric) monitoring of the combined chlorine residual and (2) adequate mixing at the point of application of sulfur dioxide.

Activated carbon Carbon adsorption for dechlorination provides complete removal of both combined and free residual chlorine [17, 26]. When activated carbon is used for dechlorination, the following reactions occur [26]:

Reactions with chlorine:

$$C + 2Cl_2 + 2H_2O \rightarrow 4HCl + CO_2 \qquad (7\text{-}62)$$

Reactions with chloramines:

$$C + 2NH_2Cl + 2H_2O \rightarrow CO_2 + 2NH_4^+ + 2Cl^- \qquad (7\text{-}63)$$

$$C + 4NHCl_2 + 2H_2O \rightarrow CO_2 + 2N_2 + 8H^+ + 8Cl^- \qquad (7\text{-}64)$$

Granular activated carbon is used in either a gravity or pressure filter bed. If carbon is to be used solely for dechlorination, it must be preceded by an activated-carbon process for the removal of other constituents susceptible to removal by activated carbon. In treatment plants where granular activated carbon is used to remove organics, either the same or separate beds can be used for dechlorination, and regeneration will be feasible.

Because granular carbon in column applications has proved to be very effective and reliable, activated carbon should be considered where dechlorination is required. However, this method is quite expensive. It is expected that the primary application of activated carbon for dechlorination will be in situations where high levels of organic removal are also required.

7-7 DISINFECTION WITH OZONE

Ozone, a strong oxidizing agent that has similar bactericidal properties to chlorine, has been found to be equal or superior to chlorine in its viricidal effects (viral inactivation).

Description

Although ozone historically has been used for disinfection primarily in the field of water treatment, recent advances in ozone generation and solution technology have made it economically more competitive than in the past for wastewater disinfection [10, 27]. Ozone can also be used in wastewater treatment for odor control and in advanced wastewater treatment for the removal of soluble refractory organics, in lieu of the carbon-adsorption process.

Ozone is produced when a high voltage is imposed across a discharge gap in the presence of a gas containing oxygen [27]. Because ozone is a relatively unstable gas, it is generated onsite from air or pure oxygen. Pure-oxygen feed can be obtained from a separate oxygen generation system that includes recycle of the oxygen from the ozone contact chamber back to the ozone generation system (see Fig. 7-19); or oxygen can be used with a pure-oxygen activated-sludge system where oxygen is vented from the ozone contactor and recycled as a source for the activated-sludge system. For many treatment plants, ozonation used in conjunction with the pure-oxygen activated-sludge process appears to be the most economical alternative for disinfection.

Analysis

When ozone is added to water, it rapidly reverts to oxygen as follows:

$$2O_3 \rightarrow 3O_2 \qquad (7\text{-}65)$$

Because of this phenomenon, no chemical residual persists in the treated effluent

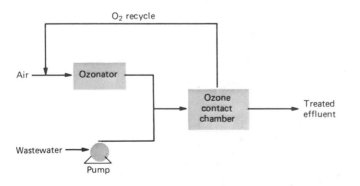

Figure 7-19 Closed-loop ozonation system with oxygen recycle.

that may require removal as is the case with chlorine residuals. Ozonation does not produce dissolved solids and is not affected by the ammonium ion or pH influent to the process. Aeration of the wastewater deriving from the use of ozone for disinfection is an added benefit. For these reasons, ozonation is considered a viable alternative to either chlorination or hypochlorination, especially where dechlorination may be required.

7-8 OTHER CHEMICAL APPLICATIONS

In addition to the major applications of chemicals discussed in this chapter, a number of other applications are occasionally encountered in the collection, treatment, and disposal of wastewater. The more important of these applications and the chemicals used are identified in Table 7-11. As shown, chlorine is by

Table 7-11 Additional chemical applications in wastewater collection, treatment, and disposal

Application	Chemicals used[a]	Remarks
Collection:		
Slime-growth control	Cl_2, H_2O_2	Control of fungi and slime-producing bacteria
Corrosion control (H_2S)	Cl_2, H_2O_2	Control brought about by destruction of H_2S in sewers
Odor control	Cl_2, H_2O_2, O_3	Especially in pumping stations and long, flat sewers
Treatment:		
Grease removal	Cl_2	Added before preaeration
BOD reduction	Cl_2, O_3	Oxidation of organic substances
Ferrous sulfate oxidation	Cl_2[b]	Production of ferric sulfate and ferric chloride
Filter-ponding control	Cl_2	Residual at filter nozzles
Filter-fly control	Cl_2	Residual at filter nozzles, used during fly season
Sludge-bulking control	Cl_2, H_2O_2, P	Temporary control measure
Digester supernatant oxidation	Cl_2	
Digester and Imhoff tank		
foaming control	Cl_2	
Ammonia oxidation	Cl_2	Conversion of ammonia to nitrogen gas (see Chap. 15)
Odor control	Cl_2, H_2O_2, O_3	
Oxidation of refractory		
organic compounds	O_3	
Disposal:		
Bacterial reduction	Cl_2, H_2O_2, O_3	Plant overflows, storm water
Odor control	Cl_2, H_2O_2, O_3	

[a] Cl_2 = chlorine, H_2O_2 = hydrogen peroxide, O_3 = ozone, P = polymer.
[b] $6FeSO_4 \cdot 7H_2O + 3Cl_2 \rightarrow 2FeCl_3 + 2Fe_2(SO_4)_3 + 42H_2O$.

far the most commonly used chemical, although hydrogen peroxide is gaining in popularity. The effectiveness of the various chemical additions is site-specific; so optimum dosage requirements are unavailable. Because chlorine has been used extensively, however, some representative dosage ranges have been established, and these are given in Chap. 8.

DISCUSSION TOPICS AND PROBLEMS

7-1 To aid sedimentation in the primary settling tank, 25 g/m^3 of ferrous sulfate (FeSO$_4 \cdot$7H$_2$O) is added to the wastewater. Determine the minimum alkalinity required to react initially with the ferrous sulfate. How many grams of lime should be added as CaO to react with the Fe(HCO$_3$)$_2$ and the dissolved oxygen in the wastewater to form insoluble Fe(OH)$_3$?

7-2 Assume that 50 kg of (a) alum (mol wt 666.7) and (b) ferrous sulfate and lime as Ca(OH)$_2$ is added per 4000 m^3 of wastewater. Also assume that all insoluble and very slightly soluble products of the reactions with the exception of 15 mg/L CaCO$_3$, are precipitated as sludge. How many kg of sludge/1000 m^3 will result in each case?

7-3 Raw wastewater is to be treated chemically for suspended-solids and phosphorus removal through coagulation and sedimentation. The wastewater characteristics are as follows: $Q = 0.75$ m^3/s; orthophosphorus = 10 mg/L as P; alkalinity = 200 mg/L expressed as CaCO$_3$ [essentially all due to the presence of Ca(HCO$_3$)$_2$]; total suspended solids = 220 mg/L.

    (a) Determine the sludge production in kg dry wt/d and m^3/d under the following conditions: (1) Alum (Al$_2$(SO$_4$)$_3 \cdot$ 14.3 H$_2$O) dosage of 150 mg/L; (2) 100 percent removal of orthophosphorus as insoluble AlPO$_4$; (3) 95 percent removal of original TSS; (4) all alum not required for reaction with phosphate reacts with alkalinity to form Al(OH)$_3$, which is 100 percent removed; (5) wet sludge has a water content of 93 percent and a specific gravity of 1.04.

    (b) Determine the sludge production in a kg dry wt/day and m^3/d under the following conditions: (1) Lime (Ca(OH)$_2$) dosage of 450 mg/L to give PH of approximately 11.2; (2) 100 percent removal of orthophosphorus as insoluble hydroxylapatite (Ca$_{10}$(PO$_4$)$_6$(OH)$_2$); (3) 95 percent removal of original TSS; (4) added lime (i) reacts with phosphate, (ii) reacts with all alkalinity to form CaCO$_3$, 20 mg/L of CaCO$_3$ is soluble and remains in solution and the rest is 100 percent removed, and (iii) remainder stays in solution; (5) wet sludge has a water content of 92 percent and a specific gravity of 1.05.

    (c) Determine the net increase in calcium hardness in mg/L as CaCO$_3$ for the treatment specified in part b. (Courtesy E. Foree.)

7-4 Compute the equilibrium carbon dioxide level in a quiescent high mountain reservoir. Assume that the atmosphere contains 0.03 percent CO$_2$ and that the temperature is 10°C.

7-5 The data in the table were obtained from a test program designed to evaluate a new diffused-air aeration system. Using these data, determine the value of $K_L a$ at 20°C and the equilibrium dissolved-oxygen concentration in the test tank. The test program was conducted using tap water at a temperature of 24°C.

C, mg/L	1.5	2.7	3.9	4.8	6.0	7.0	8.2
dC/dt, mg/L · h	8.4	7.5	5.3	4.9	4.2	2.8	2.0

7-6 If the volume of the test tank used to evaluate the aeration system in Prob. 7.5 were equal to 100,000 L and the air flowrate were equal to 2000 L/m, determine the oxygen-transfer efficiency at 20°C and 760 mm Hg.

7-7 Using the equation developed in Example 7-1, estimate the air flowrate in m^3/s required to increase the oxygen content of chlorinated effluent for zero to 4 mg/L. The effluent flowrate is equal to 0.25 m^3/s. Assume that the transfer efficiency is 6 percent and the temperature is 15°C. What is the air requirement when the temperature is 25°C?

7-8 Laboratory tests were conducted on a waste containing 50 mg/L phenol. Four jars containing 1 liter of the waste were dosed with powdered activated carbon. When equilibrium was reached, the contents of each jar were analyzed for phenol. The results are shown in the following table. Determine the constants a and b in the Langmuir equation and the dosage required to yield an effluent with a phenol concentration of 0.10 mg/L.

Jar	Carbon added, g	Equilibrium conc. of phenol, mg/L
1	0.5	6.0
2	0.64	1.0
3	1.0	0.25
4	2.0	0.08

7-9 The data in the table were obtained in a series of laboratory tests performed on an effluent from a secondary wastewater treatment process.

Chlorine dosage, mg/L	Residual fecal coliform count, no./100 ml; contact time, min.		
	15	30	60
1	10,000	2000	500
2	3,000	350	90
4	400	65	20
6	110	30	12
8	54	19	6
10	30	10	1

(*a*) Plot the data on semilog paper. Using these data, determine the value of the exponent n and the constant in Eq. 7-36 for residual coliform counts of 200/100 ml and 1000/100 ml.

(*b*) The following data apply to the wastewater treatment plant.

Item	May–October	November–April
Average flow, m^3/d	20000	26000
Peak daily flow, m^3/d	40000	52000
Maximum permissible fecal coliform count in effluent, no./100 ml	200	1000

Determine the required volume in m^3 of a chlorine contact chamber designed to provide 30-min contact at the average winter flow. Using the equations developed in part *a*, determine the minimum dosage required in mg/L to give the required kill under each of the four flow conditions given above. Assuming that the yearly chlorine requirement can be computed on the basis of the average flow for each of the two 6-month periods, determine the minimum yearly chlorine requirement in kilograms. (*Courtesy E. Foree.*)

7-10 The chlorine residuals measured when various dosages of chlorine were added to a wastewater are given below. Determine (a) the breakpoint dosage and (b) the design dosage to obtain a residual of 0.75 mg/L free available chlorine.

Dosage, mg/L	0.1	0.5	1.0	1.5	2.0	2.5	3.0
Residual, mg/L	0.0	0.4	0.8	0.4	0.4	0.9	1.4

7-11 Derive a rate expression from Eq. 7-54 that can be used to assess the efficiency of a complete mix continuous-flow stirred-tank reactor as a chlorine contact basin.

7-12 Using Eq. 7-54 and the rate expression derived in Prob. 7-11, compare the volume required for a continuous-flow stirred-tank reactor to that for a plug-flow reactor to achieve a 10^4 reduction in the coliform count of a treated effluent. Assume that in both cases the chlorine residual to be maintained is 5 mg/L.

7-13 Determine the amount of activated carbon that would be required per year to dechlorinate treated effluent containing a chlorine residual of 5 mg/L (as Cl_2) from a plant with an average flowrate of 3800 m³/d. What dosage of sulfur dioxide would be required?

7-14 Discuss the advantages of using ozone as a disinfectant. Cite a minimum of four references in your discussion.

REFERENCES

1. Allen, J. B., R. S. Joyce, and R. H. Kasch: Process Design Calculations for Adsorption from Liquid in Fixed Beds of Granular Activated Carbon, *J. Water Pollut. Control Fed.*, vol. 39, no. 2, 1967.
2. Berg, G.: The Virus Hazard in Water Supplies, *J. New England Water Works Assoc.*, vol. 78, p. 79, 1964.
3. Chick, H.: Investigation of the Laws of Disinfection, *J. Hygiene*, vol. 8, p. 92, 1908.
4. Collins, H. F.: *Effects of Initial Mixing and Residence Time Distribution on the Efficiency of the Wastewater Chlorination Process*, paper presented at the California State Department of Health Annual Symposium, Berkeley and Los Angeles, Calif. (May 1970).
5. Collins, H. F.: *Kinetics of Wastewater Chlorination*, Ph.D. thesis, Department of Civil Engineering, University of California, Berkeley, 1971.
6. Culp, G. L.: Chemical Treatment of Raw Sewage/1 and 2, *Water Wastes Eng.*, vol. 4, nos. 7, 10, 1967.
7. Danckwertz, P. V.: Significance of Liquid Film Coefficients in Gas Absorption, *J. Ind. Eng. Chem.*, vol. 43, p. 1460, 1951.
8. Eliassen, R., and J. Trump: High-Energy Electrons Offer Alternative to Chlorine, *Calif. Water Pollut. Control Assoc. Bull.*, vol. 10, no. 3, January 1974.
9. Esvelt, L. A., Kaufman, W. J., and R. E. Selleck: *Toxicity Removal from Municipal Wastewater*, University of California, Sanitary Engineering Research Laboratory Report 71-7, 1971.
10. Evans, F. L.: *Ozone for Water and Wastewater Treatment*, Ann Arbor Science Publishers, Inc., Ann Arbor, Mich., 1972.
11. Fair, G. M., et al.: The Behavior of Chlorine as a Water Disinfectant, *J. Am. Water Works Assoc.*, vol. 40, p. 1051, 1948.
12. Fair, G. M., J. C. Geyer, and D. A. Okun: *Water and Wastewater Engineering*, vol. 2, Wiley, New York, 1968.
13. Graber, S. D.: Discussion/communication by V. Kothandaraman and R. L. Evans on Hydraulic Model Studies of Chlorine Contact Tanks, *J. Water Pollut. Control Fed.*, vol. 44, no. 10, 1972.

14. Higbie, R.: The Rate of Absorption of Pure Gas into a Still Liquid during Short Periods of Exposure, *Trans. Am. Inst. Chem. Eng.*, vol. 31, p. 365, 1935.
15. Huang, J. Y.: Selecting Aerators for Wastewater Treatment, *Plant Eng.*, Dec. 24, 1975.
16. King, H. R.: Oxygen Absorption in Spiral Flow Tanks, *Sewage Ind. Wastes*, vol. 27, nos. 8–10, 1955.
17. Kovach, J. L.: Activated Carbon Dechlorination, *Ind. Water Eng.*, pp. 30–32, October/November 1973.
18. Lager, J. A., and W. G. Smith: *Urban Stormwater Management and Technology: An Assessment*, U.S. Environmental Protection Agency Report 670/2-74-040, Cincinnati, Ohio, 1974.
19. Lewis, W. K., and W. C. Whitman: Principles of Gas Adsorption, *Ind. Eng. Chem.*, vol. 16, p. 1215, 1924.
20. Metcalf, L., and H. P. Eddy: *American Sewerage Practice*, vol. III, 3d ed., McGraw-Hill, New York, 1935.
21. Metcalf & Eddy, Inc.: *Wastewater Engineering: Collection, Treatment, Disposal*, McGraw-Hill, New York, 1972.
22. Michigan Department of Natural Resources for the Environmental Protection Agency Bureau of Water Management: Chlorinated Municipal Wastes, Toxicities to Rainbow Trout and Fathead Minnows, *Water Pollut. Control. Res. Ser.*, no. 18050 GZZ, December 1971.
23. Pelczar, M. J., Jr., and R. D. Reid: *Microbiology*, 2d ed., McGraw-Hill, New York, 1965.
24. Perry, J. H.: *Chemical Engineers' Handbook*, 4th ed., McGraw-Hill, New York, 1963.
25. *Physical-Chemical Wastewater Treatment Plant Design*, U.S. Environmental Protection Agency, Technology Transfer Seminar Publication, 1973.
26. *Process Design Manual for Nitrogen Control*, U.S. Environmental Protection Agency, Technology Transfer, October 1975.
27. Rosen, H. M.: Ozone Generation and Its Economical Application in Wastewater Treatment, in F. L. Evans, III (ed.), *Ozone for Water and Wastewater Treatment*, Ann Arbor Science Publishers, Inc., Ann Arbor, Mich., 1972.
28. Sawyer, C. N., and W. O. Lynch: Effects of Detergents on Oxygen Transfer in Bubble Aeration, *J. Water Pollut. Control Fed.*, vol. 32, no. 1, 1960.
29. Schroeder, E. D.: *Water and Wastewater Treatment*, McGraw-Hill, New York, 1977.
30. Selleck, R. E., H. F. Collins, and G. C. White: *Kinetics of Wastewater Chlorination in Continuous Flow Processes*, paper presented at International Association of Water Pollution, Research Conference, San Francisco, July 29, 1970.
31. Selleck, R. E., H. F. Collins, and G. C. White: *Problems in Obtaining Adequate Disinfection in Wastewater Treatment Plants*, paper presented at Special Disinfection Symposium, Sanitary Engineers, ASCE, University of Massachusetts, Amherst, July 1970.
32. Shaw, D. J.: *Introduction to Colloid and Surface Chemistry*, Butterworth, London, 1966.
33. Stone, R. W., W. J. Kaufman, and A. J. Horne: *Long-Term Effects of Toxicants and Biostimulants on the Waters of Central San Francisco Bay*, University of California, Sanitary Engineering Research Laboratory Report 73-1, May 1973.
34. Stumm, W., and J. J. Morgan: *Aquatic Chemistry*, Wiley-Interscience, New York, 1970.
35. Stumm, W., and J. J. Morgan: Chemical Aspects of Coagulation, *J. Am. Water Works Assoc.*, vol. 54, no. 8, 1962.
36. Stumm, W., and C. R. O'Melia: Stoichiometry of Coagulation, *J. Am. Water Works Assoc.*, vol. 60, p. 514, 1968.
37. Sung, R. D.: Effects of Organic Constituents in Wastewater on the Chlorination Process, Ph.D. thesis, Department of Civil Engineering, University of California, Davis, Calif., 1974.
38. Thomas, A. W.: *Colloid Chemistry*, McGraw-Hill, New York, 1934.
39. U.S. Environmental Protection Agency: *Manual on Activated Carbon Adsorption*.
40. Weber, W. J., Jr., and J. C. Morris: Adsorption in Heterogeneous Aqueous Systems, *J. Am. Water Works Assoc.*, vol. 56, no. 4, 1964.
41. Weber, W. J., Jr.: *Physiochemical Processes for Water Quality Control*, Wiley-Interscience, New York, 1974.
42. White, G. C.: *Handbook of Chlorination*, Van Nostrand Reinhold Company, New York, 1972.

EIGHT

DESIGN OF FACILITIES FOR PHYSICAL AND CHEMICAL TREATMENT OF WASTEWATER

The purpose of this chapter is to discuss the design of the unit operations and processes that were described in Chaps. 6 and 7. The principal unit operations and processes and their functions as applied to the treatment of wastewater are reported in Table 8-1 (see also Fig. 6-1). As shown, physical operations are used for the removal of coarse solids, suspended and floating solids, and grease, and for the pumping of sludge. Chemical processes are used for the removal of suspended and colloidal solids by precipitation, disinfection of the wastewater, and control of odors.

Although each operation and process identified in Table 8-1 will be discussed in this chapter, the ones that are most commonly encountered in the design of wastewater-treatment facilities will be considered in greater detail: racks and screens, grit chambers, sedimentation, filtration, and chlorination. The processing of sludge produced from primary treatment operations and processes as well as that produced by biological and chemical secondary treatment processes, is discussed separately in Chap. 11 not only because of its importance in wastewater engineering but also because of its special handling requirements.

8-1 RACKS AND COARSE SCREENS

The first step in wastewater treatment is the removal of the coarse solids. The usual procedure is to pass the influent wastewater through racks or coarse screens. Racks are almost always used for the removal of very large objects. Where racks or coarse screens are not used, comminutors (see Sec. 8-2) can be used to grind up the coarse solids without removing them from the flow.

Table 8-1 Functions of various unit operations and processes used in wastewater-treatment flowsheets for secondary treatment

Operation or process	Function	See section
Racks and coarse screens	Removal of coarse solids by interception. Considered a pretreatment operation	8-1
Comminutors and grinders	Grinding of solids remaining after coarse screening. Considered a pretreatment operation	8-2
Grit chambers	Removal of grit, sand, and gravel, usually following comminution. Considered a pretreatment operation	8-3
Flow equalization	Equalization of flow and mass loadings of BOD and suspended solids on subsequent treatment facilities	8-4
Other pretreatment operations:		8-5
Skimming	Removal of lighter floating solids, including grease, soap, cork, wood, vegetable debris, etc.	
Flocculation	Improvement of settling characteristics of suspended solids	
Preaeration	Improvement of hydraulic distribution. Replenishment of dissolved oxygen	
Sedimentation	Removal of settleable solids and floating material. Principal operation used in the primary treatment of wastewater	8-6
Other solids-removal operations and units:		8-7
Flotation	Used as replacement for gravity sedimentation	
Fine screens	Used as replacement for gravity sedimentation	
Imhoff and septic tanks	Used for the removal of solids from small residential areas or from individual residences	
Chemical precipitation	Removal of settleable and colloidal solids and phosphorus. Used as the first step in the independent physical-chemical treatment of wastewater	8-8
Granular-medium filtration	Removal of residual suspended solids after biological or chemical secondary treatment and from stabilization ponds	8-9
Microscreening	Removal of residual suspended solids after biological or chemical secondary treatment and from stabilization ponds	8-10
Chlorination	Used principally for the disinfection of wastewater; also used for odor control	8-11
Odor control[a]	Various operations and processes used for the removal and elimination of odors emanating from various treatment facilities	8-12

[a] Not strictly defined as an operation or process.

Table 8-2 Typical design information for hand-cleaned and mechanically cleaned bar racks

Item	Hand-cleaned	Mechanically cleaned
Bar size:		
Width, mm	5–15	5–15
Depth, mm	25–75	25–75
Clear spacing between		
bars, mm	25–50	15–75
Slope from vertical, deg	30–45	0–30
Approach velocity, m/s	0.3–0.6	0.6–1.0
Allowable head loss, mm	150	150

Note: mm × 0.03937 = in
        m/s × 3.2808 = ft/s

Racks

Bar racks may be hand cleaned or mechanically cleaned. Characteristics of the two types are compared in Table 8-2. Details on each type of rack and some of the factors that must be considered in the design of rack installations are presented in the following discussion.

Hand-cleaned bar racks Hand-cleaned bar racks are frequently used ahead of pumps in small wastewater pumping stations. In the past they have been used at the head works to small wastewater-treatment plants. The tendency in recent years has been to provide mechanically cleaned racks, or comminutors, even for small installations, not only to minimize the manual labor required to clean the racks and remove and dispose of the rakings, but also to reduce flooding and overflows due to clogging.

    Where used, the length of the hand-cleaned rack should not exceed the distance that can be conveniently raked by hand. The rack bars are usually not less than 10 mm (3/8 in) thick by 50 mm (2 in) deep. They are welded to spacing bars located at the rear face, out of the way of the tines of the rake. A perforated drainage plate should be provided at the top of the rack where the rakings may be stored temporarily for drainage. A typical hand-cleaned rack is shown in Fig. 8-1.

    The rack channel should be designed to prevent the accumulation of grit and other heavy materials in the channel ahead of the rack and following it. The channel floor should be level or should slope downward through the screen without pockets to trap solids. Fillets may be desirable at the base of the sidewalls. The channel preferably should have a straight approach, perpendicular to the bar rack, to promote uniform distribution of screenable solids throughout the flow and on the rack.

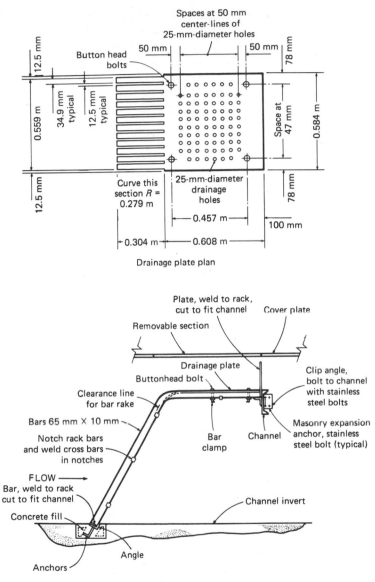

Drainage plate plan

Removable channel bar rack

Figure 8-1 Typical hand-cleaned bar rack. Note: mm × 0.03937 = in; m × 3.2808 = ft.

To provide adequate rack area for accumulation of screenings between racking operations, it is essential that the velocity of approach be limited to approximately 0.45 m/s (1.5 ft/s) at average flow. Additional area to limit the velocity may be obtained by widening the channel at the rack and by placing the rack on a flatter slope. As screenings accumulate, partially plugging the rack, the head will increase, submerging new areas for the flow to pass through. The structural design of the screen should be adequate to prevent collapse if it becomes completely plugged.

Mechanically cleaned bar racks Mechanically cleaned bar racks may be purchased from a number of manufacturers. The design engineer determines in advance the type of equipment to be used, the dimensions of the rack channel, the range of depth of flow in the channel, the clear spacing between bars, and the method of control of the rack. Either front-cleaned or back-cleaned racks may be obtained. Each type has advantages and disadvantages. Examples of front-cleaned and back-cleaned racks are shown in Fig. 8-2.

In the front-cleaned model (see Fig. 8-2a), the bars are fixed at the bottom but are supported by the traveling rake teeth above. This allows the bars to move, and where round bars are used, solids measuring up to two times the nominal clear spacing between the bars may pass through the rack.

The back-cleaned rack (see Fig. 8-2b) was developed to eliminate jamming due to obstructions at the base of the screen and has successfully overcome this problem at the expense of introducing two more. The rakes travel downward in back of the screen, free from obstructions. In one manufacturer's model they enter between the bars and project through them from behind; in another, they pass under the rack in a boot before starting up the front face of the rack. If there are large solids at the base of the rack, the rakes enter underneath them, pulling them upward and lifting them up the rack or rolling them out of the way without jamming.

Design of mechanically cleaned bar rack installations Two or more units should be installed so that one unit may be taken out of service for maintenance. Stop-log grooves should be provided ahead of, and behind, each rack so that the unit can be dewatered for painting, chain or cable replacement, replacement of teeth, removal of obstructions, straightening of bent bars, etc. If only one unit is installed, it is absolutely essential that a bypass channel with a manually cleaned rack be provided for emergency use. Flow through the bypass channel normally would be prevented by stop logs or a closed sluice gate.

The rack channel should be designed to prevent the settling and accumulation of grit and other heavy materials. The majority of racks use endless chains operating over sprockets to move the rakes. They are normally provided with "hand"-"off"-"automatic" controls. On "hand" position the rakes operate continuously. On "automatic" position they may be operated when the differential head loss increases above a certain minimum value or by a time clock. Operation by a time clock for a period, adjustable by the operator, out of a 15-minute

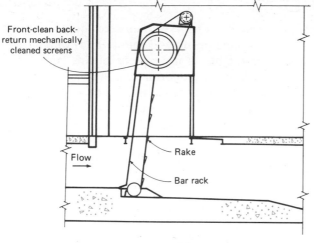

(a)

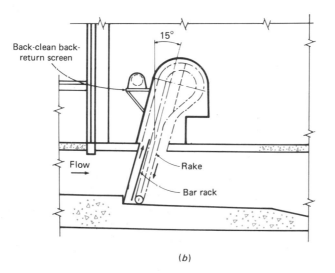

(b)

Figure 8-2 Typical mechanically cleaned bar racks. (*a*) Front-clean back-return. (*b*) Back-clean back-return.

cycle is recommended with either a high-water or high-differential contact that will place the rack in continuous operation when needed.

Sheet-metal enclosures with access doors are available for the head works of the screens above the operating flow level. They neatly enclose the mechanism and the screenings hopper, but they are a nuisance to the operator and are frequently omitted in below-ground screen chambers and areas not open to the general public. Enclosures should always be provided for racks located outdoors.

Coarse Screens

Early coarse screens were of the circular or disk type and were equipped with a perforated bronze screen plate with slotted openings 3 mm (1/8 in) wide or less. At present, comparatively few treatment plants use screens of this type. For a description of these early units, including more information on the quantity and character of screenings and data on removal efficiencies, the reader is referred to Ref. 11. Modern coarse screens are of the disk or drum type, with stainless-steel or nonferrous wire-mesh screen cloth. Typically, the openings vary from 6 to 20 mm (0.25 to 0.75 in) or more.

The disk type has a vertical circular screening surface that rotates on a horizontal shaft set slightly above the water surface. It is available in sizes from 1.2 to 5.5 m (4 to 18 ft) in diameter. The drum type revolves at about 4 r/min around a horizontal axis and operates slightly less than half submerged (see Fig. 6-3). The wastewater flows in one end of the drum and outward through the screen cloth. Drum screens are available in various sizes, from 1 to 1.5 m (39 in to 5 ft) in diameter and from 1.2 to 3.7 m (4 to 12 ft) in length.

In both the disk and drum types, the solids are raised above the liquid level by rotation of the screen and are backflushed into receiving troughs by high-pressure jets. With the finer-mesh cloth, effluent may be used for spray water.

Quantities of Screenings

The quantity of screenings collected for disposal varies, depending on the type of rack or screen used as well as on the sewer system and geographic location. For estimating purposes, the following values are suggested.

The quantity of screenings removed by bar racks usually varies from about 0.0035 to 0.0375 $m^3/10^3$ m^3 (0.5 to 5 ft^3/Mgal) of wastewater treated; the average is about 0.015 $m^3/10^3$ m^3 (2 ft^3/Mgal) [20]. The average quantity of screenings versus the size of openings between bars is shown in Fig. 8-3. In plants served by combined sewers, the quantity of screenings has been observed to increase greatly during periods of storm flow. The screenings removed by coarse screens have amounted to approximately 0.0375 to 0.225 $m^3/10^3$ m^3 (5 to 30 ft^3/Mgal) of wastewater treated, equivalent to 5 to 15 percent of suspended matter [11].

Disposal of Screenings

Means of disposal of screenings include (1) discharge to grinders or disintegrator pumps where they are ground and returned to the wastewater and (2) removal by hauling to disposal areas (landfill). In small installations, screenings may be disposed of by burial on the plant site. Alternatively, they may be disposed of with the municipal solid wastes. In large installations, incineration may be found appropriate.

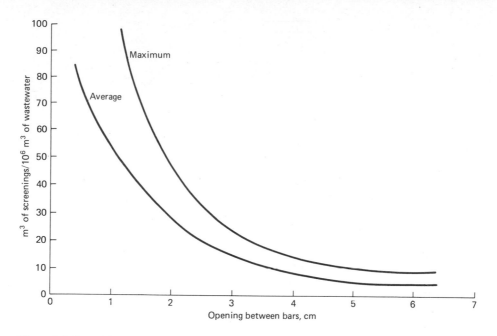

Figure 8-3 Quantities of screenings collected from mechanically cleaned bar racks. (*Adapted from FMC, Link Belt.*)

8-2 COMMINUTION

To improve the downstream operations and processes and to eliminate problems caused by the varied sizes of solids that are present in wastewater, the solids are often cut up into a smaller, more uniform size. Devices that are used to cut up (comminute) the solids in wastewater are known as comminutors.

Often, comminuting devices are used to cut up the material retained on the screens so that it may be returned to the flow stream for removal in the subsequent downstream treatment operations and processes. There is a wide divergence of views, however, on the advisability of doing this. One school of thought maintains that once material has been removed from wastewater it should not be returned, regardless of the form. The other school of thought maintains that once cut up, the solids are more easily handled in the downstream processes. The merits of each approach should be considered for every application.

Description

Although comminutors are now available from a number of manufacturers, the original device of this type was developed by the Chicago Pump Co. As shown in Fig. 8-4, the comminutor consists of a vertical revolving-drum screen with 6-mm (1/4-in) slots in small machines and 10-mm (3/8-in) slots in large machines.

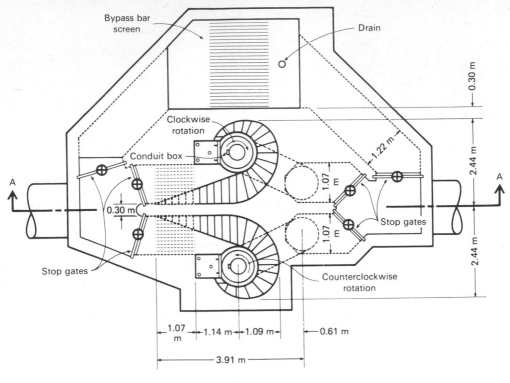

PLAN

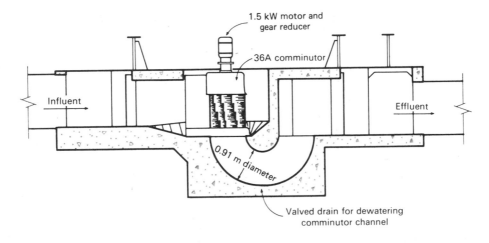

SECTION A–A

Figure 8-4 Plan and cross-sectional views of a comminutor installation. (*From FMC, Chicago Pump.*) Note: m × 3.2808 = ft; kW × 0.948 = hp.

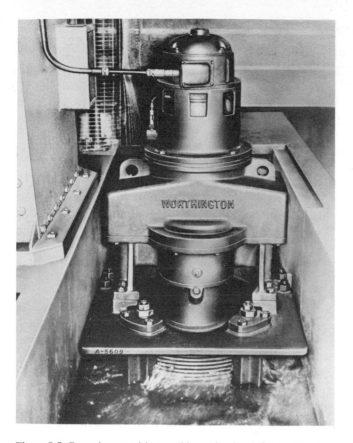

Figure 8-5 Comminutor with movable cutting head. (*From Worthington Corp.*)

Coarse material is cut by the cutting teeth and the shear bars on the revolving drum as the solids are carried past a stationary comb. The small sheared particles pass through the drum slots and out of a bottom opening through an inverted siphon and into the downstream channel.

Another type of comminuting device consists of a stationary semicircular screen grid mounted in a rectangular channel with rotating circular cutting disks (see Fig. 8-5). The grid intercepts the larger solids, while smaller solids pass through the space between the grid and cutting disks. The teeth are mounted on the rotating part, and the comb is located on the stationary screen grid.

Still another type consists of a semicircular vertical stationary stainless-steel screen with horizontal slots set in a rectangular channel concave to the flow. A motor-driven vertical arm, with cutting teeth, oscillates back and forth between the slots, conveying the screenings to the sides of the unit, where they are shredded between the oscillating cutting teeth and stationary cutter bars.

Application and Design

Comminuting devices may be preceded by grit chambers to prolong the life of the equipment and to reduce the wear on the cutting surfaces and on portions of the mechanism where there is a small clearance between moving and stationary parts. Frequently, they are installed in the wet well of pumping stations to protect the pumps against clogging by rags and large objects. They are used especially in smaller cities that are served by separate sanitary sewers carrying a minimum of grit. Provisions must be made to bypass comminutors in case flows exceed the capacity of the comminutor or in case there is a power or mechanical failure.

Because these units are complete in themselves, no detailed design is necessary. Manufacturers' data and rating tables for these units should be consulted for recommended channel dimensions, capacity ranges, upstream and downstream submergence, and power requirements.

8-3 GRIT CHAMBERS

Grit chambers are designed to remove grit, consisting of sand, gravel, cinders, or other heavy solid materials that have subsiding velocities or specific gravities substantially greater than those of the organic putrescible solids in wastewater. Grit also includes eggshells, bone chips, seeds, coffee grounds, and large organic particles, such as food wastes. Grit chambers are provided to protect moving mechanical equipment from abrasion and accompanying abnormal wear; to reduce formation of heavy deposits in pipelines, channels, and conduits; and to reduce the frequency of digester cleaning that may be required as a result of excessive accumulations of grit in such units. The removal of grit is essential ahead of centrifuges and heat-exchanger and high-pressure diaphragm pumps. On the other hand, where untreated sludge is to be dewatered on vacuum filters and incinerated, grit chambers that are far less efficient have given satisfactory service.

Grit chambers may be located ahead of all other units in treatment plants where removal of grit would facilitate operations. However, the installation of mechanically cleaned bar racks or comminutors ahead of grit chambers makes the operation of grit removal and cleaning facilities easier.

Locating grit chambers ahead of wastewater pumps, when it is desirable to do so, would normally involve placing them at considerable depth at added expense. It is therefore usually deemed more economical to pump the wastewater, including the grit, to grit chambers located at a convenient position ahead of the treatment-plant units, recognizing that the pumps may require greater maintenance.

Types

There are two general types of grit chambers: horizontal-flow and aerated. In the horizontal-flow type, the flow passes through the chamber in a horizontal direction and the straight-line velocity of flow is controlled by the dimensions of the unit or by the use of special weir sections at the effluent end. The aerated type consists of a spiral-flow aeration tank where the spiral velocity is controlled by the dimensions and the quantity of air supplied to the unit.

Horizontal-flow grit chambers In the past, most grit chambers were of the horizontal-flow, velocity-controlled type. These chambers were designed to maintain a velocity as close to 0.3 m/s (1.0 ft/s) as practical. Such a velocity will carry most organic particles through the chamber and will tend to resuspend any that settle but will permit the heavier grit to settle out.

The design of horizontal-flow grit chambers should be such that, under the most adverse conditions, the lightest particle of grit will reach the bed of the channel prior to its outlet end. Normally, grit chambers are designed to remove all grit particles that will be retained on a 65-mesh screen (0.21 mm diameter), although many chambers have been designed to remove grit particles retained on a 100-mesh screen (0.15 mm diameter). The length of channel will be governed by the depth required by the settling velocity and the control section, and the cross-sectional area will be governed by the rate of flow and by the number of channels. Allowance should be made for inlet and outlet turbulence. Representative design data for horizontal-flow grit chambers are presented in Table 8-3. The detailed design of horizontal-flow grit chambers is illustrated in the first edition of this text [12] and in Refs. 3, 11, and 20.

Table 8-3 Typical design information for horizontal-flow grit chambers

Item	Value	
	Range	Typical
Detention time, s	45–90	60
Horizontal velocity, m/s	0.25–0.4	0.3
Settling velocity for removal of:		
65-mesh material, m/min[a]	1.0–1.3	1.15
100-mesh material, m/min[a]	0.6–0.9	0.75
Head loss in control section as		
percent of depth in channel, %	30–40	36[b]
Allowance for inlet and outlet		
turbulence	$2D_m^c$–0.5 L^d	

[a] If the specific gravity of the grit is significantly less than 2.65, lower velocities should be used.
[b] For Parshall-flume control section.
[c] D_m = maximum depth in grit chamber.
[d] L = theoretical length of grit chamber.

   Note: m/s × 3.2808 = ft/s
                    m/min × 3.2808 = ft/min

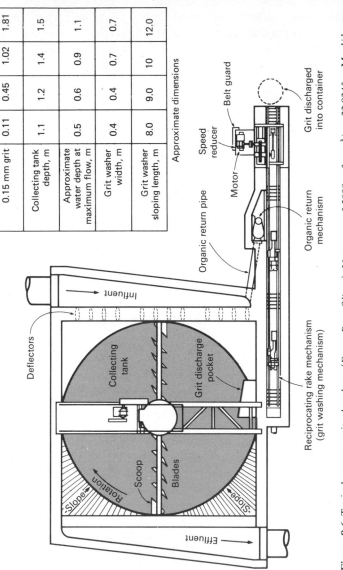

Collecting tank diameter, m	3.0	6.0	9.0	12.0
Max. flow, m³/s 0.21 mm grit	0.17	0.70	1.58	2.80
0.15 mm grit	0.11	0.45	1.02	1.81
Collecting tank depth, m	1.1	1.2	1.4	1.5
Approximate water depth at maximum flow, m	0.5	0.6	0.9	1.1
Grit washer width, m	0.4	0.4	0.7	0.7
Grit washer sloping length, m	8.0	9.0	10	12.0

Approximate dimensions

Figure 8-6 Typical square grit chamber. (*From Dorr-Oliver.*) Note: m × 3.2808 = ft; m³/s × 22.8245 = Mgal/d; mm × 0.03937 = in.

Square grit chambers Where square grit chambers, such as those shown in Fig. 8-6, are used, it is advisable to use two units. These chambers are designed on the basis of overflow rates that are dependent on particle size and the temperature of the wastewater. A typical set of design curves is shown in Fig. 8-7.

In square grit chambers, the solids are raked by a rotating mechanism to a sump at the side of the tank, from which they are moved up an incline by a reciprocating rake mechanism. While passing up the incline, organic solids are separated from the grit and flow back into the basin. By this method, a cleaner, dryer grit is obtained, comparable to the washed grit from separate grit washers.

The wear on removal equipment of the conveyor type, whether buckets, plows, or scrapers, has been considerable in the larger plants servicing areas drained by

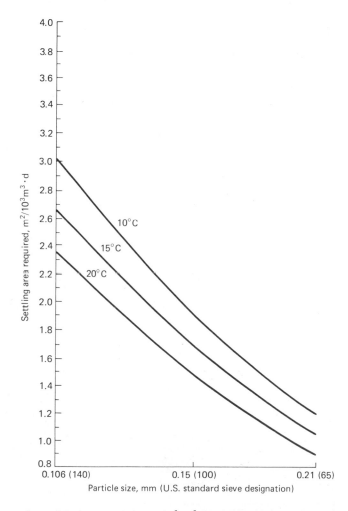

Figure 8-7 Area required per 10^3 m^3 for settling grit particles with a specific gravity of 2.65 in wastewater at indicated temperatures. (*From Dorr-Oliver.*)

combined sewers. For this reason, grab buckets of the clamshell type, operating on an overhead monorail system, have been installed at some plants for periodic cleaning as required. Chambers may or may not be dewatered during cleaning, but the flow should be shut off by closing the outlet gate.

Aerated grit chambers The discovery of grit accumulations in spiral-flow aeration tanks preceded by grit chambers led to the development of the aerated grit chamber. The excessive wear on grit-handling equipment and the necessity in most cases for separate grit-washing equipment with horizontal-flow grit chambers are two of the major factors contributing to the current popularity of the aerated grit chamber.

Aerated grit chambers are usually designed to provide detention periods of about 3 minutes at the maximum rate of flow. The cross section of the tank is similar to that provided for spiral circulation in activated-sludge aeration tanks, except that a grit hopper about 0.9 m (3 ft) deep with steeply sloping sides is located along one side of the tank under the air diffusers (see Fig. 8-8). The diffusers are located about 0.45 or 0.6 m (1.5 or 2 ft) above the normal plane of the bottom. Basic design data for aerated grit chambers are presented in Table 8-4. The design of aerated grit chambers is illustrated in Example 8-1.

The velocity of roll or agitation governs the size of particles of a given specific gravity that will be removed. If the velocity is too great, grit will be

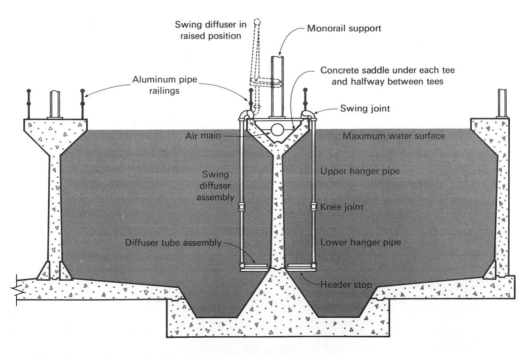

Figure 8-8 Typical section through aerated grit chamber.

Table 8-4 Typical design information for aerated grit chambers

Item	Value Range	Value Typical
Dimensions:		
Depth, m	2–5	
Length, m	7.5–20	
Width, m	2.5–7.0	
Width-depth ratio	1:1–5:1	2:1
Detention time at peak		
flow, min	2–5	3
Air supply,		
$m^3/m \cdot min$ of length	0.15–0.45	0.3
Grit and scum quantities:		
Grit, $m^3/10^3 \ m^3$	0.004–0.200	0.015

Note: $m \times 3.2808 = ft$
$m^3/m \cdot min \times 10.7639 = ft^3/ft \cdot min$
$m^3/10^3 \ m^3 \times 133.6806 = ft^3/Mgal$

carried out of the chamber; if it is too small, organic material will be removed with the grit. Fortunately, the quantity of air is easily adjusted. With proper adjustment, almost 100 percent removal will be obtained, and the grit will be well washed. Wastewater will move through the tank in a helical path (see Fig. 8-9) and will make two to three passes across the bottom of the tank at maximum flow and more at lesser flows. Wastewater should be introduced in the direction of the roll. To determine the required head loss through the chamber, the expansion in volume caused by the air must be considered.

Many aerated grit chambers have been provided with means for grit removal by grab buckets, traveling on monorails, centered over the grit collection and storage trough (see Fig. 8-10). Other installations are equipped with chain-and-bucket conveyors, running the full length of the storage troughs, which move the grit to one end of the trough and elevate it above the wastewater level in a continuous operation. Screw conveyors, jet pumps, and air lifts have also been used.

Example 8-1: Design of aerated grit chamber Design an aerated grit chamber for the treatment of municipal wastewater. The average flowrate is 0.5 m^3/s (11.4 Mgal/d). Assume that the wastewater characteristics are similar to the medium-strength wastewater described in Table 3-5, and that the sustained-peaking-factor curve given in Fig. 2-6 is applicable.

SOLUTION

1 Establish the peak flowrate for design. Assume that the aerated grit chamber will be designed for the 1-day sustained peak flowrate. From Fig. 2-6 the peaking factor is found to be 2.75, and the peak design flowrate is

$$Peak \ flowrate = 0.5 \ m^3/s \times 2.75 = 1.38 \ m^3/s$$

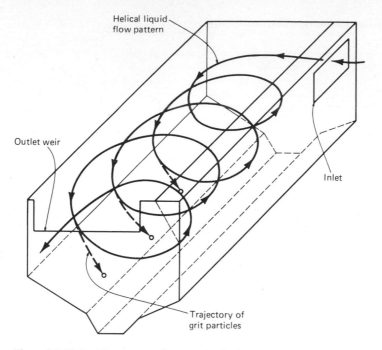

Helical liquid flow pattern

Outlet weir

Inlet

Trajectory of grit particles

Figure 8-9 Helical flow pattern in aerated grit chamber.

2 Determine the aerator volume. Because it will be necessary to drain the chamber periodically for routine maintenance, use two chambers. Assume that the average detention time at the peak flowrate is 3 min.

$$\text{Aeration chamber volume, m}^3 = (1/2)(1.38 \text{ m}^3/\text{s})3 \text{ min} \times 60 \text{ s/min}$$
$$= 124.2 \text{ m}^3$$

3 Determine the dimensions of the aeration basin. Use a depth-width ratio of $1:2$ and assume that the width is 3 m.
   a   Depth $= 1.2(3 \text{ m}) = 3.6 \text{ m}$
   b $\text{Length} = \dfrac{\text{volume}}{\text{width} \times \text{depth}} = \dfrac{124.2 \text{ m}^3}{3 \text{ m} \times 3.6 \text{ m}} = 11.5 \text{ m}$
   c   Increase the length by 15 percent to account for inlet and exit conditions.

$$\text{Adjusted length} = 11.5 \text{ m}(1.15) = 13.2 \text{ m}$$

4 Determine the air-supply requirement. Assume that 0.04 $\text{m}^3/\text{min} \cdot \text{m}$ of length will be adequate.
   a   Air required (length basis) $= 13.2 \text{ m} \times 0.04 \text{ m}^3/\text{min} \cdot \text{m}$

$$= 0.53 \text{ m}^3/\text{min}$$

5 Estimate the quantity of grit that must be handled. Assume a value of $50 \times 10^{-3} \text{ m}^3/10^3 \text{ m}$.

$$\text{Volume grit} = [(1.38 \text{ m}^3/\text{s}) \times 86,400 \text{ s/d} \times 50 \times 10^{-6} \text{ m}^3/\text{m}^3]$$
$$= 5.96 \text{ m}^3/\text{d} \ (210 \text{ ft}^3/\text{d})$$

Comment In designing aerated grit chambers, it is especially important that the size of the related grit-handling facilities be based on the sustained peak flowrate.

Figure 8-10 Grab bucket used to remove grit from aerated grit chamber.

Quantities of Grit

The quantities of grit will vary greatly from one location to another, depending on the type of sewer system, the characteristics of the drainage area, the condition of the sewers, the frequency of street sanding to counteract icing conditions, the types of industrial wastes, the number of household garbage grinders served, and the proximity and use of sandy bathing beaches. Typical values are reported in Table 8-4.

Disposal of Grit

Possibly the most common method of grit disposal is as fill, covered if necessary to prevent objectionable conditions. In some large plants, grit is incinerated with the sludge. In New York City and in some other large coastal cities, grit and screenings are barged to sea and dumped. Generally the grit must be washed before removal.

Washing The character of grit normally collected in horizontal-flow grit chambers varies widely from what might be normally considered as clean grit to grit that includes a large proportion of putrescible organic material. Unwashed grit may contain 50 percent, or more, organic material. Unless promptly disposed of, this material may attract insects and rodents. It has a distinctly disagreeable odor.

Several types of washers are available. One type relies on an inclined submerged rake or screw that provides the necessary agitation for separation of the grit from the organic materials and, at the same time, raises the washed grit to a point of discharge above water level. Another type is a separate jig that depends on an up-and-down flow of liquid through the grit bed to wash out the organic material. Its performance is excellent but it requires an additional manually supervised plant operation.

Removal from plant Grit is normally hauled to the dumping areas in trucks for which loading facilities are required. In larger plants, elevated grit storage facilities may be provided with bottom gates through which the trucks are loaded. Difficulties experienced in getting the grit to flow freely from the storage hoppers have been minimized by applying air beneath the grit and by the use of vibrators. Facilities for collection and disposal of drippings from the bottom gates are desirable. Grab buckets operating on a monorail system may also be used to load trucks directly from the grit chambers or from storage bins at grade.

In some plants, grit is successfully conveyed to grit-disposal areas by pneumatic conveyors. This system requires no elevated storage hoppers and eliminates problems in storage and trucking, but the wear on piping, especially elbows, is considerable.

8-4 FLOW EQUALIZATION

As noted in Chap. 6, both in-line and off-line flow equalization can be used to equalize the flowrate to subsequent treatment operations and processes. Where it is also desired to equalize the plant loadings, in-line equalization must be used. From a design standpoint, the principal factors that must be considered are: (1) basin construction, (2) mixing and air requirements, and (3) pump and pump control systems [14].

Basin Construction

Important considerations in the design of new equalization basins are the materials of construction, basin geometry, and operational appurtenances. If existing tanks are to be converted to equalization basins, the principal concern is with the necessary modifications. Piping and structural changes are usually most significant.

Construction materials New basins may be of earthen, concrete, or steel construction; earthen basins are generally the least expensive. Depending on local conditions, the side slopes may vary between 3:1 and 2:1. A section through a typical earthen basin is shown in Fig. 8-11. If an aerator is used, a concrete pad should be provided below the aerator to minimize erosion. If a liner is used, it should also be bonded or otherwise anchored to the pad. It should be noted that with floating aerators some minimum operating level is needed to protect the aerator. Typically, this depth will vary from 1.5 to 2 m (5 to 6 ft). The freeboard required depends on the surface area of the basin and local wind conditions. To prevent wind-induced erosion in the upper portions of the basin, it may be necessary to protect the slopes with soil cement or a partial gunite layer.

In areas of high groundwater, drainage facilities should be provided to prevent embankment failure. To further ensure a stable embankment, the tops of the dikes should be of adequate width. The use of an adequate dike width will also reduce construction costs, especially where mechanical compaction equipment is used.

If equalization basins are to be constructed of steel or concrete, standard structural design practice will control the design. Because of the large volume

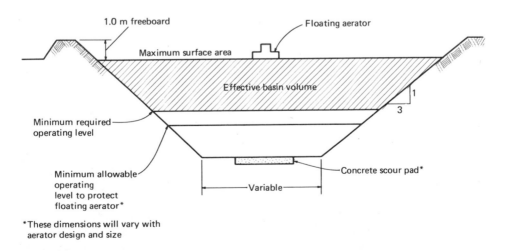

Figure 8-11 Section through earthen or lined earthen equalization basin [14].

involved, however, special attention should be given to the question of flotation, especially in areas subject to seasonally high groundwater levels.

Basin geometry The importance of basin geometry varies somewhat, depending on whether in-line or off-line equalization is used. If in-line equalization is used to dampen both the flow and the mass loadings, it is important to use a geometry that allows the basin to function as a continuous-flow stirred-tank reactor insofar as possible. Therefore, elongated designs should be avoided, and the inlet and outlet configurations should be arranged to minimize short circuiting. Discharging the influent near the mixing equipment usually minimizes short circuiting. If the geometry of the basins is controlled by the available land area and an elongated geometry must be used, it may be necessary to use multiple inlets and outlets.

Operational appurtenances Among the appurtenances that should be included in the design of equalization basins are (1) facilities for flushing any solids and grease that may tend to accumulate on the basin walls; (2) an emergency overflow in case of pump failure; (3) a high water takeoff for the removal of floating material and foam; and (4) water sprays to prevent the accumulation of foam on the sides of the basin if foam is anticipated to be a problem.

Mixing and Air Requirements

The proper operation of both in-line and off-line equalization basins requires proper mixing and aeration. Mixing equipment should be sized to blend the contents of the tank and to prevent deposition of solids in the basin. To minimize mixing requirements, grit-removal facilities should precede equalization basins where possible. Mixing requirements for blending a medium-strength municipal wastewater, having a suspended-solids concentration of approximately 220 mg/L, range from 0.004 to 0.008 kW/m^3 (0.02 to 0.04 hp/10^3 gal) of storage. Aeration is required to prevent the wastewater from becoming septic. To maintain aerobic conditions, air should be supplied at a rate of 0.01 to 0.015 m^3/m$^3 \cdot$ min (1.25 to 2.0 ft^3/10^3 gal $\cdot$ min).

One method of providing for both mixing and aeration is through the use of mechanical aerators. Baffling may be necessary to ensure proper mixing, particularly with a circular tank configuration. Minimum operating levels for floating aerators generally exceed 1.5 m (5 ft) and vary with the horsepower and design of the unit. To protect the unit, low-level shutoff controls should be provided. Because it may be necessary to dewater the equalization basins periodically, the aerators should be equipped with legs or draft tubes that allow them to come to rest on the bottom of the basin without damage.

Pumps and Pump Control

Flow equalization imposes an additional head requirement within the treatment plant. As a minimum, the required head is equal to the sum of the dynamic losses and the normal surface-level variation. Additional head may be required if the basin is to be dewatered. Depending on the local topography, it may be possible to dewater the basin upstream of the untreated wastewater pumps by gravity. Because the head requirement cannot normally be fulfilled by gravity, pumping facilities will be needed. Pumping may precede or follow equalization. In some cases pumping of both untreated and equalized flows will be required. To satisfy diurnal peak flowrates, the capacity of the influent pumps will have to be larger.

An automatically controlled flow-regulating device will be required where gravity discharge from the basin is used. Where effluent pumps are used, instrumentation should be provided to control the preselected equalization rate. Regardless of the discharge method used, a flow-measuring device should be provided on the outlet of the basin to monitor the equalized flow.

8-5 OTHER PRETREATMENT OPERATIONS

Other pretreatment operations have been used to remove material such as grease and scum from wastewater prior to primary sedimentation and to improve the treatability of wastewater [8]. Skimming, flocculation, and preaeration have been used for this purpose. Typical design information for these operations is presented in Table 8-5. Although these pretreatment operations were commonly used in the past, they are seldom used today.

Table 8-5 Typical design information for pretreatment operations

Item	Value	
	Range	Typical
Skimming tanks:		
Detention time, min	1–15	5
Flocculation:		
Detention time, min	20–60	30
Paddle-induced flocculation, maximum paddle peripheral speed, with turndown adjustment to 30% of maximum speed, m/s	0.4–1.0	0.6
Air-agitation flocculation, with porous-tube diffusers, $m^3/10^3 \ m^3$	0.6–1.2	0.8
Preaeration:		
Detention time, min	10–45	30
Tank depth, m	3–6	5
Air requirement, m^3/m^3	0.75–3.0	0.2

Note: m/s × 3.2808 = ft/s
$m^3/10^3 \ m^3$ × 133.6806 = ft³/Mgal
m × 3.2808 = ft

Skimming Tanks

A skimming tank is a chamber so arranged that floating matter rises and remains on the surface of the wastewater until removed, while the liquid flows out continuously through deep outlets or under partitions, curtain walls, or deep scum boards. This may be accomplished in a separate tank or combined with primary sedimentation, depending on the process and nature of the wastewater.

The objective of skimming tanks is the separation from the wastewater of the lighter, floating substances. The material collected on the surface of skimming tanks, whence it can be removed, includes oil, grease, soap, pieces of cork and wood, and vegetable debris and fruit skins originating in households and in industry.

Most skimming tanks are rectangular or circular and provide for a detention period of 1 to 15 minutes. The outlet, which is submerged, is opposite the inlet and at a lower elevation to assist in flotation and to remove any solids that may settle.

Flocculation

The purpose of wastewater flocculation is to form aggregates or flocs from the finely divided matter. Although not used routinely, the flocculation of wastewater by mechanical or air agitation may be worthy of consideration when it is desired to (1) increase the removal of suspended solids and BOD in primary settling facilities, (2) condition wastewater containing certain industrial wastes, and (3) improve the performance of secondary settling tanks following biological treatment processes, especially the activated-sludge process.

When used, flocculation can be accomplished (1) in separate tanks or basins specifically designed for the purpose (see Fig. 8-12), (2) in in-line facilities such as in the conduits and pipes connecting the treatment units, and (3) in combination flocculator-clarifiers. Paddles for mechanical agitation should have variable-speed drives permitting the adjustment of the top paddle speed downward to 30 percent of the top value. Similarly, where air flocculation is employed, the air-supply system should be adjustable so that the flocculation energy level can be varied throughout the tank. In both mechanical- and air-agitation flocculation systems, it is common practice to taper the energy input so that the flocs initially formed will not be broken as they leave the flocculation facilities (whether separate or in-line).

Preaeration

The objectives that are often given for aerating wastewater prior to primary sedimentation are to improve its treatability; to provide grease separation, odor control, grit removal, and flocculation; to promote uniform distribution of suspended and floating solids to treatment units; and to increase BOD removals. Of these objectives, the promotion of a more uniform distribution of suspended

(a)

(b)

Figure 8-12 Flocculators used in wastewater treatment. (a) Paddle type. (b) Turbine type. (*From Walker Process Equipment Division, Chicago Bridge & Iron Company.*)

and floating solids is probably its best application. It has been shown that short-period preaeration of 3 to 5 minutes formerly used does not significantly improve BOD or grease removal [8]. Current practice, when preaeration is used, frequently consists of increasing the detention period in aerated grit chambers. In this case, provisions for grit removal may be needed in only the first portion of the tanks.

The use of aerated channels for distributing wastewater to primary sedimentation tanks in large plants keeps the solids in suspension at all rates of flow so that the channel velocity is no longer critical. It also ensures a uniform distribution of solids to each tank, and, in conjunction with aerated grit chambers and preaeration tanks, adds dissolved oxygen to the wastewater and helps to reduce odors. The amount of air required ranges between 0.02 and 0.05 m^3/lin m · min (2 and 5 ft^3/lin ft · min) of channel. Aerated channels are often used for distributing mixed liquor to activated-sludge final settling tanks.

8-6 PRIMARY SEDIMENTATION TANKS

When a liquid containing solids in suspension is placed in a relatively quiescent state, those solids having a higher specific gravity than the liquid will tend to settle, and those with a lower specific gravity will tend to rise. These principles are used in the design of sedimentation tanks for treatment of wastewaters. The objective of treatment by sedimentation is to remove readily settleable solids and floating material and thus to reduce the suspended-solids content.

Primary sedimentation tanks may provide the principal degree of wastewater treatment, or they may be used as a preliminary step in the further processing of the wastewater. When these tanks are used as the only means of treatment, they provide for the removal of (1) settleable solids capable of forming sludge banks in the receiving waters, and (2) free oil and grease and other floating material. When they are used as a preliminary step to biological treatment, their function is to reduce the load on the biological treatment units. Efficiently designed and operated primary sedimentation tanks should remove from 50 to 70 percent of the suspended solids and from 25 to 40 percent of the BOD_5.

Primary sedimentation tanks that precede biological treatment processes may be designed to provide shorter detention periods and a higher rate of surface loading than tanks serving as the only method of treatment, except when waste-activated sludge is returned to the primary sedimentation tanks for blending with primary sludge.

Sedimentation tanks have also been used as storm-water tanks, which are designed to provide a moderate detention period (10 to 30 minutes) for overflows from either combined sewers or storm sewers. The purpose is to remove a substantial portion of the organic solids that otherwise would be discharged directly to the receiving water and that could form offensive sludge banks. Sedimentation tanks have also been used to provide sufficient detention periods for effective chlorination of such overflows.

Basis of Design

If all solids in wastewater were discrete particles of uniform size, uniform density, reasonably uniform specific gravity, and fairly uniform shape, the removal efficiency of these solids would be dependent on the surface area of the tank and time of detention. The depth of the tank would have little influence, provided that horizontal velocities would be maintained below the scouring velocity. However, the solids in most wastewaters are not of such regular character but are heterogeneous in nature, and the conditions under which they are present range from total dispersion to complete flocculation. The bulk of the finely divided solids reaching primary sedimentation tanks are incompletely flocculated but are susceptible to flocculation.

Flocculation is aided by eddying motion of the fluid within the tanks and proceeds through the coalescence of fine particles, at a rate that is a function of their concentration and of the natural ability of the particles to coalesce upon collision. As a general rule, therefore, coalescence of a suspension of solids becomes more complete as time elapses. For this reason, detention time is also a consideration in the design of sedimentation tanks. The mechanics of flocculation are such, however, that as the time of sedimentation increases, less and less coalescence of remaining particles occurs.

Detention time Normally, primary sedimentation tanks are designed to provide 90 to 150 min of detention based on the average rate of wastewater flow. Tanks that provide shorter detention periods (30 to 60 min), with less removal of suspended solids, are sometimes used for preliminary treatment ahead of biological treatment units.

Surface-loading rates Sedimentation tanks are normally designed on the basis of a surface-loading rate expressed as cubic meters per day per square meter of surface area (gallons per day per square foot of surface area). The selection of a suitable loading rate depends on the type of suspension to be separated. Typical values for various suspensions are reported in Table 8-6. Designs for municipal plants must also meet the approval of state regulatory agencies, most of which have adopted standards that must be followed.

The effect of the surface-loading rate and detention time on suspended-solids removal varies widely depending on the character of the wastewater, proportion of settleable solids, concentration of solids, and other factors. It should be emphasized that overflow rates must be set low enough to ensure satisfactory performance at peak rates of flow, which may vary from 3 times the average flow in small plants to 1.5 times the average flow in large plants (see discussion of peak flowrates in Chap. 2).

When the area of the tank has been established, the detention period in the tank is governed by water depth, as shown in Table 8-7. Overflow rates in current use result in nominal detention periods of 2 to 2.5 h, based on

Table 8-6 Typical design information for primary sedimentation tanks[a]

	Value	
Item	Range	Typical
Primary settling followed by secondary treatment:		
Detention time, h	1.5–2.5	2.0
Overflow rate, $m^3/m^2 \cdot d$		
Average flow	32–48	
Peak flow	80–120	100
Weir loading, $m^3/m \cdot d$	125–500	250
Dimensions (see Table 8-8)		
Primary settling with waste activated-sludge return:		
Detention time, h	1.5–2.5	2.0
Overflow rate, $m^3/m \cdot d$		
Average flow	24–32	
Peak flow	48–70	60
Weir loading, $m^3/m \cdot d$	125–500	250
Dimensions (See Table 8-8)		

[a] Comparable data for secondary clarifiers are presented in Table 10-7

Note: $m^3/m^2 \cdot d \times 24.5424 = gal/ft^2 \cdot d$
$m^3/m \cdot d \times 80.5196 = gal/ft \cdot d$

average design flow. As design flows in all cases are usually based on some future condition, the actual detention periods during the early years of operation are somewhat longer.

Weir rates In general, it has been found that weir loading rates have little effect on the efficiency of primary sedimentation tanks. More important is the placement of the weirs (see following discussion) and the design of the tanks. Typical weir loading rates are given in Table 8-6.

Scour velocity Scour velocity is important in sedimentation operations. Forces on settled particles are caused by the friction of water flowing over the particles. In sewers, velocities should be maintained high enough that solid particles will be kept from settling. In sedimentation basins, horizontal velocities should be kept low so that settled particles are not scoured from the bottom of the basin. The critical velocity is given by Eq. 8-1, which was developed by Camp [4] using the results from studies by Shields:

$$V_H = \left[\frac{8k(s-1)gd}{f}\right]^{1/2} \tag{8-1}$$

Table 8-7 Detention times for various surface-loading rates and tank depths for sedimentation tanks

Surface-loading rate, $m^3/m^2 \cdot d$	Detention time, h			
	3.0-m depth	3.5-m depth	4.0-m depth	5.0-m depth
24	3.0	3.5	4.0	5.0
32	2.3	2.6	3.0	3.8
48	1.5	1.8	2.0	2.5
60	1.2	1.4	1.6	2.0
80	0.9	1.1	1.2	1.5
100	0.7	0.8	1.0	1.2
120	0.6	0.7	0.8	1.0

Note: $m^3/m^2 \cdot d \times 24.5424 = gal/ft^2 \cdot d$
$m \times 3.2808 = ft$

where V_H = horizontal velocity that will just produce scour
s = specific gravity of particles
d = diameter of particles
k = constant which depends on type of material being scoured

Typical values of k are 0.04 for unigranular sand and 0.06 or more for sticky, interlocking matter. The term f is the Darcy-Weisbach friction factor, which depends on the characteristics of the surface over which flow is taking place and the Reynolds number. Typical values of f are 0.02 to 0.03. Either SI metric or U.S. customary units may be used in Eq. 8-1, so long as they are consistent, since k and f are dimensionless.

Tank Type, Size, and Shape

Almost all treatment plants of any size, except for those with Imhoff tanks, now use mechanically cleaned sedimentation tanks of standardized circular or rectangular design. The selection of the type of sedimentation unit for a given application is governed by the size of the installation, by rules and regulations of local control authorities, by local site conditions, and by the experience and judgment of the engineer and his estimate of the economics involved. In some cases, alternative bids have been taken on circular and rectangular tanks. Two or more tanks should be provided so that the process may remain in operation while one tank is out of service for maintenance and repair work. At large plants, the number of tanks is determined largely by size limitations. Typical dimensions and other data for rectangular and circular sedimentation tanks are presented in Table 8-8.

Table 8-8 Typical design information on rectangular and circular sedimentation tanks used for primary treatment of wastewater

Tank type	Value	
	Range	Typical
Rectangular:		
Depth, m	3.0–5.0	3.6
Length, m	15–90	25–40
Width, m[a]	3–24	6–10
Flight travel speed, m/min	0.6–1.2	1.0
Circular:		
Depth, m	3.0–5.0	4.5
Diameter, m	3.6–60.0	12–45
Bottom slope, mm/m	60–160	80
Flight travel speed, r/min	0.02–0.05	0.03

[a] If widths of rectangular mechanically cleaned tanks are greater than 6 m (20 ft), multiple bays with individual cleaning equipment may be used, thus permitting tank widths up to 24 m (80 ft) or more.

Note: m × 3.2808 = ft
mm/m × 0.012 = in/ft

Rectangular tanks A rectangular tank is shown in Fig. 8-13. Sludge-removal equipment for this type of tank is available from a number of manufacturers and usually consists of a pair of endless conveyor chains. Attached to the chains at regular intervals are crosspieces of wood extending the full width of the tank or bay (see Fig. 8-14). The solids settling in the tank are scraped to sludge hoppers in small tanks and to transverse troughs in large tanks. The transverse troughs are equipped with collecting mechanisms (cross collectors), usually of the same type as the longitudinal collectors, which convey solids to one or more sludge hoppers. In some recent designs, screw conveyors have been used for the cross collectors. Rectangular tanks may also be cleaned by a bridge-type mechanism which travels up and down the tank on rubber wheels or on rails supported on the sidewalls (see Fig. 8-15). One or more scraper blades are suspended from the bridge. Some of the bridge mechanisms are designed so that the scraper blades can be lifted clear of the sludge on the return travel. Bridge-type mechanisms are also available equipped with vacuum or pump systems for the withdrawal of sludge. Such systems are used in secondary settling tanks (see Chap. 10).

Where cross collectors are not provided, multiple hoppers must be installed. Their use for collection of sludge delivered by the longitudinal conveyors has introduced operating difficulties, notably sludge accumulating on the slopes and in the corners and even arching over the sludge-drawoff piping. The use of a cross collector is advisable, except possibly in the smallest tanks, because it results in the withdrawal of a more uniform and concentrated sludge.

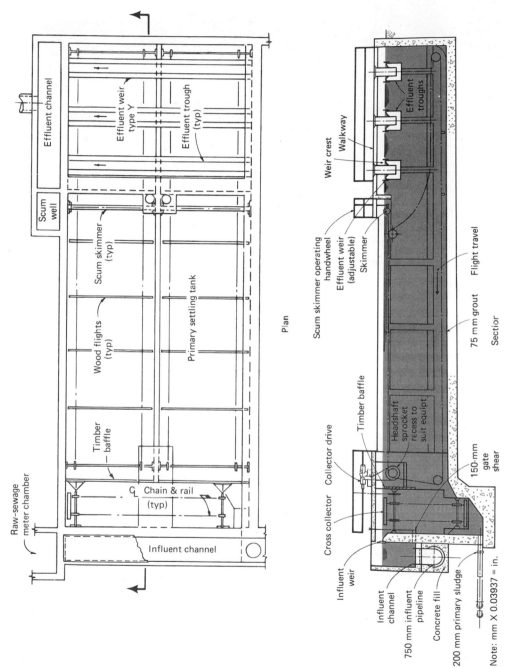

Figure 8-13 Typical rectangular primary sedimentation tank. Note: mm × 0.03937 = in.

Effluent channel

Scum well

Effluent weir type Y

Effluent trough (typ)

Scum skimmer (typ)

Wood flights (typ)

Primary settling tank

Timber baffle

Raw-sewage meter chamber

℄ Chain & rail (typ)

Influent channel

Plan

Scum skimmer operating handwheel

Weir crest

Walkway

Effluent weir (adjustable)

Effluent troughs

Skimmer

Flight travel

75 mm grout

Section

Timber baffle

Headshaft sprocket recess to suit equipt

Cross collector   Collector drive

150-mm gate shear

Influent weir

Influent channel

750 mm influent pipeline

Concrete fill

200 mm primary sludge

Note: mm X 0.03937 = in.

Figure 8-14 Empty rectangular sedimentation tank showing sludge-removal flights.

Influent channels should be provided across the inlet end of the tanks, and effluent channels should be provided across the effluent end of the tanks. It is also desirable to locate sludge-pumping facilities close to the hoppers where sludge is collected at the ends of the tanks. One sludge-pumping station can conveniently serve two or more tanks.

For large multiple installations of rectangular tanks, a pipe and operating gallery can be constructed integrally with the tanks along the influent end to contain the sludge pumps. This gallery can be extended as a service tunnel to sludge disposal, heating, and other plant units.

Scum is usually collected at the effluent end of rectangular tanks with the flights returning at the liquid surface. The scum is moved by the flights to a point where it is trapped by baffles before removal. The scum can also be moved by water sprays. The scum can be scraped manually up an inclined apron, or it can be removed hydraulically or mechanically, and for this process a number of means have been developed. For small installations, the most common scum-drawoff facility consists of a horizontal, slotted pipe that can be rotated by a lever or a screw. Except when drawing scum, the open slot is above the normal tank water level. When drawing scum, the pipe is rotated so that the open slot is submerged just below the water level, permitting the scum accumulation to flow into the pipe. Use of this equipment results in a relatively large volume of scum liquor.

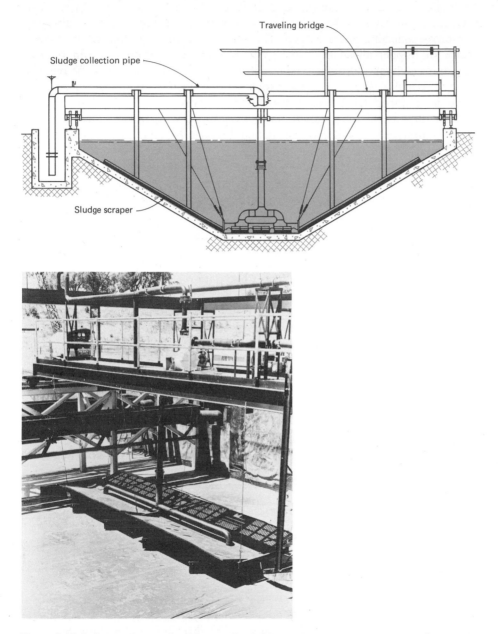

Figure 8-15 Sedimentation tank with traveling-bridge sludge-scraping mechanism. (*From Aqua Aerobics.*)

Another method for removing scum by mechanical means is a transverse rotating helical wiper attached to a shaft. By this apparatus, it is possible to draw the scum from the water surface over a short inclined apron for discharge to a cross-collecting scum trough. The scum may then be flushed to a scum ejector or hopper ahead of a scum pump. Another method of scum removal consists of a chain-and-flight type of collector that collects the scum at one side of the tank and scrapes it up a short incline for deposit in scum hoppers, whence it can be pumped to disposal units. Scum is also collected by special scum rakes in those rectangular tanks that are equipped with the carriage or bridge type of sedimentation-tank equipment. Scum is usually disposed of with the sludge produced at the plant.

Multiple rectangular tanks require less land area than multiple circular tanks and for this reason are used where ground area is at a premium. Rectangular tanks also lend themselves to nesting with preaeration tanks and aeration tanks in activated-sludge plants. They are also used generally where tank roofs or covers are required.

Circular tanks In circular tanks the flow pattern is radial (as opposed to horizontal in rectangular tanks). To achieve a radial flow pattern, the wastewater to be settled can be introduced in the center or around the periphery of the tank, as shown in Fig. 8-16. Both flow configurations have proved to be satisfactory.

In the center-feed design (see Fig. 8-16a), the wastewater is carried to the center of the tank in a pipe suspended from the bridge, or encased in concrete beneath the tank floor. At the center of the tank, the wastewater enters a circular well designed to distribute the flow equally in all directions. The sludge-removal mechanism revolves slowly and may have two or four arms equipped with scrapers. The arms also support blades for scum removal. Vacuum and pumped sludge-withdrawal mechanisms are also available (see Chap. 10). A typical center-feed circular clarifier equipped with a scraper mechanism for sludge removal is shown in Fig. 8-17.

In the rim-feed design (see Fig. 8-16b), a suspended circular baffle a short distance from the tank wall forms an annular space into which the wastewater is discharged in a tangential direction. The wastewater flows spirally around the tank and underneath the baffle, and the clarified liquid is skimmed off over weirs on both sides of a centrally located weir trough. Grease and scum are confined to the surface of the annular space.

Circular tanks 3.6 to 9 m (12 to 30 ft) in diameter have the sludge-removal equipment supported on beams spanning the tank. Tanks 10.5 m (35 ft) in diameter and larger have a central pier that supports the mechanism and is reached by a walkway or bridge (see Fig. 8-16). The bottom of the tank is sloped at about 1 in 12 (1 in/ft) to form an inverted cone, and the sludge is scraped to a relatively small hopper located near the center of the tank.

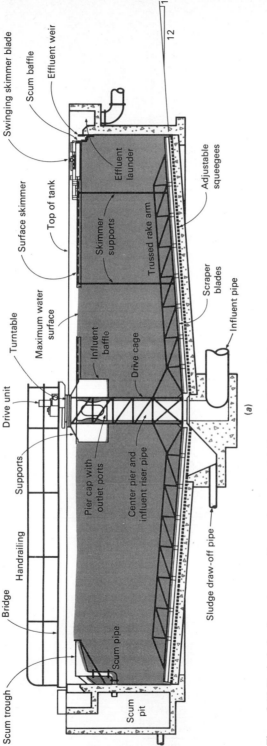

Scum trough

Bridge

Handrailing

Supports

Drive unit

Turntable

Maximum water surface

Surface skimmer

Top of tank

Swinging skimmer blade

Scum baffle

Effluent weir

Effluent launder

Skimmer supports

Influent baffle

Drive cage

Pier cap with outlet ports

Center pier and influent riser pipe

Scum pipe

Scum pit

Trussed rake arm

Adjustable squeegees

Scraper blades

Influent pipe

Sludge draw-off pipe

(a)

Figure 8-16 Typical circular primary sedimentation tanks. (a) Center-feed takeoff. (*From Infilco.*)

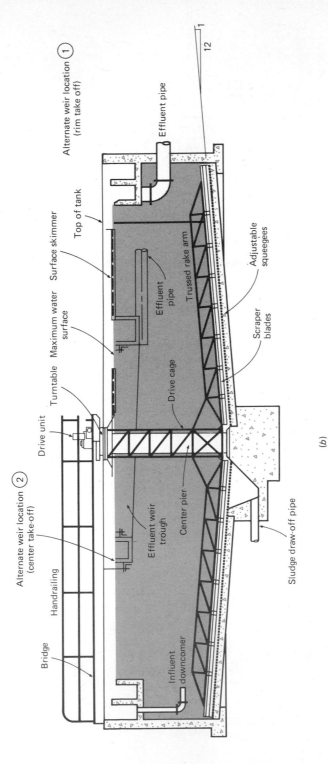

Figure 8-16 (b) Rim feed: rim one or center two takeoff. (*Adapted from Ecodyne and Clow-Yeomans.*) Note: mm × 0.03937 = in.

(b)

Figure 8-17 Empty center-feed sedimentation basin equipped with sludge scrapers.

Multiple tanks are customarily arranged in groups of two or four. The flow is divided among the tanks by a control chamber located between the tanks. Sludge is usually withdrawn to the control chamber, from which it is pumped to the sludge-disposal units.

Quantities of Sludge

The volume of sludge produced in primary settling tanks must be known or estimated so that these tanks and subsequent sludge-handling, processing, and disposal facilities can be properly designed. The sludge volume will depend on (1) the characteristics of the untreated wastewater, including strength and freshness; (2) the period of sedimentation and the degree of purification to be effected in the tanks; (3) the condition of the deposited solids, including specific gravity, water content, and changes in volume under the influence of tank depth or mechanical sludge-removal devices; and (4) the period between sludge-removal operations. Data on the specific gravity and moisture content of the sludge removed from primary sedimentation tanks are reported in Table 8-9. Example 8-2 and subsequent discussion illustrate how these factors enter into the calculation of the required storage capacity.

Table 8-9 Typical information on the specific gravity and concentration of sludge from primary sedimentation tanks

Type of sludge	Specific gravity	Solids concentration, %[a]	
		Range	Typical
Primary only:			
Sanitary[b]	1.03	4–12	6
Combined	1.05	4–12	6.5
Primary and waste-activated			
sludge	1.03	3–10	4
Primary and trickling-filter humus	1.03	4–10	5

[a] Percent dry solids.
[b] Medium-strength wastewater (see Table 3-5).

Example 8-2: Sludge-volume estimation Estimate the volume of primary sludge produced per 10^3 m^3 from a typical medium-strength wastewater. Assume that the detention time in the primary tank is 2 h and that removal efficiency of suspended solids is 60 percent.

1 Estimate the suspended-solids concentration. From Table 3-5, a medium-strength wastewater is found to contain 220 mg/L suspended solids.
2 Determine the mass of dry solids removed per 10^3 m^3.

$$\text{Dry solids} = 0.6 \times \frac{220 \text{ g/m}^3}{10^3 \text{ g/kg}} \times 10^3 \text{ m}^3 = 132 \text{ kg}$$

3 Determine the volume of sludge using the data in Table 8-9 and Eq. 11-2. If the specific gravity of the sludge is 1.03 and it contains 94 percent moisture (see Table 8-9), the volume is

$$\text{Volume}/10^3 \text{ m}^3 = \frac{132 \text{ kg}}{1.03 \times 1000 \text{ kg/m}^3 \ (0.06)}$$

$$= 2.14 \text{ m}^3/10^3 \text{ m}^3$$

Comment Because of the many problems associated with handling, treatment, and disposal of sludge, it is important to produce a sludge that is as thick as possible (e.g., minimize volume to be handled) consistent with the processing facilities. The determination of the volume of sludge when chemicals are used is illustrated in Example 8-3.

The calculation in Example 8-2 is directly applicable to the design of sludge-pumping facilities for primary sedimentation tanks. Sludge should be removed by pumping at least once per shift and more frequently in hot weather to avoid deterioration of the effluent. In large plants, sludge pumping may be controlled by a time clock providing continuous on-off operation. In primary sedimentation tanks used in activated-sludge plants, provision may be required for the excess activated sludge that may be discharged into the influent of the preliminary tanks for settlement and consolidation with the fresh sludge.

For sedimentation tanks used with trickling filters, provision may be required for the "unloading" of trickling filters and for the accumulation of sludge over longer periods than ordinarily used in primary sedimentation tanks, if mechanical equipment for sludge removal is not provided. For sedimentation tanks in the activated-sludge process, provision must be made for light, flocculant sludge of 98 to 99.5 percent moisture and for quantities of sludge ranging from 1500 to 10,000 mg/L in the influent mixed liquor. Further consideration of the requirements for sedimentation tanks following trickling filters and aeration units is included in Chap. 10, which deals with these processes.

8-7 OTHER SOLIDS-REMOVAL OPERATIONS AND UNITS

Flotation and screening are unit operations that may be used in place of primary sedimentation for removal of suspended and floating solids. Two other solids-removal units in which the removal of solids is accomplished by gravity settling are the Imhoff tank and the septic tank. Imhoff tanks are particularly suitable for small communities; septic tanks are used by individual households and small subdivisions.

Flotation

Flotation of untreated wastewater, settled wastewater, and storm-water overflows has received considerable attention recently. The process has the advantage of high surface-loading rates and high removals of grease and floatable material. For these applications, design air-solids ratios have not been well defined. From practical experience, it appears that air quantities of 2 to 3 percent by volume of the wastewater flowrate yield satisfactory results.

Recent designs have varied from that shown in Fig. 6-26b by injecting air into the retention tank and providing mixing of the air and recycle wastewater in the tank. Such designs enable 80 to 95 percent saturation compared to 50 percent for static designs. The semisaturated recycle is then piped to the flotation tanks. A backpressure valve maintains the retention-tank pressure within 28.6 to 35.5 kN/m^2 (4 to 5 lb$_f$/in^2). Turbulence or energy dissipation should be avoided in the inlet design to prevent reduction in flotation efficiency.

One of the largest wastewater installations designed for the use of dissolved-air flotation instead of conventional primary settling tanks is the Sand Island wastewater-treatment plant in Honolulu, Hawaii. In this design, six circular flotation tanks 45.7 m (150 ft) in diameter are used. The tanks have an average water depth of 4.2 m (13.7 ft), and they are covered with a steel roof to prevent the escaping of odors. The pressurization system required and all necessary sludge and dewatering pumps are housed between the tanks in a large equipment gallery.

Fine Screens

With the development of better screening materials and equipment, the use of fine screens for grit removal and as a replacement for (and a means of upgrading the performance of) primary sedimentation tanks is increasing.

Replacement for primary settling tanks Within the past 5 yr, the use of fine screens as replacements for primary settling facilities has increased steadily. The two most common types of screens used for this purpose are the inclined self-cleaning type and the rotary-drum type (see Fig. 8-18). Typical design information on these screens is presented in Table 8-10. From information on a number of full-scale installations, it appears that grit removals of 80 to 90 percent, BOD_5 removals of 15 to 25 percent, and suspended-solids removals of 15 to 30 percent can be achieved with fine screens of the types shown in Fig. 8-18. It also has been found that if the solids in the wastewater are ground up using comminutors, the BOD removals will not be as high. Typical design information on inclined, rotary-drum, and centrifugal screens is presented in Table 8-10.

Where fine screens are used as replacements for primary sedimentation facilities, the following (secondary) facilities must be sized appropriately to handle the solids and BOD_5 not removed by the screens as compared to the use of primary sedimentation facilities. Typically, fine screens can be installed for approximately 25 percent of the cost of conventional primary sedimentation

Table 8-10 Typical design information on screening devices used for the primary treatment of wastewater

Type of screen	Screening surface			Hydraulic capacity $m^3/min \cdot m^2$	Composition of waste solids, % solids by weight	Suspended-solids removal, %
	Size classification	Size range, μ	Screen material			
Inclined	Medium	250–1500	Stainless-steel wedge-wire screen	0.6–2.4	10–15	15–30
Rotary drum	Medium	250–1500	Stainless-steel wedge-wire screen	0.005–0.040	10–25	15–30
Centrifugal	Fine–medium	10–500	Stainless steel, polyester, and various other fabric cloths	0.010–0.050	0.05–0.02[a]	60–70

[a] Should be used in conjunction with primary settling facilities or with a solids thickener.
  Note: $m^3/m \cdot min \times 10.7639 = ft^3/ft \cdot min$
          $m^3/m^2 \cdot min \times 24.5424 = gal/ft^2 \cdot min$

(a)

(b)

Figure 8-18 Fine screens used for primary treatment. (*a*) Fixed-inclined screens. (*b*) Rotary-drum screen.

facilities. For this reason, it may (depending on local conditions) be more cost-effective to remove the differential solids and BOD loadings in the secondary facilities.

Upgrading existing facilities An interesting application of fine screens is in the upgrading of existing primary settling facilities that are overloaded or are, for other reasons, not performing properly. The type of fine screen that has been used for this application is the centrifugal screen shown in Fig. 6-3d. When a centrifugal screen or similar device is used to upgrade the performance of primary sedimentation facilities, the flowsheet shown in Fig. 8-19 is used. As shown, a variable portion of the flow is diverted through the screen. The underflow from the screen containing the retained solids is discharged to the existing primary sedimentation tank for thickening. The screened wastewater, from which the solids have been removed, is discharged to the effluent line from the primary sedimentation facilities. The performance of centrifugal screens used in this application is essentially the same as that of primary sedimentation facilities.

Comparing the use of the centrifugal screen with the other screens reported in Table 8-10, the following points are important: (1) the solids removed with the wedge-wire screens (inclined and rotary drum) are physically larger and much dryer than those removed with the centrifugal screen; (2) there is a significant difference in the suspended-solids-removal efficiency of the wedge-wire and centrifugal screens; and (3) some form of thickening will be required for the underflow from centrifugal or similar screens. It is for the latter reason that centrifugal screens can be used most effectively with existing sedimentation facilities.

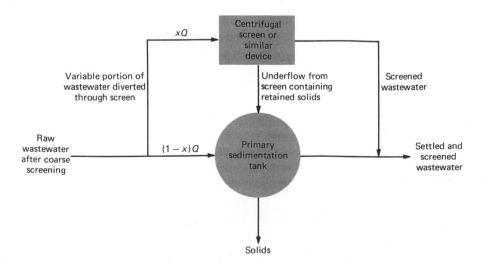

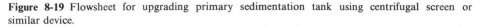

Figure 8-19 Flowsheet for upgrading primary sedimentation tank using centrifugal screen or similar device.

Imhoff and Septic Tanks

In Imhoff and septic tanks, the removal of settleable solids and the anaerobic digestion of these solids are accomplished simultaneously.

Imhoff tanks The Imhoff tank (see Fig. 8-20) consists of a two-story tank in which sedimentation is accomplished in an upper compartment and digestion is accomplished in a lower compartment. Settling solids pass through trapped slots into the unheated lower compartment for digestion. Scum accumulates in the sedimentation compartment and in vents adjoining the sedimentation compartments. Gas produced in the digestion process in the lower compartment escapes through the vents.

Before separate heated digestion tanks were developed, the Imhoff tank was widely used, but it has limited application today, mostly in relatively small plants. It is simple to operate, and there is no mechanical equipment to maintain. Operation consists of removing scum daily and discharging it into the nearest gas vent, reversing the flow of wastewater twice a month to even up the solids in the two ends of the digestion compartment, and drawing sludge periodically to the sludge beds. Recent designs developed by manufacturers for a modified form of Imhoff tank provide means of heating the sludge compartment and mechanical removal of sludge. Conventional unheated Imhoff tanks are usually rectangular, although some small circular tanks have been used. The design of Imhoff tanks is considered in detail in Ref. 11.

Septic tanks Septic tanks are used principally for the treatment of wastes from individual residences. In rural areas they are also used for establishments such as schools, summer camps, parks, trailer parks, and motels. Although single-chamber tanks are often used, two or more chambers in series are preferable (see Fig. 8-21). In a dual-chamber septic tank, the first compartment provides for sedimentation, sludge digestion, and sludge storage. The second compartment provides additional sedimentation and sludge-storage capacity and thus serves to protect against the discharge of sludge and other material that might escape the first chamber. Septic tanks designed for residential use generally have a 24-h detention period. For larger installations serving multiple families or institutions, a shorter detention period may be permissible. In either case, it is essential that adequate storage capacity be provided so that the deposited sludge remains in the tank for a sufficient length of time to undergo decomposition or digestion before being withdrawn. In general, sludge should be removed every 2 to 3 yr. Sludge pumped from septic tanks is usually discharged to a nearby wastewater-treatment plant for treatment. The characteristics of septic-tank sludge (called septage) are presented in Table 3-6.

Effluent from septic tanks is normally discharged to subsurface tile or leaching fields from where it percolates into the ground. In the past, because of a lack of understanding of the fundamental factors governing their design and operation, a number of tile-field installations have failed. However, as a result of

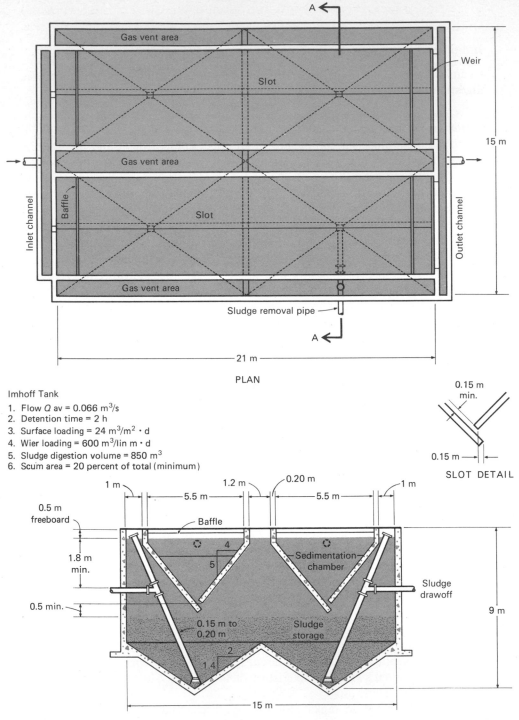

Imhoff Tank

1. Flow Q av = 0.066 m³/s
2. Detention time = 2 h
3. Surface loading = 24 m³/m² · d
4. Wier loading = 600 m³/lin m · d
5. Sludge digestion volume = 850 m³
6. Scum area = 20 percent of total (minimum)

PLAN

SLOT DETAIL

SECTION A–A

Figure 8-20 Typical Imhoff tank. Note: m × 3.2808 = ft; m³/s × 22.8245 = Mgal/d.

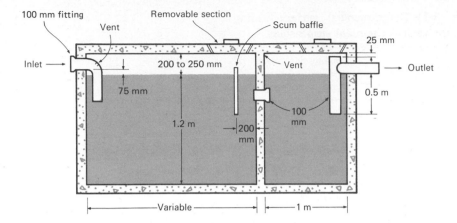

Figure 8-21 Typical septic tank. Note: mm × 0.03937 = in.

extensive studies conducted at the Richmond Field Station of the University of California, these factors are now understood more clearly, and the design of leaching fields is now a more rational undertaking. Because the details of leaching-field design are beyond the scope of this text, it is recommended that the interested reader review the many reports that have been issued by the University of California and the U.S. Public Health Service.

8-8 CHEMICAL PRECIPITATION

Chemical precipitation, discovered in 1762, was a well-established method of wastewater treatment in England as early as 1870. Lime was used as a precipitant in most cases, sometimes alone, but more often in combination with calcium chloride, magnesium chloride, alum, ferric chloride, ferrous sulfate (copperas), charcoal, or any one of a number of other substances [11]. Chemical treatment was used extensively in the United States in the 1890s and early 1900s, but with the development of biological treatment, the use of chemicals was abandoned and biological treatment was adopted. In the early 1930s, attempts were made to develop new methods of chemical treatment, and a number of plants were installed. Details on these early processes may be found in Refs. 6 and 11.

At the present time (1977), chemical precipitation is used (1) as a means of improving the performance of primary settling facilities, (2) as a basic step in the independent physical-chemical treatment of wastewater, and (3) for the removal of phosphorus. Aside from the determination of the required chemical dosages, the principal design considerations related to the use of chemical precipitation involve the analysis and design of the necessary sludge-processing facilities and the selection and design of the chemical storage, feeding, piping, and control systems.

Table 8-11 Recommended surface-loading rates for various chemical suspensions

Suspension	Loading rate, m³/m² · d	
	Range	Peak flow
Alum floc[a]	25–50	50
Iron floc[a]	25–50	50
Lime floc[a]	30–60	60
Untreated wastewater	25–50	50

[a] Mixed with the settleable suspended solids in the untreated wastewater and colloidal or other suspended solids swept out by the floc.

Note: $m^3/m^2 \cdot d \times 24.5424 = gal/ft^2 \cdot d$.

Improved Sedimentation-Tank Performance

As noted in Chap. 7, many different substances have been used as precipitants. The degree of clarification obtained when chemicals are added to untreated wastewater depends on the quantity of chemicals used and the care with which the process is monitored and controlled. With chemical precipitation it is possible to remove 80 to 90 percent of the suspended solids, 70 to 80 percent of the BOD_5, and 80 to 90 percent of the bacteria. Comparable removal values for well-designed and well-operated primary sedimentation tanks without the addition of chemicals are 50 to 70 percent of the suspended solids, 25 to 40 percent of the BOD_5, and 25 to 75 percent of the bacteria. Because of the variable characteristics of wastewater, the required chemical dosages should be determined from bench- or pilot-scale tests. Recommended surface-loading rates for various chemical suspensions to be used in the design of the sedimentation facilities are given in Table 8-11.

Independent Physical-Chemical Treatment

In many localities, the addition of industrial wastes to sewers has rendered the resulting wastewater mixture untreatable by biological means. In such situations, physical-chemical treatment is an alternative approach. The principal drawback to this method of treatment, which has limited its widespread use, is the handling and disposal of the great volumes of sludge resulting from the addition of chemicals.

A flowsheet for the physical-chemical treatment of untreated wastewater is presented in Fig. 8-22. As shown, after first-stage precipitation and pH adjustment, if required, the wastewater is passed through a granular-medium filter to remove any residual floc and then through carbon columns to remove dissolved organic compounds. Although the filter is shown as optional, its use is recommended to reduce the binding and head-loss buildup in the carbon columns. The treated

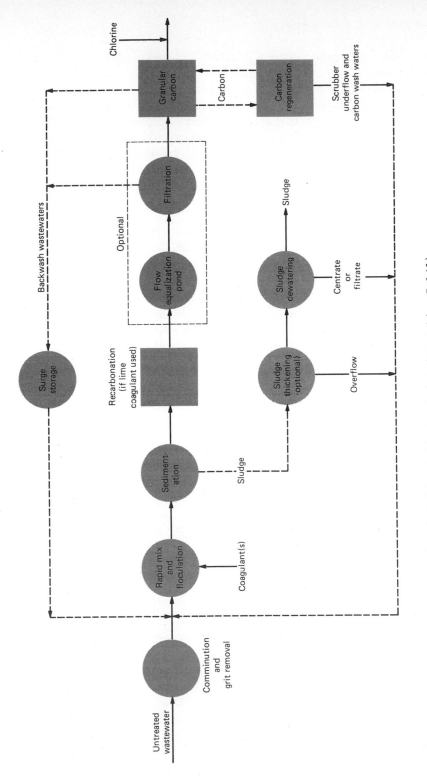

Figure 8-22 Typical flowsheet of an independent physical-chemical treatment plant. *(Adapted from Ref. 13.)*

effluent from the carbon column is usually chlorinated before discharge to the environment.

Depending on the treatment objectives, the required chemical dosages and application rates would have to be determined from bench- or pilot-scale tests. Typical chemical dosages that are used where phosphorus is also to be removed are given in Sec. 12-7. Settling velocities for chemical suspensions are presented in Table 8-11. Because one of the major considerations in the design of chemical precipitation facilities is the handling of the resulting sludge, the procedure involved in estimating the volume of chemical sludge is illustrated in Example 8-3, following the discussion of phosphorus removal.

Phosphorus Removal

Chemicals that have been used for the removal of phosphorus include lime, alum, and ferric chloride or sulfate. Polymers have also been used effectively in conjunction with lime and alum. To accomplish the removal of phosphorus, these chemicals have been added to the untreated wastewater, in biological treatment processes, and in separate facilities following the biological secondary treatment of wastewater. Each of these methods of application is considered in detail in Sec. 12-7, which deals specifically with the removal of phosphorus. Determination of the volume of sludge produced when lime is used to precipitate phosphorus from wastewater, as compared with the volume of sludge that would be expected without precipitation, is illustrated in Example 8-3.

Table 8-12 Summary of pertinent reactions required to determine the quantities of sludge produced during the precipitation of phosphorus with lime, alum, and iron Fe (III)[a]

Reaction	Chemical species in sludge	Eq. no.
Lime:		
1. $5Ca^{+2} + 3PO_4^{-3} + OH^- \rightleftharpoons Ca_5(PO_4)_3(OH)$	$Ca_5(PO_4)_3(OH)$	7-17
2. $Mg^{+2} + 2OH^- \rightleftharpoons Mg(OH)_2$	$Mg(OH)_2$	7-18
3. $Ca^{+2} + CO_3^= \rightleftharpoons CaCO_3$	$CaCO_3$	7-19
Alum:		
1. $CaO + H_2O \rightleftharpoons Ca(OH)_2$		
2. $Al^{+3} + PO_4^{-3} \rightleftharpoons AlPO_4$	$AlPO_4$	7-23
3. $Al^{+3} + 3OH^- \rightleftharpoons Al(OH)_3$	$Al(OH)_3$	7-24
Iron Fe(III):		
1. $CaO + H_2O \rightleftharpoons Ca(OH)_2$		
2. $Fe^{+3} + PO_4^{-3} \rightleftharpoons FePO_4$	$FePO_4$	7-20
3. $Fe^{+3} + 3OH^- \rightleftharpoons Fe(OH)_3$	$Fe(OH)_3$	7-21

[a] Adapted in part from Ref. 13.

Estimation of Sludge Quantities

The handling and disposal of the sludge resulting from chemical precipitation was in the past and still is one of the greatest difficulties of this method of treatment. Sludge is produced in great volume from most chemical-precipitation operations, often reaching 0.5 percent of the volume of wastewater treated. A summary of the pertinent reactions required for determining the quantity of sludge when using lime, alum, or iron for the precipitation of phosphorus is given in Table 8-12. The computational procedures involved in estimating the quantity of sludge resulting from the chemical precipitation of phosphorus are illustrated in Example 8-3.

Example 8-3: Estimation of sludge volume from chemical precipitation of untreated wastewater for the removal of phosphorus Estimate the mass and volume of sludge produced from untreated wastewater without and with the use of lime for the chemical precipitation of phosphorus. Assume that 60 percent of the suspended solids are removed in the primary settling tank without the addition of chemicals, and that the addition of 400 mg/L of lime [$Ca(OH)_2$] results in an increased removal of suspended solids to 85 percent. Also, assume that the following data apply to this problem:

Wastewater flowrate, m^3/d	1000
Wastewater suspended solids, mg/L	220
Wastewater volatile suspended solids, mg/L	150
Wastewater PO_4^{-3} as P, mg/L	10
Wastewater total hardness as $CaCO_3$	241.3
Wastewater Ca^{+2}, mg/L	80
Wastewater Mg^{+2}, mg/L	10
Effluent PO_4^{-3} as P, mg/L	0.5
Effluent Ca, mg/L	60
Effluent Mg, mg/L	0

SOLUTION

1 Compute the mass and volume of solids removed without chemicals, assuming that the sludge contains 94 percent moisture and has a specific gravity of 1.03 (see Eq. 11-2).
   a Determine the mass of suspended solids removed.

$$M_{ss} = \frac{0.6(220 \text{ g/m}^3) \times 1000 \text{ m}^3/\text{d}}{10^3 \text{ g/kg}} = 132 \text{ kg/d}$$

   b Determine the volume of sludge produced using Eq. 11-2.

$$V_2 = \frac{132 \text{ kg/d}}{1.03 \times 1000 \text{ kg/m}^3 \times (0.06)} = 2.14 \text{ m}^3/\text{d}$$

2 Using the equations summarized in Table 8-12, determine the mass of $Ca_5(PO_4)_3OH$, $Mg(OH)_2$, and $CaCO_3$ produced from the addition of 400 mg/L of lime.
   a Determine the mass of $Ca_5(PO_4)_3OH$ formed.
      i Determine the moles of P removed.

$$\text{mol P removed} = \frac{10 \text{ mg/L} - 0.5 \text{ mg/L}}{30.97 \text{ g/mol} \times 10^3 \text{ mg/g}}$$

$$= 0.307 \times 10^{-3} \text{ mol/L}$$

    ii Determine the moles of $Ca_5(PO_4)_3OH$ formed.

$$\text{mol } Ca_5(PO_4)_3OH \text{ formed} = 1/3 \times 0.307 \times 10^{-3} \, mol/L$$

$$= 0.102 \times 10^{-3} \, mol/L$$

    iii Determine the mass of $Ca_5(PO_4)_3OH$ formed.

$$\text{Mass } Ca_5(PO_4)_3OH = 0.102 \times 10^{-3} \, mol/L \times 502 \, g/mol \times 10^3 \, mg/g$$

$$= 51.3 \, mg/L$$

  b Determine the mass of $Mg(OH)_2$ formed.
    i Determine the moles of Mg removed

$$\text{mol } Mg^{+2} \text{ removed} = \frac{10 \, mg/L}{24.31 \, g/mol \times 10^3 \, mg/g}$$

$$= 0.411 \times 10^{-3} \, mol/L$$

    ii Determine the mass of $Mg(OH)_2$ formed.

$$\text{mol } Mg(OH)_2 = 0.411 \times 10^{-3} \, mol/L \times 58.3 \, g/mol \times 10^3 \, mg/g$$

$$= 24.0 \, mg/L$$

  c Determine the mass of $CaCO_3$ formed.
    i Determine the mass of Ca^{+2} in $Ca_5(PO_4)_3(OH)$

$$\text{Mass Ca in } Ca_5(PO_4)_3(OH) = (40 \, g/mol)5 \times 0.102 \times 10^{-3} \, mol/L \times 10^3 \, mg/g$$

$$= 20.4 \, mg/L$$

    ii Determine the mass of Ca added in the original dosage.

$$\text{Mass Ca in } Ca(OH)_2 = \frac{40 \, g/mol \times 400 \, mg/L}{74 \, g/mol}$$

$$= 216.2 \, mg/L$$

    iii Determine the mass of Ca present as $CaCO_3$.

$$\text{Ca in } CaCO_3 = \text{Ca in } Ca(OH)_2 + \text{Ca in influent wastewater} - \text{Ca in } Ca_5(PO_4)_3OH$$
$$- \text{Ca in effluent wastewater}$$

$$= 216.2 + 80 - 20.4 - 60$$

$$= 215.8 \, mg/L$$

    iv Determine the mass of $CaCO_3$.

$$\text{Mass } CaCO_3 = \frac{215.8 \, mg/L}{40 \, g/mol} \times 100 \, g/mol$$

$$= 540 \, mg/L$$

3 Determine the total mass of solids removed as a result of the lime dosage.
  a Suspended solids in wastewater

$$M_{ss} = \frac{0.85(220 \, g/m^3) \times 1000 \, m^3 d}{10^3 \, g/kg} = 187 \, kg/d$$

b Chemical solids

$$M_{Ca_5(PO_4)_3OH} = \frac{51.2 \text{ g/m}^3 \times 1000 \text{ m}^3/\text{d}}{10^3 \text{ g/kg}} = 51.3 \text{ kg/d}$$

$$M_{Mg(OH)_2} = \frac{24 \text{ g/m}^3 \times 1000 \text{ m}^3/\text{d}}{10^3 \text{ g/kg}} = 24 \text{ kg/d}$$

$$M_{CaCO_3} = \frac{540 \text{ g/m}^3 \times 1000 \text{ m}^3/\text{d}}{10^3 \text{ g/kg}} = 540 \text{ kg/d}$$

c Total mass of solids removed

$$M_T = (187 + 51.3 + 24 + 540) \text{ kg/d}$$

$$= 802.3 \text{ kg/d}$$

4 Determine the total volume of sludge resulting from chemical precipitation, assuming that the sludge has a specific gravity of 1.07 and a moisture content of 92.5 percent (see Chap. 11).

$$V_s = \frac{802.3 \text{ kg/d}}{1.07 \times 1000 \text{ kg/m}^3 (0.075)} = 10.0 \text{ m}^3/\text{d}$$

5 Prepare a summary table of sludge masses and volumes without and with chemical precipitation.

	Sludge	
Treatment	Mass, kg/d	Volume, m³/d
Without chemical precipitation	132.0	2.14
With chemical precipitation	802.3	10.0

Comment The magnitude of the sludge-disposal problem when chemicals are used for the removal of phosphorus is evident from a review of the data presented in the summary table given in step 5. Disposal methods that have been used for chemical sludges include (1) spreading on soil, (2) lagooning, (3) landfilling, and (4) ocean dumping. Ocean discharge, commonly used in the past and still used by some coastal cities, is no longer acceptable. Details of these methods will be found in Chap. 11. One of the advantages of using lime or alum is the possibility of recovering calcium oxide and aluminate by burning. Development of effective pure-product recovery methods would permit recycling of the precipitant, reduction of chemical costs, and minimization of the disposal problem.

Chemical Storage, Feeding, Piping, and Control Systems

The design of chemical-precipitation operations involves not only the sizing of the various unit operations and processes but also the necessary appurtenances. Because of the corrosive nature of many of the chemicals used for the chemical precipitation of wastewater and the different forms in which they are available, special attention must be given to the design and the materials of construction used for the chemical storage, feeding, piping, and control systems. A detailed discussion of these topics is beyond the scope of this text; an excellent discussion of the requirements for these appurtenances may be found in Ref. 15.

8-9 GRANULAR-MEDIUM FILTRATION

The important filtration process variables and operative particle removal mechanisms were considered in detail in Chap. 6. With that information serving as a background, the purpose of this section is to identify the major factors that should be considered in the design of effluent filtration systems. Topics that will be considered are (1) driving force, number, and size of filter units; (2) filter-bed selection; (3) filter backwashing systems; (4) filter appurtenances; (5) filter problems; and (6) filter control systems and instrumentation. Specific details on the piping and physical structures involved are not presented because they vary in each situation.

Driving Force, Number, and Size of Filter Units

One of the first decisions to be made is what type of filtration system is to be used. This decision is often based on plant-related variables, such as the space available, the time available for construction, and local costs.

Next, the number and size of filter units must be determined. Usually, the surface area required is based on the peak filtration and peak plant flowrates. The allowable peak filtration flowrate usually is established on the basis of results from pilot-plant studies. The number of units generally should be kept to a minimum to reduce the cost of piping and construction, but it should be sufficient to assure (1) that the backwash flowrates do not become excessively large and (2) that when one filter unit is taken out of service for backwashing, the transient loading on the remaining units will not be so high that material contained in the filters will be dislodged [2]. The sizes of the individual units should be consistent with the sizes of equipment available for use as underdrains, wash-water troughs, and surface washers. Typically, width-length ratios of gravity filters vary from 1:1 to 1:4. For pressure filters, it is common practice to use standard sizes that are available from manufacturers or from local fabricators if the filter is to be constructed locally.

Filter-Bed Selection

In Chap. 6, the principal types of filter beds were classified according to the number of filtering media as single-medium, dual-medium, and tri-medium (or multimedium). A further classification may be made according to stratification, as follows: (1) single-medium stratified, (2) single-medium or mixed-medium unstratified, (3) dual-medium stratified, and (4) multimedium stratified. Thus the first step in the specification of a filter bed is to select the type of bed to be used. The various classifications are considered briefly in the following discussion.

After the type of filter bed is selected, the next step is to specify the characteristics of the medium, or media if more than one is used. Typically, this involves the selection of the grain or pore size, the shape, the medium gradation, the specific gravity, the depth, and the hardness and solubility of the various materials used in the filter bed. In addition, it is necessary to determine the type of medium support to be used and the depth of submergence over the filter bed.

Single-medium stratified filter bed Although single-medium stratified beds of conventional design have been used for wastewater filtration, they are not used routinely. The principal reason is their unfavorable head-loss-buildup characteristics.

Single-medium unstratified filter bed Two types of single-medium unstratified filter beds are now in use. In the first type, a single, uniform, coarse medium (2 to 3 mm) is used in beds with depths up to 2 m (6.5 ft). It has been found that these large-medium, deep filters offer longer filter runs. Depending on the type of treatment process, these filters can also be used for the simultaneous denitrification of the wastewater, although the filtration rate will be significantly lower. The principal disadvantages are (1) the need for a uniform size of medium, (2) the high backwash velocities required to fluidize the bed for effective cleaning, and (3) the added cost for the backwashing facilities and the structure needed to contain the deep beds.

In the second type, a single medium of varying sizes is used with a combined air-water backwash (see Fig. 8-23). This type has proved to be an effective alternative to the filter with a single medium of uniform size. The combined air-water backwash scours the accumulated material from the filtering medium without the need for fluidizing the entire bed. This backwash system also eliminates the normal stratification that occurs in single-medium and multimedium beds when only a water backwash or an air-followed-by-water backwash is used.

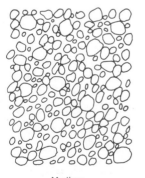

Medium   Pore size distribution

Figure 8-23 Cross section through single-medium unstratified filter bed [7].

Thus it is possible to obtain a filter bed with a more or less uniform pore-size distribution throughout its depth. From the analysis in Chap. 6, it can be concluded that the uniform pore-size distribution achieved in unstratified beds will increase the potential for the removal of suspended particles in the lower portions of the filter. By comparison, in a stratified filter bed, the potential decreases with depth because of increasing pore size.

Typically, the effective size of the medium used in unstratified filters is about the same as that used in the upper layers of a dual-medium filter bed. The depth of such filters is approximately 0.9 m (3.0 ft). Additional details on the performance of single-medium unstratified filters may be found in Ref. 7. The appurtenances used in conjunction with unstratified filters are essentially the same as those used for conventional downflow filters. A specially designed backwash water trough to separate the air and the grains of the filtering medium is described in the subsequent discussion dealing with filter backwash facilities.

Dual-medium and multimedium filter beds Some dual-medium filter beds that have been used are composed of (1) anthracite and sand, (2) activated carbon and sand, (3) resin beds and sand, and (4) resin beds and anthracite. Multimedium beds that appear to have promise are composed of (1) anthracite, sand, and garnet or ilmenite; (2) activated carbon, anthracite, and sand; (3) weighted spherical resin beads (charged and uncharged), anthracite, and sand; and (4) activated carbon, sand, and garnet or ilmenite.

Typical data on the depth and characteristics of the filtering materials most commonly used in dual-medium and multimedium filters are presented in Table 8-13. Because filter performance is related directly to the characteristics of the liquid to be filtered, it is recommended that pilot-plant studies be conducted to determine the optimum combination of filter materials. If it is not possible to conduct such studies, the data in Table 8-13 may be used as a guide.

Filtration Backwashing Systems

A filter bed can function properly only if the backwashing system used in conjunction with the filter effectively cleans the material removed within the filter. The methods commonly used for backwashing granular-medium filter beds include (1) water backwash only, (2) water backwash with auxiliary surface-water-wash agitation, (3) water backwash with auxiliary air scour, and (4) combined air-water backwashing. With the first three methods, fluidization is necessary to achieve effective cleaning of the filter bed at the end of the run. With the fourth method, fluidization is not necessary. Typical backwash velocities required to fluidize various filter beds are reported in Table 8-14.

Water backwash only The idea in the system using water backwash only is to get the grains of the filtering medium fluidized so that the shearing action of the

Table 8-13 Typical design data for dual-medium and multimedium filters[a]

Characteristic	Value	
	Range	Typical
Dual-medium:		
Anthracite:		
Depth, mm	300–600	450
Effective size, mm	0.8–2.0	1.2
Uniformity coefficient	1.3–1.8	1.6
Sand:		
Depth, mm	150–300	300
Effective size, mm	0.4–0.8	0.55
Uniformity coefficient	1.2–1.6	1.5
Filtration rate, $L/m^2 \cdot min$	80–400	200
Multimedium:		
Anthracite (top layer of quadmedium filter):		
Depth, mm	200–400	200
Effective size, mm	1.3–2.0	1.6
Uniformity coefficient	1.5–1.8	1.6
Anthracite (second layer of quadmedium filter):		
Depth, mm	100–400	200
Effective size, mm	1.0–1.6	1.2
Uniformity coefficient	1.5–1.8	1.6
Anthracite (top layer of tri-medium filter):		
Depth, mm	200–500	400
Effective size, mm	1.0–2.0	1.4
Uniformity coefficient	1.4–1.8	1.6
Sand:		
Depth, mm	200–400	250
Effective size, mm	0.4–0.8	0.5
Uniformity coefficient	1.3–1.8	1.6
Garnet or ilmenite:		
Depth, mm	50–150	100
Effective size, mm	0.2–0.6	0.3
Uniformity coefficient	1.5–1.8	1.6
Filtration rate, $L/m^2 \cdot min$	80–400	200

[a] Developed from Refs. 2 and 14.
Note: mm × 0.03937 = in
$L/m^2 \cdot min$ × 0.0245 = $gal/ft^2 \cdot min$

water as it moves past the individual grains scours any material that has accumulated on them. Cleasby and Amirtharajah [1] have found that the maximum shearing force on the grains occurs when the porosity of the fluidized bed is about 0.68 to 0.71. Although these systems have been used in water-treatment plants, it has been found that some form of auxiliary scour is required to clean

Table 8-14 Typical backwash flowrates required to fluidize various filter beds[a]

Type of filter	Size of critical granular medium	Minimum backwash velocity needed to fluidize bed[b]	
		$m^3/m^2 \cdot min$	m/h
Single-medium (sand)	2 mm	1.8–2.0	108–120
Dual-medium (anthracite and sand)	See Table 8-13	0.8–1.2	48–72
Tri-medium (anthracite, sand, and garnet or ilmenite)	See Table 8-13	0.8–1.2	48–72

[a] Adapted in part from Refs. 7, 18, and 19.
[b] Varies with size, shape, and specific gravity of the medium and the temperature of the backwash water.

Note: $m^3/m^2 \cdot min \times 24.5424 = gal/ft^2 \cdot min$
$m/h \times 3.2808 = ft/h$
$mm \times 0.03937 = in$

filters used for wastewater filtration effectively. This is especially true when filtering settled biological effluent.

Water backwash with auxiliary surface wash Surface washers (see Fig. 8-24) are often used to provide the shearing force required to clean the grains of the filtering medium used for wastewater filtration. Operationally, the surface washing cycle is started about 1 or 2 min before the water-backwashing cycle is started. Both cycles are continued for about 2 min, at which time the surface wash is terminated. Water usuage is as follows: for a single-sweep surface backwashing system, from 20 to 40 $L/m^2 \cdot min$ (0.5 to 1.0 $gal/ft^2 \cdot min$); for a dual-sweep surface backwashing system, from 60 to 80 $L/m^2 \cdot min$ (1.5 to 2.0 $gal/ft^2 \cdot min$) [2].

Water backwash with auxiliary air scour The use of air to scour the filter is common in Europe and is gaining in popularity in the United States. Operationally, air is usually applied for 3 or 4 min before the water-backwashing cycle begins. In some systems, air is also injected during the first part of the water-washing cycle. Typical air flowrates range from 10 to 16 $m^3/m^2 \cdot min$ (3 to 5 $ft^3/ft^2 \cdot min$) [2]. Because of the violent action of the air injected into the filter bed, conventional gravel underdrain systems cannot be used. Therefore, where air scour is used, the filtering medium is placed directly

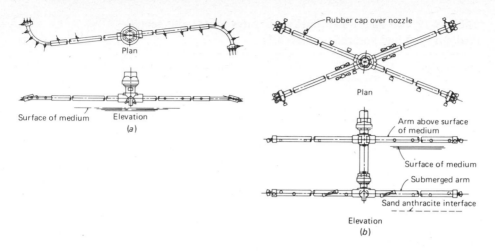

Figure 8-24 Typical surface-wash agitators. (*a*) Single-arm agitator. (*b*) Dual-arm agitators. (*From F. B. Leopold Co.*)

on a specially designed underdrain system (see discussion of underdrain systems later in this section).

Combined air-water backwash The combined air-water backwash system is used in conjunction with the single-medium unstratified filter bed. Operationally, air and water are applied simultaneously for several minutes. The specific duration of the combined backwash varies with the design of the filter bed. Ideally, during the backwash operation, the filter bed should be agitated sufficiently so that the grains of the filter medium move in a circular pattern from the top to the bottom of the filter as the air and water rise up through the bed. Some typical data on the quantity of water and air required are reported in Table 8-15. The reduced wash-water requirements for this system can be appreciated by comparing the values given in Table 8-15 with those given in Table 8-14. At the end of the combined air-water backwash, a 2- to 3-min water backwash at subfluidization velocities is used to remove any air bubbles that may remain in the filter bed [7]. This last step is required to eliminate the possibility of air binding within the filter.

Filter Appurtenances

The principal filter appurtenances are as follows: (1) the underdrain system used to support the filtering materials, collect the filtered effluent, and distribute the backwash water and air (where used); (2) the wash-water troughs used to remove the backwash water from the filter; and (3) the surface washing systems used to help remove attached material from the filtering material.

Table 8-15 Air and water backwash rates used with single-medium sand and anthracite filters [7]

	Medium characteristics		Backwash rate	
Medium	Effective size, mm	Uniformity coefficient	Water, $m^3/m^2 \cdot min$	Air, $m^3/m^2 \cdot min^a$
Sand	1.00	1.40	0.41	13.1
	1.49	1.40	0.61	19.7
	2.19	1.30	0.81	26.2
Anthracite	1.10	1.73	0.28	6.6
	1.34	1.49	0.41	13.1
	2.00	1.53	0.61	19.7

^{a} Air at 21°C (70°F), 1.0 atm.
Note: $m^3/m^2 \cdot min \times 24.5424 = gal/ft^2 \cdot min$
$m^3/m^2 \cdot min \times 3.2808 = ft^3/ft^2 \cdot min$

Underdrain systems The type of underdrain system to be used depends on the type of backwash system. In conventional water backwash filters without air scour, it is common practice to place the filtering medium on a support consisting of several layers of graded gravel. The design of a gravel support for a granular medium is delineated in the AWWA Standard for Filtering Material B100-72. When there is to be a gravel layer, an underdrain system such as the one shown in Fig. 8-25 is used. With air scour or combined air-water backwash systems, there is no gravel layer, and underdrain systems of the type shown in Fig. 8-26 must be used. In the single-medium unstratified filter bed, the grain size is larger; so the slot or screen size can also be larger.

Wash-water troughs Wash-water troughs are now constructed of plastic or sheet metal or of concrete with adjustable weir plates (see Fig. 8-25). The particular design of the trough will depend to some extent on the other equipment to be used in the design and construction of the filter.

A recent development for use in conjunction with the single-medium unstratified filter bed is shown in Fig. 8-27. As shown, a dual or single baffle is used to create a quiescent zone around the wash-water trough. Any grains of the filtering medium reaching this zone will settle and thus will not be washed out of the filter. This allows the combined air-water wash to be used until the bed is cleaned thoroughly.

Surface washers Surface washers for filters can be fixed or mounted on rotary sweeps, as shown in Fig. 8-24. According to data on a number of systems, it appears that the rotary-sweep washers are the most effective.

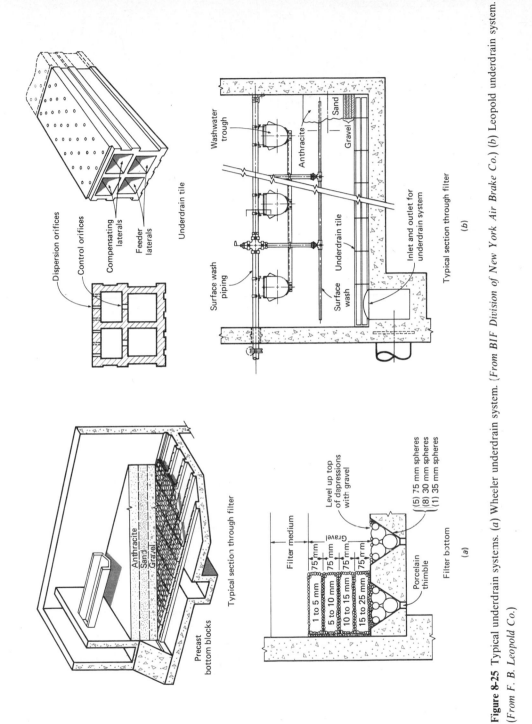

Figure 8-25 Typical underdrain systems. (*a*) Wheeler underdrain system. (*From BIF Division of New York Air Brake Co.*) (*b*) Leopold underdrain system. (*From F. B. Leopold Co.*)

Dispersion orifices
Control orifices
Compensating laterals
Feeder laterals
Underdrain tile

Washwater trough
Anthracite
Sand
Gravel
Surface wash piping
Underdrain tile
Surface wash
Inlet and outlet for underdrain system
Typical section through filter
(*b*)

Anthracite
Sand
Gravel
Typical section through filter
Precast bottom blocks

Level up top of depressions with gravel
Filter medium
Gravel
75 mm
75 mm
75 mm
75 mm
1 to 5 mm
5 to 10 mm
10 to 15 mm
15 to 25 mm
(5) 75 mm spheres
(8) 30 mm spheres
(1) 35 mm spheres
Porcelain thimble
Filter bottom
(*a*)

369

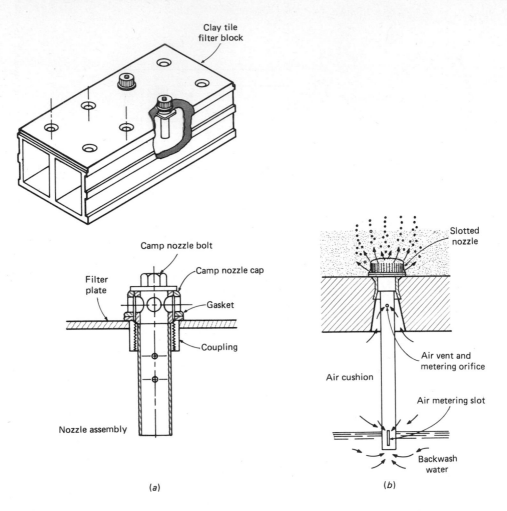

Figure 8-26 Underdrain nozzles for use in filters without gravel support for filter medium. (*a*) Camp-walker filter underdrain system. (*From Walker Process Equipment Division, Chicago Bridge & Iron Company.*) (*b*) Air-water nozzle underdrain system. (*From Infilco Degremont, Inc.*)

Filter Problems

The principal problems encountered in wastewater filtration and the control measures that have proved to be effective are reported in Table 8-16. Because these problems can affect both the performance and operation of a filter system, care should be taken in the design phase to provide the necessary facilities that will minimize their impact.

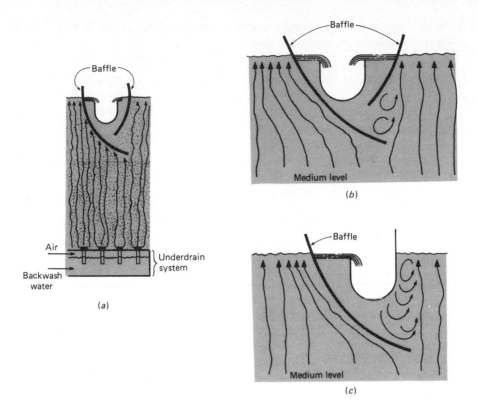

Figure 8-27 Details of baffle system developed for use with single-medium unstratified filter bed to minimize loss of filtering medium during backwash operation. (*a*) Section through filter showing flow patterns during the air-water backwash cycle. (*b*) Dual-baffle detail. (*c*) Single-baffle detail. (*From General Filter Company.*)

Filter Control Systems and Instrumentation

The selection and design of the control system and related instrumentation is as important in the development of a well-operating filtration system as is the design of the filter bed and the associated underdrain, wash-water trough, and surface wash systems. Although the specific details are beyond the scope of this book, some of the important considerations are identified in the following discussion. Additional details may be found in Refs. 18 and 19.

Filter control systems As briefly described in Chap. 6, the two principal operational methods for filters are constant rate and declining rate. Because the merits of each system and its variations depend on the specific application, each method should be considered separately in each case.

Table 8-16 Summary of commonly encountered problems in the filtration of wastewater and control measures for those problems[a]

Problem	Description/control
Turbidity breakthrough	Unacceptable levels of turbidity are recorded in the effluent from the filter, even though the terminal head loss has not been reached. To control the buildup of effluent turbidity levels, chemicals and polymers have been added ahead of the filter bed, and polymers have been added within the filter bed. The point of chemical or polymer addition must be determined locally
Mudball formation	Mudballs are agglomerations of biological floc, dirt, and the filtering medium or media. If the mudballs are not removed, they will grow into large masses that often sink into the filter bed and ultimately reduce the effectiveness of the filtering and backwashing operations. The formation of mudballs can be controlled by auxiliary washing processes, such as air scour or water surface wash concurrent with, or followed by, water wash alone
Buildup of emulsified grease	The buildup of emulsified grease within the filter bed increases the head loss and thus reduces the length of run. Both air-scour and water-surface-wash systems help control the buildup of grease. In extreme cases, it may be necessary to install a washing system for the medium and to use special solutions for the purpose
Development of cracks and contraction of filter bed	If the filter bed is not cleaned properly, the grains of the filtering medium become coated. As the filter compresses, cracks develop, especially at the sidewalls of the filter. Ultimately, mudballs may develop. This problem can be controlled by adequate backwashing and scouring
Loss of filtering medium or media (mechanical)	In time, some of the filtering medium or media may be lost during backwashing and through the underdrain system (where the gravel support has been upset or the underdrain system has been installed improperly). The mechanical loss of the filtering medium can be minimized through the proper placement of the wash-water troughs and underdrain system. Special baffles for use with wash-water troughs have also proved to be effective (see Fig. 8-27)
Loss of filtering medium or media (operational)	Depending on the characteristics of the biological floc, grains of the filtering medium can become attached to it, forming aggregates light enough to be floated away during the back-washing operations. This problem can be minimized by the addition of an auxiliary air and/or water scouring system
Gravel mounding	Gravel mounding occurs when the various layers of the support gravel are disrupted by the application of excessive rates of flow during the backwashing operation. A gravel support with an additional 50- to 74-mm layer of high-density material, such as ilmenite or garnet, can be used to overcome this problem. Gravel mounding is not a problem in uniformly graded filters

[a] Adapted in part from Ref. 2.
Note: mm × 0.03937 = in.

Filter instrumentation Apart from the specific instrumentation associated with the operation of the filter control system, the following operational parameters are normally monitored and/or controlled: (1) influent water level, (2) effluent rate of flow, (3) effluent turbidity, (4) filter head loss, and (5) backwash and surface wash-water flowrates. In some systems where polymers are added to improve the quality of the effluent, a signal from the monitors of the turbidity and effluent flowrate is used to pace the polymer feed system.

8-10 MICROSCREENING

The microscreen is a surface filtration device that has been used to remove a portion of the residual suspended solids from secondary effluents and from stabilization-pond effluents (see Chap. 10).

Description

Microstraining involves the use of variable low-speed (up to 4 r/min), continuously backwashed, rotating-drum filters operating under gravity conditions (see Fig. 8-28). The principal filtering fabrics have openings of 23 or 35 μ and are fitted on the drum periphery. The wastewater enters the open end of the drum and

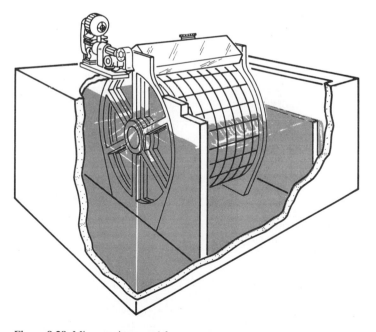

Figure 8-28 Microstrainer used for wastewater treatment.

flows outward through the rotating screening cloth. The collected solids are backwashed by high-pressure jets into a trough located within the drum at the highest point of the drum.

The typical suspended-solids removal achieved with these units is about 55 percent. The range is from about 10 to 80 percent. Problems encountered with microstrainers include incomplete solids removals and inability to handle solids fluctuations. Reducing the rotating speed of the drum and less frequent flushing of the screen have resulted in increased removal efficiencies but reduced capacity.

Functional Design

The functional design of a microscreen unit involves the following considerations: (1) the characterization of the suspended solids with respect to the concentration and degree of flocculation; (2) the selection of unit design-parameter values that will not only assure capacity to meet maximum hydraulic loadings with critical

Table 8-17 Typical design information on microscreens used for screening of secondary settled effluent

Item	Typical value	Remarks
Screen size	20–35 μ	Stainless-steel or polyester screen cloths are available in sizes ranging from 15 to 60
Hydraulic loading rate	$3–6 \times 10^{-3}$ m^3/m^2 · min	Based on submerged surface area of drum
Head loss through screen	75–150 mm	Bypass should be provided when head exceeds 200 mm
Drum submergence	70–75% of height; 60–70% of area	Varies depending on design of screen
Drum diameter	2.5–5 m	3 m (10 ft) is used most commonly; smaller sizes increase backwash requirements
Drum speed	4.5 m/min at 75 mm head loss; 35–45 m/min at 150 mm head loss	Maximum rational speed is limited to 45 m/min
Backwash requirements	2% of throughput at 350 kPa; 5% of throughput at 100 kPa	

^{a} Adapted in part from Ref. 14.

Note: m^3/m^2 · min $\times$ 24.5424 = gal/ft^2 · min

mm $\times$ 0.03937 = in

m $\times$ 3.2808 = ft

kPa $\times$ 0.1450 = lb$_f$/in^2

m/min $\times$ 3.2808 = ft/min

solids characteristics but also provide desired design performance over the expected range of hydraulic and solids loadings; and (3) the provision of backwash and clearing facilities to maintain the capacity of the screen [14]. Typical design information for microscreens is presented in Table 8-17. Because of the variable performance of these units, it is recommended that pilot-plant studies be conducted, especially if the units are to be used to remove solids from stabilization-pond effluents.

8-11 CHLORINATION

The chemistry of chlorine in water and wastewater has been discussed in Chap. 7 along with an analysis of how chlorine functions as a disinfectant. However, chlorine has been applied for a wide variety of objectives other than disinfection in the wastewater-treatment field. Therefore, the purpose of this section is to discuss briefly (1) the various uses and required dosages, (2) the chlorine compounds most commonly used, (3) the equipment and methods used in its application, and (4) the design of chlorination facilities for disinfection.

Application

To aid in the design and selection of the required chlorination facilities and equipment, it is important to know the uses, including dosage ranges, to which chlorine and its compounds have been applied.

Uses The principal uses of chlorine and its compounds in the collection, treatment, and disposal of wastewater were reported in Table 7-11. Of the many different applications, disinfection of wastewater effluents is still the most important, although serious questions are being raised about the merits of effluent chlorination in general.

Dosages The dosages for various applications, with the exception of disinfection, are reported in Table 8-18. When used for disinfection, the probable amounts of chlorine that may be required are presented in Table 8-19. A range of dosage values is given because they will vary depending on the characteristics of the wastewater. It is for this reason that laboratory chlorination studies must be conducted to determine optimum chlorine dosages.

Chlorination capacities are generally selected to meet the specific design criteria of the state or other regulatory agencies controlling the receiving body of water (see Example 8-4). In any case, where the residual in the effluent is specified or the final number of coliform bacteria is limited, the actual amount of chlorine must be determined by onsite testing. However, in the absence of more specific data, the maximum values given in Table 8-19 can be used as a guide in sizing chlorination equipment.

Table 8-18 Dosages for various chlorination applications in wastewater collection, treatment, and disposal

Application	Dosage range, mg/L
Collection:	
Corrosion control (H$_2$S)	2–9[a]
Odor control	2–9[a]
Slime-growth control	1–10
Treatment:	
BOD reduction	0.5–2[b]
Breakpoint chlorination	See Chaps. 7, 12
Digester and Imhoff-tank foaming control	2–15
Digester supernatant oxidation	20–140
Ferrous sulfate oxidation	[c]
Filter-fly control	0.1–0.5
Filter-ponding control	1–10
Grease removal	2–10
Sludge-bulking control	1–10
Disposal:	
Bacterial reduction	2–20
Disinfection	See Table 8-19

[a] Per mg/L of H$_2$S.
[b] Per mg/L of BOD$_5$ destroyed.
[c] $6FeSO_4 \cdot 7H_2O + 3Cl_2 \rightarrow 2FeCl_3 + 2Fe_2(SO_4)_3 + 42H_2O$.

Chlorine Compounds

The principal chlorine compounds used at wastewater-treatment plants are chlorine (Cl$_2$), calcium hypochlorite [Ca(OCl)], and sodium hypochlorite [Na(OCl)$_2$]. When the latter two forms are used, the chlorination process is known as hypochlorination.

Table 8-19 Typical chlorine dosages for disinfection

Effluent from	Dosage range, mg/L
Untreated wastewater (prechlorination)	6–25
Primary sedimentation	5–20
Chemical-precipitation plant	2–6
Trickling-filter plant	3–15
Activated-sludge plant	2–8
Multimedium filter following activated-sludge plant	1–5

Note: mg/L = g/m^3.

Chlorine Chlorine is supplied as a liquefied gas under high pressure in containers varying in size from 45.36-kg (100-lb) and 68.04-kg (150-lb) cylinders to 0.9072-Mg (1-ton) containers, multiunit tank cars containing fifteen 0.9072-Mg (1-ton) containers, and tank cars with capacities of 14.5, 27.2, and 49.9 Mg (16, 30, and 55 tons). Selection of the size of chlorine pressure vessel should depend on an economic study of transportation costs, demurrage, handling charges, available space, and rate of chlorine usage. Storage and handling facilities can be designed with the aid of information developed by the Chlorine Institute. Additional details may be found in Ref. 20. Although all the safety devices and precautions that must be designed into the chlorine-handling facilities are too numerous to mention, the following are fundamental:

1. Chlorine gas is both very poisonous and very corrosive. Adequate exhaust ventilation at floor level should be provided because chlorine gas is heavier than air.
2. Chlorine storage and chlorinator equipment rooms should be walled off from the rest of the plant and should be accessible only from the outdoors. A fixed glass viewing window should be included in an inside wall. Fan controls and gas masks should be located at the room entrance.
3. Dry chlorine liquid and gas can be handled in black wrought-iron piping, but chlorine solution is highly corrosive and should be handled in rubber-lined or tough plastic piping with diffusers of hard rubber.
4. Storage should be provided for a 30-d supply. Cylinders in use are set on platform scales, set flush with the floor, and the loss of weight is used as a positive record of chlorine dosage.

Calcium hypochlorite Calcium hypochlorite is available commercially in either a dry or a wet form. High-test calcium hypochlorite contains at least 70 percent available chlorine. In dry form, it is available as a powder or as granules, compressed tablets, or pellets. A wide variety of container sizes is available depending on the source. Because calcium hypochlorite granules or pellets are readily soluble in water and, under proper storage conditions, are relatively stable, they are often favored over other available forms. Because of its oxidizing potential, calcium hypochlorite should be stored in a cool, dry location away from other chemicals in corrosion-resistant containers.

Sodium hypochlorite Sodium hypochlorite solution is available in strengths from 1.5 to 15 percent with 3 percent the usual maximum strength; thus transportation costs may limit its application. The solution decomposes more readily at high concentrations and is affected by exposure to light and heat. It must therefore be stored in a cool location in a corrosion-resistant tank. However, where sodium hypochlorite is available at a reasonable cost, its use should certainly be investigated by the design engineer.

Chlorination Equipment and Dosage Control

In this section, the equipment used to inject (feed) chlorine or its related compounds into the wastewater and the methods used to control the required dosages are discussed.

Chlorine feeders Chlorine may be applied directly as a gas or in an aqueous solution. Some small-capacity chlorinators use pressure injection of the gas into the wastewater, but there are certain dangers inherent in leakages of this poisonous gas into the atmosphere of the treatment plant. Further, because of the hazards involved in handling chlorine gas, direct application is limited to large installations where adequate safety precautions can be rigidly enforced.

The most widely accepted types of chlorinators are those using vacuum-feed devices. Standard schematics for chlorination systems using (1) evaporators and 0.9072-Mg (1-ton) or larger containers, (2) 0.9072-Mg (1-ton) containers, and (3) 68.04-kg (150-lb) chlorine cylinders are shown in Figs. 8-29 and 8-30, respectively. In each of these systems, the chlorine injector is the basic component. The injector is used to create the vacuum that is used to draw chlorine gas from the storage supply through the chlorinator, which serves as a metering device, and into the injector. At the injector, the chlorine dissolves in the injector water supply to form hypochlorous acid. From the injector, the hypochlorous acid solution flows to the point where it is to be injected into the wastewater. The differential head across the chlorinator (the measuring device) may be held constant or varied in accordance with the flow by changing the degree of vacuum.

Evaporators (see Fig. 8-29) are used where the maximum rate of chlorine-gas withdrawal from a 0.9072-Mg (1-ton) container must exceed approximately 180 kg/d (400 lb/d). Although multiple-ton cylinders can be connected to provide more than 180 kg/d, the use of an evaporator conserves space. Evaporators are almost always used when the total dosage exceeds 680 kg/d (1500 lb/d) [21]. The sizing of chlorinators is considered in Example 8-4.

Example 8-4: Chlorinator selection Determine the capacity of a chlorinator for a treatment plant with an average wastewater flow of 10^3 m^3/d. The peaking factor for the treatment plant is 3.0 and the maximum required chlorine dosage (set by state regulations) is to be 20 mg/L.

SOLUTION
1 Determine the capacity of the chlorinator at peak flow.

$$Cl_2/d = \frac{20 \text{ g/m}^3 \times 10^3 \text{ m}^3/d \times 3.0}{10^3 \text{ g/kg}}$$

$$= 60 \text{ kg/d}$$

Use two 60 kg/d units with one unit serving as a spare. Although this capacity will not be required during most of the day, it must be available to meet the chlorine requirements at peak flow. Best practice calls for the availability of a standby chlorinator.

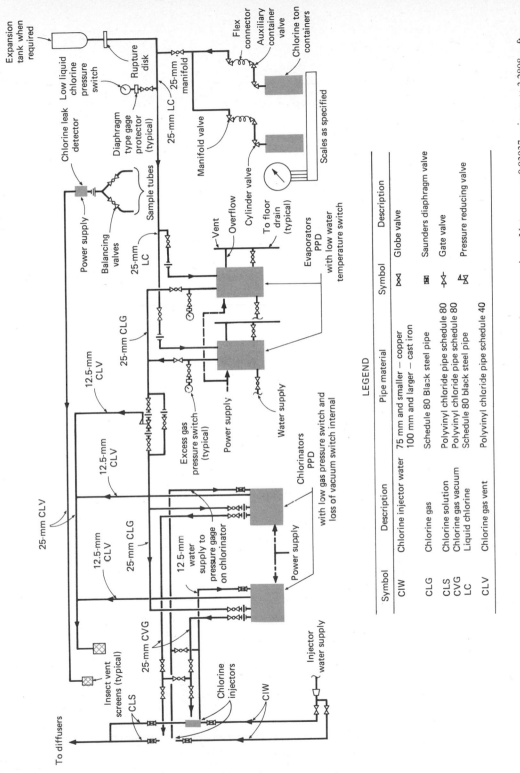

Figure 8-29 Standard chlorine schematic for chlorination system using evaporators and ton or larger containers. Note: mm × 0.03937 = in; m × 3.2808 = ft.

LEGEND

Symbol	Description	Pipe material
CIW	Chlorine injector water	75 mm and smaller — copper 100 mm and larger — cast iron
CLG	Chlorine gas	Schedule 80 Black steel pipe
CLS	Chlorine solution	Polyvinyl chloride pipe schedule 80
CVG	Chlorine gas vacuum	Polyvinyl chloride pipe schedule 80
LC	Liquid chlorine	Schedule 80 black steel pipe
CLV	Chlorine gas vent	Polyvinyl chloride pipe schedule 40

Symbol	Description
▷◁	Globe valve
▦	Saunders diaphragm valve
▷▽	Gate valve
◮▽	Pressure reducing valve

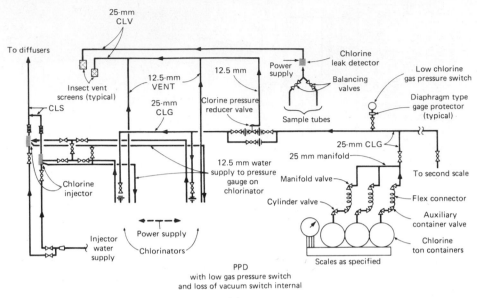

(a)

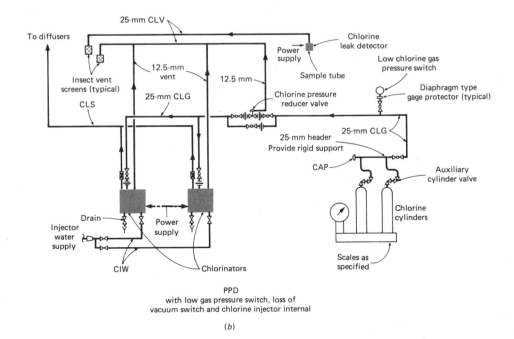

(b)

Figure 8-30 Standard chlorine schematics. (a) For chlorination system using ton containers. (b) For chlorination system using chlorine cylinders. Note: mm × 0.03937 = in.

380

2 Estimate the daily consumption of chlorine. Assume an average dosage of 10 mg/L

$$Cl_2/d = \frac{10 \text{ mg/L} \times 10^3 \text{ m}^3/\text{d}}{10^3 \text{ g/kg}} = 10 \text{ kg (83.4 lb/d)}$$

Hypochlorite feeders For very small installations serving about 10 people, it is possible to use drip-type feeders. For up to 100 people, orifice-controlled feeders using a constant-head tank fed by gravity from an overhead reservoir can be used successfully. The only difficulty is clogging of orifices so that periodic maintenance is required. The most satisfactory means of feeding sodium or calcium hypochlorite is through the use of low-capacity proportioning pumps (see Fig. 8-31).

Generally, the pumps are available in capacities up to 450 L/d (120 gal/d), with adjustable stroke for any value below this. Large capacities or multiple units are available from some of the manufacturers. The pumps can be arranged to feed at a constant rate, or they can be programmed by a time clock to start and stop at desired intervals. Such intervals can be determined by a totalizing meter, which will send electric impulses to the pump to feed for a selected length of time at a selected rate after a chosen number of cubic meters (gallons) of wastewater have flowed by the meter. When the purpose of chlorination is disinfection, this method is applicable only if the flow passes into a chamber of ample size where sufficient mixing takes place to ensure continuous chlorination. The feeding of sodium and calcium hypochlorite by this method has gained acceptance over the past 15 yr.

Dosage control Dosage may be controlled in several ways. The simplest method is manual control; the operator changes the feed rate to suit conditions. The required dosage is usually determined by measuring the residual after 15 minutes of contact time and adjusting the dosage to obtain a residual of 0.5 mg/L. A second method is to use a program control that changes the feed rate to follow a preselected pattern. This is the most economical method of obtaining automatic control. A third method is to pace the chlorine flowrate to the wastewater flowrate as measured by a primary meter such as a Parshall flume or flow tube. A fourth method is to control the chlorine dosage by automatic measurement of the chlorine residual. An automatic analyzer with signal transmitter and recorder is required. Finally, a compound system that incorporates both the third and fourth methods may be used. In a compound system, the control signals obtained from the wastewater flowmeter and from the residual recorder are superimposed to provide more precise control of chlorine dosage and residual. Additional details on these systems may be found in Ref. 21.

Chlorination Facilities

As pointed out in Chap. 7, other things being equal, the contact time and the chlorine residual are the two principal factors involved in achieving effective bacterial kill. The contact time is usually specified by the regulatory agency and

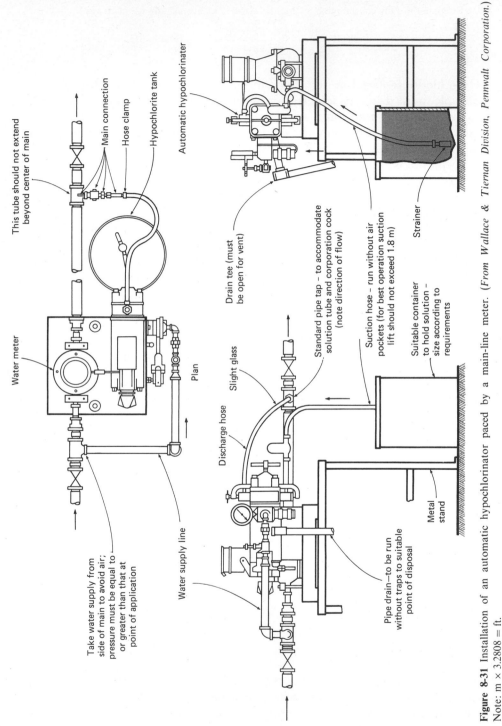

Figure 8-31 Installation of an automatic hypochlorinator paced by a main-line meter. *(From Wallace & Tiernan Division, Pennwalt Corporation.)* Note: m × 3.2808 = ft.

The labels in the figure read:

This tube should not extend beyond center of main

Main connection

Hose clamp

Hypochlorite tank

Automatic hypochlorinator

Water meter

Plan

Take water supply from side of main to avoid air; pressure must be equal to or greater than that at point of application

Water supply line

Discharge hose

Slight glass

Drain tee (must be open for vent)

Standard pipe tap – to accommodate solution tube and corporation cock (note direction of flow)

Suction hose – run without air pockets (for best operation suction lift should not exceed 1.8 m)

Suitable container to hold solution – size according to requirements

Strainer

Metal stand

Pipe drain—to be run without traps to suitable point of disposal may range from 15 to 45 min; periods of 15 min at peak flow are common. The appropriate chlorine residual that needs to be maintained should be determined from actual plant studies. In the absence of any other information, it may be estimated by using Eq. 7-59. The necessary computations are illustrated in Example 8-5.

Important practical design factors that must be considered in the design of chlorination facilities include (1) method of chlorine addition and provision for mixing, (2) the design of the chlorine contact basin, (3) maintenance of solids-transport velocity, (4) provision for bypassing the chamber or portions of it, and (5) outlet control MPN measurement. These topics are considered in the following discussion.

Example 8-5: Estimation of required chlorine residuals Estimate the chlorine residual that must be maintained to achieve a coliform count equal to or less than 200/100 mL in an effluent from an activated-sludge treatment facility assuming that the effluent contains a coliform count of $10^7/100$ mL. The specified contact time is 30 min. What will be the required residual to meet the specified effluent coliform count for a flowrate corresponding to the 1-d sustained value as given in Fig. 2-6?

SOLUTION
1   Determine the chlorine residual needed to meet the effluent-discharge requirement using Eq. 7-59.

$$\frac{N_t}{N_0} = (1 + 0.23\, C_t t)^{-3}$$

$$\frac{2 \times 10^2}{10^7} = (1 + 0.23\, C_t t)^{-3}$$

$$2 \times 10^{-5} = (1 + 0.23\, C_t t)^{-3}$$

$$1 + 0.23\, C_t t = (0.5 \times 10^5)^{1/3} = 36.84$$

$$C_t t = (36.84 - 1)/0.23 = 155.8$$

For a value of t equal to 30 min,

$$C_t = 155.8/30 = 5.2 \text{ mg/L}$$

2   Determine the residual for the 1-d peak sustained flowrate. From Fig. 2-6, the ratio of the 1-d sustained flowrate to the average flowrate is 2.75. Because the chlorine contact time will be reduced by this value, the corresponding residual is

$$C_t = 155.8/(30/2.75) = 14.3 \text{ mg/L}$$

Comment   For most small plants, the maintenance of a residual of 14.3 mg/L may be prohibitively expensive. This is one of the reasons that a statistical approach is usually employed in setting bacterial-disinfection requirements. Also, depending on the characteristics of the wastewater, the dosage required to achieve a residual value of 14.3 mg/L may be sufficient to reach the break-point, in which case the use of Eq. 7-59 is no longer valid.

Further, in applying Eq. 7-59 it has been assumed that the chlorine contact basin was either a batch reactor or an ideal plug-flow reactor and that ideal initial mixing was achieved. Because batch reactors are seldom used in other than very small plants, a plug-flow reactor would normally be used. However, from the discussion in Chap. 5, it is known that there will be axial dispersion in an actual plug-flow reactor. Therefore, to account for the effects of the inherent dispersion, it will usually be necessary to increase the value of the residual computed in step 1.

Injection and intial mixing The design of any chlorine contact system should provide for addition of chlorine solution through a diffuser, which may be a plastic or hard-rubber pipe with drilled holes through which the chlorine solution can be uniformly distributed into the path of wastewater flow, or the solution can flow directly to the propeller of a rapid mixer for instantaneous and complete diffusion. Typical diffusers are shown in Fig. 8-32. In most cases, the type of injection diffuser will depend on the means to be used to accomplish the initial mixing of the chlorine solution and the wastewater.

The effective initial mixing of the chlorine solution with the wastewater can be accomplished in a number of ways: (1) closed conduits, (2) hydraulic jumps, (3) submerged weirs, (4) Venturi flumes, (5) over-and-under baffles, and (6) mechanical means. As noted in Chap. 6, in the first five methods, mixing is accomplished as a result of the turbulence that exists in the flow regime (see Fig. 8-33). If mechanical means are used, turbulence is induced by rotating devices such as impellers, paddles, and propellers, or by gases such as in an air-lift pump. The particular method of achieving effective intial mixing will vary with each situation. If wastewater is transported to the chlorine contact basin in a closed conduit of sufficient length to allow for complete mixing, the use of in-line injection facilities should be considered. In pipelines exceeding 900 mm (36 in) in diameter, in-line mixers should be used to ensure positive initial mixing. Additional details on the effects of initial mixing on chemical reactions may be found in Ref. 5.

Chlorine contact basin design Because of the importance of contact time, careful attention should be given to design of the contact chamber so that at least 80 to 90 percent of the wastewater is retained in the basin for the specified contact time. The best way to achieve this is by using a plug-flow round-the-end type of contact chamber or a series of interconnected basins or compartments. The advantage of using a plug-flow reactor or a compartmentalized chlorine contact chamber can be evaluated using the analysis presented in Chap. 5.

Plug-flow chlorine contact basins that are built in a serpentine fashion (e.g., folded back and forth) to conserve space require special attention in their design. The reason for this is the development of dead zones with respect to flow that will reduce the hydraulic detention times. The results of model studies of various basin configurations are shown in Fig. 8-34. The use of baffles, as shown in scheme IIA, produced the best results. A full-scale chlorine contact basin using a similar baffle arrangement is shown in Fig. 8-35.

If the time of travel in the outfall sewer at the maximum design flow is sufficient to equal or exceed the required contact time, it may be possible to eliminate the chlorine contact chambers. In some small plants, chlorine contact basins have been constructed of large-diameter sewer pipe.

Maintenance of solids-transport velocity The horizontal velocity at minimum flow in the chamber should be sufficient to scour the bottom or at least to give a minimum deposition of sludge solids that may have passed through the settling tank. Horizontal velocities should be at least 2 to 4.5 m/min (6.5 to 15 ft/min).

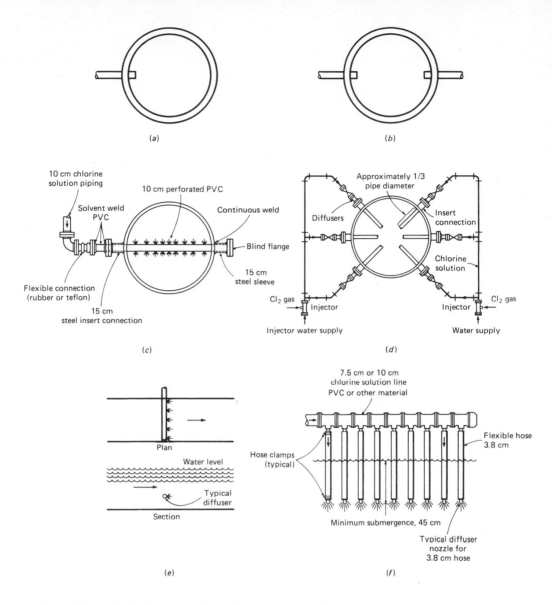

Figure 8-32 Typical diffusers used to inject chlorine solutions. (*Adapted in part from Ref. 21.*) (*a*) Single injector for small pipe. (*b*) Dual injector for small pipe. (*c*) Across-the-pipe diffuser for pipes larger than (0.9 m) in diameter. (*d*) Diffuser system for large conduits. (*e*) Single across-the-channel diffuser. (*f*) Typical hanging-nozzle-type chlorine diffuser for open channels.

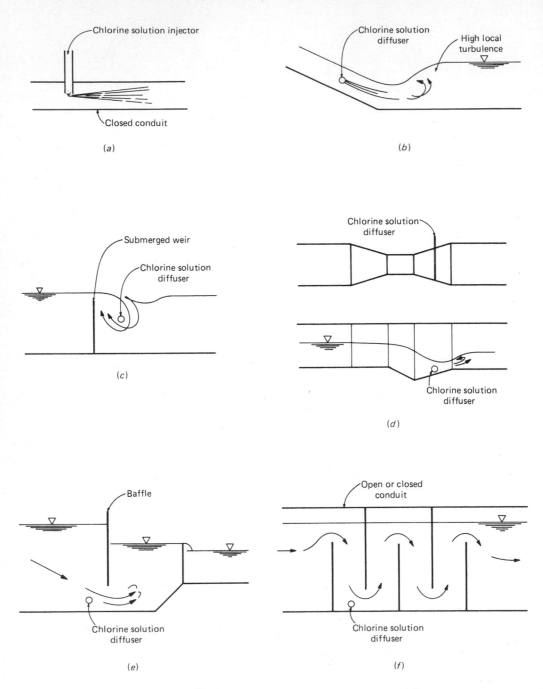

Figure 8-33 Methods of mixing chlorine solutions with wastewater by using the turbulence inherent in flow regime. (*a*) Pipeline. (*b*) Hydraulic jump. (*c*) Submerged weir. (*d*) Parshall flume. (*e*) Under baffle. (*f*) Over and under baffles.

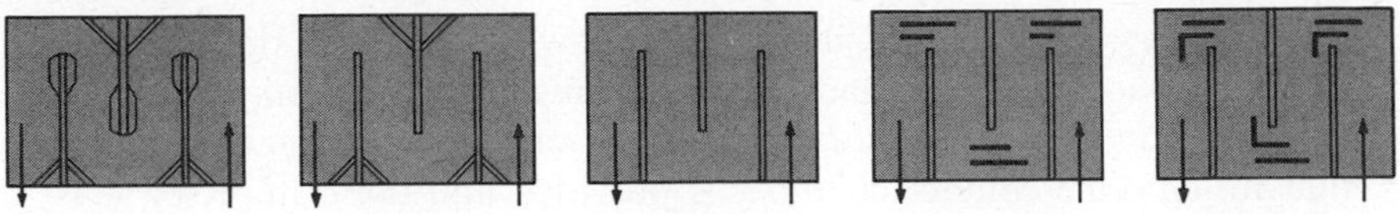

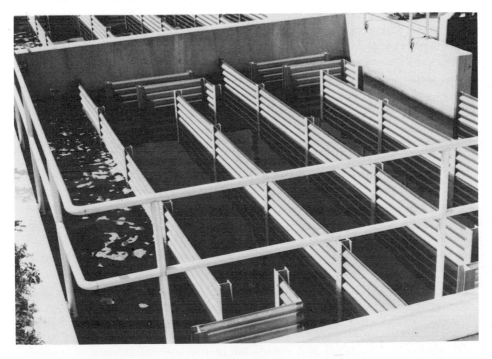

	Scheme I	Scheme IA	Scheme IB	Scheme II	Scheme IIA
Flow, m³/s	3.61	3.61	3.61	3.61	3.61
Water depth, m	4.33	4.33	4.33	4.33	4.33
Contact time, min					
Minimum	21.0	17.6	15.3	15.2	25.0
Mean	29.2	26.4	23.9	20.5	31.1
Maximum	36.5	37.8	34.8	31.9	39.5

Figure 8-34 Effect of chlorine-tank baffle design on actual detention time. (*Adapted from Ref. 10.*) Note: m³/s × 22.8245 = Mgal/d; m × 3.2808 = ft.

Chlorination basin bypass Provision should be made for dewatering the chlorine contact tank and for the removal of sludge by flushing or manual operation. If multiple-contact tanks are not used, provision should be made for a bypass and for emergency chlorination of the effluent.

Outlet control and MPN measurement The flow at the end of the contact chamber may be metered by means of a V-notch or rectangular weir. Control devices for chlorination in direct proportion to the flow may be operated from

Figure 8-35 Typical plug-flow chlorine contact basin with deflecting vanes.

these meters or from the main plant flowmeter. Final determination of the success of a chlorine contact chamber must be based on samples taken and analyzed for chlorine residual and MPN of coliform organisms. In the event that no chlorine contact chamber is provided and the outfall sewer is used for contact, the sample can be obtained at the point of chlorination, held for the theoretical detention time, and the residual determined. The sample is then dechlorinated and subsequently analyzed for bacteria by normal laboratory procedures.

8-12 ODOR CONTROL

In wastewater-treatment plants, the principal sources of odors are from (1) septic wastewater containing hydrogen sulfide and other odorous compounds on arrival at the plant, (2) industrial wastes discharged to the collection system, (3) unwashed grit, (4) scum on primary settling tanks, (5) organically overloaded biological-treatment processes, (6) sludge-thickening tanks, (7) waste-gas-burning operations where lower-than-optimum temperatures are used, (8) chemical mixing operations, (9) sludge incinerators, and (10) digested sludge in drying beds or sludge-holding basins.

With proper attention to design details, such as the use of submerged inlets and weirs, proper process loadings, the combustion of off gases at proper temperatures, and good housekeeping, the routine development of odors at treatment plants should not be a problem. It must also be recognized, however, that odors will occasionally develop. When they do, it is important that immediate steps be taken to control them. Often, this will involve the addition of some chemical, such as chlorine, hydrogen peroxide, lime, or ozone. In extreme cases, it may be necessary to cover some of the treatment facilities. Formed-in-place styrofoam domes have been used for the purpose. Where covers are used, the trapped gases must be collected and treated. The specific method of treatment will depend on the characteristics of the odorous compound.

The principal odor-control methods now used may be classified as physical, biological, and chemical. The major methods within each category are summarized in Table 8-20. Where activated carbon has been used for the control of odors, it has been found that the removal of odors depends on the concentration of the hydrocarbons in the odorous gas. It appears that the hydrocarbons are adsorbed preferentially before compounds such as H_2S are removed. The composition of the odorous gases to be treated must be defined if activated carbon is to be used (see Prob. 8-14).

The specific method that should be applied will vary with local conditions. However, because odor-control measures tend to be expensive, the cost of making process changes or modifications to the facilities to eliminate odor development should always be evaluated and compared to the cost of various alternative odor-control measures before their adoption is suggested.

Table 8-20 Methods to control odorous gases found in wastewater systems[a]

Method	Description and/or application
Physical methods:	
Combustion	Gaseous odors can be eliminated by combustion at temperatures varying from 650 to 760°C (1200 to 1400°F). The required temperature can be reduced by using catalysts. Sewer gases are often combusted in conjunction with treatment-plant solids
Adsorption, activated carbon	Odorous gases can be passed through beds of activated carbon to remove odors. Carbon regeneration can be used to reduce costs. Additional details may be found in Chap. 6
Adsorption on sand or soil	Odorous gases can be passed through sand or soil. Odorous gases from pumping stations may be vented to the surrounding soils or to specially designed beds containing sand or soils. The role of bacteria in the operation of such systems is not well understood at the present time
Oxygen injection	The injection of oxygen (either air or pure oxygen) into the wastewater to control the development of anaerobic conditions has proved to be effective
Masking agents	Perfume scents can be added to the wastewater to overpower or mask objectionable sewer-gas odors. In some cases, the odor of the masking agent is worse than the original odor
Scrubbing towers	Odorous gases can be passed through specially designed scrubbing towers to remove odors. Some type of chemical or biological agent is usually used in conjunction with the tower
Chemical methods:	
Scrubbing with various alkalies	Odorous gases can be passed through solutions of alkalies, such as calcium and sodium hydroxide, to remove odors. If the level of carbon dioxide is very high, costs may be prohibitive
Chemical oxidation	Oxidizing the odor compounds in the wastewater is one of the most common methods used to achieve odor control. Chlorine, ozone, and hydrogen peroxide are among the many oxidants that have been used. Chlorine also limits the development of a slime layer
Chemical precipitation	Chemical precipitation refers to the precipitation of sulfide with metallic salts, especially iron
Biological methods:	
Trickling filters or activated-sludge aeration tanks	Odorous gases can be passed through trickling-filter beds or injected into activated-sludge aeration tanks to remove odorous compounds
Special biological stripping towers	Specially designed towers can be used to strip odorous compounds. Typically, the towers are filled with plastic media of various types on which biological growths can be maintained

[a] Developed in part from Refs. 3, 12, and 17.

DISCUSSION TOPICS AND PROBLEMS

8-1 A vertical bar rack with 25-mm openings is to screen wastewater arriving at the treatment plant in a circular sewer with $d = 1.25$ m, $n = 0.013$, $s = 0.00064$. The maximum carrying capacity is four times the average dry-weather flow. Find the size of the steel bars that make up the rack, the number of bars in the rack, and the head loss for dry-weather flow conditions. Use rectangular bars.

8-2 Design an aerated grit chamber for a plant with an average flowrate of 16,000 m^3/d and a peak flowrate of 40,000 m^3/d. Determine the amount of air required and the pressure at the discharge of the blowers. Allow a 300-mm loss in the diffusers, and add the submergence plus 30 percent for loss in piping and valves. Determine the power required using an appropriate blower formula. Use a blower efficiency of 60 percent. Determine the monthly power bill, assuming a motor efficiency of 90 percent and a power cost of $0.03/kWh.

8-3 Discuss the advantages and disadvantages of aerated grit chambers versus horizontal-flow grit chambers.

8-4 Design a skimming chamber for a wastewater-treatment plant with a flow of 10,000 m^3/d. Use a detention time of 10 minutes and a rise velocity of 200 mm/min. Determine the surface area, surface loading, and chamber dimensions.

8-5 Design a circular radial-flow sedimentation tank for a town with a projected population of 45,000. Assume that the wastewater flow is 350 L/capita · d. Design for 2 h detention at 120 percent of the average flow. Determine the tank depth and diameter to produce an overflow rate of 35 m^3/m^2 · d for average flow. Assume standard tank dimensions to fit mechanisms that are made in diameters of whole meters and in depths of 0.2 m from 2 to 4 m.

8-6 A rectangular settling tank has an overflow rate of 30 m^3/m^2 · d and dimensions of 2.5 m deep by 6.0 m wide by 15.0 m long. Determine whether or not particles with a diameter of 0.1 mm and a specific gravity of 2.5 will be scoured from the bottom. Use $f = 0.03$ and $k = 0.04$.

8-7 Determine the percentage increase in the hydraulic and organic loading rates of the primary settling facilities of a treatment plant when 206 m^3/d of waste-settled activated sludge containing 2000 mg/L of suspended solids is discharged to the existing primary facilities for thickening. The average plant flowrate is 0.25 m^3/s, and the influent suspended-solids concentration is about 350 mg/L. The design overflow rate for the primary settling tanks without the waste sludge is 32 m^3/m^2 · d and the detention time is 2.8 h. Do you believe that the added incremental loadings will affect the performance of the primary settling facilities? Document the basis for your answer.

8-8 Prepare a table and compare the data from a minimum of six references with regard to the following primary sedimentation-tank-design parameters: (1) detention time (with and without preaeration); (2) expected BOD removal; (3) expected suspended-solids removal; (4) mean horizontal velocity; (5) surface loading (m^3/m^2 · d); (6) effluent weir overflow rate per unit length; (7) Froude number; (8) size of organic-particle removed; (9) length-to-width ratio (rectangular tanks); (10) average depth. List all references.

8-9 Contrast pressure and vacuum flotation with sedimentation discussing the following parameters:
  (*a*) Detention time
  (*b*) Surface-loading rate
  (*c*) Power input
  (*d*) Efficiency
  (*e*) Most favorable application for each type

8-10 Gravity filters are to be used to treat 0.25 m^3/s of settled effluent at a filtration rate of 200 L/m^2 · min. The filtration rate with one filter taken out of service for backwashing is not to exceed 250 L/m^2 · min. Determine the number of units and the area of each unit to satisfy these conditions. If each filter is backwashed for 5 min every 24 h at a wash rate of 60 m/h, determine the percentage of filter output used for washing if the filter is out of operation for a total of 30 min/d. What would be the total percentage of filter output used for backwashing if a surface washing system that requires 30 L/m^2 · min of filtered effluent is to be installed?

8-11 Determine the quantity of chlorine, in kilograms per day, necessary to disinfect a daily average primary effluent flow of 40,000 m^3/d. Use a dosage of 16 mg/L, and size the contact chamber for a contact time of 15 min at maximum flow, which is assumed to be two times the average flow.

8-12 You are called in as a consultant by the community of Rolling Hills to improve the performance of the chlorination facilities at their treatment plant. Their problem is that it has not been possible to achieve the bacterial kills called for in the discharge requirements. When you arrive at the treatment facilities, the mayor proudly shows you the chlorine contact basin that is designed as a CFSTR. The first thing the mayor says is, "Isn't it beautiful?" What is your answer to this statement, if any, and what long-term remedies might you propose? Assume that the disinfection process can be described adequately with first-order kinetics $(r_N = -kN)$.

8-13 The total sulfur concentration $(H_2S + HS^- + S^=)$ in a wastewater is 6 mg/L as S. Using the following expressions and data, determine the pH at which 99 percent of the total sulfur will remain in solution, assuming equilibrium conditions. If the concentration of hydrogen sulfide in the sewer atmosphere is not to exceed 2.0 ppm by volume, what pH must be maintained? In solving this problem, assume that the gas volume is equal to the liquid volume.

$$H_2S\ (gas) \rightleftharpoons H_2S\ (aq) \qquad \frac{[H_2S]}{P_{H_2S}} = 0.1$$

$$H_2S\ (aq) \rightleftharpoons H^+ + HS^- \qquad \frac{[H][HS]}{[H_2S]} = 10^{-7}$$

$$HS^- \rightleftharpoons H^+ + S^= \qquad \frac{[H][S]}{[HS]} = 10^{-15}$$

8-14 Based on the results of pilot-plant studies, it has been found that the H_2S saturation value for activated carbon is about 0.2 g H_2S/g activated carbon. The same saturation value has been found to apply to the gaseous hydrocarbons found in sewer off gasses. (a) If the density of activated carbon is 540 kg/m^3, determine the number of cubic meters of gas containing 10 ppm of H_2S by volume that can be processed per cubic meter of activated carbon. (b) How much activated carbon would be required on an annual basis if the H_2S in the air from within a pumping station is to be removed before being discharged to the atmosphere? The pump-station dry-well volume below grade is 100 m^3 and the air in the pump station contains 5 ppm of H_2S by volume and 100 ppm of hydrocarbons by volume (molecular weight = 100). Assume that 30 air changes per hour will be required [12].

REFERENCES

1. American Water Works Association: *Backwashing of Granular Filters, A Report of the Filtration Committee, AWWA*, August 1975.
2. Bishop, S. L., and B. W. Behrman: *Filtration of Wastewater Using Granular Media*, paper presented at the 1976 Thomas R. Camp Lecture Series on Wastewater Treatment and Disposal, Boston Society of Civil Engineers, Boston, 1976.
3. Camp, T. R.: Grit Chamber Design, *Sewage Works J.*, vol. 14, p. 368, 1942.
4. Camp, T. R.: Sedimentation and the Design of Settling Tanks, *Trans. ASCE*, vol. 111, 1946.
5. Collins, H. F.: *Effects of Initial Mixing and Residence Time Distribution on the Efficiency of the Wastewater Chlorination Process*, paper presented at the California State Department of Health Annual Symposium, Berkeley and Los Angeles, Calif., May 1970.
6. Culp, G. L.: Chemical Treatment of Raw Sewage/1 and 2, *Water Wastes Eng.*, vol. 4, nos. 7, 10, 1967.
7. Dahab, M. F., and J. C. Young: Unstratified-Bed Filtration of Wastewater, *J. Environ. Eng. Div., ASCE*, vol. 103, EE 12714, February 1977.

8. Eliassen, R., and D. F. Coburn: *Pretreatment—Versatility and Expandability,* presented at the ASCE Environmental Engineering Conference, Chattanooga, Tenn., 1968.

9. Kerri, K. D., and J. Brady (eds.): *Operation and Maintenance of Wastewater Collection Systems,* prepared for the U.S. Environmental Protection Agency, by Office of Water Program Operations by California State University, Department of Civil Engineering, Sacramento, 1976.

10. Louie, D., and M. Fohrman: Hydraulic Model Studies of Chlorine Mixing and Contact Chambers, *J. Water Pollut. Control Fed.,* vol. 40, no. 2, February 1968.

11. Metcalf, L., and H. P. Eddy: *American Sewerage Practice,* vol. III, 3d ed., McGraw-Hill, New York, 1935.

12. Metcalf & Eddy, Inc.: *Wastewater Engineering: Collection, Treatment, Disposal,* McGraw-Hill, New York, 1972.

13. *Physical-Chemical Wastewater Treatment Plant Design,* U.S. Environmental Protection Agency, Technology Transfer Seminar Publication, 1973.

14. *Process Design Manual for Upgrading Existing Wastewater Treatment Plants,* U.S. Environmental Protection Agency, Technology Transfer, October 1974.

15. *Process Design Manual for Phosphorus Removal,* U.S. Environmental Protection Agency, Technology Transfer, April 1976.

16. Steffensen, S. W., and N. Nash: Hypochlorination of Wastewater Effluents in New York City, *J. Water Pollut. Control Fed.,* vol. 39, no. 8, 1967.

17. Thistlethwayte, D. K. B. (ed.): *Control of Sulphides in Sewerage Systems,* Butterworth, Melbourne, Australia, 1972, and Ann Arbor Science Publishers, Ann Arbor, Mich., 1972.

18. *Wastewater Filtration—Design Considerations,* U.S. Environmental Protection Agency, Technology Transfer Seminar Publication, 1974.

19. *Wastewater Filtration—Design Considerations,* U.S. Environmental Protection Agency, Technology Transfer Report, 1977.

20. Water Pollution Control Federation: *Sewage Treatment Plant Design,* Manual of Practice 8, Washington, D.C., 1967.

21. White, G. C.: *Handbook of Chlorination,* Van Nostrand Reinhold Company, New York, 1972.

NINE

BIOLOGICAL UNIT PROCESSES

The objectives of the biological treatment of wastewater are to coagulate and remove the nonsettleable colloidal solids and to stabilize the organic matter. For domestic wastewater (i.e., sewage), the major objective is to reduce the organic content and, in many cases, the nutrients, such as nitrogen and phosphorus. For agricultural return wastewater, the objective is to remove the nutrients, specifically nitrogen and phosphorus, that are capable of stimulating the growth of aquatic plants. For industrial wastewater, the objective is to remove or reduce the concentration of organic and inorganic compounds. Because many of these compounds are toxic to microorganisms, pretreatment may be required.

With proper analysis and environmental control, almost all wastewaters can be treated biologically. Therefore, it is the responsibility of the sanitary engineer to understand the requirements of each biological process and to ensure that the proper environment is produced and effectively controlled. In view of the importance of biological treatment, it is the purpose of this chapter (1) to present some of the fundamentals of wastewater microbiology, (2) to discuss the key factors governing biological growth and waste-treatment kinetics, and (3) to illustrate the application of fundamentals and kinetics to the analysis of the biological processes used most commonly for wastewater treatment. To provide some perspective on how this information relates to the actual biological treatment processes that are now used, the processes are first identified and their applications are briefly described. The information presented in this chapter provides the background for the design of biological treatment processes discussed in Chaps. 10 through 12.

9-1 BIOLOGICAL TREATMENT: AN OVERVIEW

This brief discussion of biological treatment is necessarily qualitative because it is intended to serve only as an overview of the basic concepts underlying biological treatment. It will introduce the reader to the biological-treatment processes in common use and to the information to be presented in this chapter.

Some Useful Definitions

To understand the concepts of biological treatment, it will be helpful to know the following terms:

Aerobic processes are biological-treatment processes that occur in the presence of oxygen. Certain bacteria that can survive only in the presence of dissolved oxygen are known as obligate (restricted to a specified condition in life) aerobes.

Anaerobic processes are biological-treatment processes that occur in the absence of oxygen. Bacteria that can survive only in the absence of any dissolved oxygen are known as obligate anaerobes.

Anoxic denitrification is the process by which nitrate nitrogen is converted biologically to nitrogen gas in the absence of oxygen. This process is also known as anaerobic denitrification.

Facultative processes are biological-treatment processes in which the organisms are indifferent to the presence of dissolved oxygen. These organisms are known as facultative microorganisms.

Microaerophils are a group of microorganisms that grow best in the presence of low concentrations of oxygen.

Carbonaceous BOD removal is the biological conversion of the carbonaceous organic matter in wastewater to cell tissue and various gaseous end products. In the conversion, it is assumed that the nitrogen present in the various compounds is converted to ammonia.

Nitrification is the two-stage biological process by which ammonia is converted first to nitrite and then to nitrate.

Denitrification is the biological process by which nitrate is converted to nitrogen and other gaseous end products.

Stabilization is the biological process by which the organic matter in the sludges produced from the primary settling and biological treatment of wastewater is stabilized, usually by conversion to gases and cell tissue. Depending on whether this stabilization is carried out under anaerobic or aerobic conditions, the process is known as anaerobic or aerobic digestion.

Substrate is the term used to denote the organic matter or nutrients that are converted during biological treatment or that may be limiting in biological treatment. For example, the carbonaceous organic matter in wastewater is referred to as the substrate that is converted during biological treatment.

Suspended-growth processes are the biological-treatment processes in which the microorganisms responsible for the conversion of the organic matter or other constituents in the wastewater to gases and cell tissue are maintained in suspension within the liquid.

Attached-growth processes are the biological-treatment processes in which the microorganisms responsible for the conversion of the organic matter or other constituents in the wastewater to gases and cell tissue are attached to some inert medium, such as rocks, slag, or specially designed ceramic or plastic materials. Attached-growth treatment processes are also known as fixed-film processes.

Role of Microorganisms

In the removal of carbonaceous BOD, the coagulation of nonsettleable colloidal solids and the stabilization of organic matter are accomplished biologically using a variety of microorganisms, principally bacteria. The microorganisms are used to convert the colloidal and dissolved carbonaceous organic matter into various gases and into cell tissue. Because cell tissue has a specific gravity slightly greater than that of water, the resulting tissue can be removed from the treated liquid by gravity settling.

It is important to note that unless the cell tissue that is produced from the organic matter is removed from the solution, complete treatment has not been accomplished because the cell tissue, which itself is organic, will be measured as BOD in the effluent. If the cell tissue is not removed, the only treatment that has been achieved is that associated with the bacterial conversion of a portion of the organic matter originally present to various gaseous end products.

Treatment Processes

The major biological processes used for wastewater treatment are identified in Table 9-1. There are four major groups: aerobic processes, anoxic processes, anaerobic processes, and a combination of the aerobic/anoxic or anaerobic processes. The individual processes are further subdivided, depending on whether treatment is accomplished in suspended-growth systems, attached-growth systems, or combinations thereof.

The principal applications of these processes, also identified in Table 9-1, are for (1) the removal of the carbonaceous organic matter in wastewater, usually measured as BOD, total organic carbon (TOC), or chemical oxygen demand (COD); (2) nitrification; (3) denitrification; and (4) stabilization. In this chapter, the emphasis will be on the removal of carbonaceous material, both aerobically and anaerobically. Nitrification and denitrification are considered in greater detail in Chap. 12.

Table 9-1 Major biological treatment processes used for wastewater treatment

Type	Common name	Use[a]	See section
Aerobic processes:			
Suspended growth	Activated-sludge process		
	Conventional (plug flow)		9-5
	Continuous-flow stirred-tank		
	Step aeration		
	Pure oxygen	Carbonaceous BOD removal (nitrification)	10-1
	Modified aeration		
	Contact stabilization		
	Extended aeration		
	Oxidation ditch		
	Suspended-growth nitrification	Nitrification	9-5, 12-4
	Aerated lagoons	Carbonaceous BOD removal (nitrification)	9-5
	Aerobic digestion		9-5
	Conventional air	Stabilization, carbonaceous BOD removal	11-9
	Pure oxygen		
	High-rate aerobic algal ponds	Carbonaceous BOD removal	9-5
Attached growth	Trickling filters		9-6
	Low-rate	Carbonaceous BOD removal (nitrification)	10-3
	High-rate		
	Roughing filters	Carbonaceous BOD removal	9-6
	Rotating biological contactors	Carbonaceous BOD removal (nitrification)	9-6
	Packed-bed reactors	Nitrification	9-6, 12-4

Type	Common name	Use[a]	
Combined processes	Trickling filter, activated sludge Activated sludge, trickling filter	Carbonaceous BOD removal (nitrification)	10-4
Anoxic processes: Suspended growth Attached growth	Suspended-growth denitrification Fixed-film denitrification	Denitrification	9-7

Anaerobic processes:
Suspended growth
	Anaerobic digestion
	  Standard-rate, single-stage
	  High-rate, single-stage
	  Two-stage
		Stabilization, carbonaceous BOD removal	9-8

Attached growth	Anaerobic contact process	Carbonaceous BOD removal	11-8
	Anaerobic filter	Carbonaceous BOD removal, stabilization (denitrification)	9-8
	Anaerobic lagoons (ponds)	Carbonaceous BOD removal (stabilization)	9-9
Aerobic/anoxic or anaerobic processes: Suspended growth	Single-stage nitrification-denitrification	Carbonaceous BOD removal, nitrification, denitrification	9-9
	Nitrification-denitrification	Nitrification, denitrification	12-5
	Facultative lagoons (ponds)	Carbonaceous BOD removal	12-5
	Maturation or tertiary ponds	Carbonaceous BOD removal (nitrification)	9-10
Attached growth Combined processes	Anaerobic-facultative lagoons Anaerobic-facultative-aerobic lagoons	Carbonaceous BOD removal	9-10

[a] Major use is presented first; other uses are identified in parentheses.

9-2 SOME FUNDAMENTALS OF MICROBIOLOGY

Basic to the design of a biological-treatment process, or to the selection of the type of process to be used, is an understanding of the form, structure, and biochemical activities of the important microorganisms. In this section, the cytology and physiology of the microorganisms commonly encountered in wastewater treatment will be presented and discussed. For a more thorough discussion of the topics considered in this section, Refs. 37 and 45 are recommended.

Basic Concepts

In the past, microorganisms were commonly grouped into two kingdoms: plants and animals. Because of taxonomic difficulties, the recent trend is to group them into three kingdoms: protista, plants, and animals. The members of the kingdom protista are called protists. Summary data on the characteristics of the microorganisms in each kingdom are presented in Table 9-2. Although the most significant differences among protists, plants, and animals are shown in the table, the three kingdoms are similar in that the cell is the basic unit of life for each, regardless of the complexity of the organism.

Cell structure In general, most living cells are quite similar. As shown in Fig. 9-1, they have a cell wall, which may be either a rigid or a flexible membrane. If they are motile, they usually possess flagella or some hairlike appendages. The interior of the cell contains a colloidal suspension of proteins, carbohydrates, and other complex organic compounds, called the cytoplasm.

Table 9-2 The three kingdoms of microorganisms

Kingdom	Representative members	Characterization
Animal	Rotifers	
	Crustaceans	Multicellular,
Plant	Mosses	with tissue
	Ferns	differentiation
	Seed plants	
Protista:		
Higher[a]	Algae	
	Protozoa	Unicellular or
	Fungi	multicellular,
	Slime molds	without tissue
Lower[b]	Blue-green algae	differentiation
	Bacteria	

[a] Contain true nucleus (eucaryotic cells).
[b] Contain no nuclear membrane (procaryotic cells).

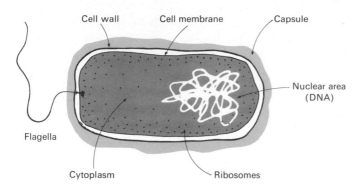

Figure 9-1 Generalized schematic of bacterial cell [20].

Each cell contains nucleic acids, the hereditary material that is vital to reproduction. The cytoplasmic area contains ribonucleic acid (RNA), whose major role is in the synthesis of proteins. Also within the cell wall is the area of the nucleus, which is rich in deoxyribonucleic acid (DNA). DNA contains all the information necessary for the reproduction of all the cell components and may be considered to be the blueprint of the cell. In some cells, the DNA is enclosed by a membrane and the nucleus is clearly defined (eucaryotic cells). In other cells, the nucleus is poorly defined (procaryotic cells). As shown in Table 9-2, bacteria and blue-green algae are examples of procaryotic cells.

Energy and carbon sources To continue to produce and function properly, an organism must have a source of energy and carbon for the synthesis of new cellular material. Inorganic elements, such as nitrogen and phosphorus, and other trace elements, such as sulfur, potassium, calcium, and magnesium, are also vital to cell synthesis. Two of the most common sources of cell carbon for microorganisms are carbon dioxide and organic matter. If an organism derives its cell carbon from carbon dioxide, it is called autotrophic; if it uses organic carbon, it is called heterotrophic.

Table 9-3 General classification of microorganisms by sources of energy and carbon

Classification	Energy source	Carbon source
Autotrophic:		
Photosynthetic	Light	CO_2
Chemosynthetic	Inorganic oxidation-reduction reaction	CO_2
Heterotrophic	Organic oxidation-reduction reaction	Organic carbon

Energy is also needed in the synthesis of new cellular material. For autotrophic organisms, the energy can be supplied by the sun, as in photosynthesis, or by an inorganic oxidation-reduction reaction. If the energy is supplied by the sun, the organism is called autotrophic photosynthetic. If the energy is supplied by an inorganic oxidation-reduction reaction, it is called autotrophic chemosynthetic. For heterotrophic organisms, the energy needed for cell synthesis is supplied by the oxidation or fermentation of organic matter. A classification of micro-organisms by sources of energy and cell carbon is presented in Table 9-3.

Aerobic and anaerobic metabolism Organisms can also be classed according to their ability to use oxygen. Aerobic organisms can exist only when there is a supply of molecular oxygen. Anaerobic organisms can exist only in an environment that is void of oxygen. Facultative organisms have the ability to survive with or without free oxygen.

Important Microorganisms

The sanitary engineer should be familiar with the characteristics of the following microorganisms because of their importance in biological-treatment processes: (1) bacteria, (2) fungi, (3) algae, (4) protozoa, (5) rotifers, (6) crustaceans, and (7) viruses.

Bacteria Bacteria are single-cell protists. They use soluble food and, in general, will be found wherever moisture and a food source are available. Their usual mode of reproduction is by binary fission, although some species reproduce sexually or by budding. Even though there are thousands of different species of bacteria, their general form falls into one of three categories: spherical, cylindrical, and helical. Bacteria vary widely in size. Representative sizes are 0.5 to 1.0 μ in diameter for the spherical, 0.5 to 1.0 μ in width by 1.5 to 3.0 μ in length for the cylindrical (rods), and 0.5 to 5 μ in width by 6 to 15 μ in length for the helical (spiral). The general structure of a cell was discussed earlier.

Tests on a number of different bacteria indicate that they are about 80 percent water and 20 percent dry material, of which 90 percent is organic and 10 percent inorganic. An approximate formula for the organic fraction is $C_5H_7O_2N$ [17]. As indicated by the formula, about 53 percent by weight of the organic fraction is carbon. The formulation $C_{60}H_{87}O_{23}N_{12}P$ has been proposed when phosphorus is also considered [29]. Compounds comprising the inorganic portion include P_2O_5 (50 percent), SO_3 (15 percent), Na_2O (11 percent), CaO (9 percent), MgO (8 percent), K_2O (6 percent), and Fe_2O_3 (1 percent). Since all these elements and compounds must be derived from the environment, a shortage of any of these substances would limit and, in some cases, alter growth.

Temperature and pH play a vital role in the life and death of bacteria, as well as in other microscopic plants and animals. It has been observed that the rate of reaction for microorganisms increases with increasing temperature, doubling with about every 10°C of rise in temperature until some limiting

Table 9-4 Some typical temperature ranges for various bacteria

Type	Temperature, °C	
	Range	Optimum
Cryophilic[a]	−2–30	12–18
Mesophilic	20–45	25–40
Thermophilic	45–75	55–65

[a] Also called psychrophilic.

Note: $1.8(°C) + 32 = °F$.

temperature is reached. According to the temperature range in which they function best, bacteria may be classified as cryophilic or psychrophilic, mesophilic, and thermophilic. Typical temperature ranges for bacteria in each of these categories are presented in Table 9-4. For a more detailed discussion of the organisms in the various temperature ranges, see Refs. 20 and 45.

The pH of a solution is also a key factor in the growth of organisms. Most organisms cannot tolerate pH levels above 9.5 or below 4.0. Generally, the optimum pH for growth lies between 6.5 and 7.5.

Metabolically, bacteria can be classified as heterotrophic or autotrophic. The most common autotrophs are chemosynthetic, but a few are able to perform photosynthesis. The purple sulfur bacteria (*Thiorhodaceae*) and the green sulfur bacteria (*Chlorobiaceae*) are representative examples of autotrophic photosynthetic bacteria. In biological wastewater treatment, the heterotrophic are usually the most important group, because of their requirement for organic compounds for cell carbon. The autotrophic and heterotrophic bacteria can be further classified as aerobic, anaerobic, or facultative, depending on their need for oxygen.

Fungi In sanitary engineering, fungi are considered to be multicellular, non-photosynthetic, heterotrophic protists. Microbiologists are less restrictive in their definition and include bacteria as a class of fungi, but in sanitary engineering bacteria are so important that they are considered separately.

Fungi are usually classed by their mode of reproduction. They reproduce sexually or asexually, by fission, budding, or spore formation. Molds or "true fungi" produce microscopic units (hyphae), which collectively form a filamentous mass called the mycelium. In sanitary engineering, the terms "fungi" and "molds" are used synonymously. Yeasts are fungi that cannot form a mycelium and are therefore unicellular.

Most fungi are strict aerobes. They have the ability to grow under low-moisture conditions and can tolerate an environment with a relatively low pH. The optimum pH for most species is 5.6; the range is 2 to 9. Fungi also have a low nitrogen requirement and need approximately one-half as much as bacteria.

The ability of the fungi to survive under low pH and nitrogen-limiting conditions makes them very important in the biological treatment of some industrial wastes and in the composting of solid organic wastes.

Algae Algae are unicellular or multicellular, autotrophic, photosynthetic protists. Algae are undesirable in water supplies because they produce bad tastes and odors. In water filtration plants, the presence of algae will shorten filter runs. The green color of most species and their ability to form mats lower the aesthetic value of the water. In oxidation ponds, algae are valuable in that they have the ability to produce oxygen through the mechanism of photosynthesis. At night, when light is no longer available for photosynthesis, they use up the oxygen in respiration. Respiration also occurs in the presence of sunlight; however, the net reaction is the production of oxygen. Equations 9-1 and 9-2 represent simplified biochemical reactions for photosynthesis and respiration.

Photosynthesis:

$$CO_2 + 2H_2O \xrightarrow{\text{light}} \underset{\substack{\text{New}\\\text{algae}\\\text{cells}}}{(CH_2O)} + O_2 + H_2O \qquad (9\text{-}1)$$

Respiration:

$$CH_2O + O_2 \rightarrow CO_2 + H_2O \qquad (9\text{-}2)$$

In an aquatic environment, it can be seen that this type of metabolic system will produce a diurnal variation in dissolved oxygen. The ability of algae to produce oxygen is vital to the ecology of the water environment. For an aerobic or facultative oxidation pond to operate effectively, algae are needed to supply oxygen to aerobic, heterotrophic bacteria. This symbiotic relationship between algae and bacteria will be expanded upon in the section on aerobic oxidation ponds at the end of this chapter.

Because algae use carbon dioxide in photosynthetic activity, high pH conditions can result. In addition, as the pH increases, the alkalinity components change, and carbonate and hydroxide alkalinity tend to predominate. If the water has a high concentration of calcium, calcium carbonate will precipitate when the carbonate and calcium-ion concentrations become great enough to exceed the solubility product. This removal of the carbonate ion by precipitation will keep the pH from continuing to increase. As is the case with dissolved oxygen, there is also a diurnal variation in pH. During the day algae use carbon dioxide, which results in a rising pH, while at night they produce carbon dioxide, which results in a falling pH.

Like other microorganisms, algae require inorganic compounds to reproduce. The principal nutrients required, other than carbon dioxide, are nitrogen and phosphorus. Other trace elements, such as iron, copper, and molybdenum, are also very noteworthy. It is noteworthy that the problem of preventing excessive algal growth in natural waters has, to date, centered around nutrient removal. Some scientists advocate the removal of nitrogen from treatment-plant effluents, while others recommend the removal of phosphorus, and still others recommend removal of both nitrogen and phosphorus. Not to be forgotten are the trace elements, which in some cases may be the limiting nutrients in the growth of algae.

Four classes of freshwater algae are of importance [44]:

1. *Green (Chlorophyta)*. The green algae are principally a freshwater species, and can be unicellular or multicellular. A distinguishing feature of this species is that the chlorophyll and other pigments are contained in chloroplasts. Chloroplasts are the sites of photosynthesis and are membrane-surrounded structures that contain chlorophyll and other pigments. Common green algae are those of the *Chlorella* group found in stabilization ponds.
2. *Motile green (Volvocales Euglenophyta)*. Colonial in nature, these algae are bright green, unicellular, and flagellated. *Euglena* is a member of this particular group of algae. *Mastigophora* that contain chlorophyll are also often included in this category.
3. *Yellow green or golden brown (Chrysophyta)*. Most forms of the *Chrysophyta* are unicellular. They are freshwater inhabitants, and their characteristic color is due to yellowish brown pigments that conceal the chlorophyll. Of this group of algae, the most important are the diatoms. They are found in both fresh and salt waters. Diatoms have shells, which are composed mainly of silica. Deposits of these shells are known as diatomaceous earth, which is used as a filter aid.
4. *Blue-green (Cyanophyta)*. The blue-green algae are of very simple form and are similar to bacteria in several respects. They are unicellular, usually enclosed in a sheath, and have no flagella. They differ from other algae in that their chlorophyll is not contained in chloroplasts but is diffused throughout the cell. The blue-green algae are of interest in water and wastewater engineering for numerous reasons. They have the ability to form large, dense mats on the surface of the water and as a result lower the aesthetic value of the water. At times they can impart undesirable tastes and odors to waters. An important characteristic is their ability to use nitrogen from the atmosphere as a nutrient in cell synthesis. Thus the removal of nitrogenous compounds from the water will not eliminate the source of nitrogen for these species of algae.

Protozoa Protozoa are motile, microscopic protists that are usually single cells. The majority of protozoa are aerobic heterotrophs, although a few are anaerobic. Protozoa are generally an order of magnitude larger than bacteria and often consume bacteria as an energy source. In effect, the protozoa act as polishers of the effluents from biological waste-treatment processes by consuming bacteria and particulate organic matter.

Protozoa are usually divided into the following five groups:

1. *Sarcodina*. The *Sarcodina* are characterized by their pseudopods, or false feet, which they use for movement and the capturing of food. *Endamoeba histolytica*, the cause of an intestinal disease in man, is a member of the *Sarcodina* group.

2. *Mastigophora.* The *Mastigophora* are characterized by their flagella, which they use for motility. Some microbiologists divide the *Mastigophora* into two groups: those that do and do not contain chlorophyll. *Euglena* and its colorless counterpart *Astasia* are examples.
3. *Sporozoa.* The *Sporozoa* are spore-forming protozoa and obligate parasites. Their only real interest to sanitary engineers is that certain *Sporozoa*, namely, four species of *Plasmodium*, cause malaria.
4. *Infusoria or Ciliata.* Movement by means of cilia is characteristic of these protozoa. The cilia are hairlike extensions from the cell membrane. Besides being responsible for the organism's movement, they are also important in assisting the protozoa to capture solid food. Sanitary engineers usually consider the *Ciliata* to be divided into two types, the free-swimming and the stalked. The free-swimming type must swim after bacteria. They require a great deal of food, because they expend so much energy in swimming. The *Paramecium* is a free-swimming ciliate that is important in wastewater treatment. The stalked ciliates are attached to something solid and must catch their food as it passes by. Because their movement is limited, they require less food for energy. The *Vorticella* is a stalked ciliate that is important in biological-treatment processes, especially in the activated-sludge process.
5. *Suctoria.* The *Suctoria* are protozoa that have long tentacles, which they use to capture other protozoa and then draw the protoplasm from these protozoa into their own bodies. During the early stages of their life cycle, the *Suctoria* have cilia, but later in the adult stage they obtain their tentacles.

Rotifers The rotifer is an aerobic, heterotrophic, and multicellular animal. Its name is derived from the fact that it has two sets of rotating cilia on its head which are used for motility and capturing food. Rotifers are very effective in consuming dispersed and flocculated bacteria and small particles of organic matter. Their presence in an effluent indicates a highly efficient aerobic biological purification process.

Crustaceans Like the rotifer, the crustacean is an aerobic, heterotrophic, and multicellular animal. Unlike the rotifer, the crustacean has a hard body, or shell. Crustaceans are an important source of food for fish and as such are normal occupants in most natural waters. Except for their occasional presence in underloaded oxidation ponds, they do not exist in biological-treatment systems to any noticeable extent. Their presence is indicative of an effluent that is low in organic matter and high in dissolved oxygen.

Viruses A virus is the smallest biological structure containing all the information necessary for its own reproduction. Viruses are so small that they can be seen only with the aid of an electron microscope. They are obligate parasites and as such require a host in which to live. Once they have a host, they redirect its complex machinery to produce new virus particles. Eventually the host cell ruptures, releasing new virus particles, which can go on to infect new cells.

Viruses are usually classified by the host they infect. Many viruses that produce diseases in man are also known to be excreted in the feces of man. Thus in wastewater treatment the sanitary engineer has the responsibility of ensuring that these viruses are effectively controlled. This is usually done by chlorination and proper disposal of the plant effluent.

Cell Physiology

Since the bacteria are the most widely occurring microorganism in biological waste treatment, the discussion of cell physiology that follows will be focused on the bacteria; however, the basic principles are applicable to all living cells.

Role of enzymes The process by which microorganisms grow and obtain energy is complex and intricate; there are many pathways and cycles. Vital to the reactions involved in these pathways and cycles are the actions of enzymes, which are organic catalysts produced by the living cell. Enzymes are proteins or proteins combined either with an inorganic molecule or with a low-molecular-weight organic molecule. As catalysts, enzymes have the capacity to increase the speed of chemical reactions greatly without altering themselves.

There are two general types of enzymes, extracellular and intracellular. When the substrate or nutrient required by the cell is unable to enter the cell wall, the extracellular enzyme converts the substrate or nutrient to a form that can then be transported into the cell. Intracellular enzymes are involved in the synthesis and energy reactions within the cell.

Enzymes are known for their high degree of efficiency in converting substrate to end products. One enzyme molecule can change many molecules of substrate per minute to end products. Enzymes are also known for their high degree of substrate specificity. This high degree of specificity means that the cell must produce a different enzyme for every substrate it uses. An enzyme reaction can be represented by the following general equation:

$$(E) \; + \; (S) \; \rightarrow (E)(S) \; \rightarrow \; (P) \; + \; (E) \tag{9-3}$$

$$\underset{\text{Enzyme}}{} \quad \underset{\text{Substrate}}{} \quad \underset{\substack{\text{Enzyme-}\\\text{substrate}\\\text{complex}}}{} \quad \underset{\text{Product}}{} \quad \underset{\text{Enzyme}}{}$$

As illustrated, the enzyme functions as a catalyst by forming a complex with the substrate, which is then converted to a product and the original enzyme. At this point, the product may be acted upon by another enzyme. In fact, a sequence of complexes and products may be formed before the final end product is produced. In a living cell, the transformation of the original substrate to the final end product is accomplished by such an enzyme system. The activity of enzymes is substantially affected by pH and temperature, as well as by substrate concentration. Each enzyme has a particular optimum pH and temperature. The optimum pH and temperature of the key enzymes in the cell are reflected in the overall temperature and pH preferences of the cell.

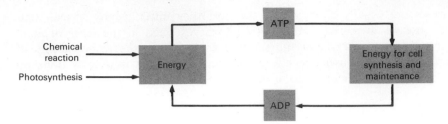

Figure 9-2 Schematic representation of ADP-ATP cellular-energy transfer system.

Dissimilatory and assimilatory processes Along with enzymes, energy is required to carry out the biochemical reactions in the cell. Energy is released in the cell by oxidizing organic or inorganic matter or by a photosynthetic reaction. The energy released is captured and stored in the cell by certain organic compounds. The most common storing compound is adenosine triphosphate (ATP). The energy captured by this compound is used for cell synthesis, maintenance, and motility. When the ATP molecule has expended its captured energy to the reactions involved in cell synthesis and maintenance, it changes to a discharged state called adenosine diphosphate (ADP). This ADP molecule can then capture the energy released in the breakdown of organic or inorganic matter. Having done this, the compound again assumes an energized state as the ATP molecule. The ADP-ATP cellular-energy system is shown schematically in Fig. 9-2. In this context, dissimilatory processes may be considered to be those processes associated with the production and/or capture of energy, whereas assimilatory processes are those associated with the production of cell tissue.

Simplified biochemical exothermic reactions (those that release energy) for heterotrophic and autotrophic bacteria are listed in Table 9-5. Although it is true that the energy they release is used to charge ADP molecules, many steps, all catalyzed by enzyme systems, are involved. A discussion of the pathways and cycles in these energy-releasing reactions is beyond the scope of this text; however, in simple terms, the overall metabolism of bacterial cells can be thought of as consisting of two biochemical reactions: energy and synthesis. The first reaction releases energy so that the second reaction of cellular synthesis can proceed. Both reactions are the result of numerous systems within the cell, and each system consists of many enzyme-catalyzed reactions. The energy released in the "energy reaction" is captured by the enzyme-catalyzed system involving ATP

Table 9-5 Typical exothermic biochemical reactions

Biochemical energy reaction	Nutrition of bacteria
$C_6H_{12}O_6 + 6O_2 \rightarrow 6CO_2 + 6H_2O$	Heterotrophic, aerobic
$C_6H_{12}O_6 \rightarrow 3CH_4 + 3CO_2$	Heterotrophic, anaerobic
$2NH_4^+ + 3O_2 \rightarrow 2NO_2^- + 2H_2O + 4H^+$	Autotrophic, chemosynthetic, aerobic
$5S + 2H_2O + 6NO_3^- \rightarrow 5SO_4^{--} + 3N_2 + 4H^+$	Autotrophic, chemosynthetic, anaerobic

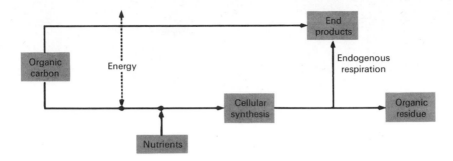

Figure 9-3 Schematic representation of heterotrophic bacterial metabolism.

and then transferred via ATP to the energy-requiring "synthesis reaction." A schematic presentation of cellular or bacterial metabolism is given in Figs. 9-3 and 9-4. These diagrams apply to aerobic, anaerobic, or facultative organisms.

It can be seen in Fig. 9-3 that, for heterotrophic bacteria, only a portion of the organic waste is converted into end products. The energy obtained from this biochemical reaction is used in the synthesis of the remaining organic matter into new cells. As the organic matter in the wastewater becomes limiting, there will be a decrease in cellular mass, because of the utilization of cellular material without replacement. If this situation continues, eventually all that will remain of the cell is a relatively stable organic residue. This overall process of a net decrease in cellular mass is termed endogenous respiration. Its place in the flow of energy and carbon for heterotrophic and autotrophic organisms is shown in Figs. 9-3 and 9-4.

When an autotrophic organism synthesizes new cellular material, the carbon source is carbon dioxide. The energy source for a cell synthesis is either light or the energy given off from an inorganic oxidation-reduction reaction. The flow of carbon and energy for autotrophic, chemosynthetic bacteria and for autotrophic, photosynthetic bacteria is shown in Figs. 9-4 and 9-5, respectively.

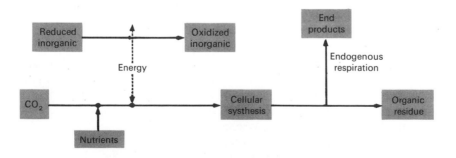

Figure 9-4 Schematic representation of chemosynthetic autotrophic bacterial metabolism.

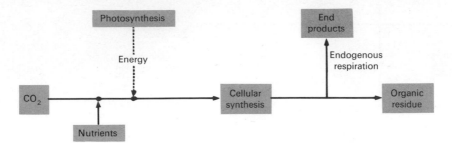

Figure 9-5 Schematic representation of photosynthetic autotrophic bacterial metabolism.

Nutrient requirement Nutrients, rather than organic or inorganic substrate in the wastewater, may at times be the limiting material for cell synthesis and growth. Bacteria, like algae, require nutrients, principally nitrogen and phosphorus, for growth. These nutrients may not always be present in sufficient quantities, as in the case of high-carbohydrate industrial wastes (e.g., sugar beets, sugar cane). Nutrient addition to the waste may be necessary for the proper growth of the bacteria and the subsequent degradation of the waste material [48]. The position of nutrient addition in the flow of energy and carbon for heterotrophic and autotrophic organisms is shown in Figs. 9-3 to 9-5.

Aerobic and anaerobic cycles in nature To the sanitary engineer, there are two very important cycles in nature involving the growth and decay of organic matter:

1. The aerobic cycle, in which oxygen is used for the decay of the organic matter
2. The anaerobic cycle, in which oxygen is not used for the decay of organic matter

These two cycles are shown in Figs. 9-6 and 9-7. The elements of nitrogen and sulfur are shown as integral parts of the cycles. These are two important elements in the synthesis and decay of organic matter, but they are certainly not the only ones. Other elements are involved, and biochemical cycles could be drawn including them.

It should be noted that the title of aerobic and anaerobic applies only to the right-hand side of Figs. 9-6 and 9-7, or to the decomposition portion of the cycles. Here, dead organic matter is first broken down into initial and intermediate products, before the final stabilized products are produced. Both heterotrophic and autotrophic bacteria are involved in the many biodegradation processes required to obtain the final stabilized products. In the aerobic systems, the final products of degradation are more fully oxidized and hence at a lower energy level than the final products of the anaerobic degradation system. This accounts for the fact that much more energy is released in aerobic than in anaerobic degradation. As a result, anaerobic degradation is a much slower process.

The left-hand side of the cycle is the same in both the aerobic and anaerobic systems. This side is involved with the building or synthesis of the organic

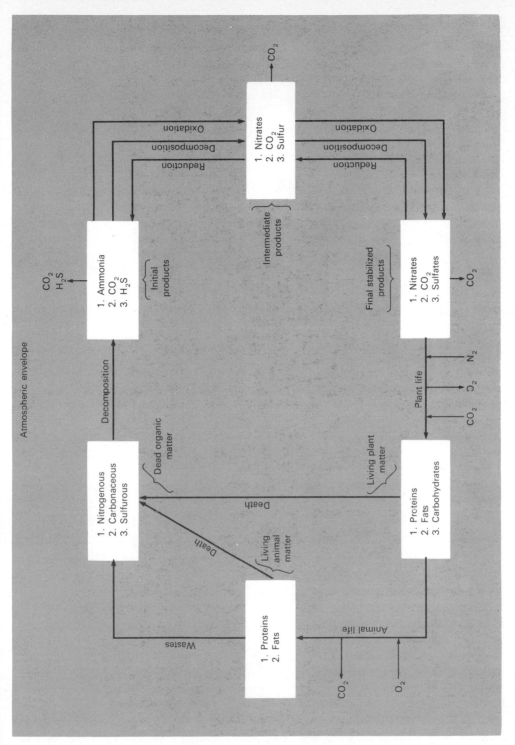

Figure 9-6 The aerobic cycle.

409

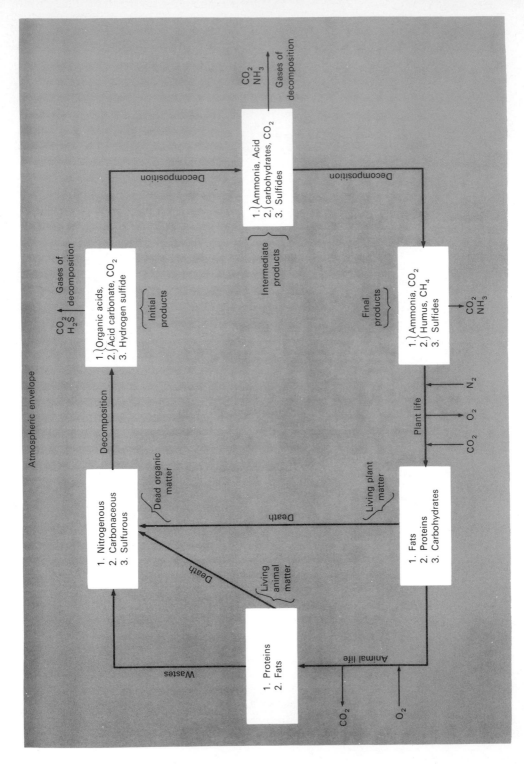

Figure 9-7 The anaerobic cycle.

matter necessary for plant and animal life. Eventually, because of death or wastes from animal life, dead organic matter is made available to the bacterial decomposers, and the cycle is again repeated.

The decomposition portion of the cycle that occurs in nature is the concern of the sanitary engineer. By controlling the environment of the microorganisms, the decomposition of wastes is speeded up. Regardless of the type of waste, the biological-treatment process consists of controlling the environment required for optimum growth of the microorganisms involved.

9-3 BACTERIAL GROWTH AND BIOLOGICAL OXIDATION

Effective environmental control in biological waste treatment is based on an understanding of the basic principles governing the growth of microorganisms. Therefore, the following discussion is concerned with the growth of bacteria, the microorganisms of primary importance in biological treatment.

General Growth Patterns in Pure Cultures

As mentioned earlier, bacteria can reproduce by binary fission, by a sexual mode, or by budding. Generally, they reproduce by binary fission, that is, by dividing; the original cell becomes two new organisms. The time required for each fission, which is termed the generation time, can vary from days to less than 20 min. For example, if the generation time is 30 min, one bacterium would yield 16,777,216 bacteria after a period of 12 h. This is a hypothetical figure, for bacteria would not continue to divide indefinitely because of various environmental limitations, such as substrate concentration, nutrient concentration, or even system size.

Growth in terms of bacterial numbers The general growth pattern of bacteria in a batch culture is shown in Fig. 9-8. Initially, a small number of organisms are inoculated into a fixed volume of culture medium, and the number of viable organisms is recorded as a function of time. The growth pattern based on the number of cells has four more or less distinct phases.

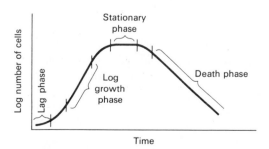

Figure **9-8** Typical bacterial growth curve.

1. The lag phase. Upon addition of an inoculum to a culture medium, the lag phase represents the time required for the organisms to acclimate to their new environment.
2. The log-growth phase. During this period the cells divide at a rate determined by their generation time and their ability to process food (constant percentage growth rate).
3. The stationary phase. Here the population remains stationary. Reasons advanced for this phenomenon are (a) that the cells have exhausted the substrate or nutrients necessary for growth, and (b) that the growth of new cells is offset by the death of old cells.
4. The log-death phase. During this phase the bacteria death rate exceeds the production of new cells. The death rate is usually a function of the viable population and environmental characteristics. In some cases, the log-death phase is the inverse of the log-growth phase.

Growth in terms of bacterial mass The growth pattern can also be discussed in terms of the variation of the mass of microorganisms with time. This growth pattern consists of the following three phases:

1. The log-growth phase. There is always an excess amount of food surrounding the microorganisms, and the rate of metabolism and growth is only a function of the ability of the microorganism to process the substrate.
2. Declining growth phase. The rate of growth and hence the mass of bacteria decrease because of limitations in the food supply.
3. Endogenous phase. The microorganisms are forced to metabolize their own protoplasm without replacement, since the concentration of available food is at a minimum. During this phase, a phenomenon known as lysis can occur in which the nutrients remaining in the dead cells diffuse out to furnish the remaining cells with food (known as cryptic growth).

Growth in Mixed Cultures

It is important to note that the preceding discussions concerned a single population of microorganisms. Often biological-treatment units are composed of complex, interrelated, mixed biological populations, with each particular microorganism in the system having its own growth curve. The position and shape of a particular growth curve in the system, on a time scale, depend on the food and nutrients available and on environmental factors, such as temperature, pH, and whether the system is aerobic or anaerobic. The variation of microorganism predominance with time in the aerobic stabilization of liquid organic waste is given in Fig. 9-9. While the bacteria are of primary importance, many other microorganisms take part in the stabilization of the organic waste. When designing or analyzing a biological-treatment process, the engineer should think in terms of an ecosystem, such as the one shown in Fig. 9-9, and not in terms of a

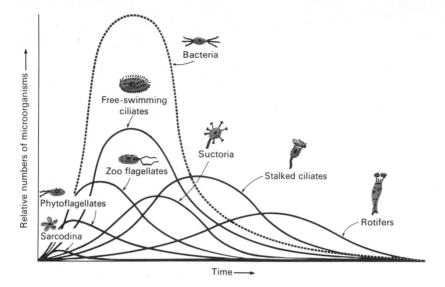

Figure 9-9 Relative growth of microorganisms stabilizing an organic waste in a liquid environment [30].

"black box" that contains mysterious microorganisms. Other ecosystems that are important in biological waste treatment will be discussed more thoroughly later in this chapter.

Bacterial Oxidation

The conversion of organic matter to gaseous end products and cell tissue can be accomplished aerobically, anaerobically, or facultatively using suspended-growth or attached-growth systems. The aerobic conversion of the organic matter in the batch culture discussed previously can be explained in terms of the sketch shown in Fig. 9-3. As shown, a portion of the organic material is oxidized to end products. It was noted previously that this process is carried out to obtain the energy necessary for the synthesis of new cell tissue. In the absence of organic matter, the cell tissue will be endogenously respired to gaseous end products and a residual to obtain energy for maintenance. In most biological-treatment systems, these three processes occur simultaneously.

Stoichiometrically, the three processes may be represented as follows for an aerobic process:

Oxidation (dissimilatory process):

$$\underset{\substack{\text{(organic} \\ \text{matter)}}}{\text{COHNS}} + O_2 + \text{bacteria} \rightarrow CO_2 + NH_3 + \underset{\substack{\text{end} \\ \text{products}}}{\text{other}} + \text{energy} \qquad (9\text{-}4)$$

Synthesis (assimilatory process):

$$COHNS + O_2 + bacteria + energy \rightarrow C_5H_7NO_2 \qquad (9\text{-}5)$$

(organic
matter)
(new
bacterial
cells)

Endogenous respiration (autoxidation):

$$C_5H_7NO_2 + 5O_2 \rightarrow 5CO_2 + NH_3 + 2H_2O + energy \qquad (9\text{-}6)$$

In these equations, COHNS represents the organic matter in wastewater. The formula $C_5H_7NO_2$, which represents cell tissue, is a generalized value obtained from experimental studies and was first suggested by Hoover and Porges in 1952 [17]. Although the endogenous respiration reaction is shown as resulting in relatively simple end products and energy, actually stable organic end products are also formed.

9-4 KINETICS OF BIOLOGICAL GROWTH

The need for a controlled environment and biological community in the design of biological waste-treatment units is stressed throughout this chapter. The classes of microorganisms of importance in wastewater treatment have been discussed, along with their cytological and metabolic characteristics and their growth patterns. Although the characteristics of the environment needed for their growth have been described, nothing has been said about how to control the environment of the microorganisms. Environmental conditions can be controlled by pH regulation, temperature regulation, nutrient or trace-element addition, oxygen addition or exclusion, and proper mixing. Control of the environmental conditions will ensure that the microorganisms have a proper medium in which to grow.

To ensure that the microorganisms will grow, they must be allowed to remain in the system long enough to reproduce. This period depends on their growth rate, which is related directly to the rate at which they metabolize or utilize the waste. Assuming that the environmental conditions are controlled properly, effective waste stabilization can be ensured by controlling the growth rate of the microorganisms.

The purpose of this section is to consider the kinetics of biological growth. The application of the kinetics developed here is documented throughout the remainder of this chapter.

Logarithmic Growth: Batch Culture

In the batch culture described in Sec. 9-3, bacteria increase in proportion to their mass in the log-growth phase (see Fig. 9-8). The rate of growth for this phase is defined by the following relationship:

$$r_g = \mu X \qquad (9\text{-}7)$$

where r_g = rate of bacterial growth, mass/unit volume · time
μ = specific growth rate, time^{-1}
X = concentration of microorganism, mass/unit volume

Growth is occurring in a batch system, and it will be remembered from Chap. 5 (see Eq. 5-28) that

$$\frac{dX}{dt} = r_g \qquad \text{for batch culture} \tag{9-8}$$

Thus the following relationship is also valid for a batch reactor:

$$\frac{dX}{dt} = \mu X \qquad \text{for batch culture} \tag{9-9}$$

Substrate Limited Growth

Again referring to the batch-culture experiment discussed in Sec. 9-3, if one of the essential requirements (substrate and nutrients) for growth were present in only limited amounts, it would be depleted first and growth would cease (see Fig. 9-8). In a continuous culture, growth is limited. Experimentally, it has been found that the effect of a limiting substrate or nutrient can be defined adequately using the following expression as proposed by Monod [33, 34]:

$$\mu = \mu_m \frac{S}{K_s + S} \tag{9-10}$$

where μ = specific growth rate, time^{-1}
μ_m = maximum specific growth rate, time^{-1}
S = concentration of growth-limiting substrate in solution, mass/unit volume
K_s = half-velocity constant, substrate concentration at one-half the maximum growth rate, mass/unit volume

The effect of substrate concentration on the specific growth rate is shown in Fig. 9-10.

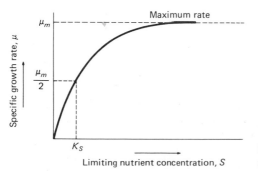

Figure 9-10 Plot showing the effects of a limiting nutrient on the specific growth rate.

If the value of μ from Eq. 9-10 is substituted in Eq. 9-7, the resulting expression for the rate of growth is

$$r_g = \frac{\mu_m XS}{K_s + S} \qquad (9\text{-}11)$$

Cell Growth and Substrate Utilization

In the batch-growth culture system described in Sec. 12-3, a portion of the substrate is converted to new cells and a portion is oxidized to inorganic and organic end products. This process is represented in general terms using the equations for bacterial oxidation and synthesis given in Sec. 9-3. Because the quantity of new cells produced has been observed to be reproducible for a given substrate, the following relationship has been developed between the rate of substrate utilization and the rate of growth:

$$r_g = -Yr_{su} \qquad (9\text{-}12)$$

where r_g = rate of bacterial growth, mass/unit volume · time
Y = maximum yield coefficient measured during any finite period of logarithmic growth [42], and defined as the ratio of the mass of cells formed to the mass of substrate consumed, mass/mass
r_{su} = substrate utilization rate, mass/unit volume · time

On the basis of laboratory studies, Ribbons [39] concluded that yield depends on (1) the oxidation-reduction state of the carbon source and nutrient elements, (2) the degree of polymerization of the substrate, (3) pathways of metabolism, (4) the growth rate, and (5) various physical parameters of cultivation.

If the value of r_g from Eq. 9-11 is substituted in Eq. 9-12, the rate of substrate utilization can be defined as follows:

$$r_{su} = -\frac{\mu_m XS}{Y(K_s + S)} \qquad (9\text{-}13)$$

In Eq. 9-13, the term μ_m/Y is often replaced by the term k, defined as the maximum rate of substrate utilization per unit mass of microorganisms:

$$k = \frac{\mu_m}{Y} \qquad (9\text{-}14)$$

If the term k is substituted for the term (μ_m/Y) in Eq. 9-13, the resulting expression is

$$r_{su} = -\frac{kXS}{K_s + S} \qquad (9\text{-}15)$$

Effects of Endogenous Metabolism

In bacterial systems used for wastewater treatment, the distribution of cell ages is such that not all the cells in the system are in the log-growth phase. Consequently, the expression for the rate of growth must be corrected to account for the energy required for cell maintenance. Other factors, such as death and predation, must also be considered. Usually, these factors are lumped together, and it is assumed that the decrease in cell mass caused by them is proportional to the concentration of organisms present. This decrease is often identified in the literature as the endogenous decay. A more theoretical discussion of the differences between the energy of maintenance and endogenous respiration may be found in Refs. 5, 6, 15, 23, and 48. The endogenous decay term can be formulated as follows:

$$r_d \text{ (endogenous decay)} = -k_d X \qquad (9\text{-}16)$$

where k_d = endogenous decay coefficient, time^{-1}
X = concentration of cells, mass/unit volume

When Eq. 9-16 is combined with Eqs. 9-11 and 9-12, the following expressions are obtained for the net rate of growth:

$$r'_g = \frac{\mu_m X S}{K_s + S} - k_d X \qquad (9\text{-}17)$$

$$r'_g = -Y r_{su} - k_d X \qquad (9\text{-}18)$$

where r'_g = net rate of bacterial growth, mass/unit volume · time

The corresponding expression for the net specific growth rate is given by Eq. 9-19, which is the same as the expression proposed by Van Uden [46]:

$$\mu' = \mu_m \frac{S}{K_s + S} - k_d \qquad (9\text{-}19)$$

where μ' = net specific growth rate, time^{-1}

The effects of endogenous respiration on the net bacterial yield are accounted for by defining an observed yield as follows:

$$Y_{obs} = -\frac{r'_g}{r_{su}} \qquad (9\text{-}20)$$

The application of these equations is illustrated in Sec. 9-5.

Effects of Temperature

The temperature dependence of the biological reaction-rate constants is very important in assessing the overall efficiency of a biological-treatment process. Temperature not only influences the metabolic activities of the microbiological

Table 9-6 Temperature coefficients for various biological processes[a]

	Value	
Process	Range	Typical
Activated sludge	1.00–1.04	1.02
Aerated lagoons	1.06–1.12	1.08
Trickling filters	1.02–1.14	1.08

[a] Adapted in part from Refs. 1 and 9.

population, but also has a profound effect on such factors as gas-transfer rates and the settling characteristics of the biological solids. The effect of temperature on the reaction rate of a biological process is usually expressed in the following form:

$$\frac{r_T}{r_{20}} = \theta^{(T-20)} \tag{9-21}$$

where r_T = reaction rate at $T°C$
r_{20} = reaction rate at 20°C
θ = temperature-activity coefficient
T = temperature, °C

Values of θ for biological processes are presented in Table 9-6. These values should not be confused with values given previously in Chap. 3 for the BOD determination.

Application of Kinetics to Biological Treatment

Before discussing the individual biological treatment processes listed in Table 9-1, the general application of the kinetics of biological growth will be explained. In this discussion, an aerobic treatment process carried out in a continuous-flow stirred-tank reactor without recycle will be considered (see Fig. 9-11). The schematic shown is the same as that for the activated-sludge process without recycle to be considered in Sec. 9-5. The purpose here is to illustrate (1) the development of microorganism and substrate balances, (2) the prediction of effluent microorganism and substrate concentrations, (3) the development of process design factors, and (4) the effects of kinetics on process design, performance, and stability.

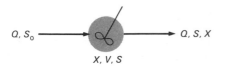

Q, S_0 Q, S, X

X, V, S

Figure 9-11 Schematic of a continuous-flow stirred-tank reactor without recycle.

Microorganism and substrate mass balances Using the procedures described in Chap. 5, a mass balance for the mass of microorganisms in the continuous-flow stirred-tank reactor shown in Fig. 9-11 can be written as follows:

1. General word statement:

$$
\begin{array}{l}
\text{Rate of accumulation} \\
\text{of microorganism} \\
\text{within the system} \\
\text{boundary}
\end{array}
=
\begin{array}{l}
\text{rate of flow of} \\
\text{microorganism} \\
\text{into the system} \\
\text{boundary}
\end{array}
-
\begin{array}{l}
\text{rate of flow of} \\
\text{microorganism} \\
\text{out of the system} \\
\text{boundary}
\end{array}
+
\begin{array}{l}
\text{net growth of} \\
\text{microorganism} \\
\text{within the} \\
\text{system boundary}
\end{array}
\tag{9-22}
$$

2. Simplified word statement:

$$\text{Accumulation} = \text{inflow} - \text{outflow} + \text{net growth} \tag{9-23}$$

3. Symbolic representation:

$$\frac{dX}{dt} V = QX_0 - QX + V(r'_g) \tag{9-24}$$

where dX/dt = rate of microorganism growth measured in terms of mass (volatile suspended solids), mass VSS/unit volume · time

V = reactor volume

Q = flowrate, volume/time

X_0 = concentration of microorganisms in influent, mass VSS/unit volume

X = concentration of microorganisms in reactor, mass VSS/unit volume

r'_g = net rate of microorganism growth, mass VSS/unit volume · time

In Eq. 9-24 and subsequent expressions derived from it, the volatile fraction of the total biological suspended solids is used as an approximation of active biological mass. The assumption is made that the volatile fraction is proportional to the activity of the microbial mass in question. Although a number of other measures, such as nitrogen, protein, DNA, and ATP content, have been used, the volatile suspended solids test is used principally because of its simplicity.

If the value of r'_g from Eq. 9-17 is substituted into Eq. 9-24, the result is

$$\frac{dX}{dt} V = QX_0 - QX + V\left(\frac{\mu_m X S}{K_s + S} - k_d X\right) \tag{9-25}$$

If it is assumed that the concentration of microorganisms in the influent can be neglected and that steady-state conditions prevail ($dX/dt = 0$), Eq. 9-25 can be simplified to yield

$$\frac{Q}{V} = \frac{1}{\theta} = \frac{\mu_m S}{K_s + S} - k_d \tag{9-26}$$

where θ = hydraulic detention time, V/Q

The substrate balance corresponding to the microorganism mass balance given in Eq. 9-25 is as follows:

$$\frac{dS}{dt} V = QS_0 - QS + V\left(-\frac{kXS}{K_s + S}\right) \qquad (9\text{-}27)$$

At steady state $(dS/dt = 0)$, the resulting equation is

$$S_0 - S - \theta \frac{kXS}{K_s + S} = 0 \qquad (9\text{-}28)$$

where $\theta = V/Q$

Prediction of effluent microorganism and substrate concentrations The effluent microorganism and substrate concentrations may be obtained as follows. If Eq. 9-26 is solved for the term $S/K_s + S$ and the resulting expression is substituted into Eq. 9-28 and simplified using Eq. 9-14, then effluent steady-state concentration is found to be given as

$$X = \frac{\mu_m(S_0 - S)}{k(1 + K_d\theta)} = \frac{Y(S_0 - S)}{1 + K_d\theta} \qquad (9\text{-}29)$$

Similarly, the effluent substrate concentration is found to be equal to

$$S = \frac{K_s(1 + \theta k_d)}{\theta(Yk - k_d) - 1} \qquad (9\text{-}30)$$

Thus, if the kinetic coefficients are known, Eqs. 9-29 and 9-30 can be used to predict effluent microorganism and substrate concentrations.

Development of process design relationships Although Eqs. 9-29 and 9-30 are useful in predicting the effects of various system changes, they are somewhat difficult to use from a design standpoint because of the many constants involved. For this reason, more usable process design relationships have been developed. The relationships to be considered in this discussion include the mean cell residence time, the specific utilization rate, the observed yield, and the food-microorganism ratio.

If the definition of the net growth rate r_g' given by Eq. 9-18 is used in Eq. 9-24, the resulting equation is

$$\frac{dX}{dt} V = QX_0 - QX + V(-Yr_{su} - k_d X) \qquad (9\text{-}31)$$

Assuming that the influent concentration equals zero and steady-state conditions, Eq. 9-31 reduces to

$$\frac{Q}{V} = \frac{1}{\theta} = -Y\frac{r_{su}}{X} - k_d \qquad (9\text{-}32)$$

Using Eqs. 9-17 through 9-19, it can be shown that μ', the net specific growth rate, may be defined as follows:

$$\mu' = \frac{Q}{V}$$ (9-33)

If Q and V are now multiplied by X, the cell concentration, the reciprocal of Eq. 9-33 is defined as the mean cell residence time and is identified by the symbol θ_c.

$$\theta_c = \frac{VX}{QX} \frac{\text{mass of cells in reactor}}{\text{mass of cells wasted per day}}$$ (9-34)

Substituting θ_c for θ in Eq. 9-32 yields

$$\frac{1}{\theta_c} = -Y\frac{r_{su}}{X} - k_d$$ (9-35)

In Eq. 9-35, the term $(-r_{su}/X)$ is known as the specific substrate utilization rate. In the following discussions, this term will be referred to as U.

$$U = -\frac{r_{su}}{X}$$ (9-36)

The term r_{su} is determined using the following expression:

$$r_{su} = -\frac{Q}{V}(S_0 - S) = -\frac{S_0 - S}{\theta}$$ (9-37)

The specific utilization rate is then calculated as follows:

$$U = \frac{S_0 - S}{\theta X} = \frac{Q}{V}\frac{S_0 - S}{X}$$ (9-38)

where $S_0 - S$ = mass concentration of substrate utilized, mg/L
θ = hydraulic detention time, d
X = mass of VSS in reactor, mg/L

If the term U is substituted for the term $(-r_{su}/X)$ in Eq. 9-35, the resulting equation is

$$\frac{1}{\theta_c} = YU - k_d$$ (9-39)

From Eq. 9-39 it can be seen that $1/\theta_c$, the net specific growth rate, and U, the specific utilization ratio, are related directly.

Using this relationship, it can also be shown that the observed yield (see Eq. 9-20) can be defined as follows:

$$Y_{obs} = \frac{YU - k_d}{U} \tag{9-40}$$

or

$$Y_{obs} = \frac{Y}{1 + \theta_c k_d} \tag{9-41}$$

A term that is closely related to the specific utilization rate U that is also commonly used in practice is known as the food-microorganism ratio (F/M), which is defined as follows:

$$F/M = \frac{S_0}{\theta X} \tag{9-42}$$

The terms U and F/M are related by the process efficiency as follows:

$$U = \frac{(F/M)E}{100} \tag{9-43}$$

where E is the process efficiency as defined by Eq. 9-44:

$$E = \frac{S_0 - S}{S_0} \times 100 \tag{9-44}$$

where E = process efficiency, percent
S_0 = influent substrate concentration
S = effluent substrate concentration

The application of these design relationships is illustrated in the discussion of the various processes given later in this chapter and in Chaps. 10 and 11.

Analysis of process performance and stability The effects of kinetics, considered previously, on the performance and stability of the system shown in Fig. 9-11, will now be examined further. It was shown in Eq. 9-39 that $1/\theta_c$, the net microorganism growth rate, and U, the specific utilization ratio, are related directly. In the following discussion, it will also be shown that because θ_c is easy to measure, it is the more desirable parameter for process control. By comparing Eq. 9-28 with Eq. 9-38, it can be shown that

$$U = \frac{kS}{K_s + S} \tag{9-45}$$

from which the following equation is obtained:

$$S = \frac{UK_s}{k - U} \tag{9-46}$$

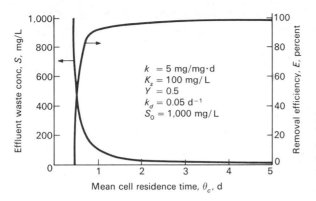

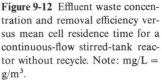

$k = 5$ mg/mg·d
$K_s = 100$ mg/L
$Y = 0.5$
$k_d = 0.05$ d^{-1}
$S_0 = 1,000$ mg/L

Figure 9-12 Effluent waste concentration and removal efficiency versus mean cell residence time for a continuous-flow stirred-tank reactor without recycle. Note: mg/L = g/m^3.

For a specified waste, a biological community, and a particular set of environmental conditions, the kinetic coefficients Y, k, K_s, and k_d are fixed. Consequently, the effluent-waste concentration S is a direct function of either θ_c or U. Setting one of these three parameters not only fixes the other two but also specifies the efficiency of biological waste stabilization.

For a continuous-flow stirred-tank system without recycle that is growth-specified (specified values of S_0, Y, K_s, k, and k_d), fixing the mean cell residence time θ_c also establishes the microorganism concentration in the reactor. This can be shown by writing Eqs. 9-29 and 9-30 in terms of θ_c and noting that the effluent waste concentration S in Eq. 9-29 is also a function of θ_c, as defined by Eq. 9-30.

Equations 9-29 and 9-30 are plotted in Fig. 9-12 for a growth-specified continuous-flow stirred-tank system without recycle. As shown, the effluent concentration S and the treatment efficiency E are related directly to θ_c, which is equal to θ. In a system such as this, there is no separate control of the microorganisms, because the mean microorganism-retention time θ_c and the liquid-retention time θ are the same.

To obtain a high treatment efficiency, θ_c must be long, which means that θ must also be long. This mode of operation is characteristic of conventional anaerobic treatment systems and some modified activated-sludge processes. The analysis presented can also be used, but not without caution, to simulate oxidation ponds and lagoons. Both of these are flow-through systems with no recycle, but they are not completely mixed. Settling is an effective unit operation in the overall efficiency of these processes, and the foregoing equations did not include the effect of settling.

It can also be seen from Fig. 9-12 that there is a certain value of θ_c below which waste stabilization does not occur. This critical value of θ_c is called the minimum mean cell residence time θ_c^M. Physically, θ_c^M is the residence time at which the cells are washed out or wasted from the system faster than they can reproduce. The minimum mean cell residence time can be calculated from

Eq. 9-26, since for this condition the influent waste concentration S_0 is equal to the effluent waste concentration S:

$$\frac{1}{\theta_c^M} = Y \frac{kS_0}{K_s + S_0} - k_d \tag{9-47}$$

In many situations encountered in waste treatment, S_0 is much greater than K_s so that Eq. 9-47 can be rewritten to yield

$$\frac{1}{\theta_c^M} \approx Yk - k_d \tag{9-48}$$

Use of Eqs. 9-47 and 9-48 determines the minimum mean cell residence time θ_c^M. Obviously, biological-treatment systems should not be designed with θ_c values equal to θ_c^M. To ensure adequate waste treatment, biological-treatment systems are usually designed and operated with a θ_c^d value from 2 to 20 times θ_c^M. In effect, the ratio of θ_c^d to θ_c^M can be considered to be a process safety factor, SF [38]:

$$SF = \frac{\theta_c^d}{\theta_c^M} \tag{9-49}$$

Determination of Kinetic Coefficients

Values for the parameters Y, k, K_s, and k_d must be available for a particular model to be used effectively. To determine these coefficients, bench-scale reactors, such as those shown in Figs. 9-13 and 9-14, or pilot-scale systems are used.

In determining these coefficients, the usual procedure is to operate the units over a range of effluent substrate concentrations; therefore, several different θ_c's (at least five) should be selected for operation ranging from 1 to 10 days. Using the data collected at steady-state conditions, mean values should be determined for Q, S_0, S, X, and r'_{su}. Equating Eq. 9-15 to the value of r_{su} given by Eq. 9-37 results in the following expression:

$$r_{su} = -\frac{kXS}{K_s + S} = -\frac{S_0 - S}{\theta} \tag{9-50}$$

Dividing by X yields

$$\frac{kS}{K_s + S} = \frac{S_0 - S}{\theta X} \tag{9-51}$$

The linearized form of Eq. 9-51, obtained by taking its inverse, is

$$\frac{X\theta}{S_0 - S} = \frac{K_s}{k} \frac{1}{S} + \frac{1}{k} \tag{9-52}$$

The values of K_s and k can be determined by plotting the term $[X\theta/(S_0 - S)]$ versus $(1/S)$. The values of Y and k_d may be determined using Eq. 9-35, by plotting $(1/\theta_c)$ versus $(-r_{su}/X)$. The slope of the straight line passed through the

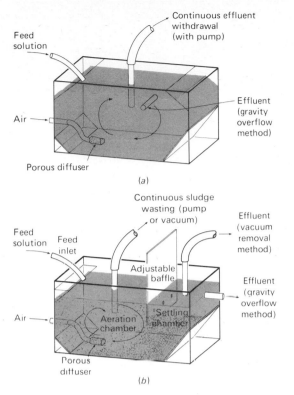

(a)

(b)

Figure 9-13 Bench-scale continuous-flow stirred-tank reactors used for the determination of kinetic coefficients (a) without solids recycle and (b) with solids recycle.

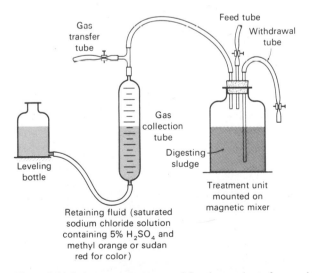

Figure 9-14 Laboratory reactor used for the conduct of anaerobic treatment studies.

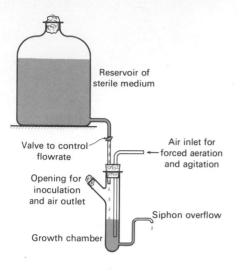

Reservoir of
sterile medium

Valve to control
flowrate

Air inlet for
forced aeration
and agitation

Opening for
inoculation
and air outlet

Siphon overflow

Growth chamber

Figure 9-15 A simplified diagram of a chemostat [45].

plotted experimental datum points is equal to Y, and the intercept is equal to k_d. The procedure is illustrated in Example 9-1.

A chemostat apparatus, such as the one depicted in Fig. 9-15, can also be used to determine kinetic coefficients. Growth rates are determined by measurements of effluent turbidity. Such units are normally used for pure culture determinations.

McCarty [28] has presented a unique method of determining the growth-yield coefficient Y, based on a thermodynamic approach to substrate utilization and cell synthesis. In general, the growth parameters k and K_s have not been evaluated as extensively as Y and k_d. A lack of values for these parameters places some limitations on the kinetic models presented in this section; however, with the data available, reasonable estimates of the growth parameters can be made. Typical values for these coefficients for the various treatment processes are presented in the following discussion.

Example 9-1: Determination of kinetic coefficients from laboratory data Determine the values of the coefficients k, K_s, μ_m, Y, and k_d using the following data derived from bench-scale activated-sludge studies using a continuous flow stirred tank reactor without recycle (see Fig. 9-13a).

Unit no.	S_0, mg/L BOD$_5$	S, mg/L BOD$_5$	$\theta = \theta_c$, d	X, mg VSS/L
1	300	7	3.2	128
2	300	13	2.0	125
3	300	18	1.6	133
4	300	30	1.1	129
5	300	41	1.1	121

SOLUTION
1 Determine the coefficients K_s and k.
    a Set up a computation table to determine the coefficients K_s and k using Eq. 9-52.

$$\frac{X\theta}{S_0 - S} = \frac{K_s}{k}\frac{1}{S} + \frac{1}{k}$$

$S_0 - S$, mg/L	$X\theta$, mg VSS/d/L	$\dfrac{X\theta}{S_0 - S}$, d	$\dfrac{1}{S}$, mg/L^{-1}
293	409.6	1.398	0.143
287	250.0	0.865	0.077
282	212.8	0.755	0.056
270	141.9	0.526	0.033
259	133.1	0.514	0.024

    b Plot the term $(X\theta/S_0 - S)$ versus $(1/S)$, as shown in Fig. 9-16.
      i From Eq. 9-52, the y intercept equals $(1/k)$.

$$\frac{1}{k} = 0.32\text{ d} \qquad k = 3.125\text{ d}^{-1}$$

      ii From Eq. 9-52, the slope of the curve in Fig. 9-16 equals K_s/k.

$$\frac{K_s}{k} = \frac{0.5\text{ d}}{0.065(\text{mg/L})^{-1}} = 7.692\text{ mg/d} \cdot \text{L}$$

$$K_s = 7.692\text{ mg/d} \cdot \text{L} \times 3.125\text{ d}^{-1}$$

$$= 24.0\text{ mg/L}$$

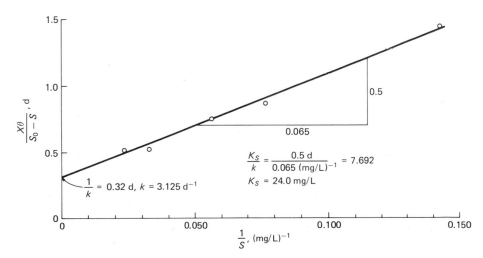

Figure 9-16 Plot of experimental data to determine the kinetic coefficients K_s and k in Example 9-1.

2 Determine the coefficients Y and k_d.

 a Set up a computation to determine the coefficients using Eq. 9-35.

$$\frac{1}{\theta_c} = -Y\frac{r_{su}}{X} - k_d$$

$$\frac{1}{\theta_c} = Y\frac{S_0 - S}{X\theta} - k_d$$

Unit no.	$\dfrac{1}{\theta_c}$, d^{-1}	$\dfrac{S_0 - S}{X\theta}$, d^{-1}
1	0.313	0.715
2	0.500	1.156
3	0.625	1.325
4	0.909	1.901
5	0.909	1.946

 b Plot the term $(1/\theta_c)$ versus $(S_0 - S/X\theta)$, as shown in Fig. 9-17.

 i From Eq. 9-35, the y intercept equals $(-k_d)$.

$$-k_d = -0.05 \ d^{-1}$$

$$k_d = 0.05 \ d^{-1}$$

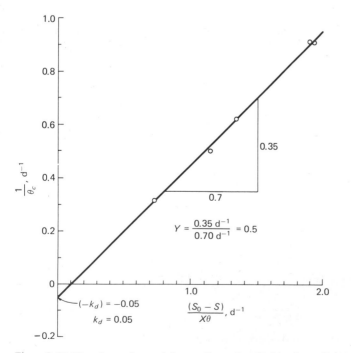

Figure 9-17 Plot of experimental data to determine the kinetic coefficients Y and k_d in Example 9-1.

ii From Eq. 9-35, the slope of the curve in Fig. 9-17 equals Y.

$$Y = \frac{0.35 \ d^{-1}}{0.70 \ d^{-1}} = 0.5$$

3 Determine the value of the coefficient μ_m using Eq. 9-14.

$$\mu_m = kY$$
$$= 3.125 \ d^{-1} \times 0.5$$
$$= 1.563 \ d^{-1}$$

Comment In this example, the kinetic coefficients were derived from data obtained using bench-scale continuous-flow stirred-tank reactors without recycle. Similar data can be obtained using continuous-flow stirred-tank reactors with recycle (see Sec. 9-5 and Fig. 9-13*b*). An advantage of using reactors with recycle is that the mean cell-residence time can be varied independently of the hydraulic detention time. A disadvantage is that small bench-scale reactors operated with solids recycle are difficult to control.

Other Rate Expressions

In reviewing the kinetic expressions used to describe the growth of microorganisms and the removal of substrate, it is very important to remember that the expressions presented are empirical and were used for the purpose of illustration, and that they are not the only expressions available. Other expressions that have been used to describe the rate of substrate utilization include the following:

$$r_{su} = -K \tag{9-53}$$

$$r_{su} = -KS \tag{9-54}$$

$$r_{su} = -KXS \tag{9-55}$$

$$r_{su} = -KX \frac{S}{S_0} \tag{9-56}$$

Expressions for the specific growth rate (see Eq. 9-10) have been proposed by a number of persons, including Monod, Teissier, Contois, and Moser.

What is fundamental in the use of any rate expression is its application in a mass-balance analysis. In this connection, it does not matter if the rate expression selected has no relationship to those used commonly as described in the literature, so long as it describes the observed phenomenon. It is equally important to remember that specific rate expressions should not be generalized to cover a broad range of situations on the basis of limited data or experience.

9-5 AEROBIC SUSPENDED-GROWTH TREATMENT PROCESSES

As noted in Table 9-1, the principal suspended-growth biological-treatment processes are (1) the activated-sludge process, (2) the suspended-growth nitrification process, (3) aerated lagoons, (4) the aerobic digestion process, and (5) high-rate oxidation ponds. Of these, the activated-sludge process is by far the one most commonly used for the secondary treatment of domestic wastewater, and for this reason it will be stressed in this section.

Activated-Sludge Process

The activated-sludge process was developed in England in 1914 by Ardern and Lockett [2] and was so named because it involved the production of an activated mass of microorganisms capable of aerobically stabilizing a waste. Many versions of the original process are in use today, but fundamentally they are all similar. The system shown in Fig. 9-18 is the continuous-flow stirred-tank activated-sludge system. Other activated-sludge systems are listed in Table 9-1 and discussed in Chap. 10.

Process description Operationally, biological waste treatment with the activated-sludge process is typically accomplished using a flowsheet such as that shown in Fig. 9-18. Organic waste is introduced into a reactor where an aerobic bacterial culture is maintained in suspension. The reactor contents are referred to as the mixed liquor. In the reactor, the bacterial culture carries out the conversion in general accordance with the stoichiometry discussed in Sec. 9-2. The aerobic environment in the reactor is achieved by the use of diffused or mechanical aeration, which also serves to maintain the mixed liquor in a completely mixed regime. After a specified period of time, the mixture of new cells and old cells is passed into a settling tank where the cells are separated from the treated wastewater. A portion of the settled cells is recycled to maintain the desired concentration of organisms in the reactor, and a portion is wasted. The portion wasted corresponds to the new growth of cell tissue, r'_g (see Eq. 9-18), associated with a particular wastewater. The level at which the biological mass in the reactor should be kept depends on the desired treatment efficiency and other considerations related to growth kinetics. Microorganism concentrations maintained in various activated-sludge treatment systems are listed in Table 11-4.

Process microbiology To design and operate an activated-sludge system efficiently, it is necessary to understand the importance of the microorganisms in the system. In nature, the key role of the bacteria is to decompose organic matter produced by other living organisms. In the activated-sludge process, the bacteria are the most important microorganisms because they are responsible for the decomposition of the organic material in the influent. In the reactor, or mixed-liquor tank, a portion of the organic waste matter is used by aerobic and facultative bacteria to obtain energy for the synthesis of the remainder of the organic material into new cells, as shown in Fig. 9-3. Only a portion of the original waste is actually oxidized to low-energy compounds, such as NO_3, SO_4, and CO_2; the remainder is synthesized into cellular material. Also, many intermediate products are formed before the final end products, shown in the right-hand side of Fig. 9-3, are produced.

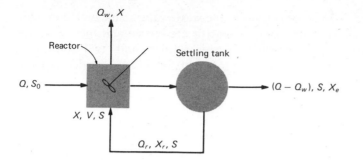

Figure 9-18 Continuous-flow stirred-tank reactor with cellular recycle.

In general, the bacteria in the activated-sludge process are gram-negative and include members of the genera *Pseudomonas, Zoogloea, Achromobacter, Flavobacterium, Nocardia, Bdellovibrio, Mycobacterium*, and the two nitrifying bacteria, *Nitrosomonas* and *Nitrobacter* [16]. Additionally, various filamentous forms, such as *Sphaerotilus, Beggiatoa, Thiothrix, Lecicothrix*, and *Geotrichum*, may also be present [16]. While the bacteria are the microorganisms that actually degrade the organic waste in the influent, the metabolic activities of other microorganisms are also important in the activated-sludge system. For example, protozoa and rotifers act as effluent polishers. Protozoa consume dispersed bacteria that have not flocculated, and rotifers consume small biological floc particles that have not settled.

Further, although it is important that bacteria decompose the organic waste as quickly as possible, it is also important that they form a satisfactory floc, which is a prerequisite for the effective separation of the biological solids in the settling unit. It has been observed that as the mean cell residence time of the cells in the system is increased, the settling characteristics of the biological floc are enhanced. The reason is that as the mean age of the cells is increased, the surface charge is reduced, and the microorganisms start to produce extracellular polymers, eventually becoming encapsulated in a slime layer (see Fig. 9-1). The presence of these polymers and the slime promotes the formation of floc particles that can be removed readily by gravity settling. For domestic wastes, mean cell residence times on the order of 3 to 4 d are required to achieve effective settling. Typical values of mean cell residence times used in the design and operation of various activated-sludge processes are shown in Table 10-4.

Even though excellent floc formation is obtained, the effluent from the system could still be high in biological solids as a result of poor design of the secondary settling unit, poor operation of the aeration units, or the presence of filamentous microorganisms, such as *Sphaerotilus, E. coli* [16, 24], and fungi. These subjects are discussed in detail in Chap. 10.

Process analysis: continuous-flow stirred-tank with recycle In this system, shown schematically in Fig. 9-18, the reactor contents are completely mixed, and it is assumed that there are no microorganisms in the wastewater influent. The system contains a unit in which the cells from the reactor are settled and then returned to the reactor. Because of the presence of this settling unit, two additional assumptions must be made in the development of the kinetic model for this system:

1. Waste stabilization by the microorganisms occurs only in the reactor unit. This assumption leads to a conservative model (in some systems there may be some waste stabilization in the settling unit).
2. The volume used in calculating the mean cell residence time for the system includes only the volume of the reactor unit.

In effect, it is assumed that the settling tank serves as a reservoir from which solids are returned to maintain a given solids level in the aeration tank. If the system is such that these assumptions do not hold true, then the model should be modified. For example, in pure-oxygen activated-sludge systems, it has been found that up to 50 percent of the total solids in the system may be present in the secondary settling tank. This subject is considered further in the following discussion and in Chap. 10.

The mean hydraulic retention time for the system θ_s is defined as

$$\theta_s = \frac{V_s}{Q} \tag{9-57}$$

where V_s = volume of reactor plus volume of settling tank
Q = influent flowrate

The mean hydraulic retention time for the reactor θ is defined as

$$\theta = \frac{V}{Q} \tag{9-58}$$

where V is the volume of the reactor.

For the system shown in Fig. 9-18, the mean cell residence time θ_c is

$$\theta_c = \frac{VX}{Q_w X + (Q - Q_w)X_e} \tag{9-59}$$

where Q_w = flowrate of liquid containing the biological cells to be wasted from the system (in this case from the reactor)
X_e = microorganism concentration in effluent from settling unit

In a system with a properly operating settling unit, the quantity of cells in the effluent is very small, and Eq. 9-59 can be simplified to yield

$$\theta_c \simeq \frac{V}{Q_w} \tag{9-60}$$

Comparing Eq. 9-59 with Eqs. 9-57 and 9-58, it can be seen that for a given reactor volume, θ_c is theoretically independent of both θ and θ_s. Practically speaking, however, θ_c cannot be completely independent of θ and θ_s. The factors relating θ_c to θ and θ_s will be discussed later.

A mass balance for the microorganisms in the entire system can be written as:

1. General word statement:

$$
\begin{array}{llll}
\text{Rate of accumulation} & \text{rate of flow of} & \text{rate of flow of} & \text{net growth of} \\
\text{of microorganism} & \text{microorganism} & \text{microorganism} & \text{microorganism} \\
\text{within the system} = & \text{into the system} - & \text{out of the system} + & \text{within the} \\
\text{boundary} & \text{boundary} & \text{boundary} & \text{system boundary}
\end{array}
\tag{9-61}
$$

2. Simplified word statement:

$$\text{Accumulation} = \text{inflow} - \text{outflow} + \text{net growth} \tag{9-62}$$

3. Symbolic representation:

$$\frac{dX}{dt} V = QX_0 - [Q_w X + (Q - Q_w)X_e] + V(r_g') \tag{9-63}$$

Substituting Eq. 9-18 for the rate of growth, and assuming that the cell concentration in the influent is zero and steady-state conditions prevail, $(dX/dt = 0)$ yields

$$\frac{Q_w X + (Q - Q_w)X_e}{VX} - - Y \frac{r_{su}}{X} - k_d \tag{9-64}$$

Making use of Eq. 9-59, Eq. 9-64 can be simplified and rearranged to yield

$$\frac{1}{\theta_c} = - Y \frac{r_{su}}{X} - k_d \tag{9-65}$$

or, by using Eq. 9-36,

$$\frac{1}{\theta_c} = YU - k_d \tag{9-66}$$

Equation 9-66 is the same as Eq. 9-39, which was developed in the general analysis for a continuous-flow stirred-tank system without recycle. In both systems, there is a direct relationship between θ_c and U. Thus, in the growth-specified recycle system, the effluent waste concentration S is directly related to θ_c or U. For both recycle and nonrecycle systems, controlling θ_c or U establishes the effluent concentration. Kinetic coefficients that can be used in the solution of Eqs. 9-66 and 9-19 are presented in Table 9-7.

In a continuous-flow stirred-tank system without recycle, both θ_c and U are direct functions of the hydraulic retention time of the reactor θ. In a recycle system, however, θ_c and U are theoretically independent of the hydraulic retention time of the reactor θ and of the system θ_s. Thus it is possible to achieve a high θ_c, and therefore good treatment efficiency, without raising θ or θ_s.

Table 9-7 Typical kinetic coefficients for the activated-sludge process[a]

Coefficient	Basis	Value[b]	
		Range	Typical
k	d^{-1}	2–10	5.0
K_s	mg/L BOD_5	25–100	60
	mg/L COD	15–70	40
Y	mg VSS/mg BOD_5[c]	0.4–0.8	0.6
	mg VSS/mg COD	0.25–0.4	0.4
k_d	d^{-1}	0.04–0.075	0.06

[a] Derived in part from Refs. 28, 41, 42, 43.
[b] Values reported are for 20°C.
[c] VSS = volatile suspended solids.
 Note: 1.8(°C) + 32 = °F.

The mass concentration of microorganisms X in the reactor can be obtained by using Eq. 9-37 in conjunction with Eq. 9-65 and solving for X:

$$X = \frac{\theta_c}{\theta} \frac{Y(S_0 - S)}{1 + k_d \theta_c} \qquad (9\text{-}67)$$

In the analysis of the foregoing equations, the predominant theme was that in a recycle system, θ_c and U were independent of θ and θ_s, whereas in a nonrecycle system, θ_c, U, and θ were directly related. Since both θ_c and U are directly related to the effluent quality of treatment efficiency, as shown by Eqs. 9-30 and 9-46, controlling either θ_c or U in a biological-treatment process will directly control the process efficiency. This can be done independently of θ or θ_s in a recycle system. The choice of which parameter to use for treatment control, θ_c or U, is a matter of ease of attainment.

To determine the specific utilization ratio U (see Eq. 9-38), the food utilized and the mass of microorganisms effective in this utilization must be known. The food utilized can be evaluated by determining the difference between the influent and the effluent COD or BOD_5. The evaluation of the active mass of microorganisms is usually what makes the use of U impractical as a control parameter. As noted earlier, the most common parameter used as a measure of the biological solids is the volatile suspended solids in the treatment unit. This parameter is not entirely satisfactory, because of the variability of volatile matter in the waste that is not related to active cellular material.

Using θ_c as a treatment control parameter, there is no need to determine the amount of active biological solids in the system, nor is there the need to evaluate the amount of food utilized. The use of θ_c is simply based on the fact that, to control the growth rate of microorganisms and hence their degree of waste stabilization, a specified percentage of the cell mass in the system must be wasted each day. Thus, if it is determined that a θ_c of 10 days is needed for a desired treatment efficiency, then 10 percent of the total cell mass is wasted from the system per day. In the continuous-flow stirred-tank system, cell wastage can be accomplished by wasting from the reactor or mixed-liquor tank. As shown in Eq. 9-60, by wasting cells directly from the reactor, only Q_w and V need to be known to determine θ_c. Wasting cells in this manner provides for a direct method of controlling and measuring θ_c.

In most biological-treatment processes, cell wastage is accomplished by drawing off from the sludge recycle line. If this were done, Eq. 9-59 would become

$$\theta_c = \frac{VX}{Q'_w X_r + (Q - Q'_w)X_e} \tag{9-68}$$

where X_r = microorganism concentration in return sludge line
Q'_w = cell wastage rate from recycle line

Assuming that X_e is very small, Eq. 9-68 can be rewritten as

$$\theta_c \simeq \frac{VX}{Q'_w X_r} \tag{9-69}$$

Thus wasting from the recycle line requires that the microorganism concentrations in both the mixed liquor and return sludge be known.

For an activated-sludge process using a continuous-flow stirred-tank with recycle, just as in the nonrecycle system, there is a minimum mean cell residence time θ_c^M below which waste stabilization cannot occur. The specific value of θ_c^M is a function of the waste concentration and the biological kinetic parameters Y, k, K_s, and k_d. The determination of these coefficients from laboratory data has been illustrated in Example 9-1. Equations 9-68 and 9-69 can be used to determine θ_c^M, with or without recycle. Values of θ_c used in the design of biological process are based on the value of θ_c^M for the particular waste.

As shown earlier, regardless of the location from which cells are wasted, θ_c is independent of θ. Practically speaking, however, in the successful operation of a wastewater-treatment plant using biological processes, a minimum hydraulic retention time θ must be met before θ_c becomes a controlling parameter. The factors involved in selecting the proper θ for a biological-treatment unit are discussed in Chap. 10. It is sufficient to note here that the most important factors tending to negate the complete independence between θ_c and θ are (1) the oxygen transfer rate in the reactor unit of an aerobic system, (2) the proper operation of the settling unit, and (3) the settling characteristics of suspended solids in the mixed liquor.

Plug flow with cellular recycle The plug-flow system with cellular recycle, shown schematically in Fig. 9-19, can be used to model certain forms of the activated-sludge process. The distinguishing feature of this recycle system is that the hydraulic regime of the reactor is of a plug-flow nature. In a true plug-flow

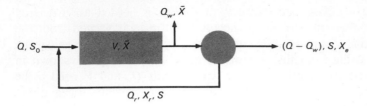

Figure 9-19 Plug-flow reactor with cellular recycle.

model, all the particles entering the reactor stay in the reactor an equal amount of time. Some particles may make more passes through the reactor because of recycle, but while they are in the tank, they all pass through in the same amount of time.

A kinetic model of the plug-flow system is mathematically difficult, but Lawrence and McCarty [21] have made two simplifying assumptions that lead to a useful kinetic model of the plug-flow reactor:

1. The concentration of microorganisms in the influent to the reactor is approximately the same as that in the effluent from the reactor. This assumption applies only if $\theta_c/\theta > 5$. The resulting average concentration of microorganisms in the reactor is symbolized as $\overline{X}$.
2. The rate of substrate utilization as the waste passes through the reactor is given by the following expression:

$$r_{su} = -\frac{kS\overline{X}}{K_s + S} \tag{9-70}$$

Integrating Eq. 9-70 over the retention time of the waste in the tank, and simplifying results in the following expression

$$\frac{1}{\theta_c} = \frac{Yk(S_0 - S)}{(S_0 - S) + (1 + \alpha)K_s \ln (S_i/S)} - k_d \tag{9-71}$$

where S_0 = influent concentration
S = effluent concentration
S_i = influent concentration to reactor after dilution with recycle flow
$$= \frac{S_0 + \alpha S}{1 + \alpha}$$

α = recycle ratio
other terms = as defined previously

Equation 9-71 is quite similar to Eq. 9-65, which applied to complete-mix systems, with or without recycle. The main difference in the two equations is that in Eq. 9-71 θ_c is also a function of the influent waste concentration S_0.

It should be noted that in Fig. 9-19 the excess microorganisms are essentially wasted from the reactor and not from the recycle line. Thus θ_c for the plug-flow–

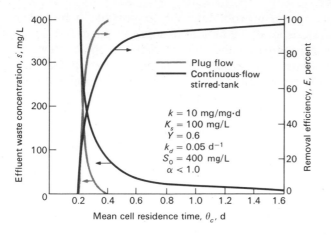

Figure 9-20 Effluent waste concentration and removal efficiency for continuous-flow stirred-tank and plug-flow reactors with recycle versus mean cell residence time [21]. Note: mg/L = g/m³.

recycle system could also be defined as in Eq. 9-69 with the same assumptions applying. The average hydraulic retention time of the waste in the reactor θ and the average hydraulic retention time of the waste in the plug-flow system θ_s can be defined using Eqs. 9-57 and 9-58. The average microorganism concentration in the reactor of the plug-flow system can be obtained using Eq. 9-67 by noting that $\bar{X}$ must be substituted for X.

The true plug-flow-recycle system is theoretically more efficient in the stabilization of most soluble wastes than is the continuous-flow stirred-tank recycle system. This is shown graphically in Fig. 9-20. In actual practice, a true plug-flow regime is difficult to obtain because of longitudinal dispersion. This difficulty, plus the fact that the plug-flow system cannot handle shock loads as well as the continuous-flow stirred-tank system, tends to reduce differences in treatment efficiency in the two models. By dividing the aeration tank into a series of reactors, it has been shown that treatment performance can be improved without a major loss in the ability of the system to handle shock loads [18]. Reactor selection is discussed further in Chap. 10.

Sedimentation facilities for the activated-sludge process Although the sedimentation tank is not often stressed, it is an integral part of the activated-sludge process. The design of the reactor cannot be considered independently of the design of the associated settling facilities. To meet discharge requirements for suspended solids and BOD associated with the volatile suspended solids in the effluent and to maintain θ_c independent of θ, it must be possible to separate the mixed-liquor solids and to return a portion to the reactor. Over the years, because of the variability observed, many myths have developed about the design of settling facilities for the activated-sludge process. These myths are discussed in a recent article by Dick [7].

Because of the variable process microbiology that is possible, it has been found that the settling characteristics of the biological solids in the mixed liquor will differ with each plant, depending on the characteristics of the wastewater and the many variables associated with process design and operation. For this reason, when settling facilities are designed for an existing or a proposed new treatment facility, column settling tests should be performed, and the design should be based on the results of these tests. If it is not possible to perform settling tests, the design should be based on the solids loadings. Both approaches are considered in Chap. 10. Schroeder [41] has also presented a unified approach to the design of the activated reactor and the settling facilities.

Suspended-Growth Nitrification

The foregoing discussion of the activated-sludge process has been limited to the aerobic biological degradation of organic carbonaceous material. Although this is the primary concern in the treatment of domestic wastewater, it is also often desirable to stabilize those inorganic compounds that can exert a BOD demand. The most important inorganic compound is ammonia, because its presence in the plant effluent can stimulate the lowering of the dissolved oxygen in the receiving stream through the biological process of nitrification. In nitrification, ammonia is oxidized biologically to nitrate. Nitrate, the final oxidation state of the nitrogen compounds, represents a stabilized product.

In practice, nitrification can be achieved either in the same reactor used in the treatment of the carbonaceous organic matter or in a separate suspended-growth reactor following a conventional activated-sludge treatment process. When carbonaceous removal and nitrification are achieved in the same reactor, the process is often identified as single-stage nitrification. When a separate facility is used for nitrification, it normally includes a reactor and settling tank of the same general design-configuration used for the activated-sludge process. The oxidation of ammonia to nitrate can be carried out with either air or pure oxygen. The details of the nitrification process are considered in Chap. 12.

Aerobic Aerated Lagoons

Aerated lagoons (or ponds) evolved from facultative stabilization ponds when surface aerators were installed to overcome the odors from organically overloaded ponds. Although a number of definitions of aerobic aerated-lagoon processes will be found in the literature, the following definition will be used in this text.

Process description The aerated-lagoon process is essentially the same as the conventional extended-aeration activated-sludge process ($\theta_c = 10$ days), except that an earthen basin is used for the reactor, and the oxygen required by the process is supplied by surface or diffused aerators. In an aerobic lagoon, all the solids are maintained in suspension. In the past, aerated lagoons were operated as flow-through activated-sludge systems without recycle, usually followed by large settling ponds. To meet secondary treatment standards of the U.S. Environmental Protection Agency (see Table 4-1), many aerated lagoons are now used in conjunction with settling facilities and incorporate the recycle of biological solids.

Process microbiology Because the aerated-lagoon process is essentially the same as the activated-sludge process, the microbiology is also similar. Some differences occur because the large surface area associated with aerated lagoons can cause more significant temperature effects than are normally encountered in the conventional activated-sludge process.

Seasonal and continuous nitrification may be achieved in aerated-lagoon systems. The degree of nitrification depends on the design and operating conditions within the system and on the wastewater temperature. Generally, with higher wastewater temperatures and lower loadings (increased sludge-retention time), higher degrees of nitrification can be achieved.

Process analysis The analysis of an aerated lagoon can be carried out using either the approach described in Sec. 9-4 for a completely mixed aerobic treatment system without recycle, or the approach described previously in this section for the activated-sludge process with recycle, depending on the method of operation to be used.

Another approach is to assume that the observed BOD_5 removal (either overall, including soluble and suspended-solids contribution, or soluble only) can be described in terms of a first-order $(r_{su} = -kS)$ or a quasi-second-order $(r_{su} = -kSX)$ removal function. The required analysis for a continuous-flow stirred-tank reactor without recycle has been outlined previously in this chapter and in Chap. 5 (see Eq. 5-38). From that analysis, the pertinent equations for a single aerated lagoon are:
For first-order kinetics:

$$\frac{S}{S_0} = \frac{1}{1 + k_1(V/Q)} \tag{9-72}$$

For quasi-second-order kinetics:

$$\frac{S}{S_0} = \frac{1}{1 + k_2 X(V/Q)} \tag{9-73}$$

where S = effluent BOD_5 concentration, mg/L
S_0 = influent BOD_5 concentration, mg/L
k_1, k_2 = observed overall BOD_5 removal rate constant, d^{-1}, L/mg · d
V = volume, m^3
Q = flowrate, m^3/d
X = mixed-liquor volatile suspended solids, mg/L

The corresponding equation derived from a consideration of soluble substrate-removal kinetics, as given by Eq. 9-15, is

$$\frac{S}{S_0} = \frac{1}{1 + [kX/(K_s + S)](V/Q)} \tag{9-74}$$

The terms in Eq. 9-74 are as defined previously. Application of these equations is considered in Prob. 9-13 and Example 10-4.

Aerobic Digestion

Aerobic digestion is an alternative method of treating the organic sludges produced from various treatment operations. Aerobic digesters may be used to treat (1) only waste-activated or trickling-filter sludge, (2) mixtures of waste-activated or trickling-filter sludge and primary sludge, or (3) waste sludge from activated-sludge treatment plants designed without primary settling. Today, two variations of the aerobic digestion process are in common use: conventional and pure oxygen. Thermophilic aerobic digestion is also under investigation. Additional details on all these processes are presented in Chap. 11.

Process description In conventional aerobic digestion, the sludge is aerated for an extended period of time in an open, unheated tank using conventional air diffusers or surface aeration equipment. The process may be operated in a continuous or batch mode. Smaller plants use the batch system in which sludge is aerated and completely mixed for an extended period of time, followed by quiescent settling and decantation [32]. In continuous systems, a separate tank is used for decantation and concentration.

Pure-oxygen aerobic digestion is a modification of the aerobic digestion process in which pure oxygen is used in lieu of air. The resultant sludge is similar to conventional aerobically digested sludge. The pure-oxygen modification is an emerging technology that currently is being investigated in a number of full-scale installations.

Thermophilic aerobic digestion represents still another refinement of the aerobic digestion process. Carried out with thermophilic bacteria at temperatures ranging from 25 to 50°C (77 to 122°F) above the ambient air temperature, this process can achieve high removals of the biodegradable fraction (up to 80 percent) at very short detention times (3 to 4 days) [19].

Process microbiology Aerobic digestion, as mentioned, is similar to the activated-sludge process. As the supply of available substrate food is depleted, the microorganisms begin to consume their own protoplasm to obtain energy for cell-maintenance reactions. When this occurs, the microorganisms are said to be in the endogenous phase. As shown in Eq. 9-6, cell tissue is aerobically oxidized to carbon dioxide, water, and ammonia. Actually, only about 75 to 80 percent of the cell tissue can be oxidized; the remainder is composed of inert components and organic compounds that are not biodegradable. The ammonia from this oxidation is subsequently oxidized to nitrate as digestion proceeds.

If activated or trickling-filter sludge is mixed with primary sludge, and the combination is to be aerobically digested, there will be both direct oxidation of the organic matter in the primary sludge and endogenous oxidation of the cell tissue. Operationally, most aerobic digesters can be considered to be arbitrary-flow reactors without recycle.

Process analysis Factors that must be considered in the analysis of aerobic digesters include: hydraulic residence time, process loading criteria, oxygen requirements, energy requirements for mixing, environmental conditions, and process operation. The design of aerobic digesters is considered in Chap. 11.

Aerobic Stabilization Ponds

In their simplest form, aerobic stabilization ponds are large, shallow earthen basins that are used for the treatment of wastewater by natural processes involving the use of both algae and bacteria [23]. Although it is common to group all pond systems together when discussing them, in this chapter they are discussed according to the classification presented in Table 9-1. In Chap. 10, their design is considered collectively.

Process description An aerobic stabilization pond contains bacteria and algae in suspension, and aerobic conditions prevail throughout its depth. There are two basic types of aerobic ponds. In the first type, the objective is to maximize the production of algae. These ponds are usually limited to a depth of about 15.2 to 45.7 cm (6 to 18 in).

In the second type, the objective is to maximize the amount of oxygen produced, and pond depths of up to 1.53 m (5 ft) are used. In both types, oxygen, in addition to that produced by algae, enters the liquid through atmospheric diffusion. To achieve best results with aerobic ponds, their contents must be mixed periodically using pumps or surface aerators.

Process microbiology In aerobic photosynthetic ponds, the oxygen is supplied by natural surface aeration and by algal photosynthesis. Except for the algal population, the biological community present in stabilization ponds is similar to that present in an activated-sludge system. The oxygen released by the algae through the process of photosynthesis is used by the bacteria in the aerobic degradation of organic matter. The nutrients and carbon dioxide released in this degradation are, in turn, used by the algae. This cyclic-symbiotic relationship is shown in Fig. 9-21. Higher animals, such as rotifers and protozoa, are also present in the pond, and their main function is to polish the effluent.

The particular algal group, animal group, or bacterial species present in any section of an aerobic pond depend on such factors as organic loading, degree

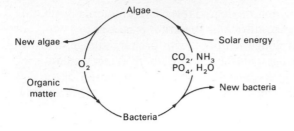

Figure 9-21 Schematic representation of the symbiotic relationship between algae and bacteria.

of pond mixing, pH, nutrients, sunlight, and temperature. Temperature has a profound effect on the operation of aerobic ponds, particularly in regions having cold winters.

Process analysis The efficiency of BOD_5 conversion in aerobic ponds is high, ranging up to 95 percent; however, it must be remembered that, although the soluble BOD_5 has been removed from the influent wastewater, the pond effluent will contain an equivalent or larger concentration of algae and bacteria that ultimately may exert a higher BOD_5 than the original waste. Various means of removing the algae from the treated wastewater are discussed in Chap. 10.

A number of theoretical approaches have been proposed for the analysis of aerobic stabilization ponds. Because of the many uncontrollable variables involved, however, the ponds are still usually designed by using appropriate loading factors derived from pilot-plant studies and observations of operating systems. The pond loading is adjusted to reflect the amount of oxygen available from photosynthesis and atmospheric reaeration.

9-6 AEROBIC ATTACHED-GROWTH TREATMENT PROCESSES

Aerobic attached-growth biological-treatment processes usually are used to remove organic matter found in wastewater. They are also used to achieve nitrification (the conversion of nitrogen in the form of ammonia to nitrate). The attached-growth processes include the trickling filter, the roughing filter, rotating biological contactor, and fixed-bed nitrification reactor. Because the trickling-filter process is used most commonly, it will be considered in greater detail than the other processes.

Trickling Filter

The trickling filter is shown in Fig. 9-22. The first trickling filter was placed in operation in England in 1893. The concept of a trickling filter grew from the use of contact filters, which were watertight basins filled with broken stones. In operation, the contact bed was filled with wastewater from the top, and the wastewater was allowed to contact the media for a short time. The bed was then drained and allowed to rest before the cycle was repeated. A typical cycle

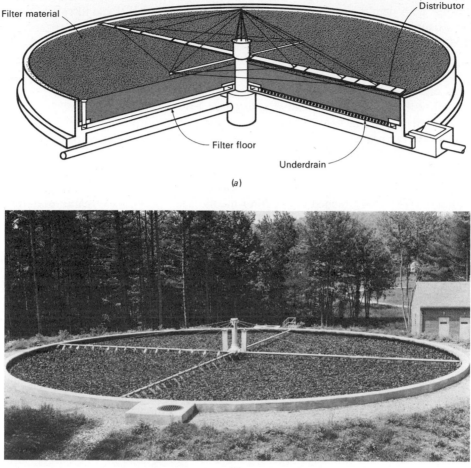

Figure 9-22 Trickling filters. (*a*) Cutaway view of a trickling filter. (*From Dorr-Oliver.*) (*b*) Conventional rock-filled type of filter.

required 12 hours (6 hours for operation and 6 hours of resting). The limitations of the contact filter included a relatively high incidence of clogging, the long rest period required, and the relatively low loading that could be used.

Process description The modern trickling filter consists of a bed of highly permeable media to which microorganisms are attached and through which wastewater is percolated or trickled—hence the name. The filter media usually consist of rocks, varying in size from 25 to 100 mm (1 to 4 in) in diameter. The depth of the rock varies with each particular design but usually ranges from 0.9 to 2.5 m (3 to 8 ft) and averages 1.8 m (6 ft). Trickling filters that use plastic media, a more recent innovation, have been built in square and other shapes with depths varying from 9 to 12 m (30 to 40 ft). Rock filter beds are now usually circular, and the liquid wastewater is distributed over the top of the bed by a rotary distributor.

Filters are constructed with an underdrain system for collecting the treated wastewater and any biological solids that have become detached from the media. This underdrain system is important both as a collection unit and as a porous structure through which air can circulate (see Fig. 9-22). The collected liquid is passed to a settling tank where the solids are separated from the treated wastewater. In practice, a portion of the liquid collected in the underdrain system or the settled effluent is recycled, usually to dilute the strength of the incoming wastewater.

The organic material present in the wastewater is degraded by a population of microorganisms attached to the filter media (see Fig. 9-23). Organic material from the liquid is adsorbed onto the biological film or slime layer. In the outer portions of the biological slime layer, the organic material is degraded by aerobic microorganisms. As the microorganisms grow, the thickness of the slime layer increases, and the diffused oxygen is consumed before it can penetrate the full depth of the slime layer. Thus an anaerobic environment is established near the surface of the media.

As the slime layer increases in thickness, the adsorbed organic matter is metabolized before it can reach the microorganisms near the media face. As a result of having no external organic source available for cell carbon, the microorganisms near the media face enter into an endogenous phase of growth and lose their ability to cling to the media surface. The liquid then washes the slime off the media, and a new slime layer starts to grow. This phenomenon of losing the slime layer is called sloughing and is primarily a function of the organic and hydraulic loading on the filter. The hydraulic loading accounts for shear velocities, and the organic loading accounts for the rate of metabolism in the slime layer. On the basis of hydraulic and organic loading rates, filters are usually divided into two classes: low-rate and high-rate. Their relative merits are discussed in Chap. 10.

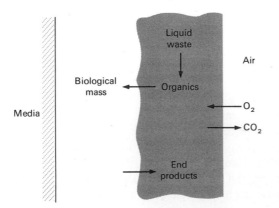

Figure 9-23 Schematic representation of the cross section of a biological slime in a trickling filter.

Process microbiology The biological community in the filter consists primarily of protists, including aerobic, anaerobic, and facultative bacteria, fungi, algae, and protozoans. Higher animals, such as worms, insect larvae, and snails are also present.

Facultative bacteria are the predominating microorganisms in the trickling filter. Along with the aerobic and anaerobic bacteria, their role is to decompose the organic material in the wastewater. *Achromobacter*, *Flavobacterium*, *Pseudomonas*, and *Alcaligenes* are among the bacterial species commonly associated with the trickling filter. Within the slime layer, where adverse conditions prevail with respect to growth, the filamentous forms *Sphaerotilus natans* and *Beggiatoa* will be found. In the lower reaches of the filter, the nitrifying bacteria *Nitrosomonas* and *Nitrobacter* will be present [16].

The fungi present are also responsible for waste stabilization, but their contribution is usually important only under low-pH conditions or with certain industrial wastes. At times, their growth can be so rapid that the filter clogs and ventilation becomes restricted. Among the fungi species that have been identified are *Fusazium*, *Mucor*, *Penicillium*, *Geotrichum*, *Sporatichum*, and various yeasts [16].

Algae can grow only in the upper reaches of the filter where sunlight is available. *Phormidium*, *Chlorella*, and *Ulothrix* are among the algae species commonly found in trickling filters [16]. Generally, algae do not take a direct part in waste degradation, but during the daylight hours they add oxygen to the percolating wastewater. From an operational standpoint, the algae are troublesome because they can cause clogging of the filter surface, which produces odors.

The protozoa in the filter are predominantly of the ciliata group, including *Vorticella*, *Opercularia*, and *Epistylis* [16]. As in the activated-sludge process, their function is not to stabilize the waste but to control the bacterial population. The higher animals, such as snails, worms, and insects, feed on the biological films in the filter and, as a result, help to keep the bacterial population in a state of high growth or rapid food utilization. The higher animal forms are not as common in high-rate tower trickling filters.

Variations in the individual population of the biological community occur throughout the filter depth with changes in organic loading, hydraulic loading, influent wastewater composition, pH, temperature, air availability, and other factors, as described in the following discussion.

Process analysis In predicting the performance of trickling filters, the organic and hydraulic loadings and the degree of purification required are among the important factors that must be considered. Over the years, a number of investigators have proposed equations to describe the removals observed, including Atkinson [3], Eckenfelder [8, 9, 10], Fairall [11], Galler and Gotaas [12], the National Research Council [35], and Velz [47].

Because of the unstable characteristics of the biological slime layer and the unpredictable hydraulic characteristics, a generalized kinetic model of the trickling filter is very difficult to develop. The problems involved can also be appreciated

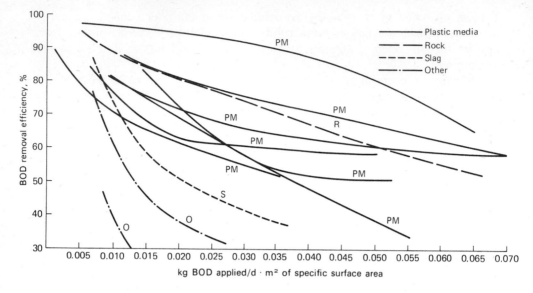

Figure 9-24 BOD removal efficiency as a function of applied loading for various trickling-filter installations. (*Data from Metcalf & Eddy files.*)

by referring to Fig. 9-24, in which performance data taken from Metcalf & Eddy files have been plotted for different plants. The scatter in these data is remarkable. It is of little wonder that a universal design equation is not available. In the discussion that follows, the equations proposed by Atkinson and Eckenfelder will be explored.

Atkinson and his coworkers [3] have proposed the following model to describe the rate of flux of organic material into the slime layer, assuming that diffusion into the slime layer controls the rate of reaction and that there is no concentration gradient across the liquid film (see Fig. 9-25).

$$r_s = -\frac{Ehk_0 \bar{S}}{K_m + \bar{S}} \tag{9-75}$$

where r_s = rate of flux of organic material into the slime layer
 E = effectiveness factor $(0 \le E \le 1)$
 h = thickness of slime layer, m
 k_0 = maximum reaction rate, d^{-1}
 $\bar{S}$ = average BOD concentration in the bulk liquid in volume element
 K_m = half-velocity constant

Because the effectiveness factor E is approximately proportional to the BOD

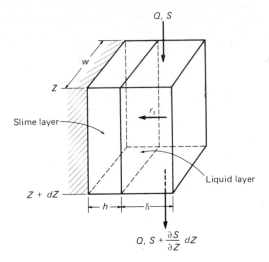

Q, S

w

Z

Slime layer

r_s

Liquid layer

$Z + dZ$

h δ

$Q, S + \dfrac{\partial S}{\partial Z} dZ$

Figure 9-25 Definition sketch for the analysis of the trickling-filter process [40].

concentration in the liquid, Eq. 9-75 can be rewritten as follows:

$$r_s = -\frac{f\,hk_0\,\bar{S}^2}{K_m + \bar{S}} \tag{9-76}$$

where f = proportionality factor

This model can be applied to the analysis of a trickling filter by performing a mass-balance analysis for the organic material contained in the liquid volume (see Fig. 9-25).

1. General word statement:

| Rate of accumulation of substrate within the volume element | = | rate of flow of substrate into the volume element | − | rate of flow of substrate out of the volume element | + | rate of substrate flux into the slime layer from the volume element | (9-77) |

2. Simplified word statement:

$$\text{Accumulation} = \text{inflow} - \text{outflow} + \text{utilization} \tag{9-78}$$

3. Symbolic representation:

$$\frac{\partial \bar{S}}{\partial t}\,dV = QS - Q\left(S + \frac{\partial S}{\partial Z}dZ\right) + dZw\left(-\frac{f\,hk_0\,\bar{S}^2}{K_m + \bar{S}}\right) \tag{9-79}$$

where Q = volumetric flowrate
w = width of section under consideration
Z = filter depth

Assuming that steady-state conditions prevail $(\partial \bar{S}/\partial t = 0)$, Eq. 9-79 can be simplified to yield

$$Q \frac{dS}{dZ} = -f k_0 \, hw \frac{\bar{S}^2}{K_m + \bar{S}} \qquad (9\text{-}80)$$

If it is now assumed that the value of the saturation coefficient K_m is small relative to the value of BOD, then Eq. 9-80 can be written as

$$\frac{dS}{dZ} = -\frac{f h k_0 \, w \bar{S}}{Q} \qquad (9\text{-}81)$$

Eq. 9-81 can now be integrated between the limits of S_e and S_i and 0 and Z to yield

$$\frac{S_e}{S_i} = \exp\left[-(f h k_0)\frac{wZ}{Q}\right] \qquad (9\text{-}82)$$

where S_e = effluent concentration, mg/L
  S_i = influent concentration resulting after the untreated incoming waste-water is mixed with recycled effluent, mg/L

The use of Eq. 9-82 involves the determination of the coefficients f, h, and k_0. The form of Eq. 9-82 is also similar to an equation proposed by Eckenfelder [10] to describe the observed performance of trickling filters:

$$\frac{S_e}{S_i} = \exp\left[-KZS_a^m\left(\frac{A}{Q}\right)^n\right] \qquad (9\text{-}83)$$

where K = observed reaction-rate constant (value usually obtained from pilot-plant studies), m/d
  Z = filter depth, m

  S_a = specific surface area of filter = $\dfrac{\text{surface area } A_s, \, \text{m}^2}{\text{unit volume } V, \, \text{m}^3}$

  A = cross-sectional area of filter, m^2
  Q = volumetric flowrate applied to filter, m^3/d
  m, n = empirical constants

In comparing Eqs. 9-82 and 9-83, it will be noted that the terms $f h k_0 \, wZ/Q$ in Eq. 9-82 correspond to the terms $KZS_a^m A/Q^n$ in Eq. 9-83. Using these similarities, the term $f h k_0$ can be estimated from an analysis of data presented by Eckenfelder [10]. To do this, Eq. 9-82 is rewritten as follows:

$$-\ln \frac{S_e}{S_i} = \left(f h k_0\right)\frac{wZ}{Q} \qquad (9\text{-}84)$$

If the term $(-\ln S_e/S_i)$ is plotted versus the term (wZ/Q), the slope of the curve obtained will be equal to the term $(f h k_0)$. To use the data reported by Eckenfelder with Eq. 9-84, it is necessary to assume m and n equal 1 and to note

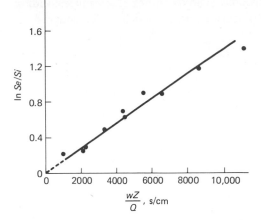

Figure 9-26 Analysis of organic data for high-rate trickling filters packed with plastic medium [40, 41].

the following relationship between terms $S_a Z(A/Q)$ in Eq. 9-83 and the term wZ/Q in Eq. 9-82:

$$\frac{S_a Z A}{Q} = \frac{A_s}{V} \frac{Z A}{Q} = \frac{A_s}{Q} = \frac{wZ}{Q} \tag{9-85}$$

After performing the necessary conversions, the data in Ref. 10 can be plotted as shown in Fig. 9-26 [41]. The slope of the curve, which represents the term $f hk_0$, is found to be equal to about 1.35×10^{-4} cm/s or 0.117 m/d. The use of this value for design is illustrated in Example 10-9.

It should be noted that Eqs. 9-82 and 9-83 can be written in terms of a volumetric loading rate (V/Q) so that a depth term does not enter into the final equation. A typical example of such an equation is given below [4].

$$\frac{S_e}{S_i} = \exp\left(-K_T S_a^a Q_v^{-b}\right) \tag{9-86}$$

where S_e = BOD of settled effluent from the filter, mg/L
S_i = BOD of wastewater applied to the filter, mg/L
K_T = experimental rate constant, m/d
S_a = specific surface area of media, m²/m³
Q_v = volumetric flowrate applied per unit volume of filter, m³/m³ · d
a, b = experimental constants

In using Eqs. 9-82, 9-83, or 9-86 for the design of trickling filters, it should be remembered that the value of the term $f hk_0$ or K_T will vary with so many local factors that coefficient values derived from the literature, such as those presented in Chap. 10, must be used with great caution. One of the most important local factors is the effect of temperature, which usually can be accounted for by correcting the reaction-rate constant using Eq. 9-21 and the data presented in Table 9-6.

Another factor over which there is a considerable amount of misunderstanding is the effect of recirculation on filter performance, which is considered in Example 9-2.

Example 9-2: Evaluation of the effects of recirculation in a trickling-filter process Examine the effect of recirculation ratios varying from 0 to 6 for a trickling-filter process with a value of 0.26 m/d for the term (fhk_0). Assume a wZ/Q value of 6.9 d/m when the recycle ratio is zero.

SOLUTION

1 Determine the fraction of the applied organic load removed using Eq. 9-82, when the recycle ratio is equal to zero.

$$\frac{S_e}{S_i} = \exp\left[-(fhk_0)\frac{wZ}{Q}\right]$$

for $fhk_0 = 0.26$ m/d

$$\frac{wZ}{Q} = 6.9 \text{ d/m}$$

$$\frac{S_e}{S_i} = \exp\left[-(0.26 \text{ m/d})(6.9 \text{ d/m})\right]$$

$$= \exp(-1.794)$$

$$= 0.166$$

Fraction removed $= 1 - \dfrac{S_e}{S_i}$

$$= 1 - 0.166 = 0.834$$

(Note that, when the recycle ratio is equal to zero, the fraction removed as computed above also corresponds to the fraction of the influent BOD removed.)

2 Determine the fraction of the applied load removed for various recycle ratios. Using Eq. 9-82 in the following form, set up a computation table.

$$\frac{S_e}{S_i} = \exp\left[-(fhk_0)\frac{wZ}{(1+\alpha)Q}\right]$$

α	0	1	2	3	4	5	6
S_e/S_i	0.166	0.408	0.550	0.639	0.699	0.742	0.774
$1 - S_e/S_i$	0.834	0.592	0.450	0.361	0.301	0.258	0.226

3 Determine the fraction of the influent load removed from the data determined in step 2. This can be accomplished by performing a materials balance on the influent to the filter, taking into account the influent and recycle flows as follows:

$$QS_0 + \alpha QS_e = (1+\alpha)QS_i$$

where $S_0 = $ BOD in the influent
$S_e = $ BOD in the recycle
$\alpha = $ recycle ratio
$S_i = $ BOD in the influent applied to the filter

Assume that $Q = 1$, and set up the foregoing expression in a format for computing the term S_e/S_0 using the data from step 2:

$$\frac{S_e}{S_0} = \left[(1 + \alpha)\frac{S_i}{S_e} - \alpha\right]$$

Compute the fraction of the influent removed using the data from step 2:

α	$1 + \alpha$	$(1 + \alpha)\dfrac{S_i}{S_e}$	$(1 + \alpha)\dfrac{S_i}{S_e} - \alpha$	$\dfrac{S_e}{S_0}$	$1 - \dfrac{S_e}{S_0}$
0	1	6.024	6.024	0.166	0.834
1	2	4.902	3.902	0.256	0.744
2	3	5.455	3.455	0.289	0.711
3	4	6.260	3.260	0.307	0.693
4	5	7.153	3.153	0.317	0.683
5	6	8.086	3.086	0.324	0.676
6	7	9.044	3.044	0.329	0.671

Comment From the preceding computation it can be seen that as the recycle ratio is increased, the degree of treatment is reduced, whether it is measured in terms of the load applied to the filter $(1 - S_e/S_i)$ or the incoming load $(1 - S_e/S_0)$. Comparing the computations presented in steps 2 and 3, it can be seen that the effect of recycle is less when the incoming load is considered. In general, this is contrary to what is commonly reported in the literature, although the minimal or even deleterious effects of increasing the recycle ratio have been identified in several references [5, 31, 42]. The use of recycle to reduce the strength of the applied load, to maintain optimum wetting of the filter, or for hydraulic control remains a valid use for recycle.

Mass-transfer limitations One of the problems encountered in the design of trickling filters is the determination of the maximum organic material that can be applied to the filter before oxygen becomes a limiting variable. Recognizing the limitations of any analytical approach because of the many variables involved, this problem can nevertheless be approached by equating the transfer of organic material from the liquid film, as defined by Eq. 9-76, to the rate of oxygen transfer using Eq. 7-19. A factor to account for the yield must also be included. This can be done by multiplying the substrate transfer-rate term by a factor such as $(1 - y)$, where y is the expected yield expressed as a decimal. From data reported in the literature, it appears that when the bulk concentrations are in the range of 400 to 500 mg/L, oxygen transfer may become a limiting factor [40]. The airflow through filters is considered in Chap. 10.

Sedimentation facilities for trickling filters As in the activated-sludge process, the settling unit is an important part of the trickling-filter process. It is needed for removal of suspended solids sloughed off during periods of unloading with low-rate filters and for removal of lesser amounts of solids sloughed off continuously by high-rate filters. If recirculation is used, some of the settled solids may be recycled and some may be wasted, but the recycle of the settled biological solids is not as important as in the activated-sludge process. In the trickling-filter process, the majority of the active microorganisms are attached to the filter media and do not pass out of the reactor as in the activated-sludge process. Although recirculation can help in seeding the filter, the primary purposes of recirculation are to dilute strong influent wastewater and to bring the filter effluent back in contact with the biological population for further treatment. Recirculation is almost always included in high-rate trickling-filter systems.

Roughing Filters

Roughing filters are used principally to reduce the organic loading on downstream processes and in seasonal nitrification applications where the purpose is to reduce the organic load so that a downstream biological process will dependably nitrify the wastewater during the summer months.

Process description Although the earliest roughing filters were shallow, stone-media systems, the present trend is toward use of synthetic media or redwood at depths of 3.7 to 12 m (12 to 40 ft). As with other biological processes, roughing-filter performance is temperature-sensitive. When roughing filters are used for the removal of a portion of the organic material present, or to enhance downstream nitrification, a drop in efficiency is not critical. In extremely cold climates, ice formation may occur on the distributor and the upper layers of the medium.

Process microbiology The biological activity in a roughing filter is essentially the same as that described for the trickling filter. Some differences will be noted in the organisms present because of the higher shearing action resulting from the higher hydraulic flowrates applied to those units.

   The biological growth is susceptible to the same heavy metals and organic substances as conventional suspended-growth systems, but the process has shown greater resistance to shock loading than suspended-growth systems. Because of the relatively short hydraulic retention time available, organics that are not readily biodegradable are not affected.

Process performance analysis Roughing filters are normally designed using loading factors developed from pilot-plant studies and data derived from full-scale installations, although the analysis presented previously for the trickling filter can be used. Appropriate design values will be found in Chap. 10.

Rotating Biological Contactors

A rotating biological contactor consists of a series of closely spaced circular disks of polystyrene or polyvinyl chloride. The disks are partially submerged in wastewater and rotated slowly through it (see Fig. 9-27).

Figure 9-27 Rotating biological contractor. (*From Autotrol Corp.*)

Process description In operation, biological growths become attached to the surfaces of the disks and eventually form a slime layer over the entire wetted surface area of the disks. The rotation of the disks alternately contacts the biomass with the organic material in the wastewater and then with the atmosphere for adsorption of oxygen. The disk rotation affects oxygen transfer and maintains the biomass in an aerobic condition. The rotation also is the mechanism for removing excess solids from the disks by the shearing forces it creates and maintaining the sloughed solids in suspension so they can be carried from the unit to a clarifier. Rotating biological contactors can be used for secondary treatment, and they can also be operated in the seasonal and continuous-nitrification modes.

Process performance analysis Rotating biological contactors are usually designed on the basis of loading factors derived from pilot-plant and full-scale installations, although their performance can be analyzed using an approach similar to that for trickling filters. Both hydraulic and organic loading-rate criteria are used in sizing units for secondary treatment. The loading rates for warm weather and year-round nitrification will be considerably lower than the corresponding rates for secondary treatment. Typical design values are presented in Chap. 10.

Rotating biological contactors generally are more reliable than other fixed-film processes because of the large amount of biological mass present (low operating F/M). This also permits them to withstand hydraulic and organic surges more effectively. The effect of staging in this plug-flow system eliminates short circuiting and damps shock loadings.

Packed-Bed Reactors

Still another attached-growth process is the packed-bed reactor used for both the removal of carbonaceous BOD and nitrification [50]. Typically, a packed-bed reactor consists of a container (reactor) that is packed with a medium to which the microorganisms can become attached. Wastewater is introduced from the bottom of the container through an appropriate underdrain system or inlet chamber. Air or pure oxygen necessary for the process is also introduced with the wastewater. This process is considered further in Chap. 12, which deals with the advanced treatment of wastewater.

9-7 ANOXIC SUSPENDED-GROWTH AND ATTACHED-GROWTH PROCESSES

The removal of nitrogen in the form of nitrate by conversion to nitrogen gas can be accomplished biologically under anoxic (without oxygen) conditions. The process is known as denitrification. In the past, the conversion process was often identified as anaerobic denitrification. However, the principal biochemical pathways are not anaerobic but rather a modification of aerobic pathways; therefore, the use of the term anoxic in place of anaerobic is considered appropriate [38]. The principal denitrification processes may be classified as suspended-growth and attached-growth processes. The microbiology and the analysis of these processes are considered in detail in Chap. 15. The information presented in this section is meant to serve only as a brief introduction.

Suspended-Growth Denitrification

Suspended-growth denitrification is usually carried out in a plug-flow type of activated-sludge system (i.e., following any process that converts ammonia and organic nitrogen to nitrates (nitrification). The anaerobic bacteria obtain energy for growth from the conversion of nitrate to nitrogen gas but require an external source of carbon for cell synthesis. Nitrified effluents are usually low in carbonaceous matter and so methanol is commonly used as a carbon source, but industrial wastes that are poor in nutrients have also been used. Because the nitrogen gas formed in the denitrification reaction hinders the settling of the mixed liquor, a nitrogen-gas stripping reactor should precede the denitrification clarifier. Removal of residual methanol-induced BOD is an added benefit of the stripping tank [32].

Fixed-Film Denitrification

Fixed-film denitrification is carried out in a column reactor containing stone or one of several synthetic media upon which the bacteria grow. Depending on the size of the medium, this process may or may not need to be followed by a clarifier. Adequate wasting of solids occurs through the low-level suspended-solids carryover in the effluent. Periodic backwashing and/or air scour is necessary to prevent solids buildup in the column that can cause excessive head loss. As in the suspended-growth denitrification process, an external carbon source is usually necessary. Most applications of this process involve the downflow mode (either gravity or pressure), but expanded-bed (upflow) techniques have also been tested [32]. Additional process modifications in terms of the physical systems used are considered in Chap. 12.

9-8 ANAEROBIC SUSPENDED-GROWTH TREATMENT PROCESSES

The two most common anaerobic suspended-growth processes used for the treatment of wastewater are (1) the anaerobic digestion process and (2) the anaerobic contact process. Because the anaerobic digestion process is of such fundamental importance in the stabilization of organic material and biological solids, it will be emphasized in the following discussion.

Anaerobic Digestion

Anaerobic digestion is one of the oldest processes used for the stabilization of sludges. It involves the decomposition of organic and inorganic matter in the absence of molecular oxygen. The major applications have been, and remain today, in the stabilization of concentrated sludges produced from the treatment of wastewater and in the treatment of some industrial wastes. It has recently been demonstrated, however, that dilute organic wastes can be treated anaerobically.

Process description In the anaerobic digestion process, the organic material in mixtures of primary settled and biological sludges under anaerobic conditions is biologically converted to methane (CH_4) and carbon dioxide (CO_2). The process is carried out in an airtight reactor. Sludges are introduced continuously or intermittently and retained in the reactor for varying periods of time. The stabilized sludge, which is withdrawn continuously or intermittently from the process, is nonputrescible, and its pathogen content is greatly reduced.

Two types of digesters are now in use, standard-rate and high-rate. In the standard-rate digestion process (see Fig. 9-28a), the contents of the digester are usually unheated and unmixed. Detention times for this process vary from 30 to 60 days. In a high-rate digestion process (see Fig. 9-28b), the contents of the digester are heated and completely mixed. The required detention time is 15 days or less. A combination of these two basic processes is known as the two-stage process (see Fig. 9-28c). The primary function of the second stage is to separate the digested solids from the supernatant liquor; however, additional digestion and gas production may occur.

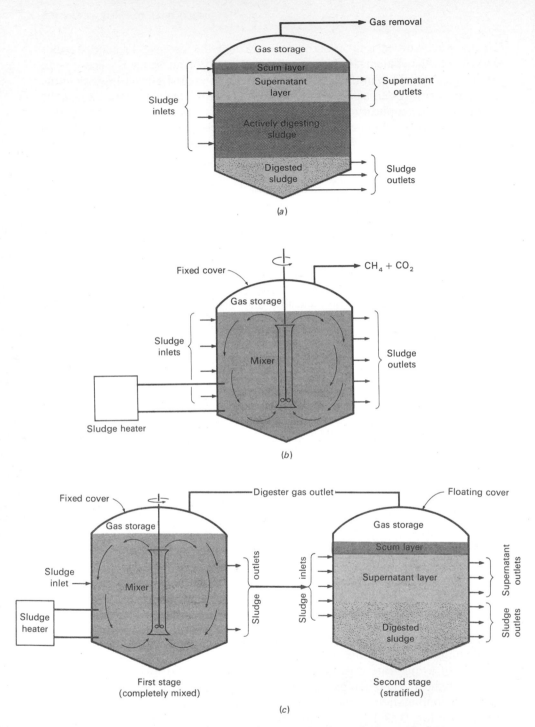

Figure 9-28 Typical anaerobic digesters. (*a*) Conventional standard-rate single-stage process. (*b*) High-rate, continuous-flow stirred tank, single-stage process. (*c*) Two-stage process.

Process microbiology The biological conversion of the organic matter in treat-ment-plant sludges is thought to occur in either two or three steps. In the three-step sequence, the first step involves the enzyme-mediated transformation (liquefaction) of higher-weight molecular compounds into compounds suitable for use as a source of energy and cell carbon. The second step involves the bacterial conversion of the compounds resulting from the first step into identifiable lower-molecular-weight intermediate compounds. The third step involves the bacterial conversion of the intermediate compounds into simpler end products, principally methane and carbon dioxide. In the two-step sequence, the first two steps described are thought to occur simultaneously and are defined as the first step.

In the two-step sequence, the microorganisms responsible for the decomposi-tion of the organic matter are commonly divided into two groups. The first group hydrolyzes and ferments complex organic compounds to simple organic acids, the most common of which are acetic and propionic acid. This group of microorganisms, described as nonmethanogenic, consists of facultative and obligate anaerobic bacteria. Collectively, these microorganisms are also identified in the literature as "acid formers." Among the nonmethanogenic bacteria that have been isolated from anaerobic digesters are *Clostridium spp., Peptococcus anaerobus, Bifidobacterium spp., Desulphovibrio spp., Corynebacterium spp., Lacto-bacillus, Actinomyces, Staphylococcus,* and *Escherichia coli.* Other physiological groups present include those producing proteolytic, lipolytic, ureolytic, or cellulytic enzymes [16].

The second group of microorganisms converts the organic acids formed by the first group to methane gas and carbon dioxide. The bacteria responsible for this conversion are strict anaerobes and are called methanogenic. Collectively, they are identified in the literature as "methane formers." Many of the methano-genic organisms identified in anaerobic digesters are similar to those found in the stomachs of ruminant animals and in organic sediments taken from lakes and rivers. The principal genera of microorganisms that have been identified include the rods (*Methanobacterium, Methanobacillus*) and spheres (*Methanococcus, Methanosarcina*) [16].

The most important bacteria of the methanogenic group are the ones that degrade acetic acid and propionic acid. They have very slow growth rates; as a result, their metabolism is usually considered rate-limiting in the anaerobic treatment of an organic waste. It is in this second step that actual waste stabilization is accomplished by the conversion of the organic acids into methane and carbon dioxide. Methane gas is highly insoluble, and its departure from solution represents actual waste stabilization.

With regard to the specific mechanisms involved in the formation of methane, it appears that two pathways are possible, depending on the nature of the starting substrate. Although there are numerous specific adaptations, methanogenic bacteria appear to be capable of using the following three categories of substrates:

1. The lower fatty acids containing six or fewer carbon atoms (formic, acetic, propionic, butyric, valeric, caproic)
2. The normal and isoalcohols containing from one to five carbon atoms (methanol, ethanol, propanol, butanol, pentanol)
3. Three inorganic gases (hydrogen, carbon monoxide, and carbon dioxide)

The two mechanisms involved can be described by considering the following equations [16]. In the first, involving substrates such as ethanol, butyrate, and hydrogen, methane is produced from substrate oxidation and from the reduction of atmospheric carbon dioxide. In the second, involving substrates such as acetate and propionate, methane is formed by the reduction of carbon dioxide that was formed during the oxidation of substrate.

Reduction of atmospheric CO_2:

$$2C_2H_5OH + CO_2 \rightarrow 2CH_3COOH + CH_4 \tag{9-87}$$

$$4H_2 + CO_2 \rightarrow CH_4 + 2H_2O \tag{9-88}$$

Reduction of CO_2 formed from reaction:

$$CO + H_2O \rightarrow CO_2 + H_2 \tag{9-89}$$

$$CO_2 + 4H_2 \rightarrow CH_4 + 2H_2O \tag{9-90}$$

$$CO + 2H_2 \rightarrow CH_4 + H_2O \tag{9-91}$$

$$4C_2H_5COOH + 8H_2O \rightarrow 4CH_3COOH + 4CO_2 + 24H \tag{9-92}$$

$$3CO_2 + 24H \rightarrow 3CH_4 + 6H_2O \tag{9-93}$$

$$4C_2H_5COOH + 2H_2O \rightarrow 4CH_3COOH + CO_2 + 3CH_4 \tag{9-94}$$

$$CH_3COOH \rightarrow CH_4 + CO_2 \tag{9-95}$$

In addition, many other groups of anaerobic and facultative bacteria use the various inorganic ions present in the sludge. *Desulfovibrio* is responsible for the reduction of the sulfate ion SO_4^- to the sulfide ion S^{--}. Other bacteria reduce nitrates NO_3^- to nitrogen gas N_2 (denitrification).

To maintain an anaerobic treatment system that will stabilize an organic waste efficiency, the nonmethanogenic and methanogenic bacteria must be in a state of dynamic equilibrium. To establish and maintain such a state, the reactor contents should be void of dissolved oxygen and free from inhibitory concentrations of such constituents as heavy metals and sulfides. Also, the pH of the aqueous environment should range from 6.6 to 7.6. Sufficient alkalinity should be present to ensure that the pH will not drop below 6.2, because the methane bacteria cannot function below this point. When digestion is proceeding satisfactorily, the alkalinity will normally range from 1000 to 5000 mg/L, and the volatile acids will be less than 250 mg/L. A sufficient amount of nutrients, such as nitrogen and phosphorus, must also be available to ensure the proper growth of the biological community. Temperature is another important environmental parameter. The optimum temperature ranges are the mesophilic, 30 to 38°C (85 to 100°F), and the thermophilic, 49 to 57°C (120 to 135°F).

Process analysis The advantages and disadvantages of the anaerobic treatment of an organic waste, as compared to aerobic treatment, stem directly from the slow growth rate of the methanogenic bacteria. Of particular interest are those bacteria responsible for the fermentation of acetic and propionic acids. These methane formers have very slow growth rates. The low growth yield signifies that only a small portion of the degradable organic waste is being synthesized into new cells. Typical kinetic coefficients for anaerobic digestion are reported in Table 9-8. With the methanogenic bacteria, most of the organic waste is converted to methane gas, which is combustible and therefore a useful end product. If sufficient quantities are produced, as is customary with muncipal wastewater sludge, the methane gas can be used to operate gas engines or heat the digesting sludge.

Because of the low cellular growth rate and the conversion of organic matter to methane gas and carbon dioxide, the resulting solid matter is reasonably well stabilized. After drying or dewatering, it frequently is suitable for disposal in sanitary landfills or on land as a soil conditioner or humuslike material. On the other hand, the sludge solids resulting from aerobic processes must either be digested, usually anaerobically, or dewatered and incinerated, on account of the large proportion of cellular organic material. A small amount, proportionately, is heat-dried and sold as fertilizer.

Table 9-8 Typical kinetic coefficients for the anaerobic digestion of various substrates[a]

	Coefficient	Basis	Value[b] Range	Value[b] Typical
Domestic sludge	Y	mg VSS/mg BOD[c]	0.040–0.100	0.06
	k_d	d^{-1}	0.020–0.040	0.03
Fatty acid	Y		0.040–0.070	0.050
	k_d	d^{-1}	0.030–0.050	0.040
Carbohydrate	Y		0.020–0.040	0.024
	k_d	d^{-1}	0.025–0.035	0.03
Protein	Y		0.050–0.090	0.075
	k_d	d^{-1}	0.010–0.020	0.014

[a] Derived in part from Refs. 21, 22, 27, and 36.
[b] Values reported are for 20°C.
[c] VSS = volatile suspended solids.
  Note: $1.8(°C) + 32 = °F$.

The high temperatures necessary to achieve adequate treatment are often listed as disadvantages of the anaerobic treatment process; however, high temperatures are necessary only when sufficiently long mean cell residence time cannot be obtained at nominal temperatures. In the anaerobic treatment systems shown in Fig. 9-28, the mean cell residence time of the microorganisms in the reactor is equivalent to the hydraulic detention time of the liquid in the reactor. As the operation temperature is increased, the minimum mean cell residence time is reduced significantly. This means that, at higher temperatures, the system can be operated at a lower mean cell residence time. Thus heating of the reactor contents lowers not only the mean cell residence time necessary to achieve adequate treatment but also the hydraulic detention time, and a smaller reactor volume can be used.

Anaerobic Contact Process

Some industrial wastes that are high in BOD can be stabilized very efficiently by anaerobic treatment. In the anaerobic contact process, untreated wastes are mixed with recycled sludge solids and then digested in a reactor sealed to the entry of air. The contents are mixed completely. After digestion, the mixture is separated in a clarifier or vacuum flotation unit, and the supernatant is discharged as effluent, usually for further treatment. Settled anaerobic sludge is then recycled to seed the incoming wastewater [13]. Because of the low synthesis rate of anaerobic microorganisms, the excess sludge that must be disposed of is minimal. This process has been used successfully for the stabilization of meat-packing and other high-strength soluble wastes.

9-9 ANAEROBIC ATTACHED-GROWTH TREATMENT PROCESSES

The most common anaerobic attached-growth treatment process is the anaerobic filter process used for the treatment of both domestic and industrial wastes. Anaerobic ponds are also discussed in this section, although they are not an attached-growth process in the strict sense.

Anaerobic Filter

The anaerobic filter, a relatively recent development in the field of wastewater treatment, is a column filled with various types of solid media used for the treatment of the carbonaceous organic matter in wastewater [51]. The waste flows upward through the column, contacting the medium on which anaerobic bacteria grow and are retained. Because the bacteria are retained on the medium and not washed off in the effluent, mean cell residence times on the order of 100 days can be obtained. Large values of θ_c can be achieved with short hydraulic retention times; so the anaerobic filter can be used for the treatment of low-strength wastes at ambient temperature.

Anaerobic Ponds

Anaerobic ponds are used for the treatment of high-strength organic wastewater that also contains a high concentration of solids. Typically, an anaerobic pond is a deep earthen pond with appropriate inlet and outlet piping. To conserve heat energy and to maintain anaerobic conditions, anaerobic ponds have been constructed with depths up to 6.1 m (10 ft). The wastes that are added to the pond settle to the bottom. The partially clarified effluent is usually discharged to another treatment process for further treatment.

Usually, these ponds are anaerobic throughout their depth, except for an extremely shallow surface zone. Stabilization is brought about by a combination of precipitation and the anaerobic conversion of organic wastes to CO_2, CH_4, other gaseous end products, organic acids, and cell tissues. BOD_5 conversion efficiencies up to 70 percent are obtainable routinely. Under optimum operating conditions, removal efficiencies up to 85 percent are possible.

9-10 COMBINED AEROBIC/ANOXIC OR ANAEROBIC TREATMENT PROCESSES

Of the various combined aerobic/anoxic or anaerobic treatment processes identified in Table 9-1, only the facultative stabilization and tertiary or maturation pond will be considered briefly in this section. The combined nitrification-denitrification processes are discussed in Chap. 12. Design details on ponds and lagoons are presented in Chap. 10.

Facultative Lagoons (Ponds)

Ponds in which the stabilization of wastes is brought about by a combination of aerobic, anaerobic, and facultative bacteria are known as facultative (aerobic-anaerobic) stabilization ponds.

Process description As shown in Fig. 9-29, three zones exist in a facultative pond: (1) a surface zone where aerobic bacteria and algae exist in a symbiotic relationship, as previously discussed; (2) an anaerobic bottom zone in which accumulated solids are actively decomposed by anaerobic bacteria; and (3) an intermediate zone that is partly aerobic and partly anaerobic, in which the decomposition of organic wastes is carried out by facultative bacteria.

In practice, oxygen is maintained in the upper layer by the presence of algae or by the use of surface aerators (see Chap. 10). If surface aerators are used, algae are not required. The advantage of using surface aerators is that a higher organic load can be applied. However, the organic load must not exceed the amount of oxygen that can be supplied by the aerators without completely mixing the pond contents, or the benefits to be derived from anaerobic decomposition will be lost.

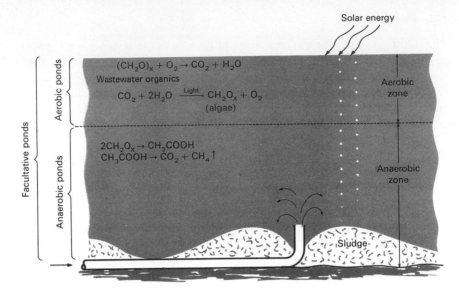

Figure 9-29 Schematic representation of a waste-stabilization pond. (*Adapted from Ref. 14.*)

Process microbiology The biological community in the upper or aerobic layer is similar to that of an aerobic pond. The microorganisms in the bottom layer of the pond are facultative and anaerobic bacteria. The metabolic activities of these bacteria were discussed earlier in this chapter.

Process analysis The amount of effort that has been devoted to the characterization of facultative ponds is staggering, and an equal amount has probably been spent trying to develop appropriate design equations. Although many design equations have been published, there is no universal equation. Part of the explanation for this is that, to a large extent, the process is undefined because of the vagaries of nature. For example, all predictive equations for effluent quality are essentially meaningless when windy conditions prevail. Under such conditions, the effluent quality will be a function of the degree of wind mixing and the quantity of the deposited solids that have been suspended. For this reason, facultative ponds are usually designed on the basis of loading factors developed from field experience.

Tertiary-Maturation Ponds

Tertiary-maturation low-rate stabilization ponds are designed to provide for secondary effluent polishing and seasonal nitrification [24, 32]. The biological mechanisms involved are similar to other aerobic suspended-growth processes. Operationally, the residual biological solids are endogenously respired, and ammonia is converted to nitrate using the oxygen supplied from surface reaeration and from algae. A detention time of 18 to 20 days has been suggested as the minimum period required to provide for complete endogenous respiration of the residual solids. To maintain aerobic conditions, the applied loadings are quite low.

As with all biological nitrification systems, the efficiency of tertiary low-rate ponds decreases with decreasing wastewater temperature. To provide a reliably nitrified effluent that is low in BOD_5 and suspended solids, an efficient and reliable effluent solids-removal process will be required.

DISCUSSION TOPICS AND PROBLEMS

9-1 A 1-L sample contains 250 mg of casein ($C_8H_{12}O_3N_2$). If 0.5 mg of bacterial cell tissue ($C_5H_7NO_2$) is synthesized per milligram of casein consumed, determine the amount of oxygen required to complete the oxidation of casein to end products and cell tissue. The end products of the oxidation are carbon dioxide (CO_2), ammonia (NH_3), and water. Assume that the nitrogen not incorporated in cell-tissue production will be converted to ammonia.

9-2 Assuming that the endogenous coefficient k_d can be neglected, develop expressions that can be used to determine the substrate and cell concentration as a function of time for a batch reactor. If the initial concentration of substrate and cell is 100 and 200 mg/L, respectively, determine the amount of substrate remaining after 1 h. If the endogenous coefficient is equal to 0.04 d^{-1}, estimate the error made by neglecting this factor. Assume that the following constants apply: $k = 2.0\,\text{h}^{-1}$; $K_s = 80$ mg/L; $Y = 0.4$ mg/mg.

9-3 If the dilution rate D is defined as Q/V and the endogenous coefficient is neglected, develop expressions that can be used to estimate the effluent substrate and cell concentration from a continuous-flow stirred-tank reactor without recycle as a function of the dilution rate. If $Y = 0.5$ mg/mg, $\mu_m = 1.0\,\text{h}^{-1}$, $K_s = 200$ mg/L, and $S_0 = 10,000$ mg/L, prepare a plot of the substrate and cell concentration versus the dilution rate. Use centimeter paper and plot the dilution rate going from zero to 1.0 h^{-1}. Use 1-cm divisions for each 1000 mg/L of substrate, and 2-cm divisions for each 1000 mg/L of cells.

9-4 A wastewater is to be treated aerobically in a continuous-flow stirred-tank reactor with no recycle. Determine θ_c^M using the following constants: $K_s = 50$ mg/L; $k = 5.0$ d^{-1}; $k_d = 0.06$ d^{-1}; and $Y = 0.60$. The initial wastewater substrate concentration is 200 mg/L.

9-5 Using a design value of $\theta_c = 2$ d and the constants given in Prob. 9-4, determine the effluent substrate concentration, the specific utilization rate $(-r_{su}/X)$, the food-to-microorganism ratio, $(S_0/X\theta)$, and the concentration of microorganisms in the reactor.

9-6 The following data were obtained using four bench-scale continuous-flow activated-sludge units to treat a food-processing waste. Using these data, determine Y and kd.

		Parameter	
Unit	X, g MLVSS	r_g', g MLVSS/d	U, g BOD_5/d · g MLVSS
1	18.81	0.88	0.17
2	7.35	1.19	0.41
3	7.65	1.42	0.40
4	2.89	1.56	1.09

9-7 Derive Eq. 9-71 for a plug-flow reactor.

9-8 It is proposed to use a continuous-flow stirred-tank reactor with recycle for the treatment of a medium-strength wastewater (see Table 3.5). Determine the amount of oxygen required for the carbonaceous oxidation of the wastewater (assume nitrification does not occur) at a mean cell-residence time of 6 d. Use the kinetic coefficients given in Table 9-7 and assume that the organic compounds in wastewater can be represented as $C_6H_{12}O_6$, the nitrogen as NH_4^+, and the phosphorus as $H_2PO_4^-$. Represent the cell tissue produced in the process as $C_{60}H_{87}O_{23}N_{12}P$. What percentage of the influent nitrogen and phosphorus will be present in the effluent?

9-9 Assuming that the waste specified in Prob. 9-8 can be nitrified completely at a mean cell-residence time of 15 d, estimate the total amount of oxygen required. How does the amount of oxygen required for carbaneous oxidation compare?

9-10 If 75 percent of the cell tissue produced during biological treatment is biodegradable, estimate the ultimate carbonaceous production of cell tissue using Eq. 9-6. If the K value (base 10) is equal to 0.1, determine the BOD_5 of the cell tissues. Express your answer in terms mg BOD_5/mg cell tissue.

9-11 Determine the kinetic coefficients k, K_s, μ_m, Y, and k_d from the following data derived using a laboratory scale complete mix reactor with solids recycle (see Fig. 9-13b).

Unit no.	S_0, mg/L	S, mg/L	θ, d	X, mg VSS/L	θ_c, d
1	400	10	0.167	3950	3.1
2	400	14.3	0.167	2865	2.1
3	400	21.0	0.167	2100	1.6
4	400	49.5	0.167	1050	0.8
5	400	101.6	0.167	660	0.6

9-12 An activated-sludge process with a short aeration time is to be used following a tower trickling filter for treatment of domestic wastewater. Using the following information and data, determine the cell concentration (MLVSS) that must be maintained in the aeration tank if both the effluent BOD_5 and suspended solids must be less than 25 mg/L. What recycle rate will be required under typical and best operating characteristics for the secondary settling tank? Assume the BOD_5 of the effluent solids is equal to 0.65 times the concentration of the solids. (1) Average flowrate = 4000 m³/d; (2) peak flowrate = 8000 m³/d; (3) detention time in aeration basin at peak flow = 0.5 h; (4) trickling filter effluent: BOD_5 = 60 mg/L, suspended solids = 60 mg MLVSS/d (see table below); (5) settling data for the MLVSS derived at a nearby location with a similar plant; (6) settled biological solids are to be recycled to head end of the aerator; (7) kinetic coefficients for the aeration process; $k = 5.0$ d⁻¹, $K_s = 60$ mg/L, $Y = 0.5$ mg/mg, $k_d = 0.06$ d⁻¹; (8) aeration tank type = CFSTR.

X, mg MLVSS/L	X_r, mg MLVSS/L	
	Typical	Best
1000	3200	6000
2000	5200	8000
3000	6600	9400
4000	8000	10,200

9-13 A complete mix aerated lagoon is to be designed with a detention time of 5 d. Using Eq. 9-74 and the data given below, determine the effluent soluble BOD_5. Estimate the total BOD_5 by considering the BOD_5 of the biological solids. What would the value of K_1 in Eq. 9-72 and K_2 in Eq. 9-73 have to be to yield the same results as obtained with Eq. 9-74. How do the values you computed compare to values reported in the literature? Cite at least three references.

1. Influent characteristics:
   Flowrate = 4000 m^3/d
   Total BOD$_5$ = 200 mg/L
   Filtered BOD$_5$ = 150 mg/L
   Suspended solids = 200 mg/L
2. Kinetic coefficients:
   $k = 4.0$ d^{-1}
   $K_s = 80$ mg/L
   $Y = 0.45$ mg/mg
   $k_d = 0.05$ d^{-1}
3. BOD$_5$ of effluent solids = 0.65 (suspended solids)
4. Assume that the BOD$_5$ associated with suspended solids is totally converted in the process

9-14 The following data were obtained from a pilot-plant study involving the treatment of a combined domestic-industrial wastewater with a tower trickling filter filled with a plastic medium. The BOD$_5$ applied to the filter after primary settling was equal to 325 mg/L. The area of the pilot filter was 0.25 m^2 and the wetted perimeter-to-area ratio for the filter medium is 0.95 cm/cm^2. The wastewater temperature at the time the tests were run was 16°C. Using these data, determine the value of the term $(f\,hk_0)$ in Eq. 9-82.

	Removal efficiency, %			
	Flowrate, m/d			
Depth, m	2	4	6	8
2	52	33	22	18
4	77	52	39	31
6	89	67	52	43
8	95	78	63	51

9-15 Using the kinetic coefficient derived in Prob. 9-14, determine the maximum rate of flow that can be applied to a 6-m tower filter that is to be designed to remove 50 percent of applied BOD$_5$ under winter conditions. The applied BOD$_5$ is equal to 325 mg/L and the critical sustained winter wastewater temperature is 7°C. If the average summer wastewater temperature is equal to 22°C, what degree of removal can be expected during the summer?

REFERENCES

1. Adams, C. E., Jr., and W. W. Eckenfelder, Jr. (eds.): *Process Design Techniques for Industrial Waste Treatment*, Enviro Press, Nashville, Tenn., 1974.
2. Ardern, E., and W. T. Lockett: Experiments on the Oxidation of Sewage without the Aid of Filters, *J. Soc. Chem. Ind.*, vol. 33, pp. 523, 1122, 1914.
3. Atkinson, B., I. J. Davies, and S. Y. How: The Overall Rate of Substrate Uptake by Microbial Films, parts I and II, *Trans. Inst. Chem. Eng.*, 1974.
4. Bruce, A. M., and J. C. Merkens: Further Studies of Partial Treatment of Sewage by High-Rate Biological Filtration, *J. Inst. Water Pollut. Contr.*, vol. 72, no. 5, London, 1973.
5. Burkhead, C. E.: *Energy Relationships in Aerobic Microbial Systems*, Ph.D. thesis, University of Kansas, Lawrence, Kans., 1966.
6. Dawes, E. A., and D. W. Ribbons: The Endogenous Metabolism of Microorganisms, *Ann. Rev. Microbiol.*, vol. 16, p. 241, 1962.

7. Dick, R. I.: Folklore in the Design of Final Settling Tanks, *J. Water Pollut. Control Fed.*, vol. 48, no. 4, April 1976.
8. Eckenfelder, W. W., Jr.: Closure to "Trickling Filtration Design and Performance," *J. Sanit. Eng. Div.*, ASCE, vol. 89, pp. 3, 65, 1963.
9. Eckenfelder, W. W., Jr.: *Industrial Water Pollution Control*, McGraw-Hill, New York, 1966.
10. Eckenfelder, W. W., Jr.: Trickling Filtration Design and Performance, *Trans. ASCE*, vol. 128, 1963.
11. Fairall, J. M.: Correlation of Trickling Filter Data, *Sewage Ind. Wastes*, vol. 28, no. 9, 1956.
12. Galler, W. S., and H. B. Gotass: Optimization Analysis for Biological Filter Design, *J. Sanit. Eng. Div.*, ASCE, vol. 92, no. SA1, 1966.
13. Gates, W. E., et al.: A Rational Model for the Anaerobic Contact Process, *J. Water Pollut. Control Fed.*, vol. 39, no. 12, 1967.
14. Gloyna, E. F., and W. W. Eckenfelder, Jr. (eds.): *Advances in Water Quality Improvement*, University of Texas Press, Austin, 1968.
15. Heukelekian, H., H. E. Orford, and R. Manganelli: Factors Affecting the Quantity of Sludge Protection in the Activated Sludge Process, *Sewage Ind. Wastes*, vol. 23, no. 8, 1951.
16. Higgins, I. J., and R. G. Burns: *The Chemistry and Microbiology of Pollution*, Academic, London, 1975.
17. Hoover, S. R., and N. Porges: Assimilation of Dairy Wastes by Activated Sludge, II: The Equation of Synthesis and Oxygen Utilization, *Sewage Ind. Wastes*, vol. 24, 1952.
18. Jenkins, D., and W. E. Garrison: Control of Activated Sludge by Mean Cell Residence Time, *J. Water Pollut. Control Fed.*, vol. 40, no. 11, part 1, 1968.
19. Jewell, W. J.: Personal communication, 1977.
20. Kimball, J. W.: *Biology*, 2d ed., Addison-Wesley, Reading, Mass., 1966.
21. Lawrence, A. W., and P. L. McCarty: A Unified Basis for Biological Treatment Design and Operation, *J. Sanit. Eng. Div.*, ASCE, vol. 96, no. SA3, 1970.
22. Lawrence, A. W., and P. L. McCarty: Kinetics of Methane Fermentation in Anaerobic Treatment, *J. Water Pollut. Control Fed.*, vol. 41, no. 2, part 2, 1969.
23. Mallette, F. M.: Validity of the Concept of Energy of Maintenance, *Ann. N.Y. Acad. Sci.*, vol. 102, p. 521, 1963.
24. Mara, D. D.: *Bacteriology for Sanitary Engineers*, Churchill Livingstone, Edinburgh, 1974.
25. McCarty, P. L.: Anaerobic Waste Treatment Fundamentals, *Public Works*, vol. 95, nos. 9–12, 1964.
26. McCarty, P. L.: Kinetics of Waste Assimilation in Anaerobic Treatment, *Developments in Industrial Microbiology*, vol. 7, American Institute of Biological Sciences, Washington, D.C., 1966.
27. McCarty, P. L.: Anaerobic Treatment of Soluble Wastes, in E. F. Gloyna and W. W. Eckenfelder (eds.), *Advances in Water Quality Improvement*, University of Texas Press, Austin, 1968.
28. McCarty, P. L.: *Energetics and Bacterial Growth*, Fifth Rudolf Research Conference, Rutgers University, New Brunswick, N.J., 1969.
29. McCarty, P. L.: *Phosphorus and Nitrogen Removal by Biological Systems*, Proceedings, Wastewater Reclamation and Reuse Workshop, Lake Tahoe, Calif., p. 226, June 25–27, 1970.
30. McKinney, R. E.: *Microbiology for Sanitary Engineers*, McGraw-Hill, New York, 1962.
31. Mehta, D. S., H. H. Davis, and R. P. Kingsburg: Oxygen Theory in Biological Treatment Plant Design, *J. Sanit. Eng. Div.*, ASCE, vol. 98, no. SA3, 1972.
32. Metcalf & Eddy, Inc.: *Report to National Commission on Water Quality on Assessment of Technologies and Costs for Publicly Owned Treatment Works*, vol. 3, prepared under Public Law 92–500, Boston, 1975.
33. Monod, J.: *Recherches sur la croissance des cultures bacteriennes*, Herman et Cie., Paris, 1942.
34. Monod, J.: The Growth of Bacterial Cultures, *Ann. Rev. Microbiol.*, vol. III, 1949.
35. National Research Council: Trickling Filters (in Sewage Treatment at Military Installations), *Sewage Works J.*, vol. 18, no. 5, 1946.
36. O'Rourke, J. T.: *Kinetics of Anaerobic Treatment at Reduced Temperatures*, Ph.D. dissertation, Stanford University, Stanford, Calif., 1968.
37. Pelczar, M. J., Jr., and R. D. Reid: *Microbiology*, 2d ed., McGraw-Hill, New York, 1965.
38. *Process Design Manual for Nitrogen Control*, U.S. Environmental Protection Agency, Office of Technology Transfer, Washington, D.C., October 1975.

39. Ribbons, D. W.: Quantitative Relationships between Growth Media Constituents and Cellular Yields and Composition, in J. W. Norris and D. W. Ribbons (eds.), *Methods in Microbiology*, vol. 3A, Academic, London, 1970.

40. Schroeder, E. D., and G. Tchobanoglous: Mass Transfer Limitations on Trickling Filter Design, *J. Water Pollut. Control Fed.*, vol. 48, no. 4, April 1976.

41. Schroeder, E. D.: *Water and Wastewater Treatment*, McGraw-Hill, New York, 1977.

42. Sherrard, J. H.: *Kinetics and Stoichiometry of Wastewater Treatment*, Department of Civil Engineering, Virginia Polytechnic Institute and State University, Blacksburg, Va., 1977.

43. Sherrard, J. H.: Personal communication.

44. Smith, G. M.: *The Fresh-Water Algae of the United States*, 2d ed., McGraw-Hill, New York, 1950.

45. Stanier, R. Y., J. L. Ingraham, and E. A. Adelberg: *The Microbial World*, 4th ed., Prentice-Hall, Englewood Cliffs, N.J., 1976.

46. Van Uden, N.: Transport-Limited Growth in the Chemostat and Its Competitive Inhibition; A Theoretical Treatment, *Arch. Mikrobiol.*, vol. 58, 1967.

47. Velz, C. J.: A Basic Law for the Performance of Biological Beds, *Sewage Works J.*, vol. 20, no. 4, 1948.

48. Weston, R. F., and W. W. Eckenfelder: Application of Biological Treatment to Industrial Wastes, I. Kinetics and Equilibria of Oxidative Treatment, *Sewage Ind. Wastes*, vol. 27, p. 802, 1955.

49. Wood, D. K., and G. Tchobanoglous: Trace Elements in Biological Waste Treatment with Specific Reference to the Activated Sludge Process, *Proc. 29th Ind. Waste Conf.*, 1974.

50. Young, J. C., E. R. Baumann, and D. J. Wall: Packed-Bed Reactors for Secondary Effluent BOD and Ammonia Removal, *J. Water Pollut. Control Fed.*, vol. 47, no. 1, January 1975.

51. Young, J. C., and P. L. McCarty: The Anaerobic Filter for Waste Treatment, *J. Water Pollut. Control Fed.*, vol. 41, 1969.

CHAPTER TEN

DESIGN OF FACILITIES FOR THE BIOLOGICAL TREATMENT OF WASTEWATER

Biological processes are used to convert the finely divided and dissolved organic matter in wastewater into flocculant settleable solids that can be removed in sedimentation tanks. Although these processes (also called secondary processes) are employed in conjunction with the physical and chemical processes that are used for the primary treatment of wastewater, as discussed in Chap. 8, they are not substitutes. Primary sedimentation is most efficient in removing coarse solids, whereas the biological processes are most efficient in removing organic substances that are either soluble or in the colloidal size range.

Four of the most commonly used biological processes are (1) the activated-sludge process, (2) aerated lagoons, (3) trickling filters, and (4) stabilization ponds. The activated-sludge process, or one of its many modifications, is most often used for large installations; stabilization ponds are most often used for small installations. Typical treatment plant flowsheets incorporating these processes are illustrated in Fig. 10-1.

The process design and the physical facilities required for the implementation of these four important processes are discussed in detail in this chapter. The use of combined aerobic biological treatment is also discussed briefly. Although not all the processes reported in Table 9-1 are discussed, the basic approach developed in this chapter can be applied to their design also.

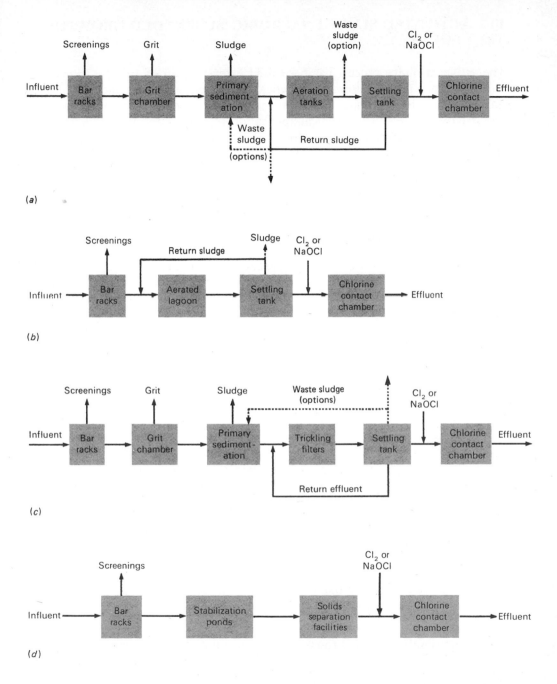

(a)

(b)

(c)

(d)

Figure 10-1 Typical (simplified) flowsheets for biological processes used for wastewater treatment. (a) Activated-sludge process. (b) Aerated lagoons. (c) Trickling filters. (d) Stabilization ponds.

10-1 ACTIVATED-SLUDGE (AEROBIC SUSPENDED-GROWTH) TREATMENT

The activated-sludge process has been used extensively in its original form as well as in many modified forms. Theoretical aspects of the process, including the microbiology, the reaction kinetics, and to some extent the operation, were discussed in Chap. 9. The practical application of this process is considered in this section. The discussion is divided into nine parts: (1) process design considerations, (2) process design, (3) types of processes and modifications, (4) facilities for diffused-air aeration, (5) mechanical aerators, (6) generation of pure oxygen, (7) design of aeration tanks and appurtenances, (8) design of solids separation facilities, and (9) operational difficulties.

Process Design Considerations

In the design of the activated-sludge process, consideration must be given to (1) loading criteria, (2) selection of the type of reactor, (3) sludge production and process control, (4) oxygen requirements and transfer, (5) nutrient requirements, (6) environmental requirements, (7) solids separation, and (8) effluent characteristics. Each of these subjects will be discussed, and then a detailed design example will be presented.

Loading criteria Over the years, a number of both empirical and rational parameters have been proposed for the design and control of the activated-sludge process. The two most commonly used parameters are (1) the food-to-microorganism ratio (F/M), and (2) the mean cell residence time θ_c (see Chap. 9).

The food-to-microorganism ratio is defined as

$$F/M = \frac{S_0}{\theta X} \qquad (9\text{-}42)$$

where F/M = food-to-microorganism ratio, d^{-1}
S_0 = influent BOD or COD concentration, g/m^3
θ = hydraulic detention time of the aeration tank
$\theta = V/Q$
V = aeration tank volume, m^3
Q = influent wastewater flowrate, m^3/d
X = concentration of volatile suspended solids in the aeration tank, g/m^3

The relationship of the food-to-microorganism ratio to the specific utilization rate U is

$$U = \frac{(F/M)E}{100} \qquad (9\text{-}43)$$

where E = process efficiency, %

Substituting Eq. 9-42 for the food-to-microorganism ratio and $[(S_0 - S)/S_0)](100)$ for the efficiency yields

$$U = \frac{S_0 - S}{\theta X} \qquad (9\text{-}38)$$

The mean cell residence time can be defined with either of the following two relationships, depending on the volume used:

Definition based on aeration tank volume

$$\theta_c = \frac{VX}{Q_w X_w + Q_e X_e} \qquad (10\text{-}1)$$

where θ_c = mean cell residence time based on the aeration-tank volume, d

V = aeration tank volume, m³

X = concentration of volatile suspended solids in the aeration tank, g/m³

Q_w = waste sludge flowrate, m³/d

X_w = concentration of volatile suspended solids in the waste stream, g/m³

Q_e = treated effluent flowrate, m³/d

X_e = concentration of volatile suspended solids in the treated effluent, g/m³

Definition based on total system volume

$$\theta_{ct} = \frac{X_t}{Q_w X_w + Q_e X_e} \qquad (10\text{-}2)$$

where θ_{ct} = mean cell residence time based on the total system

X_t = total mass of volatile suspended solids in the system, including the solids in the aeration tank, in the settling tank, and in the sludge-return facilities, g

other terms are as defined in Eq. 10-1

It is recommended that the design of the reactor be based on θ_c (Eq. 10-1) on the assumption that substantially all the substrate conversion occurs in the aeration tank. In systems where a large portion of the total solids may be present in the settling-tank and sludge-return facilities, Eq. 10-2 can be used to compute the amount of solids to be wasted. Its use is based on the assumption that the biological solids will undergo endogenous respiration regardless of where they are in the system under either aerobic or anaerobic conditions. The use of these factors is considered further in Ref. 49.

Comparing these parameters, the specific utilization rate U (F/M ratio multiplied by the efficiency) can be considered to be a measure of the rate at which substrate (BOD) is utilized by a unit mass of organisms, and θ_c can be considered to be a measure of the average residence time of the organisms in the system. The relationship between mean cell residence time θ_c, the food-to-microorganism ratio F/M, and the specific utilization rate U is

$$\frac{1}{\theta_c} = Y \frac{F}{M} \frac{E}{100} - k_d = YU - k_d \qquad (9\text{-}39)$$

Typical values for the food-to-microorganism ratio reported in the literature vary from 0.05 to 1.0. On the basis of laboratory studies and actual operating data from a number of different treatment plants throughout the United States, it has been found that mean cell residence times of about 6 to 15 d result in the production of a stable, high-quality effluent and a sludge with excellent settling characteristics.

Empirical relationships based on detention time and organic loading factors have also been used. The detention time is usually based on the influent wastewater flowrate. Typically, detention times in the aeration tank range from 4 to 8 h. Organic loadings, expressed in terms of kilograms of BOD_5 applied daily per cubic meter of aeration-tank volume vary from 0.3 to more than 3. While the concentration of the mixed liquor, the food-to-microorganism ratio, and the mean cell residence time (which may be considered operating variables as well as design parameters) are ignored when such empirical relationships are used, these relationships do have the merit of requiring a minimum aeration-tank volume that has proved to be adequate for the treatment of domestic wastewater. Problems have developed, however, when such relationships are used to design facilities for the treatment of wastewater containing industrial wastes.

Selection of reactor type One of the main steps in the design of any biological process is the selection of the type of reactor or reactors (see Chap. 5) to be used in the treatment process. Operational factors that are involved include (1) the reaction kinetics governing the treatment process, (2) oxygen-transfer requirements, (3) the nature of the wastewater to be treated, (4) the local environmental conditions, and (5) the initial construction costs and operation and maintenance costs, considered in conjunction with the secondary settling facilities. Because the relative importance of these factors will vary with each application, they should be considered separately when the type of reactor is to be selected. Their importance to the activated-sludge process is described briefly in the following discussion.

The first factor, the effect of reaction kinetics on the selection of the reactor, was illustrated in detail in Chaps. 5 and 9. The two types of reactors that are commonly used are the continuous-flow stirred-tank reactor and the plug-flow reactor. From a practical standpoint, it is interesting to note that the hydraulic detention times of many of the continuous-flow stirred-tank and plug-flow reactors in actual use are about the same. The reason is that the combined substrate (soluble and nonsoluble) removal rate for domestic wastes is approximately zero order with respect to the concentration of the substrate. It is quasi-first-order with respect to the concentration of cells.

The second factor that must be considered in the selection of reactors for the activated-sludge process is oxygen-transfer requirements. In the past, with conventional plug-flow aeration systems, it was often found that sufficient oxygen could not be supplied to meet the oxygen requirements of the head end of the reactor. This condition led to the development of the following modifications of the activated-sludge process: (1) the tapered aeration process in which an attempt was made to match the air supplied to the oxygen demand, (2) the step aeration process where the incoming waste and return solids are distributed along the length of the reactor (usually at quarter points), and (3) the continuous-flow stirred-tank process where the air supplied uniformly matches or exceeds the oxygen demand. Most of the past oxygen-transfer limitations have been overcome by better selection of process operational parameters and improvements in the design and application of aeration equipment. Pure oxygen instead of air can also be used to overcome this limitation.

The third factor that can influence the type of reactor selected is the nature of waste. For example, because the incoming waste is more or less uniformly dispersed in a continuous-flow stirred-tank reactor, it can, as compared to a plug-flow reactor, more easily withstand shock loads resulting from the slug discharge of organic and toxic materials to sewers from industrial operations. The continuous-flow stirred-tank process has been used in a number of installations for this reason. Equalization facilities can be used just as effectively for the purpose.

The fourth factor is local environmental conditions. Of these, temperature is perhaps the most important. For example, for a zero-order biological substrate-removal process where the temperature coefficient as defined by Eq. 5-13 is equal to about 1.12, if the temperature drops by 10°C (18°F), the required reactor volume will be three times as large, assuming that the solids level is not increased. In this situation, a series of continuous-flow stirred-tank reactors, or a plug-flow reactor whose length could be reduced with stop gates, could be used effectively. It should be noted, however, that in practice the conventional activated-sludge process is not significantly influenced by temperature, as reported values for the temperature coefficient vary only from 1.0 to 1.03.

The fifth factor, cost, is an extremely important consideration in selecting the type and size of reactor. Because the associated settling facilities are an integral part of the activated-sludge process, the two units must be considered together. The interrelationship of the costs of these units as a function of the concentration of the mixed-liquor volatile suspended solids is shown in Fig. 10-2. The theoretical interrelationships between the two units are considered in Ref. 46. If equalization facilities are used, the cost of the three units must be considered simultaneously.

Sludge production and process control It is important to know the quantity of sludge to be produced per day, because it will affect the design of the sludge handling and disposal facilities. The quantity of sludge that must be wasted on a daily basis can be estimated by Eq. 10-3:

$$P_x = Y_{obs} Q(S_0 - S) \times (10^3 \text{ g/kg})^{-1} \qquad (10\text{-}3)$$

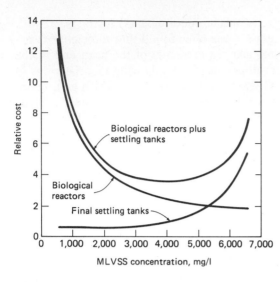

Relative cost

Biological reactors plus
settling tanks

Biological
reactors

Final settling tanks

0   1,000 2,000 3,000 4,000 5,000 6,000 7,000

MLVSS concentration, mg/l

Figure 10-2 Relative cost of activated-sludge treatment facilities as a function of the concentration of the aeration tank solids [8]. Note: $mg/L = g/m^3$.

where P_x = net waste activated sludge produced each day, measured in terms of volatile suspended solids, kg/d

Y_{obs} = observed yield, g/g

Q, S_0, S = as defined previously

The observed yield can be computed using Eq. 9-41:

$$Y_{obs} = \frac{Y}{1 + k_d(\theta_c \text{ or } \theta_{ct})} \tag{9-41}$$

The use of θ_c or θ_{ct} in Eq. 9-41 depends on whether the solids in the aeration tank or the solids in the total system are considered in the analysis. If a high percentage of the solids are retained in the settling-tank and sludge-return facilities, the use of θ_{ct} is reasonable, especially if it is assumed that endogenous respiration goes on regardless of whether the bacterial culture is in an aerobic or anaerobic environment. However, it should be noted that the value of the constant would be different from the values reported in the literature. Because no suitable value exists at the present time for a combined aerobic-anaerobic k_d, the aerobic value can be used as an estimate.

The actual amount of liquid that must be pumped to achieve process control depends on the method used and the location from which the wasting is to be accomplished. (Also, because the solids capture of the sludge-processing facilities is not 100 percent, and some solids are returned, the actual wasting rate will be higher than the theoretically determined value. This subject is considered further in Chap. 11.) For example, if the mean cell residence time is used for process control and wasting is from the aeration basin, then the rate of pumping can be estimated using Eq. 10-4:

$$\theta_c = \frac{VX}{Q_{wa}X + Q_e X_e} \tag{10-4}$$

where Q_{wa} = waste sludge flowrate from the aeration tank, m^3/d
Other terms = as defined in Eq. 10-1

If it is assumed that the concentration of solids in the effluent from the settling tank is low, then Eq. 10-4 reduces to

$$\theta_c \sim \frac{V}{Q_{wa}} \qquad (10\text{-}5)$$

or

$$Q_{wa} \sim \frac{V}{\theta_c} \qquad (10\text{-}6)$$

Thus, the process may be controlled by daily wasting a quantity of flow equal to the volume of the aeration tank divided by the mean cell residence time. If wasting is from the sludge return line, and the solids in the plant effluent are again neglected, the wasting rate can be computed using the following relationship:

$$\theta_c \sim \frac{VX}{Q_{wr} X_r} \qquad (10\text{-}7)$$

and

$$Q_{wr} \sim \frac{VX}{\theta_c X_r} \qquad (10\text{-}8)$$

where Q_{wr} = waste sludge flowrate from the sludge return line, m^3/d
X_r = concentration of sludge in the return line, g/m^3
other terms = as defined previously

To determine the waste flowrate using Eq. 10-8, both the solids concentration in the aerator and return line must be known. Because a ratio is involved, however, a small laboratory centrifuge can be used to establish the ratio of X/X_r.

If the food-to-microorganism method of control is adopted, the wasting flowrate from the return line can be determined using the following relationship:

$$P_x = Q_{wr} X_r \times (10^3 \ g/kg)^{-1} \qquad (10\text{-}9)$$

where P_x = waste activated sludge, kg/d
Q_{wr} = waste flowrate, m^3/d
X_r = solids concentration in the return line, g/m^3

In this case, the concentration of solids in the sludge return line must be analyzed. If a correlation is developed between the solids concentration and the depth of sediment in a centrifuge tube, a simple laboratory centrifuge can be used to determine X_r.

If process control is based on one of the other loading criteria, the quantity of solids to be wasted must be established by trial and error.

Oxygen requirements and transfer The theoretical oxygen requirements can be determined by knowing the BOD_5 of the waste and the amount of organisms wasted from the system per day. The reasoning is as follows. If all the BOD_5

were converted to end products, the total oxygen demand would be computed by converting BOD_5 to BOD_L, using an appropriate conversion factor. It is known that a portion of the waste is converted to new cells that are subsequently wasted from the system; therefore, if the BOD_L of the wasted cells is subtracted from the total, the remaining amount represents the amount of oxygen that must be supplied to the system. The BOD_L of a mole of cells can be estimated as follows:

$$C_5H_7NO_2 + 5O_2 \rightarrow 5CO_2 + 2H_2O + NH_3 \qquad (9\text{-}6)$$

$$\begin{array}{cc} 113 & 5(32) \\ \text{cells} & \end{array}$$

$$\frac{kg\ O_2}{kg\ cells} = \frac{160}{113} = 1.42$$

where the BOD of the cells is equal to

$$BOD_L = 1.42\ (\text{mass of cells, g/m}^3) \qquad (10\text{-}10)$$

Therefore, the theoretical oxygen requirements for the removal of the carbonaceous organic matter in wastewater for an activated-sludge system can be computed as

$$kg,\ O_2/d = \left(\begin{array}{c}\text{total mass of BOD}_L \\ \text{utilized, kg/d}\end{array}\right) - 1.42\ \left(\begin{array}{c}\text{mass of organisms} \\ \text{wasted, kg/d}\end{array}\right) \qquad (10\text{-}11)$$

In terms that have been defined previously,

$$kg,\ O_2/d = \frac{Q(S_0 - S) \times (10^3\ \text{g/kg})^{-1}}{f} - 1.42(P_x) \qquad (10\text{-}12)$$

where f = conversion factor for converting BOD_5 to BOD_L.

Then, if the oxygen-transfer efficiency of the aeration system is known or can be estimated, the actual air requirements may be determined. The air supply must be adequate to (1) satisfy the BOD of the waste, (2) satisfy the endogenous respiration by the sludge organisms, (3) provide adequate mixing, and (4) maintain a minimum dissolved-oxygen concentration of 1 to 2 mg/L throughout the aeration tank.

For food-to-microorganism ratios greater than 0.3, the air requirements for the conventional process amount to 30 to 55 m³/kg (500 to 900 ft³/lb) of BOD_5 removed. At lower food-to-microorganism ratios, endogenous respiration, nitrification, and prolonged aeration periods increase air use to 75 to 115 m³/kg (1200 to 1800 ft³/lb) of BOD_5 removed.

For diffused-air aeration, the amount of air used has commonly ranged from 3.75 to 15.0 m³/m³ (0.5 to 2.0 ft³/gal) at different plants, with 7.5 m³/m³ (1.0 ft³/gal) an early rule-of-thumb design factor. Because the air use depends on the strength of the wastewater, the amount has become a derived quantity for record-keeping purposes and is no longer a basic design criterion.

To meet the sustained peak organic loadings discussed in Chap. 3, it is recommended that the aeration equipment be designed with a safety factor of at least 2. The *Ten States Standards* [3] require the air-diffusion system to be capable of delivering 150 percent of the normal requirements, which are assumed to be $62 \ m^3/kg$ ($1000 \ ft^3/lb$) of BOD in the wastewater applied to the aeration tanks.

Nutrient requirements If a biological system is to function properly, nutrients must be available in adequate amounts. As discussed in Chaps. 3 and 9, the principal nutrients are nitrogen and phosphorus. Based on an average composition of cell tissue of $C_5H_7NO_2$, about 12.4 percent by weight of nitrogen will be required. The phosphorus requirement is usually assumed to be about one-fifth of this value. These are typical values, not fixed quantities, because it has been shown that the percentage distribution of nitrogen and phosphorus in cell tissue varies with the age of the cell and environmental conditions.

Other nutrients required by most biological systems are reported in Table 10-1. The inorganic composition of *E. coli* is shown in Table 10-2. The data in Table 10-2 can be used to estimate the concentration of trace elements required for the maintenance of proper biological growth. Because the total amount of nutrients required will depend on the net mass of organisms produced, nutrient quantities will be reduced for processes operated with long mean cell residence times. This fact can often be used to explain why two similar activated-sludge plants operated at different θ_c may not perform the same way when treating the same waste. The role of trace elements is discussed in greater detail in an article by Wood and Tchobanoglous [58].

Table 10-1 Inorganic ions necessary for most organisms [17]

Substantial quantities	Trace quantities	
Na^+ (except for plants)	Fe^{2+}	
K^+	Cu^{2+}	
Ca^{2+}	Mn^{2+}	
Mg^{2+}	Zn^{2+}	
PO_4^{3-}	B^{3+}	required by plants, certain protists
Cl		
SO_4^{2-}	Mo^+	required by plants, certain protists, and animals
HCO_3^-	V^{2+}	required by certain protists and animals
	Co^{2+}	required by certain animals, protists, and plants
	I^- Se^{2-}	required by certain animals only

Table 10-2 Inorganic composition of *E. Coli* [19]

Element	Percentage of cell dry weight
Sodium	1.3
Potassium	1.5
Calcium	1.4
Magnesium	0.54
Chloride	0.41
Iron	0.2
Manganese	0.01
Copper	0.01
Aluminum	0.01
Zinc	0.01

Environmental requirements Environmental factors of importance include temperature and pH. These factors were discussed in Chap. 9 and the reader is referred to that discussion. The effect of temperature was also discussed previously in connection with reactor selection. Although pH is not always a problem in the conventional activated-sludge process, it is important to note that in biological treatment, the alkalinity of the wastewater is reduced because of the presence of carbon dioxide as well as other conversion products. The control of pH may be required in low-alkalinity wastewaters and is often required in aerobic digestion (see Chap. 11) and in nitrification (see Chap. 12).

Solids separation Probably the most important aspect of biological waste treatment is the design of the facilities used to separate the biological solids from the treated wastewater, for it is axiomatic that if the solids cannot be separated and returned to the aeration tank, the activated-sludge process will not function properly. Unfortunately, the importance of the separation step is little appreciated and even today does not receive the attention it should. Because of its importance, a separate discussion has been devoted to this subject (see "Design of Solids-Separation Facilities" later in this section).

Effluent characteristics Organic content is a major parameter of effluent quality. The organic content of effluent from biological treatment processes is usually composed of the following three constituents:

1. Soluble biodegradable organics
   a. Organics that escaped biological treatment
   b. Organics formed as intermediate products in the biological degradation of the waste
   c. Cellular components (result of cell death or lysis)

2. Suspended organic material
   a. Biological solids produced during treatment that escaped separation in the final settling tank
   b. Colloidal organic solids in the plant influent that escaped treatment and separation
3. Nonbiodegradable organics
   a. Those originally present in the influent
   b. By-products of biological degradation

The kinetic equations developed in Chap. 9 for the effluent quality theoretically apply only to the soluble organic waste that escaped biological treatment. Clearly, this is only a portion of the organic waste concentration in the effluent. In a well-operating activated-sludge plant that is treating domestic wastes, the soluble carbonaceous BOD_5 in the effluent, determined on a filtered sample, will usually vary from 2 to 10 mg/L.

Process Design

Application of the aforementioned factors to the design of an activated-sludge treatment process is illustrated in Example 10-1. For purposes of the example, a continuous-flow stirred-tank system has been selected. Schematically, such a system would be depicted as shown in Fig. 10-3. Its distinguishing features are

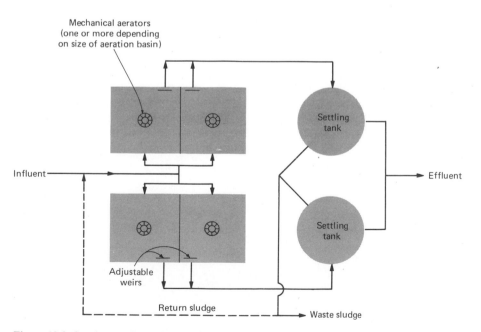

Figure 10-3 Continuous-flow stirred-tank activated-sludge process (typical schematic for four-cell process).

(1) uniform distribution of the inflow and return solids to the reactor (aeration tank) and (2) uniform withdrawal of mixed liquor from the reactor. Application of the principles discussed in this chapter and in Chap. 9 to other types of systems is covered in the problems at the end of this chapter.

Example 10-1 Design of activated-sludge process Design a continuous-flow stirred-tank activated-sludge process to treat 0.25 m³/s (5.71 Mgal/d) of settled wastewater having 250 mg/L of BOD_5. The effluent is to have 20 mg/L of BOD_5 or less. Assume that the temperature is 20°C and that the following conditions are applicable:

1. Influent volatile suspended solids to reactor are negligible.
2. Ratio of mixed-liquor volatile suspended solids (MLVSS) to mixed-liquor suspended solids (MLSS) = 0.8.
3. Return-sludge concentration = 10,000 mg/L of suspended solids (SS).
4. Mixed-liquor volatile suspended solids (MLVSS) = 3500 mg/L.
5. Design mean cell residence time $\theta_c^d = 10$ d.
6. Hydraulic regime of reactor = continuous-flow stirred-tank.
7. Effluent contains 22 mg/L of biological solids, of which 65 percent is biodegradable.
8. The value of the BOD_5 can be obtained by multiplying the value of BOD_L by a factor of 0.68 (corresponds to a K value of 0.1 d⁻¹ in the BOD equation).
9. Waste contains adequate nitrogen, phosphorus, and other trace nutrients for biological growth.
10. The 1-d sustained peak flow rate is 2.5 times the average flowrate.

SOLUTION

1. Estimate the concentration of soluble BOD_5 in the effluent using the following relationship:

Effluent BOD_5 = influent soluble BOD_5 escaping treatment + BOD_5 of effluent suspended solids

    a Determine the BOD_5 of the effluent suspended solids.
      i Biodegradable portion of effluent biological solids = 0.65(22 mg/L) = 14.3 mg/L
      ii Ultimate BOD_L of the biodegradable effluent solids (refer to Eq. 10-10) = [0.65(22 mg/L)] (1.42 mg/mg) = 20.3 mg/L
      iii BOD_5 of effluent suspended solids = 20.3 mg/L (0.68) = 13.8 mg/L
    b Solve for the influent soluble BOD_5 escaping treatment.

$$20 \text{ mg/L} = S + 13.8 \text{ mg/L}$$
$$S = 6.2 \text{ mg/L}$$

2 Determine the treatment efficiency E using Eq. 9-44.

$$E = \frac{S_0 - S}{S_0} 100$$

    a The efficiency based on soluble BOD_5 is

$$E_s = \frac{(250 - 6.2) \text{ mg/L}}{250 \text{ mg/L}} 100 = 97.5\%$$

    b The overall plant efficiency is

$$E_{overall} = \frac{(250 - 20) \text{ mg/L}}{250 \text{ mg/L}} 100 = 92\%$$

3 Compute the reactor volume. The volume of the reactor can be determined using Eq. 9-67,

$$X = \frac{\theta_c^d \, Y(S_0 - S)}{\theta(1 + k_d \theta_c^d)}$$

and Eq. 9-68,

$$\theta = \frac{V}{Q}$$

a Substituting for θ in Eq. 9-67 and solving for V yields

$$V = \frac{\theta_c^d \, QY(S_0 - S)}{X(1 + k_d \theta_c^d)}$$

b Compute the reactor volume using the following data:

$\theta_c^d = 10$ d
$Q = 0.25$ m^3/s $= 0.25$ m^3/s (86,400 s/d)
    $= 21,600$ m^3/d
$Y = 0.50$ mg/mg (assumed, see Table 9.7)
$S_0 = 250$ mg/L
$S - 6.2$ mg/L
$X = 3500$ mg/L
$k_d = 0.06$ d^{-1} (assumed, see Table 9.7)

$$V = \frac{(10 \text{ d})(21,600 \text{ m}^3/\text{d})(0.50)[(250 - 6.2) \text{ mg/L}]}{(3500 \text{ mg/L}) [(1 + 0.06 \text{ d}^{-1})(10 \text{ d})]}$$

$$= 4702 \text{ m}^3 \text{ (1.24 Mgal)}$$

4 Compute the quantity of sludge that must be wasted each day.
a Determine Y_{obs} using Eq. 9-41.

$$Y_{obs} = \frac{Y}{1 + k_d \theta_c^d} = \frac{0.5}{(1 + 0.06 \text{ d}^{-1})(10 \text{ d})} = 0.3125 \qquad (9\text{-}41)$$

b Determine the mass of volatile waste activated sludge using Eq. 10-3.

$$P_x = Y_{obs} Q(S_0 - S)(10^3 \text{ g/kg})^{-1}$$

$$= \frac{0.3125(21,600 \text{ m}^3/\text{d})(250 \text{ g/m}^3 - 6.2 \text{ g/m}^3)}{10^3 \text{ g/kg}}$$

$$= 1645.7 \text{ kg/d} \text{ (3628.0 lb/d)}$$

c Determine the total mass of sludge based on total suspended solids.

$$P_{x(SS)} = 1645.7 \text{ kg/d}/0.8$$

$$= 2057.1 \text{ kg/d} \text{ (4535.0 lb/d)}$$

Note If in step 4a it had been assumed that the additional amount of sludge in the settling tanks and sludge return lines was equal to 30 percent of the amount of sludge in the aerator, then, assuming that the values of Y and k_d are applicable, the computed value of Y_{obs} would have been equal to 0.281. The mass of sludge computed in step 4c would then have been equal to 1849.7 kg/d (4078.0 lb/d) instead of 2057.1 kg/d (4535.0 lb/d).

5 Compute the sludge-wasting rate if wasting is accomplished (a) from the aeration basin and (b) from the sludge return line. Neglecting the solids lost in the effluent, the wasting rate is:

a From the aeration basin, using Eq. 10-6,

$$Q_{wa} \sim \frac{V}{\theta_c^d}$$

$$\sim \frac{4702 \text{ m}^3}{10 \text{ d}}$$

$$\sim 470.2 \text{ m}^3/\text{d} \ (0.124 \text{ Mgal/d})$$

b From the sludge recycle line, using Eq. 10-8,

$$Q_{wr} \sim \frac{VX}{\theta_c^d X_r}$$

$$\sim \frac{(4702 \text{ m}^3)(3500 \text{ mg/L})}{(10 \text{ d})(10,000 \text{ mg/L} \times 0.8)}$$

$$\sim 205.7 \text{ m}^3/\text{d} \ (0.054 \text{ Mgal/d})$$

Note In either case the weight of sludge wasted is the same (1645.7 kg VSS/d), and either wasting method will achieve a design θ_c^d of 10 d for the system. Because a portion of the wasted solids are recycled to the process from the sludge treatment facilities, the actual wasting rate will be from 10 to 20 percent higher than the computed value (see Example 11-9).

6 Compute the recirculation ratio by writing a mass balance around the inlet to the reactor.

$$\text{Aerator VSS concentration} = 3500 \text{ mg/L}$$
$$\text{Return VSS concentration} = 8000 \text{ mg/L}$$
$$3500(Q + Q_r) = 8000(Q_r)$$

$$\frac{Q_r}{Q} = \alpha = 0.78$$

7 Compute the hydraulic retention time for the reactor.

$$\theta = \frac{V}{Q} = \frac{4702 \text{ m}^3}{(0.25 \text{ m}^3/\text{s})(3600 \text{ s/h})} = 5.2 \text{ h} = 0.217 \text{ d}$$

8 Compute the oxygen requirements based on ultimate carbonaceous demand, BOD_L.
Note Although O_2 requirements for nitrification are neglected in this example, they can be important and must be considered in the design of systems operating at mean cell residence times sufficiently high to allow nitrification to occur (see discussion in Chap. 12).

a Compute the mass of ultimate BOD_L of the incoming wastewater that is converted in the process, assuming that the BOD_5 is equal to 0.68 BOD_L.

$$\text{Mass of } BOD_L \text{ utilized} = \frac{Q(S_0 - S)(10^3 \text{ g/kg})^{-1}}{0.68}$$

$$= \frac{(21,600 \text{ m}^3/\text{d})(250 \text{ g/m}^3 - 6.2 \text{ g/m}^3)(10^3 \text{ g/kg})^{-1}}{0.68}$$

$$= 7744 \text{ kg/d}$$

b Compute the oxygen requirement using Eq. 10-12.

$$\text{kg, } O_2/\text{d} = 7744 \text{ kg/d} - 1.42(1645.7 \text{ kg/d})$$

$$= 5407.1 \text{ kg/d} \ (11,920.6 \text{ lb/d})$$

9 Check the F/M ratio and the volumetric loading factor.
   a Determine the F/M ratio using Eq. 9-42.

$$F/M = \frac{S_0}{\theta X} = \frac{250 \text{ mg/L}}{(0.217 \text{ d})(3500 \text{ mg/L})} = 0.33 \text{ d}^{-1}$$

   b Determine the volumetric loading.

$$\text{Volumetric loading} = \frac{S_0 \, Q(10^3 \text{ g/kg})^{-1}}{V}$$

$$= \frac{(250 \text{ g/m}^3)(21.600 \text{ m}^3/\text{d})(10^3 \text{ g/kg})^{-1}}{4702 \text{ m}^3}$$

$$= 1.15 \text{ kg BOD}_5/\text{m}^3$$

10 Compute the volume of air required, assuming that the oxygen-transfer efficiency for the aeration equipment to be used is 8 percent. A safety factor of 2 should be used to determine the actual design volume for sizing the blowers.
   a The theoretical air requirement, assuming that air contains 23.2 percent oxygen by weight, is

$$\frac{5407 \text{ kg/d}}{(1.201 \text{ kg/m}^3)(0.232)} = 19,406 \text{ m}^3/\text{d}$$

   b Determine the actual air requirement.

$$\frac{19,406 \text{ m}^3/\text{d}}{0.08} = 242,575 \text{ m}^3/\text{d}$$

   or

$$\frac{242,575 \text{ m}^3/\text{d}}{1440 \text{ min/d}} - 168 \text{ m}^3/\text{min}$$

   c Determine the design air requirement.

$$2(168 \text{ m}^3/\text{min}) = 336 \text{ m}^3/\text{min}$$

11 Check the air volume using the actual value determined in step 10b.
   a Air requirement per unit volume:

$$\frac{242,575 \text{ m}^3/\text{d}}{21,600 \text{ m}^3/\text{d}} = 11.2 \text{ m}^3/\text{m}^3$$

   b Air requirement per kilogram of BOD_5 removed:

$$\frac{242,575 \text{ m}^3/\text{d}}{(250 \text{ g/m}^3 - 6.2 \text{ g/m}^3)(21,600 \text{ m}^3/\text{d})(10^3 \text{ g/kg})^{-1}}$$

$$= 46.1 \text{ m}^3/\text{kg of BOD}_5 \text{ removed}$$

12 Design the required settling facilities. The necessary facilities are designed in Example 10-2 (following the discussion of the design of secondary settling facilities presented later in this section).

Types of Processes and Modifications

The activated-sludge process is very flexible and can be adapted to almost any type of biological waste treatment problem. The purpose of this section is to discuss the details of both the conventional activated processes and some of the

Table 10-3 Operational characteristics of activated-sludge processes

Process modification	Flow model	Aeration system	BOD removal efficiency, %	Remarks
Conventional	Plug flow	Diffused air, mechanical aerators	85–95	Use for low-strength domestic wastes. Process is susceptible to shock loads.
Tapered aeration	Plug flow	Diffused air	85–95	Air supply tapered to match organic loading demand.
Continuous-flow stirred-tank reactor	Continuous-flow stirred-tank reactor	Diffused air mechanical aerators	85–95	Use for general application. Process is resistant to shock loads.
Step aeration	Plug flow	Diffuse air	85–95	Use for general application to wide range of wastes.
Modified aeration	Plug flow	Diffused air	60–75	Use for intermediate degree of treatment where cell tissue in the effluent is not objectionable.
Contact stabilization	Plug flow	Diffused air, mechanical aerators	80–90	Use for expansion of existing systems, package plants. Process is flexible.
Extended aeration	Continuous-flow stirred-tank reactor	Diffused air, mechanical aerators	75–95	Use for small communities, package plants. Process is flexible.
Kraus process	Plug flow	Diffused air	85–95	Use for low-nitrogen, high-strength wastes.
High-rate aeration	Continuous-flow stirred-tank reactor	Mechanical aerators	75–90	Use for general applications with turbine aerators to transfer oxygen and control the floc size.
Pure-oxygen systems	Continuous-flow stirred-tank reactors in series	Pure oxygen with mechanical dispersion	85–95	Use for general application where limited volume is available, near economical source of oxygen (turbine or surface aerators).

Table 10-4 Design parameters for activated-sludge processes

Process modification	θ_c, d	F/M, kg BOD$_5$ applied/ kg MLVSS · d	Volumetric loading, kg BOD$_5$ applied/m^3 · d	MLSS, mg/L	V/Q, h	Q_r/Q
Conventional	5–15	0.2–0.4	0.3–0.6	1,500–3,000	4–8	0.25–0.5
Tapered aeration	5–15	0.2–0.4	0.3–0.6	1,500–3,000	4–8	0.25–0.5
Continuous-flow stirred-tank reactor	5–15	0.2–0.6	0.8–2.0	3,000–6,000	3–5	0.25–1.0
Step aeration	5–15	0.2–0.4	0.6–1.0	2,000–3,500	3–5	0.25–0.75
Modified aeration	0.2–0.5	1.5–5.0	1.2–2.4	200–500	1.5–3	0.05–0.15
Contact stabilization	5–15	0.2–0.6	1.0–1.2	(1,000–3,000)[a] (4,000–10,000)[b]	(0.5–1.0)[a] (3–6)[b]	0.25–1.0
Extended aeration	20–30	0.05–0.15	0.1–0.4	3,000–6,000	18–36	0.75–1.50
Kraus process	5–15	0.3–0.8	0.6–1.6	2,000–3,000	4–8	0.5–1.0
High-rate aeration	5–10	0.4–1.5	1.6–1.6	4,000–10,000	0.5–2	1.0–5.0
Pure-oxygen systems	8–20	0.25–1.0	1.6–3.3	6,000–8,000	1–3	0.25–0.5

[a] Contact unit.
[b] Solids stabilization unit.

Note: kg/m$^3 \cdot$ d $\times$ 62.4280 = lb/10^3 f$^3 \cdot$ d
kg/kg · d $\times$ 1.0 = lb/lb · d
mg/L = g/m^3

modifications that have become standardized. The operational characteristics and typical removal efficiencies for these processes are listed in Table 10-3, and design parameters are shown in Table 10-4.

Conventional The conventional activated-sludge process consists of an aeration tank, a secondary clarifier, and a sludge recycle line (Fig. 10-4). Sludge wasting is accomplished from the recycle or mixed-liquor line. The flow model is plug flow with cellular recycle, as described in Chap 9. Both influent settled wastewater and recycled sludge enter the tank at the head end and are aerated for a period of about 6 h. The influent wastewater and recycled sludge are mixed by the

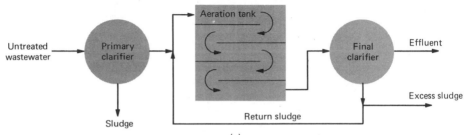

(a)

(b)

Figure 10-4 Conventional (plug-flow) activated-sludge process. (a) Schematic. (b) Aerial view of plug-flow process. (*Greenwich, Conn.*)

action of diffused or mechanical aeration, which is constant as the mixed liquor moves down the tank. During this period, adsorption, flocculation, and oxidation of the organic matter take place. The mixed liquor is settled in the activated-sludge settling tank, and sludge is returned at a rate of approximately 25 to 50 percent of the influent flowrate.

Tapered aeration The objective of tapered aeration is to match the quantity of air supplied to the demand exerted by the microorganisms, as the liquor traverses the aeration tank. Tapered aeration thus affects only the arrangement of the diffusers in the aeration tank and the amount of air consumed. It is widely used and, in a strict sense, is only a modification of the conventional process.

At the inlet of the aeration tank where fresh settled wastewater and return activated sludge first come in contact, the oxygen demand is very high. The diffusers are spaced close together to achieve a high oxygenation rate and thus satisfy the demand. As the mixed liquor traverses the aeration tank, synthesis of new cells occurs, increasing the number of microorganisms and decreasing the concentration of available food. This results in a lower food-to-microorganism ratio and a lowering of the oxygen demand. The spacing of diffusers is thus increased toward the tank outlet to reduce the oxygenation rate. Two beneficial results are obtained. Reduced oxygenation means that less air is required, reducing the size of blowers and the initial and operating costs. Avoidance of overaeration will inhibit the growth of nitrifying organisms, which can cause high oxygen demands.

Continuous-flow stirred-tank The continuous-flow stirred-tank process (Fig. 10-3) represents an attempt to duplicate the hydraulic regime of an ideal continuous-flow stirred-tank reactor. Typically, the influent settled wastewater and return-sludge flow are introduced at several points in the aeration tank. The mixed liquor is aerated as it passes through the tank. The aeration-tank effluent is collected and settled in the activated-sludge settling tank. The organic load on the aeration tank and the oxygen demand are uniform from one end to the other. As the mixed liquor passes across the aeration tank from the influent ports to the effluent channel, it is completely mixed by diffused or mechanical aeration.

Step aeration The step-aeration process is a modification of the activated-sludge process in which the settled wastewater is introduced at several points in the aeration tank to equalize the food-to-microorganism ratio, thus lowering the peak oxygen demand. The process was developed by Gould and was first applied at the Tallmans Island plant in New York City in 1939.

The aeration tank is subdivided into four or more parallel channels through the use of baffles. Each channel is a separate step, and the several steps are linked together in series. A typical flowsheet for the process is shown in Fig. 10-5. Return activated sludge enters the first step of the aeration tank along with a portion of the settled wastewater. The piping is so arranged that an increment of wastewater is introduced into the aeration tank at each step. If desired, the

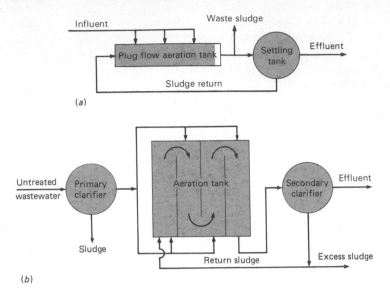

Figure 10-5 Flowsheet for step-aeration activated-sludge process. (*a*) Simplified schematic. (*b*) Typical physical configuration.

first step can be used for reaeration of the return activated sludge alone. Flexibility of operation is one of the important features of this process.

The basic theory of the step-aeration process is the same as that of the activated-sludge process. In step aeration, however, the oxygen demand is more uniformly spread over the length of the aeration tank, resulting in better utilization of the oxygen supplied. The multiple-point introduction of wastewater maintains an activated sludge with high absorptive properties, so that the soluble organics are removed within a relatively short contact period. Higher BOD loadings are therefore possible per unit of aeration-tank volume.

Modified aeration The flow diagram for the modified aeration process is identical with that of the conventional- or tapered-aeration process. The difference in the systems is that modified aeration uses shorter aeration times, usually 1.5 to 3 h, and a high food-to-microorganism ratio. As shown in Table 10-4, the mixed-liquor suspended-solids concentration is relatively low, whereas the organic loading is high. The resulting BOD removal is in the range of 60 to 75 percent; thus the process is not suitable where a high-quality effluent is desired. Some difficulties have been experienced with the process because of the poor settling characteristics of the sludge and the high suspended-solids concentration in the effluent [55].

Contact stabilization The contact-stabilization process was developed to take advantage of the absorptive properties of activated sludge. The flowsheet is shown in Fig. 10-6. In some cases, primary settling is eliminated. It has been postulated

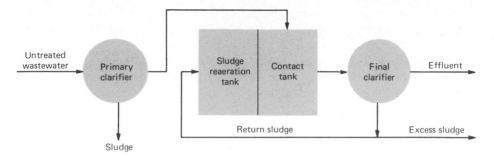

Figure 10-6 Flowsheet for contact-stabilization activated-sludge process.

that BOD removal occurs in two stages in the activated-sludge process. The first is the absorptive phase, which requires 20 to 40 min. During this phase most of the colloidal, finely suspended, and dissolved organics are absorbed in the activated sludge. The second phase, oxidation, then occurs, and the absorbed organics are assimilated metabolically. In the activated-sludge processes mentioned so far, these two phases occur in a single tank. In the contact-stabilization process, the two phases are separated and occur in different tanks.

The settled wastewater is mixed with reaerated activated sludge and aerated in a contact tank for 30 to 90 min. During this period, the organics are absorbed by the sludge floc. The sludge is then separated from the treated effluent by sedimentation, and the returned sludge is aerated from 3 to 6 h in a sludge aeration tank. During this period, the absorbed organics are used for energy and production of new cells. A portion of the return sludge is wasted prior to recycle, to maintain a constant mixed-liquor volatile-suspended-solids concentration in the tanks.

The aeration volume requirements are approximately 50 percent of those of a conventional- or tapered-aeration plant. It is thus often possible to double the plant capacity of an existing conventional plant by redesigning it to use contact stabilization. The redesign may require only changes in plant piping or relatively minor changes in the aeration system.

The contact-stabilization process has been found to work very well on domestic wastes; however, before using it on industrial wastes or mixtures of domestic and industrial wastes, laboratory tests should be performed. Its value in industrial waste treatment is limited largely to wastes in which the organic matter is not predominantly soluble.

Extended aeration The extended-aeration process operates in the endogenous respiration phase of the growth curve, which necessitates a relatively low organic loading and long aeration time. Thus it is generally applicable only to small treatment plants with capacities of less than 3800 m^3/d (1 Mgal/d).

This process is used extensively for prefabricated package plants that are provided for the treatment of wastes from housing subdivisions, isolated institutions, small communities, schools, etc. Although separate sludge wasting generally is not provided, it may be added where the discharge of the excess solids is objectionable. Operating experience has indicated that problems have developed in many plants where wasting facilities have not been provided. Aerobic digestion of the excess solids, followed by dewatering on open sand beds, usually follows separate sludge wasting. Primary sedimentation is omitted from the process to simplify the sludge treatment and disposal.

The oxidation ditch Developed for small towns in the Netherlands, the oxidation ditch is essentially an extended-aeration process. It is used in many small European towns and has found a variety of different applications in the United States. A schematic of a typical oxidation ditch is shown in Fig. 10-7a. It consists of a ring-shaped channel about 1 to 1.5 m (3 to 5 ft) deep. An aeration rotor,

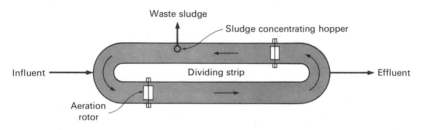

(a)

(b)

Figure 10-7 Oxidation ditch activated-sludge process. (a) Schematic of original oxidation process. (b) Aerial view of the Carrousel modification of original oxidation ditch. (*From Envirobic Systems.*)

consisting of a modified Kessener brush, is placed across the ditch to provide aeration and circulation. The screened wastewater enters the ditch, is aerated by the rotor, and circulates at about 0.3 to 0.6 m/s (1 to 2 ft/s).

Operation of the ditch shown in Fig. 10-7a is intermittent. Modifications can be made for continuous operation. For intermittent operation, the cycle consists of (1) closing the inlet valve and aerating the wastewater, (2) stopping the rotor and letting the contents settle, and (3) opening both inlet and outlet valves, thereby allowing the incoming wastewater to displace an equal volume of clarified effluent.

Carrousel process A technical modification of the original oxidation ditch concept, developed during the 1950s by Pasveer, is known as the Carrousel (see Fig. 10-7b). In this system, vertically mounted mechanical aerators are used to impart oxygen and at the same time to provide sufficient horizontal velocity to the liquid to prevent solids from settling in the aeration channels. For most applications, secondary (final) settling tanks are the only major components needed in addition to the aeration channels. Settled sludge is returned from the settling tanks to the aeration channels. Excess sludge is wasted periodically. Sand-bed drying is the most common method of handling the wasted excess sludge, although other techniques may be used. The settling-tank overflow may be disinfected and discharged into the receiving waters.

Kraus process Wastes that are deficient in nitrogen are difficult to treat by the activated-sludge process. L. S. Kraus, the operator of the treatment plant at Peoria, Ill., in about 1955, was faced with the problem of treating high-carbohydrate wastes in combination with domestic wastewater. His solution was to aerate the supernatant from the sludge digesters, digested sludge, and a portion of the return sludge in a separate reaeration tank for approximately 24 h, converting the ammonia nitrogen into nitrate, and then to mix it with the return activated sludge [18]. This accomplished two things. The nitrate in the aerated supernatant corrected the nitrogen deficiency in the high-carbohydrate waste, and the heavy solids contained in the digested sludge improved the settleability of the mixed liquor. This process was adopted at the San Jose–Santa Clara plant in California for treating the large flow of fruit-processing wastes mixed with domestic wastewater received during the canning season. The flowsheet for the process is shown in Fig. 10-8.

High-rate aeration High-rate aeration is a modification in which high concentrations of mixed-liquor suspended solids are combined with high volumetric loadings (see Table 10-4). This combination allows high food-to-microorganism ratios (0.4 to 1.5) and long mean cell residence times with hydraulic detention times of 0.5 to 2 h. Adequate mixing in the reactor to effect oxygen transfer and to control floc size is achieved through the use of turbine mixers.

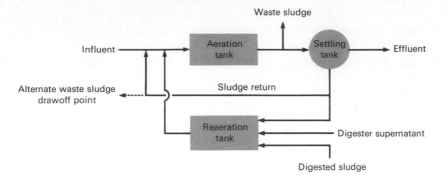

Figure 10-8 Flowsheet for Kraus activated-sludge process.

Pure oxygen Since 1970 the use of pure oxygen as a substitute for air in the activated-sludge process has gained in popularity. In application, the aeration tanks are covered, and the oxygen that is introduced into the aeration tank is recirculated. A schematic of a pure-oxygen system using a series of small continuous-flow stirred-tank reactors is shown in Fig. 10-9. Because CO_2 is released and O_2 is utilized by the microorganisms as a result of their activity, a portion of the gas must be wasted, and new oxygen must be added. Depending on the buffer capacity of the wastewater and the amount of CO_2 removed from the system, pH adjustment may be required. Using Henry's law as an approximation, if the mole fraction of the oxygen above the liquid is 0.8, then the amount of oxygen that can be put into the liquid is about four times the amount that normally could be put in with air for a given set of conditions. The generation of pure oxygen is discussed later in this section. Additional design details may be found in Ref. 38.

When the pure-oxygen process was being first promoted, a number of claims were made about and for it. On the basis of a proper analysis of data from full-scale operating plants and pilot plant data gathered over the past 8 years, many of the original claims have been disproved, especially the claim that smaller quantities of biological solids are produced as compared to those from a conventional-air activated-sludge system [48]. During this period, the process has been well proved and has been shown to provide a high standard of performance. It now appears that the use of pure oxygen is particularly applicable where (1) the available space for the construction of treatment facilities is limited, (2) wide fluctuations occur in the organic loading to the plant, and (3) strong municipal or industrial wastewaters are to be treated [39].

Diffused-Air Aeration

The two basic methods of aerating wastewater are (1) to introduce air or pure oxygen into the wastewater with submerged porous diffusers or air nozzles or (2) to agitate the wastewater mechanically so as to promote solution of air

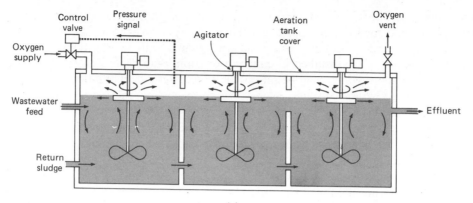

(a)

(b)

Figure 10-9 Pure-oxygen activated-sludge process (*from Union Carbide, Linde Division*). (*a*) of three-stage configuration. (*b*) Aerial view of pure-oxygen activated-sludge process.

from the atmosphere. A diffused-air system consists of diffusers that are submerged in the wastewater, header pipes, air mains, and the blowers and appurtenances through which the air passes. The following discussion covers only the selection of diffusers; the design of blowers and air piping is described in the previous edition of this text [31].

Diffusers The diffusers most commonly used in aeration systems are designed to produce fine, medium, or coarse (relatively large) bubbles. They are described in Table 10-5, and typical devices are shown schematically in Fig. 10-10.

Plate diffusers are installed in concrete or aluminum plate holders, holding six or more plates, which may be set either in recesses or on the bottom of the aeration tank. Groups of plate holders are connected to the air-supply piping at intervals along the tank length, and each group is controlled by a valve. Tube diffusers are screwed into air manifolds, which may run the length of the tank close to the bottom along one side of the tank, or short manifold headers may be mounted on movable drop pipes. With the movable drop pipes, it is possible to raise a header out of the water without interrupting the process and without dewatering the tank. The diffusers can then be removed for cleaning or replacement.

The dome diffuser shown in Fig. 10-10b consists of a porous dome 17.8 cm (7 in) in diameter, constructed of an aluminum oxide material. The dome diffuser is designed to ensure uniform permeability and to produce a flow of fine air bubbles approximately 2 mm (0.08 in) in diameter. The upward motion of the fine bubbles keeps the bottom of the tank swept clear of settled deposits and provides gentle mixing. The domes are mounted on a network of polyvinyl chloride air piping which runs the length of the aeration tank (see Fig. 10-11). Row spacing and diffuser spacing range between 300 and 760 mm (12 and 30 in), depending on the process modification. The polyvinyl chloride piping is attached to the floor at 1.58-m (5-ft) intervals by a polyvinyl chloride floor fixture and adjustable saddle. A single bolt serves as a structural fastener, air inlet, and control orifice.

With porous diffusers, it is essential that the air supplied be clean and free of dust particles that might clog the diffusers. Air filters, often consisting of viscous impingement and dry types in series ahead of the blowers, are commonly

Table 10-5 Description of diffused-air aeration devices

Size of bubble produced	Transfer efficiency	Description	See Fig.
Fine	High	1. Ceramically bonded grains of fused crystalline aluminum oxide.	10-10a, b
		2. Vitreous silicate-bonded grains of pure silica.	10-10a
		3. Resin-bonded grains of pure silica.	10-10a
Medium	Medium	1. Plastic-wrapped diffuser tubes.	10-10c
		2. Woven-fabric sock or sleeve diffusers.	10-10d
Coarse	Low	1. Various orifice devices.	10-10e
		2. Sparger air escapes from periphery of flexible or rigid disk that is displaced when the manifold pressure exceeds the head on the disk.	10-10f
		3. Slot-orifice injectors.	

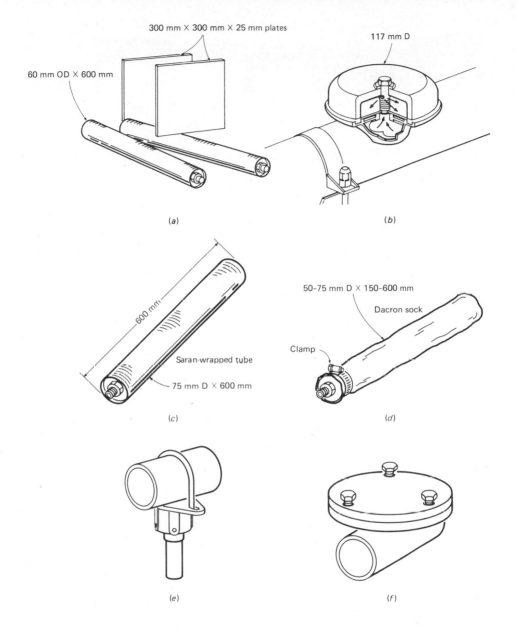

Figure 10-10 Typical diffused-air aeration devices. (*a*) Plate and tube diffusers. (*From Ferro Corp.*). (*b*) Dome diffuser. (*From Norton Co.*) (*c*) Saran-wrapped precision diffuser. (*From FMC, Chicago Pump.*) (*d*) Flexofuser. (*From FMC, Chicago Pump.*) (*e*) Monosparj. (*From Walker Process Equipment Division, Chicago Bridge & Iron Company.*) (*f*) Nonclog. (*From Envirotech, Eimoco Div.*) (*a, b*) Fine bubble; (*c, d*) medium bubble; (*e, f*) coarse bubble. Note, mm × 0.03937 = in.

Figure 10-11 Plug-flow aeration tank equipped with dome aeration devices (*from Norton Co.*).

used. Precoated bag filters and electrostatic filters have also been used. The dust problem can be alleviated by using diffusers that have large orifices.

Several types of medium- and coarse-bubble diffusers are available (see Fig. 10-10c, d, e, f). All these diffusers produce larger bubbles than porous diffusers and consequently have a slightly lower aeration efficiency; however, the advantages of lower cost, less maintenance, and the absence of stringent air-purity requirements offset the slightly lower efficiency. Water containing iron salts has caused serious clogging of porous diffusers at several plants. All diffusers will clog, however, if sufficient pressure is not maintained on the air headers at all times to prevent wastewater from accumulating in the bottom of the headers.

Diffuser performance The efficiency of oxygen transfer depends on the type and porosity of the diffuser, the size of the bubbles produced, the depth of submersion, and other factors that were explained in Chap. 7. Typical transfer efficiencies and rates for various devices are reported in Table 10-6. In general, the efficiency of porous fine-bubble diffusers varies from 10 to 30 percent or more, depending on tank depth. The diffuser shown in Fig. 10-10b can produce oxygen-transfer efficiencies in mixed liquor ranging from about 14 percent at a depth of 3.7 m (12 ft) to 25 percent at 8.0 m (26 ft). Typically, the efficiency for medium-bubble diffusers varies from about 6 to 15 percent; for coarse-bubble diffusers it is about 6 percent [4].

The amount of air used per kilogram of BOD removed varies greatly from one plant to another, and there is risk in comparing the air use at different plants, not only because of the factors mentioned above but also because of different loading rates, control criteria, and operating procedures. Extra high air flowrates applied along one side of a tank reduce the efficiency of oxygen transfer and may even reduce the net oxygen transfer by increasing circulating velocities. The result is a shorter residence time of air bubbles as well as larger bubbles with less transfer surface [24].

Table 10-6 Typical information on the oxygen-transfer capabilities of various aeration devices[a]

Aeration system	Typical[b] transfer efficiency, %	Transfer rate, kg O_2/kW · h	
		Standard[c]	Field[d]
Diffused-air system			
Fine bubble	10–30+	1.2–2.0	0.7–1.4
Medium-bubble	6–15	1.0–1.6	0.6–1.0
Coarse-bubble	4–8	0.6–1.2	0.3–0.9
Turbine-sparger system		1.2–1.4	0.7–1.0
Static-tube system	7 10	1.2–1.6	0.7–0.9
Jet	10–25	1.2–2.4	0.7–1.4
Pure-oxygen system			
Mechanical surface aeration and cryogenic generation			1.4–1.8
Mechanical surface aeration and pressure-swing adsorption generation			1.0–1.3
Turbine-sparger and cryogenic generation			1.2–1.5
Low-speed surface		1.2–2.4	0.7–1.3
Low-speed surface with draft tube		1.2–2.4	0.7–1.4
High-speed floating aerator		1.2–2.4	0.7–1.3
Rotor-brush aerator		1.2–2.4	0.7–1.3

[a] Derived in part from Refs. 38 and 42.
[b] Depends on depth.
[c] Standard conditions: tap water, 20°C; 101.325 Pa = 101.325 kN/m^2; and initial dissolved oxygen = 0 mg/L.
[d] Field conditions: wastewater, 15°C; altitude 150 m, α = 0.85, β = 0.9; operating dissolved oxygen: air systems = 2 mg/L, pure oxygen systems = 6 mg/L.

Note: kg/kW · h × 1.6440 = lb/hp · h
        kN/m^2 × 0.1450 = lb_f/in^2
            1.8(°C) + 32 = °F
              mg/L = g/m^3

Static tube aerator In this device, air is introduced at the bottom of a circular tube that can vary in height from 0.5 to 1.25 m (1.5 to 4.0 ft). Internally, the tubes are fitted with alternately placed deflection plates or turbine impellers to increase the contact of the air with the wastewater. Mixing is accomplished because the tube aerator acts as an air-lift pump. Some typical oxygen transfer data for such devices are presented in Table 10-6. It should be noted that the performance of these units is highly dependent on the liquid depth and tank geometry.

Jet aerator Oxygen transfer in jet aerators is accomplished by mixing pressurized air and water within a jet nozzle and then discharging the air-liquid mixture into the tank. The velocity of the air-liquid mixture discharged from the nozzle and the rising plume of fine air bubbles that form after discharge bring about the mixing of the wastewater within the tank. Oxygen-transfer efficiencies as high as 25 percent have been reported for these devices (see Table 10-6).

Mechanical Aerators

The two principal types of mechanical aerators that are now commonly used may be classified as surface and submerged turbine aerators. In surface aerators, oxygen is entrained from the atmosphere; in turbine aerators, oxygen is entrained both from the atmosphere and from air or pure oxygen introduced in the tank bottom. In either case, the pumping action of the aerator and that of the turbine help to keep the contents of the aeration tank or basin mixed. In the following discussion, both types will be described, along with aerator performance and the energy requirement for mixing.

Surface aerators Mechanical surface aerators are often the utmost in simplicity in an aeration system (see Fig. 10-12). They may be obtained in sizes from 0.75 to 75 kW (1 to 100 hp). They consist of submerged or partially submerged impellers that are attached to motors which are mounted on floats or on fixed structures. The impellers are fabricated from steel, cast iron, noncorrosive alloys, and fiberglass-reinforced plastic and are used to agitate the wastewater vigorously, entraining air in the wastewater and causing a rapid change in the air-water interface to facilitate solution of the air. Surface aerators may be classified according to the speed of rotation of the impeller as low- and high-speed. In low-speed aerators, the impeller is driven through a reduction gear by an electric motor. The motor and gearbox are usually mounted on a platform that is supported either by piers extending to the bottom of the tank or by beams that span the tank. They have also been mounted on floats. In high-speed aerators, the impeller is coupled directly to the rotating element of the electric motor. High-speed aerators are almost always mounted on floats. These units were originally developed for use in ponds or lagoons where the water surface elevation fluctuates, or where a rigid support would be impractical.

The Kessener brush aerator, an old device very popular in Europe, is used

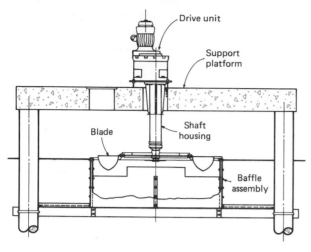

(a)

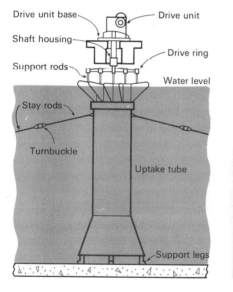

(b)

Figure 10-12 Mechanical surface aerators. (a) Low-speed turbine. *(From Clow Corporation.)* (b) Simplex cone. *(From Bird Machine Co.)*

to provide both aeration and circulation in oxidation ditches [52, 53]. It is horizontally mounted just above the water surface and consists of a cylinder with bristles of steel protruding from its perimeter into the wastewater. The drum is rotated rapidly by an electric motor drive, spraying wastewater across the tank, promoting circulation, and entraining air in the wastewater.

Submerged turbine aerators Most mechanical surface aerators are upflow types that rely on violent agitation of the surface and air entrainment for their efficiency. With turbine aerators, however, air or pure oxygen may also be introduced by diffusion into the wastewater beneath the impeller of downflow or radial aerators. The impeller is used to disperse the air bubbles and mix the tank contents (see Fig. 10-13). A draft tube may be used with either upflow or downflow models to control the flow pattern of the circulating liquid within the aeration tank. The draft tube is a cylinder with flared ends mounted concentrically with the impeller, and extending from just above the floor of the aeration tank to just beneath the impeller.

Aerator performance Surface aerators are rated in terms of their oxygen-transfer rate expressed as kilograms of oxygen per kilowatt-hour (pounds of oxygen per horsepower-hour) at standard conditions. Standard conditions exist when the temperature is 20°C, the dissolved oxygen is 0.0 mg/L, and the test liquid is tap water. Testing and rating are normally done under nonsteady-state conditions using fresh water, deaerated with sodium sulfite. Efficiencies of up to 4.25 kg O_2/kW · h (7 lb O_2/hp · h) have been claimed for various surface aerators, and high efficiencies can be obtained from small pilot plant aerators. Commercial-size surface aerators now available range in efficiency from 1.20 to 2.4 kg O_2/kW · h (2 to 4 lb O_2/hp · h). Oxygen-transfer data for various surface aerators are reported in Table 10-6. Efficiency claims should be accepted by the design engineer only when they are supported by actual test data for the actual model and size of aerator under consideration. For design purposes, the standard performance data must be adjusted to reflect anticipated field conditions. This is accomplished using the following equation. The term within the brackets represents the correction factor.

$$N = N_0 \left[\frac{\beta C_{\text{walt}} - C_L}{9.17} \right] 1.024^{T-20} \alpha \qquad (10\text{-}13)$$

where N = kg O_2/kW · h transferred under field conditions
N_0 = kg O_2/kW · h transferred in water at 20°C, and zero dissolved oxygen
β = salinity-surface tension correction factor, usually 1
C_{walt} = oxygen-saturation concentration for waste at given temperature and altitude (see Fig. 10-14), mg/L
C_L = operating oxygen concentration, mg/L
T = temperature, °C
α = oxygen-transfer correction factor for waste, usually 0.8 to 0.85 for wastewater

The application of this equation is illustrated in Sec. 10-2, which deals with the design of aerated lagoons.

Energy requirement for mixing As with diffused-air systems, the size and shape of the aeration tank is very important if good mixing is to be achieved. Aeration

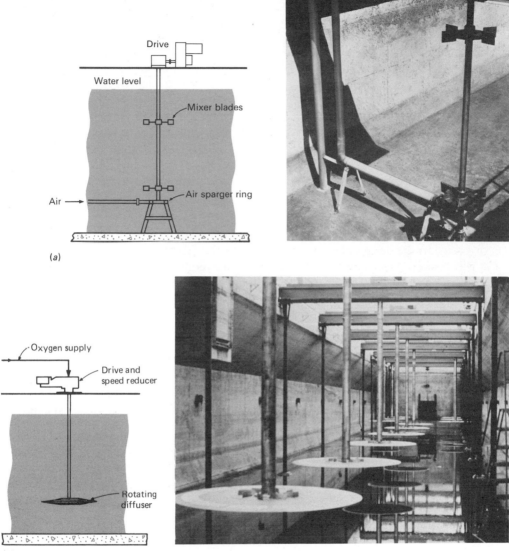

Figure 10-13 Mechanical submerged-turbine aerators. (*a*) Turbine sparger. (*From Permutit.*) (*b*) Turbine for pure oxygen. (*From FMC.*)

tanks may be square or rectangular and may contain one or more units. Water depth may vary from 1.25 to 3.75 m (4 to 12 ft) when using surface aerators. Depths up to 10.7 m (35 ft) have been used with draft-tube mixers.

In diffused-air systems, the air requirement to ensure good mixing varies from 20 to 30 m^3/10^3 m$^3 \cdot$ min (20 to 30 ft^3/10^3 ft$^3 \cdot$ min) of tank volume.

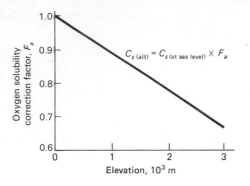

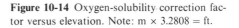

Figure 10-14 Oxygen-solubility correction factor versus elevation. Note: m × 3.2808 = ft.

Typical power requirements for maintaining a completely mixed flow regime with mechanical aerators vary from 15 to 30 $kW/10^3$ m^3 (0.6 to 1.15 $hp/10^3$ ft^3), depending on the design of the aerator and the geometry of the tank, lagoon, or basin. In the design of aerated lagoons for the treatment of domestic wastes, it is extremely important that the mixing power requirement be checked because in most instances it will be the controlling factor. With both the diffused-air systems and mechanical aerators, the power required for the liquid alone is usually less than that required for a liquid containing suspended solids.

Generation of Pure Oxygen

After the quantity of oxygen required is determined, it is necessary, where pure oxygen is to be used, to specify the type of oxygen generator that will best serve the needs of the plant. There are two basic oxygen-generator designs: (1) the traditional cryogenic air-separation process for large applications, and (2) a pressure-swing adsorption (PSA) system for the somewhat smaller and more common plant sizes.

Cryogenic air separation [38] The cryogenic air-separation process involves the liquefaction of air, followed by fractional distillation to separate it into its components (mainly nitrogen and oxygen). A schematic diagram of this process is shown in Fig. 10-15a. First, the entering air is filtered and compressed. Then it is fed to the reversing heat exchangers, which perform the dual function of cooling and removing the water vapor and carbon dioxide by freezing these mixtures out into the exchanger surfaces. This process is accomplished by periodically switching or reversing the feed air and the waste nitrogen streams through identical passes of the exchangers to regenerate their water vapor and carbon dioxide removal capacity.

   Next, the air is processed through "cold and gel traps," which are adsorbent beds that remove the final traces of carbon dioxide as well as most hydrocarbons from the feed air. It is then divided into two streams. The first stream is fed directly to the lower column of the distillation unit. The second stream

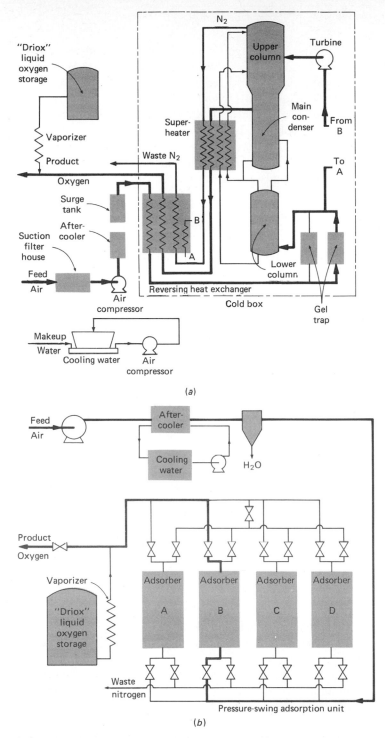

Figure 10-15 Schematics of systems for the generation of oxygen used in the pure-oxygen activated-sludge process [38]. (*a*) Cryogenic generation system. (*b*) Pressure-swing adsorption system.

is returned to the reversing heat exchangers and partially warmed to provide the required temperature difference across the exchanger. This stream is then passed through an expansion turbine and fed into the upper column of the distillation unit. An oxygen-rich liquid exits from the bottom of the lower column, and the liquid nitrogen exits from the top. Both streams are then subcooled and transferred to the upper column. In this column, the descending-liquid phase becomes progressively richer in oxygen, and the liquid that subsequently collects in the condenser reboiler is the oxygen-product stream. This oxygen is recirculated continually through an adsorption trap to remove all possible residual traces of hydrocarbons. The waste nitrogen exits from the top portion of the upper column and is heat-exchanged along with the oxygen product to recover all available refrigeration and to regenerate the reversing heat exchangers.

Pressure-swing adsorption [38] The pressure-swing adsorption system uses a multibed adsorption process to provide a continuous flow of oxygen gas. A schematic diagram of the four-bed system is shown in Fig. 10-15b. The feed air is compressed and passed through one of the adsorbers. The adsorbent removes the carbon dioxide, water, and nitrogen gas, and produces relatively high-purity oxygen. While one bed is adsorbing, the others are in various stages of regeneration.

The concept of the pressure-swing adsorption generator is that the oxygen is separated from the feed air by adsorption at high pressure, and the adsorbent is regenerated by "blowdown" to low pressure. The process operates on a repeated cycle having two basic steps, adsorption and regeneration. During the adsorption step, feed air flows through one of the adsorber vessels until the adsorbent is partially loaded with impurity. At that time the feed-air flow is switched to another adsorber, and the first adsorber is regenerated. During regeneration, the impurities are cleaned from the adsorbent so that the bed will be available again for the adsorption step. Regeneration is carried out by depressurizing to atmospheric pressure, purging with some of the oxygen, and repressurizing back to the pressure of the feed air.

Design of Aeration Tanks and Appurtenances

After the activated-sludge process and the aeration system have been selected and a preliminary design has been prepared, the next step is to design the aeration tanks and support facilities. The following discussion covers (1) aeration tanks, (2) froth-control systems, (3) return-sludge requirements, and (4) sludge wasting.

Aeration tanks Aeration tanks usually are constructed of reinforced concrete and left open to the atmosphere. A cross section of a typical aeration tank is shown in Fig. 10-16. The rectangular shape permits common-wall construction for multiple tanks. If the total capacity exceeds 140 m³ (5000 ft³), the total aeration-tank volume required should be divided among two or more units

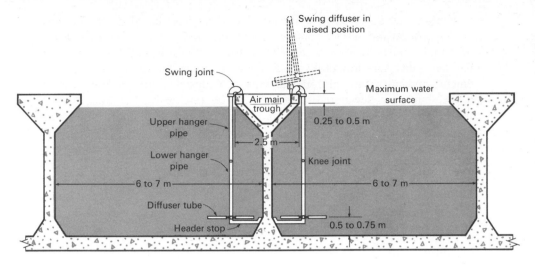

Figure 10-16 Cross section of a typical activated-sludge aeration tank with fine-bubble diffuser aeration system. Note: m × 3.2808 = ft.

capable of independent operation. The total capacity required should be determined from the biological process design. Although the air bubbles dispersed in the wastewater occupy perhaps 1 percent of the total volume, no allowance is made for this in tank sizing. The volume occupied by submerged piping is usually negligible. In Europe, to save space, steel towers varying in height from 15 to 30 m (50 to 100 ft) have also been used.

If the wastewater is to be aerated with diffused air, the geometry of the tank may significantly affect the aeration efficiency and the amount of mixing obtained. The depth of wastewater in the tank should be between 3 and 5 m (10 and 16 ft) so that the diffusers can work efficiently. Freeboard from 0.3 to 0.6 m (1 to 2 ft) above the waterline should be provided. The width of the tank in relation to its depth is important if spiral-flow mixing is used. The width-to-depth ratio for such tanks may vary from 1.0 : 1 to 2.2 : 1. This limits the width of a tank channel to 6 to 12 m (20 to 36 ft).

In large plants, the channels become quite long and sometimes exceed 150 m (500 ft) per tank. Tanks may consist of one to four channels with round-the-end flow in multiple-channel tanks. Large plants should contain not less than four tanks and preferably six to eight or more. Some of the largest plants contain from 30 to 40 tanks arranged in several groups or batteries.

For tanks with diffusers on both sides or in the center of the tank, greater widths are permissible. The important point is to restrict the width of the tank so that "dead spots" or zones of inadequate mixing are avoided. The dimensions and proportions of each independent unit should be such as to maintain adequate velocities so that deposition of solids will not occur. Triangular baffles or fillets may be placed longitudinally in the corners of the channels to eliminate dead spots and to deflect the spiral flow.

Individual tanks should have inlet and outlet gates or valves so that they may be removed from service for inspection and repair. The common walls of multiple tanks must therefore be able to withstand the full hydrostatic pressure from either side. Aeration tanks must have adequate foundations to prevent settlement; and, in saturated soil, they must be designed to prevent flotation when the tanks are dewatered. This is done by thickening the floor slab or installing relief valves connected by graded trenches beneath the floor. Drains or sumps for aeration tanks are desirable for dewatering. In large plants where tank dewatering might be more common, it may be desirable to install mud valves in the bottoms of all tanks. These should be connected to a central dewatering pump or pumping station or to a plant drain discharging to the wet well of the plant pumping station. For small plants, portable contractors' pumps are suitable for dewatering service. It should be possible to empty a tank in 16 h.

Froth-control systems Wastewater normally contains soap, detergents, and other surfactants that produce foam when the wastewater is aerated. If the concentration of mixed-liquor suspended solids is high, the foaming tendency is minimized. Large quantities of foam may be produced during startup of the process, when the mixed-liquor suspended solids are low, or whenever high concentrations of surfactants are present in the wastewater. The foaming action produces a froth that contains sludge solids, grease, and large numbers of wastewater bacteria. The wind may lift the froth off the tank surface and blow it about, contaminating whatever it touches. The froth, besides being unsightly, is a hazard to those working with it because it is very slippery, even after if collapses. In addition, once the froth has dried, it is difficult to clean off.

It is essential, therefore, to have some method for controlling froth formation. A commonly used system consists of a series of spray nozzles mounted along the top edge of the aeration tank opposite the air diffusers. Screened effluent or clear water is sprayed through these nozzles either continuously or on a time clock–controlled program, and this physically breaks down the froth as it forms. Another approach is to meter a small quantity of antifoaming chemical additive into the inlet of the aeration tank or preferably into the spray water [53].

Return-sludge requirements The purpose of the return of sludge is to maintain a sufficient concentration of activated sludge in the aeration tank so that the required degree of treatment can be obtained in the time interval desired. The return of activated sludge from the clarifier to the inlet of the aeration tank is the essential feature of the process. Ample return-sludge pump capacity should be provided and is essential if sludge solids are not to be lost in the effluent. The solids tend to form a sludge blanket in the bottom of the tank which varies in thickness from time to time and may fill the entire depth of the tank at peak flows if the return-sludge pump capacity is inadequate. Return-sludge pump capacities of 20 to 30 percent of the wastewater flow, which were formerly provided, have proved inadequate. Current designs provide capacities of 50 to

100 percent of the wastewater flow for large plants and up to 150 percent of the wastewater flow for small plants.

In general, the return-sludge pumps should be set so that the return flow is approximately equal to the percentage ratio of the volume occupied by the settleable solids from the aeration tank effluent to the volume of the clarified liquid (supernatant) after settling for 30 min in a 1000-mL graduated cylinder. This ratio should not be less than 15 percent at any time. For example, if the settleable solids occupied a volume of 150 mL after 30 min of settling, the percentage volume would be equal to 17.7 percent [(150 mL/850 mL) × 100]. If the plant flow were 2 m³/s (46 Mgal/d), the return-sludge rate should be 0.35 m³/s (0.177 × 2 m³/s) (8 Mgal/d).

Another method often used to control the rate of return-sludge pumping is based on an empirical measurement known as the sludge-volume index (SVI). This index is defined as the volume in milliliters occupied by one gram of activated-sludge mixed-liquor solids, dry weight, after settling for 30 min in a 1000-mL graduated cylinder. In practice it is taken to be the percentage volume occupied by the sludge in a mixed-liquor sample (taken at the outlet of the aeration tank) after 30 min of settling O_v, divided by the suspended-solids concentration of the mixed liquor expressed as a percentage P_w. If the sludge-volume index is known, then the percentage of return sludge, in terms of the recirculation ratio Q_r/Q required to maintain a given percentage of mixed-liquor solids concentration in the aeration tank, is $100Q_r/Q = 100/[(100/P_w SVI) - 1]$. For example, to maintain a mixed-liquor solids concentration of 0.3 percent (3000 mg/L), the percentage of sludge that must be returned when the sludge-volume index is 100 is equal to $100/[(100/0.30 × 100) - 1]$, or 43 percent.

The sludge-volume index has also been used as an indication of the settling characteristics of the sludge. However, the index value that is characteristic of a good settling sludge varies with the characteristics and concentration of the mixed-liquor solids, so observed values at a given plant should not be compared with those reported for other plants or in the literature. For example, if the solids did not settle at all but occupied the entire 1000 mL at the end of 30 min, the maximum index value would be obtained and would vary from 1000 for a mixed-liquor solids concentration of 1000 mg/L to 100 for a mixed-liquor solids concentration of 10,000 mg/L. For such conditions, the computation has no meaning other than the determination of limiting values.

In some installations, operations of the return-sludge pumps is controlled by a series of photoelectric cells or a sonic system located in the settling tank. The operation of these devices is based on the fact that the top of the sludge blanket usually forms a distinct interface with clarified liquids above it.

Sludge wasting The excess activated sludge produced each day must be wasted to maintain a given food to microorganism ratio or mean cell residence time. This can be accomplished most easily and accurately by withdrawing mixed liquor directly from the aeration tank or the aeration-tank effluent pipe where the concentration of solids is uniform. The waste mixed

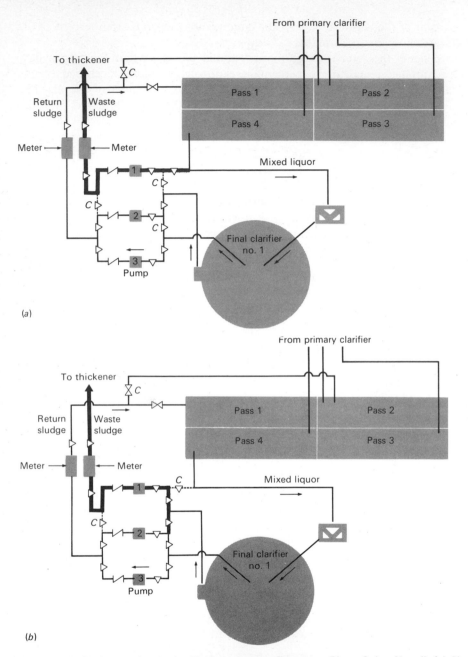

Figure 10-17 Sludge wasting at the Wahiawa Sewage Treatment Plant, Oahu, Hawaii. (*a*) Sludge wasting from mixed-liquor line. (*b*) Sludge wasting from return-sludge line.

liquor can then be discharged to a sludge-thickening tank or to the primary tanks where it mixes and settles with the untreated primary sludge (see Fig. 10-17a). At present, the most common practice is to thicken the activated sludge in the settling tanks and to waste from the return-sludge line (see Fig. 10-17b). The waste sludge is discharged to the primary tanks, to thickening tanks, or to other sludge-thickening facilities.

Design of Solids-Separation Facilities

The function of the activated-sludge settling tank is to separate the activated-sludge solids from the mixed liquor. This is the final step in the production of a well-clarified, stable effluent low in BOD and suspended solids and, as such, represents a critical link in the operation of an activated-sludge-treatment process. Although much of the information presented in Chaps. 6 and 8 in connection with the design of primary sedimentation tanks is applicable, the presence of the large volume of flocculant solids in the mixed liquor requires that special consideration be given to the design of activated-sludge settling tanks. As mentioned previously, these solids tend to form a sludge blanket in the bottom of the tank that will vary in thickness. This blanket may fill the entire depth of the tank and overflow the weirs at peak flowrates if the return-sludge pump capacity or the size of the settling tank is inadequate. Further, the mixed liquor, on entering the tank, has a tendency to flow as a density current interfering with the separation of the solids and the thickening of the sludge. To cope successfully with these characteristics, the engineer must consider the following factors in the design of these tanks: (1) type of tank to be used, (2) settling characteristics of the sludge as related to the thickening requirements for proper plant operation, (3) rates of surface loading and solids loading, (4) flow-through velocities, (5) weir placement and loading rates, and (6) scum removal.

Tank types Activated-sludge settling tanks may be either circular or rectangular (see Fig. 10-18). Square tanks are now seldom used. Circular tanks have been constructed with diameters ranging from 3 to 60 m (10 to 200 ft), although the more common range is from 10 to 30 m (~30 to 100 ft). The tank radius should preferably not exceed five times the sidewater depth. Basically, there are two types of circular tanks to choose from: the center-feed and the rim-feed clarifier. Both types use a revolving mechanism to transport and remove the sludge from the bottom of the clarifier. Mechanisms are of two types: those that scrape or plow the sludge to a center hopper similar to the types used in primary sedimentation tanks, and those that remove the sludge directly from the tank bottom through suction orifices that serve the entire bottom of the tank in each revolution. Of the latter, in one type the suction is maintained by reduced static head on the individual suction pipes (Fig. 10-19a). In another patented suction system, sludge is removed through a manifold either hydrostatically or by pumping (Fig. 10-19b).

(a)

(b)

Figure 10-18 Typical secondary settling tanks. (a) Circular. (b) Rectangular.

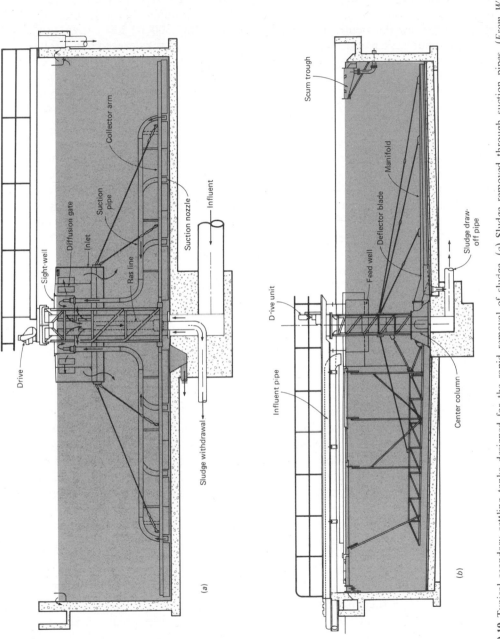

Figure 10-19 Typical secondary settling tanks designed for the rapid removal of sludge. (*a*) Sludge removed through suction pipes. (*From Walker Process Equipment Division, Chicago Bridge & Iron Company.*) (*b*) Sludge removed through manifold. (*From Envirex.*)

Rectangular tanks must be proportional so that good distribution of incoming flow is achieved and horizontal velocities are not excessive. It is recommended that, where possible, the maximum length of rectangular tanks should not exceed 10 times the depth, but lengths up to 90 m (300 ft) have been used successfully in large plants. Where widths of rectangular tanks exceed 6 m (20 ft), multiple sludge-collection mechanisms may be used to permit tank widths up to 24 m (80 ft). Regardless of tank shape, the sludge collector selected should be able to meet the following two operational conditions: (1) the collector should have a high capacity so that when a high sludge-recirculation rate is desired, channeling of the overlying liquid through the sludge will not result, and (2) the mechanism should be sufficiently rugged to be able to transport and remove the very dense sludges that could accumulate in the settling tank during periods of mechanical breakdown or power failure.

Two types of sludge collectors are commonly used: (1) traveling flights, and (2) traveling bridges. Traveling flights are similar to those used for the removal of sludge in primary settling tanks. For very long tanks, it is desirable to use two sets of chains and flights in tandem with a central hopper to receive the sludge. Tanks in which mechanisms move the sludge to the effluent end in the same direction as the density current have shown superior performance in some instances. The traveling bridge, which is similar to a traveling overhead crane, travels along the sides of the sedimentation tank or a support structure if several bridges are used. The bridge serves as the support for the sludge-removal system, which usually consists of a scraper and a suction manifold from which the sludge is pumped. The sludge is discharged to a collection trough that runs the length of the tank.

Settling characteristics of sludge Operationally, secondary settling facilities must perform two functions: (1) separation of the mixed-liquor suspended solids from the treated wastewater, which results in a clarified effluent, and (2) thickening of the return sludge [8]. Both functions must be taken into consideration if secondary settling facilities are to be designed properly. Furthermore, because both functions will be affected by the depth, adequate consideration must be given to the selection of a depth that will provide the necessary volume for both functions. For example, ample volume must be provided for storage of the solids during periods in which sustained peak plant loadings are experienced (see Chap. 3). Also, peak daily flowrate variations must be considered because they affect the sludge-removal requirements.

In general, the area required for clarification must be based on the overflow-rate equivalent of the smallest particle that is to be removed from the clarified liquid in the upper portions of the settling tank. Often, the design settling rate for clarification is taken as the interface settling rate derived from column tests for the sludge concentration at which the plant is to operate. Unfortunately, this velocity is usually much greater than the rate required to remove the light, fluffy particles that are usually found in the effluent from activated-sludge plants.

If these light particles are to be removed, adequate volume must be provided in the clarification zone of the clarifier. The time required for the adequate settling of these particles depends on whether the settling phenomena for the particles can be described as discrete or flocculant [8].

The area required for the thickening of the applied mixed liquor depends on the limiting solids flux that can be transported to the bottom of the sedimentation basin. Because this varies with the characteristics of the sludge, column settling tests should be conducted to determine the relationship between the sludge concentration and the settling rate. The required area can then be determined using the solids flux analysis procedure described in Chap. 6. The depth of the thickening portion of the sedimentation tank must be adequate enough to (1) ensure the maintenance of an adequate sludge-blanket depth so that unthickened solids are not recycled, and (2) temporarily store the excess solids that are periodically applied in excess of the transmitting capacity of the given suspension. These factors are considered in Example 10-2.

When industrial wastes are to be treated by the activated-sludge process, it is recommended that pilot plant studies be conducted to evaluate the settling characteristics of the mixed liquor. These studies are also desirable in the case of the more familiar municipal wastes where the process variables, such as the concentration of the mixed-liquor suspended solids and the mean cell residence time are outside the range of common experience. It is important that such studies be conducted over a temperature range that is representative of both the average and the coldest temperatures to be encountered.

Activated-sludge solids have a specific gravity so near that of water that the increased density and viscosity of the wastewater under winter conditions affect the settling properties of the sludge adversely. In addition, the settling properties of the sludge may vary from time to time because of changes in the amount and specific gravity of the suspended solids passing through the primary settling tanks, the character and amount of industrial wastes contained in the wastewater, and the composition of the microbial life of the floc. For these reasons, conservative design criteria are necessary if overflow of sludge solids is not to occur intermittently.

Surface and solids loading rates Although the design of settling facilities without the benefit of settling tests should be avoided, it is often necessary. When this situation develops, the engineer must resort to the use of surface and solids loading rates for the design of these facilities.

Because of the large amount of solids that may be lost in the effluent if design criteria are exceeded, effluent overflow rates should be based on peak flow conditions. The overflow rates given in Table 10-7 are typical values used for the design of biological systems. These rates are based on the cubic meter per day of wastewater flow per square meter of tank surface instead of on the mixed-liquor flow. The reason is that the overflow rate is equivalent to an upward flow velocity, whereas the return-sludge flow is drawn off the bottom of the tank and does not contribute to the upward flow velocity.

Table 10-7 Typical design information for secondary clarifiers[a, b]

Type of treatment	Overflow rate, $m^3/m^2 \cdot d$		loading, $kg/m^2 \cdot h^c$		Depth, m
	Average	Peak	Average	Peak	
Settling following trickling filtration	16–24	40–48	3.0–5.0	8.0	3–4
Settling following air-activated sludge (excluding extended aeration)	16–32	40–48	3.0–6.0	9.0	3.5–5
Settling following extended aeration	8–16	24–32	1.0–5.0	7.0	3.5–5

[a] Adapted in part from Ref. 42.

[b] The information contained in this table should not be used for the purpose of design unless settling column test data or other field data are unavailable.

[c] Allowable solids loadings are generally governed by sludge settling characteristics associated with cold weather operations.

Note: $m^3/m^2 \cdot d \times 24.5424 = gal/ft^2 \cdot d$
$kg/m^2 \cdot h \times 0.2048 = lb/ft^2 \cdot h$
$m \times 3.2808 = ft$

With pure-oxygen activated-sludge systems, it has been observed that the settling properties of the resulting sludge vary significantly with (1) the characteristics of the wastewater being treated, (2) its temperature, and (3) the degree of pretreatment. Therefore, it is recommended that pilot plant tests be conducted on the wastewater to be treated before designing the settling tanks for pure-oxygen aeration systems.

The solids loading rate on an activated-sludge settling tank may be computed by dividing the total solids applied by the surface area of the tank. The preferred units, which are the same as those used to compute the solids flux discussed previously, are kilograms per square meter per hour, although units of kilograms per square meter per day are common in the literature. The former is favored because the solids loading factor should be evaluated both at peak and average flow conditions. If peaks are of short duration, average 24-h values may govern; if peaks are of long duration, peak values should be assumed to govern to prevent the solids from overflowing the tank.

In effect, the solids loading rate represents a characteristic value for the suspension under consideration. It has been observed that in a settling tank of fixed surface area the effluent quality will deteriorate if the solids loading is increased beyond the characteristic value for the suspension [8, 23]. Typical solids loading values reported in the literature for activated-sludge mixed liquor vary from about 3.0 to 6.0 $kg/m^2 \cdot h$ (0.6 to 1.25 $lb/ft^2 \cdot h$ [40, 57]. Although higher rates may be observed under optimum conditions, they should not be used for design without extensive experimental work covering all seasons and operating variables.

Example 10-2 Design of secondary settling facilities Design the sedimentation facilities to be used in conjunction with the activated-sludge process designed in Example 10-1, using the following settling data derived from a pilot plant study:

MLSS, mg/L	1600	2500	2600	4000	5000	8000
Initial settling velocity, m/h	3.353	2.438	1.524	0.610	0.305	0.091

SOLUTION
1 Develop the gravity solids flux curve from the given data.
   a Plot the given settling column test data (see Fig. 10-20).
   b Using the curve plotted in Fig. 10-20, obtain the data necessary to develop the solids flux curve.

Solids concentration X, g/m^3	1000	1500	2000	2500	3000	4000	5000	6000	7000	8000	9000
Initial settling velocity V_i, g/m^3	4.0	3.5	2.8	1.8	1.14	0.55	0.31	0.2	0.13	0.094	0.07
Solids flux XV_i, kg/m^2 · h^a	4.00	5.25	5.60	4.50	3.42	2.20	1.55	1.20	0.91	0.75	0.63

^{a} $\dfrac{X(\text{g/m}^3)V(\text{m/h})}{10^3 \text{ g/kg}}$

   c Plot the solids flux values determined in step b versus the concentration (see Fig. 10-21).
2 Using the solids flux curve developed in step 1, determine the limiting solids flux values for underflow concentrations varying from 8000 to 12000 mg/L.
   a Using the alternative geometric construction procedure outlined in Chap. 6, draw straight lines tangent to the solids flux curve passing through the desired underflow concentration (see Fig. 10-21).
   b Prepare a summary table of the limiting solids flux values (y intercept) for the various underflow concentrations.

Underflow concentration, g/m^3	8,000	9,000	10,000	11,000	12,000
Limiting solids flux SF_L, kg/m^2 · h	4.2	3.4	2.85	2.5	2.1

3 Determine the recycle ratio necessary to maintain mixed-liquor suspended-solids concentration at 4375 mg/L = (3500 mg/L)/0.8.
   a The required recycle ratio can be determined by performing a materials balance on the influent to the reactor. The resulting expression is

$$Q(X_0) + Q_r(X_u) = (Q + Q_r) \times 4375 \text{ mg/L}$$

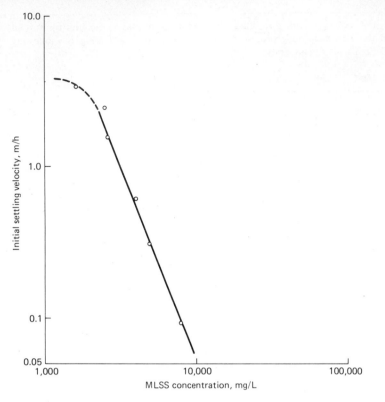

Figure 10-20 Plot of initial settling velocity versus mixed-liquor suspended-solids concentration for Example 10-2.

where Q = influent flow rate, m^3/s
Q_r = recycle flow rate, m^3/s
X_0 = influent suspended solids, g/m^3
X_u = underflow suspended solids, g/m^3

Assuming $X_0 = 0$ and $Q_r = \alpha Q$, the above expression can be written as

$$\alpha Q X_u - \alpha(4375 \text{ mg/L})Q = Q(4375 \text{ mg/L})$$

$$\alpha = \frac{4375 \text{ mg/L}}{X_u \text{ mg/L} - 4375 \text{ mg/L}}$$

where α = recycle ratio, Qr/Q

b Determine the required recycle ratios for the various underflow concentrations.

X_u, mg/L	8,000	9,000	10,000	11,000	12,000
$X_u - 4375$, mg/L	3625	4625	5625	6625	7625
α	1.21	0.95	0.78	0.66	0.57

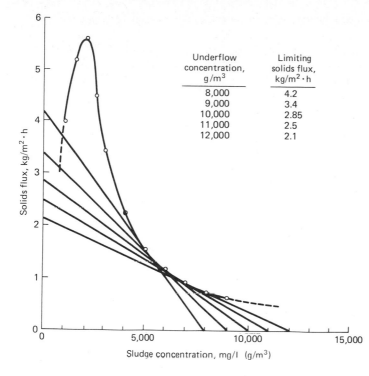

Underflow concentration, g/m³	Limiting solids flux, kg/m² · h
8,000	4.2
9,000	3.4
10,000	2.85
11,000	2.5
12,000	2.1

Figure 10-21 Solids flux curve derived from initial settling data shown in Fig. 10-20 for Example 10-2.

4   Determine the required thickening area of the clarifier for the various underflow concentrations and recycle ratios using the following modified format of Eq. 6-?.

$$SF_a = \frac{(1 + \alpha)(QX)(3600 \text{ s/h})}{A \; 10^3 \text{ g/kg}}$$

where SF_a = average applied solids flux, kg/m² · h
α = recycle ratio
Q = flowrate, m³/s
X = mixed-liquor suspended-solids concentration, g/m³
A = cross-sectional area of clarifier, m²

a   Assume that $SF_a = SF_L$, the limiting solids flux determined in step 2.
b   Set up a computation table to determine the required area.

X_u, mg/L	8,000	9,000	10,000	11,000	12,000
SF_L, kg/m² · h	4.2	3.4	2.85	2.5	2.1
α	1.21	0.95	0.78	0.66	0.57
A, m²	2072	2258	2459	2615	2944

5 Determine the overflow rates corresponding to the solids loading computed in step 4b.

X_u, mg/L	8,000	9,000	10,000	11,000	12,000
Solids loading, kg/m^2 · h^a	4.2	3.4	2.85	2.5	2.1
OF^b, m^3/m^2 · d	10.42	9.57	8.78	8.26	7.34

^{a} Corresponds to the limiting solids flux.
^{b} OF = overflow rate based on plant flow and area computed in step 4b.

6 Check the clarification requirement, assuming that the final design will be based on an underflow concentration of 10,000 g/m^3.
   a As shown in step 5, the overflow rate for an underflow concentration of 10,000 g/m^3 is equal to 8.78 m^3/m^2 · d. This is equivalent to a settling rate of 0.37 m/h.
   b Referring to the settling curve shown in Fig. 10-20, a settling rate of 0.37 m/h would correspond to a solids concentration of 4700 mg/L. Because the concentration of the solids at the interface will be below this value, the area for clarification appears to be conservative.

7 Estimate the required depth for thickening. Assume that the minimum allowable depth for the clarified zone in the sedimentation tank is to be set at 1.5 m (5 ft).
   a Estimate the required depth of the thickening zone. Assume that under normal conditions the mass of sludge retained in the secondary settling tank is equal to 30 percent of the mass in the aeration tank and that the average concentration of solids in the sludge zone is equal to approximately 7000 mg/L [(4000 + 10,000) mg/L/2].
     i Determine the mass of solids in the aeration basin.

$$\text{Aeration-tank solids} = \frac{(4702 \text{ m}^3)(4375 \text{ g/m}^3)}{10^3 \text{ g/kg}} = 20,571 \text{ kg}$$

     ii Determine the mass of solids in the sedimentation basin.

$$\text{Sedimentation-basin solids} = 0.3(20,571 \text{ kg}) = 6171 \text{ kg}$$

     iii Determine the depth of the sludge zone in the sedimentation basin using the following relationship:

$$\frac{(A, \text{ m}^2)(d, \text{ m})(7000 \text{ g/m}^3)}{10^3 \text{ g/kg}} = 6171 \text{ kg}$$

$$d = \frac{(6171 \text{ kg})(1000 \text{ g/kg})}{(7000 \text{ g/m}^3)(2459 \text{ m}^2)}$$

$$= 0.36 \text{ m (1.2 ft)}$$

   b Estimate the required storage capacity in the sludge zone assuming that excess solids must be stored in the secondary sedimentation tank at peak flow conditions because of limitations in the sludge processing facilities. Assume that the 2-d sustained peak flowrate of 2.5 Q_{avg} (see Fig. 2-5) and the 7-d sustained peak BOD loading of 1.5 BOD$_{avg}$ (see Fig. 3-6) occur simultaneously. Also assume that the excess solids must be stored in the secondary sedimentation tank at peak flow conditions because of limitations in the sludge processing facilities.
     i Estimate the solids produced under the given conditions using Eq. 10-3.

$$P_x = Y_{obs} Q(S_0 - S) \times (10^3 \text{ g/kg})^{-1}$$

$$Y_{obs} = 0.3125$$

$$Q = 2.5(21{,}600 \text{ m}^3/\text{d})$$

$$S_0 = 1.5(250 \text{ mg/L})$$

$$S = 15 \text{ mg/L} \qquad \text{(assumed under increased loading conditions)}$$

$$(P_x)_{sp} = \frac{0.3125(2.5 \times 21{,}600 \text{ m}^3/\text{d})[(375 - 15) \text{ mg/L}}{1000 \text{ g/kg}}$$

$$= 6075 \text{ kg/d}$$

  ii Because the peak flowrate is sustained for 2 d, the total solids for the 2-d period are equal to 12,150 kg.

  iii Compute the required depth for sludge storage in the sedimentation tank. Assume that the total solids in the sedimentation tank are now equal to 18,321 kg (12,150 kg + 6171 kg).

$$d = \frac{(18{,}321 \text{ kg})(1000 \text{ g/kg})}{(7000 \text{ g/m}^3)(2459 \text{ m}^2)}$$

$$= 1.06 \text{ m} \ (3.49 \text{ ft})$$

  c Estimate the total required depth

$$\text{Depth} = (1.5 + 0.36 + 1.06)\text{m}$$

$$= 2.92 \text{ m, use } 3.0 \text{ m} \ (9.8 \text{ ft})$$

8 Check the surface overflow rate at peak flow.
  a The peak flow is

$$Q_p = 2.5(21{,}600 \text{ m}^3/\text{d})$$

$$= 54{,}000 \text{ m}^3/\text{d}$$

  b The surface overflow rate at peak flow is

$$\text{Peak overflow rate} = \frac{54{,}000 \text{ m}^3/\text{d}}{2459 \text{ m}^2}$$

$$= 22 \text{ m}^3/\text{m}^2 \cdot \text{d}$$

This value is well below the peak value in Table 10-7.

9 Prepare a summary table of the sedimentation-tank design data.

Item	Value	
	SI units	U.S. customary units
Surface area	2459 m²	26,468 ft²
Depth[a]	3.0 m	9.8 ft
Detention time (avg)	8.2 h	8.2 ft
Mixed-liquor suspended solids	4375 mg/L	4375 mg/L
Limiting solids flux	2.85 kg/m² · h	14.0 lb/ft² · d
Overflow rate		
Average	8.78 m³/m² · d	215.9 gal/ft² · d
Peak	22 m³/m² · d	541.0 gal/ft² · d

  [a] Does not include freeboard.

Flow-through velocity To avoid troubles due to density currents and scouring of sludge already deposited, horizontal velocities should be limited. For rectangular tanks, horizontal flow-through velocities based on the maximum mixed-liquor flow should not exceed a value of about 30.5 m/h (100 ft/h). In a circular center-feed tank, the inlet baffle should have a diameter of 15 to 20 percent of the tank diameter and should not extend more than 1 m (3 ft) below the surface to avoid scouring of deposited sludge from the sludge drawoff sump.

Weir placement and loading Activated-sludge mixed liquor entering the tank will flow along the tank bottom as a density current until it encounters a counter-current pattern or an end wall. Unless this is designed against, solids may be discharged over the effluent weir [44]. Anderson's experimental work at Chicago on tanks 38.4 m (126 ft) in diameter indicated that a circular weir trough placed at two-thirds to three-fourths of the radial distance from the center was in the optimum position to intercept well-clarified effluent [44]. With low surface loadings and weir rates, the placement of the weirs in small tanks does not significantly affect the performance of the clarifier. Circular clarifiers are manufactured with overflow weirs located near both the center and the perimeter of the tank. The minimum water depth below effluent weirs so located should be 3.1 m (10 ft) to prevent overflow of density currents. If weirs are located at the tank perimeter or at end walls in rectangular tanks, the minimum depth should be 3.7 m (12 ft).

Weir loading rates in large tanks should preferably not exceed 375 m³/lin m · d (30,000 gal/lin ft · d) of weir at maximum flow when located away from the upturn zone of the density current, or 250 m³/lin m · d (20,000 gal/lin ft · d) when located within the upturn zone. In small tanks, the weir loading rate should not exceed 125 m³/lin m · d (10,000 gal/lin ft · d) at average flow or 250 m³/lin m · d at maximum flow. The upflow velocity in the immediate vicinity of the weir should be limited to about 3.7 to 7.3 m/h (12 to 24 ft/h). This can be used to determine the spacing of multiple weirs in rectangular tanks.

Scum removal Whether or not to provide for skimming the final tanks depends on the characteristics of the incoming wastewater, the extent of primary treatment, and the type of treatment process. With the modified aeration process in which the primary settling tanks are usually omitted, skimming of the final tanks is essential. Skimming facilities are now required on all federally funded projects.

Operational Difficulties

Two of the most common problems encountered in the operation of an activated-sludge plant are rising sludge and bulking sludge. Because few plants have escaped these problems, it is appropriate to discuss their nature and methods for their control.

Rising sludge Occasionally sludge that has good settling characteristics will be observed to rise or float to the surface after a relatively short settling period. The cause of this phenomenon is denitrification, in which the nitrites and nitrates in the wastewater are converted to nitrogen gas (see Chap. 12). As nitrogen gas is formed in the sludge layer, much of it is trapped in the sludge mass. If enough gas is formed, the sludge mass becomes buoyant and rises or floats to the surface. Rising sludge can be differentiated from bulking sludge by noting the presence of small gas bubbles attached to the floating solids.

Rising-sludge problems can be overcome by (1) increasing the rate of return activated-sludge pumping from the activated-sludge settling tank, (2) decreasing the rate of flow of aeration liquor into the offending tank if the sludge depth cannot be reduced by increasing the return activated-sludge withdrawal rate, (3) where possible, increasing the speed of the sludge-collecting mechanism in the settling tanks, and (4) decreasing the mean cell residence time by increasing the sludge-wasting rate.

Bulking sludge A bulked sludge is one that has poor settling characteristics and poor compactability. The two principal types of sludge bulking problems have been identified. One is caused by the growth of filamentous organisms or organisms that can grow in a filamentous form under adverse conditions, and the other is caused by bound water in which the bacterial cells composing the floc swell through the addition of water to the extent that their density is reduced and they will not settle. The causes of sludge bulking that are most commonly cited in the literature are related to (1) the physical and chemical characteristics of the wastewater, (2) treatment plant design limitations, and (3) plant operation.

Wastewater characteristics that can affect sludge bulking include fluctuations in flow and strength, pH, temperature, staleness, nutrient content, and the nature of the waste components. Design limitations include air-supply capacity, clarifier design, return-sludge pumping-capacity limitations, and short circuiting or poor mixing. Operational causes include low dissolved oxygen in the aeration tank, organic waste overloading of aeration tanks, and final clarifier operation. In almost all cases, all the aforementioned conditions represent some sort of adverse operating condition.

In the control of bulking, where a number of variables are possible causes, it is good to have a checklist of things to investigate. The following are recommended: (1) wastewater characteristics, (2) dissolved-oxygen content, (3) process loading, (4) return-sludge pumping rate, (5) internal plant overloading, and (6) clarifier operation.

The nature of the components found in wastewater or the absence of certain components, such as trace elements, can lead to the development of a bulked sludge [58]. If it is known that industrial wastes are being introduced into the system either intermittently or continuously, the quantity of nitrogen and phosphorus in the wastewater should be checked first, because limitations of both or either are known to favor bulking. Wide fluctuations in pH are also known to be deterimental in plants of conventional design. Wide fluctuations in organic waste loads due to batch-type operations can also lead to bulking and should be checked.

Limited dissolved oxygen has been noted more frequently than any other cause of bulking. If the problem is due to limited oxygen, it can usually be confirmed by operating the air blowers at full capacity. Under these conditions, the blowers should have adequate capacity to maintain at least 2 mg/L of dissolved oxygen in the aeration tank, a generally accepted value. If this level of oxygen cannot be maintained, solution to the problem may require the installation of additional blowers.

The food-to-microorganism ratio should be checked to make sure that it is within the range of generally accepted values (see Table 10-4). When plant operation is controlled by the mean cell residence time parameter, the food-to-microorganism ratio need not be checked. However, the mean cell residence time should be checked to make sure it is within the range normally found to provide efficient treatment (see Table 10-4). If it is not within the range given in this table, the sludge-wasting rate should be adjusted as discussed previously.

The recommendations presented previously should be followed with regard to the return-sludge pumping rate.

To avoid internal plant overloading, it should be determined that the liquid removed from the sludge during mechanical sludge dewatering, or other similar operations, is not being returned to the plant flow during times of peak hydraulic and organic loading.

If bulking conditions continue to persist after all the aforementioned factors have been checked, a critical study of clarifier behavior should be made. This is particularly true of center-fed circular tanks where sludge is removed from the tank directly under the point where the mixed liquor enters the tank. Explorations in the sludge blanket may show that a large part of the sludge is actually retained in the tank for many hours rather than the desired 30 min. If this is the case, then the design is at fault, and changes must be made in facilities.

In an emergency situation or while the aforementioned factors are being investigated, chlorine and hydrogen peroxide may be used to provide temporary help. Chlorination of wastewater or of return sludge has been practiced quite extensively as a means of controlling bulking. Although chlorination is effective in controlling bulking caused by filamentous growths, it is ineffective when bulking is due to light floc containing bound water. Chlorination of return sludge should be based on its dry solids content. A reasonable range is between 0.2 and 1.0 percent by weight. Chlorination normally results in the production of a turbid effluent until such time as the sludge is freed of the filamentous forms. Chlorination of a nitrifying sludge will also produce a turbid effluent, because of the death of the nitrifying organisms. Hydrogen peroxide has also been used in the control of filamentous organisms in bulking sludge. Dosage of hydrogen peroxide and treatment time depend on the extent of the filamentous development. Reasons cited in favour of this control method include a lesser effect on the normal activated-sludge microorganisms [7].

10-2 AERATED-LAGOON (AEROBIC SUSPENDED-GROWTH) TREATMENT

An aerated lagoon is a basin in which wastewater is treated either on a flow-through basis or with solids recycle. Oxygen is usually supplied by means of surface aerators or diffused-air aeration units. As noted in Chap. 9, aerated lagoons evolved from facultative stabilization ponds after aeration devices were added to overcome the odors that developed when they became organically overloaded. The action of the aerators and that of the rising air bubbles from the diffuser are used to keep the contents of the basin in suspension.

The contents of an aerobic lagoon are mixed completely, and neither the incoming solids nor the biological solids produced from waste conversion settle out. In effect, the essential function of this type of lagoon is waste conversion. Depending on the detention time, the effluent contains about one-third to one-half the value of the incoming BOD in the form of cell tissue. Before the effluent can be discharged, however, the solids must be removed by settling (a settling tank or basin is a normal component of most lagoon systems). If the solids are returned to the lagoon, there is no difference between this process and a modified activated-sludge process. Aerial views of aerated lagoons are shown in Fig. 10-22. A schematic of an aerated lagoon with solids recycle used for the treatment of the wastewater from a small cannery is shown in Fig. 10-23. Solids return is often used to improve performance during winter months.

Process Design Considerations

Factors that must be considered in the process design of aerated lagoons include (1) BOD removal, (2) effluent characteristics, (3) oxygen requirements, (4) temperature effects, (5) energy requirement for mixing, and (6) solids separation. The first four factors are considered in the following discussion, and their application is illustrated in Example 10-4. The energy required for mixing was discussed previously (see "Mechanical Aerators"). Solids separation is discussed at the end of this section.

BOD removal Because an aerated lagoon can be considered to be a continuous-flow stirred-tank reactor without recycle, the basis of design can be the mean cell residence time, as outlined in Chap. 9. The mean cell residence time should be selected to assure (1) that the suspended microorganisms will bioflocculate for easy removal by sedimentation and (2) that an adequate safety factor is provided when compared to the mean cell residence time of washout. Typical design values of θ_c^d for aerated lagoons used for treating domestic wastes vary from about 3 to 6 d. Once the value of θ_c^d has been selected, the soluble substrate concentration of the effluent can be estimated, and the removal efficiency can then be computed using the equations given in Chap. 9.

An alternative approach is to assume that the observed BOD_5 removal

(a)

(b)

Figure 10-22 Aerial views of aerated lagoons.

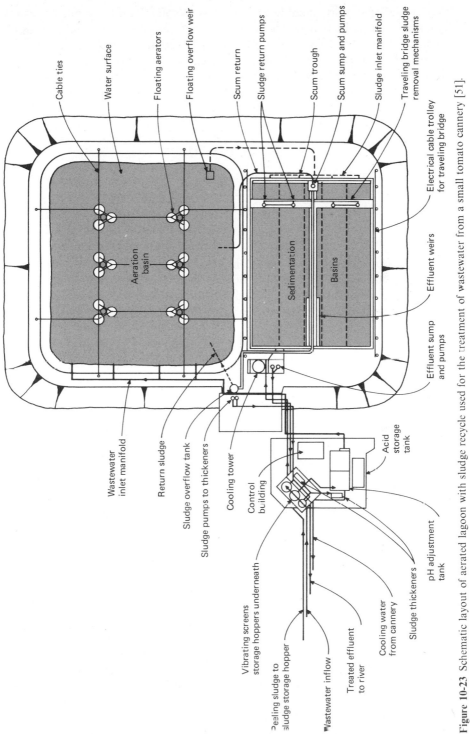

Figure 10-23 Schematic layout of aerated lagoon with sludge recycle used for the treatment of wastewater from a small tomato cannery [51].

Cable ties

Water surface

Floating aerators

Floating overflow weir

Scum return

Sludge return pumps

Scum trough

Scum sump and pumps

Sludge inlet manifold

Traveling bridge sludge removal mechanisms

Electrical cable trolley for traveling bridge

Effluent weirs

Effluent sump and pumps

Aeration basin

Sedimentation

Basins

Wastewater inlet manifold

Return sludge

Sludge overflow tank

Sludge pumps to thickeners

Cooling tower

Control building

Acid storage tank

Vibrating screens storage hoppers underneath

Peeling sludge to sludge storage hopper

Wastewater inflow

Treated effluent to river

Cooling water from cannery

Sludge thickeners

pH adjustment tank

(either overall, including soluble and suspended-solids contribution, or soluble only) can be described in terms of a first-order removal function. The BOD_5 removal is measured between the influent and lagoon outlet (not the outlet of the sedimentation facilities following the lagoon). The required analysis for a continuous-flow stirred-tank reactor has been outlined previously in Chap. 5 in connection with the selection of reactor types for the activated-sludge process. Based on that analysis, the pertinent equation for a single aerated lagoon is

$$\frac{S}{S_0} = \frac{1}{1 + k(V/Q)} \tag{5-64}$$

where S = effluent BOD_5 concentration, mg/L
  S_0 = influent BOD_5 concentration, mg/L
  k = overal first-order BOD_5 removal-rate constant, d^{-1}
  V = volume, m^3 (Mgal)
  Q = flowrate, m^3/d (Mgal/d)

Reported overall k values vary from 0.25 to 1.0. Removal rates for soluble BOD_5 would be higher. Application of this equation is illustrated in Example 10-3 presented later in this section. Additional details on the analysis of aerated lagoons may be found in Refs. 2 and 3.

Effluent characteristics The important characteristics of the effluent from an aerated lagoon include the BOD_5 and the suspended-solids concentration. The effluent BOD_5 will be made up of those components previously discussed in connection with the activated-sludge process and occasionally may contain the contribution of small amounts of algae. The solids in the effluent are composed of a portion of the incoming suspended solids, the biological solids produced from waste conversion, and occasionally small amounts of algae. The solids produced from the conversion of soluble organic wastes can be estimated using Eq. 9.

Oxygen requirement The oxygen requirement is computed as previously outlined in Sec. 10-1, which deals with the activated-sludge process design. Based on operating results obtained from a number of industrial and domestic installations, the amount of oxygen required has been found to vary from 0.7 to 1.4 times the amount of BOD_5 removed.

Temperature Because aerated lagoons are installed and operated in locations with widely varying climatic conditions, the effects of temperature change must be considered in their design. The two most important effects of temperature are (1) reduced biological activity and treatment efficiency and (2) the formation of ice.

The effect of temperature on biological activity was described in Chap. 9. From a consideration of the influent wastewater temperature, air temperature, surface area of the pond, and wastewater flowrate, the resulting temperature in the aerated lagoon can be estimated using the following equation developed by Mancini and Barnhart [20]:

$$(T_i - T_w) = \frac{(T_w - T_a)fA}{Q}$$ (10-14)

where T_i = influent waste temperature, °C (°F)
$\quad T_w$ = lagoon water temperature, °C (°F)
$\quad T_a$ = ambient air temperature, °C (°F)
$\quad f$ = proportionality factor, 0.5
$\quad A$ = surface area, m² (ft²)
$\quad Q$ = wastewater flowrate, m³/d (Mgal/d)

The proportionality factor incorporates the appropriate heat-transfer coefficients and includes the effect of surface area increase due to aeration, wind, and humidity. A typical value for the eastern United States is 0.5 in SI units (12×10^{-6} in U.S. customary units). To compute the lagoon temperature, Eq. 10-14 is rewritten as

$$T_w = \frac{Af\,T_a + QT_i}{Af + Q}$$ (10-15)

Alternatively, if climatological data are available, the average temperature of the lagoon may be determined from a heat-budget analysis by assuming the lagoon is completely mixed.

Where icing may be a problem, its effects on the operation of lagoons may be minimized by increasing the depth of the lagoon or by altering the method of operation. The effect of reducing the surface area is illustrated in Example 10-3. As computed, reducing the area by one-half increases the temperature about 3.8°C (6.9°F), which corresponds roughly to about a 50 percent increase in the rate of biological activity. As the depth of the lagoon is increased, maintenance of a completely mixed flow regime becomes difficult. If the depth is increased much beyond about 3.7 m (12 ft), draft-tube aerators must be used.

Example 10-3 Effect of pond surface area on liquid temperature Determine the effect of reducing the surface area of an aerated lagoon from 10,000 to 5000 m² (107,640 to 53,820 ft²) by doubling the depth for the following conditions:

1. Wastewater flowrate = 3800 m³/d (1 Mgal/d)
2. Wastewater temperature = 15.6°C (60°F)
3. Air temperature = −6.7°C (20°F)
4. Proportionality constant = 0.5

SOLUTION
1 Determine the lagoon water temperature for a surface area of 10,000 m² using Eq. 10-15 and the following data:

$$T_w = \frac{Af\,T_a + QT_i}{Af + Q}$$

where $A = 10{,}000 \text{ m}^2$
$\quad f = 0.5$
$\quad T_a = -6.7°C$
$\quad Q = 3800 \text{ m}^3/d$
$\quad T_i = 15.6°C$

$$T_w = \frac{10{,}000(0.5)(-6.7) + 3800(15.6)}{10{,}000(0.5) + 3800} = 2.93°C \ (37.3°F)$$

2 Determine the lagoon water temperature for a surface of 5000 m².

$$T_w = \frac{5000(0.5)(-6.7) + 3800(15.6)}{5000(0.5) + 3800} = 6.75°C \ (44.2°F)$$

Another approach that can be used to mitigate the effects of cold weather is to modify the operation pattern during the winter months. To do this, two aerated lagoons are required. During the summer months, the lagoons are operated in parallel. As the cold weather begins, parallel operation is discontinued and the lagoons are operated in series [45]. This method of operation conserves heat. Both aerators are operated until ice formation forces the shutdown of the second aerator. Thus the second lagoon functions as an anaerobic pond during the winter months. Because it is covered with ice, odor problems are minimized. In spring when the ice melts, the parallel method of operation is again adopted. With this method of operation it is possible to achieve a 60 to 70 percent removal of BOD_5 even during the coldest winter months [45]. Still another method that can be used to improve performance during the winter months is to recycle a portion of the solids removed by settling.

Process Design

The design of an aerated lagoon is illustrated in Example 10-4.

Example 10-4 Design of an aerated lagoon Design a flow-through aerated lagoon to treat a wastewater flow of 3800 m³/d (1 Mgal/d), including the number of surface aerators and their horsepower rating. The treated liquid is to be held in a settling basin with a 2-d detention time before discharge. Assume that the following conditions and requirements apply:

1. Influent suspended solids = 200 mg/L.
2. Influent suspended solids are not biologically degraded.
3. Influent soluble BOD_5 = 200 mg/L.
4. Effluent soluble BOD_5 = 20 mg/L.
5. Effluent suspended solids after settling = 20 mg/L.
6. Growth constants: $Y = 0.65$, $K_s = 100$ mg/L, $k = 6.0 \text{ d}^{-1}$, $k_d = 0.07 \text{ d}^{-1}$.
7. Total solids produced are equal to computed volatile suspended solids divided by 0.80.
8. First-order soluble BOD_5 removal-rate constant $k_{20} = 2.5 \text{ d}^{-1}$ at 20°C.
9. Summer air temperature = 30°C (86°F).
10. Winter air temperature = 10°C (50°F).

11. Wastewater temperature $= 15.6°C \ (60°F)$.
12. Temperature coefficient: $\theta = 1.06$.
13. Aeration constants: $\alpha = 0.85$, $\beta = 1.0$.
14. Elevation $= 610 \ m \ (2000 \ ft)$.
15. Oxygen concentration to be maintained in liquid $= 1.5 \ mg/L$.
16. Lagoon depth $= 3 \ m \ (10 \ ft)$.
17. Design mean cell residence time $\theta_c^d = 4 \ d$.

SOLUTION

1 On the basis of a mean cell residence time of 4 d, determine the surface area of the lagoon.

$$\text{Volume } V = Q\theta_c^d = (3800 \ m^3/d) \ 4d = 15,200 \ m^3 \ (537,000 \ ft^3)$$

$$\text{Surface area} = \frac{15,200 \ m^3}{3 \ m} = 5067 \ m^2 \ (1.25 \ acres)$$

2 Estimate the summer and winter liquid temperatures using Eq. 10-15.
Summer:

$$T_w = \frac{5067(0.5)(30) + 3800(15.6)}{5067(0.5) + 3800}$$

$$= 21.4°C \ (70.5°F)$$

Winter:

$$T_w = \frac{5067(0.5)(10) + 3800(15.6)}{5067(0.5) + 3800}$$

$$= 13.4°C \ (56.1°F)$$

3 Estimate the soluble effluent BOD_5 measured at the lagoon outlet during the summer using Eq. 9-30.

$$S = \frac{K_s(1 + \theta k_d)}{\theta(Yk - kd) - 1} \qquad (\textit{Note: } \theta = \theta_c^d \text{ for aerated lagoon without recycle.})$$

$$= \frac{100[1 + 0.07(4)]}{4[0.65(6) - 0.07] - 1}$$

$$= 8.9 \ mg/L \qquad (\textit{Note: } \text{The value in the effluent from the settling facilities will be essentially the same.})$$

This value was computed using kinetic-growth constants derived for the temperature in the range from 20 to 25°C. Thus, during the summer months, the effluent requirement of 20 mg/L or less will be met easily. Because there is no reliable information on how to correct these constants for the winter temperature of 13.3°C, an estimate of the effect of temperature can be obtained using the first-order soluble BOD_5 removal-rate constant.

4 Estimate the effluent BOD_5.

a Correct the removal-rate constant for temperature effects using Eq. 5-13.

$$\frac{k_T}{k_{20}} = \theta^{T-20}$$

Summer (21.4°C): $\quad k_{21.4} = 2.5(1.06)^{21.4-20} = 2.71$

Winter (13.4°C): $\quad k_{13.4} = 2.5(1.06)^{13.4-20} = 1.7$

b Determine the effluent BOD_5 using Eq. 5-64.

$$\frac{S}{S_0} = \frac{1}{1 + k\theta}$$

Summer (21.4°C):

$$\frac{S}{200} = \frac{1}{1 + 2.71(4)}$$

$$S = 16.9 \text{ mg/L}$$

Winter (13.4°C)

$$\frac{S}{200} = \frac{1}{1 + 1.7(4)}$$

$$S = 25.6 \text{ mg/L}$$

$$\text{Ratio of } \frac{S_{winter}}{S_{summer}} = \frac{25.6}{16.9} = 1.5$$

Applying the ratio to the soluble effluent BOD_5 computed using the kinetic-growth constants yields a value of about 13.4 mg/L. Using the ratio of the removal-rate constants yields approximately the same value.

The foregoing calculations were presented only to illustrate the method. The value of the removal-rate constant must be evaluated for the waste in question, in a bench- or pilot-scale test program as outlined in Chap. 9.

5 Estimate the concentration of biological solids produced using Eq. 9-29.

$$X = \frac{Y(S_0 - S)}{1 + k_d \theta} = \frac{0.65(200 - 8.9)}{1 + 0.07(4)} = 97 \text{ mg/L VSS}$$

An approximate estimate of the biological solids produced can be obtained by multiplying the assumed growth-yield constant (BOD_5 basis) by the BOD_5 removed.

6 Estimate the suspended solids in the lagoon effluent before settling.

$$SS = 200 \text{ mg/L} + \frac{97 \text{ mg/L}}{0.80} = 321 \text{ mg/L}$$

With the extremely low overflow rate provided in a holding basin with a detention time of 2 d, an effluent containing less than 20 mg/L of suspended solids should be attainable.

7 Estimate the oxygen requirements using Eq. 10-12.

$$O_2 \text{ kg/d} = \frac{Q(S_0 - S) \times (10^3 \text{ g/kg})^{-1}}{f} - 1.42P_x$$

a Determine P_x, the amount of biological solids wasted per day.

$$P_x = (97 \text{ g/m}^3)(3800 \text{ m}^3/\text{d}) \times (10^3 \text{ g/kg})^{-1} = 368.6 \text{ kg/d}$$

b Assuming the conversion factor for BOD_5 to BOD_L is 0.68, determine the oxygen requirements.

$$O_2 \text{ kg/d} = \frac{(3800 \text{ m}^3/\text{d})[(200 - 8.9) \text{ g/m}^3]}{0.68(1000 \text{ g/kg})} - 1.42(368.6 \text{ kg/d}) = 544 \text{ kg/d} \ (1199 \text{ lb/d})$$

8 Compute the ratio of oxygen required to BOD_5 removed.

$$\frac{O_2 \text{ required}}{BOD_5 \text{ removed}} = \frac{544 \text{ kg/d}}{[(200 - 8.9) \text{ g/m}^3](3800 \text{ m}^3/\text{d})(10^3 \text{ g/kg})^{-1}}$$

$$= 0.75$$

9 Determine the surface-aerator power requirements, assuming that the aerators to be used are conservatively rated at 2.0 kg $O_2/kW \cdot h$ (3.3 lb $O_2/hp \cdot h$).

   a Determine the correction factor for surface aerators for summer conditions using Eq. 10-13.

Oxygen-saturation concentration at $21.4°C = 8.93$ mg/L   (see Appendix C)

Oxygen-saturation concentration at $21.4°C$ corrected for altitude $= 8.93 \times 0.94 = 8.39$ mg/L (see Fig. 10-14)

$$\text{Correction factor} = \left[\frac{\beta C_{walt} - C_L}{9.17}\right] 1.024^{T-20} \alpha$$

$$= \frac{8.39 - 1.5}{9.17} (1.024^{21.4-20} 0.85$$

$$= 0.66$$

   b The field-transfer rate N is equal to

$$N = N_0(0.65) = 2.0 \text{ kg } O_2/kW \cdot h \text{ } (0.65)$$

$$= 1.3 \text{ kg } O_2/kW \cdot h \text{ } (2.1 \text{ lb } O_2/hp \cdot h)$$

The amount of O_2 transferred per day per unit is equal to 31.2 kg $O_2/kW \cdot d$. The total power required to meet the oxygen requirements is

$$kW = \frac{544 \text{ kg } O_2/d}{31.2 \text{ kg } O_2/kW \cdot d} = 17.4 \text{ kW } (23.4 \text{ hp})$$

10 Check the energy requirements for mixing. Assume that, for a completely mixed flow regime, the power requirement is 15 $kW/10^3$ m.

Lagoon volume $= 15,200$ m^3

Power required $= 15(15.2) = 228$ kW (305 hp)

Use eight 30-kW (40-hp) surface aerators.

Comment For installations designed to treat domestic wastewater, the energy requirement for mixing usually is the controlling factor in sizing the aerator. The energy needed to meet the oxygen requirement is generally the controlling factor in sizing the aerators where industrial wastes are to be treated.

Solids Separation

If the effluent from aerated lagoons is to meet the requirements for secondary treatment as defined by the U.S. Environmental Protection Agency (see Table 4-1), it will be necessary to provide some type of settling facilities. Usually, sedimentation is accomplished in a large, shallow earthen basin used expressly for the purpose, or in more conventional settling facilities. Where large earthen basins are used, the following requirements must be considered carefully: (1) the detention time must be adequate to achieve the desired degree of suspended-solids removal; (2) sufficient volume must be provided for sludge storage; (3) algal growth must be minimized; and (4) odors that may develop as a result of the anaerobic decomposition of the accumulated sludge must be controlled [1]. In some cases, because of local conditions, these requirements may be in conflict with each other.

In most cases, a minimum detention time of 1 d is required to achieve solids separation [1]. If a 1-d detention time is used, adequate provision must be made for sludge storage so that the accumulated solids will not reduce the actual liquid detention time. Further, if all the solids become deposited in localized patterns, it may be necessary to increase the detention time to counteract the effects of poor hydraulic distribution that may result. Under anaerobic conditions, about 40 to 60 percent of the deposited volatile suspended solids will be degraded each year. Assuming that first-order removal kinetics apply, the following expression can be used to estimate the decay of volatile suspended solids [1].

$$W_t = W_0 \, e^{-k_d t} \qquad (10\text{-}16)$$

where W_t = mass of volatile suspended solids that have not degraded after time
$\quad$ t, kg
$\quad$ W_0 = mass of solids deposited initially, kg
$\quad$ k_d = decay of coefficient, d^{-1} or yr^{-1}
$\quad$ t = time, d or yr

Two problems that are often encountered with the use of settling basins are the growth of algae and the production of odors. Algal growths can usually be controlled by limiting the hydraulic detention time to 2 d or less. If longer detention times must be used, the algal content may be reduced by using either a rock filter (see Sec. 10-4) or a microstrainer. Odors arising from anaerobic decomposition can generally be controlled by maintaining a minimum water depth of 1 m (3 ft). In extremely warm areas, depths up to 2 m (6.5 ft) have been needed to eliminate odors, especially those of hydrogen sulfide.

If space for large settling basins is unavailable, conventional settling facilities can be used. To reduce the construction costs associated with conventional concrete and steel settling tanks, lined earthen basins can be used [51]. An example of such a settling basin is shown in Fig. 10-23. A traveling bridge mechanism equipped with a pumped sludge-withdrawal system is used to remove the sludge from the bottom of the settling tank. The bridge mechanism can be programmed to make either complete or partial sweeps of the basin with appropriate stops at each end. Because low solids loading rates can be achieved inexpensively, higher-than-normal underflow concentrations are usually obtained [51]. The design of a large, unlined earthen sedimentation basin for an aerated lagoon is illustrated in Example 10-5.

Example 10-5 Design of a large, unlined earthen sedimentation basin for an aerated lagoon Design an earthen sedimentation basin for the aerated lagoon designed in Example 10-4. Assume that the hydraulic detention time is to be 2 d and that the liquid level above the sludge layer at its maximum level of accumulation is to be 1.5 m (5 ft). For the purposes of this example, assume that 70 percent of the total solids discharged to the sedimentation basin are volatile. Also assume that the sedimentation pond is cleaned after 4 yr.

SOLUTION

1 Determine the mass of sludge that must be accumulated in the basin each year without anaerobic decomposition.

$$\text{Mass} = (SS_i - SS_e) \times Q \times (10^3 \text{ g/kg})^{-1} \times 365 \text{ d/yr}$$

where SS_i = suspended solids in the influent to the sedimentation basin, g/m^3
SS_e = suspended solids in the effluent from the sedimentation basin, g/m^3
Q = flowrate, m^3/d

a Compute the total mass of solids added per year.

$$\text{Mass} = [(321 - 20) \text{ g/m}^3](3800 \text{ m}^3/\text{d}) \times (10^3 \text{ g/kg})^{-1}(365 \text{ d/yr})$$

$$= 417,487 \text{ kg/yr}$$

b Compute the mass of volatile and fixed solids added per year, assuming that VSS = 0.77 SS.
   i Volatile solids:

$$(\text{Mass})_{VSS} = (417,487 \text{ kg/yr})(0.7)$$

$$= 292,241 \text{ kg/yr}$$

   ii Fixed solids:

$$(\text{Mass})_{FS} = (417,487 - 292,241) \text{ kg/yr}$$

$$= 125,246 \text{ kg/yr}$$

2 Determine the amount of sludge that will accumulate at the end of 4 yr. Assume that the maximum volatile-solids reduction that will occur is equal to 60 percent and that it will occur within 1 yr. To simplify the problem, assume that the deposited volatile suspended solids undergo a linear decomposition. Because the volatile solids will decompose to the maximum extent within 1 yr, the following relationship can be used to determine the maximum amount of volatile solids available at the end of each year of operation:

$$(\text{VSS})_t = [0.7 + 0.4(t - 1)](292,241 \text{ kg/yr})$$

where $(\text{VSS})_t$ = mass of volatile suspended solids at the end of t yr, kg
t = time, yr

a Mass of volatile suspended solids accumulated at the end of 4 yr:

$$\text{VSS}_t = [0.7 + 0.4(4 - 1)](292,241 \text{ kg/yr})$$

$$= 555,258 \text{ kg}$$

b Total mass of solids accumulated at the end of 4 yr:

$$\text{SS}_t = 555,258 \text{ kg} + 4 \text{ yr}(125,246 \text{ kg/yr})$$

$$= 1,056,242 \text{ kg}$$

3 Determine the required liquid volume and the dimensions for the sedimentation basin.
a Volume of sedimentation basin:

$$V = (2 \text{ d})(3800 \text{ m}^3/\text{d}) = 7600 \text{ m}^3$$

b Surface area of sedimentation basin:

$$A_s = \frac{7600 \text{ m}^3}{1.5 \text{ m}} = 5067 \text{ m}^2$$

The aspect ratio for the surface area of the sedimentation basin (ratio of width to length) depends on the geometry of the available site.

4 Determine the depth required for the storage of sludge.

   a Determine the mass of accumulated sludge per square meter.

$$\text{Accumulated mass of sludge} = 1{,}056{,}242 \text{ kg}$$

$$\text{Mass per unit area} = \frac{1{,}056{,}242 \text{ kg}}{5067 \text{ m}^2} = 208.5 \text{ kg/m}^2$$

   b Determine the required depth, assuming that the deposited solids will compact to an average value of 15 percent and that the density of the accumulated solids is equal to 1.06.

$$\frac{208.5 \text{ kg/m}^2}{d, \text{ m}} = (1.06)(0.15)(1000 \text{ kg/m}^3)$$

$$d = \frac{208.5 \text{ kg/m}^2}{(1.06)(0.15)(1000 \text{ kg/m}^3)}$$

$$= 1.31 \text{ m } (4.3 \text{ ft})$$

Because it may be difficult to provide a total depth of 2.81 m (1.5 m + 1.31 m), it may be necessary either to increase the detention time or to clean the sedimentation basins more frequently.

10-3 TRICKLING-FILTER (AEROBIC ATTACHED-GROWTH) TREATMENT

Starting with the first municipal installation in the United States in 1908 at Reading, Pa., trickling filters have been widely used to provide biological waste-water treatment. The process microbiology and theoretical analysis of trickling filters were described in Chap. 9. The discussion in this section covers the classification of filters, the process design, and the design of physical facilities.

Filter Classification

Trickling filters are classified by hydraulic or organic loading rates as low-rate, intermediate-rate, high-rate, and super-rate. The range of loadings normally encountered and other operational characteristics are shown in Table 10-8.

Low-rate filters A low-rate filter is a relatively simple, highly dependable device that produces an effluent of consistent quality with an influent of varying strength. Generally, a constant hydraulic loading is maintained, not by recirculation, but by suction-level-controlled pumps or a dosing siphon. Dosing tanks are small, usually with only a 2-min detention time based on twice the average design flow, so that intermittent dosing is minimized. Even so, at small plants, low nighttime flows may result in intermittent dosing [42].

    If the interval between dosing is longer than 1 or 2 h, the efficiency of the progress deteriorates because the character of the biological slime is altered by a lack of moisture. In most low-rate filters, only the top 0.6 to 1.2 m (2 to 4 ft)

Table 10-8 Typical design information for trickling filters[a]

Item	Low-rate filter	Intermediate-rate filter	High-rate filter	Super-rate (roughing) filter
Hydraulic, loading, $m^3/m^2 \cdot d$	1–4	4–10	10–40	40–200
Organic loading, $kg/m^3 \cdot d$	0.08–0.32	0.24–0.48	0.32–1.0	0.80–6.0
Depth, m	1.5–3.0	1.25–2.5	1.0–2.0	4.5–12
Recirculation ratio	0	0–1	1–3; 2–1	1–4
Filter media[b]	Rock, slag, etc.	Rock, slag, etc.	Rock, slag, synthetic materials	Synthetic materials, redwood
Power requirements, $kW/10^3 \ m^3$	2–4	2–8	6–10	10–20
Filter flies	Many	Intermediate	Few, larvae are washed away	Few or none
Sloughing	Intermittent	Intermittent	Continuous	Continuous
Dosing intervals	Not more than 5 min. (generally intermittent)	15 to 60 s (continuous)	Not more than 15 s (continuous)	Continuous
Effluent	Usually fully nitrified	Partially nitrified	Nitrified at low loadings	Nitrified at low loadings

[a] Adapted in part from Ref. 42.
[b] See Table 10-9 for the physical characteristics of the various filter media.

Note: $m \times 3.2808 = ft$
$m^3/m^2 \cdot d \times 1.0691 = mgal/acre \cdot d$
$m^3/m^2 \cdot d \times 0.0170 = gal/ft^2 \cdot min$
$kg/m^3 \cdot d \times 62.4280 = lb/10^3 \ ft^3 \cdot d$
$kW \times 1.3410 = hp$

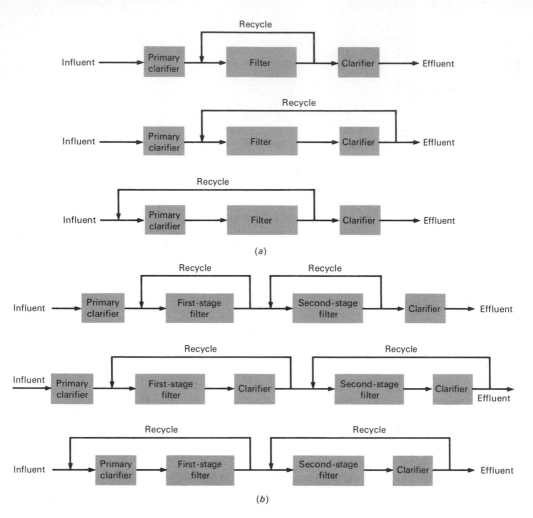

Figure 10-24 Intermediate-rate and high-rate trickling-filter flowsheets with various recirculation patterns. (*a*) Single-stage filters; (*b*) two-stage filters.

of the filter medium will have appreciable biological slime. As a result, the lower portions of the filter may be populated by autotrophic nitrifying bacteria which oxidize ammonia nitrogen to nitrite and nitrate forms. If the nitrifying population is sufficiently well established, and if climatic conditions and wastewater characteristics are favorable, a well-operated low-rate filter can provide not only good BOD removal but also a highly nitrified effluent.

Head loss through the filter may be 1.5 to 3 m (5 to 10 ft), which may be objectionable if the site is too flat to permit gravity flow. With a favorable gradient, the ability to use gravity flow is a distinct advantage.

Low-rate filters are not without drawbacks. Odors are a common problem, especially if the wastewater is stale or septic, or if the weather is warm. Filters should not be located where the odors would create a nuisance. Filter flies (*Psychoda*) may breed in the filters unless control measures are used.

Intermediate-rate and high-rate filters In intermediate-rate and high-rate filters, recirculation of the filter effluent or final effluent permits higher organic loadings. Flow diagrams for various high-rate configurations are shown in Fig. 10-24.

Recirculation of effluent from the trickling-filter clarifier permits the high-rate filter to achieve the same removal efficiency as the low-rate or intermediate-rate filter. Recirculation of filter effluent around the filter (first flowsheet in Fig. 10-24a and b) results in the return of viable organisms. It has been observed that this method of operation often improves treatment efficiency. Recirculation also helps to prevent ponding in the filter and to reduce the nuisance from odors and flies [42].

Super-rate filters Super-rate trickling filters have evolved as a result of the development of various types of synthetic and wood packing media. The major applications of super-rate filters have been for high-strength wastes and as roughing units (see Fig. 10-25). Because of their high surface area per unit

Figure 10-25 Super-rate tower filters for the treatment of combined domestic and cannery wastewater.

volume, synthetic media filters can perform as well as high-rate filters at lower loadings. At extremely low loading rates, synthetic media filters have been used to achieve the nitrification of treated effluent [42].

Process Design

As noted in Chap. 9, a universal equation is not available for the design of trickling filters. However, Eq.. 9-82 and the equations proposed by Eckenfelder [10] and by Bruce and Mertens [5] have proved to be adequate in terms of describing the observed removals in trickling filters. These equations are:

$$\frac{S_e}{S_i} = \exp\left[-(f\,hk_0)\,\frac{wZ}{Q}\right] \qquad (9\text{-}82)$$

$$\frac{S_e}{S_i} = \exp\left[-KZS_a^m\left(\frac{A}{Q}\right)^n\right] \qquad (9\text{-}83)$$

$$\frac{S_e}{S_i} = \exp\left(-K_T S_a^a Q_v^{-b}\right) \qquad (9\text{-}86)$$

where S_e = BOD$_5$ of settled effluent from the filter, mg/L
$\quad S_i$ = BOD$_5$ of wastewater applied to the filter, mg/L
$\quad K, K_T$ = observed removal-rate constant, m/d
$\quad wZ$ = surface area of filter medium, m^2 (see Chap. 9)
$\quad Z$ = depth of filter, m
$\quad S_a$ = specific surface area per unit volume, m^2/m^3
$\quad\quad = A_s/V$
$\quad A_s$ = surface area, m^2
$\quad V$ = volume, m^3
$\quad A$ = cross-sectional area of filter surface, m^2
$\quad Q$ = volumetric flowrate, m^3/d
$\quad Q_v$ = volumetric flowrate applied per unit volume, m^3/m$^3 \cdot$ d
m, n, a, b = empirical constants

If the constants are set equal to 1 and the terms within the brackets are equated, it will be noted that all these equations are essentially the same. It has been shown [26] that they are the same as the basic equation proposed by the National Research Council [31]:

$$\frac{S_a}{Q_v} = \frac{S_a ZA}{Q} = \frac{A_s}{V}\frac{ZA}{Q} = \frac{A_s}{Q}\frac{wZ}{Q} \qquad (9\text{-}85)$$

Therefore, only the application of Eq. 9-83 will be illustrated.

In an extensive study conducted in England, Bruce and Merkens [5] found that the value of the removal-rate constant for wastewater K_T at 15°C varied from about 0.025 to 0.060 m/d, and the value of the temperature coefficient θ was 1.08, so that

$$K_T = K_{15}(1.08^{T-15}) \qquad (5\text{-}13)$$

If these values are corrected to a temperature of about 22°C, they can be compared to the value of the removal-rate constant found by analyzing Eckenfelder's data, which was equal to 0.117 m/d (see Fig. 9-28). The corrected values vary from 0.0428 to 0.1028 m/d. Typical values in the literature from treatment systems in the United States vary from about 0.06 to 0.120 m/d. From this analysis, it appears that a value of 0.10 m/d is representative for a temperature of 20°C.

The value of influent applied BOD_5, S_i, can be determined by setting up a materials balance around the filter as follows:

$$QS_0 + \alpha QS_e = (1 + \alpha)(QS_i)$$

where Q = flowrate, m^3/d
S_0 = BOD in influent after primary settling (if primary settling is used), mg/L
α = recycle ratio = Q_r/Q
S_e = BOD in settled effluent
S_i = BOD in influent applied to filter

Solving for S_i yields the following expression:

$$S_i = \frac{S_0 + \alpha S_e}{1 + \alpha} \tag{10-17}$$

In the theoretical discussion of the trickling-filter process in Chap. 9, it was demonstrated that as the degree of recycle is increased, the treatment efficiency is reduced. This finding also helps to explain, in part, the observation that as the applied loading is increased, the performance is decreased, because the rate of recirculation is usually increased when the plant load is increased. The design of a trickling filter is illustrated in Example 10-6.

Example 10-6 Trickling-filter design Design a plastic-medium tower filter to replace some existing facultative ponds that are now used to treat wastes from a rural community in which a small vegetable cannery is located. The following information and data apply. Insofar as possible, the data were derived from local records and field tests.

1. Average year-round domestic wastewater flowrate = 10,000 m^3/d (2.64 Mgal/d).
2. Sustained-peak seasonal cannery flowrate = 5000 m^3/d (1.32 Mgal/d).
3. The canning season is May through October.
4. Average year-round domestic BOD_5 = 220 mg/L.
5. Sustained-peak combined domestic and cannery BOD_5 = 550 mg/L.
6. Critical temperature data
   a. Sustained low for May and October = 20°C.
   b. Sustained low for January = 0°C.
7. Effluent BOD_5 requirement = 30 mg/L.
8. BOD_5 removal-rate constant for tower filter derived from pilot plant studies conducted during the summer when the average temperature was 25°C = 0.10 m/d.
9. Temperature-correction coefficient = 1.08.

10. Specific area of filter packing material $= 85$ m^2/m^3.
11. Maximum allowable filter height because of local site restrictions $= 10$ m.
12. The value of the coefficients m and n in Eq. 9-83 were found to be equal to 1.

SOLUTION

1 Determine the bulk volume of filter medium required during the cannery season using Eq. 9-83.

$$\frac{S_e}{S_i} = \exp\left(-KZS_a\frac{A}{Q}\right)$$

a Correct the observed BOD reaction-rate coefficient for the sustained temperatures observed during May and October.

$K_{20} = K_{25}\,\theta^{T-25}$

$K_{20} = (0.10$ m/d$)(1.08^{20-25})$

$\quad = 0.068$ m/d

b To avoid oxygen-transfer limitations, assume that the influent BOD$_5$ will be diluted with recycle flow to a value of about 350 mg/L. Determine the required recycle ratio for this influent value:

$$550Q + 30Q_r = 350(Q + Q_r)$$

$$\frac{Q_r}{Q} = \frac{550 - 350}{350 - 30}$$

$$\alpha = 0.63$$

c Compute the required bulk volume ZA using the following data:

$S_e = 30$ g/m^3
$S_i = 350$ g/m^3
$K = 0.068$ m/d
$S_a = 85$ m^2/m^3
$Q = (1 + 0.63)(15{,}000$ m^3/d$) = 24{,}450$ m^3/d

$$-\ln\frac{S_e}{S_i} = KZS_a\left(\frac{A}{Q}\right)$$

$$-\left(\ln\frac{30}{350}\right) = (0.068 \text{ m/d})(85 \text{ m}^2/\text{m}^3)\,\frac{ZA}{24{,}450 \text{ m}^3/\text{d}}$$

$$ZA = \frac{2.46(24{,}450 \text{ m}^3/\text{d})}{(0.068 \text{ m/d})(85 \text{ m}^2/\text{m}^3)} = 10{,}406 \text{ m}^3$$

2 Determine the bulk volume of filter medium needed during the winter to meet the effluent requirements.
  a Correct the BOD reaction-rate coefficient for the low sustained temperature value observed during January.

$$K_0 = (0.10 \text{ m/d})(1.08^{0-25}) = 0.015$$

b Compute the required bulk volume using the following data:

$$S_e = 30 \text{ mg/L}$$
$$S_i = 220 \text{ mg/L}$$
$$K = 0.015 \text{ m/d}$$
$$S_a = 85 \text{ m}^2/\text{m}^3$$
$$Q = 10,000 \text{ m}^3/\text{d}$$

$$-\ln \frac{S_e}{S_i} = KZS_a \frac{A}{Q}$$

$$-\left(\ln \frac{30}{220}\right) = (0.015 \text{ m/d})(85 \text{ m}^2/\text{m}^3)\frac{ZA}{10,000 \text{ m}^3/\text{d}}$$

$$ZA = \frac{1.99(10,000 \text{ m}^3/\text{d})}{(0.015 \text{ m/d})(85 \text{ m}^2/\text{m}^3)}$$

$$= 15,608 \text{ m}^3$$

Therefore, winter conditions control.

3 Determine the area and depth. Because an infinite number of combinations are possible, a detailed analysis of the cost trade-off would have to be performed. For the purpose of this example, assume that the tower filter is to function like a high-rate filter, so the applied rate will be set at 40 m³/m² · d. Using this value, the area and depth required are:

a Area of filter using a hydraulic loading of 40 m³/m² · d:

$$A = \frac{24,450 \text{ m}^3/\text{d}}{40 \text{ m}^3/\text{m}^2 \cdot \text{d}}$$

$$= 611.3 \text{ m}^2$$

b Depth of filter at a loading of 40 m³/m² · d:

$$d = \frac{15,608 \text{ m}^3}{611.3 \text{ m}^2} = 25.53 \text{ m} \qquad \text{(This value is obviously unacceptable.)}$$

c Determine the required surface area assuming a tower depth of 10 m.

$$A = \frac{15,608 \text{ m}^3}{10 \text{ m}}$$

$$= 1561 \text{ m}^2$$

Use two filters with a diameter equal to 32 m (105 ft). The corresponding area is equal to 1608 m² (2 × 804 m²).

d Check the hydraulic loadings.

i Summer conditions = $\dfrac{24,450 \text{ m}^3/\text{d}}{1608 \text{ m}^2}$ = 15 m³/m² · d

ii Winter conditions = $\dfrac{10,000 \text{ m}^3/\text{d}}{1608 \text{ m}^2}$ = 6.2 m³/m² · d

Comment The hydraulic loading values determined in step 3*d* are generally lower than those reported in Table 10-8 for a high-rate filter. The reasons are (1) the high organic loading during the canning season, (2) the low winter temperatures, and (3) the high degree of treatment required. In this situation, the cost effectiveness of covering the filters to increase the operating temperature could be explored.

Design of Physical Facilities

Factors that must be considered in the design of trickling filters include (1) the type and dosing characteristics of the distribution system; (2) the type and physical characteristics of filter medium to be used; (3) the configuration of the underdrain system; (4) provision for adequate ventilation, either natural or forced air; and (5) the design of the required settling tanks.

Distribution systems The rotary distributor for trickling filtration has become a standard for the process because it is reliable and easy to maintain. It consists of two or more arms that are mounted on a pivot in the center of the filter and revolve in a horizontal plane. The arms are hollow and contain nozzles through which the wastewater is discharged over the filter bed. The distributor assembly may be driven either by the dynamic reaction of the wastewater discharging from the nozzles or by an electric motor. The speed of revolution varies with the flowrate for the reaction-driven unit, but it should be in the range of one revolution in 10 min or less for a two-arm distributor. Clearance of 150 to 225 mm (6 to 9 in) should be allowed between the bottom of the distributor arm and the top of the bed. This permits the wastewater streams from the nozzles to spread out and cover the bed uniformly, and it prevents the ice accumulations from interfering with the distributor motion during freezing weather.

Distributor arms may be of constant cross section for small units, or they may be tapered to maintain minimum transport velocity. Nozzles are spaced unevenly so that greater flow per unit of length is achieved near the periphery than at the center. For uniform distribution over the area of the filter, the flowrate per unit of length should be proportional to the radius from the center. The head loss through the distributor is in the range of 0.6 to 1.5 m (2 to 5 ft). Important features that should be considered in selecting a distributor are the ruggedness of construction, ease of cleaning, ability to handle large variations in flow while maintaining adequate rotational speed, and corrosion resistance of the material and its coating system. Distributors are manufactured for beds with diameters up to 60 m (200 ft).

Dosing tanks providing intermittent operation or recirculation by pumping may be used to ensure that the minimum flow is sufficient to rotate the distributor and discharge wastes from all nozzles. Four-arm distributors may be provided with weir boxes confining the flow to two arms at minimum flowrates.

Fixed-nozzle distribution systems consist of a series of spray nozzles located at the points of equilateral triangles covering the filter bed. A system of pipes placed in the filter distributes the wastewater uniformly to the nozzles. Special nozzles having a flat spray pattern are used, and the head is varied systematically so that the spray falls first at a maximum distance from the nozzle and then at a decreasing distance as the head slowly drops. In this way, a uniform dose is applied over the whole area of the bed. Half-spray nozzles are used along the sides of the filter. Twin dosing tanks, with sloping bottoms that provide more volume at the higher head (required by the greater spray area), supply the nozzles by discharging through automatic siphons and are arranged to fill and dose alternately. At maximum flow, there should be a minimum rest period of 30 s between doses; the rest period increases as the rate of flow decreases. The head required, measured from the surface of the filter to the maximum water level in the dosing tank, is normally 2.4 to 3 m (8 to 10 ft). For a complete description of trickling filters with fixed nozzles and the hydraulic computations involved, the reader is referred to Metcalf and Eddy [28].

Filter media The ideal filter medium is a material that has a high surface area per unit of volume, is low in cost, has a high durability, and does not clog easily. Typical packing media are shown in Fig. 10-26. The physical characteristics of commonly used filter media, including those shown in Fig. 10-26, are reported in Table 10-9. The most suitable material is generally a locally available river rock or gravel graded to a uniform size within the range of 25 to 75 mm (1 to 3 in). Trap rock is particularly satisfactory. Other materials, such as slag, cinders, or hard coal, have also been used. Stones less than 25 mm (1 in) in diameter do not provide sufficient pore space between the stones to permit free flow of wastewater and sloughed solids. Plugging of the medium and ponding inside the filter or at the surface will result. Large-diameter stones avoid the ponding problem but have a relatively small surface area per unit volume; thus they cannot support as large a biological population. The specification of size uniformity is a way of ensuring adequate pore space.

Important characteristics of filter media are strength and durability. Durability may be determined by the sodium sulfate test, which is used to test the soundness of concrete aggregates. Detailed specifications for filter media can be found in the ASCE Manual 13, *Filtering Materials for Sewage Treatment Plants*.

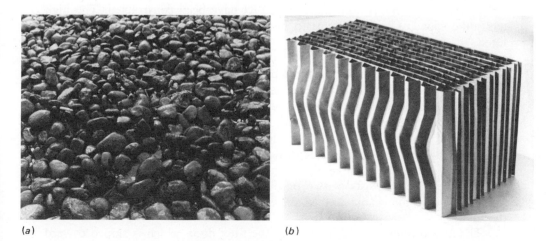

(a)   (b)

Figure 10-26 Typical packing mediums for trickling filters. (a) River rock used in conventional filters. (b) Plastic used in tower trickling filters.

Table 10-9 Physical properties of trickling-filter media[a]

Medium	Nominal size, mm	Mass/unit volume, kg/m^3	Specific surface area, m^2/m^3	Void space, percent
River rock				
Small	25–65	1250–1450	55–70	40–50
Large	100–120	800–1000	40–50	50–60
Blast-furnace slag				
Small	50–80	900–1200	55–70	40–50
Large	75–125	800–1000	45–60	50–60
Plastic				
Conventional	600 × 600 × 1200[b]	30–100	80–100	94–97
High-specific surface	600 × 600 × 1200[b]	30–100	100–200	94–97
Redwood	1200 × 1200 × 500[b]	150–175	40–50	70–80

[a] Derived in part from Refs. 28 and 41.
[b] Module size.

Note: mm × 0.03937 = in
kg/m^3 × 62.4280 = lb/10^3 ft^3
m^2/m^3 × 0.3048 = ft^2/ft^3

One type of synthetic medium consists of interlocking sheets of plastic, which are arranged like a honeycomb to produce highly porous and clog-resistant medium. The sheets are shipped unassembled, requiring very little space in comparison with the final volume of the assembled medium. The sheets are corrugated, so that when the medium is assembled, a strong lightweight grid is formed. Subassemblies of the medium form modules that can be arranged to fit any filter configuration. Filters as deep as 12 m (40 ft) have been constructed. The high hydraulic capacity and resistance to plugging can best be used in a high-rate filter.

Underdrains A trickling filter has three primary systems: distribution, treatment (the medium), and collection. The collection system catches the filtered wastewater and solids discharged from the filter medium and conveys them to a conduit leading to the final sedimentation tank. It consists of the filter floor, collection channel, and the underdrains [15]. A typical underdrain system for a tower filter is shown in Fig. 10-27. The underdrains are specially designed vitrified-clay blocks with slotted tops that admit the wastewater and support the medium. The body of the block consists of two or three channels with curved inverts that form the underdrain channels when laid end to end and cover the entire floor of the filter.

The underdrains are laid directly on the filter floor, which is sloped toward the collection channel at a 1 to 2 percent gradient. Underdrains may be open

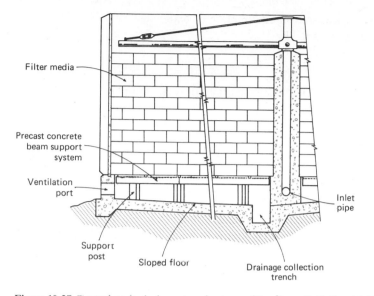

Figure 10-27 Typical underdrain system for tower filter [from B. F. Goodrich].

at both ends, so that they may be inspected easily and flushed out if they become plugged. The underdrains also ventilate the filter, providing the air for the micro-organisms that live in the filter slime, and they should at least be open to a circumferential channel for ventilation at the wall as well as to the central collection channel.

Air-flow-through filters An adequate flow of air is of fundamental importance to the successful operation of a trickling filter. The principal factors responsible for airflow in an open-top filter are natural draft and wind forces. In the case of natural draft, the driving force for airflow is the temperature difference between the ambient air and the air inside the pores. If the wastewater is colder than the ambient air, the pore air will be cold and the direction of flow will be downward. If the ambient air is colder than the wastewater, the flow will be upward. The latter is less desirable from a mass-transfer point of view because the partial pressure of oxygen, and thus the oxygen-transfer rate, is lowest in the region of highest oxygen demand. In many areas of the country, there are periods, especially during the summer, when essentially no airflow occurs through the trickling filter because temperature differentials are negligible.

Draft, which is the pressure head resulting from the temperature difference, may be determined from Eq. 10-18 [47]:

$$D_{air} = 0.353 \left(\frac{1}{T_c} - \frac{1}{T_n} \right) Z \qquad (10\text{-}18)$$

where D_{air} = air draft measured in centimeters of water
  T_c = cold temperature, °K
  T_n = hot temperature, °K
  Z = height of the filter, cm

A more conservative estimate of the average pore air temperature is obtained by using the log mean temperature, T_m:

$$T_m = \frac{T_2 - T_1}{\ln (T_2/T_1)} \tag{10-19}$$

where T_1 = warmer temperature, °K
  T_2 = colder temperature, °K

The volumetric air flowrate may be estimated by setting the draft equal to the sum of the head losses that result from the passage of air through the filter and underdrain system.

Natural draft has proved adequate for trickling filters, provided that the following precautions are taken [57]:

1. Underdrains and collecting channels should be designed to flow no more than half full to provide a passageway for the air.
2. Ventilating manholes with open-grating types of covers should be installed at both ends of the central collection channel.
3. Large-diameter filters should have branch collecting channels with ventilating manholes or vent stacks installed at the filter periphery.
4. The open area of the slots in the top of the underdrain blocks should not be less than 15 percent of the area of the filter.
5. One square foot gross area of open grating in ventilating manholes and vent stacks should be provided for each 23 m² (250 ft²) of filter area.

In extremely deep or heavily loaded filters, there may be some advantage in forced ventilation if it is properly designed, installed, and operated. Such a design should provide for a minimum air flow of 0.3 m³/m² · min (1 ft³/ft² · min) of filter area in either direction. It may be necessary during periods of extremely low air temperature to restrict the flow of air through the filter to keep it from freezing.

Trickling-filter settling tanks The function of settling tanks that follow trickling filters is to produce a clarified effluent. They differ from activated-sludge settling tanks in that sludge recirculation, which is essential to the activated-sludge process, is lacking. All the sludge from trickling-filter settling tanks is removed to sludge-processing facilities. The design of these tanks is similar to the design of primary settling tanks, except that the surface loading rate is based on the plant flow plus the recycle flow (see Fig. 10-1) minus the underflow (often neglected). The overflow rate at peak flow should not exceed 48 m³/m² · d (1200 gal/ft² · d).

10-4 COMBINED AEROBIC TREATMENT PROCESSES

The number of treatment flowsheets that can be derived by combining various aerobic processes discussed previously is endless. Two of the more common flowsheets now in use will be considered in this section: (1) a trickling-filter process followed by an activated-sludge process, and (2) an activated-sludge process followed by a trickling-filter process. Such combination processes have been used successfully for a number of years for the treatment of all types of wastewater, especially combined domestic and industrial wastewater [43]. In this application, the first process in the series can be considered as a "roughing process" whose function is to reduce the loading on the following process to a level that will allow it to function in an optimum manner. Another application of these combined flowsheets is in the upgrading of existing processes to meet the new secondary treatment requirements of the U.S. Environmental Protection Agency (see Table 4-1) and other more stringent requirements that may be adopted in the future.

Series Trickling-Filter and Activated-Sludge Processes

The most common of the combined aerobic treatment processes is the trickling-filter process followed by an activated-sludge process (see Fig. 10-28). This process flowsheet is usually used to reduce the strength of wastewater where industrial and domestic wastewater are treated in common treatment facilities. It is also often used to upgrade an existing activated-sludge system by adding an upstream trickling filter. An alternative arrangement involves the addition of an activated-sludge process downstream from an existing trickling filter.

The process microbiology for these combined processes is essentially the same

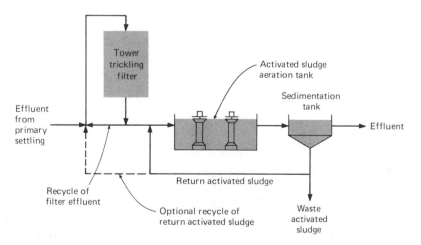

Figure 10-28 Combination treatment process (roughing filter followed by an activated-sludge process).

as for the individual processes described previously in Chap. 9. Some microorganism population shifts occur in the trickling filter because of the high hydraulic loadings that are normally used. Also, many of the microorganisms associated with conventional rock or slag trickling filters are not present because tower filters are normally used.

The performance of the series processes considered in this section is usually evaluated on the basis of loading factors, although here again the kinetic expressions developed previously for the activated-sludge process and trickling filters can theoretically be applied. When the loading-factor approach is used, the BOD removed $(kg/m^3 \cdot d)$ is plotted against the applied BOD $(kg/m^3 \cdot d)$ and a regression line is fitted to the experimental data.

An interesting aspect associated with the use of the series trickling-filter and activated-sludge processes is the additional treatment of the biological solids that is achieved. Biological solids that enter the activated-sludge process from the trickling filter are partially stabilized and also flocculated [43]. This further treatment is not well modeled by the kinetic expressions presented in Sec. 9-4, so volumetric loading factors are often used to describe the process flowsheet.

Series Activated-Sludge and Trickling-Filter Processes

A trickling filter downstream from an activated-sludge process can be used to achieve seasonal nitrification or to upgrade an overloaded activated-sludge process. On the basis of an experimental study, Roskopf et al. [43] have reported that the series activated-sludge and trickling-filter processes appear to have good stability. The process microbiology and performance analysis for this series of processes are quite similar to those described previously, so they are not repeated here.

10-5 STABILIZATION PONDS

A stabilization pond (or lagoon) is a relatively shallow body of water contained in an earthen basin of controlled shape, which is designed for the purpose of treating wastewater. The term oxidation pond, often used, is synonymous. Ponds have become very popular in small communities because their low construction and operating costs offer a significant financial advantage over other recognized treatment methods. Ponds are also used extensively for the treatment of industrial wastewater and mixtures of industrial and domestic wastewater that are amenable to biological treatment. Installations are now serving such industries as oil refineries, slaughterhouses, dairies, poultry-processing plants, and rendering plants [37].

Because of their popularity, stabilization ponds are discussed in detail in this section. The purpose is to describe (1) the various types of ponds that are used, (2) their application, (3) the process design, (4) solids separation techniques that can be used with ponds to meet secondary treatment requirements of the U.S. Environmental Protection Agency, and (5) the design of the physical facilities.

Pond Classification

Stabilization ponds are usually classified according to the nature of the biological activity that is taking place as aerobic, anaerobic, or aerobic-anaerobic. This general scheme was used in Table 9-1, where the various pond processes were also classified according to whether they were suspended-growth, attached-growth, or combination-growth processes. The principal types of stabilization ponds that are in common use are identified in Table 10-10. Other classification schemes that have been used are based on the type of influent (untreated, screened, or settled-wastewater or activated-sludge effluent); the pond-overflow condition (nonexistent, intermittent, or continuous); and the method of oxygenation (photosynthesis, atmospheric surface reaeration, or mechanical aerators). In addition, as described in the following discussion, a variety of different pond systems have been developed and applied to meet specific treatment objectives.

Application

Stabilization ponds have been used singly or in various combinations to treat both domestic and industrial wastes. Typical applications are reported in Table 10-10. As shown, aerobic ponds are used primarily for the treatment of soluble organic wastes and effluents from wastewater treatment plants. The aerobic-anaerobic ponds are the most common type and have been used to treat domestic wastewater and a wide variety of industrial wastes (see Fig. 10-29). Anaerobic ponds are especially effective in bringing about the rapid stabilization of strong organic wastes. Usually, anaerobic ponds are used in series with aerobic-anaerobic ponds to provide complete treatment.

Where ponds are commonly used, most states have regulations governing their design, installation, and management (operation). A minimum of a 60-d detention is often required for flow-through facultative ponds receiving untreated wastewater. Even higher detention times (90 to 120 d) have been specified frequently. A high degree of coliform removal is assured even with a 30-d detention. When a pond 1.8 m (6 ft) deep is designed on the basis of 100 persons per acre per day, the detention period is approximately 200 d.

Process Design and Analysis

The design of stabilization ponds is perhaps the least well defined of all the designs of biological treatment processes. Numerous methods have been proposed in the literature, yet when the results are correlated, a wide variance is usually found. Typical design parameters for the different types of ponds previously discussed are reported in Tables 10-11 and 10-12; data for aerated lagoons are

Figure 10-29 Aerial view of facultative stabilization ponds (*from John Carollo Engineers, Walnut Creek, California*).

also included for comparison. Most of the data were derived from operating experience with a wide variety of individual ponds and pond systems. Some methods that have been proposed for their design, including a consideration of sludge buildup, are described and illustrated in the following discussion.

Aerobic high-rate pond Because of the difficulties of separating the algae from the treated wastewater, the aerobic high-rate pond is not used routinely for the treatment of domestic wastewater. It is used, however, where the algae are harvested to recover protein in the form of cell tissue that can be fed to livestock of all types.

One of the more successful pilot plant studies involving the use of an aerobic high-rate pond has been carried out for the past 2 yr at the farm of the Melbourne and Metropolitan Board of Works at Werribee, Victoria, Australia [6, 9]. A schematic of the pilot plant facilities is shown in Fig. 10-30. The pond was of shallow design and was equipped with pumps for the purpose of intermittently circulating the liquid contents. The total area was about 0.28 ha (0.69 acres), and the operating depths varied from 170 to 200 mm (6.7 to 7.9 in). The influent to the pond was primary treated wastewater. During the study, loading

Table 10-10 Types and applications of stabilization ponds in common use

Type of pond or pond system	Common name	Identifying characteristic	Application
Aerobic (150–450 mm)	High-rate aerobic pond	Designed to optimize the production of algae cell tissue and achieve high yields of harvestable protein.	Nutrient removal, treatment of soluble organic wastes, conversion of wastes
Aerobic	a. Low-rate aerobic pond	Designed to maintain aerobic conditions throughout the liquid depth.	Treatment of soluble organic wastes and secondary effluents
	b. Maturation or tertiary pond	Similar to low-rate aerobic ponds but very lightly loaded.	Used for polishing (upgrading) effluents from conventional secondary treatment processes, such as trickling filter or activated sludge [29]
Aerobic-anaerobic (oxygen source: algae)	Facultative pond	Deeper than a high-rate pond. Photosynthesis and surface reaeration provide oxygen for aerobic stabilization in upper layers. Lower layers are facultative. Bottom layer of solids undergoes anaerobic digestion.	Treatment of untreated screened or primary settled wastewater and industrial wastes
Aerobic-anaerobic (oxygen source: surface aerators)	Facultative pond with mechanical surface aeration	As above, but small mechanical aerators used to provide oxygen for aerobic stabilization.	Treatment of untreated screened or primary settled wastewater and industrial wastes
Anaerobic	Anaerobic lagoon (pond), anaerobic pretreatment ponds	Anaerobic conditions prevail throughout, usually followed by aerobic or facultative ponds.	Treatment of domestic and industrial wastes
Anaerobic + aerobic-anaerobic with recirculation from aerobic-anaerobic to anaerobic	Pond system	Used to achieve specific treatment objectives	Complete treatment of wastewater and industrial wastes
Anaerobic + aerobic-anaerobic + aerobic pond system with recirculation from aerobic to anaerobic	Pond system	Used to achieve specific treatment objectives	Complete treatment of wastewater and industrial wastes with high bacterial removals

Table 10-11 Typical design parameters for aerobic stabilization ponds

Parameter	Aerobic (high-rate) pond	Aerobic pond[a]	Aerobic (maturation) pond	Aerated lagoons
Flow regime	Intermittently mixed	Intermittently mixed	Intermittently mixed	Completely mixed
Pond size, ha	0.25–1	<4 multiples	1–4	1–4 multiples
Operation[b]	Series	Series or parallel	Series or parallel	Series or parallel
Detention time, d[b]	4–6	10–40	5–20	3–10
Depth, m	0.30–0.45	1–1.5	1–1.5	2–6
pH	6.5–10.5	6.5–10.5	6.5–10.5	6.5–8.0
Temperature range, °C	5–30	0–30	0–30	0–30
Optimum temperature, °C	20	20	20	20
BOD_5 loading, kg/ha · d[c]	80–160	40–120	≤15	
BOD_5 conversion	80–95	80–95	60–80	80–95
Principal conversion products	Algae, CO_2, bacterial cell tissue	Algae, CO_2, bacterial cell tissue	Algae, CO_2, bacterial cell tissue, NO_3	CO_2, bacterial cell tissue
Algal concentration, mg/L	100–260	40–100	5–10	
Effluent suspended solids, mg/L[d]	150–300	80–140	10–30	80–250

[a] Conventional aerobic ponds designed to maximize the amount of oxygen produced rather than the amount of algae produced.
[b] Depends on climatic conditions.
[c] Typical values (much higher values have been applied at various locations). Loading values are often specified by state control agencies.
[d] Includes algae, microorganisms, and residual influent suspended solids. Values are based on an influent soluble BOD_5 of 200 mg/L and, with the exception of the aerobic ponds, an influent suspended-solids concentration of 200 mg/L.

Note: ha × 2.4711 = acre
  m × 3.2808 = ft
  kg/hd · d × 0.8922 = lb/acre · d

552

Table 10-12 Typical design parameters for anaerobic and facultative stabilization ponds

Parameter	Aerobic-anaerobic (facultative) pond	Aerobic-anaerobic (facultative) pond	Anaerobic pond	Aerated lagoons
Flow regime	·········	Mixed surface layer	·········	Completely mixed
Pond size, ha	1–4 multiples	1–4 multiples	0.2–1 multiples	1–4 multiples
Operation[a]	Series or parallel	Series or parallel	Series	Series or parallel
Detention time, d[a]	7–30	7–20	20–50	3–10
Depth, m	1–2	1–2.5	2.5–5	2–6
pH	6.5–9.0	6.5–8.5	6.8–7.2	6.5–8.0
Temperature range, °C	0–50	0–50	6–50	0–40
Optimum temperature, °C	20	20	30	20
BOD_5 loading, kg/ha · d[b]	15–80	50–200	200–500	
BOD_5 conversion	80–95	80–95	50–85	80–95
Principal conversion products	Algae, CO_2, CH_4, bacterial cell tissue	Algae, CO_2, CH_4, bacterial cell tissue	CO_2, CH_4, bacterial cell tissue	CO_2, bacterial cell tissue
Algal concentration, mg/L	20–80	5–20	0–5	
Effluent suspended solids, mg/L[c]	40–100	40–60	80–160	80–250

[a] Depends on climatic conditions.

[b] Typical values (much higher values have been applied at various locations). Loading values are often specified by state control agencies.

[c] Includes algae, microorganisms, and residual influent suspended solids. Values are based on an influent soluble BOD_5 of 200 mg/L and, with the exception of the aerobic ponds, an influent suspended-solids concentration of 200 mg/L.

Note: ha × 2.4711 = acre
m × 3.2808 = ft
kg/ha · d × 0.8922 = lb/acre · d
mg/L = g/m^3

553

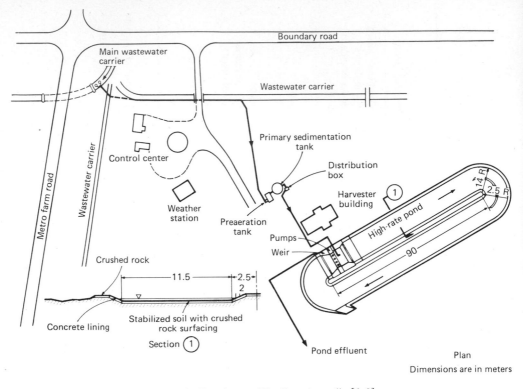

Figure 10-30 High-rate aerobic pond pilot plant at Werribee, Australia [6, 9].

rates varied from 80 kg BOD_5/ha · d (70 lb BOD_5/acre · d) in the winter period to 160 kg BOD_5/ha · d (140 lb BOD_5/acre · d) in the summer period.

On the basis of an extended period of operation, the soluble BOD_5 was reduced from an average value of 276 to 9 mg/L, thus achieving a 97 percent conversion (treatment) efficiency. As would be expected, the effluent contained significant amounts of algal cell tissue measured in the form of volatile suspended solids. The average concentration of volatile suspended solids in the pond effluent was 585 mg/L as compared to an influent value of 160 mg/L [6, 9]. Clearly, such an effluent is unacceptable from the standpoint of secondary treatment requirements of the U.S. Environmental Protection Agency. Even with the use of extensive settling ponds, it is doubtful that an effluent with an acceptable suspended-solids concentration could be achieved. Thus, if such high-rate ponds are to be used, some positive means must be available for the separation of the algal cell tissue. This subject is considered further later in this section (see "Solids Separation").

Aerobic ponds Although a number of different approaches have been used to design aerobic stabilization ponds [12, 13, 21, 22, 25], no universally accepted approach is available. The most common procedure is based on the use of appropriate loading factors, such as those reported in Table 10-11. Where oxygen production is the main objective, depths up to 1.5 m (5 ft) have been used. From a study of a number of different ponds, Oswald [36] has developed the curve shown in Fig. 10-31 in which the BOD loading in kilograms per hectare per day is plotted versus the depth to which aerobic conditions were maintained. Developed for fall conditions, this curve can be used as a guide in selecting appropriate loading factors.

An alternative method of arriving at an appropriate loading factor to maintain aerobic conditions, developed by Oswald and his coworkers at the University of California [34, 35, 36], is to equate the oxygen resources of the pond to the applied organic loading. The principal source of oxygen in an aerobic stabilization pond is photosynthesis, which is governed by solar energy. This method of analysis was delineated in the first edition of this text.

Because the bacteria in an aerobic pond are responsible for the removal (conversion) of BOD_5, the principles developed in Chap. 9 can be applied. Since kinetic values are not now available, a first-order removal-rate constant as given

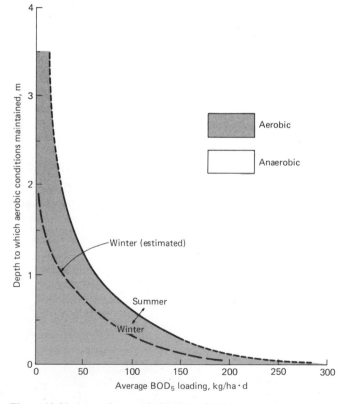

Figure 10-31 Approximate relationship of BOD loading to depth of aerobic zone in stabilization ponds for fall conditions [adapted from Ref. 36].

in Eq. 5-64 is frequently assumed. The rate constant must be related functionally to the temperature, solar radiation intensity, organic loading, and nature of the wastewater. Further, on the basis of the first-order removal equation developed by Wehner and Wilhelm for a reactor with an arbitrary flow-through pattern (between a continuous-flow stirred-tank pattern and a plug-flow pattern), the term kt must be related to the hydraulic characteristics of the pond. Because the contents of aerobic ponds must be mixed to achieve best performance, it is estimated that a typical value for the pond dispersion factor would be about 1.0. Typical values for the overall first-order BOD_5 removal-rate constant k vary from about 0.05 to 1.0 per day, depending on the operational and hydraulic characteristics of the pond [54]. One of the problems in comparing the values reported in the literature is that the basis (for example, soluble BOD_5 conversion, total BOD_5 including the contribution of suspended solids) on which they were derived often is not clearly stated.

The design of an aerobic stabilization pond is illustrated in Example 10-7.

Example 10-7 Design of an aerobic stabilization pond Design an aerobic stabilization pond to treat an industrial wastewater flow of 3800 m³/d (1.0 Mgal/d) with a soluble BOD_5 of 100 mg/L. Assume that the following conditions apply:

1. Influent suspended solids = negligible
2. BOD_5 (conversion) = 90 percent
3. First-order soluble BOD_5 removal-rate constant = 0.25 d⁻¹ at 20°C
4. Temperature coefficient θ = 1.06 at 20°C
5. Pond temperature in summer = 32°C
6. Pond temperature in winter = 10°C
7. Maximum individual pond area = 4 ha (10 acres)
8. Maximum pond depth = 1 m (3.3 ft)
9. Pond dispersion factor (assumed) = 1.0

SOLUTION

1 From Fig. 5-12, determine the value of kt for the pond for a dispersion factor of 1.0 and a removal efficiency of 90 percent.

$$kt = 5$$

2 Determine the detention time.
    a Winter conditions:

$$k_{10} = 0.25(1.06)^{10-20} = 0.14$$

$$0.14t = 5$$

$$t = 35.7d$$

    b Summer conditions:

$$k_{32} = 0.25(1.06)^{32-20} = 0.5$$

$$0.5t = 5$$

$$t = 10d$$

3   Determine the pond surface area requirements.

$$\text{Surface area} = \frac{(3800 \text{ m}^3/\text{d})(35.7 \text{ d})}{1 \text{ m}}$$

$$= 135,660 \text{ m}^2$$

$$= 13.6 \text{ ha} (33.6 \text{ acres})$$

4   Check the design using the curve given in Fig. 10-31.
  a   From Fig. 10-31, assume that an appropriate loading factor for a pond with a depth of 1.0 m for winter conditions would be about 30 kg/ha · d.
  b   Using the loading factor obtained from Fig. 10-31, determine the detention time required using the following relationship:

$$\theta = \frac{d(\text{BOD}_5)(10^4 \text{ m}^2/\text{ha})}{LR(10^3 \text{ g/kg})}$$

where θ = hydraulic detention time, d
    d = pond depth, m
    BOD_5 = applied BOD_5, g/m³
    LR = loading rate, kg/ha · d

$$\theta = \frac{(1.0 \text{ m})(100 \text{ g/m}^3)(10^4 \text{ m}^2/\text{ha})}{(30 \text{ kg/ha} \cdot \text{d})(10^3 \text{ g/kg})}$$

   = 33.3 d   (checks with detention time value determined in step 2)

Aerobic-anaerobic (facultative) ponds The design of aerobic-anaerobic ponds closely follows that of the method outlined in connection with the design of aerobic ponds. Because of the method of operation (for example, maintenance of quiescent conditions to promote the removal of suspended solids by sedimentation), it is anticipated that dispersion factors for such ponds would vary from 0.3 to 1.0.

Another factor that must be considered is sludge accumulation, which is important in terms of the oxygen resources and the overall performance of the pond. For example, in cold climates, a portion of the incoming BOD_5 is stored in the accumulated sludge during the winter months. As the temperature increases in the spring and summer, the accumulated BOD_5 is anaerobically converted, and the oxygen demand of gases and acids produced may exceed the oxygen resources of the aerobic surface layer of the pond. When it is anticipated that BOD_5 storage will be a problem, surface aerators are recommended. If the design is based on BOD_5, the aerators should have a capacity adequate to satisfy from 175 to 225 percent of the incoming BOD_5. Another problem caused by the accumulation of sludge is a reduction in performance of the pond as measured by the suspended-solids content of the effluent. The design of an aerobic-anaerobic pond using surface aerators is illustrated in Example 10-8.

Example 10-8 Aerobic-anaerobic stabilization-pond design Design an aerobic-anaerobic stabilization pond to treat a wastewater flow of 3800 m^3/d (1.0 Mgal/d). Because the ponds are to be installed near a residential area, surface aerators will be used to maintain oxygen in the upper layers of the pond. Assume that the following conditions apply:

1. Influent suspended solids = 200 mg/L
2. Influent BOD_5 = 200 mg/L
3. Summer liquid temperature = 25°C (77°F)
4. Winter liquid temperature = 15°C (59°F)
5. Overall first-order BOD_5 removal-rate constant = 0.25 d^{-1} at 20°C
6. Temperature coefficient θ = 1.06
7. Pond depth = 1.8 m (6 ft)
8. Pond dispersion factor = 0.5
9. Overall BOD_5 removal efficiency = 80 percent

SOLUTION

1 From Fig. 5-12, determine the value of kt for a dispersion factor of 0.5 and a BOD_5 removal efficiency of 80 percent.

$$kt = 2.4$$

2 Determine the temperature coefficient for summer and winter conditions.
   a Winter:

$$k_{15} = (0.25 \ d^{-1})[(1.06)^{15-20}] = 0.187 \ d^{-1}$$

   b Summer:

$$k_{25} = (0.25 \ d^{-1})[(1.06)^{25-20}] = 0.335 \ d^{-1}$$

3 Determine the detention time for winter and summer conditions.
   a Winter:

$$(0.187 \ d^{-1})(t) = 2.4$$

$$t = 12.8 \ d$$

   b Summer:

$$(0.335 \ d^{-1})(t) = 2.4$$

$$t = 7.2 \ d$$

4 Determine the pond volumes and surface requirements.
   a Winter:

$$\text{Volume} = (3800 \ m^3/d)(12.8)$$
$$= 48,640 \ m^3 \ (1.72 \times 10^6 \ ft^3)$$

$$\text{Surface area} = \frac{48,640 \ m^3}{1 \ m}$$

$$= 48,640 \ m^2 = 4.86 \ ha \ (12.0 \ acre)$$

   b Summer:

$$\text{Volume} = 27,360 \ m^3 \ (9.66 \times 10^5 \ ft^3)$$

$$\text{Surface area} = 27,360 \ m^2 = 2.74 \ ha \ (6.8 \ acre)$$

Therefore, winter conditions control the design.

5 Determine the surface loading.

$$kg\ BOD_5/ha \cdot d = \frac{(3800\ m^3/d)(200\ g/m^3)(10^3\ g/kg)^{-1}}{4.86\ ha}$$

$$= 156\ kg/ha \cdot d$$

6 Determine the power requirements for the surface aerators. Assume that the capacity of the aerators in terms of oxygen transferred is to be twice the value of the BOD_5 applied per day and that a typical aerator will transfer about 24 kg $O_2/kW \cdot d$.

$$kg\ O_2/d\ required = 2(3800\ m^3/d)(200\ g/m^3) \times (10^3\ g/kg)^{-1}$$

$$= 1520\ kg/d\ (3351\ lb/d)$$

$$kW = \frac{1520\ kg/d}{24\ kg/kW \cdot d} = 63.3\ kW\ (84.9\ hp)$$

Use six 10-kW units.

7 Check the power input to determine if mixing will occur.

$$kW/10^3\ m^3 = \frac{60\ kW}{48.6 \times 10^3\ m^3} = 1.23\ kW/10^3\ m^3$$

Regardless of how the ponds are operated (for example, series, parallel), the power required to keep the surface aerated will not be sufficient to mix the pond contents (about 3 to $kW/10^3\ m^3$ is about the absolute minimum required).

Anaerobic ponds The design of anaerobic stabilization ponds follows the principles presented in Chap. 9 and previously in this chapter, so a design example will not be presented. Because anaerobic ponds are similar to anaerobic digesters, with the exception of mixing, the process design methods outlined in Chap. 11 should be reviewed.

Pond systems Pond systems such as those previously discussed by applying the aforementioned equations sequentially, taking into account recirculation where it is used. Stabilization ponds may be used in parallel or series arrangement to achieve special objectives. Series operation is beneficial where a high level of BOD or coliform removal is important. The effluent from aerobic-anaerobic ponds in series operation has a much lower algal concentration than that obtained in parallel operation, with a resultant decrease in color and turbidity. Many serially operated multiple-unit installations have been designed to provide complete treatment or complete retention of the wastewater, the liquid being evaporated into the atmosphere or percolated into the ground. Parallel units provide better distribution of settled solids. Smaller units are conducive to better circulation and have less wave action. The additional cost of equipping units for both series and parallel operation is usually nominal. In some instances, actual savings can be demonstrated because of the lesser volume of earthmoving needed to adapt two or more smaller units to the topography.

Recirculation Recirculation of pond effluent has been used effectively to improve the performance of pond systems operated in series. Occasionally, internal recirculation is used. If three aerobic-anaerobic ponds are used in series, the normal mode of operation involves recirculating effluent from either the second or third pond to the first pond. The same situation applies if an anaerobic pond is substituted for the first aerobic-anaerobic pond. Recirculation rates varying from 0.5 to 2.0 Q (plant flow) have been used. If recirculation is to be considered, it is recommended that the pumps have a capacity of at least Q.

Solids Separation

To meet the prescribed requirements for secondary treatment of the U.S. Environmental Protection Agency (see Table 4-1), some type of solids removal facilities must be used in conjunction with most stabilization ponds [29, 56]. The principal means that have been used or that are under investigation and development are (1) conventional and earthen settling basins, (2) chemical precipitation, (3) dissolved-air flotation, (4) fine screening, (5) intermittent sand filters, (6) rock filters, and (7) a specialized algal harvesting device.

Conventional and earthen settling basins Both conventional and earthen settling basins can be used in conjunction with stabilization ponds. Where conventional facilities are contemplated, the use of an earthen-lined sedimentation basin, such as shown in Fig. 10-23, should be investigated. If space is available, large earthen settling basins can be used. Actually, these basins serve as maturation ponds, as discussed previously. Where such basins are used, adequate provision must be made for sludge storage.

Chemical precipitation In general, adding chemicals to the entire flow from a stabilization pond using conventional facilities is not especially cost-effective. To overcome the difficulties associated with sludge handling from conventional facilities, Friedman (11) has suggested an alternative approach which takes advantage of the cyclic pH variation that occurs in a pond. From experimental studies, he found that flocculation of the algal cell tissue is greatly enhanced at pH values above 9. Using this finding, he has proposed adding magnesium hydroxide [$Mg(OH)_2$] to the contents of a pond late in the afternoon, when pH values between 9 and 11 are reached under natural conditions, to form a $Mg(OH)_2$ precipitate.

To achieve effective separation of the algal cells, a two-pond system is required. Late in the afternoon when the pH is highest, supernatant is drawn off from the first "active" pond and directed to a second pond used for settling. As the effluent flows into the second pond, $Mg(OH)_2$ is added to raise the pH and to bring about the precipitation of $Mg(OH)_2$, which in turn clarifies the liquid by its sweep action as it settles. The required $Mg(OH)_2$ dosage appears to be no less than 10 mg/L. It seems that this system could be used in conjunction with one of the other solids separation systems, such as fine screening, to develop a workable year-round process for the removal of solids from stabilization ponds.

Flotation The clarification of algal cells from lagoon effluents with flotation may be considered a natual extension of the attachment of minute bubbles of internally produced oxygen. Chemicals such as alum, ferric sulfate, and polyelectrolytes are generally required to create floc structures amenable to flotation.

The separation of suspended materials by flocculation and flotation is closely related to chemical dosage. However, the relationship between chemical requirements and suspended-solids removal is extremely variable for different wastewater effluents. Typical dosages of 50 to 200 mg/L of alum are required for the removal of 70 to 90 percent of the suspended solids. The effective chemical dose for flotation may be a function of wastewater characteristics other than algal concentration.

The essential parameters controlling process design and performance are the air-to-solids ratio (defined as the kilograms of air released per kilogram of solids in the effluent), hydraulic loading, aeration pressure, and detention time. Typical operating conditions are, respectively: 0.01 to 0.08 kg air/kg solids; 50 to 150 $m^3/m^2 \cdot d$ (1200 to 3600 gal/ft$^2 \cdot$ d); 70 to 345 kN/m^2 (10 to 50 lb/in^2); and 20 to 30 min. Optimum conditions have been reported with 25 percent pressurized recycle flows.

The float-solids concentration may vary according to chemical dose, aeration pressure, detention time, and the operation efficiency of the skimming device. Typical float solids concentrations are 1 to 2 percent. Further dewatering of float solids has been obtained by dewatering the algal slurry on a sand bed.

Fine screening A logical application of fine screening, involving the use of microscreens and centrifugal screens, would appear to be the removal of algae from stabilization pond effluents. Unfortunately, the results obtained with these devices have been variable, depending on the algal community in the pond effluent. The results of a study of the facultative ponds at Davis, Calif., are summarized in Table 10-13 [50]. As shown, the shifts in the algal community are quite dramatic.

From the results of this study and others, it appears that fine screens could be coupled with one of the other separation methods to provide a very effective treatment system. For example, additional removals could be obtained during periods when the performance of the screens is not adequate by using slow sand filters. Another advantage of fine screens is that the screened solids can be recirculated to improve the performance of the stabilization pond(s).

Intermittent sand filtration The slow sand-filtration process has been used for domestic wastewater treatment for over a century and a half. The process may be more correctly described as the modification of intermittent application of wastewater and now is more commonly referred to as intermittent sand filtration. Long-term experience with the process has resulted in standard operating practices

Table 10-13 Identification of algal species in facultative pond effluent, Davis, Calif., 1975 [50]

	Jan. 8	Jan. 29	Feb. 7	Mar. 4	Mar. 31	Apr. 18	Apr. 21	Apr. 30	May 20	June 12	Jul. 23	Sept 4	Sept. 29	Oct. 15	Oct. 22
Algal genera															
Aphanizomenon	12,400[a]	12,700	12,000	11,800	11,000	8,400	3,300	1,200	600	300	0	200	5,200	25,600	15,600
(size)	130[b]	130	130	160	160	160	160	160	50	25	...	10	20	25	25
Anacystis	0	0	0	0	0	200	2,100	12,700	11,500	9,400	7,900	8,300	900	100	0
(size)	...	...	...	...	...	5	5	5	25	60	60	70	c[c]	c[c]	0
Ankistrodesmus	200	300	400	1200	400	100	0	0	0	0	0	100	1,200	600	100
(size)	d[d]	d[d]	d[d]	d[d]	d[d]	5	5	0	25	0	0	10	20	25	15
Chlamydomonas	0	0	0	0	0	0	0	0	0	0	100	400	1,100	800	600
(size)	...	...	...	...	...	...	...	...	...	...	3	3	3	3	3
Chlorella	0	0	0	0	0	0	0	0	0	0	Trace	200	47,000	1,400	0
(size)	...	...	...	...	...	...	...	...	...	...	3	5	5	5	3
Golenkinia	Trace	400	700	200	200	800	500	Trace	Trace	0	0	0	0	0	0
(size)	...	5	5	7	7	10	10	7	5	...	...	5	5	5	...
Scenedesmus	400	500	700	700	1,100	2,100	3,000	1,600	0	0	0	200	200	300	300
(size)	3 × 7	3 × 7	5 × 10	5 × 10	5 × 10	7 × 15	7 × 15	7 × 15	...	...	...	3 × 7	3 × 7	3 × 7	3 × 7
Pond quality															
COD, mg/L	128	108	111	134	146	210	243	302	327	315	285	240	193	194	185
SS, mg/L	54	61	51	50	55	130	131	166	178	161	150	142	88	81	66

[a] Colonies or cells per mL.
[b] Average size, μ.
[c] Very irregular shape.
[d] Fusiform shape.

Note: mg/L = g/m³

for filter surface and media cleansing. Intermittent sand filters are operated by applying pond effluent on a periodic or intermittent basis until a predetermined head loss limited to the available freeboard is reached. At this point, the bed is drained and cleaned. Most designs call for adequate underdraining and collection, and for subsequent disinfection prior to disposal. Removal and replacement of the top layer of clogged sand has been shown to provide the most effective means of restoring the filtering ability of the sand.

The efficiency of the process is primarily a function of the algal growths on the filter surface. Slow sand infiltration, in effect, is a combined biological and filtration process. Effluent quality from the treatment of typical domestic wastewater has been well above that required by the secondary treatment standards of the U.S. Environmental Protection Agency. BOD and suspended-solids values of about 5 mg/L have been reported for filter effluent [16]. The use of intermittent sand filters for nitrification and filtration applications also appears to be a very competitive alternative as a tertiary treatment option where land costs are not high.

Hydraulic loading rates depend on the effective medium size. The frequency of cleansing depends on the solids content and characteristics of the influent wastewater to the process. The minimum desirable filter bed depth has been shown to be 45 cm (18 in). Recent applications of intermittent sand filtration have shown that alternative loading criteria are required, depending on the upstream treatment processes used. For more information on the performance of intermittent sand filters, references [16] and [30] are recommended.

Rock filters Rock or coarse-medium filtration has been developed primarily for the removal of algae cells from lagoon effluents (see Fig. 10-32). The preferred mode of operation generally provides for the horizontal flow of lagoon effluent through a slime-layered rock bed (typically a modified tertiary pond dike). Final effluent is collected with a horizontal pipe containing evenly spaced holes located at the lagoon water surface elevation [32]. Adaptations of the concept are now used at a number of locations [29].

The principal design concept involves the buildup, clogging, and sloughing of suspended solids on the filter.

Consideration must be given to the potential production of hydrogen sulfide and other odorous compounds and to a net increase in the concentration of ammonia nitrogen. It appears that odors are more likely to develop where the total alkalinity of the wastewater is below about 200 mg/L [33]. Effluent performance data at several locations where rock filters are used to polish lagoon effluents have been encouraging (effluent concentrations of about 30 mg/L BOD_5 and 30 mg/L total suspended solids). The potential usefulness of this technique to upgrade lagoon performance to meet secondary treatment requirements appears highly favorable. In lagoons with a very large algal population, the submerged rock filter can be used as a roughing device to reduce the load on subsequent solids separation facilities [33].

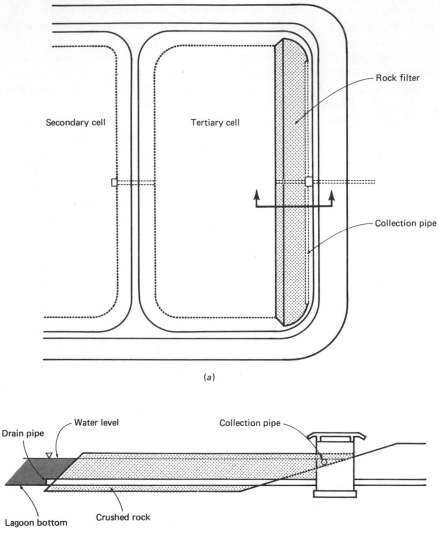

(a)

(b)

Figure 10-32 Rock filter for the separation of solids from effluent from facultative pond [32]. (a) Plan of rock filter. (b) Section through rock filter.

Algal harvester An ingenious device that is similar to a paper making machine has been developed to remove the algae from high-rate ponds. Essentially, a thin layer of paper precoat is formed on a fabric belt through which the algae-laden water is filtered. The wet paper mat containing the filtered algae is subsequently dried and removed from the fabric belt. The product can then be processed and used as an animal feedstock [9].

Combination facilities From a careful review of all the methods used to separate solids from stabilization pond effluents, it appears that the major limitation is that each one is evaluated on an all-or-nothing basis. Many feasible systems derived from combining one or more of the processes are thereby often overlooked. For example, fine screens that work effectively for 8 to 10 months of the year are rejected because they do not work for a 2-month period. Clearly, what is required is a more integrated approach to the evaluation of the various methods that are now available for the separation of solids from stabilization pond effluents.

Design of Physical Facilities

Although the process design for ponds is imprecise, careful attention must be given to the design of the physical facilities to ensure optimum performance. Factors that should be considered include (1) location of influent lines, (2) outlet structure design, (3) dike construction, (4) liquid depth, (5) treatment of lagoon bottom, and (6) control of surface runoff.

Influent lines For small ponds, a center inlet is preferred. For ponds of 4 ha (10 acres) or more, the inlet can be installed 120 m (400 ft) from the dike. For large aerobic-anaerobic ponds, multiple inlets are desirable to distribute the settleable solids over a larger area. For increased flexibility, movable inlets are being used more frequently.

Outlet structure(s) The outlet structure(s) should permit lowering the water level at a rate of less than 0.3 m/week (1 ft/week) while the facility is receiving its normal load. It should be large enough to permit easy access for normal maintenance. Provision for complete draining of the pond is desirable. During ice-free periods, discharge should be taken just below the water surface. This releases effluent of the highest quality and ensures retention of floating solids. For flow-through ponds, the maximum rate of effluent discharge is considerably less than the rate of peak wastewater flow, because of pond losses and the leveling out of peak flows. Overflow structures generally comparable to a sewer manhole are most frequently used, and selected level discharge is facilitated through valved piping or other adjustable overflow devices. Overflow lines should be vented if the design would otherwise permit siphoning.

Dikes Dikes should be constructed in a way that minimizes seepage. Compaction afforded by the use of conventional construction equipment is usually adequate. Vegetation should be removed, and the area upon which the embankment is to be placed should be scarified. It is generally unnecessary to key the dikes into impervious subsoil, but this precaution may be advisable for sandy topsoils.

The dike should be wide enough to accommodate mowing machines and other maintenance equipment. A width of 2.5 m (8 ft) is generally considered adequate, and narrower dikes may be satisfactory for small installations. Slopes are influenced by the nature of the soil and the size of the installation. For outer slopes, a 3 horizontal to 1 vertical is satisfactory. Inner slopes are generally from 1 vertical to 3 to 4 horizontal, although slopes exceeding 1 to 5 for larger installations and 1 to 3 for smaller installations are sometimes specified.

The freeboard is to some extent influenced by the size and shape of the installation, because wave heights are greater on larger bodies of water. One meter (3 ft) above maximum liquid level is usually specified as minimum freeboard, but 0.6 m (2 ft) is considered adequate by some states, particularly for installations of 2.5 ha (6 acres) or less not exposed to severe winds.

Liquid depth Optimum liquid depth for adequate circulation is influenced to some extent by the pond area; greater depth is allowed for large units. Shallow ponds encourage the growth of emergent vegetation and consequently may foster mosquito breeding.

There is a distinct advantage for facilities that permit operation at selected depths up to 1.5 m (5 ft), and provision for additional depth may be desirable for large installations. Facilities for adjusting pond levels can be provided at small cost. For ponds 12.0 ha (30 acres) or larger, provision for periodic operation at depths greater than 1.5 m may be advantageous.

Pond bottom The bottom of aerobic and most aerobic-anaerobic ponds should be made as level as possible except around the inlet. The finished elevation should not vary more than 15 cm (6 in) from the average elevation of the bottom. An exception to this is where the bottom of an aerobic-anaerobic pond is designed specifically to retain the settleable solids in hoppered compartments or cells. The bottom should be well compacted to avoid excessive seepage. Where excessive percolation is a problem, increased hydraulic loading or partial sealing may merit consideration to maintain a satisfactory water level in the lagoon. Porous areas, such as gravel pockets, should receive particular attention.

Surface-runoff control Ponds should not receive significant amounts of surface runoff. If necessary, provision should be made for diverting surface water around the ponds. For new installations, where maintenance of a satisfactory water depth is a problem, the diversion structure may be designed to admit surface runoff to the lagoon when necessary.

DISCUSSION TOPICS AND PROBLEMS

10-1 In Example 10-1, compute the required quantities of nitrogen and phosphorus if the nitrogen requirement is $0.12P_x$ and the phosphorus requirement is one-fifth of the nitrogen requirement. In what forms should these nutrients be added?

10-2 Prepare a one-page abstract of the following article: R. E. McKinney and W. J. O'Brien: Activated Sludge—Basic Design Concepts, *J. WPCF*, vol. 40, no. 11, part 1, 1968. Prepare a one-page summary of the more important design criteria given in the article.

10-3 Prepare a one-page abstract of the following article: M. T. Garrett, Jr.: Hydraulic Control of Activated Sludge Growth Rate, *Sew. Ind. Wastes*, vol. 39, no. 3, 1958.

10-4 A conventional activated-sludge plant is to treat 4000 m^3/d of wastewater having a BOD_5 of 200 mg/L after settling. The process loading is 0.30 kg BOD/d · kg MLVSS. The detention time is 6 h and the recirculation ratio is 0.33. Determine the value of MLVSS.

10-5 A conventional activated-sludge plant is operated at a mean cell residence time of 10 d. The reactor volume is 8000 m^3, and the MLSS concentration is 3000 mg/L. Determine (*a*) the sludge production rate, (*b*) the sludge-wasting flowrate when wasting from the reactor, and (*c*) the sludge-wasting flowrate when wasting from the recycle line. Assume that the concentration of suspended solids in the recycle is equal to 10,000 mg/L.

10-6 The step-aeration activated-sludge system shown in Fig. 10-5 is to be analyzed as a series of continuous-flow stirred-tank reactors (see Fig. 10-33). Using the design parameters given below, determine the MLVSS concentration in each tank.

$$V = 240 \text{ m}^3$$
$$S_1 = 4 \text{ mg/L} \qquad S_2 = 6 \text{ mg/L}$$
$$S_3 = 8 \text{ mg/L} \qquad S_4 = 10 \text{ mg/L}$$
$$S_0 = 250 \text{ mg/L}$$
$$Q_0 = 4000 \text{ m}^3/\text{d} \qquad Q_r = 800 \text{ m}^3/\text{d}$$
$$X_r = 10,000 \text{ mg/L}$$
$$Y = 0.65 \qquad k_d = 0.05$$

10-7 The following data were obtained from an aerator test performed in tap water at a temperature of 7.5°C. Determine the value of $K_l a$ expressed in terms of h^{-1} using both Eqs. 7-19 and Eq. 7-21.

Time, min	C, mg/L
0	0
5	1.8
10	3.2
15	4.4
20	5.5
25	6.4
30	7.2
35	7.9
40	8.4

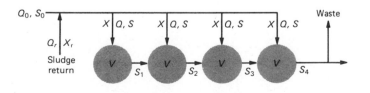

Figure 10-33 Flowsheet for step-aeration system used in Prob. 10-6.

10-8 Using the $K_L a$ value determined in Prob. 10-7, estimate the maximum strength of waste that could be treated in a complete-mix activated-sludge process if the observed value of the yield is equal to 0.35 mg/mg.

10-9 Determine the temperature of the wastewater in a 32-ha aerated lagoon. Wastewater is discharged to the lagoon at a rate of 2000 m^3/d. Use a typical f value of 2.5 for the midwestern United States. The temperature of the air is 10°C, and the temperature of the incoming wastewater is 20°C.

10-10 Design an aerated lagoon to treat 10,000 m^3/d of wastewater under the following conditions:
  (a) Influent soluble BOD_5 and suspended solids = 150 mg/L
  (b) Overall first-order BOD_5 removal-rate constant = 2.0 d^{-1} at 20°C
  (c) Summer temperature = 27°C
  (d) Winter temperature = 7°C
  (e) Wastewater temperature = 15°C
  (f) Temperature coefficient = 1.07
  (g) $\alpha = 0.85$, $\beta = 1.0$
  (h) Elevation = 1250 m
  (i) Oxygen concentration to be maintained = 2.0 mg/L
  (j) Lagoon depth = 2.5 m
  (k) Hydraulic residence time = 10 d
  (l) Temperature proportionality constant $f = 0.5$

Determine the surface area, summer and winter temperatures in the lagoon, and the effluent BOD_5 in summer and winter. If the growth yield is approximately 0.5 (BOD_5 basis), determine the biological-solids concentration in the lagoon, the oxygen requirements, and the power requirements for summer and winter conditions. Use surface aerators rated at 1.5 kg $O_2/kW \cdot h$.

10-11 Prepare a plot of the natural draft available in a 6.5-m tower trickling filter for air flow in the upward and downward direction between the ambient air and the air within the filter. Use (1) the wastewater temperature and (2) the log mean temperature as estimates of the air temperature within the filter. Assume that the maximum temperature difference between the ambient air and the wastewater is ±15°C and the temperature of the wastewater is 26°C.

10-12 A tower trickling filter 6.5 m in height is to be used to treat a combined domestic-industrial waste. The temperature of the wastewater is 26°C. Using the following temperature data taken at the plant site and the plot developed in Prob. 10.10, determine the air draft available throughout the day in centimeters of water. Use the log-mean estimate for the air temperature within the filter.

Time	Air temperature, °C
Midnight	22.0
2	18.4
4	16.6
6	17.3
8	22.0
10	32.0
Noon	40.8
2	42.0
4	41.0
6	38.5
8	32.0
10	26.0
Midnight	22.0

If a bulk-air flowrate through the filter of 0.1 $m^3/m^2 \cdot min$ is needed to meet the oxygen requirements of the waste, determine the number of hours during each day that the air flow will be
insufficient. Assume that the area of the vent openings in the bottom of the filter is equal to 5
percent of the filter surface area and that the headloss through the filter can be approximated
using the following expression:

$$h_F = 0.030 \frac{V_v^2}{2g} \frac{\rho_a}{\rho_w}$$

where h_F = headloss through the filter, cm of H_2O
$\quad$ V_v = air flow velocity through the vent, m/min
$\quad$ g = acceleration due to gravity = 9.8 m/s^2
$\quad$ ρ_a = density of air, kg/m^3
$\quad$ ρ_w = density of water, kg/m^3

10-13 An industrial waste is to be treated with a tower trickling filter followed by an activated sludge
process. Primary settling will not be used. The packing medium in the tower trickling filter is to be
plastic and the operational mean cell residence time for the activated sludge process is to be 5 d during
the critical summer period and vary from 5 to 15 d during the winter. The lowest average sustained
winter temperature (at least two weeks) is 5°C and the highest average sustained summer temperature is 26°C. The characteristics of the industrial waste, data derived from pilot plant studies, and
related design data are presented below. Using these data, size the units and determine the concentration of mixed liquor suspended solids to be maintained during summer and winter operation;
the recycle rates around the filter and activated sludge process; the quantity of sludge to be disposed;
and the quantity of nutrients that must be added. Assume the flowrate, which is 20,000 m^3/d, has been
equalized.

Wastewater characteristics:
$\quad$ BOD_5 = 1200 mg/L
$\quad\quad$ SS = 100 mg/L
$\quad\quad$ VSS = 0 mg/L
$\quad$ Total nitrogen as N = 10 mg/L
$\quad$ Total phosphorus as P = 4 mg/L
$\quad$ Total iron as Fe = 0.15 mg/L

Trickling filter pilot plant data:
$\quad$ $K_{20°C}$ = 0.075 m/d
$\quad$ Y (BOD_5) = 0.70 mg/mg
$\quad\quad$ θ = 1.06

Activated sludge pilot plant data:
$\quad$ Y (BOD_5) = 0.8 mg/mg
$\quad\quad$ k_d = 0.1 d^{-1}
$\quad\quad$ k = 6.0 d^{-1}
$\quad\quad$ K_s = 90 mg/L
$\quad\quad$ θ = 1.035

Design parameters:
$\quad$ w (for trickling filter) = 1.0 cm/cm^2 of trickling filter cross-sectional area
$\quad\quad\quad$ θ_c^d = 5 d (critical summer period)
$\quad\quad\quad$ θ_c^d = 5 – 15 d (winter)

10-14 Review at least four articles dealing with the design of rotating biological contactors and
prepare a summary table of the design parameters presented in the articles. Cite the articles reviewed.

10-15 Using the design parameters summarized in Prob. 10-14, design a treatment process using
rotating biological contactors to treat a wastewater with the characteristics given in Example 10-1.
Compare and contrast your process design to that given in Example 10-1.

10-16 Design an aerobic stabilization pond to treat 10,000 m^3/d of wastewater with a BOD_5 removal efficiency of 90 percent under the following conditions:

    (a) Influent $BOD_5 = 250$ mg/L

    (b) Overall first-order BOD_5 removal-rate constant $= 0.2$ d^{-1} at 20°C

    (c) Pond temperature in summer $= 30$°C

    (d) Pond temperature in winter $= 12$°C

    (e) Temperature coefficient $= 1.06$

    (f) Maximum pond area $= 4.0$ ha

    (g) Maximum pond depth $= 1.5$ m

    (h) Pond dispersion factor $= 0.5$

Determine the detention times and area requirements for summer and winter conditions.

REFERENCES

1. Adams, C. E. Jr., and W. W. Eckenfelder, Jr. (eds.): *Process Design Techniques for Industrial Waste Treatment*, Enviro, Nashville, 1974.
2. Bartsch, E. H., and C. J. Randall: Aerated Lagoons—A Report on the State of the Art, *J. WPCF*, vol. 43, no. 4, April 1971.
3. Benjes, Henry: Theory of Aerated Lagoons, *Second International Symposium for Waste Treatment Lagoons*, Kansas City, June 1970.
4. Bewtra, J. K., and W. R. Nicholas: Oxygenation from Diffused Air in Aeration Tanks, *J. WPCF*, vol. 36, no. 10, 1964.
5. Bruce, A. M., and J. C. Merkens: "Further Studies of Partial Treatment of Sewage by High-Rate Biological Filtration," *J. Inst. Water Pollut. Contr.*, vol. 72, no. 5, London, 1973.
6. Caldwell Connell Engineers: *Algae Harvesting from Sewage*, Environmental Study Report 1, Bureau of Environmental Studies, Department of Environment, Housing and Community Development, Canberra, Australia, 1976.
7. Caropreso, F., C. W. Raleigh, and J. C. Warner: *Attack with H_2O_2 and Microscope*, California Water Pollution Control Association Bulletin, July 1974.
8. Dick, R. I.: Folklore in the Design of Final Settling Tanks, *J. WPCF*, vol. 48, no. 4, April 1976.
9. Dodd, J. C.: "Harvesting Algae with a Paper Precoated Belt-type Filter with Integral Dewatering and Drying," Ph.D. dissertation, University of California, Davis, 1972.
10. Eckenfelder, W. W., Jr.: *Industrial Water Pollution Control*, McGraw-Hill, New York, 1966.
11. Friedman, A. A.: Seminar on Algae Removal from Oxidation Ponds, University of California, Davis, June 28, 1975.
12. Gloyna, E. F.: Basis for Waste Stabilization in Pond Designs, in E. F. Gloyna and W. W. Eckenfelder, Jr. (eds.), *Advances in Water Quality Improvement*, University of Texas Press, Austin, 1968.
13. Gloyna, E. F.: *Waste Stabilization Ponds*, World Health Organization Monograph Series No. 60, Geneva, 1971.
14. Great Lakes-Upper Mississippi River Board of State Sanitary Engineers: *Recommended Standards for Sewage Works (Ten State Standards)*, 1960.
15. *Handbook of Trickling Filter Design*, Public Works Journal Corporation, Ridgewood, N. J., 1968.
16. Harris, S. E., et al.: Intermittent Sand Filtration for Upgrading Waste Stabilization Pond Effluents, *J. WPCF*, vol. 49, no. 83, January 1977.
17. Kimball, J. W.: *Biology*, 2d ed., Addison-Wesley, Reading, Mass., 1968.
18. Kraus, L. S.: A Rugged Activated Sludge Process, paper presented at the Annual Meeting of the Ohio Sewage and Industrial Waste Treatment Conference, 1955.
19. Luria, S. E.: The Bacterial Protoplasm: Composition and Organization, in I. C. Gunsalus and R. Y. Stanier (eds.), *The Bacteria I*, Academic, New York, 1960.
20. Mancini, J. L., and E. L. Barnhart: Industrial Waste Treatment in Aerated Lagoons, in E. F. Gloyna and W. W. Eckenfelder, Jr. (eds.), *Advances in Water Quality Improvement*, University of Texas Press, Austin, 1968.

21. Marais, G. V. R., and V. A. Shaw: A Rational Theory for the Design of Sewage Stabilization Ponds in Central and South Africa, *Trans. Civ. Eng. S.A.*, Johannesburg, vol. 3, p. 205, 1961.
22. Marais, G. R., and Capri, M. J.: A Simplified Kinetic Theory for Aerated Lagoons, *Second International Symposium for Waste Treatment Lagoons*, Kansas City, June 1970.
23. McCabe, B. J., and W. W. Eckenfelder, Jr.: *Biological Treatment of Sewage and Industrial Wastes*, vol. 2, Reinhold, New York, 1958.
24. McKinney, R. E., and W. J. O'Brien: Activated Sludge—Basic Design Concepts, *J. WPCF*, vol. 40, no. 11, part 1, 1968.
25. McKinney, R. E.: *Waste Treatment Lagoons—State of the Art*, U.S. Environmental Protection Agency Project No. 17090 EXH, 1971.
26. Mehta, D. S., H. H. Davis, and R. P. Kingsburg: Oxygen Theory in Biological Treatment Plant Design, *J. San. Eng. Div.*, ASCE, vol. 98, no. SA3, June 1972.
27. Metcalf & Eddy, Inc.: *Wastewater Engineering: Collection, Treatment, Disposal*, McGraw-Hill, 1972.
28. Metcalf, L., and H. P. Eddy: *American Sewerage Practice*, 3d ed., vol. III, McGraw-Hill, New York, 1935.
29. Middlebrooks, E. J., et al., (eds.): *Upgrading Wastewater Stabilization Ponds to Meet New Discharge Standards*, Symposium Proceedings, Utah State University, Logan, Utah, 1974.
30. Middlebrooks, E. J., J. H. Reynolds, and C. II. Middlebrooks: *Performance and Upgrading of Wastewater Stabilization Ponds*, U.S. Environmental Protection Agency, Technology Transfer Design Seminar for Small Wastewater Treatment Systems.
31. National Research Council: Trickling Filters (in Sewage Treatment at Military Installations), *Sewage Works J.*, vol. 18, no. 5, 1946.
32. O'Brien, W. J.: Algae Removal by Rock Filtration, in *Transactions Twenty-Fifth Annual Conference on Sanitary Engineering*, University of Kansas, 1975.
33. O'Brien, W. J.: Personal communication, 1976.
34. Oswald, W. J., and H. B. Gotaas: Photosynthesis in Sewage Treatment, *Trans. ASCE*, vol. 122, 1957.
35. Oswald, W. J.: Fundamental Factors in Stabilization Pond Design, *Advance in Waste Treatment*, Pergamon, New York, 1963.
36. Oswald, W. J.: Advances in Anaerobic Pond System Design, in E. F. Gloyna and W. W. Eckenfelder, Jr. (eds.), *Advances in Water Quality Improvement*, University of Texas Press, Austin, 1968.
37. Otte, G. B.: "Design of Waste Stabilization Ponds." Masters Thesis, Department of Civil Engineering, University of California, Davis, 1974.
38. *Oxygen Activated Sludge Wastewater Treatment Systems: Design Criteria and Operating Experience*, U.S. Environmental Protection Agency, Technology Transfer Seminar Publication, rev. ed. January 1974.
39. Parker, D. S., and M. S. Merrill: Oxygen and Air Activated Sludge: Another View, *J. WPCF*, vol. 48, no. 11, November 1976.
40. Pflanz, P.: Performance of (Activated Sludge) Secondary Sedimentation Basins, in *Proceedings of the Fourth International Conference, International Association on Water Pollution Research*, Prague, 1969, p. 569.
41. *Process Design Manual for Nitrogen Control*, U.S. Environmental Protection Agency, Office of Technology Transfer, Washington, D.C., October 1974.
42. *Process Design Manual for Upgrading Existing Wastewater Treatment Plants*, U.S. Environmental Protection Agency, Office of Technology Transfer, Washington, D.C., October 1974.
43. Roskopf, R. F., J. C. Young, and E. R. Baumann: Trickling Filter–Activated Sludge Combinations, *J. Environ. Eng. Div.*, ASCE, vol. 102, no. EE5, October 1976.
44. Sawyer, C. N.: Final Clarifiers and Clarifier Mechanism, in B. J. McCabe and W. W. Eckenfelder, Jr. (eds.), *Biological Treatment of Sewage and Industrial Wastes*, vol. 1, Reinhold, New York, 1956.
45. Sawyer, C. N.: New Concepts in Aerated Lagoon Design and Operation, in E. F. Gloyna and W. W. Eckenfelder, Jr. (eds.), *Advances in Water Quality Improvement*, University of Texas Press, Austin, 1968.

46. Schroeder, E. D.: *Water and Wastewater Treatment*, McGraw-Hill, New York, 1977.
47. Schroeder, E. D., and G. Tchobanoglous: Mass Transfer Limitations on Trickling Filter Design, *J. WPCF*, vol. 48, no. 4, April 1976.
48. Stahl, J. F.: Pure Oxygen and Diffused Air Activated Sludge at Hyperion, paper presented at the 48th Annual Conference of the California Water Pollution Control Association, South Lake Tahoe, Calif., 1976.
49. Stall, T. R., and Sherrad, J. H.: Evaluation of Control Parameters for the Activated Sludge Process, *J. WPCF*, vol. 50, no. 3, March 1978.
50. Stowell, R.: A Study of the Screening of Algae from Stabilization Ponds Masters Thesis, Department of Civil Engineering, University of California, 1976.
51. Tchobanoglous, G. and B. Ostertag: Waste Management at Hickmott Foods, Inc.—A Case Study of a Small Tomato Cannery. *Proceedings of the 30th Industrial Waste Conference*, 1975, Ann Arbor Science Publishers Inc., Ann Arbor, Mich., 1977.
52. Technical Practice Committee, Subcommittee on Aeration in Wastewater Treatment: *Aeration in Wastewater Treatment*, WCPF Manual of Practice 5, Washington, D.C., 1971.
53. *The Sewerage Manual*, published biennially by *Public Works*.
54. Thirumurthi, D.: Design of Waste Stabilization Ponds, *J. San. Eng. Div.*, ASCE, vol. 95, no. SA2, 1969.
55. Torpey, W. N., and A. H. Chasick: Principles of Activated Sludge Operation, in B. J. McCabe and W. W. Eckenfelder, Jr. (eds.), *Biological Treatment of Sewage and Industrial Wastes*, vol. 1, Reinhold, New York, 1956.
56. *Upgrading Lagoons*, U.S. Environmental Protection Agency, Technology Transfer Seminar Publication, 1973.
57. Water Pollution Control Federation: *Wastewater Treatment Plant Design*, Manual of Practice No. 8, Water Pollution Control Federation, Washington, D.C., 1977.
58. Wood, D. K., and G. Tchobanoglous: Trace Elements in Biological Waste Treatment, *J. WPCF*, vol. 47, no. 7, July 1975.

CHAPTER ELEVEN

DESIGN OF FACILITIES FOR THE
TREATMENT AND DISPOSAL OF SLUDGE

The constituents removed by a variety of methods in wastewater-treatment plants are important by-products of the treatment process. They include screenings, grit, scum, and sludge. The sludge resulting from wastewater-treatment operations and processes is usually in the form of a liquid or semisolid liquid, which typically contains from 0.25 to 12 percent solids, depending on the operations and processes used. Of the constituents removed by treatment, sludge is by far the largest in volume, and its processing and disposal is perhaps the most complex problem facing the engineer in the field of wastewater treatment. For this reason, a separate chapter has been devoted to this subject. The disposal of grit and screenings was discussed in Chap. 8, where it was noted that screenings are often ground up and disposed of together with the sludge.

The problems of dealing with sludge are complex because (1) it is composed largely of the substances responsible for the offensive character of untreated wastewater; (2) the portion of sludge produced from biological treatment requiring disposal is composed of the organic matter contained in the wastewater but in another form, and it, too, will decompose and become offensive; and (3) only a small part of the sludge is solid matter. Therefore, the main purpose of this chapter is to describe the operations and processes that are used to reduce the water and organic content of sludges.

The principal methods now used to process and dispose of sludge are listed in Table 11-1. Thickening (concentration), conditioning, dewatering, and drying are used primarily to remove moisture from sludge; digestion, incineration, and wet oxidation are used primarily to treat the organic material in the sludge. To make the study of these operations and processes more meaningful, the first two sections of this chapter are devoted to a presentation of representative sludge-treatment flowsheets and a discussion of the sources, characteristics, and quantities of sludge. Because the pumping of sludge between processes is of fundamental importance, a separate discussion (Sec. 11-3) is devoted to sludge and scum pumping. The various operations and processes are then discussed in Secs. 11-4 through 11-14. Section 11-15 deals with the preparation of solids balances for treatment facilities, and Sec. 11-16 deals with the conveyance and ultimate disposal of the sludge or ash after processing and treatment.

Table 11-1 Sludge-processing and -disposal methods[a]

Unit operation, unit process, or treatment method	Function	See Sec.
Preliminary operations		
Sludge grinding	Size reduction	11-4
Sludge degritting	Grit removal	11-4
Sludge blending	Blending	11-4
Sludge storage	Storage	11-4
Thickening		
Gravity thickening	Volume reduction	11-5
Flotation thickening	Volume reduction	11-5
Centrifugation	Volume reduction	11-5
Stabilization		
Chlorine oxidation	Stabilization	11-6
Lime stabilization	Stabilization	11-6
Heat treatment	Stabilization	11-6
Anaerobic digestion	Stabilization, mass reduction	11-7
Aerobic digestion	Stabilization, mass reduction	11-8
Conditioning		
Chemical conditioning	Sludge conditioning	11-9
Elutriation	Leaching	11-9
Heat treatment	Sludge conditioning	11-9
Disinfection		
Disinfection	Disinfection	11-10
Dewatering		
Vacuum filter	Volume reduction	11-11
Filter press	Volume reduction	11-11
Horizontal belt filter	Volume reduction	11-11
Centrifuge	Volume reduction	11-11
Drying bed	Volume reduction	11-11
Lagoon	Storage, volume reduction	11-11

Table 11-1 (*continued*)

Unit operation, unit process, or treatment method	Function	See Sec.
Drying		
Flash dryer	Weight reduction, volume reduction	11-12
Spray dryer	Weight reduction, volume reduction	11-12
Rotary dryer	Weight reduction, volume reduction	11-12
Multiple-hearth dryer	Weight reduction, volume reduction	11-12
Oil emersion dehydration	Weight reduction, volume reduction	11-12
Composting		
Composting (sludge only)	Product recovery, volume reduction	11-13
Co-composting with solid wastes	Product recovery, volume reduction	11-13
Thermal reduction		
Multiple-hearth incineration	Volume reduction, resource recovery	11-14
Fluidized-bed incineration	Volume reduction	11-14
Flash combustion	Volume reduction	11-14
Coincineration with solid wastes	Volume reduction	11-14
Co-pyrolysis with solid wastes	Volume reduction, resource recovery	11-14
Wet air oxidation	Volume reduction	11-14
Ultimate disposal		
Landfill	Final disposal	11-16
Land application	Final disposal	11-16
Reclamation	Final disposal, land reclamation	11-16
Reuse	Final disposal, resource recovery	11-16

a Developed in part from Ref. 22.

11-1 SLUDGE-TREATMENT FLOWSHEETS

A generalized flowsheet incorporating the unit operations and processes to be discussed in this chapter is presented in Fig. 11-1. As shown, an almost infinite number of combinations are possible. In practice, the most commonly used process flowsheets for sludge treatment may be divided into two general categories, depending on whether or not biological treatment is involved. Therefore, it is important that the sanitary engineer be familiar with representative flowsheets in each category.

Typical flowsheets incorporating biological processing are presented in Fig. 11-2. Depending on the source of the sludge, either gravity or air flotation thickeners are used; in some cases, both may be used in the same plant. Following biological digestion, any of the three methods shown (vacuum filtration, centrifugation, drying beds) may be used to dewater the sludge, the choice depending on local conditions.

Because the presence of industrial and other toxic wastes has presented problems in the operation of biological digesters, a number of plants have been designed with other means for sludge treatment. Three representative process flowsheets without biological treatment are shown in Fig. 11-3.

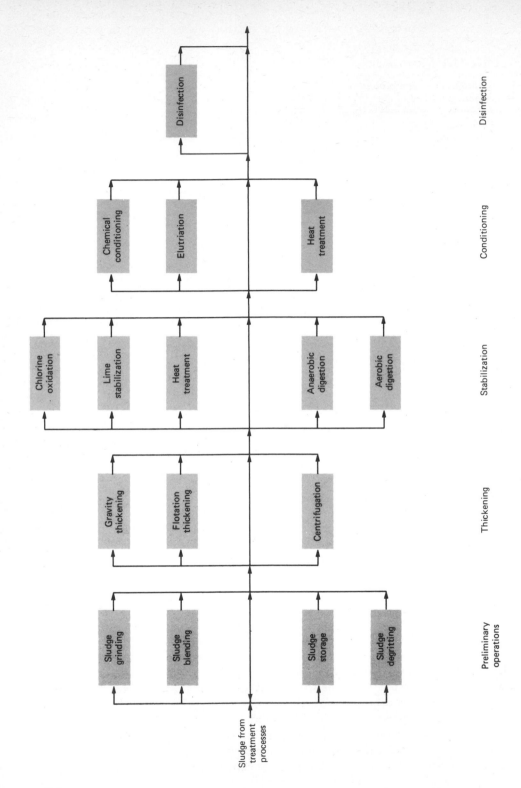

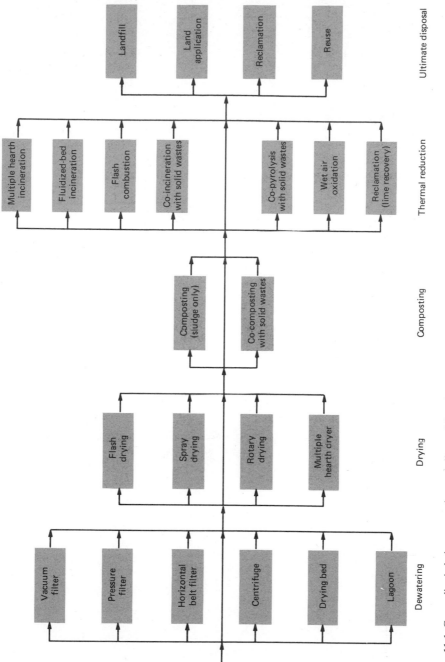

Figure 11-1 Generalized sludge-processing and disposal flowsheet.

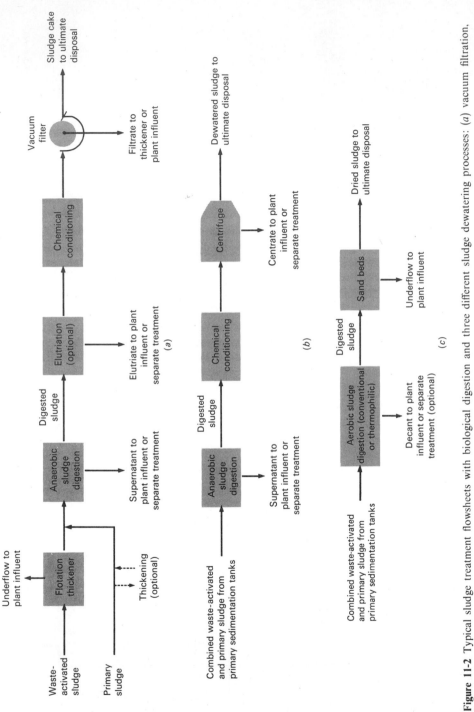

Figure 11-2 Typical sludge treatment flowsheets with biological digestion and three different sludge dewatering processes: (*a*) vacuum filtration, (*b*) centrifugation, and (*c*) drying beds.

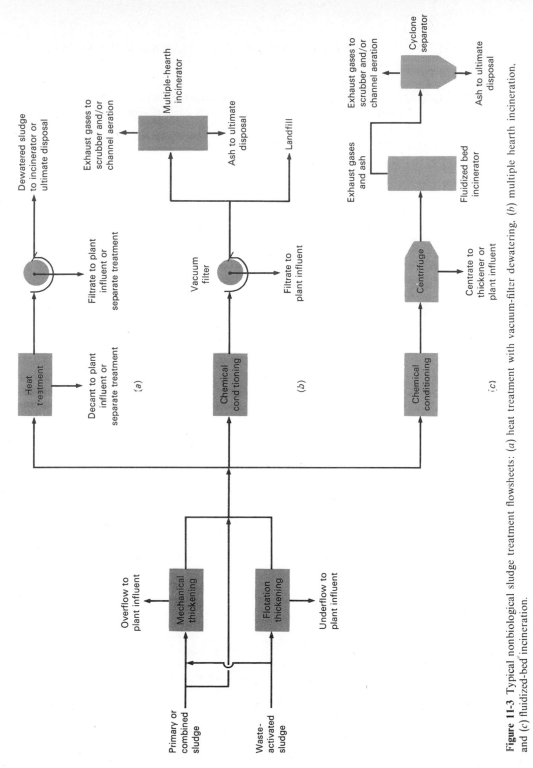

Figure 11-3 Typical nonbiological sludge treatment flowsheets: (*a*) heat treatment with vacuum-filter dewatering, (*b*) multiple hearth incineration, and (*c*) fluidized-bed incineration.

579

11-2 SOLIDS AND SLUDGE SOURCES, CHARACTERISTICS, AND QUANTITIES

To design sludge-processing, -treatment, and -disposal facilities properly, the sources, characteristics, and quantities of the solids and sludge to be handled must be known. Therefore, the purpose of this section is to present background data and information on these topics, which will serve as a basis for the material to be presented in the subsequent sections of this chapter.

Sources

The sources of solids in a treatment plant vary according to the type of plant and its method of operation. The principal sources of solids and sludge and the types generated are reported in Table 11-2. For example, in a continuous-flow stirred-tank activated-sludge process, if sludge wasting is accomplished from the mixed-liquor line or aeration chamber, the activated-sludge settling tank is not a source of sludge. On the other hand, if wasting is accomplished from the solids return line, the activated-sludge settling tank constitutes a sludge source. If the sludge from the mixed-liquor line or aeration chamber is returned to the primary

Table 11-2 Sources of solids and sludge from a conventional wastewater-treatment facility

Unit operation or process	Type of solids or sludge	Remarks
Screening	Coarse solids	Course solids are often comminuted and returned to the wastewater for removal in subsequent treatment facilities.
Grit removal	Grit and scum	Scum removal facilities are often omitted on grit removal facilities.
Preaeration	Scum	In some plants, scum removal facilities are not provided in preaeration tanks.
Primary sedimentation	Primary sludge and scum	The quantities of both sludge and scum depend on the nature of the collection system and whether industrial wastes are discharged to the system.
Aeration tank	Suspended solids	Suspended solids are produced from the conversion of BOD. If wasting is from the aeration tank, flotation thickening is normally used to thicken the waste activated sludge.
Secondary sedimentation	Secondary sludge and scum	Provision for scum removal on secondary settling tanks is now a requirement of the U.S. Environmental Protection Agency.
Sludge-processing facilities	Sludge and ashes	The characteristics and moisture content of the sludge and ashes depend on the operations and processes that are used.

settling tank for thickening, this obviates the need for a thickener and therefore reduces by one the number of independent sludge sources in the treatment plant. Processes used for thickening, digesting, conditioning, and filtering the sludge produced from primary and activated-sludge settling tanks also constitute sludge sources.

Characteristics

To treat and dispose of the sludge produced from wastewater-treatment plants in the most effective manner, it is important to know the characteristics of the solids and sludge that will be processed. The characteristics vary depending on the origin of the solids and sludge, the amount of aging that has taken place, and the type of processing to which they have been subjected. Some of the physical characteristics are summarized in Table 11-3.

Typical data on the chemical composition of untreated and digested sludges are reported in Table 11-4. Many of the chemical constituents, including nutrients, are important in considering the ultimate disposal of the processed sludge and the liquid removed from the sludge during processing. The fertilizer value of

Table 11-3 Characteristics of sludge produced during wastewater treatment

Solids or sludge	Description
Screenings	Screenings include all types of organic and inorganic materials large enough to be removed on bar racks. The organic content varies, depending on the nature of the system and the season of the year.
Grit	Grit is usually made up of the heavier inorganic solids that settle with relatively high velocities. Depending on the operating velocities, grit may also contain significant amounts of organic matter, specifically fats and grease.
Scum	Scum consists of the floatable materials skimmed from the surface of primary and secondary settling tanks. It may include grease, vegetable and mineral oils, animal fats, waxes, soaps, food wastes, vegetable skins, hair, paper and cotton, cigarette tips, plastic suppositories, rubber prophylactics, grit particles, and similar materials. The specific gravity of scum is less than 1.0 and usually around 0.95.
Primary sludge	Sludge from primary sedimentation tanks is usually gray and slimy and, in most cases, has an extremely offensive odor. It can be readily digested under suitable conditions of operation.
Chemical-precipitation sludge	Sludge from chemical precipitation tanks is usually dark in color, though its surface may be red if it contains much iron. Its odor may be objectionable, but not as bad as odor from primary sedimentation sludge. While it is somewhat slimy, the hydrate of iron or aluminum in it makes it gelatinous. If it is left in the tank, it undergoes decomposition like the sludge from primary sedimentation but at a slower rate. It gives off gas in substantial quantities and its density is increased by standing.

(continued)

Table 11-3 (*continued*)

Solids or sludge	Description
Activated sludge	Activated sludge generally has a brown flocculant appearance. If the color is quite dark, it may be approaching a septic condition. If the color is lighter than usual, there may have been underaeration with a tendency for the solids to settle slowly. Sludge in good condition has an inoffensive characteristic odor. It tends to become septic rather rapidly and then has a disagreeable odor of putrefaction. It will digest readily alone or mixed with fresh wastewater solids.
Trickling-filter sludge	Trickling-filter humus is brownish, flocculant, and relatively inoffensive when fresh. It generally undergoes decomposition more slowly than other undigested sludges, but when it contains many worms it may become offensive quickly. It is readily digested.
Digested sludge (aerobic)	Aerobically digested sludge is brown to dark brown and has a flocculant appearance. The odor of aerobically digested sludge is not offensive; it is often characterized as musty. Well-digested aerobic sludge dewaters easily, and the resulting dry solids are inoffensive.
Digested sludge (anaerobic)	Anaerobically digested sludge is dark brown to black and contains an exceptionally large quantity of gas. When thoroughly digested, it is not offensive, its odor being relatively faint and like that of hot tar, burnt rubber, or sealing wax. When drawn off on porous beds in thin layers, the solids first are carried to the surface by the entrained gases, leaving a sheet of comparatively clear water below them which drains off rapidly and allows the solids to sink down slowly on to the bed. As the sludge dries, the gases escape, leaving a well-cracked surface with an odor resembling that of garden loam.
Septage	Sludge from septic tanks is black. Unless well digested by long storage, it is offensive because of the hydrogen sulfide and other gases it gives off. The sludge can be dried on porous beds if spread out in thin layers, but objectionable odors are to be expected while it is draining unless it has been well digested.

sludge, which should be evaluated where the sludge is to be used as a soil conditioner, is based primarily on the content of nitrogen, phosphorus, and potash. The measurement of pH, alkalinity, and organic acid content is important in process control of anaerobic digestion.

The thermal content of sludge is important where incineration or some other combustion process is considered, and accurate bomb-calorimeter tests should be conducted so that a heat balance can be made for the combustion system. The thermal content of untreated primary sludge is the highest, especially if it contains appreciable amounts of grease and skimmings. Where kitchen food grinders are used, the volatile and thermal content of the sludge will also be high. The thermal content reported in Table 11-4 was calculated using Eq. 11-16, given in Sec. 11-14.

Table 11-4 Typical chemical composition of untreated and digested sludge

Item	Untreated primary sludge Range	Untreated primary sludge Typical	Digested sludge Range	Digested sludge Typical
Total dry solids (TS), %	2.0–8.0	5.0	6.0–12.0	10.0
Volatile solids (% of TS)	60–80	65	30–60	40.0
Grease and fats				
(ether-soluble, % of TS)	6.0–30.0	...	5.0–20.0	...
Protein (% of TS)	20–30	25	15–20	18
Nitrogen (N, % of TS)	1.5–6.0	4.0	1.6–6.0	4.0
Phosphorus (P_2O_5, % of TS)	0.8–3.0	2.0	1.5–4.0	2.5
Potash (K_2O, % of TS)	0–1.0	0.4	0.0–3.0	1.0
Cellulose (% of TS)	8.0–15.0	10.0	8.0–15.0	10.0
Iron (not as sulfide)	2.0–4.0	2.5	3.0–8.0	4.0
Silica (SiO_2, % of TS)	15.0–20.0	...	10.0–20.0	...
pH	5.0–8.0	6.0	6.5–7.5	7.0
Alkalinity (mg/L as $CaCO_3$)	500 1500	600	2500–3500	3000
Organic acids (mg/L as HAc)	200–2000	500	100–600	200
Thermal content (MJ/kg)	14–23	16.5[a]	6–14	9[b]

[a] Based on 65 percent volatile matter.
[b] Based on 40 percent volatile matter.
 Note: MJ/kg × 429.92 = Btu/lb

Table 11-5 Typical data on the physical characteristics and quantities of sludge produced from various wastewater-treatment processes

Treatment process	Specific gravity of sludge solids	Specific gravity of sludge	Dry solids, kg/10^3 m^3 Range	Dry solids, kg/10^3 m^3 Typical
Primary sedimentation	1.4	1.02	110–170	150
Activated sludge (waste sludge)	1.25	1.005	70–100	85
Trickling filtration (waste sludge)	1.45	1.025	55–90	70
Extended aeration (waste sludge)	1.30	1.015	80–120	100[a]
Aerated lagoon (waste sludge)	1.30	1.010	80–120	100[a]
Filtration	1.20	1.005	10–20	15
Algae removal	1.20	1.005	10–25	15
Chemical addition to primary clarifiers for phosphorus removal				
Low lime (350–500 mg/L)	1.9	1.04	250–400	300[b]
High lime (800–1600 mg/L)	2.2	1.05	600–1280	800[b]
Suspended-growth nitrification	...	...	...	...[c]
Suspended-growth denitrification	1.20	1.005	10–30	16
Roughing filters	1.28	1.020	...	...[d]

[a] Assuming no primary treatment.
[b] Sludge in addition to that normally removed by primary sedimentation.
[c] Negligible.
[d] Included in sludge production from biological secondary treatment processes.
 Note: kg/10^3 m^3 × 0.0083 = lb/10^3 gal

Quantities

Data on the quantities of sludge produced from various processes and operations are presented in Table 11-5. Corresponding data on the sludge concentrations to be expected from various processes are given in Table 11-6. Although the data in Table 11-5 are useful as presented, it should be noted that the quantity of sludge produced will vary widely.

Quantity variations The quantity of solids entering the wastewater-treatment plant daily may be expected to fluctuate over a wide range. To ensure capacity

Table 11-6 Expected sludge concentrations from various treatment operations and processes

Operation or process application	Sludge solids concentration, % dry solids	
	Range	Typical
Primary settling tank		
Primary sludge	4.00–12.0	5.00
Primary sludge to a cyclone[a]	0.50–3.00	1.50
Primary and waste activated sludge	3.00–10.0	4.00
Primary and trickling-filter humus	4.00–10.0	5.00
Primary sludge with iron addition for phosphorus removal	5.00–14.0	7.50
Primary sludge with low lime addition for phosphorus removal	2.00–8.00	4.00
Primary sludge with high lime addition for phosphorus removal	4.00–16.0	10.00
Secondary settling tank		
Waste activated sludge, with primary settling	0.50–1.50	0.75
Waste activated sludge, without primary settling	0.75–2.50	1.25
Pure-oxygen activated sludge, with primary settling	1.25–3.00	2.00
Pure-oxygen activated sludge, without primary settling	1.50–4.00	2.50
Trickling-filter humus sludge	1.00–3.00	1.50
Scum	3.00–10.0	5.00
Gravity thickener		
Primary sludge only	6.00–12.0	8.00
Primary and waste activated	3.00–10.0	4.00
Primary and trickling-filter humus	4.00–10.0	5.00
Flotation thickener		
Waste activated sludge only	3.00–6.00	4.00
Anaerobic digester		
Primary sludge only	5.00–10.0	7.00
Primary and waste activated sludge	2.50–7.00	3.50
Primary and trickling-filter humus	3.00–8.00	4.00
Aerobic digester		
Waste activated sludge only	0.75–2.50	1.25
Waste activated and primary sludge	1.50–4.00	2.50
Primary sludge only	2.50–7.00	3.50

[a] Arrangements must be provided to dilute a more concentrated sludge.

capable of handling these variations, the designer of sludge-disposal facilities should consider (1) the average and maximum rates of sludge production, and (2) the potential storage capacity of the treatment units within the plant. The variation in daily quantity that may be expected in large cities is shown in Fig. 11-4. The curve is characteristic of large cities having a number of large sewers laid on flat slopes [14]; even greater variations may be expected at small plants.

A limited quantity of solids may be stored temporarily in the sedimentation and aeration tanks. This storage provides capacity for equalizing short-term peak loads. Where digestion tanks are used, their large storage capacity provides a substantial dampening effect on peak digested-sludge loads. Sludge-disposal systems that use digestion tanks may be designed on the basis of maximum monthly loadings; those that do not use digestion tanks should be capable of handling the solids production of the maximum week. The preparation of solids balances for treatment plants is considered in Sec. 11-15.

Volume-mass relationships The volume of sludge depends mainly on its water content and only slightly on the character of the solid matter. A 10 percent sludge, for example, contains 90 percent water by mass. If the solid matter is composed of fixed (mineral) solids and volatile (organic) solids, the specific gravity of all of the solid matter can be computed using Eq. 11-1.

$$\frac{M_s}{S_s \rho_w} = \frac{M_f}{S_f \rho_w} + \frac{M_v}{S_v \rho_w} \tag{11-1}$$

where M_s = mass of solids
S_s = specific gravity of solids
ρ_w = density of water
M_f = mass of fixed solids (mineral matter)
S_f = specific gravity of fixed solids
M_v = mass of volatile solids
S_v = specific gravity of volatile solids

Therefore, if one-third of the solid matter in a sludge containing 90 percent water is composed of fixed mineral solids with a specific gravity of 2.5, and two-

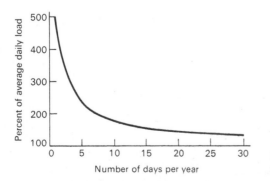

Figure 11-4 Peak sludge load as a function of the average daily load [14].

thirds is composed of volatile solids with a specific gravity of 1.0, then the specific gravity of all solids S_s would be equal to 1.25, as follows:

$$\frac{1}{S_s} = \frac{0.33}{2.5} + \frac{0.67}{1} = 0.802$$

$$S_s = \frac{1}{0.802} = 1.25$$

If the specific gravity of the water is taken to be 1.0, as it can be without appreciable error, the specific gravity of the sludge is 1.02, as follows:

$$\frac{1}{S} = \frac{0.1}{1.25} + \frac{0.9}{1.0} = 0.98$$

$$S = \frac{1}{0.98} = 1.02$$

The volume of a sludge may be computed with the following expression:

$$V = \frac{M_s}{\rho_w S_{sl} P_s} \tag{11-2}$$

where M_s = mass of dry solids, kg
ρ_w = density of water, 10^3 kg/m^3 (at 5°C)
S_{sl} = specific gravity of the sludge
P_s = percent solids expressed as a decimal

For approximate calculations for a given solids content, it is simple to remember that the volume varies inversely with the percent of solid matter contained in the sludge as given by

$$\frac{V_1}{V_2} = \frac{P_2}{P_1} \quad \text{(approximate)}$$

where V_1, V_2 = sludge volumes
P_1, P_2 = percent of solid matter

The application of these mass and volume relationships is illustrated in Example 11-1.

Example 11-1 Volume of untreated and digested sludge Determine the liquid volume before and after digestion and the percent reduction for 500 kg (dry basis) of primary sludge with the following characteristics:

	Primary	Digested
Solids, %	5	10
Volatile matter, %	60	60 (destroyed)
Specific gravity of		
fixed solids	2.5	2.5
Specific gravity of		
volatile solids	≈ 1.0	≈ 1.0

SOLUTION

1 Compute the average specific gravity of all the solids in the primary sludge using Eq. 11-1.

$$\frac{1}{S_s} = \frac{0.4}{2.5} + \frac{0.6}{1.0} = 0.76$$

$$S_s = \frac{1}{0.76} = 1.32 \qquad \text{(primary solids)}$$

2 Compute the specific gravity of the primary sludge.

$$\frac{1}{S_{sl}} = \frac{0.05}{1.32} + \frac{0.95}{1}$$

$$S_{sl} = \frac{1}{0.99} = 1.01$$

3 Compute the volume of the primary sludge using Eq. 11-2.

$$V = \frac{500 \text{ kg}}{(1000 \text{ kg/m}^3)(1.01)(0.05)}$$

$$= 9.9 \text{ m}^3 \ (349 \text{ ft}^3)$$

4 Compute the percentage of volatile matter after digestion.

$$\% \text{ volatile matter} = \frac{0.4(0.6 \times 500)(100)}{200 + 0.4(300)} = 37.5$$

5 Compute the average specific gravity of all the solids in the digested sludge using Eq. 11-1.

$$\frac{1}{S_s} = \frac{0.625}{2.5} + \frac{0.375}{1} = 0.625$$

$$S_s = \frac{1}{0.625} = 1.6 \qquad \text{(digested solids)}$$

6 Compute the specific gravity of the digested sludge.

$$\frac{1}{S_{dsl}} = \frac{0.1}{1.6} + \frac{0.90}{1} = 0.96$$

$$S_{dsl} = \frac{1}{0.96} = 1.04$$

7 Compute the volume of digested sludge using Eq. 11-2.

$$V = \frac{200 + 0.4(300)}{(1000 \text{ kg/m}^3)(1.04)(0.10)}$$

$$= 3.1 \text{ m}^3 \ (109 \text{ ft}^3)$$

8 Determine the percentage reduction in the sludge volume after digestion.

$$\% \text{ reduction} = \frac{(9.9 - 3.1) \text{ m}^3}{9.9 \text{ m}^3} \ 100 = 68.7$$

11-3 SLUDGE AND SCUM PUMPING

Sludge produced in wastewater-treatment plants must be conveyed from point to point in the plant in conditions ranging from a watery sludge or scum to a thick sludge. For each type of sludge, a different type of pump may be needed.

Pumps

Pumps used to convey sludge include the plunger, progressing-cavity, centrifugal, and torque-flow types. Diaphragm pumps may be used for pumping scum.

Plunger pumps Plunger pumps (see Fig. 11-5a) have been used frequently, and if rugged enough for the service, have proved to be quite satisfactory. The advantages of plunger pumps are as follows:

1. Pulsating action of simplex and also duplex pumps tends to concentrate the sludge in the hoppers ahead of the pumps and resuspend solids in pipelines when pumping at low velocities.
2. They are suitable for suction lifts up to 3 m (10 ft), and they are self-priming.
3. Low pumping rates can be used with large port openings.
4. Positive delivery is provided unless some object prevents the ball check valves from seating.
5. They have constant but adjustable capacity, regardless of large variations in pumping head.
6. Large discharge heads may be provided for.
7. Heavy-solids concentrations may be pumped if the equipment is designed for the load conditions.

   Plunger pumps come in simplex, duplex, and triplex models with capacities of 2.5 to 3.8 L/s or 0.0025 to 0.0038 m^3/s (40 to 60 gal/min) per plunger, and larger models are available. Pump speeds should be between 40 and 50 r/min. The pumps should be designed for a minimum head of 24 m (80 ft) in small plants and 35 m (115 ft) or more in large plants, because grease accumulations in sludge lines cause a progressive increase in head with use. Pumps are available with heads up to 70 m (230 ft) and should be considered for large plants. Capacity is decreased by shortening the stroke of the plunger; however, the pumps seem to operate more satisfactorily at or near full stroke. For this reason, many pumps are provided with variable-pitch V-belt drives for speed control of capacity.

Progressing-cavity pumps The Moyno (trade name) progressing-cavity pump (see Fig. 11-5b) has been used successfully, particularly on concentrated sludge. The pump is composed of a single-threaded rotor that operates with a minimum of clearance in a double-threaded helix of rubber. It is self-priming at suction lifts up to 8.5 m (28 ft), but it must not be operated dry or it will burn out the rubber stator. It is available in capacities up to 44 L/s or 0.044 m^3/s (700 gal/min). It is recommended that pumps handling primary and concentrated sludges be selected for a discharge pressure of not less than 276 kN/m^2 (40 lb$_f$/in^2). For primary sludges, a grinder normally precedes these pumps.

Centrifugal pumps With centrifugal pumps, the problem is to obtain a large enough pump to pass the solids without clogging and a small enough capacity to avoid pumping a sludge diluted by large quantities of the overlying wastewater.

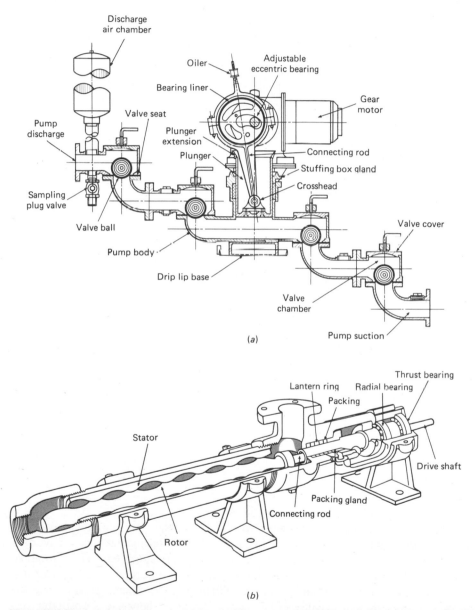

(a)

(b)

Figure 11-5 Typical sludge and scum pumps used in wastewater treatment plants. (*a*) Plunger. (*From Komline-Sanderson.*) (*b*) Progressive cavity. (*From Robbins & Myers.*)

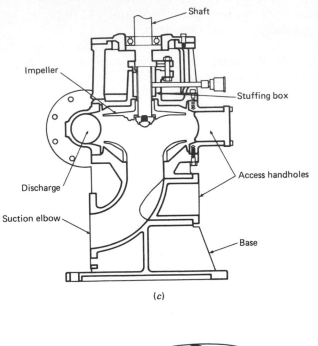

(c)

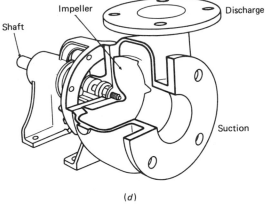

(d)

Figure 11-5 (*continued*) (*c*) Nonclog centrifugal. (*From Worthington.*) (*d*) Torque flow. (*From Envirotech, Wemco Div.*)

Centrifugal pumps of special design (screw feed, bladeless, torque flow) have been used for pumping primary sludge in large plants. Because the capacity of a centrifugal pump varies with the head, which is usually specified great enough so that the pumps may assist in dewatering the tanks, the pumps have considerable excess capacity under normal conditions. Throttling the discharge to reduce the capacity is impractical because of frequent stoppages; hence it is absolutely essential that these pumps be equipped with variable-speed drives. The superiority of centrifugal screw-feed pumps to handle primary sludge without clogging is well documented.

Centrifugal pumps with bladeless impellers have been used to some extent and in some cases are preferable to either plunger or screw-feed pumps. Bladeless pumps have approximately one-half the capacity of conventional nonclog pumps of the same nominal size and consequently approach the hydraulic requirements more closely. The design of the pump makes clogging at the suction of the impeller almost impossible.

Torque-flow pumps (see Fig. 11-5d) have fully recessed impellers and are very effective in conveying sludge. The size of particles that can be handled is limited only by the diameter of the suction or discharge openings. The rotating impeller develops a vortex in the sludge so that the main propulsive force is the liquid itself.

Slow-speed centrifugal and mixed-flow pumps are commonly used for returning activated sludge to the aeration tanks. Screw pumps are also being used for this service.

Application of Pumps to Types of Sludge

Types of sludge that are pumped include primary, chemical, and trickling-filter sludges and activated, elutriated, thickened, and concentrated sludges. Scum that accumulates at various points in a treatment plant must also be pumped. The application of pumps to types of sludge is summarized in Table 11-7.

Head-Loss Determination

The head loss encountered in the pumping of sludge depends on the type of sludge, its solids content, and the flow velocity. It has been observed that head losses increase with increased solids content, increased volatile content, and lower temperatures. When the percent volatile matter multiplied by the percent solids exceeds 600, difficulties may be encountered in pumping sludge.

The head loss in pumping unconcentrated activated and trickling-filter sludges may be from 10 to 25 percent greater than for water. Primary, digested, and concentrated sludges at low velocities may exhibit a plastic flow phenomenon in which a definite pressure is required to overcome resistance and start flow. The resistance then increases approximately with the first power of the velocity throughout the laminar range of flow, which extends to about 1.1 m/s (3.5 ft/s), the so-called lower critical velocity [5]. Above the higher critical velocity at about 1.4 m/s (4.5 ft/s), the flow may be considered turbulent. In the turbulent range, the losses for well-digested sludge may be from two to three times the losses for water. The losses for primary and dense concentrated sludges may be considerably greater. To determine the head loss, the factor k is obtained from Fig. 11-6 for a given solids content and type of sludge. The head loss when pumping sludge is determined by multiplying the head loss for water determined using the Darcy-Weisbach, Hazen Williams, or Mannings equation by k.

Table 11-7 Application of pumps to types of sludge

Type of sludge	Applicable pump	Comment
Primary sludge	Plunger; centrifugal pump (screw-feed, bladeless, or torque-flow type); positive-displacement and progressing-cavity pumps	Ordinarily, it is desirable to obtain as concentrated a sludge as practicable from primary sedimentation tanks, usually by collecting the sludge in hoppers and pumping intermittently, allowing the sludge to collect and consolidate between pumping periods. The character of primary untreated sludge will vary considerably, depending on the characteristics of the solids in the wastewater, the types of units and their efficiency, and, where biological treatment follows, the quantity of solids added from (1) overflow liquors from digestion tanks, (2) waste activated sludge, (3) humus sludge from settling tanks following trickling filters, and (4) overflow liquors from sludge elutriation tanks. The character of the sludge is such that conventional nonclog pumps cannot be used satisfactorily.
Chemical precipitation	Same as for primary sludge	
Digested sludge	Plunger; torque-flow centrifugal type; positive-displacement and progressing-cavity pumps	Well-digested sludge is homogenous, containing 5 to 8 % solids and a quantity of gas bubbles, but may contain up to 12 % solids. Poorly digested sludge may be difficult to handle. At least one positive-displacement pump should be available.
Trickling-filter Sludge	Plunger, progressing-cavity pump, nonclog centrifugal pump, and torque-flow pump	Sludge is usually of homogeneous character and can be easily pumped.
Return activated	Nonclog or mixed-flow centrifugal pump (see Fig. 11-5c)	Sludge is dilute and contains only fine solids so that it may be readily pumped with this type of pump, which operates at slow speed because the head is low and the flocculant character of the sludge should not be broken up.
Elutriated, thickened, and concentrated	Plunger	Plunger pumps are frequently used for concentrated sludge to accommodate the high friction-head losses in pump discharge lines.

Table 11-7 (*continued*)

Type of sludge	Applicable pump	Comment
	Positive-displacement and progressing-cavity pumps	This type of pump has been used successfully for dense sludges containing up to 20% solids. Because these pumps have limited clearances, it is necessary to reduce all solids to small size.
Scum	Positive-displacement or progressing-cavity pumps; plunger or diaphragm; pneumatic ejectors; centrifugal pump (screw-feed, bladeless, or torque-flow types)	Scum is often pumped by the sludge pumps; valves are manipulated in the scum and sludge lines to permit this. In larger plants, separate scum pumps are used.

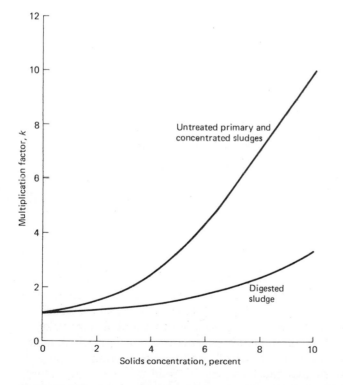

Figure 11-6 Head-loss multiplication factor for different sludge types.

Usually, the consistency of untreated primary sludge changes during pumping. At first, the most concentrated sludge is pumped. When most of the sludge has been pumped, the pump must handle a dilute sludge that has essentially the same hydraulic characteristics as water. This change in characteristics causes a centrifugal pump to operate farther out on its curve. The pump motor should be sized for the additional load, and a variable-speed drive should be supplied to reduce the flow under these conditions. If the pump motor is not sized for the maximum load obtainable when pumping water at top speed, it is likely to go out on overload or be damaged if the overload devices do not function or are set too high.

To determine the operating speeds and motor power required for a centrifugal pump handling sludge, system curves should be computed (1) for the most dense sludge anticipated, (2) for average conditions, and (3) for water. These should be plotted on a graph of the pump curves for a range of available speeds. The maximum and minimum speeds required of a particular pump are obtained from the intersection of the pump head-capacity curves with the system curves at the desired capacity. Where the maximum-speed head-capacity curve intersects the system curve for water determines the power required. In constructing the system curves for sludge, it is recommended that, for velocities from 0 to 1.1 m/s (3.5 ft/s), the head be assumed constant at the figure computed for 1.1 m/s. The intersection of the pump curves with the system curve for average conditions can be used to estimate hours of operations, average speed, and power costs.

In the plastic and laminar range, the usual flow formulas cannot be used, so the engineer must rely on judgment and experience. In this range, capacities will be small, and positive-displacement plunger pumps or progressing-cavity pumps should be used with ample head and capacity as recommended previously.

A method of computing the head loss for laminar-flow conditions was derived by Babbitt and Caldwell [1], based on the results of experimental and theoretical studies. The head loss when pumping sludge is considered to depend on the plastic properties of a non-Newtonian fluid. Although this method is theoretically sound, the selection of suitable parameters for insertion in the equations presented is difficult, based on the available data. For further information on this method, the original work [1], the paper by Chou [5], and the literature should be consulted.

Sludge Piping

In treatment plants, conventional sludge piping should not be smaller than 150 mm (6 in) in diameter (smaller-diameter glass-lined pipe has been used successfully). It need not be larger than 200 mm (8 in), unless the velocity exceeds 1.5 to 1.8 m/s (5 to 6 ft/s), in which case the pipe is sized to maintain that velocity. Gravity sludge-withdrawal lines should not be less than 200 mm (8 in) in diameter. It is common practice to install a number of cleanouts in the form of plugged crosses instead of elbows so that the lines can be rodded if necessary. Pump connections should not be smaller than 100 mm (4 in) in diameter.

A liberal number of hose gates should be installed in the piping, and an ample supply of flushing water under high pressure should be available for clearing stoppages. The flushing water should be plant effluent supplied by a centrifugal pump. The pump should have a capacity of not less than 0.010 m³/s (160 gal/min) at 500 kN/m² (72 lb_f/in^2). In large plants with larger piping, a greater capacity should be available, and the available pressure should be increased to 700 kN/m² (100 lb_f/in^2).

Grease has a tendency to coat the inside of piping used for transporting primary sludge and scum. This is more of a problem in large plants than in small ones. The coating results in a decrease in effective diameter and a large increase in pumping head. For this reason, low-capacity positive-displacement pumps are designed for heads greatly in excess of the theoretical head. Centrifugal pumps, with their larger capacity, usually pump a more dilute sludge, often containing some wastewater, and head buildup due to grease accumulations appears to occur more slowly. In some plants, provisions have been made for melting out the grease by circulating hot water, steam, or digester supernatant through the main sludge lines.

In treatment plants, friction losses are low, and there is no difficulty in providing an ample safety factor. In the design of long sludge lines that transport solids from one plant to another for further treatment, total friction losses may exceed 700 kN/m² (100 lb/in²) and should be estimated carefully. For solving such problems, the engineer should refer to the literature for design and operating data on long sludge lines [28].

11-4 PRELIMINARY OPERATIONS

Sludge grinding, degritting, blending, and storage are necessary to provide a relatively constant, homogeneous feed to sludge processing facilities. Blending and storage can be accomplished either in a single unit designed to do both or separately in other plant components.

Sludge Grinding

Sludge grinding is a process in which large material contained in sludge is cut or sheared into small particles. Some of the processes that must be preceded by sludge grinders and the purposes of grinding are reported in Table 11-8.

Sludge grinders use one of two techniques: hammermill pulverizing or cutting. This type of unit required usually depends on the specific application. Selection is generally based on the recommendation of the manufacturer of the sludge process for which grinding is a prerequisite.

Sludge Degritting

In some plants where separate grit removal facilities are not used ahead of the primary sedimentation tanks, or where the grit removal facilities are not adequate to handle peak flows and peak grit loads, it may be necessary to remove the

Table 11-8 Processes requiring the grinding of sludge [22]

Process	Purpose of grinding
Heat treatment	To prevent clogging of high-pressure pumps and heat exchangers.
Nozzle-disk and solid-bowl centrifuges	To prevent clogging in nozzles and between disks. Nozzle-disk units may also require fine screens.
Chlorine oxidation	To enhance chlorine contact with sludge particles.
Pumping with progressing-cavity pumps	To prevent clogging and reduce heat.

grit before further processing of the sludge. Where further thickening of the primary sludge is desired, it is practical to consider degritting the primary sludge. The most effective method of degritting sludge is through the application of centrifugal forces in a flowing system to achieve separation of the grit particles from the organic sludge. Such separation is achieved through the use of hydroclones, which have no moving parts. The sludge is applied tangentially to a cylindrical feed section, thus imparting a centrifugal force. The heavier grit particles move to the outside of the cylinder section and are discharged through a conical feed section. The organic sludge is discharged through a separate outlet.

The efficiency of the hydroclone is affected by pressure and by the concentration of the organics in the sludge. To obtain effective grit separation, the sludge must be relatively dilute. As the sludge concentration increases, the particle size that can be removed decreases. The general relationship between sludge concentration and effectiveness of removal for primary sludges is shown in Table 11-9.

Sludge Blending

Blending will become a more important consideration as more plants provide secondary and advanced waste treatment. Blending sludges to produce a uniform mixture is particularly important ahead of sludge stabilization and dewatering

Table 11-9 Hydroclone grit-removal efficiency for primary sludges[a]

Primary sludge concentration, % total solids	Mesh of removal[b]
1	150
2	100
3	65
4	28–35

[a] For a 30.5-cm (12-in) hydroclone at 42 kN/m^2 (6 lb/in^2 gage) at 12.9 L/s (205 gal/min).
[b] About 95 percent or more of indicated particle size is removed.

processes and incineration. Sludge from primary, secondary, and advanced processes can be blended in several ways:

1. *In primary settling tanks.* Secondary or tertiary sludges can be returned to the primary settling tanks where they will settle and mix with the primary sludge.
2. *In pipes.* This procedure requires careful control of sludge sources and feed rates to ensure the proper blend.
3. *In sludge-processing facilities with long detention times.* Aerobic and anaerobic digesters (continuous-flow stirred-tank type) can blend the feed sludges uniformly.
4. *In a separate blending tank.* This practice provides the best opportunity to control the quality of the blended sludges.

    In small treatment plants of less than 0.045 m^3/s (1 Mgal/d), blending is usually accomplished in the primary settling tanks. In large facilities, optimum efficiency is achieved by separately thickening and sludges before blending. A typical blending tank with mixing facilities is shown in Fig. 11-7.

Sludge Storage

Sludge storage must be provided to smooth out fluctuations in the rate of sludge production and to allow sludge to accumulate during periods when subsequent sludge-processing facilities are not operating (i.e., night shifts, weekends, and periods of unscheduled equipment downtime). Sludge storage to provide a

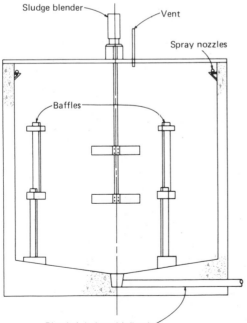

Sludge blender

Vent

Spray nozzles

Baffles

Blended sludge withdrawl

Figure 11-7 Typical sludge mixer and blender used in conjunction with sludge storage idea [*from South Essex Sewer District*].

uniform feed rate is particularly important ahead of the following processes: chlorine oxidation, lime stabilization, heat treatment, mechanical dewatering, drying, and thermal reduction.

Short-term sludge storage may be accomplished in wastewater settling tanks or in sludge-thickening tanks. Long-term sludge storage may be accomplished in sludge-stabilization processes with long detention times (i.e., aerobic and anaerobic digestion) or in specially designed separate tanks. In small installations, sludge is usually stored in the settling tanks and digesters. In large installations that do not use aerobic and anaerobic digestion, sludge is often stored in separate blending and storage tanks. Such tanks may be sized to retain the sludge for a period of several hours to several days. The determination of the required storage volume is illustrated in Example 11-2. Sludge is often aerated to prevent septicity and to promote mixing. Mechanical mixing may be necessary to assure complete blending of the sludge. Chlorine and hydrogen peroxide are often used to arrest septicity and to control the odors from sludge storage and blend tanks.

Example 11-2 Determination of volume required for sludge storage Assume that the yearly average rate of sludge production from an activated-sludge treatment plant is 12,000 kg/d (26,500 lb/d). Develop a curve of sustained sludge mass loading rates that can be used to determine the size of sludge-storage facilities required with various downstream sludge-processing units. Then, using this curve, determine the volume required for sludge storage, assuming that sludge accumulated for 7 d is to be processed in 5 working days, and that sludge accumulated for 14 d is to be processed in 10 working days. Note that the 5- and 10-d work periods correspond to 1 and 2 weeks, respectively, assuming that certain sludge-processing facilities, such as vacuum filters, will not be operated on the weekends.

SOLUTION

1 Develop a curve of sustained sludge mass loadings.
   a Because no information is specified, it will be assumed that the sustained sludge production will mirror the sustained BOD plant loadings given in Fig. 3-6a and used in Example 3-1.
   b Set up an appropriate computation table and compute the values necessary to plot the curve.

(1) Length of sustained peak, d	(2) Peaking factor[a]	(3) Peak solids mass loading, kg/d	(4) Total sustained loading, kg[b]
1	2.4	28,800	28,800
2	2.1	25,200	50,400
3	1.9	22,800	68,400
4	1.8	21,600	86,400
5	1.7	20,400	102,000
10	1.4	16,800	168,000
15	1.3	15,600	234,000
365	1.0	12,000	

[a] From Fig. 3-6a.
[b] Total mass produced for the corresponding sustained period given in col. 1.

   c Plot the sustained solids loading curve (see Fig. 11-8).

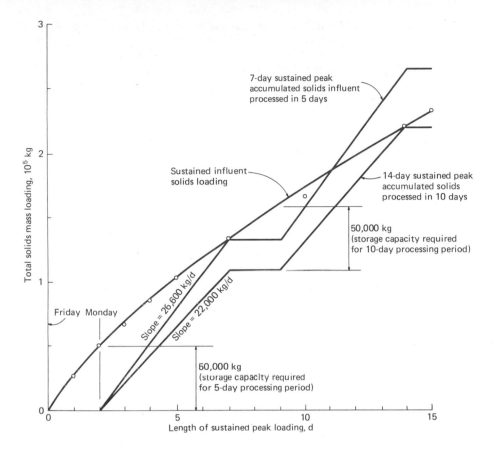

Figure 11-8 Sustained peak solids loading curve for determining volume of sludge storage facilities in Example 11-2. Note: kg × 2.2046 = lb.

2   Determine the sludge storage volume required for the stated operating conditions.

   a   Determine the daily rate at which sludge must be processed to handle the 7-d sustained peak (from Fig. 11-8) in 5 working days.

$$kg/d = \frac{133,000 \text{ kg}}{5 \text{ d}} = 26,600 \text{ kg/d}$$

   b   Determine the daily rate at which sludge must be processed to handle the 14-d sustained peak (from Fig. 11-8) in 10 working days.

$$kg/d = \frac{220,000 \text{ kg}}{10 \text{ d}} = 22,000 \text{ kg/d}$$

   c   Assuming that the sludge-storage facilities are empty on Friday just before the weekend, plot on Fig. 11-8 the average daily rate at which sludge must be processed during the 5- and 10-d periods.

   d   From Fig. 11-8, the required storage capacity in kilograms of solids is:

     i   Capacity based on 5 working days = 50,000 kg

     ii   Capacity based on 10 working days = 50,000 kg

Comment The downstream processing equipment can now be sized using the daily rate at which sludge must be processed. For example, if the number of kilograms per square meter per hour that can be processed with a vacuum filter is known, then the required surface area can be computed from the number of shifts to be used per day and the assumed value of the actual working hours per shift. In sizing equipment, a trade-off analysis should always be performed between the cost of storage and processing facilities versus labor costs (for both one shift and two shifts) to determine the most cost-effective combination.

11-5 CONCENTRATION (THICKENING)

The solids content of primary, activated, trickling-filter, or mixed sludge (i.e., primary plus activated) varies considerably, depending on the characteristics of the sludge, the sludge-removal and -pumping facilities, and the method of operation. Representative values of percent total solids are shown in Table 11-6. Thickening is a procedure used to increase the solids content of sludge by removing a portion of the liquid fraction. To illustrate, if waste activated sludge, which is typically pumped from secondary settling tanks with a content of 0.8 percent solids, can be thickened to a content of 4 percent solids, then a fivefold decrease in sludge volume is achieved. Thickening is generally accomplished by physical means, including gravity settling, flotation, and centrifugation. Some of the possibilities for separate sludge thickening are presented in Table 11-10.

Table 11-10 Occurrence of thickening methods in sludge processing[a]

Method	Type of sludge	Frequency of use and relative success
Gravity	Untreated primary	Increasing; excellent results
Gravity	Untreated primary and waste activated sludge	Often used, especially for small plants; satisfactory results with sludge concentrations in the range of 4 to 6 percent
Gravity	Untreated primary and waste activated sludge	Some new installations; marginal results
Gravity	Waste activated sludge	Essentially never used; poor results
Gravity (elutriation)	Digested primary and waste activated sludge mixture	Used to a lesser extent in new plants; used to reduce the amount of conditioning chemicals prior to mechanical dewatering
Dissolved-air flotation	Untreated primary and waste activated sludge	Some limited use
Dissolved-air flotation	Waste activated sludge	Increasing; good results
Solid-bowl conveyor centrifuge	Waste activated sludge	Some limited use; solids capture problem
Disk centrifuge	Waste activated sludge	Some limited use; data now being accumulated

[a] Adapted from Ref. 24.

Application

The volume reduction obtained by sludge concentration is beneficial to subsequent treatment processes, such as digestion, dewatering, drying, and combustion, from the following standpoints: (1) capacity of tanks and equipment required, (2) quantity of chemicals required for sludge conditioning, and (3) amount of heat required by digesters and amount of auxiliary fuel required for heat drying or incineration, or both.

On large projects when sludge must be transported a significant distance, such as to a separate plant for processing, a reduction in sludge volume may result in a reduction of pipe size and pumping costs. On small projects, the requirements of a minimum practicable pipe size and minimum velocity may necessitate the pumping of significant volumes of wastewater in addition to sludge, which diminishes the value of volume reduction. Volume reduction is very desirable when liquid sludge is transported by tank trucks for direct application to land as a solid conditioner.

Sludge thickening is achieved at all wastewater treatment plants in some manner—in the primary clarifiers, in sludge-digestion facilities, or in specially designed separate units. If separate units are used, the recycled flows are normally returned to the wastewater-treatment facilities. In small treatment facilities (less than 0.045 m³/s or 1 Mgal/d), separate sludge thickening is seldom practiced. Rather, gravity thickening is accomplished in the primary settling tank or in the sludge-digestion units, or both. In larger treatment facilities, the additional costs of separate sludge thickening are often justified by the improved control over the thickening process and the higher concentrations attainable.

Separate sludge concentration can be quite beneficial to the operation of activated-sludge–treatment plants, because it makes feasible the direct removal of aeration-tank mixed liquor for excess sludge wasting (the more usual practice is to waste the more concentrated return sludge). By removing a given volume of the mixed liquor each day for concentration and disposal, the sludge age or solids retention time, upon which the efficiency and operational characteristics of the activated-sludge process depend, can be closely maintained. It should be noted, however, that when this method of wasting is used, the size of the required sludge-thickening facilities will be considerably larger than those required where sludge wasting is from the return line. Consequently, the adoption of this method is not widespread.

Description of Thickening Equipment

The following discussion is intended to introduce the reader to the equipment used for the thickening of sludges. Most of the equipment is mechanical, and therefore the sanitary engineer usually is more concerned with its proper application to meet a given treatment objective than with the theory of its design.

Gravity thickening Gravity thickening is accomplished in a tank similar in design to a conventional sedimentation tank. Normally, a circular tank is used. Dilute sludge is fed to a center feed well. The feed sludge is allowed to settle and compact, and the thickened sludge is withdrawn from the bottom of the tank. Conventional sludge-collecting mechanisms with deep trusses (see Fig. 11-9) or vertical pickets stir the sludge gently, thereby opening up channels for water to escape and promoting densification. The continuous supernatant flow that results is returned to the primary settling tank. The thickened sludge that collects on the bottom of the tank is pumped to the digesters or dewatering equipment as required; thus, storage space must be provided for the sludge. As indicated in Table 11-10, gravity thickening is most effective on untreated primary sludge.

Flotation thickening As described in Chap. 6, there are four basic variations of the flotation-thickening operation: dissolved-air flotation; vacuum flotation; dispersed-air flotation; and biological flotation. However, only dissolved-air flotation is extensively used for sludge thickening in the United States. In dissolved-air flotation, air is introduced into a solution that is being held at an elevated pressure. A typical unit used for thickening waste activated sludge is shown in Fig. 11-10. When the solution is depressurized, the dissolved air is released as finely divided bubbles carrying the sludge to the top where it is removed. In locations where freezing is a problem, flotation thickeners are normally enclosed in a heated building.

Flotation thickening is most efficiently used for waste sludges from suspended-growth biological treatment processes, such as the activated-sludge process or the suspended-growth nitrification process. In a recently completed study, it was found that the degree of thickening achieved depended on the initial concentration of the sludge. Greater final concentrations were achieved when starting with more dilute sludges [34]. Also, it appears that the ability to thicken waste activated sludge will vary depending on the mean cell residence time at which the plant is being operated [34].

Centrifugal thickening Centrifuges are used both to thicken and to dewater sludges. As indicated in Table 11-10, their application in thickening is normally limited to waste activated sludge. Thickening by centrifugation involves the settling of sludge particles under the influence of centrifugal forces. The three basic types of centrifuges currently available for sludge thickening are nozzle-disk, solid-bowl, and basket centrifuges (see Fig. 11-11).

The operation of the nozzle-disk centrifuge is continuous. The centrifuge consists of a vertically mounted unit containing a number of stacked conical disks. Each disk acts as a separate low-capacity centrifuge. The liquid flows upward between the disks toward the center shaft, becoming gradually clarified. The solids are concentrated in the periphery of the bowl and are discharged through nozzles. Because of the small nozzle openings, these units must be preceded by sludge grinding and screening equipment to prevent clogging.

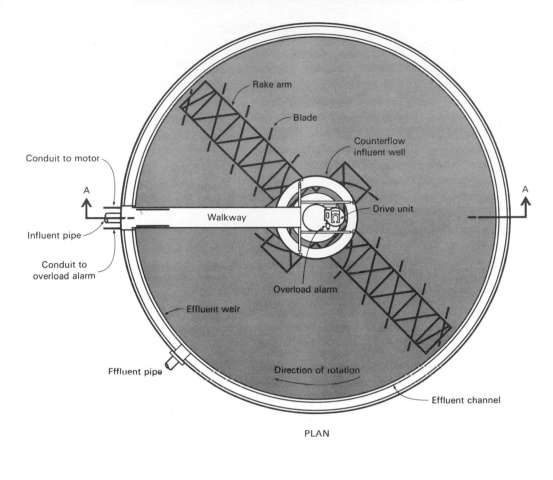

PLAN

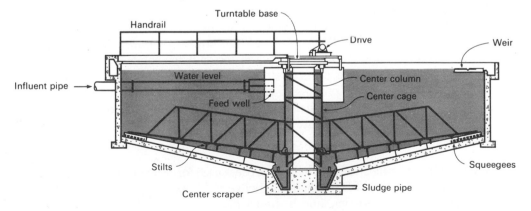

SECTION A–A

Figure 11-9 Schematic of a mechanical thickener [*from Dorr-Oliver*].

Figure 11-10 Dissolved-air flotation unit used for thickening waste activated sludge [*from Envirex*].

The operation of the solid-bowl centrifuge is also continuous. It consists of a long bowl, normally mounted horizontally and tapered at one end. Sludge is introduced into the unit continuously, and the solids concentrate on the periphery. A helical scroll, spinning at a slightly different speed, moves the accumulated sludge toward the tapered end where additional solids concentration occurs. The sludge is then discharged.

The basket centrifuge operates on a batch basis. The liquid sludge is introduced into a vertically mounted spinning bowl. The solids accumulate against the wall of the bowl. The centrate is decanted. When the solids-holding capacity of the machine has been achieved, the bowl decelerates and a scraper is positioned in the bowl to help remove the accumulated solids.

Maintenance and power costs for the centrifugal thickening process can be substantial. Therefore, the process is usually attractive only at large facilities (over 0.2 m^3/s or 5 Mgal/d) when space is limited and skilled operators are available, or for sludges that are difficult to thicken by more conventional means.

Classification Classification is actually a process used for recovering calcium from lime-precipitated sludges. The chemical sludge is fed at a high rate to a solid-bowl centrifuge. The heavier lime is discharged as a cake. The lighter organic fraction is discharged as centrate, which may be combined with biological sludges for further processing. The lime sludge is recalcined in an incinerator and then reused.

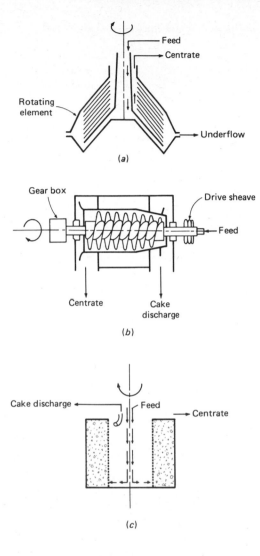

Feed
Centrate
Rotating element
Underflow
(a)

Gear box
Drive sheave
Feed
Centrate
Cake discharge
(b)

Cake discharge
Feed
Centrate
(c)

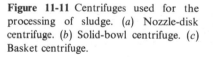

Figure 11-11 Centrifuges used for the processing of sludge. (a) Nozzle-disk centrifuge. (b) Solid-bowl centrifuge. (c) Basket centrifuge.

Design of Thickeners

Three types of thickeners are commonly used to thicken sludges from various sources in wastewater-treatment plants: gravity, dissolved-air flotation, and centrifugal. In designing thickening facilities, it is important to (1) provide adequate capacity to meet peak demands and (2) prevent septicity, with its attendant odor problems, during the thickening process. To reduce the size of the units, the use of sludge storage facilities should be evaluated.

Gravity thickeners Gravity thickeners are designed on the basis of hydraulic surface loading and solids loading. The principles that apply are the same as those used in designing sedimentation tanks, discussed in Chaps. 6, 8, and 9

Table 11-11 Typical concentrations of unthickened and thickened sludges and solids loadings for gravity thickeners

Type of sludge	Sludge concentration, %		Solids loading for gravity thickeners, $kg/m^2 \cdot d$
	Unthickened	Thickened	
Separate			
Primary sludge	4.0–12.0	6.0–12.0	100–150
Trickling-filter sludge	1.0–3.0[a]	4.0–10.0	40–50
Activated sludge	0.5–2.5	1.5–4.0	20–40
Pure-oxygen sludge	1.25–4.0	2.0–5.0	25–50
Combined			
Primary and trickling-filter sludge	4.0–10.0	4.0–10.0	60–100
Primary and modified-aeration sludge	3.0–10.0	3.0–10.0	60–100
Primary and air-activated sludge	3.0–10.0	3.0–10.0	40–80

[a] Values as high as 6.5 have been reported where the sludge is removed once per day.

Note: $kg/m^2 \cdot d \times 0.2048 = lb/ft^2 \cdot d$

Typical surface-loading rates are 16 to 36 $m^3/m^2 \cdot d$ (400 to 900 $gal/ft^2 \cdot d$). Typical solids loadings are listed in Table 11-11. To maintain aerobic conditions in gravity thickeners, provisions should be made for adding aerated mixed liquor or final effluent to the thickening tank.

In operation, a sludge blanket is maintained on the bottom of the thickener to aid in concentrating the sludge. An operating variable is the sludge volume ratio, which is the volume of the sludge blanket held in the thickener divided by the volume of the thickened sludge removed daily. Values of the sludge volume ratio normally range between 0.5 and 20 d; the lower values are required during warm weather.

Flotation thickeners Higher loadings can be used with flotation thickeners than are permissible with gravity thickeners, because of the rapid separation of solids from the wastewater. Flotation thickeners may be operated at the solids loadings given in Table 11-12. For design, the minimum loadings should be used. The higher solids loadings generally result in lower concentrations of thickened sludge. The pressure requirements and air-solids ratio requirements for dissolved-air flotation thickeners were discussed in Chap. 6.

Primary-tank effluent or plant effluent is recommended as the source of air-charged water rather than flotation-tank effluent, except when chemical aids are used, because of the possibility of fouling the air-pressure system with solids. The use of polyelectrolytes as flotation aids may or may not be effective in increasing solids loadings and the concentration of thickened sludge, but it does appear to be effective in increasing the solids recovery in the floated sludge from 85 to 98 or 99 percent.

Table 11-12 Solids loadings for dissolved-air flotation units[a]

Type of sludge	Loading, kg/m² · d
Air-activated sludge (mixed-liquor)	25–75
Air-activated sludge (settled)	50–100
Pure-oxygen activated sludge (settled)	60–150
50% primary + 50% activated (settled)	100–200
Primary only	to 260

[a] Derived in part from Ref. 15.

Note: kg/m² · d × 0.2048 = lb/ft² · d

Centrifuge thickeners The performance of a centrifuge is often measured in terms of a quantity called the percent capture, which is defined by Eq. 11-3.

$$\text{Percent capture} = \left[1 - \frac{C_r(C_c - C_s)}{C_s(C_c - C_r)} \right] 100 \qquad (11\text{-}3)$$

where C_r = concentration of solids in reject wastewater (centrate), mg/L, %
C_c = concentration of solids in the cake, mg/L, %
C_s = concentration of solids in sludge feed, mg/L, %

Thus, for a constant feed concentration, the percent capture increases as the concentration of solids in the centrate decreases. In concentrating sludge solids, capture is important if a minimum amount of solids is to be returned to the treatment process.

The principal operational variables include the following: (1) characteristics of the sludge (its water-holding structure being most important); (2) rotational speed; (3) hydraulic loading rate; (4) depth of the liquid pool in the bowl (solid-bowl machines); (5) differential for speed of the screw convey (solid-bowl machines); and (6) use of polyelectrolytes to improve the performance. Because the interrelationships of these variables will be different in each location, specific design recommendations are not available; in fact, pilot plant studies are recommended.

11-6 STABILIZATION: CHEMICAL AND THERMAL PROCESSES

Sludges are stabilized to (1) reduce pathogens, (2) eliminate offensive odors, and (3) inhibit, reduce, or eliminate the potential for putrefaction. The success in achieving these objectives is related to the effects of the stabilization operation or process on the volatile or organic fraction of the sludge. Survival of pathogens, release of odors, and putrefaction occur when microorganisms are allowed to flourish in the organic fraction of the sludge. There are four means to eliminate these nuisance conditions through stabilization. They are: (1) the biological reduction of volatile content, (2) the chemical oxidation of volatile matter, (3) the addition of chemicals to the sludge to render it unsuitable for the survival of microorganisms, and (4) the application of heat to disinfect or sterilize the sludge [22].

The technologies available for sludge stabilization include (1) chlorine oxidation, (2) lime stabilization, (3) heat treatment, (4) anaerobic digestion, and (5) aerobic digestion. The first three processes are discussed in this section. Because of the importance of the anaerobic and aerobic digestion processes, they are discussed separately in Secs. 11-7 and 11-8, respectively.

Chlorine Oxidation

The chlorine-oxidation process involves the chemical oxidation of the sludge with high doses of chlorine gas, which is generally applied directly to the sludge in an enclosed reactor for a short period of time. The process is followed by dewatering. Sand bed drying is an effective means. Belt-filter-press dewatering following conditioning with the addition of polyelectrolytes is also used.

Most chlorine-oxidation units are of a prefabricated modular design, completely self-contained and skid-mounted. Application of chlorine oxidation as an exclusive means of sludge stabilization has been limited to small plants on the order of 0.2 m³/s (5 Mgal/d) and less. The process may be used for treating any biological sludge, for treating septage, and as an auxiliary means of stabilization to supplement existing overtaxed facilities. The sludge should be ground to ensure proper contact.

Because of the reaction of chlorine gas with the sludge, significant quantities of hydrochloric acid are formed. The acid can also solubilize heavy metals. Consequently, supernatants and filtrates from chlorine-oxidized sludges may contain a high concentration of heavy metals. It has been reported that the release of heavy metals is dependent on pH, sludge metal content, and the species of metal found in the sludge. Supernatant and filtrate from the process may also contain high concentrations of chloramines [24].

Implementation of chlorine oxidation requires the installation of chlorinators to feed chlorine to the process. Other chemical requirements may include sodium hydroxide and polyelectrolytes to condition the sludge prior to dewatering.

Lime Stabilization

In the lime-stabilization process, lime is added to untreated sludge in sufficient quantity to raise the pH to 12 or higher. The high pH creates an environment that is not conducive to the survival of microorganisms. Consequently, the sludge will not putrefy, create odors, or pose a health hazard so long as the pH is maintained at this level.

Lime addition to untreated sludge has been practiced for many years as a conditioning process to facilitate dewatering; however, the use of lime as a stabilization agent has only recently gained recognition. Lime stabilization requires more lime per unit weight of sludge processed than that necessary for dewatering. The higher lime dose is required to attain a higher pH. In addition, sufficient contact time must be provided before dewatering to effect a high level of pathogen kill. Lime treatment at a pH higher than 12 for a period of 3 h has been reported to achieve pathogen reduction beyond that attainable with anaerobic digestion [23].

Because lime stabilization does not destroy the organics necessary for bacterial growth, the sludge must be disposed of before the pH drops significantly, or it can become reinfested and putrefy.

Heat Treatment

Heat treatment is a continuous process in which sludge is heated in a pressure vessel to temperatures up to 260°C (500°F) at pressures up to 2.75 MN/m^2 (400 lb$_f$/in^2 gage) for short periods of time. It essentially serves as both a stabilization process and a conditioning process. It conditions the sludge by rendering the solids capable of being dewatered without the use of chemicals. When the sludge is subjected to the high temperatures and pressures, the thermal activity releases bound water and results in the coagulation of solids. In addition, hydrolysis of proteinaceous materials occurs, resulting in cell destruction and release of soluble organic compounds and ammonia nitrogen. This method of treatment is considered in greater detail in Sec. 11-9, which deals with sludge conditioning.

11-7 STABILIZATION: ANAEROBIC SLUDGE-DIGESTION PROCESS

The history of sludge digestion and its precursors can be traced from the 1850s with the development of the first tank designed to separate and retain solids. The first unit used to treat settled wastewater solids was known as the Mouras automatic scavenger. It was developed by Louis H. Mouras of Vesoul, France, in about 1860 after it had been observed that if the solids were kept in a closed vault (cesspool) they were converted to a liquid state [18].

The first person to recognize that a combustible gas containing methane was produced when wastewater solids were liquefied was Donald Cameron, who built the first septic tank for the city of Exeter, England, in 1895. He collected and used the gas for lighting in the vicinity of the plant. In 1904, the first dual-purpose tank incorporating sedimentation and sludge treatment was installed at Hampton, England. It was known as the Travis hydrolytic tank and continued in operation until 1936 [21]. Experiments on a similar unit, called a Biolytic tank, were carried out in the United States between 1909 and 1912.

In 1904, a patent was issued to Dr. Karl Imhoff in Germany for a dual-purpose tank now commonly known as the Imhoff tank. One of the first installations in the United States using separate digestion tanks was the wastewater treatment plant in Baltimore, Md. Three rectangular digestion tanks were built as part of the original plant in 1911, 16 circular digestion tanks were added in 1914, and an additional rectangular tank was added in 1921 [30].

In the period from 1920 to 1935, the anaerobic digestion process was studied extensively. Heat was applied to separate digestion tanks, and major improvements were made in the design of the tanks and associated appurtenances. It is interesting to note that the same practice is being followed today, some 40 years later, but great progress has been made in the fundamental understanding and control of the process, the sizing of tanks, and the design and application of equipment. At the same time, engineers have become aware of its limitations and are learning when not to use it.

Process Description

The microbiology of anaerobic digestion and the optimum environmental conditions for the microorganisms involved are discussed in Chap. 9 (see Sec. 9-8). The operation and physical facilities for anaerobic digestion in standard-rate, high-rate, and two-stage digesters are described in this section.

Standard-rate digestion Standard-rate (conventional) sludge digestion is usually carried out as a single-stage process (see Fig. 9-28a). The functions of digestion, sludge thickening, and supernatant formation are carried out simultaneously. A cross section of a typical standard-rate digester is shown in Fig. 11-12. Operationally, in a single-stage process, untreated sludge is added in the zone where the sludge is actively digesting and the gas is being released. The sludge is heated by means of an external heat exchanger. As gas rises to the surface, it lifts sludge particles and other materials, such as grease, oils, and fats, ultimately giving rise to the formation of a scum layer (see Fig. 9-28a).

As a result of digestion, the sludge becomes more mineralized (for example, the percentage of fixed solids increases), and it thickens because of gravity. In turn, this leads to the formation of a supernatant layer above the digesting sludge. (The biochemistry of the reactions taking place in the digesting zone is described in Chap. 9.) As a result of the stratification and the lack of intimate mixing, not more than 50 percent of the volume of a standard-rate single-stage digester is used. Because of these limitations, the standard-rate process is used for small installations.

High-rate digestion The high-rate-digestion process differs from the conventional single-stage process in that the solids loading rate is much greater (see "Process Design"). The sludge is intimately mixed by gas recirculation, pumping, or draft-tube mixers (separation of scum and supernatant does not take place), and it is heated to achieve optimum digestion rates (see Fig. 11-13). With the exception of higher loading rates and improved mixing, there are only a few differences between the primary digester in a conventional two-stage process and a high-rate digester. The mixing equipment should have greater capacity and should reach to the bottom of the tank, the gas piping will be somewhat larger, fewer

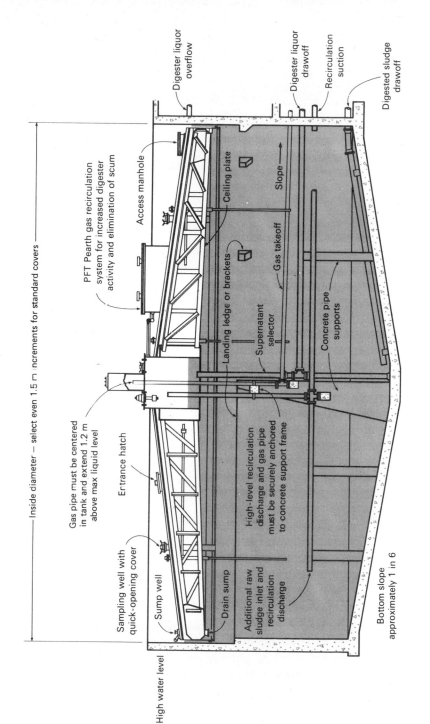

High water level

Sampling well with
quick-opening cover

Sump well

Drain sump

Additional raw
sludge inlet and
recirculation
discharge

High-level recirculation
discharge and gas pipe
must be securely anchored
to concrete support frame

Bottom slope
approximately 1 in 6

Gas pipe must be centered
in tank and extend 1.2 m
above max liquid level

Entrance hatch

PFT Pearth gas recirculation
system for increased digester
activity and elimination of scum

Access manhole

Inside diameter — select even 1.5 m increments for standard covers

Ceiling plate

Landing ledge or brackets

Supernatant
selector

Gas takeoff

Slope

Concrete pipe
supports

Digester liquor
overflow

Digester liquor
drawoff

Recirculation
suction

Digested sludge
drawoff

Figure 11-12 Cross section through a typical standard-rate digester [*from PFT*].

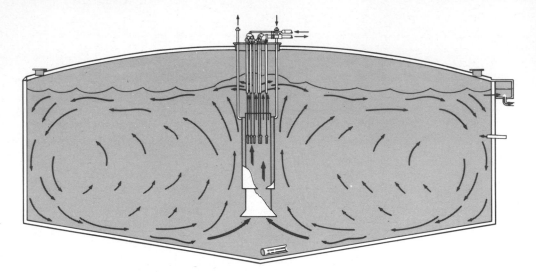

Figure 11-13 Section through gas-mixed high-rate sludge digestion tank [*from Walker Process Equipment Division, Chicago Bridge & Iron Company*].

multiple sludge drawoffs replace the supernatant drawoffs, and the tank should be deeper, if practicable, to aid the mixing process for the high-rate digester.

Sludge should be pumped to the digester continuously or by time clock on a 30-min to 2-h cycle. The incoming sludge displaces digested sludge either to a holding tank (its capacity determined by subsequent disposal methods) or to a second digester for supernatant separation and residual-gas extraction. Because there is no supernatant separation in the high-rate digester, and the total solids are reduced by 45 to 50 percent and given off as gas, the digested sludge is about half as concentrated as the untreated sludge feed. A large digester installation is shown in Fig. 11-14.

Two-stage digestion In the two-stage digestion process, the first tank is used for digestion. It is heated and equipped with mixing facilities consisting of one or more of the following: (1) sludge-recirculation pumps; (2) gas recirculation using short mixing tubes, one or more deep-draft tubes, or bottom-mounted diffusers; (3) mechanical draft-tube mixers, and (4) turbine and propeller mixers. The second tank is used for storage and concentration of digested sludge and for formation of a relatively clear supernatant. Frequently, the tanks are made identical, in which case either one may be the primary. In other cases, the second tank may be an open tank, an unheated tank, or a sludge lagoon. Tanks may have fixed roofs or floating covers. Any or all of the floating covers may be of the gas holder type. Alternatively, gas may be stored in a separate gas holder or compressed and stored under pressure. Tanks are usually circular and seldom less than 6 m (20 ft) or more than 35 m (115 ft) in diameter. They should have a water depth

Figure 11-14 Aerial view of several large anaerobic digesters (Toronto, Canada).

of not less than 7.5 m (25 ft) at the center and may be as deep as 14 m (45 ft) or more. The bottom should slope to the sludge drawoff in the center, with a minimum slope of 1 vertical to 4 horizontal.

Process Design

Ideally, the design of anaerobic sludge-digestion processes should be based on an understanding of the fundamental principles of biochemistry and microbiology discussed in Chap. 9. Because these principles have not been appreciated fully in the past, a number of empirical methods have also been used in the design of digesters. Therefore, the purpose of this discussion is to illustrate the various methods that have been used to design digesters in terms of size. These methods are based on (1) the concept of mean cell residence time, (2) the use of volumetric loading factors, (3) observed volume reduction, and (4) loading factors based on population.

Mean cell residence time Digester design based on mean cell residence time involves application of the principles discussed in Chap. 9. To review briefly, the respiration and oxidation end products of anaerobic digestion are methane gas and carbon dioxide. The quantity of methane gas can be calculated using Eq. 11-4 [20]:

$$V_{CH_4} = (0.35 \text{ m}^3/\text{kg})\{[EQS_0(10^3 \text{ g/kg})^{-1}] - 1.42(P_x)\} \qquad (11\text{-}4)$$

where V_{CH_4} = volume of methane produced, m^3/d
   0.35 = theoretical conversion factor for the amount of methane produced from the conversion of 1 kg of BOD_L
   E = efficiency of waste utilization
   Q = flowrate, m^3/d
   S_0 = ultimate BOD_L of influent, g/m^3
   1.42 = conversion factor for cell tissue to BOD_L
   P_x = net mass of cell tissue produced per day, kg/d

The derivation of the constant for converting the quantity of BOD_L converted to methane is derived in Example 11-3. The efficiency of waste utilization E normally ranges from 0.6 to 0.9 under satisfactory operating conditions.

Example 11-3 Conversion of BOD_L to methane gas Determine the amount of methane produced per kilogram of ultimate BOD_L stabilized. Assume that the starting compound is glucose $(C_6H_{12}O_6)$.

SOLUTION
1   Write a balanced equation for the conversion of glucose to CO_2 and CH_4 under anaerobic conditions.

$$C_6H_{12}O_6 \rightarrow 3CO_2 + 3CH_4$$
$$180 \qquad 132 \qquad 48$$

Note that although the glucose has been converted, the methane has an oxygen requirement for complete conversion to carbon dioxide and water.

2   Write a balanced equation for the oxidation of methane to CO_2 and H_2O, and determine the kilograms of methane formed per kilogram of BOD_L.

$$3CH_4 + 6O_2 \rightarrow 3CO_2 + 6H_2O$$
$$48 \qquad 192$$

Using this and the previous equation, the ultimate BOD_L per kilogram of glucose is (192/180) kg, and 1.0 kg of glucose yields (48/180) kg of methane, so that the ratio of the amount of methane produced per kilogram of BOD_L converted is

$$\frac{kg\ CH_4}{kg\ BOD_L} = \frac{48/180}{192/180} = 0.25$$

Therefore, for each kilogram of BOD_L converted, 0.25 kg of methane is formed.

3   Determine the volume equivalent of the 0.25 kg of methane produced from the stabilization of 1.0 kg of BOD_L.

$$V_{CH_4} = (0.25\ kg)(10^3\ g/kg)\frac{1\ mol}{16\ g}\frac{22.4\ L}{mol}(10^3\ L/m^3)^{-1}$$

$$= 0.35\ m^3\ CH_4 \text{(at standard conditions)}$$

Therefore, 0.35 m^3 of methane is produced per kilogram of ultimate BOD_L converted.

Table 11-13 Suggested mean cell residence times for use in the design of continuous-flow stirred-tank digesters [19, 20]

Operating temperature, °C	θ_c^M, d	θ_c^d, d suggested for design
18	11	28
24	8	20
30	6	14
35	4	10
40	4	10

Note: $1.8(°C) + 32 = °F$

For a continuous-flow stirred-tank high-rate digester without recycle, the mass of biological solids synthesized daily P_x can be estimated using Eq. 11-5, which is a combination of Eqs. 9-41 and 11-3.

$$P_x = \frac{YQ(ES_0) \times (10^3 \text{ g/kg})^{-1}}{1 + k_d \theta_c} \tag{11-5}$$

where P_x = net mass of cell tissue produced, kg/d
Y = yield coefficient, mg/mg or g/g
Q = flowrate, m³/d
E = efficiency of waste stabilization
S_0 = ultimate BOD$_L$ of influent, g/m³
k_d = endogenous coefficient, d⁻¹
θ_c = mean cell residence time, d

(Note that for a continuous-flow stirred-tank flow-through digester, θ_c is the same as the hydraulic retention time θ.) Values for Y and k_d as found for various types of waste are given in Table 9-7. Typical values for θ_c for various temperatures are reported in Table 11-13. The application of Eqs. 11-4 and 11-5 in the process design of a high-rate digester is illustrated in Example 11-4.

Example 11-4 Estimation of digester volume and performance Estimate the size of digester required to treat the sludge from a preliminary treatment plant designed to treat 30,000 m³/d (10 Mgal/d) of wastewater. Check the volumetric loading, and estimate the percent stabilization and the amount of gas produced per capita. For the wastewater to be treated, it has been found that the quantity of dry solids and BOD$_L$ removed is 0.15 kg/m³ (1250 lb/Mgal) and 0.14 kg/m³ (1165 lb/Mgal), respectively. Assume that the sludge contains about 95 percent moisture and has a specific gravity of 1.02. Other pertinent design assumptions are as follows:

SOLUTION

1. The hydraulic regime of the reactor is continuous-flow stirred-tank.
2. $\theta_c = 10$ days at 35°C (see Table 11-13).
3. Efficiency of waste utilization $E = 0.80$.
4. The waste contains adequate nitrogen and phosphorus for biological growth.
5. $Y = (0.05\text{-kg cells})/(\text{kg BOD utilized})$ and $k_d = 0.03$ d^{-1}.
6. Constants are for a temperature of 35°C.

1 Compute the daily sludge volume and BOD$_L$ loading.

$$\text{Sludge volume} = \frac{(0.15 \text{ kg/m}^3)(38,000 \text{ m}^3/\text{d})}{1.02(1000 \text{ kg/m}^3)(0.05)}$$

$$= 111.8 \text{ m}^3/\text{d} \ (3950 \text{ ft}^3/\text{d})$$

$$\text{BOD}_L \text{ loading} = (0.14 \text{ kg/m}^3)(38,000 \text{ m}^3/\text{d})$$

$$= 5320 \text{ kg/d} \ (11,729 \text{ lb/d})$$

2 Compute the digester volume.

$$\frac{V}{Q} = \theta_c$$

where Q = sludge flowrate
$$V = Q\theta_c$$
$$= (111.8 \text{ m}^3/\text{d})(10 \text{ d})$$
$$= 1118 \text{ m}^3 \ (39,481 \text{ ft}^3) \quad \text{(also volume of first-stage digester in two-stage process)}$$

3 Compute the volumetric loading.

$$\text{kg BOD}_L/\text{m}^3 \cdot \text{d} = \frac{5320 \text{ kg/d}}{1118 \text{ m}^3} = 4.76 \text{ kg/m}^3 \cdot \text{d}$$

4 Compute the quantity of volatile solids produced per day using Eq. 11-5.

$$P_x = \frac{YQ(ES_0) \times (10^3 \text{ g/kg})^{-1}}{1 + k_d\theta_c}$$

$$QES_0 \times (10^3 \text{ g/kg})^{-1} = (0.8)(5320 \text{ kg/d}) = 4256 \text{ kg/d}$$

$$P_x = \frac{(0.05)(4256 \text{ kg/d})}{1 + (0.03 \text{ d}^{-1})(10 \text{ d})}$$

$$= 163.7 \text{ kg/d} \ (361 \text{ lb/d})$$

5 Compute the percent stabilization.

$$\text{Percent stabilization} = \frac{QES_0 \times (10^3 \text{ g/kg})^{-1} - 1.42(P_x)}{QS_0 \times (10^3 \text{ g/kg})^{-1}} 100$$

$$= \frac{4256 \text{ kg/d} - 1.42(163.7 \text{ kg/d})}{5320 \text{ kg/d}} 100$$

$$= 76\%$$

6 Compute the volume of methane produced per day using Eq. 11-4.

$$V_{CH_4} = (0.35 \text{ m}^3/\text{kg})[(EQS_0 \times (10^3 \text{ g/kg})^{-1} - 1.42(P_x)]$$

$$= (0.35 \text{ m}^3/\text{kg})(4256 \text{ kg/d} - 232.5 \text{ kg/d})$$

$$= 1408 \text{ m}^3/\text{d} (49{,}730 \text{ ft}^3/\text{d})$$

7 Estimate the total gas production. Because digester gas is about two-thirds methane, the total volume of gas produced is:

$$\text{Total gas volume} = \frac{1408 \text{ m}^3/\text{d}}{0.67}$$

$$= 2102 \text{ m}^3/\text{d} (74{,}210 \text{ ft}^3/\text{d})$$

Loading factors One of the most common methods used to size digesters is to determine the required volume on the basis of a loading factor. Although a number of different factors have been proposed, the two that seem most favored are based on (1) the kilograms of volatile solids added per day per cubic meter of digester capacity and (2) the kilograms of volatile solids added per day per kilogram of volatile solids in the digester. From the information presented in Chap. 9, the similarity between these loading factors and the food-to-microorganism ratio is apparent. In applying these loading factors, another factor that should also be checked is the hydraulic detention time, because of its relationship to organism growth and washout (see Table 11-13) and to the type of digester used (for example, only 50 percent or less of the capacity of a conventional standard-rate single-stage digester is effective).

Ideally, the conventional single-stage digestion tank is stratified into three layers with the supernatant at the top, the active digestion zone in the middle, and the thickened sludge at the bottom. Because of the storage requirements for the digested sludge and the supernatant, and the excess capacity provided for daily fluctuations in sludge loading, the volumetric loading for standard-rate digesters is low. Detention times based on cubic meters of untreated sludge pumped vary from 30 to more than 90 d for this type of tank. The recommended solids loadings for standard-rate digesters are from 0.5 to 1.6 $\text{kg/m}^3 \cdot \text{d}$ (0.03 to 0.10 $\text{lb/ft}^3 \cdot \text{d}$) of volatile solids.

For high-rate digesters, loading rates of 1.6 to 6.4 $\text{kg/m}^3 \cdot \text{d}$ (0.10 to 0.40 $\text{lb/ft}^3 \cdot \text{d}$) of volatile solids and hydraulic detention periods of 10 to 20 d are practicable. The six high-rate digestion tanks at the Newtown Creek Plant of New York City are designed for a volatile-solids loading of 3.43 $\text{kg/m}^3 \cdot \text{d}$ (0.214 $\text{lb/ft}^3 \cdot \text{d}$) and a detention period of 17.6 d with an untreated sludge concentration of 8 percent solids. The tanks are also designed so that four tanks can handle the entire load and the other two can be used for storage and residual gas extraction. Under these conditions, the volatile-solids loading becomes 5.13 $\text{kg/m}^3 \cdot \text{d}$ (0.32 $\text{lb/ft}^3 \cdot \text{d}$) and the detention period 11.7 d [9]. Four draft-tube mixers in each tank are designed to turn over the entire tank contents in 30 min. The effect of sludge concentration and hydraulic detention time on the volatile-solids loading factor is reported in Table 11-14.

Table 11-14 Effect of sludge concentration and hydraulic detention time on volatile-solids loading factor[a]

Sludge concentration, %	Volatile-solids loading factor, kg/m³ · d			
	10 d[b]	12 d	15 d	20 d
4	3.06	2.55	2.04	1.53
5	3.83	3.19	2.55	1.91
6	4.59	3.83	3.06	2.30
7	5.36	4.46	3.57	2.68
8[c]	6.12	5.10	4.08	3.06
9	6.89	5.74	4.59	3.44
10	7.65	6.38	5.10	3.83

[a] Based on 75 percent volatile content of sludge, and a sludge specific gravity of 1.02 (concentration effects neglected).
[b] Hydraulic detention time, d.
[c] The volatile-solids loading factor at 11.7 and 17.6 d is 5.23 and 3.48 kg/m³ · d (see text discussion).
Note: kg/m³ · d × 0.0624 = lb/ft³ · d

The degree of stabilization obtained is also often measured by the percent reduction in volatile solids. This can be related either to the mean cell residence time or to the detention time based on the untreated sludge feed. Since the untreated sludge feed can be measured easily, this method is more commonly used. In plant operation, this calculation should be made routinely as a matter of record whenever sludge is drawn to processing equipment or drying beds. The results of this calculation can also be used as a guide for scheduling the withdrawal of sludge.

In calculating the volatile-solids reduction, it is assumed that the ash content of the sludge is conservative; that is, the number of pounds of ash going into the digester is equal to that being removed. A typical calculation of volatile-solids reduction is presented in Example 11-5.

Example 11-5 Determination of volatile-solids reduction Determine the total volatile-solids reduction achieved during digestion from the following analysis of untreated and digested sludge. It is assumed that (1) the weight of fixed solids in the digested sludge equals the weight of fixed solids in the untreated sludge and (2) the volatile solids are the only constituent of the untreated sludge lost during digestion.

	Volatile solids, %	Fixed solids, %
Untreated sludge	70	30
Digested sludge	50	50

SOLUTION

1 Determine the weight of the digested solids. Because the quantity of fixed solids remains the same, the weight of the digested solids based on 1.0 kg of dry untreated sludge, as computed below, is 0.6 kg.

$$\text{Fixed-solids untreated sludge, } 30\% = \frac{(0.3 \text{ kg})(100)}{0.3 \text{ kg} + 0.7 \text{ kg}}$$

Let x equal the weight of volatile solids after digestion. Then

$$\text{Fixed-solids digested sludge, } 50\% = \frac{(0.3 \text{ kg})(100)}{0.3 \text{ kg} + x}$$

$$x = \frac{(0.3 \text{ kg})(100)}{50} - 0.3 \text{ kg} = 0.3 \text{ kg to volatile solids}$$

$$\text{Weight of digested solids} = 0.3 \text{ kg} + x = 0.6 \text{ kg}$$

2 Determine the percent reduction in total and volatile solids.

a Percent reduction of total solids

$$R_{TSS} = \frac{(1.0 - 0.6 \text{ kg})(100)}{1.0 \text{ kg}} = 40\%$$

b Percent reduction in volatile solids

$$R_{VSS} = \frac{(0.7 - 0.3 \text{ kg})(100)}{0.7 \text{ kg}} = 57.1\%$$

Volume reduction It has been observed that as digestion proceeds, if the supernatant is withdrawn and returned to the head end of the treatment plant, the volume of the remaining sludge decreases approximately exponentially. If a plot is prepared of the remaining volume versus time, the required volume of the digester is represented by the area under the curve. It can be computed using Eq. 11-6:

$$V = [V_f - \tfrac{2}{3}(V_f - V_d)]t \tag{11-6}$$

where V = volume of digester, m³ (ft³)
V_f = volume of fresh sludge added per day, m³/d (ft³/d)
V_d = volume of digested sludge removed per day, m³/d (ft³/d)
t = digestion time, d

Population basis Digestion tanks are also designed on a volumetric basis by allowing a certain number of cubic meters per capita. Detention times of 35 to 45 d are recommended for design based on total tank volume, plus additional storage volume if sludge is dried on beds and weekly sludge drawings are curtailed because of inclement weather.

Based on 120 g (0.27 lb) of suspended solids per capita in the untreated wastewater (see Table 3-9), these requirements translate into the number of cubic meters per capita shown in Table 11-15. The table also includes the requirements of the *Ten States Standards* for comparison. These requirements are for heated

Table 11-15 Digestion-tank-capacity requirements[a]

	Wet sludge			Volume required	
Type of plant	Dry solids, g/capita · d	Percent solids	m³/10³ capita · d	35- to 45-d detention, m³/10³ capita	*Ten States Standards.* m³/10³ capita
Primary	72	5	1.44	50–65	56–85
Primary and trickling filter	108	4	2.70	95–122	113–142
Primary and activated sludge	114	3	3.8	133–171	113–170

[a] Based on 120 g (0.20 lb) of suspended solids per capita per day in untreated wastewater.
Note: g × 0.0022 = lb
      m³ × 35.3147 = ft³

tanks and are applied where analyses and volumes of sludge to be digested are not available. For unheated tanks, capacities must be increased, depending on local climatic conditions and the storage volume required. The capacities shown in Table 11-15 should be increased 60 percent in a municipality where the use of garbage grinders is universal and should be increased on a population-equivalent basis to allow for the effect of industrial wastes.

Gas Production, Collection, and Utilization

Wastewater gas contains about 65 to 70 percent CH_4 by volume, 25 to 30 percent CO_2, and small amounts of N_2, H_2, and other gases. It has a specific gravity of approximately 0.86 referred to air. Because production of gas is one of the best measures of the progress of digestion and because it can be used as fuel, the designer should be familiar with its production, collection, and utilization.

Gas production The volume of methane gas produced during the digestion process can be estimated using Eq. 11-4, which has been discussed previously. For example, the volume of methane gas produced from 1 kg (2.2 lb) of cells is equal to about 0.5 m^3 (17.5 ft^3). This value is obtained by multiplying the kilograms of cells by 1.42 to convert to ultimate BOD and then by multiplying by 0.35 to obtain the cubic meters of methane (see Example 11-4).

Total gas production is usually estimated from the volatile-solids loading of the digester or from the percentage of volatile-solids reduction. Typical values are from 0.5 to 0.75 m^3/kg (8 to 12 ft^3/lb) of volatile solids added, and from 0.75 to 1.12 m^3/kg (12 to 18 ft^3/lb) of volatile solids destroyed. Gas production can fluctuate over a wide range, depending on the volatile-solids content of the sludge feed and the biological activity in the digester. Excessive gas-production rates sometimes occur during startup and may cause foaming and escape of foam and gas from around the edges of floating digester covers. If stable operating conditions have been achieved and the foregoing gas-production rates are being maintained, the operator can be assured that the result will be a well-digested sludge.

Gas production can also be crudely estimated on a per capita basis. The normal yield is 15 to 22 m^3/10^3 persons · d (0.6 to 0.8 ft^3/person · d) in primary plants treating normal domestic wastewater. In secondary treatment plants, this is increased to about 28 m^3/10^3 persons · d (1.0 ft^3/person · d).

Gas collection Floating covers fit on the surface of the digester contents and allow the volume of the digester to change without allowing air to enter the digester. Gas and air must not be allowed to mix, or an explosive mixture may result. Explosions have occurred in wastewater treatment plants. Gas piping and pressure-relief valves must include adequate flame traps. The covers may also be installed to act as gas holders that store a small quantity of gas under pressure and act as reservoirs. This type of cover can be used for single-stage digesters or in the second stage of two-stage digesters.

Fixed covers provide a free space between the roof of the digester and the liquid surface. Gas storage must be provided so that (1) when the liquid volume is changed, gas, and not air, will be drawn into the digester, and (2) gas will not be lost by displacement. Gas can be stored either at low pressure in gas holders that use floating covers or at high pressure if gas compressors are used. Gas not used should be burned in a flare. Gas meters should be installed to measure gas produced and gas used or wasted.

Gas utilization One cubic meter of methane at standard temperature and pressure has a net heating value of 35,800 kJ/m^3 (960 Btu/ft^3). Since digester gas is only 65 percent methane, the low heating value of digester gas is approximately 22,400 kJ/m^3 (600 Btu/ft^3). By comparison, natural gas, which is a mixture of methane, propane, and butane, has a low heating value of approximately 37,300 kJ/m^3 (1000 Btu/ft^3).

In large plants, digester gas may be used as fuel for boiler and internal combustion engines which are, in turn, used for pumping wastewater, operating blowers, and generating electricity. Hot water from heating boilers or from engine jackets and exhaust-heat boilers may be used for sludge heating and for building heating, or gas-fired sludge-heating boilers may be used.

Digester Mixing

Proper mixing is one of the most important considerations in achieving optimum process performance. Various systems for mixing the contents of the digester have been used. The most common ones involve the use of (1) single- or multiple-draft tubes through which the sludge is circulated by a turbine mixer located within the tube (see Fig. 11-15a), and (2) gas recirculated through diffusers in the base of the digester or by means of drop pipes (see Fig. 11-15b).

Digester Heating

The heat requirements of digesters consist of the amount needed (1) to raise the incoming sludge to digestion-tank temperatures, (2) to compensate for the heat losses through walls, floor, and roof of the digester, and (3) to make up the losses that might occur in the piping between the source of heat and the tank. The sludge in digestion tanks is heated by pumping the sludge and supernatant through external heat exchangers and back to the tank (see Fig. 11-16).

Analysis of heat requirements In computing the energy required to heat the incoming sludge to the temperature of the digester, it is assumed that the specific heat of most sludges is essentially the same as that of water. This assumption has proved to be quite acceptable for engineering computations. The

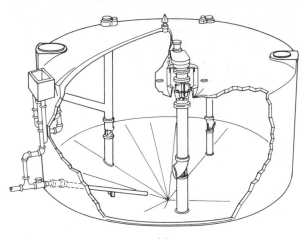

(a)

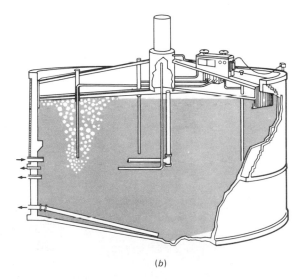

(b)

Figure 11-15 Devices used for mixing contents of anaerobic high-rate digesters. (a) Draft tube mixers. (*From Dorr-Oliver.*) (b) Gas recirculation mixers. (*From PFT-Pearth.*)

heat loss through the digester sides, top, and bottom are computed using the following expression:

$$q = UA \, \Delta T \qquad (11\text{-}7)$$

where q = heat loss, W (Btu/h)
U = overall coefficient of heat transfer, W/m$^2 \cdot$ °C (Btu/h $\cdot$ ft$^2 \cdot$ °F)
A = cross-sectional area through which the heat loss is occurring, m^2 (ft^2)
ΔT = temperature drop across the surface in question, °C (°F)

In computing the heat losses from a digester using Eq. 11-7, it is common practice to consider the characteristics of the various heat-transfer surfaces

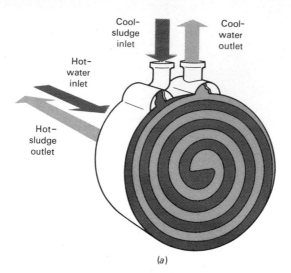

Cool-
sludge
inlet

Cool-
water
outlet

Hot-
water
inlet

Hot-
sludge
outlet

(a)

(b)

Figure 11-16 Heat exchangers used for heating digesting sludge. (a) Schematic of countercurrent spiral heat exchanger. (*From Dorr-Oliver.*) (b) Countercurrent tube and shell type heat exchanger.

separately and to develop transfer coefficients for each one. The application of Eq. 11-7 in the computation of digester heating requirements is illustrated in Example 11-7.

Heat-transfer coefficients Typical overall heat-transfer coefficients are reported in Table 11-16. As shown, separate entries are included for the walls, bottom, and top of the digester. Data are also presented on the transfer from pipe coils.

Table 11-16 Typical values for the overall coefficients of heat transfer for computing digester heat losses

Item	U, W/m$^2 \cdot$ °C
Plain concrete walls (above ground)	
300-mm-thick wall with air space plus facing	1.8–2.4
300-mm-thick wall with insulation	0.6–0.8
Plain concrete walls (below ground)	
Surrounded by dry earth	0.57–0.68
Surrounded by moist earth	1.1–1.4
Plain concrete floors in contact with moist earth	0.68–0.85
Floating covers	
With 35-mm wood deck, built-up roofing, and no insulation	1.8–2.0
With 25-mm insulating board installed under the roofing	0.9–1.0
Fixed concrete covers	
100 mm thick and covered with built-up roofing, not insulated	4.0–5.0
100 mm thick and covered, but insulated with 25-mm-thick insulating board	1.2–1.6

Note: W/m$^2 \cdot$ C × 0.1761 = Btu/h $\cdot$ ft$^2 \cdot$ °F

Digestion-tank walls may be surrounded by earth embankments that serve as insulation, or they may be of compound construction consisting of approximately 300 mm (12 in) of concrete, corkboard insulation, or an insulating air space, plus brick facing or corrugated aluminum facing over rigid insulation. The heat transfer from plain concrete walls below ground level and from floors depends on whether they are below the groundwater level. If the groundwater level is not known, it may be assumed that the sides of the tank are surrounded by dry earth and that the bottom is saturated earth. Since the heat losses from the tank warm up the adjacent earth, it is assumed that the earth forms an insulating blanket 1.5 to 3 m (5 to 10 ft) thick before stable ambient earth temperatures are reached. In northern climates, frost may penetrate to a depth of 1.2 m (4 ft). Therefore, the ground temperature can be assumed to be 0°C (32°F) at this depth and to vary uniformly above this depth to the design air temperatures at the surface, and below this depth to normal winter ground temperatures, which are 5 to 10°C (10 to 20°F) higher at the base of the wall. Alternatively, an average temperature may be assumed for the entire wall below grade.

The loss through the roof depends on the type of construction, the absence or presence of insulation and its thickness, the presence of air space (as in floating covers between the skin plate and the roofing), and whether the underside of the roof is in contact with sludge liquor or gas. Typical values of U are reported in Table 11-16.

Radiation from roofs and above-ground walls also contributes to heat losses. At the temperatures involved, the effect is small and is included in the coefficients normally used, such as those given in the foregoing discussion. For the theory of

radiant-heat transmission, the reader is referred to McAdams [17]. Heat requirements for a digester are determined in Example 11-6.

When external heaters are installed, the sludge is pumped at high velocity
through the tubes while water circulates at high velocity around the tubes. This
promotes high turbulence on both sides of the heat-transfer surface and results
in higher heat-transfer coefficients and better heat transfer. Another advantage
of external heaters is that untreated cold sludge on its way into the digesters can
be warmed, intimately blended, and seeded with sludge liquor before entering the
tank.

Example 11-6 Digester heating requirements A digester with a capacity of 45,000 kg/d (100,000 lb/d)
of sludge is to be heated by circulation of sludge through an external hot-water heat exchanger.
Assuming that the following conditions apply, find the heat required to maintain the required
digester temperature. If all heat were shut off for 24 h, what would be the average drop in
temperature of the tank contents?

1. Concrete digester dimensions:
   Diameter $= 18$ m (60 ft)
   Side depth $= 6$ m (20 ft)
   Mid-depth $= 9$ m (30 ft)
2. Heat-transfer coefficients:
   Dry earth embanked for entire depth, $U = 0.68$ W/m² · °C
   Floor of digester in groundwater, $U = 0.85$ W/m² · °C
   Roof exposed to air, $U = 0.91$ W/m² · °C
3. Temperatures:
   Air, -5°C (23°F)
   Earth next to wall, 0°C (32°F)
   Incoming sludge, 10°C (50°F)
   Earth below floor, 5°C (42°F)
   Sludge contents in digester, 32°C (90°F)
4. Specific heat of sludge $= 4200$ J/kg · °C (1.0 Btu/lb · °F)

SOLUTION
1   Compute the heat requirement for the sludge.

$$q = 45,000(32 - 10)(4200) = 4.16 \times 10^9 \text{ J/d } (3,940,000 \text{ Btu/d})$$

2   Compute the area of the walls, roof, and floor.

$$\text{Wall area} = \pi(18)(6) = 339.3 \text{ m}^2 \ (3652 \text{ ft}^2)$$

$$\text{Floor area} = \pi(9)(9^2 + 3^2)^{1/2} = 268.2 \text{ m}^2 \ (2887 \text{ ft}^2)$$

$$\text{Roof area} = \pi(9^2) = 254.5 \text{ m}^2 \ (2739 \text{ ft}^2)$$

3   Compute the heat losses by conduction using Eq. 11-7.

$$q = UA \, \Delta T$$

   a   Walls:

$$q = 0.68(339.3)(32 - 0)(24 \times 60 \times 60)$$

$$= 6.38 \times 10^8 \text{ J/d } (25,200 \text{ Btu/h})$$

b Floor:

$$q = 0.85(268.2)(32 - 5)(24 \times 60 \times 60)$$

$$= 5.32 \times 10^8 \text{ J/d (21,000 Btu/h)}$$

c Roof:

$$q = 0.91(254.5)(32 + 5)(24 \times 60 \times 60)$$

$$= 7.40 \times 10^8 \text{ J/d (29,200 Btu/h)}$$

d Total losses:

$$q_t = (6.38 + 5.32 + 7.4)(10^8)$$

$$= 19.1 \times 10^8 \text{ J/d (74,400 Btu/h)}$$

4 Compute the required heat-exchanger capacity.

Capacity = heat required for sludge and heat required for digester

$$= (41.6 + 19.1) \times 10^8 \text{ J/d}$$

$$= 60.7 \times 10^8 \text{ J/d (240,000 Btu/h)}$$

5 Determine the effect of heat shutoff.

a Digester volume $= \pi(9)^2(6 + \frac{2}{3})$

$$= 1781.3 \text{ m}^3 \text{ (471,000 gal)}$$

b Weight of sludge $= 1781.3 \times 1000$

$$= 1.7813 \times 10^6 \text{ kg (3,927,100 lb)}$$

c Drop in temperature $= \dfrac{(60.7 \times 10^8 \text{ J/d})(1.0 \text{ d})}{(1.7813 \times 10^6 \text{ kg})(4200 \text{ J/kg} \cdot {}^\circ\text{C})}$

$$= 0.81{}^\circ\text{C (1.46}{}^\circ\text{F)}$$

11-8 STABILIZATION: AEROBIC SLUDGE-DIGESTION PROCESS

Aerobic digestion is an alternative process for stabilizing organic sludges produced from various treatment operations. It may be used to treat only (1) waste activated sludge, (2) mixtures of waste activated sludge or trickling-filter sludge and primary sludge, or (3) waste sludge from activated-sludge-treatment plants designed without primary settling. To date, aerobic digestion has been used primarily in small plants, particularly those using extended aeration and contact stabilization.

Advantages claimed for aerobic digestion as compared to anaerobic digestion are as follows: (1) volatile-solids reduction is approximately equal to that obtained anaerobically; (2) lower BOD concentrations in supernatant liquor; (3) production of an odorless, humuslike, biologically stable end product that can be disposed of easily; (4) production of a sludge with excellent dewatering characteristics; (5) recovery of more of the basic fertilizer values in the sludge;

(6) fewer operational problems; and (7) lower capital cost [2]. The major disadvantage of the aerobic digestion process appears to be the higher power cost associated with supplying the required oxygen, but some recent process developments may invalidate this objection. That a useful by-product such as methane is not recovered may also be a disadvantage. Comparing the advantages and disadvantages, it appears that aerobic digestion will increase in popularity as more reliable information on process kinetics and economics is developed.

Process Description

Aerobic digestion is similar to the activated-sludge process. As the supply of available substrate (food) is depleted, the microorganisms begin to consume their own protoplasm to obtain energy for cell-maintenance reactions. When this occurs, the microorganisms are said to be in the endogenous phase. As shown in Eq. 9-6, cell tissue is oxidized aerobically to carbon dioxide, water, and ammonia. In actuality, only about 75 to 80 percent of the cell tissue can be oxidized; the remaining 20 to 25 percent is composed of inert components and organic compounds that are not biodegradable. The ammonia from this oxidation is subsequently oxidized to nitrate as digestion proceeds. The resulting overall reaction is given by the following equation:

$$C_5H_7NO_2 + 7O_2 \rightarrow 5CO_2 + NO_3^- + 3H_2O + H^+ \qquad (11\text{-}8)$$

As shown by Eq. 11-8, a pH drop can occur when ammonia is oxidized to nitrate if the alkalinity of the wastewater is insufficient to buffer the solution. Theoretically, about 7.1 kg of alkalinity, expressed as $CaCO_3$, are destroyed per kilogram of ammonia oxidized. In situations where the buffer capacity is insufficient, it may be necessary to install chemical feed equipment to maintain the desired pH.

Where activated or trickling-filter sludge is mixed with primary sludge and the combination is to be aerobically digested, there will be both direct oxidation of the organic matter in the primary sludge and endogenous oxidation of the cell tissue. Aerobic digestors can be operated as batch or continuous flow reactors (see Fig. 11-17).

At present, two proven variations of the process are available: (1) conventional aerobic digestion, and (2) pure-oxygen aerobic digestion. A third variation, thermophilic aerobic digestion, is currently under intensive investigation. These three variations are described in the following discussion. Aerobic digestion accomplished with air is the most commonly used process, so it is considered in greater detail.

Conventional-Air Aerobic Digestion

Factors that must be considered in designing aerobic digesters include hydraulic residence time, process loading criteria, oxygen requirements, energy requirements for mixing, environmental conditions, and process operation.

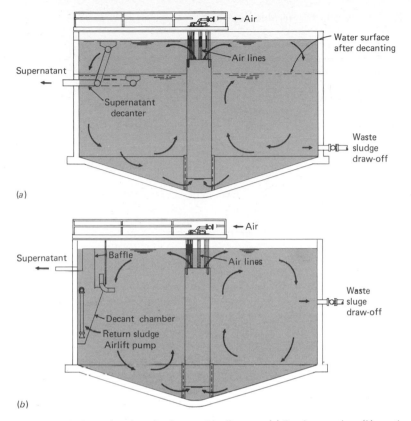

(a)

(b)

Figure 11-17 Operational modes for aerobic digesters. (*a*) Batch operation. (*b*) continuous operation.

Hydraulic residence time The amount of volatile solids in sludge is reduced more or less linearly up to a value of about 40 percent at a hydraulic detention time of about 10 to 12 d [10]. Although volatile-solids removal continues with increasing detention time, the rate of removal is reduced considerably. Depending on the temperature, the maximum reduction ranges between 45 and 70 percent. The required time and degree of volatile-solids removal also varies with the characteristics of the sludge. Typically, volatile-solids reductions vary from about 35 to 45 percent in 10 to 12 d at temperatures equal to or above 20°C (68°F). Based on these findings, recommended hydraulic detention times for aerobic digesters are given in Table 11-17. Temperature effects are discussed under "Environmental Conditions."

Loading criteria Limited information is currently available on appropriate loading criteria to use for this process. Typical values in terms of kilograms of volatile solids per cubic meter per day are given in Table 11-17. Because the hydraulic and mean cell residence times are nominally equivalent for this process (see "Process Operation"), loading criteria based on mean cell residence time

Table 11-17 Design criteria for aerobic digesters

Parameter	Value
Hydraulic detention time, days at 20°C[a]	
Waste activated sludge only	10–15
Activated sludge from plant operated without primary settling	12–18
Primary plus activated or trickling-filter sludge[b]	15–20
Solids loading, kg volatile solids/m$^3 \cdot$ d	1.6–4.8
Oxygen requirements, kg/kg destroyed	
Cell tissue[c]	~2.3
BOD$_5$ in primary sludge	1.6–1.9
Energy requirements for mixing	
Mechanical aerators, kW/10^3 m^3	20–40
Air mixing, m^3/10^3 m$^3 \cdot$ min	20–40
Dissolved oxygen level in liquid, mg/L	1–2

[a] Detention times should be increased for temperatures below 20°C (68°F). If sludge cannot be withdrawn during certain periods (e.g., weekends, rainy weather) additional storage capacity should be provided.

[b] Similar detention times are used for primary only.

[c] Ammonia produced during carbonaceous oxidation oxidized to nitrate (see Eq. 11-8).

> *Note:*   kg/m$^3 \cdot$ d $\times$ 0.0624 = lb/ft$^3 \cdot$ d
>   kW/10^3 m^3 $\times$ 0.0380 = hp/10^3 ft^3
>   m^3/10^3 m^3 = ft^3/10^3 ft^3
>   1.8(°C) + 32 = °F
>   mg/L = g/m^3

would appear to be most satisfactory. The maximum solids concentration would be governed by oxygen-transfer and mixing requirements.

If only waste activated sludge is to be aerobically digested, the mean cell residence time required to achieve a given volatile-solids reduction can be estimated using Eq. 5-64. Representative values for the decay coefficient k may be about 0.05 to 0.07/d.

Oxygen requirements The oxygen requirements that must be satisfied during aerobic digestion are those of the cell tissue and, with mixed sludges, the BOD$_5$ in the primary sludge. The oxygen requirement for the complete oxidation of cell tissue, computed using Eq. 11-8, is equal to 7 mol/mol of cells, or about 2 kg/kg of cells. The oxygen requirement for the complete oxidation of the BOD$_5$ contained in primary sludge varies from about 1.7 to 1.9 kg/kg destroyed.

On the basis of operating experience, it has been found that if the dissolved-oxygen concentration in the digester is maintained at 1 to 2 mg/L and the detention time is greater than 10 d, the sludge dewaters well [10].

Energy requirements for mixing To ensure proper operation, the contents of the aerobic digester should be well mixed. In general, because of the amount of air that must be supplied to meet the oxygen requirement, mixing should be achieved; nevertheless, power mixing requirements should be checked (see Table 11-17).

Environmental conditions Of the many environmental factors considered in Chaps. 9 and 10, temperature and pH play an important role in the operation of aerobic digesters. It has been observed that the operation of aerobic digesters is temperature-dependent, especially at temperatures below 20°C (68°F). On the basis of extremely limited data, it appears that a temperature coefficient in the range of 1.08 to 1.10 might be appropriate for adjusting the hydraulic detention time for temperatures below 20°C for hydraulic residence times on the order of 15 d. As the hydraulic detention time is increased to about 60 d, the effect of temperature is negligible. In extremely cold climates, consideration should be given to heating the sludge or the air supply, covering the tanks, or both.

Depending on the buffering capacity of the system, the pH may drop to a rather low value (5.5±) at long hydraulic detention times. Reasons advanced for this include the increased presence of nitrate ions in solution and the lowering of the buffering capacity due to air stripping [10]. Although this does not seem to inhibit the process, the pH should be checked periodically and adjusted if found to be excessively low.

Process operation In the past, aerobic digestion has normally been conducted in unheated tanks similar to those used in the activated-sludge process (see Fig. 11-18). However, as understanding of the thermophilic aerobic digestion

Figure 11-18 Overhead view of aerobic digester.

process increases, it is anticipated that more use will be made of well-insulated or even partially heated tanks. In some cases, existing anaerobic digesters have been converted and are being used as aerobic digesters. Aerobic digesters should be equipped with decanting facilities so that they may also be used to thicken the digested solids before discharging them to subsequent thickening facilities or sludge-drying beds. If the digester is operated so that the incoming sludge is used to displace supernatant and the solids are allowed to build up, the mean cell residence time will not be equal to the hydraulic residence time. The following example illustrates the design of an aerobic digester.

Example 11-7 Aerobic-digester design Design an aerobic digester to treat the waste sludge produced by the activated-sludge treatment plant in Example 10-1. Assume that the waste sludge is concentrated to 3 percent, using a gravity thickener. Also assume that the temperature is 20°C, and use a hydraulic detention time of 15 d.

SOLUTION

1   Compute the volume of sludge to be disposed of per day assuming a specific gravity of 1.03.

$$V_s = \frac{2057.1 \text{ kg/d}}{(1000 \text{ kg/m}^3)(1.03)(0.03)}$$

$$= 67 \text{ m}^3 \ (2331 \text{ ft}^3)$$

2   Determine the volume of the aerobic digester.

$$V = 67 \text{ m}^3 \times 15 \text{ d} = 1005 \text{ m}^3 \ (35,490 \text{ ft}^3)$$

3   Check the solids loading.

$$\text{kg volatile solids/m}^3 \cdot \text{d} = \frac{(2057.1 \text{ kg/d})(0.8)}{1005 \text{ m}^3}$$

$$= 1.64 \text{ kg/m}^3 \cdot \text{d} \ (0.01 \text{ lb/ft}^3 \cdot \text{d})$$

4   Determine the oxygen requirement, assuming that 40 percent of the cell tissue is oxidized completely (see Table 11-17 for oxygen requirements).

$$\text{kg O}_2/\text{d} = 2057.1(0.8)(0.4)(2.3) = 1514 \text{ kg/d} \ (3338 \text{ lb/d})$$

5   Compute the volume of air required at standard conditions (see Example 10-1).

$$\frac{1514 \text{ kg/d}}{(1.201 \text{ kg/m}^3)(0.232)} = 5434 \text{ m}^3/\text{d} \ (191,900 \text{ ft}^3/\text{d})$$

Assuming that the oxygen-transfer efficiency is 10 percent, the air requirement is

$$\frac{5434 \text{ m}^3/\text{d}}{0.10(1440 \text{ min/d})} = 37.7 \text{ m}^3/\text{min} \ (1333 \text{ ft}^3/\text{min})$$

6   Compute the air requirement per 1000 m³ of digester volume.

$$\frac{37.7 \text{ m}^3/\text{min}}{1005 \text{ m}^3} = 37.5 \text{ m}^3/10^3 \text{ m}^3 \ (37.5 \text{ ft}^3/10^3 \text{ ft}^3)$$

7   Check the mixing requirement. Because the air requirement computed in step 6 is approximately the same as the maximum value given in Table 11-19, adequate mixing should prevail.

Pure-Oxygen Aerobic Digestion

Pure-oxygen aerobic digestion is a modification of the aerobic digestion process in which pure oxygen is used in lieu of air. The resultant sludge is similar to sludge from conventional-air aerobic digestion. Influent sludge concentrations vary from 2 to 4 percent. Recycle flows are similar to those achieved by conventional-air aerobic digestion. Pure-oxygen aerobic digestion is particularly applicable in cold weather climates because of its relative insensitivity to ambient air temperatures. This modification is an emerging technology that is currently being investigated in several full-scale installations.

In one configuration being investigated, a covered aeration tank is used. In this variation, a high-purity-oxygen atmosphere is maintained above the liquid surface, and oxygen is transferred into the sludge via mechanical aerators. In another variation, an open aeration tank is used. Oxygen is introduced to the liquid sludge by a special diffuser that produces minute oxygen bubbles. The bubbles totally dissolve before reaching the air-liquid interface.

The aerobic process is exothermic. Normally, the heat generated is not retained in conventional-air aerobic digesters because of the vigorous aeration in the open tank. Temperatures of the digesting sludges were maintained above ambient air temperature in the pure-oxygen investigations because the increased gas transfer allows the heat to be retained and the sludge temperatures to rise. This effect is more pronounced in the covered tank. Maintenance of these higher temperatures in the digester results in a significant increase in the rate of volatile suspended solids destruction [22].

Pure-oxygen aerobic digestion can be used only by large installations when the incremental cost of oxygen-generation equipment is offset by the savings obtained by reduced reactor volumes and lower energy requirements for dissolution equipment. In this regard, the process is most compatible with a treatment facility using the pure-oxygen activated-sludge process, because of the potential for oxygen availability on an incremental basis.

Thermophilic Aerobic Digestion

Thermophilic aerobic digestion represents a refinement of both the conventional-air and pure-oxygen aerobic digestion. Although there are presently no full-scale installations in the United States, it has been shown in large-scale pilot studies that thermophilic aerobic digestion can be used to achieve high removals of the biodegradable fraction (up to 70 percent) at very short detention times (3 to 4 d). Thermophilic digestion without external heat input can be achieved by using the heat released during microbial oxidation of organic matter to heat the sludge. It has been estimated that more than 25 kcal/L of heat energy are released in the aerobic digestion of primary and secondary sludges (between 2 and 5 percent solids) [8]. It has also been demonstrated that this quantity of heat is sufficient to heat wet slurries containing from 95 to 97 percent water to the thermophilic range 45°C (113°F) if sufficiently high oxygen-transfer efficiencies can be obtained so that air or oxygen stripping of the heat does not occur [8].

Although it would appear that pure oxygen would be best in this application, it has been shown that thermophilic digestion can be achieved with a simple air aeration system. Typically, the process operates from 25 to 50°C (77 to 122°F) above the ambient air temperature [8]. Because of the high operating temperatures, the digested sludge is pasteurized as well. Ideally, the feed sludge should contain more than 4 percent solids to optimally support thermophilic digestion.

11-9 CONDITIONING

Sludge is conditioned expressly to improve its dewatering characteristics. The two methods most commonly used involve the addition of chemicals and heat treatment. Freezing and irradiation have also been investigated. Elutriation, a physical washing operation, is used to reduce chemical-conditioning requirements.

Chemical Conditioning

The use of chemicals to condition sludge for dewatering is economical because of the increased yields and greater flexibility obtained. Chemical conditioning results in coagulation of the solids and release of the absorbed water. Conditioning is used in advance of vacuum filtration and centrifugation. Chemicals used include ferric chloride, lime, alum, and organic polymers.

Chemicals are most easily applied and metered in the liquid form. Dissolving tanks are needed if the chemicals are received as dry powder. These tanks should be large enough for at least one day's supply of chemicals and should be furnished in duplicate. They must be fabricated or lined with corrosion-resistant material. Polyvinyl chloride, polyethylene, and rubber are suitable materials for tank and pipe linings for acid solutions. Metering pumps must be corrosion resistant. These pumps are generally of the positive-displacement type with variable-speed or variable-stroke drives to control the flowrate. Another metering system consists of a constant-head tank supplied by a centrifugal pump. A rotameter and throttling valve are used to meter the flow.

Dosage The chemical dosage required for any sludge is determined in the laboratory. Filter-leaf test kits are used to determine chemical doses, filter yields, and the suitability of various filtering media. These kits have several advantages over the Büchner funnel procedure [11]. In general, it has been observed that the type of sludge has the greatest impact on the quantity of chemical required. Difficult-to-dewater sludges require larger doses of chemicals and generally do not yield as dry a cake. Sludge types, listed in the approximate order of increasing chemical requirements for conditioning, are as follows:

1. Untreated (raw) primary sludge
2. Untreated mixed primary and trickling-filter sludge

3. Untreated mixed primary and waste activated sludge
4. Anaerobically digested primary sludge
5. Anaerobically digested mixed primary and waste activated sludge
6. Aerobically digested sludge (normally dewatered on drying beds without the use of chemicals for conditioning)

Typical chemical-conditioning requirements for various types of sludges ahead of vacuum-filter dewatering are reported in Table 11-18. Actual dosages in any given case may vary considerably from the indicated values.

Sludge mixing Intimate admixing of sludge and coagulant is essential for proper conditioning. The mixing must not break the floc after it has formed, and the detention is kept to a minimum so that sludge reaches the filter as soon after conditioning as possible. Mixing tanks are generally of the vertical type for small plants and of the horizontal type for large plants. They are ordinarily built of welded steel and lined with rubber or other acid-proof coating. A typical layout for a mixing or conditioning tank has a horizontal agitator driven by a variable-speed motor to provide a shaft speed of 4 to 10 r/min. Overflow from the tank is adjustable to vary the detention period. Vertical cylindrical tanks with propeller mixers are also used.

Elutriation

Elutriation is a unit operation in which a solid or a solid-liquid mixture is intimately mixed with a liquid for the purpose of transferring certain components to the liquid. A typical example is the washing of digested wastewater sludge before chemical conditioning to remove certain soluble organic and inorganic components that would consume large amounts of chemicals. The cost of washing the sludge is, in general, more than compensated for by the savings that result from a lower demand for conditioning chemicals. Although this operation was

Table 11-18 Dosage of chemicals for various types of sludges for vacuum filtration

(Conditioners in Percentage of Dry Sludge)

Type of sludge	Fresh solids		Digested		Elutriated, digested	
	$FeCl_3$	CaO	$FeCl_3$	CaO	$FeCl_3$	CaO
Primary	1–2	6–8	1.5–3.5	6–10	2–4	
Primary and trickling filter	2–3	6–8	1.5–3.5	6–10	2–4	
Primary and activated	1.5–2.5	7–9	1.5–4	6–12	2–4	
Activated (alone)	4–6					

used commonly in the past, it has fallen into disfavor because of the concern that the finely divided solids washed out of the sludge may not be fully captured in the main wastewater treatment facilities. In fact, the U.S. Environmental Protection Agency has stated that sludge elutriation is not considered desirable and its use will not be approved without adequate safeguards.

The usual leaching operation consists of two steps: (1) a thorough mixing of the solid or solid-liquid mixture with the leaching liquid and (2) separation of the leaching liquid. Each combination of mixing and washing is called a stage. A stage is said to be ideal if the concentration of the component being leached is the same in the separating liquid as it is in the liquid that remains with the solids. Mixing and separating can be carried out either in the same tank or in separate tanks. In sanitary engineering, separate tanks are usually used for each stage. The mixing and thickening operations should be designed in accordance with the principles previously discussed in this chapter.

Since alkalinity is usually present in high concentrations in digested sludge, it is commonly used to measure leaching efficiency. A decrease in the quantity of chemicals required to condition sludge has been correlated, by Genter in McCabe and Eckenfelder [18], with the decrease in alkalinity that results from elutriation. Additional details may be found in the previous edition of this text.

Heat Treatment

As noted previously, heat treatment is both a stabilization and a conditioning process that involves heating the sludge for short periods of time under pressure. The treatment coagulates the solids, breaks down the gel structure, and reduces the water affinity of sludge solids. As a result the sludge is sterilized, practically deodorized, and is dewatered readily on vacuum filters or filter presses without the addition of chemicals. The heat-treatment process is most applicable to biological sludges that may be difficult to stabilize or condition by other means. The high capital costs of equipment generally limit its use to large plants (more than 0.2 m^3/s or 5 Mgal/d) or facilities where space may be limited. Processes used for heat treatment include the Porteus and the low-pressure Zimpro (see Fig. 11-19).

In the Porteus process, sludge is preheated by being passed through a heat exchanger before it enters the reactor vessel. Steam is injected into the vessel to bring the temperature to within the range of 140 to 200°C (290 to 390°F) under pressures of 1.0 to 1.4 MN/m^2 (150 to 200 lb$_f$/in^2). After a 30-min detention in the reactor, the sludge is discharged through the heat exchanger and into a decant tank. The thickened sludge can be filtered to a solids content of 30 to 50 percent. Filter yields up to 100 kg/m$^2 \cdot$ h (20 lb/ft$^2 \cdot$ h) have been obtained.

In the low-pressure Zimpro process, the sludge is treated as in the Porteus process except that air is injected into the reactor vessel with the sludge. The reactor vessel is heated by steam to temperatures varying from 1.0 to 2.0 MN/m^2 (150 to 300 lb$_f$/in^2). Heat released during oxidation increases the operating temperature to a range of 180 to 315°C (350 to 600°F). The partially

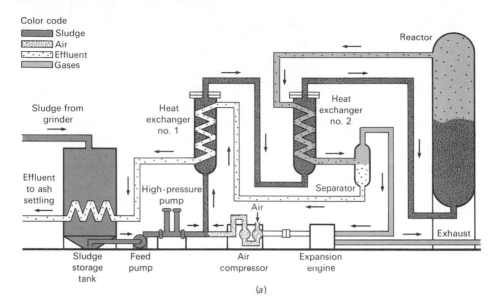

Color code
- ▬ Sludge
- ▨ Air
- ⋯ Effluent
- ▭ Gases

(a)

Figure 11-19 Zimmerman wet oxidation process [*from Zimpro*]. (a) Schematic. (b) Typical installation.

oxidized sludge may be dewatered by filtration, centrifugation, or drainage on beds. The solids content of the dewatered sludge can range from 30 to 50 percent, depending on the degree of oxidation desired. Essentially complete oxidation of volatile solids (approximately 90 percent reduction) can be accomplished with higher pressures and temperatures (see "Multiple Hearth Incineration," "Fluidized-Bed Incineration," and "Wet Oxidation" in Sec. 11-14).

A major disadvantage associated with heat treatment results from the very high strength of the supernatant and filtrate. The recycle liquor is composed mostly of organic acids, sugars, polysaccharides, amino acids, ammonia, etc. The exact composition is not well defined. Although the recycle flow represents less than 1 percent of the total wastewater stream, it is reported that the recycled BOD can account for 30 to 50 percent of the loading to the aeration system [24]. It has been shown that the recycled liquor is highly polluted and contains a high proportion of nonbiodegradable matter [24]. As a result, separate treatment of the recycle flow may be required.

Waste sludges containing a high concentration of chloride (greater than 500 mg/L) require that special metal (usually titanium) be used to construct the heat-treatment unit. This significantly increases the capital cost of the process. Feed sludges with a high hardness may result in rapid buildup of scale within the exchangers and reactor, and this increases maintenance costs. Also, the heat-treatment system and subsequent dewatering equipment will, in almost all cases, require odor-control facilities.

Other Processes

Freezing and irradiation have also been investigated as sludge conditioning methods. Laboratory investigations indicate that freezing of sludge is more effective than chemical conditioning in improving sludge filterability. Much remains to be done, however, before this method can be applied effectively. Although irradiation has been shown to be effective in improving sludge filterability, it is not considered economically competitive at present (1977).

11-10 DISINFECTION

Sludge disinfection is becoming an important consideration now that the use of sludge is gaining wider acceptance. When sludges are applied to the land, protection of public health requires that contact with pathogenic organisms be minimized.

There are many ways to destroy pathogens in liquid and dewatered sludges. The following methods have been used to achieve pathogen reduction beyond that attained by stabilization [22]:

1. Pasteurization for 30 min at 70°C (158°F).
2. High pH treatment, typically with lime, at a pH higher than 12.0 for 3 h.

3. Long-term storage of liquid digested sludge [60 days at 20°C (68°F) or 120 d at 4°C (39°F)].
4. Complete composting at temperatures above 55°C (131°F) and curing in a stockpile for at least 30 d. (Composting is discussed in Sec. 11-13.)
5. Addition of chlorine to stabilize and disinfect sludge.
6. Disinfection with other chemicals.
7. Disinfection by high-energy radiation (see Ref. 8 in Chap. 7).

As indicated in Sec. 11-6, some stabilization processes will also provide disinfection. These processes include chlorine oxidation, lime stabilization, heat treatment, and thermophilic aerobic digestion.

Anaerobic and aerobic digestion (excluding thermophilic aerobic digestion) will not disinfect the sludge, but will greatly reduce the number of pathogenic organisms. Disinfection of liquid aerobic and anaerobic digested sludges is best accomplished by pasteurization or long-term storage. Long-term storage and composting are probably the most effective means of disinfecting dewatered aerobic and anaerobic digested sludges.

Pasteurization

Pasteurization of sludge is not presently practiced in this country at full-scale treatment facilities, but it is used in Europe to disinfect sludges spread on pastures during the summer growing season [22].

The two methods that are used for pasteurizing liquid sludges involve (1) the direct injection of steam, and (2) indirect heat exchange. Because heat exchangers tend to scale or become fouled with organic matter, it appears that direct steam injection is the most feasible method. Equipment presently used for sludge pasteurization may not be cost-effective for plants with capacities of less than 0.2 m³/s (5 Mgal/d) because of the high capital costs. Pasteurization in small plants may be achieved by direct steam injection into the tank trucks that transport the sludge to the disposal site.

Long-Term Storage

Liquid digested sludge is normally stored in earthen lagoons. Storage requires that sufficient land be available. Storage is often necessary in land-application systems to retain sludge during periods when it cannot be applied because of weather or crop considerations. In this case, the storage facilities can perform a dual function by providing disinfection as well. Because of the potential contamination effects of the stored sludge, special attention must be devoted to the design of these lagoons with respect to limiting percolation and the development of odors. Additional details on sludge storage are presented in Sec. 11-16.

11-11 DEWATERING

Dewatering is a physical (mechanical) unit operation used to reduce the moisture content of sludge for one or more of the following reasons:

1. The costs for trucking sludge to the ultimate disposal site become substantially lower when sludge volume is reduced by dewatering.
2. Dewatered sludge is generally easier to handle than thickened or liquid sludge. In most cases, dewatered sludge may be shoveled, moved about with tractors fitted with buckets and blades, and transported by belt conveyors.
3. Dewatering is normally required prior to the incineration of the sludge to increase the calorific value by removal of excess moisture.
4. In some cases, removal of the excess moisture may be required to render the sludge totally odorless and nonputrescible. This is especially true for sludges stabilized by processes that create high-strength recycle flows.
5. Sludge dewatering is commonly required prior to landfilling to reduce leachate production at the landfill site.

Dewatering devices use a number of techniques for removing moisture. Some rely on natural evaporation and percolation to dewater the solids. Mechanical dewatering devices use mechanically assisted physical means to dewater the sludge more quickly. They include filtration, squeezing, capillary action, vacuum withdrawal, and centrifugal settling and compaction.

The selection of the dewatering device is determined by the type of sludge to be dewatered and the space available. For smaller plants where land availability is not a problem, drying beds or lagoons are most frequently selected. Conversely, for facilities situated on constricted sites, mechanical dewatering devices are generally chosen.

Some sludges, particularly aerobically digested sludges, are not amenable to mechanical dewatering. These sludges can be dewatered on sand beds with good results. When a particular sludge must be dewatered mechanically, it is often difficult or impossible to select the optimum dewatering device without conducting bench-scale or pilot studies.

The available dewatering processes include vacuum filters, centrifuges, filter presses, horizontal belt filters, drying beds, and lagoons.

Vacuum Filtration

The function of the unit operation of vacuum filtration is to reduce the water content of sludge, whether it is untreated, digested, or elutriated, so that the proportion of solids increases from the range of 5 to 10 percent to the range of 20 to 30 percent. At this higher percentage, wastewater sludge is a moist, easily handled cake. To visualize the amount of water to be removed, consider 1 Mg

(1000 kg or 2205 lb) of sludge with 5 percent solids, or 50 kg (110 lb) of dry solids and 950 kg (2094 lb) of water. After filtration to 30 percent solids, the 50 kg of solids would be associated with 117 kg (258 lb) of water in 167 kg (368 lb) of sludge. Thus, 833 kg (1837 lb) of water would have been extracted by the vacuum filter. This represents an 83 percent reduction in the weight of sludge to be disposed of from the treatment process.

Theory The basic theory of the filtration process was developed by Carmen [4] and extended by Coakley [6, 18] for conditions of streamline flow by application of Poiseuille's and Darcy's laws. The basic filtration equation is

$$\frac{dV}{dt} = \frac{PA^2}{\mu(rWV + R_m A)} \tag{11-9}$$

where V = volume of filtrate, m^3 (ft^3)
  t = time, s
  P = pressure, N/m^2 (lb$_f$/ft^2)
  A = area, m^2 (ft^2)
  μ = viscosity of filtrate, N · s/m^2 (lb$_f$ · s/ft^2)
  r = specific resistance of sludge cake, m/kg (ft/lb)
  W = mass of dry solids per unit volume of filtrate, kg/m^3 (lb$_m$/ft^3)
  R_m = resistance of filter medium, m^{-1} (ft^{-1})

For constant pressure, integration of Eq. 11-9 yields

$$\frac{t}{V} = \frac{\mu rWV}{2PA^2} + \frac{\mu R_m}{PA} \tag{11-10}$$

The specific resistance of a sludge can be determined from laboratory data obtained using a Büchner funnel (see Fig. 11-20) or other filters specifically designed for the purpose [6] and plotting t/V versus V. If the value of the slope of the line passed through the data points is m, then from Eq. 11-10 the specific resistance is

$$r = \frac{2PA^2 m}{\mu W} \tag{11-11}$$

Typical specific resistance values for various biological sludges are reported in Table 11-19. As a point of caution, it should be noted that the literature on vacuum filtration is confusing in terms of unit inconsistencies and the reader is advised to check units before using the various equations.

Where the resistance of the filter medium is small compared with the resistance of the filter cake, an expression that can be used to estimate the filter yield L can be derived from Eq. 11-10 by noting that $L = WV/tA$. The resulting expression is

$$L = 484.8 \left(\frac{xPW}{\mu r\theta}\right)^{1/2} \tag{11-12}$$

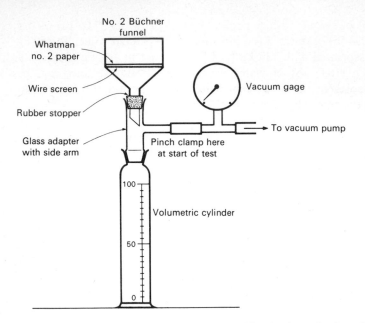

Figure 11-20 Buchner funnel test apparatus used for the determination of the specific resistance of sludge [11].

where L = filter yield, kg dry solids/m² · h (lb$_m$/ft² · h)

   W = mass of dry solids per unit volume of filtrate, kg/m³ (lb$_m$/ft³)

   P = pressure, N/m² (lb$_f$/ft²)

   x = form time per cycle time (see Fig. 11-21a)

   μ = viscosity of filtrate, N · s/m² (lb$_f$ · s/ft²)

   r = specific resistance, m/kg (ft/lb$_m$)

   θ = cycle time = time of one revolution of the drum, min

Table 11-19 Typical specific resistance values for various sludges

Sludge	Specific resistance r, m/kg[a]
Primary	1.5–5.0 × 10¹⁴
Activated	1–10 × 10¹³
Digested	1–6 × 10¹⁴
Digested and coagulant	3–40 × 10¹¹

[a] Specific resistance is often expressed as s²/g, but this is not dimensionally correct for use in Eqs. 11-9 and 11-10. To compare the values in Table 11-19, expressed in m/kg to those reported in the literature in terms of s²/g, the values in Table 11-19 must be divided by 9.81 × 10³.

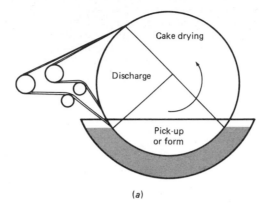

(a)

(b)

Figure 11-21 Typical vacuum filter. (a) Diagrammatic cross section. (b) Typical filter in operation.

The yield of the filter in kilograms (pounds) of dry solids per unit time may be changed by varying the suction, the speed of rotation, the portion of the cycle time during which suction takes place, and the permeability of the filter cake. The latter is controlled by the addition of sludge-conditioning chemicals. The calculation of filter yield is illustrated in Example 11-8.

Example 11-8 Vacuum-filter yield A conditioned, digested sludge with a specific resistance r of 3.1×10^7 s^2/g is to be dewatered on a rotary drum filter under a vacuum of 625 mm (25 in) Hg. The mass of filtered solids per unit volume of filtrate is to be 64 kg/m^3 (4 lb/ft^3). The filtration temperature of the sludge is 25°C (77°F) and the cycle time, half of which is form time, is 6 min. Calculate the filter yield.

SOLUTION

1 Compute the vacuum:

$$P = 625 \text{ mm Hg} = 83,300 \text{ N/m}^2 \ (1740 \text{ lb}_f/\text{ft}^2)$$

2 Substitute the following values in Eq. 11-12 and compute the filter yield:

$$L = 484.8 \left(\frac{xPW}{\mu r \theta} \right)^{1/2}$$

where $x = 0.5$
$\qquad W = 64 \text{ kg/m}^3$
$\qquad \theta = 6 \text{ min}$
$\qquad \mu = 8.90 \times 10^{-4} \text{ N} \cdot \text{s/m}^2$
$\qquad r = 3.0 \times 10^{11} \text{ m/kg}$

$$L = 484.8 \left[\frac{0.5(83,300)(64)}{(8.90 \times 10^{-4})(3.0 \times 10^{11})(6)} \right]^{1/2}$$

$$= 19.8 \text{ kg/m}^2 \cdot \text{h} \ (4.05 \text{ lb}_m/\text{ft}^2 \cdot \text{h})$$

Operation description In wastewater-treatment plants, vacuum filtration is a continuous operation that is generally accomplished on cylindrical drum filters. These drums have a filter medium which may be a cloth of natural or synthetic fibers, coil springs, or a wire-mesh fabric. The drum is suspended above and dips into a vat of sludge. As the drum rotates slowly, part of its circumference is subject to an internal vacuum that draws sludge to the filter medium. Water is drawn through the porous filter cake for that sector of the circumference, as shown in Fig. 11-21. The piping arrangement within the filter permits the suction to be maintained until the release point, at which time compressed air is blown through the medium to release the cake to a scraper for discharge. The filter medium may be washed in the small sector before suction begins again. The flowsheet of a vacuum-filtration system is shown in Fig. 11-22.

A recent innovation in vacuum-filter design consists of the top-feed filter. In this variation, the thickened sludge is distributed to the filter near the top of the drum, thereby using gravity to assist in depositing sludge solids on the drum surface.

Vacuum filters with surface areas ranging from 5 to about 60 m^2 (50 to 600 ft^2) are made by several manufacturers and can be equipped with various types of filtering cloth. Filter cloths made of cotton, wool, nylon, dacron, and other synthetic materials are available in a variety of weaves having different porosities. Cloths made of woven stainless-steel wire are sometimes used. A unique filter medium manufactured by the Komline-Sanderson Co. consists of stainless-steel coil springs that are wrapped around the filter drum in a double layer.

The performance of vacuum filters is affected by the type and age of the

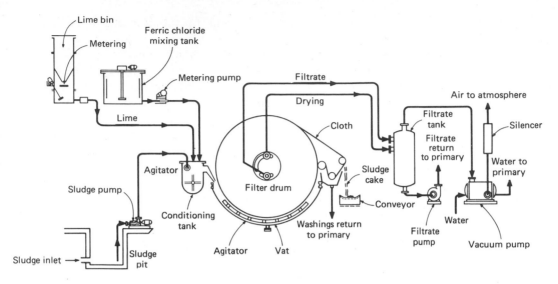

Figure 11-22 Typical vacuum-filtration system.

sludge, prior sludge processing, filter medium selected, and sludge feed temperature. Although some generalities regarding vacuum-filter performance can be made, the results obtainable in practice are extremely variable. In general, conditioning of wet sludges is necessary to achieve satisfactory yields from vacuum filters. Conditioning coagulates the sludge particles and allows the water to drain freely. As a result, a thicker filter cake is produced, and the drum can be rotated at a higher speed.

The number and size of filters are based on the type of sludge to be filtered and the number of hours of operation. At small plants, 30 h/week may be assumed; at large plants, 20 h/d may be necessary. The additional hours are used for conditioning, cleanup, and possible delays. A plant may be designed for one-shift operation initially, and for two- to three-shift operation of the same filters when it is expanded to provide for future or ultimate conditions.

The results obtained vary greatly with the characteristics of the sludge being filtered. The solids content of the sludge, among other parameters, is very important. Chemical conditioning of the sludge prior to filtration is usually practiced to control the solids content and to improve the dewatering characteristics. The optimum solids content for filtration is about 8 to 10 percent. Higher solids content makes the sludge difficult to distribute and to condition for dewatering; lower solids content requires the use of larger-than-necessary vacuum filters. Chemicals that are commonly used for conditioning sludge are lime, ferric chloride, and polyelectrolytes. Sludge from primary settling tanks, in general, requires lesser amounts of conditioning chemical than sludge from biological-waste-treatment processes. Elutriation of digested sludge, as previously discussed, reduces the chemical requirements. Laboratory tests and engineering and cost studies are necessary to determine the optimum process of sludge handling to achieve minimum construction and operating costs, including labor, power, and chemicals.

The performance of a vacuum filter is measured in terms of the yield of solids on a dry weight basis expressed as kilograms per square meter per hour (pounds per square foot per hour). The quality of the filter cake is measured by its moisture content on a wet-weight basis expressed as a percent. Filters are operated to obtain the maximum production consistent with the desired cake quality. Where the cake is to be heat-dried or incinerated, the moisture content is a critical parameter, since all the water remaining in the cake must be evaporated to steam. If the cake is conveyed into a truck and hauled to a disposal site, moisture content is not as important, although it does affect the mass that must be hauled. In such cases, the drum can be operated at the highest speed that will produce a cake that will separate easily from the filter. Moisture content normally varies from 70 to 80 percent, but after heat treatment of the sludge, filters may be operated to produce a cake of 60 to 70 percent moisture where the cake is to be heat dried or incinerated. Typical yields are shown in Table 11-20. A design rate of 17.0 kg/m² · h (3.5 lb/ft² · h) is frequently used when the quality of the sludge must be estimated and the type of filter to be furnished is based on an open specification.

Centrifugation

The centrifugation process is widely used in industry for separating liquids of different density, thickening slurries, or removing solids. The process is applicable to the dewatering of wastewater sludges and has been used with varying degrees of success in both the United States and Europe.

Table 11-20 Expected performance of vacuum filters handling properly conditioned sludge[a]

Type of sludge	Yield, kg/m² · h	Cake solids, %
Fresh solids		
Primary	20–60	20–40
Primary and trickling filter	20–40	20–30
Primary and air-activated	20–25	16–25
Primary and oxygen-activated	25–30	20–30
Air-activated (alone)	12.5–17.5	15–25
Pure-oxygen-activated (alone)	15–20	15–25
Digested solids (with or without elutriation)		
Primary	20–40	20–30
Primary and trickling filter	20–25	15–28
Primary and air-activated	20–25	12–25
Primary and oxygen-activated	25–30	15–25

[a] Adapted in part from Refs. 23 and 24.
Note: kg/m² · h × 0.2048 = lb/ft² · h

Equipment description The centrifugal devices used for thickening sludge (solid-bowl, basket, and nozzle-disk centrifuges) were discussed in Sec. 11-5. Sludge dewatering may be accomplished by solid-bowl and basket centrifuges. In the solid-bowl machine (see Fig. 11-23), sludge is fed at a constant flowrate into the rotating bowl where it separates into a dense cake containing the solids and a dilute stream called centrate. The centrate contains fine, low-density solids and is returned to the untreated-sludge thickener or primary clarifier. The sludge cake, which contains approximately 75 to 80 percent moisture, is discharged from the bowl by a screw feeder into a hopper or onto a conveyor belt. Depending on the type of sludge, solids concentration in the cake varies from 10 to 40 percent, but reductions below 25 percent are not usually feasible economically. The cake can then be disposed of by incineration or by hauling to a sanitary landfill.

Solid-bowl centrifuges are generally suitable in the same applications as vacuum filters. Their performance is governed by the same factors that affect vacuum filters: type and age of sludge, prior sludge processing, etc. The units can be used to dewater sludges with no prior chemical conditioning, but the solids

Figure 11-23 Typical solid-bowl centrifuge used for dewatering sludge [*from Bird Machine Co.*].

capture and centrate quality are considerably improved when solids are conditioned with polyelectrolytes. Chemicals for conditioning are added to the sludge within the bowl of the centrifuge. Dosage rates for conditioning with polyelectrolytes vary from 1.0 to 7.5 kg/10^3 kg (2 to 15 lb/ton) of sludge (dry solids). Typical expected performance data for solid-bowl centrifuges are reported in Table 11-21.

Basket centrifuges have been used for partial dewatering at small plants. They can be used to concentrate and dewater waste activated sludge, with no chemical conditioning, at solids capture rates up to 90 percent. To date, the use of basket centrifuges for dewatering has been limited. The combined use of basket centrifuges with solid-bowl centrifuges is also being investigated.

Design considerations The operation of centrifuges is simple, clean, relatively inexpensive, and normally does not require chemical conditioning. Special consideration must be given to providing sturdy foundations and soundproofing because of the vibration and noise that result from centrifuge operation. Adequate electric power must also be provided because large motors are required.

The major difficulty encountered in the operation of centrifuges has been the disposal of the centrate, which is relatively high in suspended, nonsettling solids. The return of these solids to the wastewater treatment units could result in a large recirculating load of these fine solids through the sludge and primary settling system and in reduced effluent quality. Two methods can be used to control the fine-solids discharge and to increase the capture. Longer residence of the liquid stream is accomplished by reducing the feed rate or by using a centrifuge with a larger bowl volume. Particle size can be increased by coagulating the

Table 11-21 Expected performance data for solid-bowl centrifuges

Type of sludge	Cake solids, %	Solids capture, % Without chemical	Solids capture, % With chemical
Untreated			
Primary	25–35	75–90	90+
Primary and trickling filter	20–25	60–80	90+
Primary and air activated	12–20	55–65	90+
Waste sludge			
Trickling filter	10–20	60–80	90+
Air activated	5–15	60–80	90+
Pure-oxygen activated	10–20	60–80	90+
Digested			
Primary	25–35	75–90	90+
Primary and trickling filter	18–25	60–75	90+
Primary and air activated	15–20	50–65	90+

sludge prior to centrifugation with ferric chloride and lime or organic polymers. Solids capture may be increased from a range of 50 to 80 percent to a range of 80 to 95 percent of influent solids.

The addition of lime will also aid in the control of odors that may develop when centrifuging untreated sludge. Untreated primary sludge can usually be de-watered to a lower moisture content than digested sludge, because it has not been subjected to the liquefying action of the digestion process, which reduces particle size. Some type of chemical conditioning is usually desirable when dewatering combined primary and waste activated sludge, regardless of whether or not it has been digested.

The area required for a centrifuge installation is less than that required for a vacuum filter of equal capacity, and the initial cost is lower. Higher power costs will partially offset the lower initial cost.

Selection of units for plant design is dependent on manufacturer's rating and performance data. Several manufacturers have portable pilot plant units, which can be used for field testing if sludge is available. Wastewater sludges from supposedly similar treatment processes but in different localities may differ markedly from each other. For this reason, pilot plant tests should be run whenever possible before final design decisions are made.

Filter Presses

In a filter press, dewatering is achieved by forcing the water from the sludge under high pressure. Advantages cited for the filter press include (1) high concentrations of cake solids, (2) filtrate clarity, (3) solids capture, and (4) chemical consumption. Disadvantages include high labor costs and limitations on filter cloth life.

Various types of filter presses have been used to dewater sludge. One type consists of a series of rectangular plates, recessed on both sides, that are sup-ported face to face in a vertical position on a frame with a fixed and movable head (see Fig. 11-24). A filter cloth is hung or fitted over each plate. The plates are held together with sufficient force to seal them to withstand the pressure applied during the filtration process. Hydraulic rams or powered screws are used to hold the plates together.

In operation, chemically conditioned sludge is pumped into the space between the plates, and pressure of 40 to 150 N/cm^2 (60 to 225 lb_f/in^2) is applied and maintained for 1 to 3 h, forcing the liquid through the filter cloth and plate outlet ports. The plates are then separated and the sludge is removed. The filtrate normally is returned to the headworks of the treatment plant. The sludge cake thickness varies from about 2.5 to 3.8 cm (1 to 1.5 in), and the moisture content varies from 55 to 70 percent. The filtration cycle time varies from 2 to 5 h and includes the time required to fill the press, the time the press is under pressure, the time to open the press, the time required to wash and discharge the cake, and the time required to close the press. To reduce the amount of labor to a minimum, most modern presses are mechanized.

The most significant costs associated with this method of dewatering are those

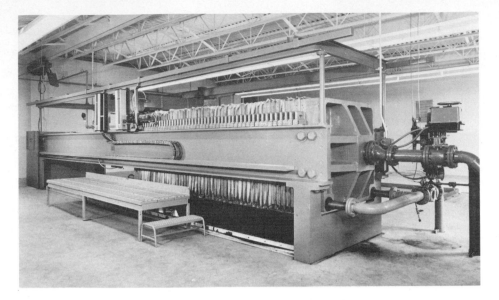

Figure 11-24 Plate and frame filter press used for dewatering sludge [*from Zimpro*].

for chemical conditioning and maintenance and replacement of filter cloths. Although filter presses are not commonly used in the United States, they have been used extensively in England and other parts of the world. The limited acceptance in the municipal wastewater treatment field in the United States has been due to the filter media which, in the past, have clogged rapidly. However, development of new types of filter media and other innovations are leading to increased acceptance.

Horizontal Belt Filters

Several new mechanical dewatering systems have been introduced in the past few years. Four new systems that appear to be viable technologies have been loosely categorized as horizontal belt filters. They are the moving-screen concentrator, belt pressure filter, capillary dewatering system, and rotating-gravity concentration. All four systems use horizontally mounted continuous belts on which the sludge is conveyed and dewatered, and all four systems appear to be designed to compete with vacuum filters. Operating complexity and energy requirements are similar. Solids capture and cake moisture content are very close to those achieved by vacuum filters.

Moving-screen concentrator In the moving-screen concentrator, thickened and polymer-treated sludge is distributed on a two-stage variable-speed moving screen (see Fig. 11-25a). On the first screen, gravity is the major means of dewatering. When the sludge is passed onto the second screen, a compression process is implemented for final dewatering. Sludge is passed under compression rollers of

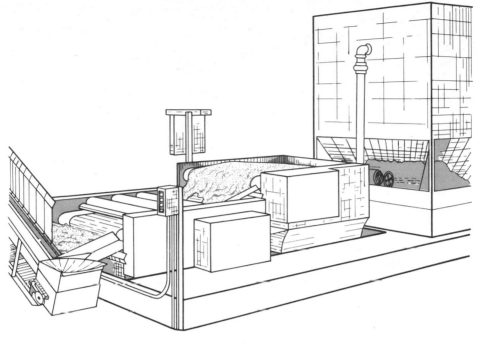

(a)

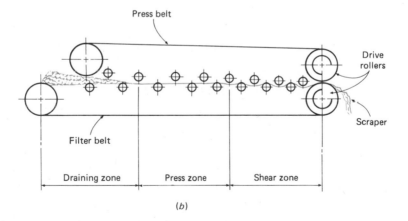

(b)

Figure 11-25 Horizontal belt filters used for dewatering sludge. (a) Moving screen concentrator system. (b) Belt pressure filter.

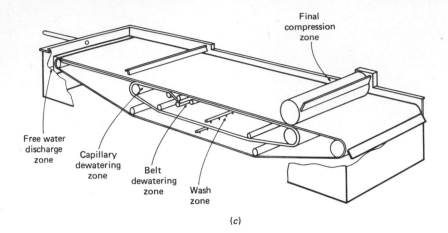

(c)

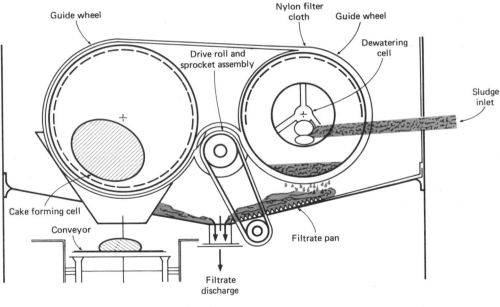

(d)

Figure 11-25 (c) Capillary dewatering unit. (d) Rotating gravity concentrator.

increasingly higher pressure. At this point, relatively little free water remains in the sludge cake, and it passes out of the secondary unit and into a disposal receptacle.

Belt pressure filter The belt pressure filter consists of two continuous belts set one above the other (see Fig. 11-25b). Conditioned sludge is fed in between the two belts. Three process zones exist. First, the sludge passes through the drainage zone where dewatering is effected by the force of gravity. Then, the sludge passes into the pressure zone where pressure is applied to the sludge by means of rollers in contact with the top belt. Finally, the sludge is passed to the shear zone where shear forces are used to bring about the final dewatering. The dewatered sludge is then removed by a scraper.

Capillary dewatering system In the capillary dewatering scheme, chemically conditioned sludge is distributed evenly over the screen where free water is released and the solids concentration is increased by 25 percent. Next, the screen carrier comes in contact with a capillary belt (see Fig. 11-25c). Here the force for dewatering comes from the capillary action of the belt. At various stations in this zone, the filtrate is removed from the belt. Next, the sludge is carried to a final compression zone where final dewatering takes place. The sludge cake is then removed from the screen by a blade. The screen is washed and the cycle begins again.

Rotating-gravity concentration The rotating-gravity-concentration process consists of two independent cells formed by a fine-mesh nylon filter cloth (see Fig. 11-25d). Dewatering occurs in the first cell, cake formation in the second. In the first cell, liquid drains from the sludge and the sludge is carried into the second cell. Here the sludge is continuously rolled into a cake of low moisture content. When the cake is large enough, excess sludge cake is discharged over the rim and onto a conveyor belt for disposal. Operation is continuous and dewatering is entirely by gravity. When more complete dewatering is required, a multiroll press is provided. The multiroll press consists of dual endless belts. The sludge cake from the rotating gravity concentrator is fed between the belts and graduated pressure is applied by rollers.

Sludge-Drying Beds

Sludge-drying beds are used to dewater digested sludge. Sludge is placed on the beds in a 200- to 300-mm (8- to 12-in) layer and allowed to dry. After drying, the sludge is removed and either disposed of in a landfill or ground for use as a fertilizer. A typical sludge-drying bed is shown in Fig. 11-26. The economical use of sludge-drying beds is generally limited to small- and medium-sized communities. For cities with populations over 20,000, consideration should be given to alternative means of sludge dewatering. In large municipalities, the initial cost, the cost of removing the sludge and replacing sand, and the large area requirements preclude the use of sludge-drying beds.

The drying area is partitioned into individual beds, approximately 6 m wide by 6 to 30 m long (20 ft wide by 20 to 100 ft long), or a convenient size so that one or two beds will be filled by a normal withdrawal of sludge from the digesters. The interior partitions commonly consist of two or three creosoted planks, one on top of the other, to a height of 380 to 460 mm (15 to 18 in), stretching between slots in precast concrete posts. The outer boundaries may be

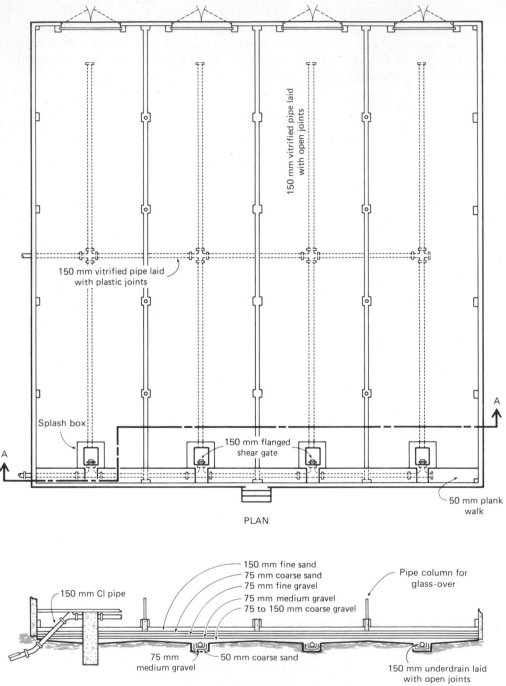

PLAN

SECTION A–A

Figure 11-26 Plan and section of a typical sludge drying bed.

of similar construction or earthen embankments for open beds, but concrete foundation walls are required if the beds are to be covered.

Open beds are used where adequate area is available and sufficiently isolated to avoid complaints caused by occasional odors. Covered beds with greenhouse types of enclosures are used where it is necessary to dewater sludge continuously throughout the year regardless of the weather, and where sufficient isolation does not exist for the installation of open beds. Well-digested sludge discharged to drying beds should present no odor problem, but to avoid nuisance from poorly digested sludge, sludge beds should be located at least 100 m (328 ft) from dwellings.

Sludge-bed loadings are computed on a per capita basis or on a unit loading of kilograms of dry solids per square meter per year. Typical data for various types of sludge are shown in Table 11-22. With covered drying beds, more sludge drawings per year can be accommodated because of the protection from rain and snow.

Sludge dewaters by drainage through the sludge mass and supporting sand and by evaporation from the surface exposed to the air. Most of the water leaves the sludge by drainage; thus the provision of an adequate underdrainage system is essential. Drying beds are equipped with lateral drainage tiles (vitrified-clay pipe laid with open joints) spaced 2.5 to 6 m (8 to 20 ft) apart. The tiles should be adequately supported and covered with coarse gravel or crushed stone (see Fig. 11-26). The sand layer should be from 230 to 300 mm (9 to 12 in) deep with an allowance for some loss from cleaning operations. Deep sand layers retard the draining process. Sand should have a uniformity coefficient of not over 4.0 and an effective size of 0.3 to 0.75 mm.

Piping to the sludge beds should drain to the beds and should be designed for a velocity of at least 0.75 m/s (2.5 ft/s). Cast-iron pipe is frequently used. Arrangements should be made to flush the lines if necessary and to prevent their freezing in cold climates. Distribution boxes are required to divert the sludge flow into the bed selected. Splash plates are placed in front of the sludge outlets to spread the sludge over the bed and to prevent erosion of the sand.

Table 11-22 Typical area requirements for open sludge-drying beds

Type of sludge	Area, $m^2/10^3$ persons[a]	Sludge loading rate, kg dry solids/$m^2 \cdot$ yr
Primary digested	90–140	120–200
Primary and humus digested	110–160	100–160
Primary and activated digested	160–275	60–100
Primary and chemically precipitated digested	185–230	100–160

[a] Corresponding area requirements for covered beds vary from 70 to 75 percent of those for the open beds.

Note: $m^2 \times 10.7639 = ft^2$
$kg/m^2 \cdot yr \times 0.2048 = lb/ft^2 \cdot yr$

Sludge can be removed from the drying bed after it has drained and dried sufficiently to be spadable. Dried sludge has a coarse, cracked surface and is black or dark brown. The moisture content is approximately 60 percent after 10 to 15 d under favorable conditions. Sludge removal is accomplished by manual shoveling into wheelbarrows or trucks or by a scraper or front-end loader. Provisions should be made for driving a truck onto or along the bed to facilitate loading.

Lagoons

Drying lagoons may be used as a substitute for drying beds for the dewatering of digested sludge. Lagoons are not suitable for dewatering untreated sludges, limed sludges, or sludges with a high-strength supernatant because of their odor and nuisance potential. The performance of lagoons, like that of drying beds, is affected by climate; precipitation and low temperatures inhibit dewatering. Lagoons are most applicable in areas with high evaporation rates.

Unconditioned digested sludge is discharged to the lagoon in a manner suitable to accomplish an even distribution of sludge. Sludge depths usually range from 0.75 to 1.25 m (2.5 to 4 ft). Evaporation is the prime mechanism for dewatering. Facilities for decanting of supernatant are usually provided, and the liquid is recycled to the treatment facility. Sludge is removed mechanically, usually at a moisture content of about 70 percent. The cycle time for lagoons varies from several months to several years. Typically, sludge is pumped to the lagoon for 18 months, and then the lagoon is rested for 6 months.

11-12 HEAT DRYING

Sludge drying is a unit operation that involves reducing water content by vaporization of water to the air. In sludge-drying beds, vapor-pressure differences account for evaporation to the atmosphere. In mechanical drying apparatuses, auxiliary heat is provided to increase the vapor-holding capacity of the ambient air and to provide the latent heat of evaporation. The purpose of heat drying is to remove the moisture from the wet sludge so that it can be incinerated efficiently or processed into fertilizer. Drying is necessary in fertilizer manufacture to permit grinding of the sludge, to reduce its weight, and to prevent continued biological action. The moisture content of the dried sludge is less than 10 percent.

Theory

Under equilibrium conditions of constant-rate drying, mass transfer is proportional to (1) the area of wetted surface exposed, (2) the difference between water content of the drying air and saturation humidity at the wet-bulb temperature of the sludge-air interface, and (3) other factors, such as velocity and turbulence of drying air expressed as a mass-transfer coefficient. The pertinent equation is

$$W = k_y(H_s - H_a)A \qquad (11\text{-}13)$$

where W = evaporation rate, kg/h (lb/h)

k_y = mass-transfer coefficient of gas phase, kg/m$^2 \cdot$ h (lb mass/ft$^2 \cdot$ h) per unit of humidity difference (ΔH)

H_s = saturation humidity of air at sludge-air interface, kg water vapor/kg dry air (lb/lb)

H_a = humidity of drying air, kg water vapor/kg dry air (lb/lb)

The sludge-air interface temperature may be taken as equal to the wet-bulb temperature of the bulk volume of drying air or hot gases, provided the temperature of the air and the walls of the container are approximately the same. For extension of the theory and its application to specific types of drying equipment, refer to McAdams [17].

It is evident that drying may be accomplished most rapidly on a finely divided sludge by exposing new areas to the drying air stream. Furthermore, maximum contact between dry air and wet sludge should be obtained to assure a maximum value of ΔH. These factors must be considered in the selection of drying apparatuses for sludge disposal in a wastewater treatment plant.

Heat-Drying Options

Five mechanical processes may be used for drying sludge: (1) flash dryers, (2) spray dryers, (3) rotary dryers, (4) multiple hearth dryers, and (5) the Carver-Greenfield process. Most systems can be made to dry or incinerate [2]. Sludge dryers are normally preceded by dewatering. Flash dryers are the most common type in use at wastewater-treatment plants.

Flash dryers Flash drying involves pulverizing the sludge in a cage mill or by an atomized suspension technique in the presence of hot gases. The equipment should be designed so that the particles remain in contact with the turbulent hot gases long enough to accomplish mass transfer of moisture from sludge to the gases.

One operation involves a cage mill that receives a mixture of wet sludge or sludge cake and recycled dried sludge. The mixture contains approximately 50 percent moisture. The hot gases and sludge are forced up a duct in which most of the drying takes place and to a cyclone, which separates the vapor and solids. It is possible to achieve a moisture content of 8 percent in this operation. A schematic drawing of the entire drying process is shown on Fig. 11-27. The dried sludge may be used or sold as fertilizer or it may be incinerated in the furnace in any proportion up to 100 percent of production.

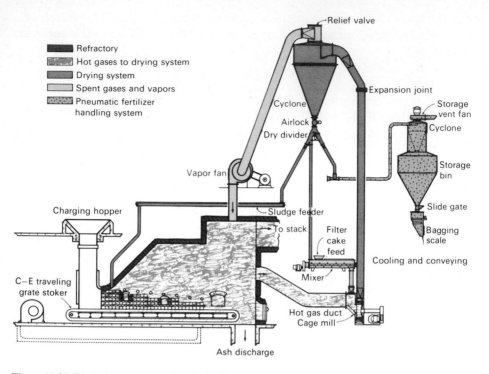

Refractory
Hot gases to drying system
Drying system
Spent gases and vapors
Pneumatic fertilizer handling system

Relief valve

Expansion joint

Cyclone

Airlock

Dry divider

Vapor fan

Storage vent fan

Cyclone

Storage bin

Charging hopper

Sludge feeder

Slide gate

C–E traveling grate stoker

To stack

Filter cake feed

Bagging scale

Cooling and conveying

Mixer

Hot gas duct

Cage mill

Ash discharge

Figure 11-27 Flash-drying system for sludge [*from Combustion Engineering*].

Spray dryers A spray dryer uses a high-speed centrifugal bowl into which liquid sludge is fed. Centrifugal force serves to atomize the sludge into fine particles and to spray them into the top of the drying chamber where steady transfer of moisture to the hot gases takes place. This is shown in Fig. 11-28. A nozzle may be used in place of the bowl if the design prevents clogging of the nozzle.

Rotary dryers Rotary kiln dryers have been used in several plants for the drying of sludge and for the drying and burning of municipal solid wastes and industrial wastes (see Fig. 11-29). Many different dryers have been developed for industrial processes [17]. In direct-heating dryers, the material being dried is in contact with the hot gases. In indirect-heating dryers, the hot gases are separated from the drying material by steel shells. In indirect-direct dryers, the hottest gases surround a central shell containing the material but return through it at reduced temperatures. Coal, oil, gas, municipal solid wastes, or the dried sludge may be used as fuel. Plows or louvers may be installed for lifting and agitating the material as the drum revolves.

Multiple hearth dryer A multiple hearth incinerator is frequently used to dry and burn sludges that have been partially dried by vacuum filtration. This is a counter-

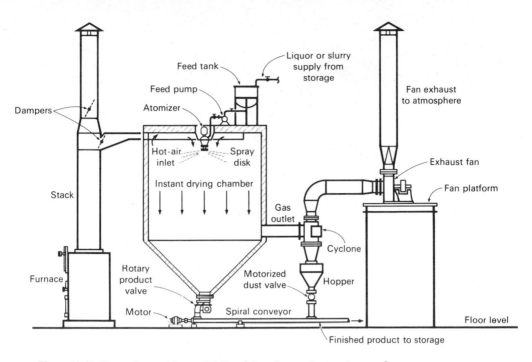

Figure 11-28 Spray dryer with parallel flow [*from Instant Drying Company*].

current operation in which heated air and products of combustion pass by finely pulverized sludge that is continually raked to expose fresh surfaces. A more detailed discussion of multiple hearth incineration is presented in Sec. 11-14.

Oil-emersion dehydration The drying of sludge can also be accomplished using a patented oil-emersion process known as the Carver-Greenfield process. Operationally, a light oil is mixed with dewatered sludge. The oil-sludge mixture, which can be pumped easily and is effective in reducing scaling and corrosion, is then passed through a four-stage falling film evaporator. Water is removed because it has a lower boiling point than the oil carrier. After evaporation, what remains is essentially a mixture of oil and dry sludge. The solids are removed from the oil with a centrifuge. The remaining oil can be separated into a light-oil and heavy-oil residue by exposing it to superheated steam.

The dry solids are suitable for further processing (e.g., pelletizing as a fuel source) or disposal. The recovery of energy and heat from the dried sludge using an incinerator-pyrolysis reactor, or gasifier, is an option that should be investigated when the Carver-Greenfield process is being evaluated. The heat recovered from the dried sludge could be used to supply the energy requirements of the process. If the solids are to be disposed of in a landfill, they can be mixed with the solids from a centrifuge or vacuum filter to reduce dewatering costs.

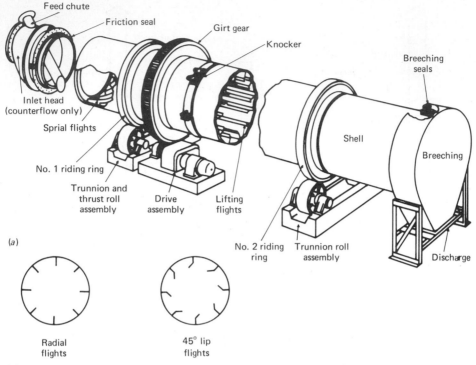

Feed chute

Friction seal

Girt gear

Knocker

Breeching seals

Inlet head (counterflow only)

Sprial flights

Shell

Breeching

No. 1 riding ring

Trunnion and thrust roll assembly

Drive assembly

Lifting flights

No. 2 riding ring

Trunnion roll assembly

Discharge

(a)

Radial flights

45° lip flights

(b)

Figure 11-29 Rotary dryer used for drying sludge. (*a*) Isometric view. (*b*) Alternative flight arrangements.

Air-pollution and odor control

The two most important control measures associated with heat drying of sludge are fly-ash collection and odor control. Cyclone separators having efficiencies of 75 to 80 percent are suitable for vent-gas temperatures up to 340 or 370°C (650 or 700°F). Wet scrubbers have higher efficiencies and will condense some of the organic matter in the vent gas, but may carry over water droplets.

Sludge drying occurs at temperatures of approximately 370°C, whereas 650 to 760°C (1200 to 1400°F) is required for complete incineration. To achieve destruction of odors, the exhaust gases must reach approximately 730°C (1350°F). Thus, if the gases evolved in the drying process are reheated in an incinerator to a minimum of 730°C, odors will be eliminated. At lower temperatures, partial oxidation of odor-producing compounds may occur, resulting in an increase in the intensity or disagreeable character of odor produced.

11-13 COMPOSTING

Composting is a process in which organic material undergoes biological degradation to a stable end product. Sludge that has been composted properly is a sanitary, nuisance-free, humuslike material. Approximately 20 to 30 percent of the volatile solids are converted to carbon dioxide and water. In addition, because the sludge is usually processed in the thermophilic temperature range, the product sludge is essentially pasteurized. The composted sludge may be used as a soil conditioner. Although the process works well, the primary problem has been the lack of a market for the stabilized end product [7]. However, currently there is renewed interest in composting, based primarily on work carried out at Beltsville, Md., by the Agricultural Research Service and the Economic Research Service of the U.S. Department of Agriculture.

Process Description

Most composting operations consist of three basic steps: (1) preparation of the wastes to be composted, (2) decomposition of the prepared wastes, and (3) preparation and marketing of the product. Receiving, sorting, separation, size reduction, and addition of moisture and nutrients are part of the preparation step. To accomplish the decomposition step, several techniques have been developed. In windrow composting, the prepared wastes are placed in windrows in an open field. The windrows are turned once or twice a week for a composting period of about 5 weeks. The material is usually cured for an additional 2 to 4 weeks to ensure stabilization. As an alternative to windrow composting, several mechanical systems have been developed, including the aerated pile process [7]. By controlling the operation carefully in a mechanical system, it is possible to produce a humus within 5 to 7 d. Often, the composted material is removed, screened, and cured for an additional period of about 3 to 4 weeks. Once the compost has been cured, it is ready for the third step, which is product preparation and marketing. This step may include fine grinding, blending with various additives, granulating, bagging, storing, shipping, and, in some cases, direct marketing.

An enclosed mechanical compost system usually is preferred to open composting in excessively humid or cold areas because conditions can be better controlled [12]. Additional details on the process may be found in Refs. 29 and 33.

Co-composting Options

Sludge may be composted either separately, as has been discussed, or in combination with wood chips or other solid wastes (co-composting).

Co-composting with wood chips Co-composting of sludge with wood chips usually requires that the sludge be dewatered initially. In addition, it must be blended with a bulking material, such as wood chips. The most attractive of the wood-chip co-composting processes appears to be the aerated pile process (see Fig. 11-30). In this process, sludge is first mixed with wood chips. The mixed material is then placed in a pile and covered with a 300-mm (12-in) layer of screened compost for insulation and odor control. Oxygen is supplied by forced aeration. After about 21 d, plus 2 d of drying, the wood chips are postsorted and recycled.

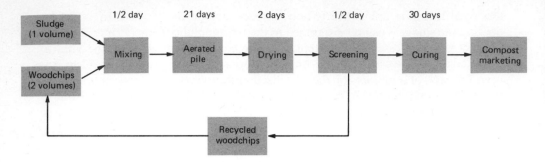

Figure 11-30 Flow diagram of Beltsville aerated pile method for composting raw sludge [7].

After curing for an additional 30 d, the compost is ready for product preparation and marketing [7]. Equipment requirements for composting in piles may be found in Ref. 33.

Feed sludge may consist of digested or untreated sludges. Digested sludge is reported to compost more slowly than untreated sludge, particularly during wet, cold periods, possibly because of the lack of sufficient digestible energy material for rapid biological oxidation [12]. On the other hand, compost systems using untreated sludges are often more susceptible to odor problems. Odor has not been a problem with the aerated pile process.

Co-composting with solid wastes Co-composting of sludges and municipal solid wastes usually does not require sludge dewatering. Feed sludges may have a solids content ranging from 5 to 12 percent. A 2 to 1 mixture of solid wastes to sludge is recommended; in fact, any amount of sludge can be mixed with solid wastes for composting, provided that the sludge is dewatered adequately. The solid wastes should be presorted and pulverized in a hammermill prior to mixing with sludge. Additional details may be found in Refs. 12 and 29.

11-14 THERMAL REDUCTION

Thermal reduction of sludge involves (1) the total or partial conversion of organic solids to oxidized end products, primarily carbon dioxide and water, by incineration or wet-air oxidation, or (2) the partial oxidation and volatilization of organic solids by pyrolysis to end products with calorific value. The major objective of thermal reduction is to reduce the quantities of solid material for disposal. Such processes are used most commonly by large plants with limited solids-disposal facilities.

Sludges processed by thermal reduction are usually dewatered untreated sludges. It is normally unnecessary to stabilize sludge before incineration. In fact, such practice may be detrimental because stabilization, specifically aerobic and anaerobic digestion, decreases the volatile content of the sludge and consequently increases the requirement for an auxiliary fuel. An exception is the implementation of heat treatment ahead of incineration. Heat-treated sludges dewater extremely well. Therefore, the sludge is normally rendered auto-combustible, i.e., no auxiliary fuel is required to sustain the burning process. Sludges may be subjected to thermal reduction separately or in combination with municipal solid wastes.

The thermal reduction processes considered in the following discussion include multiple hearth incineration, fluidized-bed incineration, flash combustion, coincineration, co-pyrolysis, wet-air oxidation, and recalcination. Before dis-cussing these processes, it will be helpful to review some fundamental aspects of thermal reduction.

Process Fundamentals

Complete combustion, incomplete combustion, and pyrolysis processes are used in the thermal reduction of sludge, as described in the following discussion.

Complete combustion Incineration of sludge implies complete combustion of all the organic substances present. The predominant elements in the carbohydrates, fats, and proteins composing the volatile matter of sludge are carbon, oxygen, hydrogen, and nitrogen (C–O–H–N). The approximate percentages of these may be determined in the laboratory by a technique known as ultimate analysis.

Oxygen requirements for complete combustion may be determined from a knowledge of the constituents, assuming that carbon and hydrogen are oxidized to the ultimate end products CO_2 and H_2O. The formula becomes

$$C_aO_bH_cN_d + (a + 0.25c - 0.5b)O_2 \rightarrow aCO_2 + 0.5cH_2O + 0.5dN_2 \quad (11\text{-}14)$$

The theoretical quantity of air will be 4.35 times the calculated quantity of oxygen, because air is composed of 23 percent oxygen on a weight basis. To ensure complete combustion, excess air amounting to about 50 percent of the theoretical amount will be required. A materials balance must be made to include the above compounds and the inorganic substances in the sludge, such as the inert material and moisture, and the moisture in the air. The specific heat of each of these substances and of the products of combustion must be taken into account in determining the heat required for the incineration process.

Heat requirements will include the sensible heat Q_s in the ash, plus the sensible heat required to raise the temperature of the flue gases to 760°C (1400°F) or whatever higher temperature of operation is selected for complete oxidation and elimination of odors, less the heat recovered in preheaters or recuperators. Latent heat Q_e must also be furnished to evaporate all of the moisture in the sludge. Total heat required Q may be expressed as

$$Q = \sum Q_s + Q_e = \sum C_p W_s(T_2 - T_1) + W_w \lambda \quad (11\text{-}15)$$

where C_p = specific heat for each category of substance in ash and flue gases
W_s = mass of each substance
T_1, T_2 = initial and final temperatures
λ = latent heat of evaporation per kilogram (pound)

It should be obvious that reduction of moisture content of the sludge is the principal way to lower heat requirements and may determine whether additional fuel will be needed to support combustion.

The heating value of a sludge may be determined by the conventional bomb-calorimeter test. An empirical formula, based on a statistical study of fuel values of vacuum-filtered sludges of different types, taking into account the amount of coagulant added before filtration, is as follows [13]:

$$Q = a \left[\frac{P_v(100)}{100 - P_c} - b \right] \frac{100 - P_c}{100} \qquad (11\text{-}16)$$

where Q = fuel value, J/kg dry solids (Btu/lb dry solids)
a = coefficient (3×10^5 for primary sludge, untreated or digested; 2.5×10^5 for fresh activated)
b = coefficient (10 for primary sludge; 5 for activated sludge)
P_v = percent of volatile solids in sludge
P_c = percent of coagulating solids added to the sludge

The fuel value of untreated sludge ranges from 11 to 23 MJ/kg (4800 to 10,000 Btu/lb) dry solids, depending on the type of sludge and the volatile content. This is equivalent to some of the lower grades of coal. Digested sludge has a heat content ranging from 6 to 13 MJ/kg (2500 to 5500 Btu/lb).

To design an incinerator for sludge volume reduction, a detailed heat balance must be prepared. Such a balance must include heat losses through the walls and pertinent equipment of the incinerator, as well as losses in the stack gases and ash. Approximately 2.0 to 2.5 MJ (1800 to 2500 Btu) are required to evaporate each 0.5 kg (1 lb) of water in the sludge. Heat is obtained from the combustion of the volatile matter in the sludge and from the burning of auxiliary fuels. For untreated primary sludge incineration, the auxiliary fuel is needed only for warming up the incinerator and maintaining the desired temperature when the volatile content of the sludge is low. The design should include provisions for auxiliary heat for startup and for assuring complete oxidation at the desired temperature under all conditions. Fuels such as oil, natural gas, or excess digester gas are suitable.

Incomplete combustion Organic substances may be oxidized under high pressures at elevated temperatures with the sludge in a liquid state by feeding compressed air into the pressure vessel. The process, known as wet combustion, was developed in Norway for pulp-mill wastes, but has been revised for the oxidation of untreated wastewater sludges pumped directly from the primary settling tank or thickener. Combustion is not complete; the average is 80 to 90 percent completion. Thus, some organic matter, plus ammonia, will be observed in the end products. For this incomplete combustion reaction, Rich [26] has suggested that

$$C_a H_b O_c N_d + 0.5(ny + 2s + r - c)O_2 \rightarrow$$

$$nC_w H_x O_y N_z + sCO_2 + rH_2O + (d - nz)NH_3 \qquad (11\text{-}17)$$

where $r = 0.5[b - nx - 3(d - nz)]$
$s = a - nw$

The results obtained from this equation can also be approximated by the COD of the sludge, which is approximately equal to the oxygen required in combustion. The range of heat released per 0.5 kg (1 lb) of air required has been found to be from 1.3 to 1.5 MJ (1200 to 1400 Btu). Maximum operating temperatures for the system vary from 175 to 315°C (350 to 600°F) with design operating pressures ranging from 1 to 20 MN/m^2 (150 to 3000 lb$_f$/in^2 gage). A flowsheet of the wet oxidation process, known as the Zimmerman process, is shown in Fig. 11-19.

Pyrolysis Because most organic substances are thermally unstable, they can, upon heating in an oxygen-free atmosphere, be split through a combination of thermal cracking and condensation reactions into gaseous, liquid, and solid fractions. Pyrolysis is the term used to describe the process. In contrast to the combustion process, which is highly exothermic, the pyrolytic process is highly endothermic. For this reason, the term destructive distillation is often used as an alternative term for pyrolysis [29].

The characteristics of the three major component fractions resulting from the pyrolysis are:

1. A gas stream containing primarily hydrogen, methane, carbon monoxide, carbon dioxide, and various other gases, depending on the organic characteristics of the material being pyrolyzed
2. A fraction that consists of a tar and/or oil stream that is liquid at room temperatures and has been found to contain chemicals such as acetic acid, acetone, and methanol
3. A char consisting of almost pure carbon plus any inert material that may have entered the process

For cellulose ($C_6H_{10}O_5$), the following expression has been suggested as being representative of the pyrolysis reaction:

$$3(C_6H_{10}O_5) \rightarrow 8H_2O + C_6H_8O + 2CO + 2CO_2 + CH_4 + H_2 + 7C \qquad (11\text{-}18)$$

In Eq. 11-18, the liquid tar and/or oil compounds normally obtained are represented by the expression C_6H_8O. It has been found that distribution of the product fractions varies dramatically with the temperature at which the pyrolysis is carried out. Additional details may be found in Ref. 16.

Thermal Reduction Processes

Applications of six types of thermal reduction processes will be described in the discussion that follows: multiple hearth incineration, fluidized-bed incineration, flash combustion, coincineration, copyrolysis, and wet air oxidation.

Multiple hearth incineration [22] The multiple hearth process converts dewatered sludge cake to an inert ash. In 1970, about 120 of these units were being used for wastewater sludge incineration. Because the process is complex and requires specially trained operators, multiple hearth furnaces are normally used only in plants larger than 0.2 m³/s (5 Mgal/d). They have been used at facilities with lower flows where land for the disposal of sludges is limited. They are also used in chemical treatment plants for the recalcining of lime sludges.

As shown in Fig. 11-31, sludge cake is fed onto the top hearth and is slowly raked to the center. From the center, it drops to the second hearth where the rakes move it to the periphery. Here it drops to the third hearth and is again raked to the center. The hottest temperatures are on the middle hearths where the sludge burns and where auxiliary fuel is also burned as necessary to warm up the furnace and to sustain combustion. Preheated air is admitted to the lowest hearth and is further heated by the sludge as it rises past the middle hearths where combustion is occurring. The air then cools as it gives up its heat to dry the incoming sludge on the top hearths.

The highest moisture content of the air is found on the top hearths where sludge with the highest moisture is heated and some water is vaporized. This air passes twice through the furnace. Cooling air is initially blown into the central column and hollow rabble arms to keep them from burning up. A large portion of this air, after passing out the central column at the top, is recirculated to the lowest hearth.

This furnace may also be designed as a dryer only. In this case, a furnace is needed to provide hot gases, and the sludge and gases both proceed downward through the furnace in parallel flow. Parallel flow of product and hot gases is frequently used in drying operations to prevent burning or scorching a heat-sensitive material.

Feed sludge must contain more than 15 percent solids because of limitations on the maximum evaporating capacity of the furnace. Auxiliary fuel is usually required when the feed sludge contains between 15 and 30 percent solids. Feed sludge containing more than 50 percent solids may create temperatures in excess of the refractory and metallurgical limits of standard furnaces. Average loading rates of wet cake are approximately 40 kg/m² · h (8 lb/ft² · h) of effective hearth area, but may range from 25 to 75 kg/m² · h (5 to 15 lb/ft² · h).

In addition to dewatering, required ancillary processes include ash-handling systems and some type of wet scrubber to meet air-pollution requirements. Recycle flows consist of scrubber water (see Fig. 11-31). Scrubber water comes in contact with, and removes, most of the particulate matter in the exhaust gases. The recycle BOD and COD is nil, and the suspended-solids content is a function

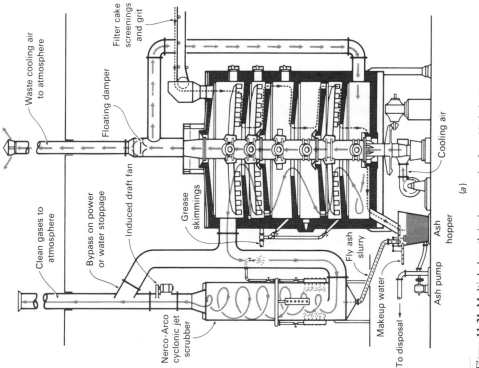

Figure 11-31 Multiple hearth incinerator. (*a, from Nichols Engineering and Research; b, from Envirotech, BSP Div.*)

Waste cooling air to atmosphere

Filter cake screenings and grit

Floating damper

Clean gases to atmosphere

Bypass on power or water stoppage

Induced draft fan

Grease skimmings

Cooling air

Ash hopper

Nerco-Arco cyclonic jet scrubber

Fly ash slurry

Makeup water

Ash pump

To disposal

(a)

(b)

of the particulates captured in the scrubber. Under proper operating conditions, particulate discharges to the air are less than $0.65 \text{ kg}/10^3 \text{ kg}$ (1.3 lb/ton) of dry sludge input. Other air pollutants are not normally considered a problem in most states.

Ash handling may be either wet or dry. In the wet system, the ash falls into an ash hopper located beneath the furnace where it is slurried with water from the exhaust gas scrubber. After agitation, the ash slurry is pumped to a lagoon or is dewatered mechanically. In the dry system, the ash is conveyed mechanically by a bucket elevator to a storage hopper for discharge into a truck for eventual disposal as fill material. The ash is usually conditioned with water. Ash density is about 5.6 kg/m^3 (0.35 lb/ft^3) dry and 880 kg/m^3 (55 lb/ft^3) wet.

Fluidized-Bed Incineration The Dorr-Oliver FS (Fluo-Solids) system uses a fluidized bed of sand as a heat reservoir to promote uniform combustion of sludge solids. The sludge is dewatered with a centrifuge or with a vacuum filter before it enters the fluidized-bed reactor. The fluidized bed must be preheated, using fuel oil or gas, to approximately 650°C (1200°F) before the sludge is introduced. In the fluidized-bed reactors, the combustion of sludge takes place in a hot suspended bed of sand at a temperature of 815°C (1500°F). Air is blown into the incinerator below the sand bed, causing it to expand and fluidize. The sludge is injected directly into the suspended sand. Most of the ash residue is swept out of the combustion area with the flue gas.

The resultant combustion gases, ash, and water vapor exit through a wet scrubber, where the ash is removed, and are then exhausted through a stack. The ash is separated from the scrubber in a cyclone separator. Recycle flows consist of scrubber water produced at a rate of approximately 25 to 40 L/kg (3 to 5 gal/lb) of dry solids feed to the fluidized bed. Most of the ash (99 percent) is captured in the scrubber water, and the suspended-solids content is approximately 20 to 30 percent of the dry solids feed. The recycle flow is normally directed to an ash lagoon. BOD and COD are nil. Particulates and other air emissions are comparable to those from the multiple hearth incinerator.

The combustion process is controlled by varying the sludge feed rate and the air flow to the reactor to oxidize completely all the organic material and to eliminate the need for sludge digestion. If the process is operated continuously or with shutdowns of short duration on untreated sludge, there is no need for auxiliary fuel after startup.

Like the multiple hearth, the fluidized bed, though very reliable, is complex and requires the use of trained personnel. For this reason, fluidized-bed incinerators are normally used in large plants, but they may be used in plants with lower flow ranges where land for the disposal of sludges is limited.

Dorr-Oliver has also developed a fuel-fired horizontal spiral-flow incinerator especially adapted to the needs of small wastewater-treatment plants. In this process, thickened sludge is atomized as it enters the combustion chamber by a strong blast of compressed air. The sludge particles are oxidized rapidly and exit as ash with the combustion gases. The ash is then separated and disposed of in the same manner as with the Fluo-Solids process.

Flash combustion [22] In flash combustion, a variation of flash drying, part or all of the dried sludge is incinerated. Part of the sludge may be incinerated to decrease fuel requirements when drying the sludge for fertilizer. All the sludge is incinerated when there is no market for the fertilizer. Flash combustion for total sludge incineration is not competitive with multiple hearth or fluidized-bed incineration, and future use for that application is not expected.

Coincineration [22] Coincineration is the process of incinerating wastewater sludges with municipal solid wastes. The major objective is to reduce the combined costs of incinerating sludge and solid wastes. At present, coincineration is not widely practiced. The process has the advantages of producing the heat energy necessary to evaporate water from sludges, supporting combustion of solid wastes and sludge, and providing an excess of heat for steam generation, if desired, without the use of auxiliary fossil fuels. In properly designed systems, the hot gases from the process can be used to remove moisture from sludges to a content of 10 to 15 percent. It has been found that the direct feeding of sludge filter cake containing 70 to 80 percent moisture over solid wastes on traveling or reciprocating grates is ineffective.

An excellent example of coincineration is the system at Waterbury, Connecticut, where the municipal solid waste incinerator is located at the wastewater-treatment plant. This plant has been in operation for a number of years. A flash-drying system, using hot effluent gases at 705°C (1300°F) from the furnaces is used to dry the sludge to a 10 to 15 percent moisture content before introducing it in suspension over the burning solid waste bed.

Coincineration of industrial wastewater sludges with solid wastes using flash drying has been done successfully in a large eastern industrial complex for approximately 3 yr. A water-filled boiler serves as the furnace for these fuels. The steam output from the boiler is used to generate in-plant electric power. An electrostatic precipitator is used to clean the exhaust gases.

For systems operating without heat recovery, a disposal ratio of 0.5 kg (1 lb) of dry wastewater solids to 2.25 kg (5 lb) of solid wastes is fired in normal operation. In the case of the water-walled boiler with heat recovery, the ratio is approximately 0.5 kg (1 lb) of dry (industrial plant) solids to 4.0 kg (8 lb) of solid wastes.

In accordance with the history of practice in municipal solid waste disposal, it is anticipated that the construction of coincineration plants will proceed very slowly, despite the advantages to the community in combining the two waste disposal functions.

Co-pyrolysis [22] Pyrolysis is the destructive distillation and decomposition of organic solids at temperatures ranging from 370 to 870°C (700 to 1600°F) in the absence of air or other gases which support combustion. Like incineration, pyrolysis reduces the volume of solid wastes and produces a sterile end product. Unlike incineration, it offers the potential advantages of eliminating air pollution and producing useful by-products.

At present, considerably more data are available for the pyrolysis of municipal solid wastes than for the co-pyrolysis process with wastewater sludge. The development of operating parameters for the optimum production of pyrolysis products, such as fuel gases, oils, and char, is necessary to make this process economically viable. Operation of pyrolysis equipment at the commercial scale is required to determine such parameters as operating temperature, operating pressure, and residence time in the pyrolyzer. A typical flowsheet for the co-pyrolysis of solid wastes and sludge is shown in Fig. 11-32.

Aside from the residual chars and wastewater streams, treatment of effluent combustion gases from the processes is required. Gas scrubbers using water as the scrubbing medium are proposed for these systems. If the residual chars or solid material from the pyrolysis process are not used as fuels or as aggregate materials in road construction, landfilling is the recommended disposal method. Because of the inert nature of the chars or ash from pyrolysis reactors and because of the relatively small quantity of ash from these processes, effluent gases comprise the major pollutant to be treated.

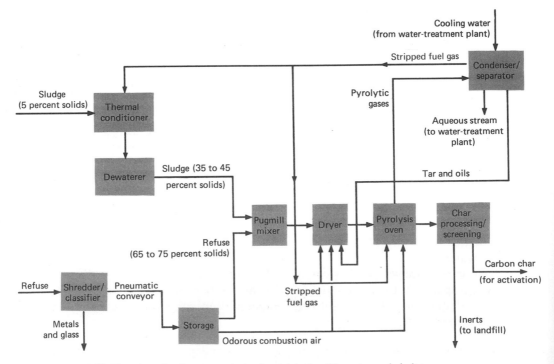

Figure 11-32 Flowsheet for the co-pyrolysis of municipal solid wastes and sludge.

The use of the co-pyrolysis process is expected to proceed rather slowly and will probably not be attempted on a large scale for another 2 to 5 yr.

Wet-air oxidation The Zimmerman process (see Fig. 11-19) involves wet oxidation of untreated sludge at an elevated temperature and pressure. The process is the same as that discussed under heat treatment, except that higher pressures and temperatures are required to oxidize the volatile solids more completely. Untreated sludge is ground and mixed with a specified quantity of compressed air. The mixture is pumped through a series of heat exchangers and then enters a reactor, which is pressurized to keep the water in the liquid phase at the reactor operating temperature of 175 to 315°C (350 to 600°F). High-pressure units can be designed to operate at pressures up to 20 MN/m^2 (3000 lb$_f$/in^2 gage). A mixture of gases, liquid, and ash leave the reactor.

The liquid and ash are returned through heat exchangers to heat the incoming sludge and then pass out of the system through a pressure-reducing valve. Gases released by the pressure drop are separated in a cyclone and released to the atmosphere. In large installations, it may be economical to expand the gases through a turbine to recover power. The liquid and stabilized solids are cooled by passing through a heat exchanger and are then separated in a lagoon or settling tank or on sand beds. The liquid is returned to the primary settling tank and the solids are disposed of by landfill. The process can be designed to be thermally self-sufficient when untreated sludge is used. When additional heat is needed, steam is injected into the reactor vessel.

A major disadvantage associated with this process is the high-strength recycle liquor produced. The liquors represent a considerable organic load on the treatment system. The BOD content of the liquor may be as high as 40 to 50 percent of that of the unprocessed sludge; the COD typically ranges from 7000 to 10,000 mg/L [24]. It is reported that a fraction of the COD is refractory and has been found difficult to remove by physical or chemical processes such as coagulation and activated carbon adsorption [24].

Wet-air oxidation has been implemented in only a limited number of installations since its introduction in the early 1960s. Some of these units have subsequently been taken out of service. Recent innovations may render this process more acceptable. In one system, the feed sludge is preconditioned by the addition of acid to lower the pH to 3. As a result, the process may be operated at lower pressures and temperatures of 230°C and 4 MN/m^2 (450°F and 600 lb$_f$/in^2 gage). In general, the wet-air oxidation system will not be applicable to treatment plants smaller than 0.2 m^3/s (5 Mgal/d).

11-15 PREPARATION OF SOLIDS MASS BALANCES

Sludge-processing facilities normally include additional solids-liquid separation steps, such as thickening or vacuum filtration. The use of such facilities results in flows that must be recycled to the treatment process or to treatment facilities designed specifically for the purpose. When the flows are recycled to the treatment process, they should be recycled to the head end of the plant and blended with the plant flow following the preliminary treatment. In many cases, equalization facilities are provided for the recycled flows so that their reinjection into the plant flow will not cause a shock loading on the subsequent treatment processes.

The recycled flows impose an incremental solids, hydraulic, and organic load on the wastewater treatment facilities which must be considered in the plant design. Hydraulic loadings are also required for the design of the plant piping. To predict these incremental values, it is necessary to perform a materials mass balance on the entire treatment plant flowsheet. The preparation of a materials mass balance is illustrated in Example 11-9. Although all the computational details are given in the example and they are, for the most part, self-explanatory, the following discussion is provided to further explain the general methodology involved in preparing materials mass balances.

Basis for Preparation of Mass Balances

Typically, a materials mass balance is computed on the basis of average flow and average BOD and suspended-solids concentrations. To properly size certain facilities, such as sludge storage tanks and plant piping, it is also important to perform a materials mass balance for the maximum expected concentration of BOD and suspended solids in the untreated wastewater. However, the maximum concentrations will not usually result in a proportional increase in the recycled BOD and suspended solids. The principal reason is that the storage capacity in the wastewater- and sludge-handling facilities tends to dampen peak solids loads to the plant. For example, for a maximum suspended-solids concentration equal to twice the average value, the resulting peak solids loading to a vacuum filter may be only 1.5 times the average loading. Further, it has been shown that periods of maximum hydraulic loading typically do not correlate with periods of maximum BOD and suspended solids. Therefore, coincident maximum hydraulic loadings should not be used in the preparation of a materials mass balance for maximum organic loadings.

Performance Data for Sludge-Processing Facilities

To prepare a materials mass balance, it will be necessary to have information on the operational performance and efficiency of the various unit operations and processes that are used for the processing of waste sludge. Some representative data on the solids capture and expected solids concentrations for the most commonly used operations are reported in Tables 11-23 and 11-24. These data were derived from an analysis of the records from a number of installations throughout the United States. However, local conditions have a significant effect on such data, so the reported values should be used as a guide only if local data are unavailable.

Table 11-23 Typical solids-capture values for various sludge-processing facilities

Operation	Solids capture, %	
	Range	Typical
Gravity thickeners		
Primary sludge and		
waste activated sludge	80–90	85
Primary sludge only	85–92	90
Flotation thickeners		
With chemicals	90–98	92
Without chemicals	80–95	90
Elutriation	60–85	80
Vacuum filtration, with		
chemicals	90–98	96
Pressure filtration, with		
chemicals	90–98	96
Centrifuging		
With chemicals	80–98	92
Without chemicals	50–90	80

Table 11-24 Typical solids-concentration values for the sludge-processing facilities given in Table 11-23

Operation	Solids concentration, %	
	Range	Typical
Gravity thickeners		
Primary sludge and		
waste activated sludge	3–6	4
Primary sludge only	4–10	6
Flotation thickeners		
With chemicals	3–7	5
Without chemicals	3–6	4
Vacuum filtration, with		
chemicals	15–30	20
Pressure filtration, with		
chemicals	20–50	36
Centrifuging		
With chemicals	10–35	18
Without chemicals	10–30	18

Table 11-25 Typical BOD and suspended-solids concentrations in the recycled flows from various sludge-processing facilities

Operation	BOD, mg/L		Suspended solids, mg/L	
	Range	Typical	Range	Typical
Gravity thickening				
Primary sludge and waste				
activated sludge	60–400	300	100–350	250
Primary sludge	100–400	250	80–300	200
Flotation thickening	50–400	250	100–600	300
Anaerobic digestion				
Standard-rate type	500–5,000	2,000	3,000–8,000	5,000
High-rate type	3,000–10,000	5,000	4,000–15,000	10,000
Aerobic digestion				
Mixed-liquor type	200–1,000	500	5,000–30,000	20,000
Decant or filtrate type	50–150	50	40–1,000	200
Heat treatment, top-liquor or				
filtrate[a]	3,000–15,000	7,000	1,000–5,000	2,000
Vacuum filtration:[a]				
Undigested sludge	500–5,000	1,000	1,000–5,000	2,000
Digested sludge	500–5,000	2,000	1,000–20,000	4,000
Centrifugation:[a]				
Undigested sludge	1,000–10,000	5,000	2,000–10,000	5,000
Digested sludge	1,000–10,000	5,000	2,000–15,000	5,000

[a] Very limited data available.

In addition to data on expected solids capture and concentrations, data on the expected concentrations of BOD and suspended solids in the return flows must also be available for the preparation of materials mass balances. Some representative data for the most commonly used operations and processes are reported in Table 11-25. The wide variation that can occur in the reported values is apparent. It is therefore stressed that the values in Table 11-25 should be used only if local data are unavailable. As a check on the data given in this table, it is often possible to compute the expected BOD and suspended solids using the data given in Tables 11-23 and 11-24. For example, in the case of a flotation thickener for which the solids capture ratio and concentration are specified, the concentration of suspended solids in the return flow can be determined. Using this value, the BOD_5 in the return flow can be estimated by making an assumption on the amount of the suspended solids that will be biodegradable. This procedure is illustrated in Example 11-9.

Example 11-9 Preparation of solids balance for treatment facility Prepare a solids balance for the treatment flowsheet shown in Fig. 11-33, using an iterative computational procedure. Assume that the design of the biological treatment process is the same as that presented in Example 10-1. Also assume for the purposes of this example that the following data apply:

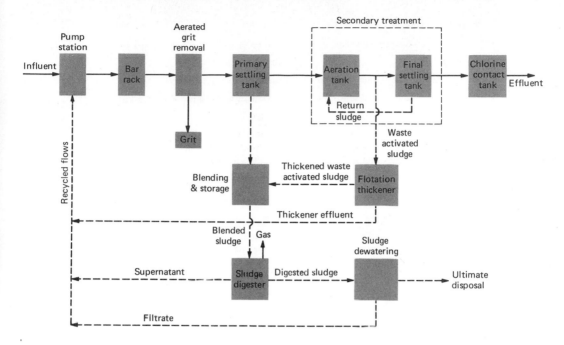

Figure 11-33 Flowsheet for treatment plant to be used for preparation of solids mass balance in Example 11-9.

1. Wastewater flowrates
   a. Average dry weather flow = 21,600 m³/d (5.7 Mgal/d).
   b. Peak dry weather flow = 2.5(21,600 m³/d) = 54,000 m³/d (14.3 Mgal/d).
2. Influent characteristics
   a. BOD_5 = 375 mg/L.
   b. Suspended solids = 400 mg/L.
   c. Suspended solids after grit removal = 360 mg/L.
3. Grit characteristics
   a. Quantity = 0.0357 m³/10³ m³ (5 ft³/Mgal).
   b. Unit mass = 1,120 kg/m³ (70 lb/ft³).
4. Solids characteristics
   a. Concentration of primary solids = 6%.
   b. Concentration of thickened waste activated sludge = 4%.
   c. Total solids in digested sludge = 5%.
   d. For the purposes of this example, assume that the specific gravity of the solids from the primary sedimentation tank and the flotation thickener is equal to 1.0.
5. Effluent characteristics
   a. BOD_5 = 20 mg/L.
   b. Suspended solids = 22 mg/L.

SOLUTION
1 Convert the given constituent quantities to daily mass values.
   a BOD_5 in influent:

$$BOD_5, kg/d = 21,600 \ m^3/d \times 375 \ g/m^3 \times (10^3 \ g/kg)^{-1}$$

$$= 8100 \ kg/d$$

b Suspended solids in influent:

$$\text{Suspended solids, kg/d} = 21{,}600 \text{ m}^3/\text{d} \times 400 \text{ g/m}^3 \times (10^3 \text{ g/kg})^{-1}$$

$$= 8640 \text{ kg/d}$$

c Suspended solids in influent after grit removal:

$$\text{Suspended solids, kg/d} = 21{,}600 \text{ m}^3/\text{d} \times 360 \text{ g/m}^3 \times 10^{-3} \text{ g/kg}$$

$$= 7776 \text{ kg/d}$$

2 Prepare the first iteration of the solids balance. (In the first iteration, the influent wastewater suspended solids and the biological solids generated in the process are distributed among the unit operations and processes that make up the process flowsheet.)

a Primary settling

  i Operating parameters:

  BOD_5 removed $= 33\%$

  Suspended solids removed $= 70\%$ (typical value)

  ii BOD_5 removed $= 0.33(8100 \text{ kg/d}) = 2700 \text{ kg/d}$

  iii BOD_5 to secondary $= (8100 - 2700) \text{ kg/d} = 5400 \text{ kg/d}$

  iv Suspended solids removed $= 0.7(7776 \text{ kg/d}) = 5443.2 \text{ kg/d}$

  v Suspended solids to secondary $= (7776 - 5443.2) \text{ kg/d} = 2332.8 \text{ kg/d}$

b Determine the volatile fraction of primary sludge.

  i Operating parameters

  Volatile fraction of suspended solids in influent prior to grit removal $= 67\%$

  Volatile fraction of grit $= 10\%$

  Volatile fraction of incoming suspended solids discharged to the secondary process $= 85\%$

  ii Volatile suspended solids in influent prior to grit removal, kg/d $= 0.67(8640 \text{ kg/d}) = 5788.8 \text{ kg/d}$

  iii Volatile suspended solids removed in grit chamber, kg/d $= 0.10(8640 - 7776) \text{ kg/d} = 86.4 \text{ kg/d}$

  iv Volatile suspended solids in secondary influent, kg/d $= 0.85(2332.8 \text{ kg/d}) = 1982.9 \text{ kg/d}$

  v Volatile suspended solids in primary sludge, kg/d $= (5788.8 - 86.4 - 1982.9) \text{ kg/d} = 3709.5 \text{ kg/d}$

  vi Volatile fraction in primary sludge

  Volatile fraction, $\%$ $= (3709.5 \text{ kg/d}/5443.2 \text{ kg/d})(100\%) = 68.2\%$

c Secondary process

  i Operating parameters (from Example 10-1):

  Mixed-liquor suspended solids $= 4375 \text{ mg/L}$

  (The recycled BOD_5 and suspended solids from the sludge processing facilities were not considered in Example 10-1 in computing the concentration of the mixed-liquor suspended solids.)

  Volatile fraction of mixed-liquor suspended solids $= 0.80$

  $Y_{obs} = 0.3125$

  ii Determine the effluent mass quantities.

$$BOD_5, \text{ kg/d} = (21{,}600 \text{ m}^3/\text{d})(20 \text{ g/m}^3)(10^{-3} \text{ g/kg})$$

$$= 432.0 \text{ kg/d}$$

$$\text{Suspended solids, kg/d} = (21{,}600 \text{ m}^3/\text{d})(22 \text{ g/m}^3)(10^{-3} \text{ g/kg})$$

$$= 475.2 \text{ kg/d}$$

(Note that the actual effluent flow will be less than the influent flowrate by the amount of liquid lost by evaporation, taken up in the production of digester gas, and that contained in the sludge cake. The actual BOD_5 and suspended solids mass values will be within 5 percent of these maximum values.)

iii   Estimate the mass of volatile solids produced in the activated-sludge process that must be wasted. The required value was computed using Eq. 10-3.

$$P_{x(VSS)} = Y_{obs} Q(S_0 - S)(10^3 \text{ g/kg})^{-1}$$

$$= \frac{0.3125(21,600 \text{ m}^3/\text{d})(250 - 6.2) \text{ g/m}^3}{10^3 \text{ g/kg}}$$

$$= 1645.7 \text{ kg/d}$$

(Note that the actual flowrate will be the primary influent less the flowrate of the primary underflow. However, the primary underflow is normally small and can be neglected. If the underflow is significant, the actual flowrate should be used to determine the volatile solids production.)

iv   Estimate the total mass of suspended solids that must be wasted assuming that the volatile fraction represents 0.80 of the total solids.

$$\text{Suspended solids, kg/d} = 1645.7/0.80 = 2057.1 \text{ kg/d}$$

Note: If it is assumed that the fixed solids portion of the influent suspended solids equals 0.15, the mass of fixed solids in the input from the primary settling facilities is equal to 349.9 kg/d (0.15 × 2332.8 kg/d). This value can then be compared with the fixed solids determined in the above computation, which is equal to 411.4 kg/d (2057.1 kg/d − 1645.7 kg/d). The ratio of these values is 1.18 (411.4 kg/d to 349.9 kg/d). Values that have been observed for this ratio vary from about 1.0 to 1.3; a value of 1.15 is considered to be the most representative.

v   Estimate the waste quantities discharged to the thickener. (It is assumed in this example that wasting is from the biological reactor.)

$$\text{Suspended solids, kg/d} = (2057.1 - 475.2) \text{ kg/d}$$

$$= 1581.9 \text{ kg/d}$$

$$\text{Flowrate, m}^3/\text{d} = \frac{(1581.9 \text{ kg/d})(10^3 \text{ g/kg})}{4375 \text{ g/m}^3}$$

$$= 361.6 \text{ m}^3/\text{d}$$

{From Example 10-1, the concentration of mixed-liquor suspended solids in the aerator is 4375 g/m³ [(3500 g/m³)/0.8]. This value will increase when the recycled BOD_5 and suspended solids are taken into consideration in the second and subsequent iterations. The computed flowrate is less than the value given in Example 10-1 (470.2 m³/d), because the solids lost in the effluent are taken into consideration.}

d Flotation thickeners

i   Operating parameters:
Concentration of thickened sludge = 4%
Assumed solids recovery = 90%
Assumed specific gravity of feed and thickened sludge = 1.0

ii   Determine the flowrate of the thickened sludge.

$$\text{Flowrate, m}^3/\text{d} = \frac{(1581.9 \text{ kg/d})(0.9)}{(1000 \text{ kg/m}^3)(0.04)} = 35.6 \text{ m}^3/\text{d}$$

iii   Determine the flowrate recycled to the plant headworks.

$$\text{Recycled flowrate} = (361.6 - 35.6) \text{ m}^3/\text{d} = 326.0 \text{ m}^3/\text{d}$$

iv   Determine the suspended solids to the digester.

$$\text{Suspended solids, kg/d} = (1581.9 \text{ kg/d})(0.9) = 1423.7 \text{ kg/d}$$

v   Determine the suspended solids recycled to the plant headworks.

$$\text{Suspended solids, kg/d} = (1581.9 - 1423.7) \text{ kg/d} = 158.2 \text{ kg/d}$$

vi Determine the recycled BOD_5.

$$\text{Suspended solids in recycle flow} = \frac{(158.2 \text{ kg/d})(10^3 \text{ g/kg})}{326.0 \text{ m}^3/\text{d}} = 485.3 \text{ mg/L}$$

$$BOD_5 \text{ of suspended solids} = (485.3 \text{ mg/L})(0.65)(1.42)(0.68) = 304.6 \text{ mg/L}$$

Note: The BOD_5 of the suspended solids was estimated using the same procedure that was applied in Example 10-1.

$$\text{Soluble } BOD_5 \text{ escaping treatment} = 6.2 \text{ mg/L (from Example 10-1)}$$

$$\text{Total } BOD_5 \text{ concentration} = (304.6 + 6.2) \text{ mg/L} = 310.8 \text{ mg/L}$$

$$BOD_5, \text{kg/d} = (310.8 \text{ g/m}^3)(326.0 \text{ m}^3/\text{d})(10^3 \text{ g/kg})^{-1}$$
$$= 101.3 \text{ kg/d}$$

e Sludge digestion
   i   Operating parameters:
   $\theta = 10$ d
   Volatile-solids destruction during digestion = 50%
   Gas production = 1.12 m³/kg of volatile solids destroyed
   BOD in digester supernatant = 5000 mg/L (0.5%)
   Total solids in digester supernatant = 5000 mg/L (0.5%)
   Total solids in digested sludge = 5%
   ii Determine the total solids fed to the digester and the corresponding flowrate.

$$\text{Total solids} = \text{solids from primary settling plus waste solids from thickener}$$

$$\text{Total solids, kg/d} = 5443.2 \text{ kg/d} + 1423.7 \text{ kg/d} = 6866.9 \text{ kg/d}$$

$$\text{Total flowrate} = \frac{5443.2 \text{ kg/d}}{0.06(10^3 \text{ kg/m}^3)} + \frac{1423.7}{0.04(10^3 \text{ kg/m}^3)}$$
$$= 90.7 \text{ m}^3/\text{d} + 35.6 \text{ m}^3/\text{d}$$
$$= 126.3 \text{ m}^3/\text{d}$$

   iii Determine the total volatile solids fed to the digester.

$$\text{Total volatile solids, kg/d} = 0.682(5443.2 \text{ kg/d}) + 0.80(1423.7 \text{ kg/d})$$
$$= 3712.3 \text{ kg/d} + 1139.0 \text{ kg/d}$$
$$= 4851.3 \text{ kg/d}$$

$$\text{Percent volatile solids in sludge mixture fed to digester} = \frac{4851.3}{6866.9} \text{ kg/d } 100\% = 70.6\%$$

   iv Determine the volatile solids destroyed.

$$\text{Volatile solids destroyed, kg/d} = 0.5(4851.3 \text{ kg/d})$$
$$= 2425.6 \text{ kg/d}$$

v Determine the mass flow to the digester.

Primary sludge at 6% solids:

$$\text{Mass flow, kg/d} = \frac{5443.2 \text{ kg/d}}{0.06}$$

$$= 90{,}720 \text{ kg/d}$$

Thickened waste activated sludge at 4% solids:

$$\text{Mass flow, kg/d} = \frac{1423.7 \text{ kg/d}}{0.04}$$

$$= 35{,}592.5 \text{ kg/d}$$

$$\text{Total mass flow} = (90.720 + 35{,}592.5) \text{ kg/d} = 126{,}312.5 \text{ kg/d}$$

Note: The total mass flow can also be computed by multiplying the total flowrate to the digester by the density of the combined sludge, if known.

vi Determine the mass quantities of gas and sludge after digestion. Assume that the total mass of fixed solids does not change during digestion and that 50 percent of the volatile solids are destroyed.

$$\text{Fixed solids} = \text{total solids} - \text{volatile solids} = (6866.9 - 4851.3) \text{ kg/d} = 2015.6 \text{ kg/d}$$

$$\text{Total solids in digested sludge} = 2015.6 \text{ kg/d} + 0.5(4851.3) \text{ kg/d} = 4441.2 \text{ kg/d}$$

Gas production, assuming that the density of digester gas is equal to 0.86 times that of air (1.202 kg/m³):

$$\text{Gas, kg/d} = (1.12 \text{ m}^3/\text{kg})(0.5)(4851.3 \text{ kg/d})(0.86)(1.202 \text{ kg/m}^3) = 2808.4 \text{ kg/d}$$

Mass balance for digester output:

$$\text{Mass input} = 126{,}312.5 \text{ kg/d}$$

$$\text{Less gas} = -2808.4 \text{ kg/d}$$

$$\text{Mass output} = 123{,}504.1 \text{ kg/d} \qquad \text{(solids and liquid)}$$

vii Determine the flowrate distribution between the supernatant at 5000 mg/L and digested sludge at 5 percent solids. Let $S = $ kg/d of supernatant suspended solids.

$$\frac{S}{0.005} + \frac{4441.2 - S}{0.05} = 123{,}504.1 \text{ kg/d}$$

$$S + 444.1 - 0.1S = 617.5$$

$$0.9S = 173.4$$

$$S = 192.7 \text{ kg/d}$$

$$\text{Digested solids} = (4441.2 - 192.7) \text{ kg/d} = 4248.5 \text{ kg/d}$$

$$\text{Supernatant flowrate} = \frac{192.7 \text{ kg/d}}{0.005(10^3 \text{ kg/m}^3)} = 38.54 \text{ m}^3/\text{d}$$

$$\text{Digested-sludge flowrate} = \frac{4248.5 \text{ kg/d}}{0.05(10^3 \text{ kg/m}^3)} = 85.0 \text{ m}^3/\text{d}$$

viii   Establish the characteristics of the recycle flow to the plant headworks.

$$\text{Flowrate} = 38.5 \text{ m}^3/\text{d}$$

$$BOD_5 = (38.54 \text{ m}^3/\text{d})(5000 \text{ g/m}^3)(10^3 \text{ g/kg})^{-1}$$

$$= 192.7 \text{ kg/d}$$

$$\text{Suspended solids} = (38.54 \text{ m}^3/\text{d})(5000 \text{ g/m}^3)(10^3 \text{ g/kg})^{-1}$$

$$= 192.7 \text{ kg/d}$$

f Sludge dewatering. (*Note:* In the analysis that follows, the weight of the polymer or other sludge conditioning chemicals that may be added were not considered. In some cases, their contribution can be significant and must be considered.)
  i   Operating parameters:
  Sludge cake = 20% solids
  Specific gravity of sludge = 1.06
  Solids capture = 95%
  Filtrate BOD_5 = 1,500 mg/L
  ii   Determine the sludge-cake characteristics.

$$\text{Solids} = (4248.5 \text{ kg/d})(0.95) = 4036.1 \text{ kg/d}$$

$$\text{Volume} = \frac{4036.1 \text{ kg/d}}{1.06(0.20)(10^3 \text{ kg/m}^3)} = 19.0 \text{ m}^3/\text{d}$$

  iii   Determine the filtrate characteristics.

$$\text{Flow} = (85.0 - 19.0) \text{ m}^3/\text{d} = 66.0 \text{ m}^3/\text{d}$$

$$BOD_5 \text{ at 1500 mg/L} = (1500 \text{ g/m}^3)(66.0 \text{ m}^3/\text{d})(10^3 \text{ g/kg})^{-1}$$

$$= 99.0 \text{ kg/d}$$

$$\text{Suspended solids} = 4248.5 \text{ kg/d})(0.05)$$

$$= 212.4 \text{ kg/d}$$

g Prepare a summary table of the recycle flows and waste characteristics for the first iteration.

Operation	Flow, m^3/d	BOD$_5$, kg/d	Suspended solids, kg/d
Flotation thickener	326.0	101.3	158.2
Digester supernatant	38.5	192.7	192.7
Dewatering filtrate	66.0	99.0	212.4
Total	430.5	393.0	563.3[a]

[a] The volatile fraction of the returned suspended solids will typically vary from 50 to 75 percent. A value of 60 percent will be used for the computation in the second iteration.

3 Prepare the second iteration of the solids balance. (*Note:* In preparing the second iteration, it is assumed that none of the recycled BOD and suspended solids are removed in the grit-removal facilities.)
  a Primary settling
    i   Operating parameters = same as those in the first iteration.

ii Total suspended solids and BOD_5 entering the primary tanks.

Total BOD_5 = influent BOD_5 + recycled BOD_5 = 8100 kg/d + 393.0 kg/d = 8493.0 kg/d

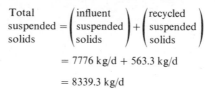

$$= 7776 \text{ kg/d} + 563.3 \text{ kg/d}$$

$$= 8339.3 \text{ kg/d}$$

iii BOD_5 removed = 0.33(8493.0 kg/d) = 2831.0 kg/d
iv BOD_5 to secondary = (8493.0 − 2831.0) kg/d = 5662.0 kg/d
v Suspended solids removed = 0.7(8339.3 kg/d) = 5837.5 kg/d
vi Suspended solids to secondary = (8339.3 − 5837.5) kg/d = 2501.8 kg/d

b Determine the volatile fraction of the primary sludge and effluent suspended solids.
   i Operating parameters
     Incoming wastewater = same as those for the first iteration
     Volatile fraction of solids in recycle returned to pump station = 60%
   ii Although the computations are not shown, the computed change in the volatile fractions determined in the first iteration are slight and, therefore, the values determined previously are used for the second iteration. If the volatile fraction of the return is less than about 50 percent, the volatile fractions should be recomputed.

c Secondary process
   i Operating parameters = same as those for the first iteration
   ii Determine the BOD_5 in the influent to the aeration tank

$$BOD_5, \text{mg/L} = \frac{5662.0 \text{ kg/d } (10^3 \text{ g/kg})}{(21,600 + 430.5) \text{ m}^3/\text{d}}$$

$$= 257 \text{ g/m}^3 = 257 \text{ mg/L}$$

iii Determine the new concentration of mixed-liquor suspended solids. The volatile suspended solids can be computed using Eq. 9-67, which was also used to determine the volume in Example 10-1. The difference in the following computation is that the volume is now fixed (4702 m³ from Example 10-1).

$$X_{vss} = \frac{\theta_c^d Q Y (S_0 - S)}{V(1 + k_d \theta_c^d)}$$

$$X_{vss} = \frac{(10 \text{ d})(22,030.5 \text{ m}^3/\text{d})0.5(257 - 6.2) \text{ mg/L}}{4702 \text{ m}^3(1 + 0.06 \text{ d}^{-1}(10 \text{ d}))}$$

$$= 3672.1 \text{ mg/l}$$

iv Determine the mixed-liquor suspended solids

$$X_{ss} = \frac{X_{vss}}{0.8}$$

$$= 3672.1 \frac{\text{mg/L}}{0.8}$$

$$= 4590.2 \text{ mg/L}$$

v Determine the cell growth using Eq. 10-3.

$$P_X = Y_{obs}\,Q(S_0 - S)(10^3\text{ g/kg})^{-1}$$

$$= 0.3125(22{,}030.5\text{ m}^3/\text{d})(257 - 6.2)\text{ g/m}^3(10^3\text{ g/kg})^{-1}$$

$$= 1726.6\text{ kg/d}$$

$$P_{X(SS)} = 1726.6\,\frac{\text{kg/d}}{0.8}$$

$$= 2158.3\text{ kg/d}$$

vi Determine the waste quantities discharged to the thickener.

Effluent suspended solids, kg/d = 475.2 m³/d (specified in the first iteration)

Total suspended solids to be wasted to the thickener, kg/d = (2158.3 − 475.2) kg/d

$$= 1683.1\text{ kg/d}$$

$$\text{Flowrate, m}^3/\text{d} = \frac{1683.1\text{ kg/d} \times (10^3\text{ g/kg})}{4590.2\text{ g/m}^3} = 366.7\text{ m}^3/\text{d}$$

d Complete the remainder of the second iteration in the same manner as the first iteration. Computations are not shown, but the resultant values for the recycle flows and characteristics are presented in the following table. The incremental change in the recycle flows and waste characteristics from the previous iteration is also reported.

				Incremental change from previous iteration		
Operation/process	Flow, m³/d	BOD₅, kg/d	Suspended solids, kg/d	Flow, m³/d	BOD₅, kg/d	Suspended solids, kg/d
Flotation thickener	328.8	107.7	168.3	2.8	6.4	10.1
Digester supernatant	41.2	205.8	205.8	2.7	13.1	13.1
Dewatering filter	70.6	105.9	227.5	4.6	6.9	15.1
Total	440.6	419.4	601.6	10.1	26.4	38.3

4 Prepare the third iteration of the solids balance. This cycle is computed in the same manner as the second iteration. Computations again are not shown, but the resultant values for the recycle flows, waste characteristics, and incremental values are presented in the following table. This is the final iteration since the incremental change in the return quantities is less than 5 percent. The flow, suspended solids, and BOD₅ values for the various processes are presented in Fig. 11-34.

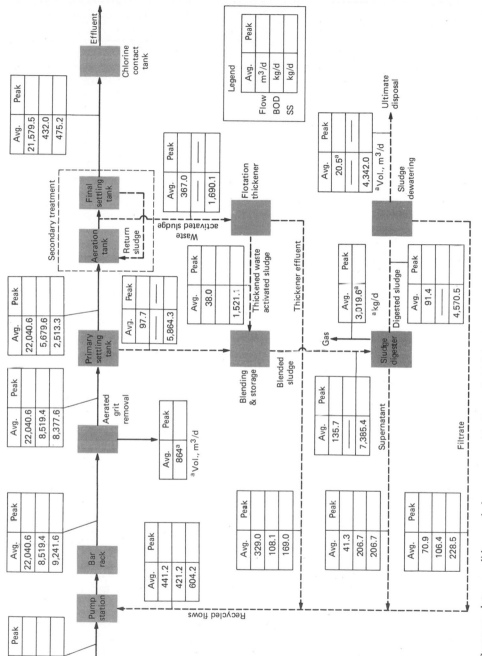

Figure 11-34 Summary data on solids mass balance prepared in Example 11-9 for the treatment plant flowsheet shown in Fig. 11-33.

				Incremental change from previous iteration		
Operation/process	Flow, m³/d	BOD₅, kg/d	Suspended solids, kg/d	Flow, m³/d	BOD₅, kg/d	Suspended solids, kg/d
Flotation thickener	329.0	108.1	169.0	0.2	0.4	0.7
Digester supernatant	41.3	206.7	206.7	0.1	0.9	0.9
Dewatering filter	70.9	106.4	228.5	0.3	0.5	1.0
Total	441.2	421.2	604.2	0.6	1.8	2.6

Comment. In this example, after three iterations the incremental change in the return quantities was less than 1 percent. In general, the iterative computational procedure should be carried out until the incremental change in all of the return quantities from the previous iteration is equal to or less than 5 percent.

11-16 FINAL SLUDGE AND SOLIDS CONVEYANCE, STORAGE AND DISPOSAL

The solids removed as sludge from preliminary and biological treatment processes are concentrated and stabilized by biological and thermal means and are reduced in volume in preparation for final disposal. Because the methods of conveyance and final disposal often determine the type of stabilization required and the amount of volume reduction that is needed, they are considered briefly in the following discussion.

Conveyance Methods [22]

Sludge may be transported long distances by (1) pipeline, (2) truck, (3) barge, (4) rail, or any combination of these four modes (see Fig. 11-35). To minimize the danger of spills, odors, and dissemination of pathogens to the air, liquid sludges should be transported in closed vessels, such as tank trucks, covered or tank barges, or railroad tank cars. Stabilized, dewatered sludges can be transferred in open vessels, such as dump trucks, or in railroad gondolas if they are covered.

The method of transportation chosen and its costs are dependent on a number of factors, including (1) the nature, consistency, and quantity of sludge to be transported; (2) the distance from origin to destination; (3) the availability and proximity of the transit modes to both origin and destination; (4) the degree of flexibility required in the transportation method chosen; and (5) the estimated useful life of the ultimate disposal facility.

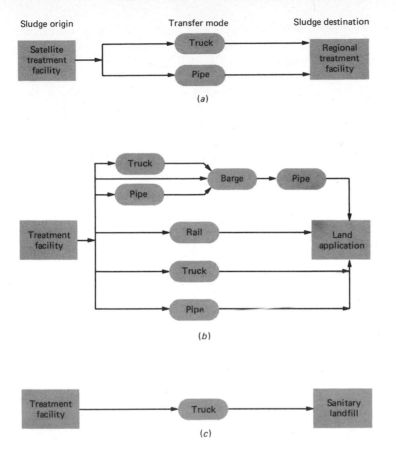

Figure 11-35 Conveyance methods for processed sludge and solids [22]. (*a*) Transfer to further treatment. (*b*) Transfer to land application and reclamation. (*c*) Transfer to sanitary landfill.

Pipeline In general, the energy requirements for long-distance transportation of untreated sludges with a solids concentration of more than 6 percent usually are prohibitive. Also, during sludge pumping, grease tends to build up in unlined pipes and corrosion problems may arise. During low flow conditions, grit tends to accumulate, thereby increasing pipe roughness, and septicity may become a problem. Most of these problems can be reduced or eliminated by maintaining high flowrates in large-diameter pipes. Transport of digested sludge is somewhat easier because the sludge is more homogeneous and has a lower grease content. The head-loss factor for digested sludge is also lower, as previously discussed in Sec. 11-3 and shown in Fig. 11-6. For the reasons cited, the piping of sludge for long distances cannot be economically justified unless a large and relatively constant volume of liquid sludge is to be transported. Further, because pipelines require a large capital investment and because their routes are fixed, the ultimate disposal sites must be large and long-lasting to justify the expenditure.

Pressure piping of sludges with a solids content of less than 10 percent from small treatment plants (0.4 m³/s or 10 Mgal/d or smaller) is economically justifiable for distances under about 16 km (10 mi). This method is commonly used by satellite plants from which effluent sludges are pumped to larger regional treatment facilities for further treatment. Sludges that are to be pumped to central or regional treatment facilities should not be prethickened. It is generally more economical to pump unthickened sludges short distances. Moreover, pumping creates turbulent conditions that break up sludge particles, thus requiring further chemical conditioning and thickening at the destination point.

Truck Trucking is the most flexible and most widely used method for transporting sludge. Either liquid or dewatered sludge may be hauled by truck to diverse destinations. It has been found that trucking dewatered sludge is the most economical method for small- to medium-sized treatment facilities where the sludge must be disposed of in sanitary landfills, or where it must be transported seasonally to several different locations. The capital investment is relatively small and the operation is not complicated. If daily sludge volumes are small, one tank truck may be able to serve two or more plants.

Long-haul trucking of thickened sludges to land application sites is somewhat more expensive than the other transportation options. However, trucking does have the advantage that the sludge may be applied directly to the ground at the site, thus eliminating the need for a distribution system at the site.

Barge Many different sizes and types of barges are available for transporting sludge. Generally, double-hulled vessels are used to reduce the possibility of spills in congested harbor areas. Barges may be either towed or self-propelled and may use either gravity or pumped discharges. Barge transport generally is economical only for large facilities treating wastewater flows in excess of 4.38 m³/s (100 Mgal/d). However, it might also be economical for smaller facilities if the effluent sludge can be either trucked or piped to a central barging transfer station where the sludge from several plants can be loaded onto a single barge.

Rail Rail transportation of sludge is not widely used in the United States today, but it may increase in the future. Rail may be used to transport sludges of any consistency, but those with high solids content are transported most economically. A denser material could be formed by mixing digested sludge with solid wastes. Such a material would also have a greater land-reclamation potential and would be more economically transported than the lower-density solid wastes. The use of rail transportation for small quantities of sludges, or for transportation of sludges over short distances, is not economically justifiable at the present time.

Environmental considerations Each transportation method contributes a minor air-pollutant load, either directly or indirectly. A certain amount of air pollution is produced from the facility that generates the electricity necessary for sludge pumping. The engines that move barges, trucks, and railroad cars also produce some air pollutants. On a mass (tonnage) basis, the transportation mode that contributes the lowest pollutant load is piping. Next, in sequence, are barging and unit train rail transportation. The highest pollutant load is from trucking. Other factors of environmental concern include traffic, noise, and construction disturbance.

Sludge Storage

It is often necessary to store sludge that has been anaerobically digested before it is recycled or disposed of ultimately. This can be accomplished in sludge storage basins (see Fig. 11-36). In addition to concentrating the sludge, it is further stabilized by continued anaerobic biological activity.

Depth of the sludge storage basins will vary from 3 to 5 m (10 to 16 ft). Solids loading rates vary from about 0.1 to 0.25 kg VSS/d $\cdot$ m^2 of surface area (20 to 50 lb VSS/d $\cdot$ 10^3 ft^2). If the basins are not loaded too heavily (≤ 0.1 kg VSS/d $\cdot$ m^2), it is possible to maintain an aerobic surface layer through the growth

Figure 11-36 Sludge storage basin with floating aerator in foreground and sludge pumping rig (mud cat) in background.

of algae and by atmospheric reaeration [27]. Alternatively, surface aerators can be used to maintain aerobic conditions in the upper layers.

The number of basins to be used should be sufficient to allow each basin to be out of service for a period of about six months. Stabilized and thickened sludge can be removed from the basins using a mud pump mounted on a floating platform. Sludge concentrations as high as 25 percent solids have been achieved in the bottom layers of these basins [27].

Final Disposal

Final disposal for the sludge and solids from treatment facilities usually involves some form of land disposal. Ocean disposal of sludge by the major coastal cities of the United States is being phased out because of changes in water pollution control regulations. A discussion of sludge-management alternatives for coastal cities is presented in Ref. 3. Reuse of wastewater sludge as a soil conditioner or fertilizer has had some success, as described in Chap. 13. The major problem in disposing of sludge in this manner lies in the economical marketing of the product. The most common methods of land disposal include spreading on land, lagooning, dumping, and landfilling. These are considered briefly in the following discussion; spreading on land is discussed in greater detail in Chap. 13.

Spreading on land Dewatered and composted sludge may be disposed of by spreading over farmlands (Fig. 11-37a) and plowing under after it has dried. Wet dewatered sludge can be incorporated into the soil directly by injection (11-37b). The humus in the sludge conditions the soil and improves its moisture retentiveness. The practice of land spreading has been used in more than 20 communities in Great Britain. Application rates have ranged between 2.3 and 5.6 cm/yr (0.9 and 2.2 in/yr) of sludge having a concentration of 2 to 5 percent solids. Studies of sludge applied at a rate of approximately 5.1 cm/yr (2 in/yr) have been made by the Metropolitan Sanitary District of Chicago. Studies are also underway in California to determine the long-term effects of applying sludge to land. Thus, much remains to be learned about land spreading of sludge.

Lagooning Lagooning of sludge is another popular disposal method because it is simple and economical if the treatment plant is in a remote location. A lagoon is an earth basin into which untreated or digested sludge is deposited. Untreated-sludge lagoons stabilize the organic solids by anaerobic and aerobic decomposition, which may give rise to objectionable odors. The stabilized solids settle to the bottom of the lagoon and accumulate. Excess liquid from the lagoon, if there is any, is returned to the plant for treatment. Lagoons should be located away from highways and dwellings to minimize possible nuisance conditions and should be fenced to keep out unauthorized persons. They should be relatively shallow, 1 to 1.5 m (4 to 5 ft), if they are to be cleaned by

(a)

(b)

Figure 11-37 Application of sludge on land. (*a*) Surface spreading of dewatered sludge. (*b*) Subsurface injection of liquid sludge.

scraping. If the lagoon is used only for digested sludge, the nuisances mentioned should not be a problem. Sludge may be stored indefinitely in a lagoon, or it may be removed periodically after draining and drying.

Dumping Dumping, such as in an abandoned mine quarry, is a suitable disposal method only for sludges and solids that have been stabilized so that no decomposition or nuisance conditions will result. Digested sludge, clean grit, and incinerator residue can be disposed of safely by this method.

Landfilling If a suitable site is convenient, a sanitary landfill can be used for disposal of sludge, grease, grit, and other solids whether it is stabilized or not. The economics of hauling sludge usually indicate that dewatering for volume reduction will result in justifiable savings. The sanitary landfill method is most suitable if it is also used for disposal of the other solid wastes of the community. In a true sanitary landfill, the wastes are deposited in a designated area, compacted in place with a tractor or roller, and covered with a 30-cm (12-in) layer of clean soil. With daily coverage of the newly deposited wastes, nuisance conditions, such as odors and flies, are minimized.

In selecting a land disposal site, consideration must be given to the nuisance and health hazards that may be caused. Trucks carrying wet sludge and grit should be able to reach the site without passing through heavily populated areas or business districts. Drainage from the site that would cause pollution of ground-water supplies or surface streams must also be guarded against. After several years' time, during which the wastes are decomposed and compacted, the land can be used for recreational or other purposes for which gradual subsidence would not be objectionable.

DISCUSSION TOPICS AND PROBLEMS

11-1 The water content of a sludge is reduced from 98 to 95 percent. What is the percent reduction in volume by the approximate method and by the more exact method, assuming that the solids contain 70 percent organic matter of specific gravity 1.00 and 30 percent mineral matter of specific gravity 2.00? What is the specific gravity of the 98 and the 95 percent sludge?

11-2 Consider an activated sludge treatment plant with a capacity of 40,000 m^3/d. The untreated wastewater contains 200 mg/L suspended solids. The plant provides 60 percent removal of the suspended solids in the primary settling tank. If the primary sludge alone is pumped, it will contain 5 percent solids. Assume that 400 m^3/d of waste activated sludge containing 0.5 percent solids is to be wasted to the digester. If the waste activated sludge is thickened in the primary settling tank, the resulting mixture will contain 3.5 percent solids. Calculate the reduction in daily volume of sludge pumped to the digester that can be achieved by thickening the waste activated sludge in the primary settling tank as compared with discharging the primary and waste activated sludge directly to the digester. Assume complete capture of the waste activated sludge in the primary settling tank.

11-3 Sludge is to be withdrawn by gravity from a primary settling tank for heat treatment. The available head is equal to 3 m, and 90 m of 150-mm pipe is to be used to interconnect the units. Determine the flowrate and velocity, assuming that the solids content of the sludge is 6 percent. Assume that the f value for water in the Darcy Weisbach equation is 0.025 and that the minor losses are equal to 0.6 m.

11-4 Determine the required digester volume for the treatment of the sludge quantities specified in Example 11-4 using the (a) volatile solids loading factor, (b) volume reduction, and (c) volumetric per capita allowance methods. Set up a comparison table to display the results obtained using the four different procedures for sizing digesters (three in this problem and one in Example 11-4). Assume the following data apply:
1. Volatile solid loading method
   - a. Solids concentration = 5%
   - b. Detention time = 10 d
   - c. Loading factor = 3.83 kgVSS/$m^3 \cdot$ d (see Table 11-14)
2. Volume reduction method
   - a. Initial volatile solids = 75%
   - b. Volatile solids destroyed = 60%
   - c. Final sludge concentration = 8%
   - d. Final sludge specific gravity = 1.04
3. Volumetric loading
   - a. Per capita contribution = 72 g/capita $\cdot$ d
   - b. Volume required = 50 $m^3/10^3$ capita

11-5 A preliminary wastewater treatment plant providing for separate sludge digestion receives an influent wastewater with the following characteristics:

$$\text{Average flow} = 8000 \text{ m}^3/\text{d}$$
$$\text{Suspended solids removed by primary sedimentation} = 200 \text{ mg/L}$$
$$\text{Volatile matter in settled solids} = 75\%$$
$$\text{Water in untreated sludge} = 96\%$$
$$\text{Specific gravity of mineral solids} = 2.60$$
$$\text{Specific gravity of organic solids} = 1.30$$

(a) Determine the required digester volume using a mean cell residence time of 12 d.

(b) Determine the minimum digester capacity using the recommended loading parameters of kilograms of volatile matter per cubic meter per day and cubic meters per 1000 persons.

(c) Assuming 90 percent moisture in the digested sludge and a 60 percent reduction in volatile matter during digestion at 32°C, determine the minimum theoretical digester capacity for this plant based on parabolic reduction in sludge volume during digestion and a digestion period of 25 d.

11-6 Consider an industrial waste consisting mainly of carbohydrates in solution. Pilot plant experiments using a continuous-flow stirred-tank anaerobic digester without recycle yielded the following data:

Run	BOD_L influent, kg/d	X_T reactor, kg	P_x effluent, kg/d
1	1000	428	85.7
2	500	115	46

Assuming a waste-utilization efficiency of 80 percent, estimate the percentage of added BOD_L that can be stabilized when treating a waste load of 5000 kg/d. Assume that the design sludge retention time (θ_c) is 10 d.

11-7 A digester is loaded at a rate of 300 kg BOD_L/d. Using a waste-utilization efficiency of 75 percent, what is the volume of gas produced when $\theta_c = 40$ d? $Y = 0.10$ and $k_d = 0.02$ d^{-1}.

11-8 Volatile acid concentration, pH, or alkalinity should not be used alone to control a digester. How should they be correlated to predict most effectively how close to failure a digester is at any time?

11-9 Prepare a one-page abstract of each of the following four articles: P. L. McCarty: Anaerobic Waste Treatment Fundamentals, *Public Works*, vol. 95, nos. 9, 10, 11, and 12, 1964.

11-10 A digester is to be heated by circulation of sludge through an external hot-water heat exchanger. Using the following data, find the heat required to maintain the required digester temperature:

(a) U_x = overall heat-transfer coefficient, W/m$^2 \cdot$ K.

(b) $U_{air} = 0.85$, $U_{ground} = 0.68$, $U_{cover} = 1.1$.

(c) Digester is a concrete tank with floating steel cover; diameter = 11 m and side-wall depth = 8 m, 4 m of which is above the ground surface.

(d) Sludge fed to digester = 15 m^3/d at 14°C.

(e) Outside temperature = -15°C.

(f) Average ground temperature = 5°C.

(g) Sludge in tank is to be maintained at 35°C.

(h) Assume a specific heat of the sludge = 4000 J/kg $\cdot$ K.

(i) Sludge contains 4 percent solids.

(j) Assume a cone-shaped cover with center 0.6 m above digester top, and a bottom with center 1.2 m below bottom edge.

11-11 Determine the specific resistance of an activated sludge sample from the following data that were obtained using a small laboratory test filter.

Time, s	Volume of filtrate, ml
100	2.5
200	4.1
400	7.0
600	9.1
800	11.0
1000	12.5

Vacuum = $90.0 \ kN/m^2$
Solids content = 4%
Filtrate viscosity = $1.015 \times 10^{-3} \ N \cdot s/m^2$

11-12 The ultimate elemental analysis of a dried sludge yields the following data:

Carbon	52.1%
Oxygen	38.3%
Hydrogen	2.7%
Nitrogen	6.9%
Total	100.0%

How many kilograms of air will be required per kilogram of sludge for its complete oxidation?

11-13 Compute the fuel value of the sludge from a primary settling tank (a) if no chemicals are added and (b) if the coagulating solids amount to 10 percent by weight of the dry sludge. The amount of volatile solids is 75 percent.

11-14 Assume that a community of 5000 persons has asked you to serve as a consultant on their sludge disposal problems. Specifically you have been asked to determine if it is feasible to compost the sludge from the primary clarifier with the communities solid waste. If this plan is not feasible you have been asked to recommend a feasible solution. Currently the waste solids from the communities biological process are thickened in the primary clarifier. Assume the following data are applicable:

Solid waste data:
 Waste production = 2.0 kg/person · d
 Compostable fraction = 55%
 Moisture content of compostable fraction = 22%
Sludge production:
 Net sludge production = 0.12 kg/person · d
 Concentration of sludge in underflow from primary clarifier = 5%
 Specific gravity of underflow solids = 1.08
Compost:
 Final moisture content of sludge-solid waste mixture = 55%

11-15 Prepare a solids balance for the peak loading condition for the treatment plant used in Example 11-9. Assume the following data apply. Enter your final values on Fig. 11-34 in your text.

Peak flow = $54,000 \ m^3/d$
Average BOD_5 at peak flow = 340 mg/L
Average suspended solids at peak flow = 350 mg/L
Suspended solids after grit removal = 325 mg/L

Use data given in Example 11-9 for other parameters.

11-16 Prepare a solids balance, using the iterative technique delineated in Example 11-9, for the flowsheet shown in Fig. 11-38. Also determine the effluent flowrate and suspended solids concentration. Assume the following data are applicable:

Influent characteristics
  Flowrate = 4000 m^3/d
  Suspended solids = 1000 mg/L
Sedimentation tank
  Removal efficiency = 85%
  Concentration of solids in underflow = 7%
  Specific gravity of sludge = 1.1
Alum addition
  Dosage = 10 mg/L of filter influent
  Chemical solution = 0.5 kg alum/L of solution
Filter
  Removal efficiency = 90%
  Washwater solids concentration = 6%
  Specific gravity of backwash = 1.08
Thickener
  Effluent solids concentration = 500 mg/L
  Concentration of solids in underflow = 12%
  Specific gravity of sludge = 1.25
Ferric chloride addition
  Dosage = 1% of underflow solids from thickener
  Specific gravity of chemical solution = 2.0 kg/L
Filter press
  Concentration of solids in filtrate = 200 mg/L
  Concentration of thickened solids = 40%
  Specific gravity of thickened sludge = 1.6

In preparing the solids balance assume that all of the unit operations respond linearly such that the removal efficiency for recycled solids is the same as that for the solids in the influent wastewater. Also assume that the distribution of the chemicals added to improve the performance of the filter and filter press is proportional to the total solids in the returns and the effluent solids.

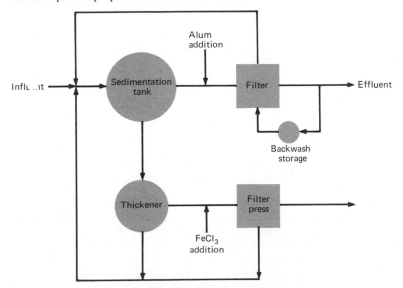

Figure 11-38 Treatment plant flowsheet for Problem 11-16.

REFERENCES

1. Babbitt, H. E., and D. H. Caldwell: *Laminar Flow of Sludge in Pipes*, University of Illinois Bulletin 319, 1939.
2. Burd, R. S.: *A Study of Sludge Handling and Disposal*, U.S. Department of the Interior Publication WP-20-4, 1968.
3. Bursztynsky, T., and J. Davis: Sludge Management Alternatives for Coastal Cities, in *Proceedings of the National Conference on Municipal Sludge Management and Disposal*, Information Transfer, Inc., Rockville, Md., 1975.
4. Carman, P. C.: A Study of the Mechanism of Filtration, parts I–III, *J. Soc. Chem. Ind.*, vols. 52, 53, London, 1933, 1934.
5. Chou, T. L.: Resistance of Sewage Sludge to Flow in Pipes, *J. San. Eng. Div.*, ASCE, September 1958.
6. Coackley, P., and B. R. S. Jones: Vacuum Sludge Filtration, *Sewage and Industrial Wastes*, vol. 28, no. 6, 1956.
7. Colacicco, D., et al.: Costs of Sludge Composting, ARS-Ne-79, Agricultural Research Service, U.S. Department of Agriculture, Bettsville, Md., 1977.
8. Cummings, R. J., and W. J. Jewell: Thermophilic Aerobic Digestion of Dairy Waste, 9th Cornell University Waste Management Conference, Syracuse, N.Y., Apr. 28, 1977.
9. Cunetta, J., and R. Feuer: Design of the Newtown Creek Water Pollution Control Project, *J. WPCF*, vol. 40, no. 4, 1968.
10. Drier, D. E., and C. A. Obma: *Aerobic Digestion of Solids*, Walker Process Equipment Co. Bulletin No. 26-S-18194, Aurora, Ill., 1963.
11. Eckenfelder, W. W., Jr.: *Industrial Water Pollution Control*, McGraw-Hill, New York, 1966.
12. Epstein, E., and G. B. Wilson: Composting Raw Sludge, in *Proceedings of the National Conference on Municipal Sludge Management and Disposal*, Information Transfer, Inc., Rockville, Md., 1975.
13. Fair, G. M., and E. W. Moore: Heat and Energy Relations in the Digestion of Sewage Solids, *Sewage Works J.*, vol. 4, pp. 242, 428, 589, and 728, 1932.
14. Fisher, A. J.: The Economics of Various Methods of Sludge Disposal, *Sewage Works J.*, vol. 8, no. 2, 1936.
15. Katz, W. J., and A. Geinopolos: Sludge Thickening by Dissolved-Air Flotation, *J. WPCF*, vol. 39, no. 6, 1967.
16. Lewis, M. F.: Sludge Pyrolysis for Energy Recovery on Pollution Control, in *Proceedings of the National Conference on Municipal Sludge Management and Disposal*, Information Transfer, Inc., Rockville, Md., 1975.
17. McAdams, W. H.: *Heat Transmission*, 2d ed., McGraw-Hill, New York, 1954.
18. McCabe, B. J., and W. W. Eckenfelder, Jr.: *Biological Treatment of Sewage and Industrial Wastes*, vol. 2, Reinhold, New York, 1958.
19. McCarty, P. L.: Anaerobic Treatment of Soluble Wastes, in E. F. Gloyna and W. W. Eckenfelder, Jr. (eds.), *Advances in Water Quality Improvement*, University of Texas Press, Austin, 1968.
20. McCarty, P. L.: Anaerobic Waste Treatment Fundamentals, *Public Works*, vol. 95, nos. 9–12, 1964.
21. Metcalf, L., and H. P. Eddy: *American Sewerage Practice*, vol. III, 3d ed., McGraw-Hill, New York, 1935.
22. Metcalf & Eddy, Inc.: *Report to National Commission on Water Quality on Assessment of Technologies and Costs for Publicly Owned Treatment Works*, vol. 2, prepared under Public Law 92-500, Boston, 1975.
23. Paulsrud, B., and A. S. Eikum: Lime Stabilization of Sewage Sludges, *Water Res.*, vol. 9, 1975.
24. *Process Design Manual for Sludge Treatment and Disposal*, U.S. Environmental Protection Agency, Office of Technology Transfer, Washington, D.C., 1974.
25. *Process Design Manual for Upgrading Existing Wastewater Treatment Plants*, U.S. Environmental Protection Agency, Office of Technology Transfer, Washington, D.C., 1974.
26. Rich, L. G.: *Unit Operations of Sanitary Engineering*, Wiley, New York, 1963.

27. Sacramento Area Consultants: *Study of Wastewater Solids Processing and Disposal,* Report prepared for Sacramento Regional County Sanitation District, Sacramento, California, June, 1975.

28. Sparr, A. E.: Pumping Sludge Long Distances, *J. WPCF,* vol. 43, no. 8, 1971.

29. Tchobanoglous, G., H. Theisen, and R. Eliassen: *Solid Wastes: Engineering Principles and Management Issues,* McGraw-Hill, New York, 1977.

30. Wagenhals, H. H., E. J. Theriault, and H. B. Hommon: *Sewage Treatment in the United States,* U.S. Public Health Bulletin 132, 1925.

31. Water Pollution Control Federation: *Sludge Dewatering,* Manual of Practice 20, 1969.

32. Weismantel, G. E.: Sludge Pyrolysis Schemes Now Head for Tryouts, *Chem. Eng.,* Dec. 8, 1975.

33. Willson, G. B.: Equipment for Composting Sewage Sludge in Windrows and in Piles, in *Composting of Municipal Residues and Sludges,* Information Transfer, Inc., Rockville, Md., 1978.

34. Wyckoff, J.: "Effects of Initial Concentration on Flotation Thickening Efficiency," unpublished Master's Thesis, Department of Civil Engineering, University of California, Davis, 1977.

TWELVE

ADVANCED WASTEWATER TREATMENT

Many of the substances found in wastewater are not affected or are little affected by conventional treatment operations and processes. These substances range from relatively simple inorganic ions, such as calcium, potassium, sulfate, nitrate, and phosphate, to an ever-increasing number of highly complex synthetic organic compounds. As the effects of these substances on the environment become more clearly understood, it is anticipated that treatment requirements will become more stringent in terms of the allowable concentration of many of these substances in the effluent from wastewater treatment plants. In turn, this will require advanced wastewater treatment facilities, which are not used extensively. An alternative to the addition of advanced wastewater treatment is the development of totally different process flowsheets.

Within the past few years, the subject of advanced wastewater treatment has become greatly expanded, and a great deal of literature has been written about it, especially with respect to the removal of nitrogen and phosphorus. The purpose of this chapter is not to report on all these developments, but rather to present an overview of this subject with respect to the removal of specific constituents of concern. The chapter contains a brief summary of some of the effects of substances that may be found in wastewater; an overview of the available types of unit operations and processes and some representative flowsheets used for the removal of the contaminants of concern identified in Chaps. 3 and 4; and a more detailed review of the more important of these operations and processes as applied to the specific constituents. Brief mention is also made of the ultimate disposal of contaminants.

12-1 EFFECTS OF CHEMICAL CONSTITUENTS IN WASTEWATER

The typical composition of domestic wastewater was reported in Table 3-5. Most domestic wastewaters also contain a wide variety of trace compounds and elements, although they are not measured routinely. If industrial wastewater is discharged to domestic sewers, the distribution of the constituents will vary considerably from that reported in Table 3-5. Some of the substances found in wastewater and the concentrations that may cause problems when discharged to the environment are reported in Table 12-1. This list is not meant to be exhaustive; rather, it is meant to highlight the fact that a wide variety of substances must be considered and that they will vary with each treatment application.

Table 12-1 Typical chemical constituents that may be found in wastewater and their effects

Constituent	Effect	Critical concentration, mg/L
Inorganic		
Ammonia	Increases chlorine demand; toxic to fish; can be converted to nitrates and, in the process, can deplete oxygen resources; with phosphorus, can lead to the development of undesirable aquatic growths	Any amount Variable[a] Any amount
Calcium and magnesium	Increase hardness and total dissolved solids	
Chloride	Imparts salty taste; interferes with agricultural and industrial processes	250 75–200
Mercury	Toxic to humans and aquatic life	0.00005
Nitrate	Stimulates algal and aquatic growth; can cause methemoglobinemia in infants (blue babies)	0.3^b 10^c
Phosphate	Stimulates algal and aquatic growth; interferes with coagulation; interferes with lime-soda softening	0.015^b 0.2–0.4 0.3
Sulfate	Cathartic action	600–1,000
Organic		
DDT	Toxic to fish and other aquatic life	0.001
Hexachloride	May be related to the development of cancer; also	0.02
Petrochemicals	may cause taste and odor problems in water	0.005–0.1
Phenolic compounds		0.0005–0.001
Surfactants	Cause foaming and may interfere with coagulation	1.0–3.0

[a] Depends on pH and temperature.
[b] For quiescent lakes [12].
[c] U.S. Environmental Protection Agency: *Part 141—National Interim Primary Drinking Water Regulations, Federal Register*, vol. 40, no. 248, December 24, 1975.

Compounds containing available nitrogen and phosphorus have received considerable attention since the mid-1960s because of their actual and suspected importance in promoting aquatic growth. In some cases, the culprit may not be either one of these substances, but some trace element, such as cobalt, molybdenum, or vanadium.

12-2 UNIT OPERATIONS AND PROCESSES AND TREATMENT FLOWSHEETS

In this section, an overview is presented of the treatment methods that have been applied and studied for the removal of the constituents of concern reported in Table 3-2, as well as other compounds and substances (see Table 12-1).

Classification

Unit operations and processes that have been applied to the further treatment of wastewater may be classified as physical, chemical, and biological. To facilitate a general comparison of the various operations and processes, information on (1) the types of wastewater to be treated, (2) the major types of constituents affected, (3) the form of the ultimate wastes to be disposed, and (4) the section and chapter where each one is considered is reported in Table 12-2. As will be noted in reviewing Table 12-2, many of the operations and processes have already been discussed and analyzed in detail in Chaps. 6 through 11.

Although the processes and operations are listed individually in Table 12-2, almost all of them are used in conjunction with other unit processes and operations. For example, when ion exchange is used to remove nitrogen or phosphorus from treated wastewater, the exchange unit is normally preceded by some form of filtration.

PROCESS SELECTION AND DEVELOPMENT OF TREATMENT FLOWSHEETS

Selection of a given operation, process, or combination thereof depends on (1) the use to be made of the treated effluent, (2) the nature of the wastewater, (3) the compatibility of the various operations and processes, (4) the available means to dispose of the ultimate contaminants, and (5) the economic feasibility of the various combinations. In some cases, because of extreme conditions, economic feasibility may not be a controlling factor in the design of an advanced wastewater treatment system.

Two very complete flowsheets in which state-of-the-art technology is used for the advanced treatment of wastewater are shown in Figs. 12-1 and 12-2. In Fig. 12-1, physical and chemical means are used to achieve nitrogen removal. In Fig. 12-2, biological means are used for nitrogen removal. In both flowsheets, breakpoint chlorination is used for disinfection and to ensure that any residual ammonia will not be discharged in the effluent; chemical precipitation is used as the principal means for the removal of phosphorus; and flow equalization is used to optimize the performance of the downstream units, specifically the multimedia filters and the breakpoint chlorination process. Although the flowsheets shown in Figs. 12-1 and 12-2 are quite different, the effluent concentrations of the major quality parameters are essentially the same.

Typical Process Performance Data

The performance of the advanced wastewater treatment operations and processes depends on the concentration and characteristics of the wastewater. The expected residual pollutant concentrations at various points in the process for the flowsheets shown in Figs. 12-1 and 12-2 are also reported in the figures. Information on other treatment processes used in conjunction with various secondary treatment processes for advanced treatment of wastewater are reported in Table 12-3. For purposes of comparison, the expected effluent qualities for an activated-sludge process and a trickling-filter process are also reported in Table 12-3.

12-3 NITROGEN CONVERSION AND REMOVAL

Before discussing the major nitrogen conversion and removal processes in detail, it will be helpful to review the forms in which nitrogen can exist in the environment, the principal sources of nitrogen in wastewater, and the various means that have been used or proposed for the removal of nitrogen in its alternative forms.

Forms of Nitrogen

Nitrogen in wastewater can exist in four forms: organic nitrogen, ammonia nitrogen, nitrite nitrogen, and nitrate nitrogen. As noted in Table 3-5, organic nitrogen and ammonia nitrogen are the principal forms in untreated wastewater. In nature, in the nitrogen cycle (see Fig. 3-18), organic nitrogen and ammonia nitrogen are converted first to nitrite and then to nitrate. The overall reaction starting with ammonia is:

$$NH_4^+ + 2O_2 \rightarrow NO_3^- + H_2O + 2H^+ \qquad (12\text{-}1)$$

For this reaction to go to completion, 4.57 g of oxygen are required per g of ammonia nitrogen. This oxygen demand is often identified as the nitrogenous oxygen demand (NOD). It should also be noted that 7.1 g of alkalinity (as $CaCO_3$) will be required per gram of ammonia nitrogen.

The problem with nitrogen in wastewater is related primarily to the oxygen demand that can be exerted if ammonia nitrogen is discharged to the environment.

Table 12-2 Advanced wastewater-treatment operations and processes

Description	Type of waste-water treated[a]	Principal or major use	Waste for ultimate disposal	See Sec.
Physical unit operations				
Air stripping or ammonia	EST	Removal of ammonia nitrogen	None	12-6
Filtration:				
multimedian	EST	Removal of suspended solids	Liquid and sludge	6-7
Diatomite bed	EST	Removal of suspended solids	Sludge	
Microstrainers	EBT	Removal of suspended solids	Sludge	
Distillation	EST nitrified + filtration	Removal of dissolved solids	Liquid	
Electrodialysis	EST + filtration + carbon adsorption	Removal of dissolved solids	Liquid	12-9
Flotation	EPT, EST	Removal of suspended solids	Sludge	
Foam fractionation	EST	Removal of refractory organics, surfactants, and metals	Liquid	
Freezing	EST + filtration	Removal of dissolved solids	Liquid	
Gas-phase separation	EST	Removal of ammonia nitrogen	None	
Land application	EPT, EST	Nitrification, denitrification, removal of ammonia nitrogen and phosphorus	None	Chap. 13
Reverse osmosis	EST + filtration	Removal of dissolved solids	Liquid	12-9
Sorption	EBT	DDIS	Liquid and sludge	

Chemical unit processes				
Breakpoint chorination	EST (filtration)	Removal of ammonia nitrogen	Liquid	12-6
Carbon adsorption	EPT, EST (filtration)[b]	Removal of dissolved organics, heavy metals, and chlorine	Liquid	12-8
Chemical precipitation	EBT	Phosphorus precipitation, removal of heavy metals, removal of colloidal solids	Sludge	12-7
Chemical precipitation in activated sludge	EPT	Removal of phosphorus	Sludge	12-7
Ion exchange	EST + filtration	Removal of ammonia and nitrate nitrogen	Liquid	12-6
Electrochemical treatment	Untreated	Removal of dissolved solids	Liquid and sludge	12-9
Oxidation	EST	Removal of refractory organics	None	12-8
Biological unit processes				
Bacterial assimilation	EPT	Removal of ammonia nitrogen	Sludge	12-5
Denitrification	Agricultural return water	Nitrate reduction	None	12-4
Harvesting of algae	EBT	Removal of ammonia nitrogen	Algae	12-4
Nitrification	EPT, EBT	Ammonia oxidation	Sludge	12-4
Nitrification-denitrification	EPT, EBT	Total nitrogen removal	Sludge	12-5

[a] EPT = effluent from primary treatment; EBT = effluent from biological treatment; EST = effluent after secondary treatment.

[b] Optional.

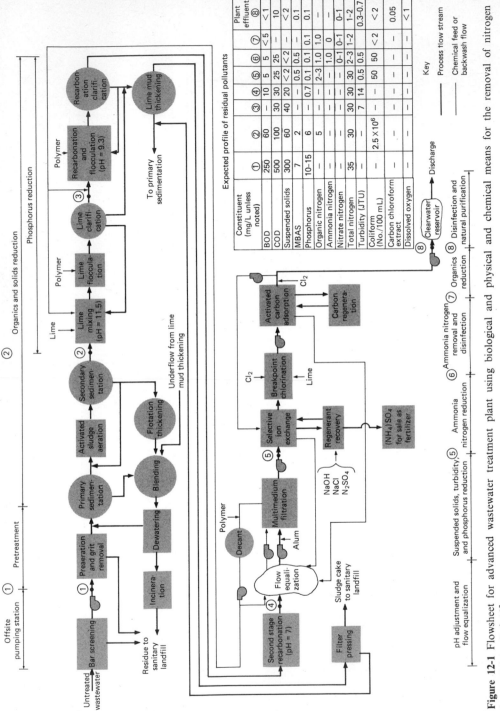

Expected profile of residual pollutants

Constituent (mg/L unless noted)	①	②	③	④	⑤	⑥	⑦	Plant effluent ⑧
BOD	250	60	–	10	5	5	<5	<1
COD	500	100	30	30	25	25	–	10
Suspended solids	300	60	40	20	<2	<2	–	<2
MBAS	7	2	–	–	0.5	0.5	–	0.1
Phosphorus	10–15	6	–	0.7	0.1	0.1	0.1	0.1
Organic nitrogen	–	5	–	–	2–3	1.0	1.0	–
Ammonia nitrogen	–	–	–	–	–	1.0	–	–
Nitrate nitrogen	–	–	–	–	–	–	0–1	0–1
Total nitrogen	35	30	30	30	30	2–3	1–2	1–2
Turbidity (JTU)	–	–	7	14	0.5	0.5	<2	0.3–0.7
Coliform (No./100 mL)	–	2.5×10^6	–	–	–	50	<2	<2
Carbon chloroform extract	–	–	–	–	50	–	–	0.05
Dissolved oxygen	–	–	–	–	–	–	–	<1

Key
— Process flow stream
— Chemical feed or backwash flow

Figure 12-1 Flowsheet for advanced wastewater treatment plant using biological and physical and chemical means for the removal of nitrogen [adapted from Ref. 18].

702

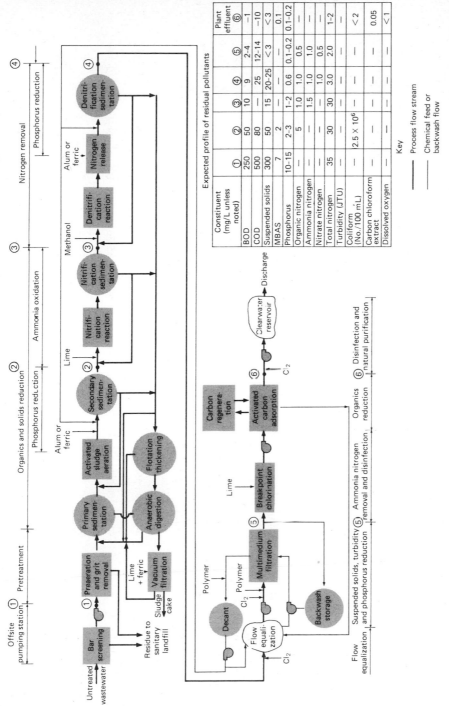

Expected profile of residual pollutants

Constituent (mg/L unless noted)	①	②	③	④	⑤	Plant effluent ⑥
BOD	250	50	10	9	2–4	~1
COD	500	80	—	25	12–14	~10
Suspended solids	300	50	15	20–25	<3	<3
MBAS	7	2	—	—	—	0.1
Phosphorus	10–15	2–3	1–2	0.6	0.1–0.2	0.1–0.2
Organic nitrogen	—	5	1.0	1.0	0.5	—
Ammonia nitrogen	—	—	1.5	1.0	1.0	—
Nitrate nitrogen	—	—	1.0	1.0	0.5	—
Total nitrogen	35	30	30	3.0	2.0	1–2
Turbidity (JTU)	—	—	—	—	—	<2
Coliform (No./100 mL)	—	2.5×10^6	—	—	—	—
Carbon chloroform extract	—	—	—	—	—	0.05
Dissolved oxygen	—	—	—	—	—	<1

Key
——— Process flow stream
- - - Chemical feed or backwash flow

Figure 12-2 Flowsheet for advanced wastewater treatment plant using three-stage biological process for the removal of nitrogen [*adapted from Ref. 18*].

703

Table 12-3 Treatment levels achievable with various operations and processes used for advanced wastewater treatment[a]

Secondary treatment[b]	Additional treatment	Suspended solids, mg/L	BOD, mg/L	COD, mg/L	Total N, mg/L	PO₄ as P, mg/L	Turbidity, mg/L	Color units,
Activated-sludge process (suspended-growth process)	None (secondary effluent)	20–30	15–25	40–80	20–60	6–15	5–15	15–80
	Granular-medium filtration	<5–10	<5–10	30–70	15–35	4–12	0.3–5	15–60
	Granular-medium filtration, carbon column	<3	<1	5–15	15–30	4–12	0.3–3	5
	Coagulation plus settling	<5	<5–10	40–70	15–30	1–2	<10	10–30
	Coagulation plus settling and granular-medium filtration	<1	<5	30–60	15–30	0.1–1.0[c]	0.1–1.0	10–30
	Coagulation, settling, granular-medium filtration, ammonia stripping	<1	<5	30–60	2–10[d]	0.1–1.0[c]	0.1–1.0	10–30

Typical effluent quality

704

Process	Additional treatment						
Coagulation, settling, granular-medium filtration, ammonia stripping, carbon columns		<1	<1	1–15	2–10[d]	0.1–1.0[c]	0.1–1.0	<5
Land treatment	Irrigation[e]; rapid infiltration[f]; overland flow[g]	<1	<2	....	3	0.3		
		2	2	....	10	3		
		10	10	....	3	12		
Trickling-filter process	None (secondary effluent); granular-medium filtration	20–40	15–35	40–100	20–60	6–15	5–15	15–80
Rotating biological contactor (attached-growth processes)		10–20	10–20	30–70	15–35	6–15	<10	15–60
	Aeration, settling, granular-medium filtration	<5–10	<5–10	30–60	15–35	4–12	0.5–5	15–60

[a] Adapted in part from Ref. 21.
[b] To meet U.S. Environmental Protection Agency effluent standards, the use of filters has now become accepted as standard practice and they are considered to be included in the definition of conventional secondary treatment.
[c] Reduction of PO_4 to this level will typically require 200 ppm of alum or 400 ppm of lime; if greater PO_4 concentrations can be tolerated, coagulant dosage is decreased.
[d] Requires elevating the pH to over 10.5 to convert nitrogen to ammonia.
[e] Percolation of primary or secondary effluent through 1.5 m of soil.
[f] Percolation of primary or secondary effluent through 4.5 m of soil.
[g] Runoff of comminuted municipal wastewater over about 45 m of slope.

Table 12-4 Effect of various treatment processes on nitrogen compounds[a]

Treatment operations or processes	Nitrogen compound			Removal of total nitrogen entering process, %[b]
	Organic nitrogen	$NH_3\text{-}NH_4^+$	NO_3^-	
Conventional treatment				
1. Primary	10–20% removed	No effect	No effect	5–10
2. Secondary	15–50% removed[c] urea → $NH_3\text{-}NH_4^{+d}$	<10% removed	Slight	10–30
Biological processes				
1. Bacterial assimilation	No effect	40–70% removed	Slight	30–70
2. Denitrification[e]	No effect	No effect → Cells	80–90% removed → Cells	70–95
3. Harvesting of algae	Partial transformation to $NH_3\text{-}NH_4^+$	→ Cells		50–80
4. Nitrification[e]	10–50	→ NO_3^-	No effect	5–20
5. Oxidation ponds	Partial transformation to $NH_3\text{-}NH_4^+$	Partial removal by stripping	Partial removal by nitrification-denitrification	20–90
Chemical processes				
1. Breakpoint chlorination[e]	Uncertain	90–100% removed	No effect	80–95
2. Chemical coagulation[f]	50–90% removed	Slight	Slight	20–30
3. Carbon sorption	30–50% removed	Slight	Slight	10–20
4. Selective ion exchange for ammonium[e]	Slight, uncertain	80–97% removed	No effect	70–95
5. Selective ion exchange for nitrate	Slight	Slight	75–90% removed	70–90

706

Physical operations				
1. Ammonia stripping[e]	No effect	$60-95\%$ removed	No effect	50–90
2. Electrodialysis	100% of suspended organic N removed	$30-50\%$ removed	$30-50\%$ removed	80–90
3. Filtration	$30-100\%$ of suspended organic N removed	Slight	Slight	20–40
4. Reverse osmosis		$60-90\%$ removed	$60-90\%$ removed	80–90
Land application				
1. Irrigation	$\rightarrow NH_3\text{–}NH_4^+$	$\rightarrow NO_3^-$, plant N	$\rightarrow N_2$, plant N	60–90
2. Rapid infiltration	$\rightarrow NH_3\text{–}NH_4^+$	$\rightarrow NO_3^-$	$\rightarrow N_2$	30–80
3. Overland flow	$\rightarrow NH_3\text{–}NH_4^+$	$\rightarrow NO_3$, plant N	$\rightarrow N_2$, plant N	70–90

[a] Adapted from Ref. 22.

[b] Depends on the fraction of influent nitrogen for which the process is effective, which may depend on other processes in the treatment plant.

[c] Soluble organic nitrogen, in the form of urea and amino acids, is substantially reduced by secondary treatment.

[d] Arrow denotes "conversion to."

[e] Principal methods now used for the control of nitrogen wastewater.

This problem can be mitigated either by converting the ammonia and organic nitrogen to nitrate or by eliminating the nitrogen from the wastewater. These methods are discussed in Secs. 12-3 through 12-5.

Sources of Nitrogen

Nitrogen enters the aquatic environment from both natural and manmade sources. Natural sources include precipitation, dustfall, nonurban runoff, and biological fixation. As a result of the activities of human beings, the quantities of nitrogen contained in precipitation, dustfall, and nonurban runoff have all increased. Other sources deriving from human activities include runoff from urban areas, municipal wastewaters, drainage from agricultural lands and feedlots, industrial wastes, and septic tank leachate. As shown in Table 3-5, the total nitrogen concentration in untreated wastewater varies from 20 to 85 mg/L. Of this total, about 40 percent is organic nitrogen and 60 percent is ammonia nitrogen. The forms of nitrogen in treated wastewater depend on both the type and the degree of treatment [22].

Operations and Processes for the Control of Nitrogen

Over the past 50 years, a number of operations and processes have been used for the conversion and removal of nitrogen from water and wastewater. Those applied to wastewater are reported in Table 12-4. There are five major categories, and the effects of each method on the organic, ammonia, and nitrate nitrogen in wastewater are listed.

   Conventional treatment processes were discussed in Chap. 9 and land disposal is discussed in depth in Chap. 13. The principal nitrogen-conversion and removal processes from the remaining three categories are considered in this chapter: nitrification in Sec. 12-4; denitrification in Sec. 12-5; and air stripping, breakpoint chlorination, and selective ion exchange in Sec. 12-6. These operations and processes were selected for detailed discussion because they have been used most often for the control of nitrogen. Details on the others may be found in the literature and in Refs. 3, 22, and 23.

12-4 NITRIFICATION

As noted in Sec. 12-3, nitrogen can exist in four forms in the aquatic environment. Most of the nitrogen in treated wastewater is in the form of ammonia. Consequently, when wastewater containing ammonia is discharged to the environment, depletion of receiving-water oxygen resources can occur as the ammonia is oxidized to nitrate. This depletion of oxygen can be eliminated if the ammonia is first oxidized to nitrate before it is discharged. Nitrification is the process used to accomplish this objective. The nitrification processes as applied to the conversion of ammonia in wastewater are considered in this section.

Nitrification Processes

In Chap. 9, the biological processes used for nitrification were identified as aerobic *suspended growth* and aerobic *attached growth*. In this section, a further distinction is made based on whether nitrification is accomplished in the same reactor used for the removal of carbonaceous BOD or in a separate reactor.

Combined carbon oxidation nitrification process Nitrifying organisms are present in almost all aerobic biological treatment processes, but usually their numbers are limited. As early as 1940, Sawyer noted that the ability of various activated sludges to nitrify was correlated to the BOD_5/NH_3 ratio [24]. More recently it has been shown that the BOD_5/TKN (total Kjeldahl nitrogen) ratio can be used as a more reliable measure [22]. For BOD_5/TKN ratios between 1 and 3, which roughly correspond to the values encountered in separate-stage nitrification systems, the fraction of nitrifying organisms is estimated to vary from 0.21 at a ratio of 1, to 0.083 at a ratio of 3 (see Table 12-5) [22]. In most activated-sludge processes, the fraction of nitrifying organisms would therefore be considerably less than the 0.083 value. In fact, it has been found that when the BOD_5/TKN ratio is greater than about 5, the process can be classified as a combined carbon oxidation nitrification process, and when the ratio is less than 3, it can be classified as a separate-stage nitrification process [22].

Nitrification can be accomplished in any of the suspended-growth activated-sludge processes identified in Table 9-1. All that is required is the maintenance of conditions suitable for the growth of nitrifying organisms. For example, in most warm climates, increased nitrification can be brought about simply by increasing the mean cell residence time and the supply of air. This technique is often used to achieve seasonal nitrification.

The two attached-growth processes that can be used for combined carbon oxidation nitrification processes are the trickling filter and the rotating biological contactor (see Table 9-1). As with the suspended-growth processes, nitrification in the attached-growth process can also be brought about or encouraged by suitable adjustment of the operating parameters. This can usually be accomplished by reducing the applied loading rate, as discussed later in this section (see "Process Analysis").

Table 12-5 Relationship between the fraction of nitrifying organisms and the BOD_5/TKN ratio [23]

BOD_5/TKN ratio	Nitrifier fraction	BOD_5/TKN ratio	Nitrifier fraction
0.5	0.35	5	0.054
1	0.21	6	0.043
2	0.12	7	0.037
3	0.083	8	0.033
4	0.064	9	0.029

Separate-stage nitrification Both suspended-growth and attached-growth processes are used to achieve separate-stage nitrification. Originally, the separate-stage suspended-growth nitrification process was developed so that its operation could be optimized in a manner similar to that of the carbon oxidation stage. In most details, separate-stage suspended-growth nitrification processes are similar in design to the activated-sludge process. Both continuous-flow stirred-tank staged-flow reactors and plug-flow reactors have been used. When very low ammonia concentrations are desired, the staged-flow or plug-flow reactors are favored.

Three different types of attached-growth processes have been used for separate-stage nitrification. They are the trickling-filter process, the rotating biological contactor, and the packed-bed reactor. Typically, a packed-bed reactor, as shown in Fig. 12-3, consists of a container (reactor) that is packed with a medium to which nitrifying microorganisms can become attached. Wastewater is introduced from the bottom of the reactor through an appropriate underdrain inlet chamber. Air or pure oxygen necessary for the process is also introduced with the wastewater.

Comparison of processes The advantages and disadvantages of the various nitrification processes are reported in Table 12-6. The selection of a particular flowsheet depends on a number of factors, including (1) whether nitrification is being incorporated into an existing treatment plant or a new treatment plant

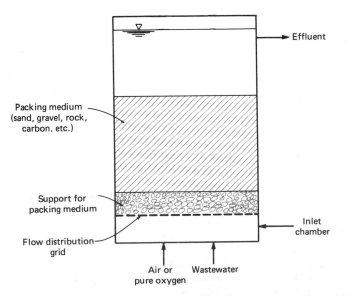

Figure 12-3 Schematic of packed-bed reactor used for the biological nitrification of wastewater [32].

Table 12-6 Comparison of nitrification alternatives

System type	Advantages	Disadvantages
Combined carbon oxidation nitrification process		
Suspended growth	Combined treatment of carbon and ammonia in a single stage; low effluent ammonia possible; inventory control of mixed liquor stable due to high BOD_5/TKN ratio	No protection against toxicants; only moderate stability of operation; stability linked to operation of secondary clarifier for biomass return; large reactors required in cold weather
Attached growth	Combined treatment of carbon and ammonia in a single stage; stability not linked to secondary clarifier as organisms on media	No protection against toxicants; only moderate stability of operation; effluent ammonia normally 1–3 mg/L (except rotating biological disk); cold weather operation impractical in most cases
Separate-stage nitrification		
Suspended growth	Good protection against most toxicants; stable operation; low effluent ammonia possible	Sludge inventory requires careful control when low BOD_5/TKN ratio; stability of operation linked to operation of secondary clarifier for biomass return; greater number of unit processes required than for combined carbon oxidation-nitrification
Attached growth	Good protection against most toxicants; stable operation; stability not linked to secondary clarifier as organisms on media	Effluent ammonia normally 1–3 mg/L Greater number of unit processes required than for combined carbon oxidation-nitrification

[a] Adapted from Ref. 22.

flowsheet is being developed, (2) whether seasonal or year-round standards are to be met, (3) operating temperatures, (4) desired effluent ammonia concentration, (5) effluent standards for other parameters, and (6) local costs.

Nitrification Stoichiometry

Nitrogen in the form of the ammonium ion is converted to nitrate in two steps by nitrifying autotrophic bacteria, as summarized by the following reactions.

Energy reaction, step 1:

$$NH_4^+ + \tfrac{3}{2}O_2 \xrightarrow{\text{Nitrosomonas}} NO_2^- + 2H^+ + H_2O \qquad (12\text{-}2)$$

Energy reaction, step 2:

$$NO_2^- + \tfrac{1}{2}O_2 \xrightarrow{\text{Nitrobacter}} NO_3^- \qquad (12\text{-}3)$$

Overall energy reaction:

$$NH_4^+ + 2O_2 \rightarrow NO_3^- + 2H^+ + H_2O \qquad (12\text{-}4)$$

Along with obtaining energy, however, some of the ammonium ion is assimilated into cell tissue. A representative synthesis reaction for this auto-trophic assimilation is as follows [14].

Synthesis:

$$4CO_2 + HCO_3^- + NH_4^+ + H_2O \rightarrow C_5H_7NO_2 + 5O_2 \qquad (12\text{-}5)$$

On the basis of the results of both laboratory studies and theoretical calculations, the following overall reaction is typical of the reactions that have been proposed to describe the autotrophic conversion of ammonium ion to nitrate [14]:

Overall ammonia conversion:

$$22NH_4^+ + 37O_2 + 4CO_2 + HCO_3^- \rightarrow C_5H_7NO_2 + 21NO_3^- + 2OH_2O + 42H^+ \qquad (12\text{-}6)$$

As noted in Chaps. 9 and 10, nitrification will occur in most aerobic biological treatment processes when the operating and environmental conditions are suitable. In the activated-sludge process, one of the important controlling variables is the mean cell residence time θ_c.

Process Analysis (Nitrification)

Although the kinetics of the nitrification process are reasonably well defined for suspended-growth systems, they are not so well defined for attached-growth systems. In the following discussion, the approach followed is similar to that used in Ref. 22. This is done because this reference has been widely disseminated. It contains an excellent review and analysis of both the nitrification and denitrification processes and the available data in the literature up to its date of publication (1975).

Suspended-growth process In general, it has been found that the kinetics expressions developed in Chap. 9 are applicable to the nitrification process, subject to environmental constraints. To avoid the necessity of having to refer to Chap. 9, the pertinent kinetic expressions used in the following analysis are summarized in Table 12-7. Details on the development of these equations may be found in Chap. 9 (Secs. 9-4 and 9-5).

From a detailed review of the pertinent nitrification literature [22], it has been found that the following factors have a significant effect on the nitrification process: ammonia nitrite concentration, BOD$_5$/TKN ratio, dissolved oxygen concentration, temperature, and pH. The impact of these variables on the nitrification process and the approach developed to account for them are

Table 12-7 Summary of kinetic expressions used for the analysis of suspended-growth nitrification and denitrification processes

Equation	Number	Definition of terms
$\mu = \dfrac{\mu_m S}{K_s + S}$	9-10	μ = specific growth rate, time^{-1} r_{su} = substrate utilization rate, mass/unit volume $\cdot$ time
$r_{su} = -\dfrac{\mu_m SX}{Y(K_s + S)}$	9-13	μ_m = maximum specific growth rate, time^{-1} S = concentration of growth limiting substrate in solution, mass/unit volume
$k = \dfrac{\mu_m}{Y}$	9-14	X = concentration of microorganisms, mass/unit volume
$r_{su} = -\dfrac{kSX}{K_s + S}$	9-15	Y = maximum yield coefficient measured during a finite period of logarithmic growth, mass of cells formed per mass of substrate consumed
$U = -\dfrac{r_{su}}{X}$	9-36	K_s = half velocity constant, mass/unit volume k = maximum rate of substrate utilization, time^{-1}
$U = \dfrac{S_0 - S}{\theta X}$	9-38	U = substrate utilization rate, time^{-1} θ = hydraulic detention time, time
$U = \dfrac{kS}{K_s + S}$	9-45	θ_c^d = design mean cell residence time, time
$\dfrac{1}{\theta_c^d} = YU - k_d$	9-39	θ_c^M = minimum mean cell residence time before washout, time SF = safety factor
$\dfrac{1}{\theta_c^M} \approx Yk - k_d$	9-48	S_0 = influent concentration, mass/unit volume
$SF = \dfrac{\theta_c^d}{\theta_c^M}$	9-49	

reported in Table 12-8 and illustrated in Fig. 12-4. Representative kinetic coefficients for the suspended-growth nitrification process are given in Table 12-9.

To apply the kinetic approach as delineated in Ref. 22 to the analysis of the suspended-growth nitrification process in a continuous-flow stirred-tank reactor, it will be necessary to take the following steps:

1. Select an appropriate safety factor to handle peak, diurnal, and transient loadings.
2. Estimate the maximum growth rate of nitrifying organisms consistent with the most adverse temperature, dissolved oxygen, and pH conditions.
3. Determine the minimum mean cell residence time based on the adjusted growth rate determined in step 2.
4. Determine the design mean cell residence time using the safety factor determined in step 1.
5. Determine the effluent nitrogen concentration.
6. Determine the hydraulic retention time to achieve the necessary effluent nitrogen concentration.

Table 12-8 Effects of the major operational and environmental variables on the suspended-growth nitrification process[a]

Factor	Description of effect
Ammonia-nitrite concentration	It has been observed that the concentration of ammonia and nitrate will affect the maximum growth rate of *Nitrosomonas* and *Nitrobacter*. The effect of either constituent can be made using a Monod-type kinetic expression $$\mu = \mu_m \frac{S}{K_s + S}$$ Because it has been found that the growth rate of *Nitrobacter* is considerably greater than that of *Nitrosomonas*, the rate of nitrification is usually modeled using the conversion of ammonia to nitrate as the rate-limiting step.
BOD$_5$/TKN	The fraction of nitrifying organisms present in the mixed liquor of a single-state carbon oxidation nitrification process has been found to be reasonably well related to the BOD$_5$/TKN ratio. For ratios greater than 5, the fraction of nitrifying organisms decreases from a value of about 0.054 (see Table 12-5).
Dissolved-oxygen concentration	The *DO* level has been found to affect the maximum specific growth rate μ_m of the nitrifying organisms. The effect has been modeled with the following relationship: $$\mu'_{m_n} = \mu_{m_n} \frac{DO}{K_{O_2} + DO}$$ Based on limited information, a value of 1.3 can be used for K_{O_2}.
Temperature	Temperature has a significant effect on nitrification rate constants. The overall nitrification rate decreases with decreasing temperature and is accounted for with the following two relationships: $$\mu'_{m_n} = \mu_m\, e^{0.098(T-15)}$$ $$K_n = 10^{0.051T - 1.158}$$ where $T = $ °C
pH	It has been observed that the maximum rate of nitrification occurs between pH values of about 7.2 and 9.0. For combined carbon oxidation nitrification systems, the effect of pH can be accounted for using the following relationship: $$\mu'_{m_n} = \mu(1 - 0.833)\,(7.2 - pH)$$

[a] Developed from information contained in Ref. 22.

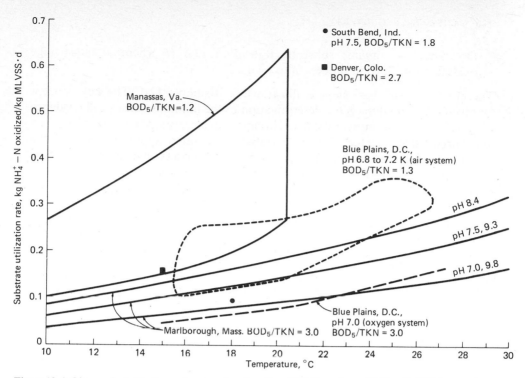

Figure 12-4 Observed nitrification rates at various locations [22]. Note: kg × 2.205 = lb. TKN = total Kjeldahl nitrogen.

Table 12-9 Typical kinetic coefficients for the suspended-growth nitrification process (pure culture values)[a,b]

Coefficient	Basis	Value	
		Range	Typical[c]
Nitrosomonas			
μ_m	d^{-1}	0.3–2.0	0.7
K_s	NH_4^+–N, mg/L	0.2–2.0	0.6
Nitrobacter			
μ_m	d^{-1}	0.4–3.0	1.0
K_s	NO_2^-–N, mg/L	0.2–5.0	1.4
Overall			
μ_m	d^{-1}	0.3–3.0	1.0
K_s	NH_4^+–N, mg/L	0.2–0.5	1.4
Y	NH_4^+–N[d], mg VSS/mg	0.1–0.3	0.2
k_d	d^{-1}	0.03–0.06	0.05

[a] Derived in part from Refs. 22 and 25.
[b] Values for nitrifying organisms in activated sludge will be considerably lower than the values reported in this table.
[c] Values reported are for 20°C.
[d] VSS = volatile suspended solids.

Note: mg/L = g/m³
1.8(°C) + 32 = °F

7. Determine the organic substrate utilization rate U where a single-stage oxidation nitrification process is to be used.

The application of these steps is illustrated in Example 12-1. The key concept involved in this analysis is the determination of the minimum mean cell residence time subject to the most critical environmental constraints and the use of an appropriate safety factor. This is essentially the same approach used in the design of the suspended-growth activated-sludge process in a continuous-flow stirred-tank reactor. The analysis of the plug-flow suspended-growth process is essentially as outlined in Sec. 9.5.

Example 12.1 Design of single-stage suspended-growth carbon oxidation nitrification process Determine the design criteria for an activated-sludge process to achieve essentially complete nitrification when treating domestic wastewater. The influent flowrate is 3400 m^3/d, the BOD_5 after primary settling is 200 mg/L, and the TKN after primary settling is 40 mg/L (see Table 3-5). Use the kinetic coefficients given in Table 12-9 except let $\mu_m = 0.5 d^{-1}$. Assume that (1) the minimum sustained temperature is 15°C, (2) the dissolved oxygen to be maintained in the reactor is 2.5 mg/L, and (3) the buffer capacity of the wastewater is adequate to maintain the pH at or above a value of 7.2.

SOLUTION
1 Estimate the safety factor to be used in the design based on the peak nitrogen loading. From a review of Fig. 3-6c, it appears that a safety factor of 2.5 should be adequate.
2 Determine the maximum growth rate for the nitrifying organisms under the stated operating conditions.
   a The following expression from Ref. 22 can be used:

$$\mu'_m = \mu_m e^{0.008(T-15)} \frac{DO}{K_{O_2} + DO} [1 - 0.833(7.2 - pH)]$$

$$\underset{\substack{\text{temperature} \\ \text{correction} \\ \text{factor}}}{} \quad \underset{\substack{\text{dissolved-} \\ \text{oxygen} \\ \\ \text{factor}}}{} \quad \underset{\substack{\text{pH} \\ \text{correction} \\ \text{factor}}}{}$$

   where μ'_m = maximum growth rate under the stated conditions of temperature, dissolved oxygen, and pH
          μ_m = maximum specific growth rate
          T = temperature
          DO = dissolved oxygen
          K_{O_2} = dissolved-oxygen half-velocity constant = 1.3
          pH = operating pH, the numerical value of the pH term is taken as 1 for the above values

   b Substitute the known values and determine μ'_m.

$$\mu_m = 0.5 \ d^{-1}$$
$$T = 15°C$$
$$DO = 2.5 \ mg/L$$
$$K_{O_2} = 1.3$$
$$pH = 7.2$$
$$\mu_m = (0.5 \ d^{-1})[e^{0.098(T-15)}]\left(\frac{2.5}{1.3 + 2.5}\right)[1 - 0.833(7.2 - 7.2)]$$
$$= 0.5 \ d^{-1} \frac{2.5}{1.3 + 2.5}$$
$$= 0.33 \ d^{-1}$$

3 Determine the maximum rate of substrate utilization k using Eq. 9-14 (see Table 12-7).

$$k' = \frac{\mu'_m}{Y}$$

$\mu'_m = 0.33 \text{ d}^{-1}$ (from step 2b above)

$Y = 0.2$ (from Table 9-8)

$$k' = \frac{0.33 \text{ d}^{-1}}{0.2} = 1.65 \text{ d}^{-1}$$

4 Determine the minimum and design mean cell residence times.

a Minimum θ_c^M:

$$\frac{1}{\theta_c^M} \sim Yk' - k_d$$

$Y = 0.2$

$k' = 1.65 \text{ d}^{-1}$ (from step 3)

$k_d = 0.05$ (from Table 12-9)

$$\frac{1}{\theta_c^M} = 0.2(1.65 \text{ d}^{-1}) - 0.05$$

$$= 0.28 \text{ d}^{-1}$$

$$\theta_c^M = \frac{1}{0.28} = 3.57 \text{ d}$$

b Design θ_c^d:

$$\theta_c^d = SF(\theta_c^M) = 2.5(3.57 \text{ d}) = 8.93 \text{ d}$$

5 Determine the design substrate-utilization factor U.

$$\frac{1}{\theta_c^d} = YU - k_d$$

$$U = \left(\frac{1}{\theta_c^d} + k_d\right)\frac{1}{Y}$$

$$= \left(\frac{1}{8.93} + 0.05\right)\frac{1}{0.2} = 0.81 \text{ d}^{-1}$$

6 Determine the concentration of ammonia in the effluent using Eq. 9-45.

$$U = \frac{kN}{K_N + N}$$

$$= 0.810 \text{ d}^{-1}$$

$k = 1.65 \text{ d}^{-1}$

$T = 15°C$

$N =$ effluent NH_4^+-N concentration, mg/L

$$K_n = 10^{0.051T - 1.158} = 0.40 \text{ mg/L (see Table 12-8)}$$

$$0.810 = \frac{1.65 N}{0.40 + N}$$

$$N = \frac{1.65 N}{0.810} - 0.4$$

$$N\left(\frac{1 - 1.65}{0.810}\right) = -0.40$$

$$N = 0.39 \text{ mg/L}$$

7   Determine the substrate-removal rate for the activated-sludge process using Eq. 9-39.

$$\frac{1}{\theta_c^d} = YU - k_d$$

$\theta_c^d = 8.93$ d   (step 4 above)

$Y = 0.5$ mg VSS/mg BOD$_5$ (from Table 9.7)

$k_d = 0.06$ d^{-1} (from Table 9.7)

$$U = \left(\frac{1}{8.93} + 0.06\right)\frac{1}{0.5}$$

$U = 0.34$ kg BOD$_5$ removed/kg MLVSS · d

If it is assumed that the process efficiency is 90 percent, the corresponding value of the food-to-microorganism ratio be equal to 0.38 kg BOD$_5$ applied per kg MLVSS · d.

8   Determine the required hydraulic detention time for BOD oxidation and nitrification using Eq. 9-38.

$$U = \frac{S_0 - S}{\theta X}$$

a   BOD$_5$ oxidation:

$$\theta = \frac{S_0 - S}{UX}$$

$S_0 = 200$ mg/L   (from problem specification)

$S = 20$ mg/L   (assumed value)

$U = 0.34$ d^{-1}(from step 7)
$X = $ MLVSS, mg/L (assume $X = 2000$ mg/L

$$\theta = \frac{(200 - 20)\ \text{mg/L}}{0.34\ \text{d}^{-1}(2000\ \text{mg/L})} = 0.26\ \text{d} = 6.4\ \text{h}$$

b   Nitrification:

$$\theta = \frac{N_0 - N}{UX}$$

$N_0 = 40$ mg/L   (from problem specification)

$N = 0.39$ mg/L   (from step 6)

$U = 0.81$ d^{-1}

$X = 2000$ mg/L $\times$ 0.08   (assumed fraction of nitrifiers)
  $= 160$ mg/L

$$\theta = \frac{(40 - 0.38)\ \text{mg/L}}{0.81\ \text{d}^{-1}(2000\ \text{mg/L} \times 0.08)} = 0.31\ \text{d} = 7.3\ \text{h}$$

Conclusion: Nitrification process controls the required hydraulic detention time.

9   Determine the required aeration-tank volume.

$$V = Q\theta = (3400\ \text{m}^3/\text{d})\ (0.31\ \text{d}) = 1054\ \text{m}^3$$

10 Determine the total amount of oxygen required.

a   The total amount of oxygen required based on average conditions can be estimated using the following expression:

$$O_2 \text{ kg/d} = \frac{Q(S_0 - S) \times (10^3 \text{ g/kg})^{-1}}{f} - 1.42(P_x) + 4.57Q(N_0 - N) \times (10^3 \text{ g/kg})^{-1}$$

where Q = flowrate, m^3/d
S_0 = influent BOD_5, g/m^3
S = effluent BOD, g/m^3
f = factor to convert BOD_5 value to BOD_L, 0.68
P_x = net mass of volatile solids (cells) produced
1.42 = conversion factor for cell tissue to BOD_L
N_0 = influent TKN, mg/L
N = effluent TKN, mg/L
4.57 = conversion factor for amount of oxygen required for complete oxidation of TKN

b   Alternatively, the following expression can be used as a rough estimate:

$$O_2 \text{ kg/d} = Q(S_0 + 4.57\text{TKN}) \times (10^3 \text{ g/kg})^{-1}$$

c   Using the expression given in step *10b*, the total oxygen required per day, with a factor of safety of 2.5, is equal to

$$O_2 \text{ kg/d} = (3400 \text{ m}^3/\text{d})[200 \text{ g/m}^3 + 4.6(40 \text{ g/m}^3)] \times (10^3 \text{ g/kg})^{-1}(2.5)$$
$$= 3264 \text{ kg/d} (7196 \text{ lb/d})$$

Comment   In addition to these computations, the alkalinity requirements should be checked. If the natural alkalinity of the wastewater is insufficient, it may be necessary to install a pH control system.

Attached-growth processes As noted previously, the principal attached-growth processes are the trickling filter, rotating biological contactor, and the packed-bed reactor. The trickling filter and the rotating biological contactor can be used in a single-stage process, but the packed-bed reactor is usually applied as a separate stage. To date, the most common approach to describe the performance of the attached-growth processes has been to use loading factors. Typical loading data to achieve nitrification with these processes are reported in Table 12-10 and in Fig. 12-5. Because the amount of ammonia that can be oxidized depends on the surface area of the filter medium, an alternative approach is to relate the observed oxidation of ammonia to the surface area. Additional details on this approach may be found in Refs. 8 and 32.

Process Applications

Assuming that sufficient air can be supplied, nitrification generally can be assured at moderate temperatures in a conventional activated-sludge system. If nitrification is to be accomplished in an activated-sludge system, certain operational adjustments must be made beyond those necessary for the stabilization of the organic matter:

1. Additional oxygen must be provided for the nitrification process.
2. A longer mean cell residence time must be used. Because the bacteria that are responsible for nitrification are strict autotrophs, they are distinctly different

Table 12-10 Typical loading rates for attached-growth processes to achieve nitrification[a]

Process	Percent nitrification	Loading rate, kg BOD$_5$/m^3 · d
Trickling filter,[b] rock medium	75–85	0.16–0.10
	85–95	0.10–0.05
Tower filter,[b] plastic medium	75–85	0.30–0.20
	85–95	0.20–0.10
Rotating biological contactor[b]	(see Fig. 12-5)	
Packed-bed reactor[c]	75–85	0.40–0.20[d]
	85–95	0.20–0.10[d]

[a] Developed in part from Refs. 22 and 25.
[b] Single-stage process.
[c] Separate-stage process (air or pure oxygen) with various packing media, including sand, gravel, coal, and plastic.
[d] kg NH$_3$–N/m^3 · d.
Note: kg/m^3 · d × 62.4280 = lb/10^3 ft^3 · d

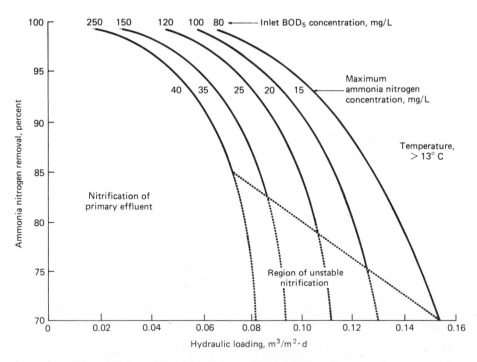

Figure 12-5 Effect of BOD$_5$ concentration and hydraulic load on nitrification in the rotating biological contactor [22]. Note: m^3/m^2 · d × 24.5424 = gal/ft^2 · d.

from the heterotrophic bacteria that are responsible for the degradation of the organic matter. Nitrifying bacteria have a growth rate that is much slower than that of the heterotrophic bacteria, so they require a longer mean cell residence time to be effective. Typical kinetic coefficients for the nitrification process are presented in Table 12-9.
3. Because the microbial conversion causes a drop in pH, provision should be made for lime or caustic addition with low-alkalinity wastewaters.

In some cases where nitrification has been accomplished in a conventional activated-sludge reactor, a problem can develop with the operation of the settling tank due to rising sludge (see Chap. 10). To avoid this problem, the nitrification step may be carried out in a separate reactor. This method of operation allows greater process flexibility, and each process (carbonaceous oxidation and nitrification) can be operated independently to achieve optimum performance. A flowsheet and aerial view of a typical treatment plant with separate stages for carbon oxidation and nitrification are shown in Fig. 12-6.

In many locations where only seasonal nitrification is required, the use of a single-stage process will probably be most effective. The decision to use a single stage or a separate stage for nitrification will also depend on whether phosphorus removal and denitrification are to be incorporated in the overall process flowsheet (see Fig. 12-2).

12-5 DENITRIFICATION AND NITRIFICATION-DENITRIFICATION

Of the methods proposed for the removal of nitrogen, the nitrification-denitrification process is perhaps the best for the following reasons: (1) high potential removal efficiency, (2) high process stability and reliability, (3) easy process control, (4) low land-area requirements, and (5) moderate cost. The removal of nitrogen with this process is carried out in either one or two steps, depending on the nature of the wastewater. If the wastewater to be treated contains nitrogen in the form of ammonia, two steps are required. In the first step, the ammonia is aerobically converted to the nitrate NO_3^- form (nitrification). In the second step, the nitrates are converted to nitrogen gas (denitrification). If the nitrogen in the wastewater is already in the form of nitrate, as in the case of irrigation return water, only the denitrification step is required. Although both denitrification and nitrification-denitrification are considered in this section, the principal focus is on the denitrification processes.

Denitrification Processes

As with the nitrification processes discussed previously, the denitrification processes in Chap. 9 were identified as being anoxic suspended growth and anoxic attached growth. (It was noted in Chap. 9 that the term anoxic is used

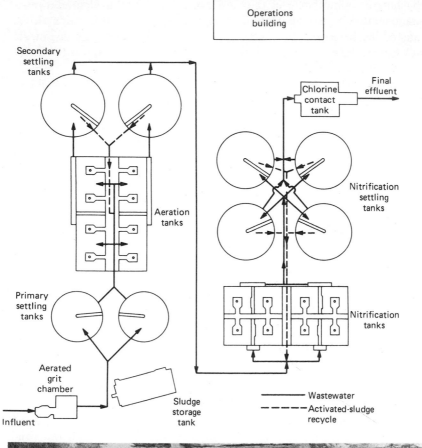

Operations building

Secondary settling tanks

Final effluent

Chlorine contact tank

Nitrification settling tanks

Aeration tanks

Nitrification tanks

Primary settling tanks

Aerated grit chamber

Sludge storage tank

Influent

———— Wastewater
— — — — Activated-sludge recycle

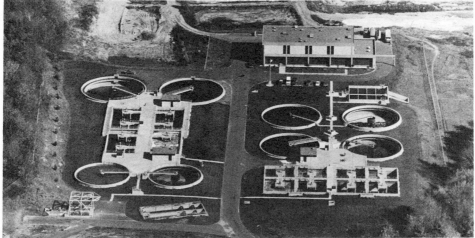

Figure 12-6 Flowsheet and aerial view of treatment plant with separate stages for carbon oxidation and nitrification (Marlborough, Mass.).

in preference to the term anaerobic when describing the denitrification process because the principal biochemical pathways are not anaerobic, but only modifications of aerobic pathways [22]). In the following discussion, an additional distinction will be made in describing the application of biological denitrification processes. This distinction is based on whether denitrification is accomplished (1) in separate reactors using methanol or some other suitable material source of organic carbon, or (2) in combined carbon oxidation nitrification–denitrification systems using wastewater or endogenous carbon sources.

Separate systems with external carbon source In these systems, an external carbon source is added to remove the nitrate. The analog to this in conventional biological waste treatment is the addition of oxygen for the removal of organic carbon. Because any excess carbon that is added over that required for the process will be measured in the effluent BOD, careful attention must be devoted to this aspect of design.

The design of suspended-growth denitrification systems is similar, in many respects, to the design of the activated-sludge systems used to remove organic carbon. A reactor is used to keep the bacterial mass in suspension; both continuous-flow stirred-tank and plug-flow reactors have been used. Because the nitrogen gas released during the denitrification process often becomes attached to the biological solids, a nitrogen release step is included between the reactor and the sedimentation facilities that are used to separate the biological solids (see Fig. 12-2). The removal of the attached nitrogen gas bubbles can be accomplished either in aerated channels that can be used to connect the biological reactor and the settling facilities or in a separate tank in which the solids are aerated for a short period of time (30 to 60 min). The stoichiometry and design of this process are described after a brief discussion of the attached-growth processes.

Since 1970, a number of different attached-growth denitrification processes, many of them proprietary, have been developed. The principal ones are identified in Table 12-11. Additional details on these processes may be found in Ref. 22.

Combined carbon oxidation nitrification–denitrification processes Because of the high cost of most organic carbon sources, a number of processes have been developed or are currently under development in which the carbon oxidation nitrification–denitrification processes are combined into a single process without any intermediate steps. These processes appear to hold promise as an effective means for nitrogen removal. Specific advantages include: (1) reduction in the volume of air needed to achieve nitrification and BOD_5 removal; (2) potential elimination of the supplemental organic carbon sources (e.g., methanol) required for complete denitrification, and (3) elimination of intermediate clarifiers required in a staged nitrification–denitrification system.

In these combined processes, either the endogenous decay of the organisms or the carbon in the wastewater is used to achieve denitrification. The flowsheet

Table 12-11 Description of attached-growth separate denitrification systems

Classification	Description	Typical removal rates at 20°C, kg N removed/$m^3 \cdot d^a$
Packed-bed reactor		
Gas-filled	The reactor shown in Fig. 12-7 is covered and filled with nitrogen gas, which eliminates the necessity of having to submerge the medium to maintain anoxic conditions.	1.6–1.8
Liquid-filled		
High-porosity medium	With both high- and low-porosity	0.1–0.12
Low-porosity medium	liquid-filled packed-bed reactors (see Fig. 12-8) backwashing of the packing medium is usually required to control the biomass.	0.2–0.4
Fluidized-bed reactor		
High-porosity medium, fine sand	Porosity is varied by adjusting the density of the medium and the flow-rate.	12–16
High-porosity medium, activated carbon		5–6

^{a} Data are reported for comparative purposes only. If any of these process are to be applied, pilot plant testing should be used to verify reported removal rates.

Note: $kg/m^3 \cdot d \times 62.4280 = lb/10^3 ft^3 \cdot d$
$$1.8(°C) + 32 = °F$$

for endogenous decay denitrification is shown in Fig. 12-9. The procedure that has been used to achieve denitrification by using the carbon in the wastewater involves a series of alternating aerobic and anoxic stages without intermediate settling. This can be accomplished, for example, in oxidation ditches by controlling the rotor oxygenation levels, or in an alternating contact process in which two separate tanks are used.

Another sequential process, termed the "Bardenpho" process by its developer, is shown in Fig. 12-10. Here, separate reactors are used for the oxidation nitrification step and the anoxic denitrification steps. Although this process must still be considered experimental, data from pilot scale testing indicate that high removals can be achieved.

Comparison of processes A general comparison of the various denitrification processes is presented in Table 12-12. From the preceding discussion it can be concluded that the number of denitrification processes that will be available in the future will continue to increase as research in this area continues. Further, because many of the processes that have been described have not been fully tested, great caution should be exercised in recommending their use. In almost

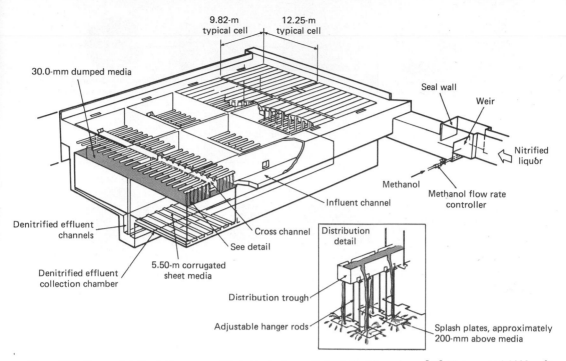

Figure 12-7 Design details of nitrogen gas-filled attached-growth denitrification column [22]. Note: m × 3.2808 = ft; mm × 0.03937 = in.

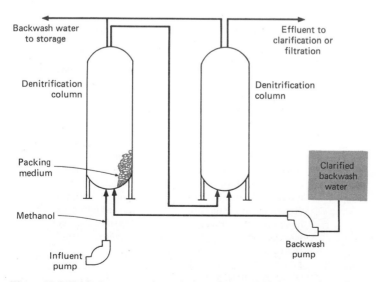

Figure 12-8 Typical process schematic for submerged high-porosity-media columns [22].

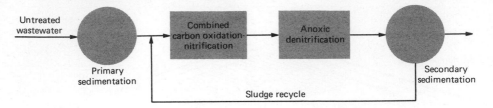

Figure 12-9 Sequential carbon oxidation-nitrification-denitrification [22].

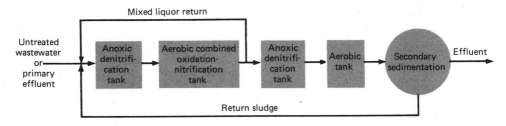

Figure 12-10 The Bardenpho system—sequential utilization of wastewater carbon and endogenous carbon [22].

Table 12-12 Comparison of denitrification alternatives[a]

System type	Advantages	Disadvantages
Suspended growth using methanol following a nitrification stage	Denitrification rapid, small structures required; demonstrated stability of operation; few limitations in treatment sequence options; excess methanol oxidation step can be easily incorporated; each process in the system can be separately optimized; high degree of nitrogen removal possible	Methanol required; stability of operation linked to clarifier for biomass return; greater number of unit processes required for nitrification-denitrification than in combined systems
Attached-growth (column) using methanol following a nitrification stage	Denitrification rapid, small structures required; demonstrated stability of operation; stability not linked to clarifier as organisms on media; few limitations in treatment sequence options; high degree of nitrogen removal possible; each process in the system can be separately optimized	Methanol required; excess methanol oxidation process not easily incorporated; greater number of unit processes required for nitrification-denitrification than in combined system

Table 12-12 Comparison of denitrification alternatives[a] (continued)

System type	Advantages	Disadvantages
Combined carbon oxidation nitrification–denitrification in suspended-growth reactor using endogenous carbon source	No methanol required; lesser number of unit processes required	Denitrification rates very low; very large structures required; lower nitrogen removal than in methanol-based system; stability of operation linked to clarifier for biomass return; treatment sequence options limited when both N and P removal required; no protection provided for nitrifiers against toxicants; difficult to optimize nitrification and denitrification separately
Combined carbon oxidation nitrification–denitrification in suspended-growth reactor using wastewater carbon source	No methanol required; lesser number of unit processes required	Denitrification rates low, large structures required; lower nitrogen removal than in methanol-based system; stability of operation linked to clarifier for biomass return; tendency for development of sludge bulking; treatment sequence options limited when both N and P removal required; no protection provided for nitrifiers against toxicants; difficult to optimize nitrification and denitrification separately

[a] From Ref. 22.

all cases, pilot plant studies are recommended. Where such studies cannot be conducted, the use of conservative design criteria is recommended.

Denitrification Stoichiometry

As early as 1860, it was observed that nitrate, nitrous oxide, and nitrogen gas were commonly produced in biological fermentations taking place in the presence of nitrates. By 1909 it was recognized that the reduction of nitrate (denitrification) involved the use of the nitrate radical as a hydrogen acceptor and that the requirements for this reaction included a source of combined hydrogen and the lack of free oxygen (anaerobic conditions) [28]. Although some autotrophic bacteria are able to reduce nitrate by using it as an electron acceptor, most of the nitrate-reducing bacteria are facultative anaerobic heterotrophs. The principal genera are *Pseudomonas, Micrococcus, Achromobacter,* and *Bacillus* [9].

Using methanol as the carbon source, the energy reaction may be represented by the following equations:

Energy reaction, step 1:

$$6NO_3^- + 2CH_3OH \rightarrow 6NO_2^- + 2CO_2 + 4H_2O \qquad (12\text{-}7)$$

Energy reaction, step 2:

$$6NO_2^- + 3CH_3OH \rightarrow 3N_2 + 3CO_2 + 3H_2O + 6OH^- \qquad (12\text{-}8)$$

Overall energy reaction:

$$6NO_3^- + 5CH_3OH \rightarrow 5CO_2 + 3N_2 + 7H_2O + 6OH^- \qquad (12\text{-}9)$$

A typical synthesis reaction as given by McCarty [15] is:

Synthesis:

$$3NO_3^- + 14CH_3OH + CO_2 + 3H^+ \rightarrow 3C_5H_7O_2N + H_2O \qquad (12\text{-}10)$$

In practice 25 to 30 percent of the amount of methanol required for energy is required for synthesis. On the basis of experimental laboratory studies, McCarty [15] developed the following empirical equation to describe the overall nitrate-removal reaction.

Overall nitrate removal:

$$NO_3^- + 1.08CH_3OH + H^+ \rightarrow 0.065C_5H_7O_2N + 0.47N_2 + 0.76CO_2 + 2.44H_2O$$

$$(12\text{-}11)$$

If all the nitrogen is in the form of nitrate, the overall methanol requirement can be determined using Eq. 12-11. However, biologically processed wastewater that is to be denitrified may contain some nitrite and dissolved oxygen. Where nitrate, nitrite, and dissolved oxygen are present, the methanol requirement can be computed using the following empirically derived equation [15]:

$$C_m = 2.47N_0 + 1.53N_1 + 0.87D_0 \qquad (12\text{-}12)$$

where C_m = required methanol concentration, mg/L
N_0 = initial nitrate-nitrogen concentration, mg/L
N_1 = initial nitrite-nitrogen concentration, mg/L
D_0 = initial dissolved-oxygen concentration, mg/L

Kinetic coefficients for the denitrification process are summarized in Table 12-13.

Table 12-13 Typical kinetic coefficients for the denitrification process[a]

Coefficient	Basis		Value[b]	
			Range	Typical
μ_m	d^{-1}		0.3–0.9	0.3
K_s	mg/L NO_3^-–N		0.06–0.20	0.10
Y	mg VSS/mg NO_3^-–N[c]		0.4–0.9	0.8
k_d	d^{-1}		0.04–0.08	0.04

[a] Derived in part from Refs. 22 and 25.
[b] Values reported are for 20°C.
[c] VSS = volatile suspended solids.
   Note: 1.8(°C) + 32 = °F

Process Analysis (Denitrification)

As with the nitrification process, the kinetics of the suspended-growth denitrification process are better defined than those of the attached-growth processes. In the following discussion, both systems will be considered.

Suspended-growth processes The effects of the major operational and environmental variables are reported in Table 12-14. The kinetic expressions used to analyze the denitrification process for a continuous-flow stirred-tank reactor are as reported in Table 12-7. The application of the kinetic approach, using the information in Table 12-14 and the various expressions in Table 12-7 to analyze the suspended-growth denitrification process, is as follows:

1. Using the kinetic data given in Table 12-13 and Fig. 12-11 and Eq. 9-14, determine the minimum mean cell residence time θ_c^M for denitrification. The kinetic coefficients must be corrected for the operating temperature using the expression given in Table 12-14.
2. Using Eq. 9-49 and an assumed safety factor, determine the design mean cell residence time, θ_c^d.
3. Using the design mean cell residence time determined in step 2 and Eq. 9-39, determine the substrate utilization rate U.
4. Using the substrate utilization rate determined in step 3, determine the effluent substrate concentration using Eq. 9-45.
5. Determine the hydraulic retention time using Eq. 9-38.
6. Determine the sludge-wasting rate using the standard definition given in Chap. 9.

Attached-growth processes To date, the approach most commonly used in assessing the performance of the attached-growth denitrification processes has involved the use of loading parameters. Although some application data are

Table 12-14 Effect of the major operational and environmental variables on the denitrification process[a]

Factor	Description of effect
Nitrate concentration	It has been observed that the concentration of nitrate will affect the maximum growth of the organisms responsible for denitrification. The effect of the nitrate concentration has been modeled using the following expression: $$\mu_D' = \mu_{m_D} \frac{C_N}{K_{SN} + C_N}$$
Carbon concentration	The effect of the carbon concentration has also been modeled using a Monod-type expression. The relationship using methanol as the carbon source is $$\mu_D' = \mu_{m_D} \frac{M}{K_M + M}$$ where M = methanol concentration, mg/L K_M = half-saturation constant for methanol, mg/L
Temperature	The effect of temperature is significant. It can be estimated using the following expression: $$P = 0.25 T^2$$ where P = percent of denitrification growth rate at 20°C T = temperature, °C
pH	From available evidence, it appears that the optimum pH range is between about 6.5 and 7.5, and the optimum condition is around 7.0.

[a] Developed from information contained in Ref. 22.
  Note: mg/L = g/m^3
         1.8(°C) + 32 = °F

reported in Table 12-10, it is recommended that pilot plant tests be conducted when any of these processes are considered for use. In the future, as additional information becomes available, the use of generalized application data may be appropriate.

Process Application

Usually, denitrification is used in conjunction with some type of nitrification process. One of the flowsheets that incorporates biological means and that has been used for the removal of nitrogen from domestic wastewater is shown schematically in Fig. 12-12. Typical design parameters for each of the processes shown in Fig. 12-12 are given in Table 12-15. This flowsheet can also be used for the removal of phosphorus by adding alum to precipitate the phosphorus in the activated-sludge settling tank. In addition to removing phosphorus, this technique also can be used to overcome the difficulties of separating organisms growing

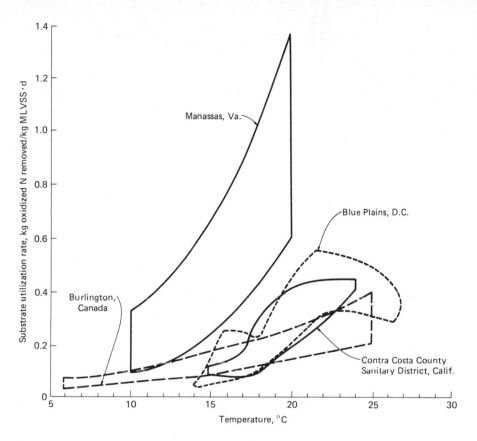

Figure 12-11 Observed denitrification rates for suspended-growth systems using methanol [22]. Note: kg × 2.2046 = lb.

in the dispersed-growth phase. Effluent from the denitrification step can be filtered, or it can be precipitated with alum for the removal of residual phosphorus and suspended solids and then filtered.

As shown in Fig. 12-12 and reported in Table 12-15 a continuous-flow stirred-tank reactor is used for the activated-sludge process, and plug-flow–mixed reactors are used for nitrification and denitrification. Mixing in the denitrification reactor is accomplished using submerged paddles. Because of the effect of temperature on the rate of nitrification and denitrification, special attention must be devoted to the design of these tanks and appurtenant facilities. Additional details on the design of the required facilities may be found in Ref. 22.

The mixed-liquor volatile suspended solids in the nitrification reactors are composed of those organisms responsible for the conversion of organic carbon (BOD) and those responsible for nitrification. The distribution of the two varies with each installation. The total mixed-liquor suspended solids in the nitrification reactor are normally 50 to 100 percent higher than the mixed-liquor volatile

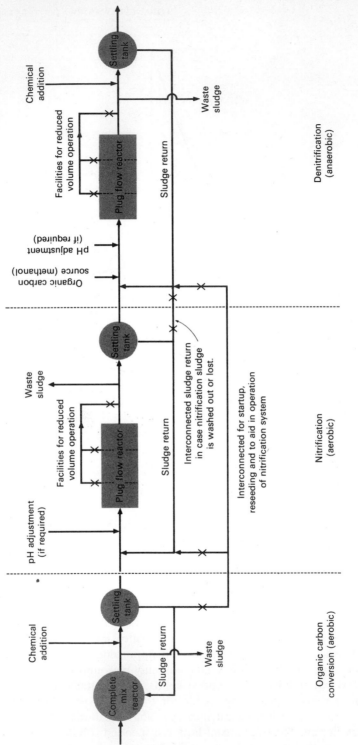

Figure 12-12 Flowsheet for a three-stage biological treatment process for nitrogen removal.

Chemical addition

Settling tank

Facilities for reduced volume operation

Plug flow reactor

Sludge return

Waste sludge

Denitrification (anaerobic)

pH adjustment (if required)

Organic carbon source (methanol)

Waste sludge

Settling tank

Facilities for reduced volume operation

Plug flow reactor

Sludge return

Interconnected sludge return in case nitrification sludge is washed out or lost.

Interconnected for startup, reseeding and to aid in operation of nitrification system

Nitrification (aerobic)

pH adjustment (if required)

Settling tank

Chemical addition

Complete mix reactor

Sludge return

Waste sludge

Organic carbon conversion (aerobic)

Table 12-15 Design parameters for a three-stage biological treatment process to remove nitrogen from domestic wastewater

Treatment				Design parameter				
Process	Description	Type of reactor	Aeration system	θ_c, d[a]	θ, h[a]	MLVSS, mg/L	pH	Temperature coefficient[b]
Organic carbon conversion (modified biological treatment)	Aerobic conversion of organic carbon to end products and cell tissue	Continuous-flow stirred-tank	Air or pure oxygen	2–5	1–3	See Table 10-4	6.5–8.0	1.00–1.03
Nitrification	Aerobic conversion of ammonia and nitrite to nitrate	Plug-flow	Air or pure oxygen[c]	10–20	0.5–3	1000–2000[d]	7.0–8.0[e]	1.08–1.10[f]
Denitrification[g]	Anoxic conversion of nitrate to nitrogen gas	Plug-flow[h]	....	1–5	0.2–2	1000–2000[d]	6.5–7	1.14–1.16[i]

[a] Indicated values for θ_c and θ are for 20°C.

[b] Temperature coefficient to be used in the equation $K_T = K_{20}\theta^{T-20}$.

[c] The theoretical oxygen requirement for nitrification can be estimated using Eq. 12-6 and adding to it the value obtained by multiplying the applied BOD by a factor of 1.5.

[d] Higher values may be observed depending on the degree of solids carryover.

[e] Lower pH values have been reported.

[f] Estimated from data developed by Wild et al. [31] and Mulbarger [20].

[g] Methanol requirement can be computed using Eq. 12-12.

[h] Covered reactors may be used to minimize air-liquid contact.

[i] Estimated from data developed by Mulbarger [20].

Note: mg/L = g/m³

suspended solids and may include residual chemical precipitates if alum precipitation is used for phosphorus removal. In denitrification reactors, the mixed-liquor volatile suspended solids have been observed to be about 40 to 70 percent of the mixed-liquor suspended solids.

Temperature will also significantly affect performance. It cannot be overstressed that unless temperature is properly considered in the design of both nitrification and denitrification systems, effluent quality (measured in terms of the amount of ammonia or nitrate in the effluent) will deteriorate at low temperatures. For example, using a temperature coefficient of 1.12 (see Table 12-15) the reactor volume at 10°C (50°F) would be approximately three times the volume required at 20°C (68°F). From a design standpoint, the additional volume could be provided by using a plug-flow–mixed reactor whose length could be reduced. Alternatively, the solids in the system could be increased to accommodate cold weather operation.

12-6 NITROGEN REMOVAL BY PHYSICAL AND CHEMICAL PROCESSES

The principal physical and chemical processes used for nitrogen removal, identified in Table 12-4, are air stripping, breakpoint chlorination, and selective ion exchange.

Air Stripping of Ammonia

The removal of ammonia from wastewaters by air stripping has received considerable attention as part of the work that has been carried out in the field of advanced waste treatment under the sponsorship of the federal government. Perhaps the best-known example is the work conducted at Lake Tahoe, Calif. [3, 4, 26]. This process also has received considerable attention in the industrial waste treatment field.

Theory The air stripping of ammonia from wastewater is a modification of the aeration process used for the removal of gases dissolved in water. Ammonium ions in wastewater exist in equilibrium with ammonia, as shown in Eq. 12-13:

$$NH_3 + H_2O \rightleftharpoons NH_4^+ + OH^- \tag{12-13}$$

As the pH of the wastewater is increased above 7, the equilibrium is shifted to the left; and the ammonium ion is converted to ammonia, which may be removed as a gas by agitating the wastewater in the presence of air. This is usually accomplished in a packed tray tower equipped with an air blower. An analysis of this process, including the data required for design, is presented in the following discussion taken from Tchobanoglous [27].

Because the reaction in Eq. 12-13 is pH dependent, the percentage distribution of ammonia and ammonium ion can be computed using the following equation:

$$NH_3, \% = \frac{NH_3 \times 100}{NH_3 + NH_4^+} = \frac{100}{1 + NH_4^+/NH_3} = \frac{100}{1 + K_b[H]/K_w} \qquad (12\text{-}14)$$

For example, at 25°C and pH 10, the percentage distribution of NH_3 is

$$\frac{100}{1 + (1.8 \times 10^{-5} \times 10^{-10})/10^{-14}} = 85\%$$

Knowing that at 25°C and pH 11 the percentage distribution of ammonia is about 98 percent, the amount of base that must be added to a wastewater can be established. Typical data on the amount of lime $[CA(OH)_2]$ required to raise the pH to 11 as a function of the wastewater alkalinity are presented in Fig. 12-13.

The amount of air required to remove the ammonia from the liquid in a stripping tower is determined as follows. A process definition sketch is presented in Fig. 12-14. An overall steady-state materials balance equation for a stripping tower is given by

$$G(Y_2 - Y_1) = L(X_2 - X_1) \qquad (12\text{-}15)$$

where G = moles of incoming gas per unit time
  L = moles of incoming liquid per unit time
  Y_1 = concentration of solute in gas at bottom of tower, moles of solute per mole of solute-free gas
  Y_2 = concentration at top
  X_1 = concentration of solute in liquid at bottom of tower, moles of solute per mole of liquid
  X_2 = concentration at top

Because Eq. 12-15 was derived solely from a consideration of the equality of input and output, it holds regardless of the internal equilibria that may control

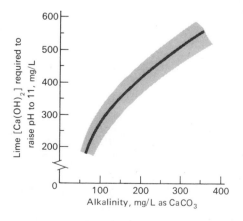

Figure 12-13 Lime dosage required to raise the pH to 11 as a function of untreated wastewater alkalinity.

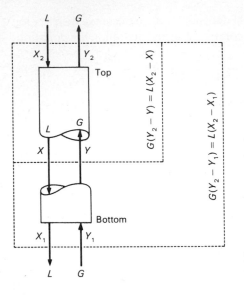

Figure 12-14 Definition sketch for an ammonia-stripping column.

the process. For a more complete analysis of this operation and other mass-transfer operations, see Ref. 29.

Equilibrium curves for ammonia in water for various temperatures and 1 atm of pressure are given in Fig. 12-15. Equilibrium curves are also often presented in terms of mole fractions [mol NH_3/(mol H_2O + mol NH_3)]. The theoretical amount of air required per cubic meter of wastewater can be computed using Eq. 12-15 and Fig. 12-15. If it is assumed that the liquid leaving and the air entering the bottom of the tower contain no ammonia, Eq. 12-15 can be written as

$$\frac{G}{L} = \frac{X_2}{Y_2} \tag{12-16}$$

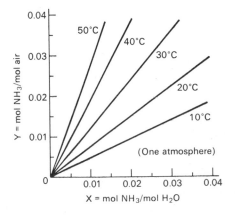

Figure 12-15 Equilibrium curves for ammonia in water.

where G/L is the ratio of the gas (air) to liquid (wastewater) required for stripping the ammonia from the wastewater.

The theoretical minimum amount of air would be required if the ammonia in the air leaving the tower were in equilibrium with the ammonia in the incoming liquid. The minimum air-to-liquid ratio in moles of air per mole of water is given by the reciprocal of the slope of the equilibrium curves in Fig. 12-15. Calculation of this ratio is illustrated in Example 12-2.

Example 12-2 Theoretical air requirements for ammonia stripping Determine the minimum air-to-liquid ratio required at 20°C for complete stripping.

SOLUTION
1 Determine the ratio X/Y at 20°C using Fig. 12-15.

$$\frac{X}{Y} = \frac{\text{mol NH}_3/\text{mol H}_2\text{O}}{\text{mol NH}_3/\text{mol air}} = \frac{0.02 \text{ mol air}}{0.015 \text{ mol H}_2\text{O}} = \frac{1.33 \text{ mol air}}{\text{mol H}_2\text{O}}$$

2 Convert the moles of air to liters of air.

$$1.33 \text{ mol} \times 22.4 \text{ L/mol} = 29.8 \text{ L}$$

3 Convert the moles of water to liters of water.

$$1.0 \text{ mol} \frac{18 \text{ g}}{\text{mol}} \frac{\text{kg}}{1000 \text{ g}} \frac{\text{L}}{\text{kg}} = 0.018 \text{ L}$$

4 Determine the required air-to-liquid ratio.

$$\frac{G}{L} = \frac{29.8 \text{ L}}{0.018 \text{ L}} = 1656 \text{ L/L} = 1656 \text{ m}^3/\text{m}^3$$

The required air-to-liquid ratio for various temperatures for stripping ammonia is given in Fig. 12-16. The theoretical ratio is derived by assuming the

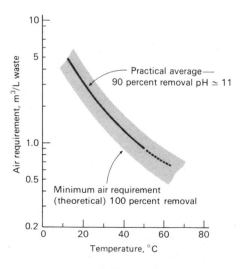

Figure 12-16 Effect of temperature on air requirements for ammonia stripping [27]. Note: $\text{m}^3/\text{L} \times 133.6806 = \text{ft}^3/\text{gal}$.

process to be 100 percent efficient with a stripping tower of infinite height—obviously unachievable in practice. Therefore, in computing the actual amount of air required, the theoretical value is often multiplied by a factor varying from 1.5 to 2.0. Practical air-to-liquid ratios are reported in Fig. 12-16.

The foregoing analysis would be reasonable even if the concentration of ammonia in the liquid leaving the tower were on the order of 1 to 2 mg/L. In this case, the value of X_1 would be very close to zero and, for all practical purposes, the minimum air requirement would be defined by the slope of the equilibrium line. In all practical cases, the ammonia concentration in the air leaving the tower will be less than the equilibrium value, and the air-to-liquid ratio will plot below the equilibrium curves in Fig. 12-15. If practically complete removal of ammonia is obtained under these conditions, air in excess of the theoretical amount is being applied as shown in Fig. 12-16. In general, removal efficiency depends on the temperature, size, and proportions of the facility, and the efficiency of the air-water contact. If the removal of ammonia is unsatisfactory, the tower has not been designed correctly or is overloaded. In this case, additional air volume may improve operation.

Application The application of air stripping in the treatment of untreated wastewater and digester supernatant is shown in Figs. 12-17 and 12-18, respectively. In the flowsheet shown in Fig. 12-17, lime is added to the untreated wastewater to precipitate phosphorus. Next, the settled wastewater is passed through a countercurrent ammonia-stripping tower. The stripped liquid is then recarbonated and resettled. A lime-recovery system is also included. This flowsheet is similar to the one used at Lake Tahoe for the treatment of secondary effluent.

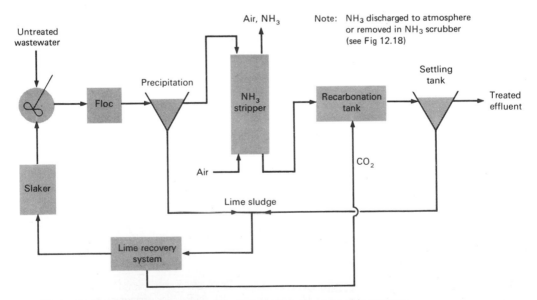

Figure 12-17 Typical flowsheet for ammonia stripping of untreated wastewater.

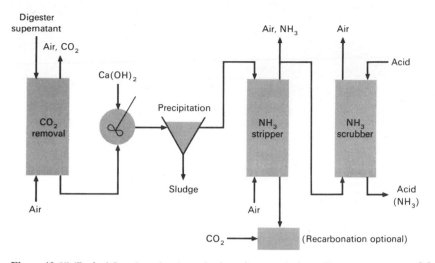

Figure 12-18 Typical flowsheet for the stripping of ammonia from digester supernatant [5].

The flowsheet for the treatment of digester supernatant (see Fig. 12-18) is similar to those discussed previously, except that CO_2 is stripped before adding lime, and an acid ammonia scrubber has been added where the stripped ammonia cannot be discharged to the atmosphere.

A complete system for the elimination of nitrogen in which ammonia stripping is used is shown in Fig. 12-19. The stripped ammonia gas is converted to nitrate in a biological packed-bed tower reactor. Biological denitrification is used to convert the nitrate to nitrogen gas.

In most cases where ammonia stripping has been applied, a number of problems have developed, such as calcium carbonate scaling within the tower and feed lines and poor performance during cold weather operation. The amount and nature (soft to extremely hard) of the calcium carbonate scale formed varies with the characteristics of the wastewater and local environmental conditions. As the temperature decreases, the amount of air required increases significantly for the same degree of removal (see Fig. 12-16). Under conditions of icing, the liquid-air contact geometry in the tower is altered, which further reduces the overall efficiency.

Apart from operational difficulties, the discharge of ammonia to the atmosphere near large bodies of water or snow-covered areas can create serious nitrogen pollution problems. In such situations, use of an acid scrubber or the adoption of an alternative method of nitrogen removal may be necessary.

Breakpoint Chlorination

Breakpoint chlorination, in which a sufficient amount of chlorine is added to oxidize the ammonia nitrogen in solution to nitrogen gas and other stable compounds, is an alternative method of achieving nitrogen control (see Table 12-4).

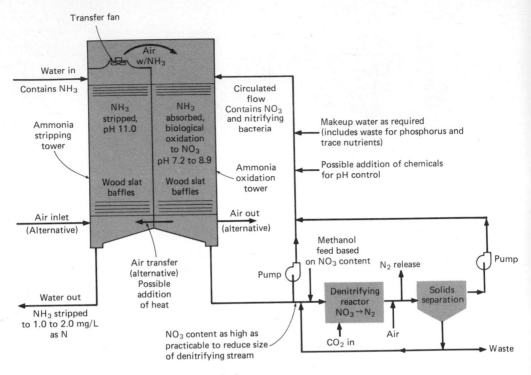

Figure 12-19 Ammonia elimination system [22].

Perhaps the most important advantage of this process is that, with proper control and flow equalization, all the ammonia nitrogen in the wastewater can be reduced to zero. An added advantage is that disinfection of the wastewater is achieved at the same time. Breakpoint chlorination (as well as flow equalization) is used in both flowsheets shown in Figs. 12-1 and 12-2.

Theory The theory of breakpoint chlorination was described in Sec. 7-5. From that discussion, a representative equation that can be used to describe the overall reaction is

$$2NH_3 + 3HOCL \rightarrow N_2 + 3H_2O + 3HCl \qquad (12\text{-}17)$$

The stoichiometric mass ratio of chlorine as Cl_2 to ammonia as N, as computed in Example 7-4, is 7.6 : 1. In practice, the ratio has been found to vary from about 8 : 1 to 10 : 1.

From both laboratory and full-scale testing programs, it has been found that the optimum pH operating range for breakpoint chlorination is between 6 and 7. If breakpoint chlorination is accomplished outside this range, it has been observed that the chlorine dosage required to reach the breakpoint increases significantly and that the rate of reaction is slower. The temperature, on the other hand, does not appear to have a major effect on the process in the ranges normally encountered in wastewater treatment. The effect of interfering substances on the process is discussed in Sec. 7.5.

Application The breakpoint chlorination process can be used for the removal of ammonia nitrogen from treatment plant effluents either alone or in combination with other processes. Usually, to avoid the large chlorine dosages required when used alone, it is used in conjuction with other nitrogen removal processes, as shown in Figs. 12-1 and 12-2.

To optimize the performance of this process and to minimize equipment and facility costs, flow equalization is usually required (see Figs. 12-1 and 12-2). Also, because of the potential toxicity problems that may develop if chlorinated compounds are discharged to the environment (see Sec. 7.6), it is usually necessary to dechlorinate the effluent. Activated carbon columns are used for this purpose in the flowsheets shown in Figs. 12-1 and 12-2. The use of the breakpoint chlorination process for seasonal nitrogen control is considered in Example 12-3.

Example 12-3 Analysis of breakpoint chlorination process used for seasonal control of nitrogen Estimate the daily required chlorine dosage and the resulting buildup of total dissolved solids when breakpoint chlorination is used for the seasonal control of nitrogen. Assume that the following data apply to this problem:

1   Plant flowrate = 3800 m^3/d (1.0 Mgal/d)
2   Effluent characteristics
   a BOD_5 = 20 mg/L
   b   Suspended solids = 25 mg/L
   c NH_3–N = 23 mg/L
3   Required effluent NH_3–N concentration = 1.0 mg/L

SOLUTION

1   Estimate the required Cl_2 dosage. Assume that the required mass ratio of chlorine to ammonia is 9 : 1.

$$\text{kg } Cl_2/d = (3800 \text{ m}^3/d)[(23 - 1) \text{ g/m}^3](9.0)(10^3 \text{ g/kg})^{-1}$$
$$= 752.4 \text{ kg/d}$$

2   Determine the increment of total dissolved solids added to the wastewater. Using the data reported in Table 7-10, the total dissolved-solids increase per mg/L of ammonia consumed is equal to 6.2.

$$\text{Total dissolved-solids increment} = 6.2(23 - 1) \text{ mg/L} = 136.4 \text{ mg/L}$$

   Comment   In this example, it was assumed that the acid produced from the breakpoint reaction would not require the addition of a neutralizing agent such as NaOH (sodium hydroxide). If the addition of NaOH were required, the total dissolved solids increase would have been significantly large. It is noted that although breakpoint chlorination can be used to control nitrogen, it may be counterproductive if in the process the treated effluent is rendered unusable for other applications because of the buildup of total dissolved solids.

Ion Exchange

Ion exchange is a unit process by which ions of a given species are displaced from an insoluble exchange material by ions of a different species in solution. It may be operated in either a batch or a continuous mode. In a batch process, the resin is simply stirred with the water to be treated in a reactor until the reaction is complete. The spent resin is removed by settling and subsequently is regenerated and reused. In a continuous process, the exchange material is placed in a bed or a packed column, and the water to be treated is passed through it. Theoretical and operational aspects may be found in Chapman [2].

Theory The chemistry of the ion-exchange process may be represented by the following equilibrium equations.

Reaction:

$$RH + Na^+ \rightleftharpoons RNa + H^+ \tag{12-18}$$
$$RNa_2 + Ca^{2+} \rightleftharpoons RCa + 2Na^+ \tag{12-19}$$

Regeneration:

$$RNa + HCl \rightleftharpoons RH + NaCl \tag{12-20}$$
$$RCa + 2NaCl \rightleftharpoons RNa_2 + CaCl_2 \tag{12-21}$$

where R represents the resin.

Equations 12-18 and 12-19 represent the reactions involved in the removal of sodium and calcium ions from water using a synthetic cationic exchange resin. Equations 12-20 and 12-21 represent the reactions involved in the regeneration of the exhausted resin. The extent of completion of the removal reactions shown depends on the equilibrium that is established between the ions in the aqueous phase and those in the solid phase. For the removal of sodium, this equilibrium is defined by the following expression:

$$\frac{[H] X_{RNa}}{[Na] X_{RH}} = K_{H \to Na} \tag{12-22}$$

where $K_{H \to Na}$ = selectivity coefficient
$\quad$ [ ] = concentration in solution phase
$\quad X_{RH}$ = mole fraction of hydrogen on exchange resin
$\quad X_{RNa}$ = mole fraction of sodium on exchange resin

The selectivity coefficient depends primarily on the nature and valence of the ion, the type of resin and its saturation, and the ion concentration in wastewater. In fact, for a given series of similar ions, exchange resins have been found to exhibit an order of selectivity or affinity for the ions. For synthetic cationic and anionic exchange resins, typical series are as follows:

$$Li^+ < H^+ < Na^+ < K^+ < Rb^+ < Ag^+$$
$$Mg^{2+} < Zn^{++} < Cu^{2+} < Co^{++} < Ca^{2+} < Sr^{2+} < Ba^{2+}$$
$$OH^- < F^- < HCO_3^- < Cl^- < Br^- < I^- < NO_3^- < ClO_4^-$$

In practice, the selectivity coefficients are determined by measurement in the laboratory and are valid only for the conditions under which they were measured. At low concentrations, the value of the selectivity coefficient for the exchange of monovalent ions by divalent ions is, in general, larger than the exchange of monovalent ions by monovalent ions. This fact has, in many cases, limited the use of synthetic resins for the removal of certain substances in wastewater, such as ammonia in the form of the ammonium ion. There are, however, certain natural zeolites that favor NH_4^+ or Cu^{2+}.

Representative listings of cationic and anionic exchange resins produced by the major manufacturers in the United States may be found in Applebaum [1]. Reported exchange capacities vary with the type and concentration of regenerant used to restore the resin. Exchange capacities for resins often are expressed in terms of grams as $CaCO_3$ per cubic meter of resin or gram equivalents per cubic meter. Conversion between these two units is accomplished using the following expression:

$$\frac{1 \text{ g equiv}}{m^3} = \frac{(1 \text{ g equiv})(50 \text{ g CaCO}_3)/(\text{g equiv})}{m^3} = 50 \text{ g CaCO}_3/m^3 \quad (12\text{-}23)$$

Calculation of the required resin volume for the ion-exchange process is illustrated in Example 12-4.

Example 12-4 Ion-exchange-resin volume requirements Determine the volume of a cationic resin with an exchange capacity of 100,000 g $CaCO_3/m^3$ required to treat 4000 m³ of water containing 18 mg/L of ammonium ion.

SOLUTION

1   Determine the total ammonium ion expressed as $CaCO_3$. In terms of $CaCO_3$, 18 mg/L of ammonium ion are equal to 50 mg/L. Therefore, the required exchange capacity is equal to

$$(50 \text{ g/m}^3)(4000 \text{ m}^3) = 200,000 \text{ g as CaCO}_3$$

2   Compute the volume of resin required.

$$\frac{200,000 \text{ g as CaCO}_3}{100,000 \text{ g as CaCO}_3/m^3} = 2 \text{ m}^3 \text{ (70.6 ft}^3\text{)}$$

3   In practice, because of leakage and other operational and design limitations, the required volume of resin usually is about 1.1 to 1.4 times that computed on the basis of exchange capacity.

Application Although both natural and synthetic ion-exchange resins are available, synthetic resins are used more widely because of their durability. Nevertheless, some natural resins (zeolites) have found application in the removal of ammonia from wastewater. Of the natural zeolites that have been investigated, Hector Clinoptilolite has proved to be one of the most effective [17]. One of the novel features of this zeolite is the regeneration system employed. Upon exhaustion, the zeolite is regenerated with lime $Ca(OH)_2$. The ammonium ion removed from the zeolite is converted to ammonia because of the high pH. At this point, the regenerating solution is passed through a stripping tower for removal of the ammonia. The stripped liquid is collected in a storage tank for subsequent reuse

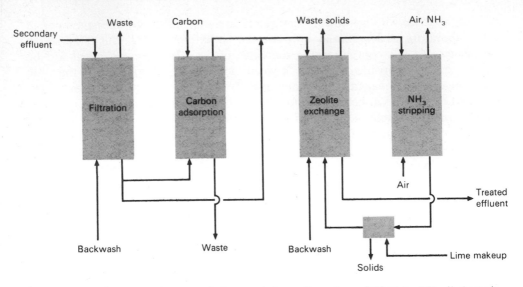

Figure 12-20 Flowsheet for the removal of ammonia by zeolite exchange [17]. Note: NH_3 discharged to atmosphere or removed in NH_3 scrubber (see Fig. 12.18).

[17]. The advantage of this system is that there is no process waste containing ammonia for which ultimate disposal must be provided. A flowsheet for this process is shown in Fig. 12-20.

A pressing problem that must be solved is the formation of calcium carbonate precipitates within the zeolite exchange bed and in the stripping tower and piping appurtenances. As indicated in Fig. 12-20 the zeolite bed is equipped with backwash facilities to remove the carbonate deposits that form within the filter.

A serious problem associated with application of ion exchange to the treatment of wastewater effluents is resin binding caused by the residual organic matter found in effluent from biological treatment. This problem has been solved partially by prefiltering the wastewater (see Fig. 12-20) or by using scavenger exchange resins before application to the exchange column [1].

To make ion exchange economical for advanced wastewater treatment, it would be desirable to use regenerants and restorants that would remove both the inorganic anions and the organic material from the spent resin. Chemical and physical restorants found to be successful in the removal of organic material from resins include sodium hydroxide, hydrochloric acid, methanol, and bentonite [6, 7].

12-7 PHOSPHORUS REMOVAL

Domestic wastewater and agricultural return water have been identified as the principal sources of phosphorus, which has often been cited as the culprit responsible for the stimulation of aquatic plants and for contributing to eutrophication in general. In this section, only the treatment of phosphorus in domestic wastewater will be considered. Additional details beyond what is presented in this section may be found in Ref. 23.

Forms of Phosphorus

Phosphorus in wastewater may be present in three forms: orthophosphate, polyphosphate, and organic phosphorus. Typically, the phosphorus enters the wastewater from human body wastes, from food wastes discharged to the sewers from kitchen grinders, and from the condensed inorganic phosphate compounds used in various household detergents. Commercial washing and cleaning compounds are also a source of phosphates.

Operations and Processes for Phosphorus Removal

With most wastewaters, approximately 10 percent of the phosphorus corresponding to the portion that is insoluble is normally removed by primary settling. Except for the amount taken up for incorporation into cell tissue, the additional removal achieved in conventional biological treatment is minimal because almost all the phosphorus present after primary sedimentation is soluble.

None of the forms of phosphorus present in wastewater are gaseous at normal temperatures and pressures, so removal must be accomplished by the formation of an insoluble precipitate that can be removed by gravity settling [11]. The principal chemicals used for this purpose are lime, alum, and ferric chloride or sulfate. Polymers have also been used effectively in conjunction with lime and alum. The chemistry of the precipitation reactions involved was described in Sec. 7-1. Factors affecting the choice of chemical to use for phosphorus removal are reported in Table 12-16.

Chemicals can be added at a variety of different points in the treatment process (see Fig. 12-21), but because polyphosphates and organic phosphorus are less easily removed than orthophosphorus, adding aluminum or iron salts after secondary treatment (where organic phosphorus and polyphosphorus are transformed into orthophosphorus) usually results in the best removal. Some

Table 12-16 Factors affecting choice of chemical for phosphorus removal [11]

1. Influent phosphorus level
2. Wastewater suspended solids and alkalinity
3. Chemical cost including transportation
4. Reliability of chemical supply
5. Sludge-handling facilities
6. Ultimate disposal methods
7. Compatability with other treatment processes in plant
8. Potential adverse environmental effects of chemical used

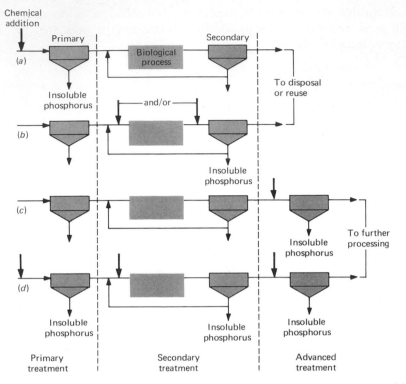

Figure 12-21 Alternative points of chemical addition for phosphorus removal: (*a*) before primary sedimentation, (*b*) before and/or following biological treatment, (*c*) following secondary treatment, and (*d*) at several locations in a process known as split treatment.

additional nitrogen removal occurs because of better settling, but essentially no ammonia is removed unless chemical additions to primary treatment reduce BOD loadings to the point where nitrification can occur. An increase in total dissolved solids can be expected. Still another method that has been proposed involves the removal of phosphorus by contact filtration [10].

Chemical addition to primary sedimentation facilities In the simplest terms, when aluminum or iron salts are added to untreated wastewater, they react with the soluble orthophosphate to produce a precipitate. When lime is used, both the calcium and the hydroxide react with the orthophosphorus to form an insoluble hydroxyaptite. Organic phosphorus and polyphosphate are removed by more complex reactions and by adsorption onto floc particles. The insolubilized phosphorus, as well as considerable quantities of BOD and suspended solids, are removed from the system as primary sludge. Adequate mixing and floccula- tion are necessary upstream of primary facilities, whether separate basins are provided or existing facilities are modified to provide these functions. Additions of polymer may be required to aid in settling. A base is sometimes necessary

Table 12-17 Typical alum dosage requirements for various levels of phosphorus removal[a]

| Phosphorus reduction, % | mol ratio, Al : P | |
	Range	Typical
75	1.25 : 1–1.5 : 1	1.4 : 1
85	1.6 : 1–1.9 : 1	1.7 : 1
95	2.1 : 1–2.6 : 1	2.3 : 1

[a] Developed in part from Ref. 23.

in low-alkalinity waters to keep pH in the 5 to 7 range with mineral addition. Mineral salts generally are applied in the range of a 1 to 3 metal ion/J phosphorus molar ratio (see Table 12-17 and Fig. 12-22). The exact application rate is determined by on-site testing. It varies with the characteristics of the wastewater and the desired phosphorus, BOD, or suspended solids removals.

Both low-lime and high-lime treatment can be used to precipitate a portion of the phosphorus (usually about 65 to 80 percent) at pH values equal to or less than 10. In the trickling-filter process, recarbonation is generally required before biological treatment. In the activated-sludge process, the carbon dioxide generated during treatment is usually sufficient to lower the pH without recarbonation. The residual phosphorus level of 1.0 mg/L can be readily achieved with the addition of effluent filtration facilities to which chemicals can be added. In the high-lime system, sufficient lime is added to raise the pH to about 11

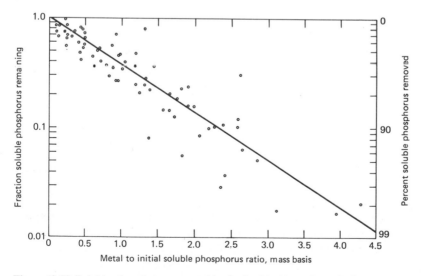

Figure 12-22 Soluble phosphorus removal by ferric chloride addition [23].

(see Fig. 12-13). After precipitation, the effluent must be recarbonated before biological treatment.

When lime is used, the principal variables controlling the dosage are the degree of removal required and the alkalinity of the wastewater. The operating dosage must usually be determined by on-site testing. Although lime recalcination lowers chemical costs, it is a feasible alternative only for large plants.

The additional BOD and suspended-solids removals afforded by chemical additions to primary treatment may solve overloading problems on downstream biological systems, or may allow nitrification (either seasonal or year-round, depending on biological system designs).

Phosphorus removal in biological waste treatment Phosphorus is removed in biological treatment by means of incorporation into cell tissue. The total amount removed depends on the net solids produced as determined using Eq. 10-9. It can be estimated by assuming that the phosphorus content of the cell tissue is about one-fifth of the nitrogen content. The actual phosphorus content may vary from one-seventh to one-third of the nitrogen value, depending on specific environmental conditions. It has been observed that the degree of phosphorus removal at some activated-sludge treatment plants is considerably higher than would be predicted on the basis of the requirements for organism growth. Two different theories have been proposed to account for this observation.

The first theory is that the removal of phosphate is brought about by chemical precipitation, as described by Menar and Jenkins [16]. The required conditions are as follows: (1) hydrolysis of complex phosphate to orthophosphates; (2) decreasing carbon dioxide production as the waste passes through a plug-flow reactor; (3) an increase in pH because less carbon dioxide is being produced and more is being removed by aeration; and (4) the development of conditions favoring the precipitation of calcium phosphate. As noted from these conditions, a long plug-flow reactor would be required.

The second theory is that the removal is accomplished by biological means. It is believed that, under certain ideal conditions, the microorganisms in the activated-sludge mixed liquor are able to remove an excess amount of phosphorus over that required for growth [13]. This phenomenon has been termed "luxury uptake." It is not clear if the phosphorus is incorporated (stored) within the cell or adsorbed on the bacterial cells, or a combination of both.

Metal-salt addition to secondary treatment Metal salts can be added to the untreated wastewater, in the activated-sludge aeration tank, or the final clarifier influent channel. In trickling-filter systems, the salts are added to the untreated wastewater or to the filter effluent. Multipoint additions have also been used. Phosphorus is removed from the liquid phase through a combination of precipitation, adsorption, exchange, and agglomeration, and it is wasted with either the primary or secondary sludges, or both. Theoretically, the minimum solubility of $AlPO_4$ occurs at pH 6.3, and that of $FePO_4$ occurs at pH 5.3; however, practical applications have yielded good phosphorus removal anywhere in the range of pH 5.5 to 7.0, which is compatible with mixed-liquor organisms.

The use of lime or ferrous salts is limited because they produce low phosphorus levels only at high pH values. In low-alkalinity waters, either sodium aluminate and alum or ferric plus lime, or both, can be used to maintain the pH higher than 5.5. Improved settling and lower effluent BOD result from chemical addition, particularly if polymer is also added to the final clarifier. Dosages generally fall in the range of a 1 to 3 metal ion–phosphorus molar ratio.

Chemical polymer addition to secondary clarifiers In certain cases, such as trickling-filtration and extended-aeration activated-sludge processes, solids may not flocculate and settle well in the secondary clarifier. This problem may become acute in plants that are overloaded. The addition of aluminum or iron salts will cause the precipitation of metallic hydroxides or phosphates, or both. Aluminum and iron salts, along with certain organic polymers, can also be used to destabilize colloidal particles. The resultant destabilized colloids and precipitates will settle readily in the secondary clarifier, reducing the suspended solids in the effluent and effecting phosphorus removal. Dosages of aluminum and iron salts usually fall in the range of a 1 to 3 metal ion–phosphorus molar ratio.

Tertiary lime coagulation filtration Lime can be added to the waste stream after biological treatment to reduce the level of phosphorus and suspended solids (see Fig. 12-21). Single-stage process and two-stage-process flowsheets are shown in Figs. 12-23 and 12-24, respectively. In the first-stage clarifier of the two-stage process (see Fig. 12-24), sufficient lime is added to raise the pH above 11 to precipitate the soluble phosphorus as basic calcium phosphate (apatite). The

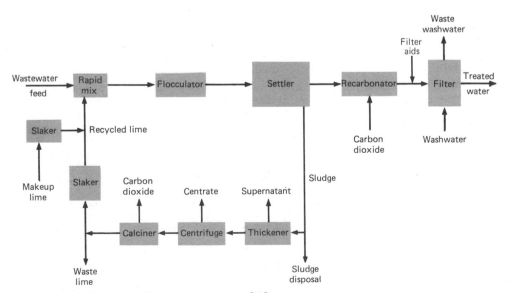

Figure 12-23 Single-stage lime treatment system [23].

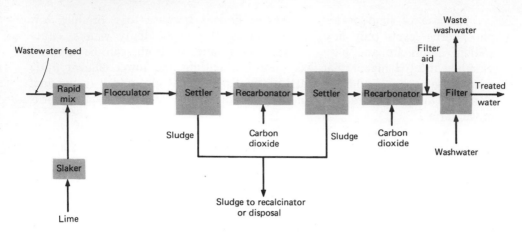

Figure 12-24 Two-stage lime treatment system [23].

calcium carbonate precipitate formed in the process acts as a coagulant for suspended-solids removal. The excess soluble calcium is removed in the second-stage clarifier as a calcium carbonate precipitate by adding carbon dioxide gas to reduce the pH to about 10. Generally, there is a second injection of carbon dioxide to the second-stage effluent. This reduces the pH to prevent scaling. To

Table 12-18 Advantages and disadvantages of phosphorus removal in various sections of a treatment plant[a]

Level of treatment	Advantages	Disadvantages
Primary	Applicable to most plants; increased BOD and suspended-solids removal; lowest degree of metal leakage; lime recovery demonstrated	Least efficient use of metal; polymer may be required for flocculation; sludge more difficult to dewater than primary sludge
Secondary	Lowest cost; lower chemical dosage than primary; improved stability of activated sludge; polymer not required	Overdose of metal may cause low-pH toxicity; with low-alkalinity wastewaters, a pH control system may be necessary; cannot use lime because of excessive pH; inert solids added to activated-sludge mixed liquor, reducing the percentage of volatile solids
Advanced (tertiary)	Lowest phosphorus in effluent; most efficient metal use; lime recovery demonstrated	Highest capital cost; highest metal leakage

[a] Adapted from Ref. 11.

remove the residual levels of suspended solids and phosphorus, the secondary clarifier effluent is passed through a multimedia filter.

In most cases, a lime-recovery system is required for a cost-effective operation. This includes a thermal regeneration facility, which converts the calcium carbonate in the sludge to lime by heating at $982°C$ ($1800°F$). The carbon dioxide from this process or other on-site stack gas (containing 10 to 15 percent carbon dioxide) is generally used as the source of carbon dioxide gas.

Comparison of Processes

The advantages and disadvantages of the removal of phosphorus by the addition of chemicals at various points in a treatment process flowsheet are summarized in Table 12-18. It is recommended that each alternative point of application be evaluated when phosphorus is to be removed.

12-8 REMOVAL OF REFRACTORY ORGANICS

Removal of the refractory organic compounds that are not converted during conventional biological treatment can be accomplished with carbon adsorption, oxidation, and by means of land application. Carbon adsorption and ozone oxidation are considered briefly in the following discussion. Removal by land application is discussed in Chap. 13.

Carbon Adsorption

To date, carbon adsorption is the method that has most often been used for the removal of the refractory organic compounds in wastewater. Either granular or powdered activated carbon can be used. The details of activated carbon adsorption were considered in Sec. 7-3. A typical flowsheet using granular carbon is shown in Fig. 12-25.

Both granular and powdered carbon appear to have a low adsorption affinity for low-molecular-weight polar organic species. If biological activity is low in the carbon contactor or in other biological unit processes, these species are difficult to remove with activated carbon. Under normal conditions, after treatment with carbon the effluent BOD ranges from 2 to 7 mg/L, and the effluent COD ranges from 10 to 20 mg/L. Under optimum conditions, it appears that the effluent COD can be reduced to about 10 mg/L.

Chemical Oxidation

In advanced wastewater-treatment applications, chemical oxidation can be used to remove ammonia, to reduce the concentration of residual organics, and to reduce the bacterial and viral content of wastewaters. The use of chlorine for the oxidation of ammonia was discussed in Sec. 12-5. Both chlorine and ozone can

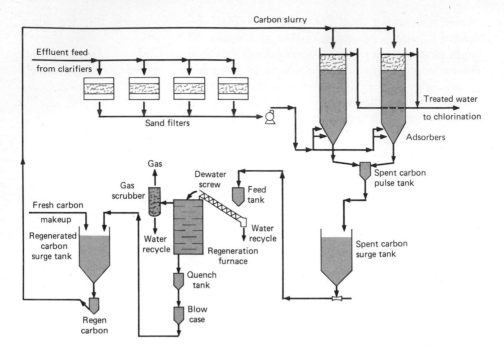

Figure 12-25 Flowsheet of activated-carbon facilities.

be used to reduce the residual organics in wastewater. When these chemicals are used for this purpose, disinfection of the wastewater is usually an added benefit. A further benefit of using ozone is the removal of color.

Typical chemical dosages for both chlorine and ozone for the oxidation of the organics in wastewater are reported in Table 12-19. The dosages increase with the degree of treatment, which is reasonable when it is considered that the

Table 12-19 Typical chemical dosages for the oxidation of organics in wastewater[a]

Chemical	Use	Dosage, kg/kg destroyed	
		Range	Typical
Chlorine	BOD$_5$ reduction	0.5–2.5	1.75[b]
		1.0–3.0	2.0[c]
Ozone	COD reduction	2.0–4.0	3.0[b]
	COD reduction	3.0–8.0	6.0[c]

[a] Derived in part from Ref. 30.
[b] Typical values for settled wastewater.
[c] In secondary effluent.

organic compounds that remain after biological treatment are typically composed of low-molecular-weight polar organic compounds and complex organic compounds built around the benzene ring structure.

Until more data become available on dosages required and on the critical operating parameters, it is recommended that pilot plant studies be conducted when either chlorine or ozone is to be used for the oxidation of organics. Because ozone can be generated conveniently at treatment plants that use the pure-oxygen activated-sludge process, it is anticipated that its use will become more common at these locations in the future. Also, as the cost of generating ozone is reduced by the development of improved generating equipment, the widespread use of ozone will become inevitable.

12-9 REMOVAL OF DISSOLVED INORGANIC SUBSTANCES

As reported in Table 12-2, a number of different unit operations and processes have been investigated in various advanced wastewater treatment applications. Although many of them have proved to be technically feasible, other factors, such as cost, operational requirements, and aesthetic considerations, have not been favorable in some cases. Nevertheless, it is important that sanitary engineers be familiar with the more important operations and processes so that in any given situation they can consider all treatment possibilities. These important operations and processes are chemical precipitation, ion exchange, reverse osmosis, and electrodialysis. The following discussion of these operations and processes has been adapted from Ref. 19.

Chemical Precipitation

As discussed in Chaps. 7 and 8 and in Sec. 12-7, precipitation of phosphorus in wastewater is usually accomplished by the addition of coagulants, such as alum, lime, or iron salts, and polyelectrolytes. Coincidently with the addition of these chemicals for the removal of phosphorus is the removal that occurs of various inorganic ions, principally some of the heavy metals. Where both industrial and domestic wastes are treated together, it may be necessary to add chemicals to the primary settling facilities, especially if on-site pretreatment measures prove to be ineffective. When this is done, the use of anaerobic digestion for sludge stabilization will usually be impossible because of the toxicity of the precipitated heavy metals. As noted in Chap. 7, one of the disadvantages of chemical precipitation is that it usually results in a net increase in the total dissolved solids of the wastewater that is being treated.

Ion Exchange

Ion exchange is a unit process by which ions of a given species are displaced from an insoluble exchange material by ions of a different species in solution. The most widespread use of this process is in domestic water softening, where sodium ions from a cationic exchange resin replace the calcium and magnesium ions in the treated water, thus reducing the hardness. For the reduction of the total dissolved solids, both anionic and cationic exchange resins must be used. The wastewater is first passed through a cation exchanger where the positively charged ions are replaced by hydrogen ions. The cation exchanger effluent is then passed over an anionic exchange resin where the anions are replaced by hydroxide ions. Thus, the dissolved solids are replaced by hydrogen and hydroxide ions that react to form water molecules.

Ion exchangers are usually of the downflow, packed-bed column type. Wastewater enters the top of the column under pressure, passes downward through the resin bed, and is removed at the bottom. When the resin capacity is exhausted, the column is backwashed to remove trapped solids and then regenerated. The cationic exchange resin is regenerated with a strong acid, such as sulfuric or hydrochloric. Sodium hydroxide is the commonly used regenerant for the anion exchange resin.

This dimineralization process can take place in separate exchange columns arranged in series, or both resins can be mixed in a single reactor. Wastewater-application rates range from 0.20 to 0.40 $m^3/m^2 \cdot min$ (5 to 10 $gal/ft^2 \cdot min$). Typical bed depths are 0.75 to 2.0 m (2 to 6.5 ft). Total exchange capacities of commercially available resins are about 50,000 to 80,000 g/m^3 (as $CaCO_3$).

High concentrations to influent suspended solids can plug the ion-exchange beds, causing high head losses and inefficient operation. Resin binding can be caused by residual organics found in biological treatment effluents. Not all dissolved ions are removed equally; each resin is characterized by a selectivity series, and some dissolved ions at the end of the series are only partially removed.

Total dissolved-solids removals of 90 to 99 percent have been reported for the treatment of water supplies with a conventional two-stage exchanger system. Even higher removals are possible with mixed-bed exchangers. In reuse applications, treatment of a portion of the wastewater flow, followed by blending with the untreated flow, would possibly reduce dissolved solids levels to those of the carriage water.

Reverse Osmosis (Hyperfiltration)

Reverse osmosis is one of several demineralization techniques applicable to the production of a water suitable for reuse (see Fig. 12-26). This process has the added benefit of removing dissolved organics, which are less selectively removed by other demineralization techniques. The primary limitations of reverse osmosis are its high cost and a general lack of operating experience in the treatment of domestic wastewaters.

Reverse osmosis is a process in which water is separated from dissolved salts in solution by filtering through a semipermeable membrane at a pressure greater than the osmotic pressure caused by the dissolved salts in the wastewater. With existing membranes and equipment, operating pressures vary from atmospheric to 10,000 kN/m^2 (1000 lb_f/in^2).

Figure 12-26 Reverse osmosis process.

The basic components of a reverse osmosis unit are the membrane, a membrane support structure, a containing vessel, and a high-pressure pump. Cellulose acetate and nylon have been used as membrane materials. Four types of membrane support configurations are now in use: spiral-wound, tubular, multiple-plant, and hollow-fiber configurations. The tubular configuration is recommended for use with domestic wastewater effluents (see Fig. 12-26). Reverse osmosis units can be arranged either in parallel to provide adequate hydraulic capacity or in series to effect the desired degree of demineralization.

A very high-quality feed is required for efficient operation of a reverse osmosis unit. Pretreatment of a secondary effluent with filtration and carbon adsorption is usually necessary. The removal of iron and manganese also is sometimes necessary to decrease scaling potential. The pH of the feed should be adjusted to a range of 4.0 to 7.5 to inhibit scale formation.

Electrodialysis

In the electrodialysis process, ionic components of a solution are separated through the use of semipermeable ion-selective membranes. Application of an electrical potential between the two electrodes causes an electric current to pass through the solution, which, in turn causes a migration of cations toward the negative electrode and a migration of anions toward the positive electrode. Because of the alternate spacing of cation- and anion-permeable membranes, cells of concentrated and dilute salts are formed.

Wastewater is pumped through the membranes which are separated by spacers and assembled into stacks. The wastewater is usually retained for about 10 to 20 s in a single stack or stage. Dissolved solids removals vary with the (1) wastewater temperature, (2) amounts of electrical current passed, (3) type and amount of ions, (4) permselectivity of the membrane, (5) fouling and scaling potential of the wastewater, (6) wastewater flowrates, and (7) number and configuration of stages.

This process may be operated in either a continuous or a batch mode. The units can be arranged either in parallel to provide the necessary hydraulic capacity or in series to effect the desired degree of demineralization. Makeup water, usually about 10 percent of the feed volume, is required to wash the membranes continuously. A portion of the concentrate stream is recycled to maintain nearly equal flowrates and pressures on both sides of each membrane. Sulfuric acid is fed to the concentrate stream to maintain a low pH and thus minimize scaling.

Problems associated with the electrodialysis process for wastewater renovation include chemical precipitation of salts with low solubility on the membrane surface and clogging of the membrane by the residual colloidal organic matter in wastewater treatment plant effluents. To reduce membrane fouling, activated-carbon pretreatment, possibly preceded by chemical precipitation and some form of multimedia filtration, may be necessary.

12-10 ULTIMATE DISPOSAL OF CONTAMINANTS

Various methods of ultimate disposal for the waste from advanced treatment facilities have been studied, and a summary of the more important ones is given in Table 12-20. From this table, it can be concluded that there are three general possibilities: (1) land or ocean disposal, or both; (2) conversion and land or ocean disposal; (3) conversion, product recovery, and land or ocean disposal. Of these possibilities, the last, involving conversion and product recovery, is the most desirable but at present the least obtainable. Further, it should be noted that ocean discharge will not be allowed in the future.

The cost of ultimate disposal for the methods reported in Table 12-20 varies over an extremely wide range, depending on the geographic location and local conditions. Because this cost component may be the controlling factor governing the overall feasibility of an advanced wastewater treatment plan, ultimate disposal methods must be considered carefully.

DISCUSSION TOPICS AND PROBLEMS

12-1 A wastewater contains 10 mg/L of ammonia nitrogen and no organic carbon. If the plant flowrate is 10,000 m^3/d, estimate the methanol requirement and cell production in kilograms per day for the complete bacterial assimilation of ammonia.

12-2 A conventional activated-sludge plant treating 4000 m^3/d of wastewater is to be operated to produce a nitrified effluent. How would this be done? Assuming a nitrified effluent is produced

Table 12-20 Methods for the ultimate disposal of concentrated contaminants resulting from advanced wastewater treatment

Disposal method	Remarks
Liquid	
Evaporation ponds	Provisions must be made to prevent groundwater contamination.
Land application	Provisions must be made to prevent groundwater contamination.
Shallow-well injection	Provisions must be made to prevent groundwater contamination.
Deep-well injection	Porous strata and natural or artificial cavities should be available.
Landfill	Liquid can be used as a wetting agent to increase compaction.
Controlled evaporation	Applicability depends on liquid volume, power costs, and local conditions.
Ocean discharge	Transportation by truck, rail, or pipeline is required. Ocean discharge will not be allowed in the future.
Sludge	
Land application	Sludge may be pretreated to aid dewatering or to remove objectionable components.
Lagooning	Provisions must be made to prevent groundwater contamination.
Landfill	Sludge can be used as a wetting agent to increase compaction.
Recovery of products	Recovery depends on sludge characteristics, recovery technology, and costs.
Wet combustion	Heat value may be recovered for use. Disposal of ash is required.
Incineration	Concentration of sludge and disposal of ash is required.
Ocean discharge	Ocean discharge will not be allowed in the future.
Ash	
Landfill	Ash can be mixed with solid wastes to increase compacted density of landfill.
Soil conditioner	Applicability depends on waste characteristics.

containing 15 mg/L of nitrate nitrogen, 1.5 mg/L of nitrite nitrogen, and 2.0 mg/L of dissolved oxygen, compute the methanol requirement for denitrification. How will the activated-sludge effluent BOD affect the methanol requirement?

12-3 Based on a review of at least four articles dealing with the use of trickling filters for nitrification, recommend an appropriate loading factor or design approach to achieve complete nitrification using a trickling filter following an activated-sludge process. Assume that the activated-sludge process is designed to remove only the carbonaceous organic matter (after secondary treatment soluble $BOD_5 = 2.0$ mg/L and NH_3 as $N = 40$ mg/L). Cite the literature references reviewed.

12-4 A wastewater contains 40 mg/L of nitrate nitrogen (177 mg/L NO_3^-) and has a flowrate of 10,000 m^3/d. Effluent requirements have been set at 2 mg/L total nitrogen. Using a mean cell residence time of 15 d and a mixed-liquor concentration of 1500 mg/L, determine the volume of a complete mix reactor that will be required to provide the necessary treatment. Use the kinetic coefficients reported in Table 12-13. Also determine the rate of cell production and the methanol utilization rate assuming that the influent DO is equal to 5 mg/L. Assume that the final clarifier will produce an effluent with 10 mg/L suspended solids.

12-5 Design a three-stage biological treatment process to remove nitrogen by nitrification followed by denitrification. After preliminary treatment, 20,000 m^3/d of wastewater containing 150 mg/L of BOD_5 and 30 mg/L of ammonia nitrogen is to be treated. Give the detention times and volumes for the three reactors. Determine the oxygen and methanol requirements for nitrification and denitrification. Determine the sludge produced in each reactor.

12-6 Prepare a distribution diagram of the relative amounts of NH_3 and NH_4^+ (expressed as a percent) that would be present in a water sample at 25°C as a function of pH.

12-7 A treated wastewater contains 25 mg/L of ammonia nitrogen which is completely stripped from the wastewater with 5500 m^3 of air per cubic meter of wastewater in a stripping tower. The temperature of operation is 15°C. Evaluate the efficiency of this operation in terms of air-to-liquid ratio. What changes could be made to reduce the air-to-liquid ratio?

12-8 An anionic resin is to be used to remove nitrate ions by ion exchange. If the resin has an exchange capacity of 50 kg as $CaCO_3/m^3$, determine the theoretical and design volumes of resin to treat 1000 m^3/d of wastewater containing 20 mg/L of nitrate nitrogen. Assume that the leakage factor is 1.3.

12-9 Phosphorus is to be removed from a secondary effluent. The plant discharge requirements have been set at 1.0 mg/L. If the soluble phosphorus in the effluent is equal to 10 mg/L, estimate the alum dosage that will be required to achieve the desired degree of removal. If the concentration of the settled alum sludge is 6 percent and the specific gravity is 1.05, estimate the volume of sludge that must be disposed of per day if the plant flowrate is 40,000 m^3/d.

12-10 A wastewater is to be treated with activated carbon to remove residual COD. The following data were obtained from a laboratory adsorption study in which 1 g of activated carbon was added to a beaker containing 1 L of wastewater at selected COD values. Using these data, determine the more suitable isotherm (Langmuir or Fruendlich) to describe the data.

Initial COD, mg/L	Equilibrium COD, mg/L
140	10
250	30
300	50
340	70
370	90
400	110
450	150

12-11 Using the results from Prob. 12-10, determine the amount of activated carbon that would be required to treat a flow of 5000 m^3/d to a final COD concentration of 20 mg/L if the COD concentration after secondary treatment is equal to 120 mg/L.

12-12 A quantity of sodium-form ion-exchange resin (5 g) is added to a water containing 2 meq of potassium chloride and 0.5 meq of sodium chloride. Calculate the residual concentration of potassium if the exchange capacity of the resin is 4.0 meq/g of dry weight and the selectivity coefficient is equal to 1.46.

REFERENCES

1. Applebaum, S. B.: *Demineralization by Ion Exchange*, Academic, New York, 1968.
2. Chapman, R. F. (ed.): *Separation Processes in Practice*, Reinhold, New York, 1961.
3. Culp, R. L., and G. L. Culp: *Advanced Wastewater Treatment*, Van Nostrand Reinhold, New York, 1971.
4. Culp, G. L., and A. Slechta: *Nitrogen Removal from Sewage*, Final Progress Report, USPHS Demonstration Grant-26-01, 1966.
5. *Development of a Process to Remove Carbonaceous, Nitrogenous, and Phosphorus Materials from Anaerobic Digester Supernatant and Related Process Streams*, Progress Report: Phase 1, prepared for Federal Water Pollution Control Administration, Environmental Engineering Laboratories, FMC Corporation, Santa Clara, 1969.

6. Eliassen, R., and G. E. Bennett: Anion Exchange and Filtration Techniques for Wastewater Renovation, *J. WPCF*, vol. 39, no. 10, part 2, 1967.
7. Eliassen, R., B. M. Wyckoff, and C. D. Tonkin: Ion Exchange for Reclamation of Reusable Supplies, *J. AWWA*, vol. 57, no. 9, 1965.
8. Haug, R. T., and P. L. McCarty: Nitrification with the Submerged Filter, *J. WPCF*, vol. 44, no. 10, 1972.
9. Higgins, I. J., and R. G. Burns: The Chemistry and Microbiology of Pollution, Academic, London, 1975.
10. Kavanaugh, M., et al.: Contact Filtration for Phosphorus Removal, *J. WPCF* (in press).
11. Kugelman, I. J.: Status of Advanced Waste Treatment, in H. W. Gehm and J. I. Bregman (eds.), *Handbook of Water Resources and Pollution Control*, Van Nostrand, New York, 1976.
12. Lackey, J. B., and C. N. Sawyer: Plankton Productivity of Certain Southeastern Wisconsin Lakes as Related to Fertilization, *Sewage Works J., WPCF*, vol. 17, p. 573, 1945.
13. Levin, G. V., and J. Shapiro: Metabolic Uptake of Phosphorus by Wastewater Organisms, *J. WPCF*, vol. 37, no. 6, 1965.
14. McCarty, P. L.: Biological Processes for Nitrogen Removal: Theory and Application, *Proceedings Twelfth Sanitary Engineering Conference*, University of Illinois, Urbana, 1970.
15. McCarty, P. L., L. Beck, and P. St. Amant: Biological Denitrification of Wastewaters by Addition of Organic Materials, *Proceedings of the 24th Purdue Industrial Waste Conference*, Lafayette, Ind., 1969.
16. Menar, A. B., and D. Jenkins: The Fate of Phosphorus in Waste Treatment Processes: The Enhanced Removal of Phosphate by Activated Sludge, *Proceedings of the 24th Purdue Industrial Waste Conference*, Lafayette, Ind., 1969.
17. Mercer, B. W., et al.: Ammonia Removal from Secondary Effluents by Selective Ion Exchange, *J. WPCF*, vol. 42, no. 10, 1969.
18. Metcalf & Eddy, Inc.: *Advanced Wastewater Treatment for the Dulles/Herndon/Reston Area, Fairfax County, Virginia, Report to the County of Fairfax*, Va., Boston, 1975.
19. Metcalf & Eddy, Inc.: *Report to the National Commission on Water Quality on Assessment of Technologies and Costs for Publicly Owned Treatment Works*, vol. 3, prepared under Public Law 92–500, Boston, 1975.
20. Mulbarger, M. C.: Nitrification and Denitrifiration in Activated Sludge Systems, *J. WPCF*, vol. 43, no. 10, 1971.
21. Neptune Microfloc, Inc.: Product Literature, Corvallis, Ore., 1972.
22. *Process Design Manual for Nitrogen Control*, U.S. Environmental Protection Agency, Office of Technology Transfer, Washington, D.C., October 1975.
23. *Process Design Manual for Phosphorus Removal*, U.S. Environmental Protection Agency, Office of Technology Transfer, Washington, D.C., April 1976.
24. Sawyer, C. N.: Activated Sludge Oxidation, V: The Influence of Nutrition in Determining Activated Sludge Characteristics, *Sewage Works J.*, vol. 12, no. 1, 1940.
25. Schroeder, E. D.: *Water and Wastewater Treatment*, McGraw-Hill, New York, 1977.
26. Slechta, A. F., and G. L. Culp: Water Reclamation Studies at the South Tahoe Public Utility District, *J. WPCF*, vol. 39, no. 5, 1967.
27. Tchobanoglous, G.: Physical and Chemical Processes for Nitrogen Removal—Theory and Application, *Proceedings Twelfth Sanitary Engineering Conference*, University of Illinois, Urbana, 1970.
28. Thimann, K. V.: *The Life of Bacteria*, 2d ed., Macmillan, New York, 1963.
29. Treybal, R. E.: *Mass-Transfer Operations*, McGraw-Hill, 2d ed., New York, 1967.
30. White, G. C.: *Handbook of Chlorination*, Van Nostrand Reinhold, New York, 1972.
31. Wild, H. E., C. N. Sawyer, and T. C. McMahon: Factors Affecting Nitrification Kinetics, *J. WPCF*, vol. 43, no. 9, 1971.
32. Young, J. C., E. R. Baumann, and D. J. Wall: Packed-Bed Reactors for Secondary Effluent BOD and Ammonia Removal, *J. WPCF*, vol, 47, no. 1, January 1975.

CHAPTER

THIRTEEN

LAND-TREATMENT SYSTEMS

Land treatment of wastewater involves the use of plants, the soil surface, and the soil matrix for wastewater treatment. Although land treatment of wastewater has been practiced for centuries, its full potential has only recently been recognized in the field of wastewater engineering. The principal reason for this is the widespread research and development activity generated as a result of the emphasis placed in Public Law 92-500 on water reuse, nutrient recycling, and the use of wastewater for crop production.

The specific topics covered in this chapter are: (1) the development of land-treatment systems, (2) fundamental considerations in land-treatment systems, (3) irrigation systems, (4) rapid-infiltration systems, (5) overland-flow systems, (6) other systems, and (7) land application of sludge. Disposal of sludge on land in sanitary landfills was discussed in Chap. 11. Information on costs of land treatment of wastewater is available in Ref. 46. A design manual on land treatment was published by the U.S. Environmental Protection Agency in 1977 [49].

13-1 DEVELOPMENT OF LAND-TREATMENT SYSTEMS

An overview of land-treatment systems is provided in this section. The historical practice is traced, and current land-treatment practice is described.

Historical Practice

Evidence of land-treatment systems in Western civilization extends back as far as ancient Athens [41]. A wastewater irrigation system in Bunzlaw, Germany, is reported to have been in operation for over 300 years, beginning in 1559 [23].

The greatest proliferation of land-treatment systems occurred in Europe in the second half of the nineteenth century. Pollution of many rivers had reached unacceptable levels, and disposal of untreated wastewater and sludge on the land was the only feasible means of treatment available at the time. "Sewage farming," the practice of transporting untreated wastewater into rural areas for irrigation and disposal, was commonly used by many European cities, including some of those shown in Table 13-1. In the 1870s, the practice was recognized in England as an acceptable form of treatment [33]. As urban areas expanded and more intensive treatment processes became available, many of these older systems were abandoned, usually because of pressures to develop the land.

Land-treatment systems in the United States also date from the 1870s [50]. As in Europe, sewage farming became relatively common as a first attempt to control water pollution. In the first half of the twentieth century, these systems were generally replaced either by in-plant treatment systems or by (1) managed farms where treated wastewater was used for crop production, (2) landscape irrigation sites, or (3) groundwater recharge sites. These newer land-treatment systems tended to predominate in the West where the resource value of wastewater was an added advantage.

The number of United States municipalities using land treatment increased from 304 in 1940 to 571 (serving a population of 6.6 million) in 1972, but this total still represents only a small percentage of the estimated 15,000 total municipal treatment facilities [65].

Table 13-1 Selected early land-treatment systems

Location	Date started	Type of system	Area, ha	Flow, m³/s
International				
Berlin, Germany	1874	Sewage farm	2,720	
Braunschweig, Germany	1896	Sewage farm	4,400	0.7
Croydon-Beddington, England	1860	Sewage farm	252	0.2
Leamington, England	1870	Sewage farm	160	0.04
Melbourne, Australia	1893	Irrigation	4,160	2.19
Mexico City, Mexico	1900	Irrigation	44,800	24.97
Paris, France	1869	Irrigation	640	3.46
Wroclaw, Poland	1882	Sewage farm	800	1.23
United States				
Calumet City, Mich.	1888	Rapid infiltration	4.8	0.05
Ely, Nev.	1908	Irrigation	160.0	0.07
Fresno, Calif.	1891	Irrigation	1,600.0	1.14
San Antonio, Tex.	1895	Irrigation	1,600.0	0.88
Vineland, N.J.	1901	Rapid infiltration	5.6	0.04
Woodland, Calif.	1889	Irrigation	96.0	0.18

Note: ha × 2.4711 = acre
m³/s × 22.8245 = Mgal/d

In recent years, much effort has been spent on developing land-treatment technology and improving methods of control. The various types of land-treatment systems have become accepted as wastewater-management techniques that should be considered equally with any others.

Land Treatment of Wastewater

The three principal processes of land treatment of wastewater, shown schematically in Fig. 13-1, are irrigation, rapid infiltration, and overland flow. Other processes, which are less widely used and generally less adaptable to large-scale use, include

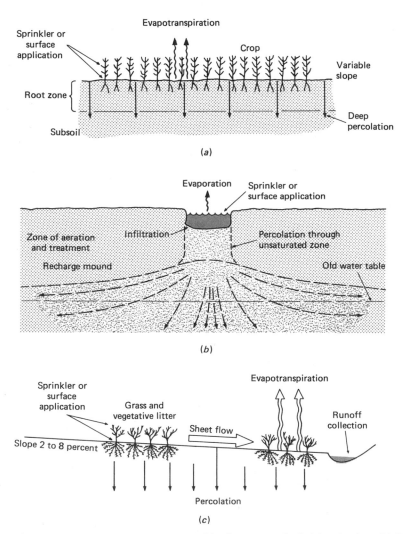

Figure 13-1 Three principal processes of land treatment [45]. (*a*) Irrigation. (*b*) Rapid infiltration. (*c*) Overland flow.

Table 13-2 Comparison of design features for alternative land-treatment processes

Feature	Irrigation	Rapid infiltration	Overland flow	Wetland application	Subsurface application
Application techniques	Sprinkler or surface[a]	Usually surface	Sprinkler or surface	Sprinkler or surface	Subsurface piping
Annual application rate, m	0.6–6.0	6–120	3–20	1–30	2–25
Field area required, ha[b]	22–225	1–22	10–44	4–113	5–56
Typical weekly application rate, cm	2.5–10	10–210	6–15[c] 15–40[d]	2.5–60	5–50
Minimum preapplication treatment provided	Primary sedimentation[e]	Primary sedimentation	Screening and grit removal	Primary sedimentation	Primary sedimentation
Disposition of applied wastewater	Evapotranspiration and percolation	Mainly percolation	Surface runoff and evapotranspiration with some percolation	Evapotranspiration, percolation, and runoff	Percolation with some evapotranspiration
Need for vegetation	Required	Optional	Required	Required	Optional

[a] Includes ridge and furrow and border strip.
[b] Field area in hectares not including buffer area, roads, or ditches for 0.044 m³/s (1 Mgal/d) flow.
[c] Range for application of screened wastewater.
[d] Range for application of lagoon and secondary effluent.
[e] Depends on the use of the effluent and the type of crop.

Note: cm × 0.3937 = in
m × 3.2808 = ft
ha × 2.4711 = acre

Table 13-3 Comparison of site characteristics for land-treatment processes

Characteristics	Irrigation	Rapid infiltration	Overland flow	Wetland application
Climatic restrictions	Storage often needed for cold weather and precipitation	None (possibly modify operation in cold weather)	Storage often needed for cold weather	Storage may be needed for cold weather
Depth to groundwater, m	0.6–0.9 (minimum)	3.0 (lesser depths acceptable where underdrainage provided)	Not critical	Not critical
Slope	Less than 20% on cultivated land; less than 40% on noncultivated land	Not critical; excessive slopes require much earthwork	Finish slopes 2–8%	Usually less than 5%
Soil permeability	Moderately slow to moderately rapid	Rapid (sands, loamy sands)	Slow (clays, silts, and soils with impermeable barriers)	Slow to moderate

Note: m × 3.2808 = ft

wetland application, subsurface application, and aquaculture. Comparisons of typical design features, major site characteristics, and the expected quality of treated water are presented in Tables 13-2, 13-3, and 13-4, respectively.

Irrigation Irrigation, the predominant land-treatment process in use today, involves the application of effluent to the land for treatment and for meeting the growth needs of plants. The applied effluent is treated by physical, chemical,

Table 13-4 Comparison of expected quality of treated water from land-treatment processes, mg/L

Constituent	Irrigation[a]		Rapid infiltration[b]		Overland flow[c]	
	Average	Maximum	Average	Maximum	Average	Maximum
BOD	<2	<5	2	<5	10	<15
Suspended solids	<1	<5	2	<5	10	<20
Ammonia nitrogen as N	<0.5	<2	0.5	<2	0.8	<2
Total nitrogen as N	3	<8	10	<20	3	<5
Total phosphorus as P	<0.1	<0.3	1	<5	4	<6

[a] Percolation of primary or secondary effluent through 1.5 m of soil.
[b] Percolation of primary or secondary effluent through 4.5 m of soil.
[c] Runoff of comminuted municipal wastewater over about 45 m of slope.
Note: m × 3.2808 = ft

and biological means as it seeps into the soil. Effluent can be applied to crops or vegetation (including forestland) either by sprinkling (Fig. 13-2) or by surface techniques for purposes such as avoidance of surface discharge of nutrients; economic return from use of water and nutrients to produce marketable crops; water conservation by exchange when lawns, parks, or golf courses are irrigated; and preservation and enlargement of greenbelts and open space.

Where water for irrigation is valuable, crops can be irrigated at consumptive use rates of 2.5 to 7.5 cm/week (1 to 3 in/week), depending on the crop, and the economic return from the sale of the crop can be balanced against the increased cost of the land and distribution system. On the other hand, where water for irrigation is of little value, hydraulic loadings can be maximized (provided that renovated water quality criteria are met), thereby minimizing system costs. Crops grown under high-rate irrigation rates of 6 to 10 cm/week (2.4 to 4 in/week) are usually water-tolerant grasses with lower potential for economic return but with high nutrient-uptake capacities.

Rapid infiltration In rapid-infiltration systems, effluent is applied to the soil at high rates (10 to 210 cm/week or 4 to 84 in/week) by spreading in basins or by sprinkling. Treatment occurs as the water passes through the soil matrix. System objectives can include (1) groundwater recharge, (2) natural treatment followed by pumped withdrawal or underdrains for recovery, and (3) natural treatment with renovated water moving vertically and laterally in the soil and recharging a surface watercourse.

Where groundwater quality is being degraded by salinity intrusion, groundwater recharge can be used to reverse the hydraulic gradient and protect the existing groundwater. Where existing groundwater quality is not compatible with

Figure 13-2 Sprinkler irrigation at Pleasanton, Calif.

expected renovated water quality, or where existing water rights control the discharge location, a return of renovated water to surface water can be designed, using pumped withdrawal, underdrains, or natural drainage.

Overland flow Overland flow is essentially a biological treatment process in which wastewater is applied over the upper reaches of sloped terraces and allowed to flow across the vegetated surface to runoff collection ditches. Renovation is accomplished by physical, chemical, and biological means as the wastewater flows in a thin sheet down the relatively impervious slope.

Overland flow can be used either as a secondary treatment process, where discharge of a nitrified effluent low in BOD is acceptable, or as an advanced wastewater treatment process. The latter objective will allow higher rates of application—15 to 40 cm/week (6 to 16 in/week)—depending on the degree of treatment required. Where a surface discharge is prohibited, runoff can be recycled or applied to the land in irrigation or rapid-infiltration systems.

Other systems The use of wetlands and aquaculture for wastewater treatment has received much attention recently because of the large areas of existing marshlands and wetlands and because of their ability to influence water quality. Both artificial and existing wetlands have been studied using untreated as well as secondary effluents [56, 62]. The use of water hyacinths [4] and various combinations of plants and fish [9] to effect improvements in water quality has also been reported.

Land Application of Sludge

Residue from treatment of municipal wastewater can be applied to the land to serve as a fertilizer and a soil conditioner. The practice is widespread in rural and noncoastal communities [10]. Loading rates are generally limited by nitrogen loadings. Long-term use of the selected site may be limited by the trace-element content of the sludge.

13-2 FUNDAMENTAL CONSIDERATIONS IN LAND-TREATMENT SYSTEMS

Knowledge of wastewater characteristics, treatment mechanisms, vegetation, and public health requirements is fundamental to the successful design and operation of land treatment systems.

Wastewater Characteristics and Treatment Mechanisms

The soil surface and profile can provide physical and chemical treatment of wastewater and a habitat for microorganisms that can provide biological treatment. The capability of land treatment systems to remove organics, nitrogen, phosphorus, exchangeable cations, trace elements, and microorganisms from applied wastewater depends on a variety of factors.

Organic matter For organic matter the soil is a highly efficient biological treatment system. Organic matter is filtered by grass, litter, and topsoil and is reduced by biological oxidation. Because high organic loadings may create anaerobic conditions in the soil matrix and result in the production of odors, an intermittent loading schedule is used. This allows air to penetrate the soil and supply oxygen to the bacteria that oxidize the organic matter.

Nitrogen Nitrogen can be removed in land treatment by crop uptake and harvest and by denitrification. In irrigation systems, crop uptake is the main mechanism, and the quantities of uptake vary widely according to the crop planted. In overland flow, denitrification is most important, but crop uptake is also significant. In rapid infiltration, the only significant nitrogen removal process is denitrification, which must be managed by using relatively long flooding cycles, or wastewater with ample BOD, to create anaerobic conditions in the soil profile. The most important chemical and biological reactions that occur when nitrogen is applied to a land-treatment system are described in Fig. 13-3.

    The relationship between organic matter, as measured by BOD, and nitrogen in wastewater is important to the removal of both constituents. Typical ratios for various types of wastewaters and their effects on carbon assimilation and nitrogen availability to soil and plants are given in Table 13-5.

Phosphorus The major phosphorus-removal processes in land-treatment systems are chemical precipitation and adsorption, although plants do take up some amounts. The phosphorus, which occurs mainly in the form of orthophosphates, is adsorbed by clay minerals and certain organic soil fractions in the soil matrix. Chemical precipitation with calcium (at neutral to alkaline pH values) and iron or aluminum (at acid pH values) occurs at a slower rate than adsorption, but

Table 13-5 Typical ratios of BOD to nitrogen for various types of wastewaters

Type of wastewater	BOD/N ratio	Effects on biological decomposition
Many food-processing wastewaters	80 to 100	Carbon assimilation relatively slow; available soil nitrogen immobilized and made less available to plants
Livestock manure	20	Carbon assimilation not limited; nitrogen available to plants
Untreated municipal wastewater	5	Nitrogen in excess of bacterial needs for carbon assimilation; released to soil and plants
Secondary effluent	1 to 2	Carbon limited for denitrification; nitrogen released to soil and plants

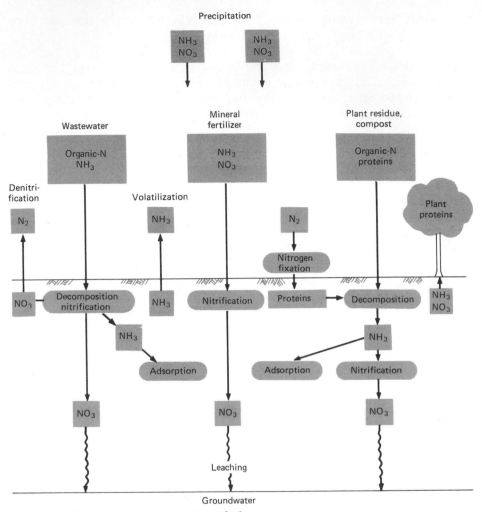

Figure 13-3 Nitrogen transformations in soil [45].

it is equally important. Adsorbed phosphorus can be held tightly and is generally resistant to leaching. Phosphorus removal depends on the type of land treatment system used.

Exchangeable cations Exchangeable cations, particularly sodium, calcium, and magnesium ions, deserve special consideration, because high sodium concentrations in clay-bearing soils disperse soil particles and decrease soil permcability. To determine the sodium hazard, the sodium adsorption ratio (SAR) was developed by the U.S. Department of Agriculture Salinity Laboratory, and is defined as follows:

$$SAR = \frac{Na}{\sqrt{(Ca + Mg)/2}} \qquad (13\text{-}1)$$

where Na = sodium, meq/L
    Ca = calcium, meq/L
    Mg = magnesium, meq/L

High SAR values (>9) may adversely affect the permeability of fine-textured soils. Occasionally, high sodium concentrations in soil can be toxic to plants, although the effects on permeability generally occur first.

Trace elements Although many trace elements are essential for plant growth, some become toxic at higher levels to both plant life and microorganisms. Retention of trace elements, especially heavy metals, in the soil matrix occurs mainly through sorption (the term includes adsorption and precipitation reactions) and ion exchange. The retention capacity for most metals in most soils is generally high, especially for pH values above 7. Under low pH conditions, some metals can be leached out of soils.

Microorganisms Bacterial removal mechanisms common to most methods of land treatment include straining, die-off, sedimentation, entrapment, and adsorption. In wastewater sprinkling systems, some bacteria are intercepted by vegetation where desiccation, die-off, and predators eliminate them.

Vegetation

Plants in land-treatment systems are used to (1) take up nitrogen and phosphorus from the applied wastewater, (2) maintain and increase water-intake rates and soil permeability, (3) reduce erosion, and (4) serve as a medium for micro-organisms (overland flow). Their principal use in irrigation systems is for nutrient removal. Some common crops and typical uptake rates for nitrogen and phosphorus are presented in Table 13-6. For rapid-infiltration systems, the primary requirement is for a water-tolerant plant that will help to maintain high infiltration rates. For overland-flow systems, the need is for a mixture of warm- and cool-season perennial grasses with high water tolerances.

Public Health

Aspects of public health that are related to land treatment include (1) bacterio-logical agents and the possible transmission of disease to higher biological forms, including humans; (2) chemicals that may reach groundwater and pose risks to health if ingested; and (3) crop quality when crops are irrigated with wastewater effluents.

Bacteriological agents The survival of pathogenic bacteria and viruses in sprayed aerosol droplets, on and in the soil, and the effects on workers have received considerable attention [55, 61]. It is important to realize that any connection between pathogens applied to land through wastewater and the contraction of disease in animals or humans would require a long and complex path of

Table 13-6 Nitrogen and phosphorus removal by crops [24, 45]

Type of vegetation	Description	Dry matter yield, kg/ha	Nitrogen, kg/ha	Phosphorus, kg/ha
Forage crops				
Alfalfa	Perennial legume	15,770	504	39
Bromegrass	Cool-season perennial	9856	186	32
Coastal Bermuda grass	Warm-season perennial	19,712	560	80
Reed canary grass	Cool-season perennial	13,000	350	36
Ryegrass	Cool-season annual or perennial	8030	235	67
Sweet clover	Biennial legume	7375	177	18
Tall fescue	Cool-season perennial	6854	133	30
Field crops				
Barley	Winter annual	......	70	17
Corn	Summer annual	......	174	19
Cotton	Summer annual	......	74	13
Milomaize	Summer annual	......	91	16
Soybeans	Summer annual legume	......	105	12

Note: kg/ha × 0.8922 = lb/acre

epidemiological events. Nevertheless, questions have been raised, concern exists, and precautions should be taken in dealing with possible disease transmission.

Sprinklers used to apply effluents produce a mist that may be transported by wind currents. Mist droplets that are extremely small in both dimension and mass are referred to as aerosols. Aerosols are tiny airborne colloidal-like droplets of liquid (0.01 to 50 μ in diameter). Aerosols generated in connection with inadequately disinfected wastewater may contain active bacteria and viruses. However, Sorber reported that aerosolization occurred for only about 0.3 percent of the wastewater being sprinkled, as determined by fluorescein tracer tests [61].

Studies of aerosol and mist travel have been conducted using untreated wastewater as well as disinfected secondary effluent [55]. Although bacteria traveled farther in aerosols from undisinfected wastewater, the reported maximum distances ranged from 30 to 200 m (100 to 600 ft). Generally, the wind travel of bacteria increases with increases in relative humidity and wind speed, and with decreases in ultraviolet radiation [55].

The need for buffer zones or disinfection to minimize public health risks from aerosols should be assessed on a case-by-case basis considering (1) the degree of public access to the site, (2) the size of the irrigated area, (3) the feasibility of providing buffer zones or plantings of trees or shrubs, and (4) the prevailing climatic conditions. Buffer zones of 20 to 60 m (65 to 200 ft) from roads, property lines, and buildings are typical. Alternatives to buffer zones include plantings of trees; use of sprinklers that spray downward or at low trajectories; and ceasing sprinkling, or at least sprinkling on interior portions of the site, during high winds.

Groundwater quality The U.S. Environmental Protection Agency regulations on interim primary drinking-water standards are listed in Table 13-7. Because nitrate has been demonstrated to be the causative agent of methemoglobinemia in children, its concentration in drinking water is limited in the standards to 10 mg/L as nitrate nitrogen [60]. As indicated, nitrogen can be managed in land-treatment systems by promoting crop uptake or denitrification to maintain the nitrate nitrogen below 10 mg/L in the groundwater.

The trace metals are usually removed from the percolating water by adsorption or chemical precipitation within the first few meters of soil. In most cases, however, the concentrations of trace elements in the untreated wastewater prior to application are usually below those limits set forth in Table 13-7.

Bacterial removal from effluents passing through fine soils is quite complete; it may also be extensive in the coarse, sandy soil used for rapid infiltration systems. Fractured rock or limestone cavities may provide a passage for bacteria that can travel several hundred feet from the point of application. This situation can be avoided by proper geological investigations during site selection.

Table 13-7 Proposed regulations of the U.S. Environmental Protection Agency on interim primary drinking-water standards, 1975 [73]

Constituent or characteristic	Value	Reason for standard
Physical Turbidity, units	1[a]	Aesthetic
Chemical, mg/L		
Arsenic	0.05	Health
Barium	1.0	Health
Cadmium	0.01	Health
Chromium	0.05	Health
Fluoride	1.4 2.4[b]	Health
Lead	0.05	Health
Mercury	0.002	Health
Nitrates as N	10	Health
Selenium	0.01	Health
Silver	0.05	Cosmetic
Total coliform, per 100 mL	1	Disease
Pesticides, mg/L		
Endrin	0.0002	Health
Lindane	0.004	Health
Methoxychlor	0.1	Health
Toxaphene	0.005	Health
2,4–D	0.1	Health
2,4,5–TP	0.01	Health

[a] Five mg/L of suspended solids may be substituted if it can be demonstrated that it does not interfere with disinfection

[b] Dependent upon temperature; higher limits for lower temperatures.

Note: mg/L = g/m^3

Table 13-8 Crop quality compared to suggested tolerance levels for heavy metals at site in Melbourne, Australia

| Constituent | Grasses at Melbourne[a] | | Suggested tolerance levels, ppm[b] |
	Control sample, ppm	Wastewater, irrigated, ppm	
Cadmium	0.77	0.89	3
Cobalt	<0.64	0.64	5
Copper	6.5	12	150
Iron	970	361	750
Manganese	149	49	300
Nickel	2.7	4.9	3
Lead	<2.5	<2.5	10
Zinc	50	63	300

[a] From Ref. 27.
[b] From Ref. 40.

Crop quality Few data exist on the quality of crops grown with wastewater effluents compared to the quality of crops grown with water and nutrients obtained from other sources. Data on crops grown using sludge are not usually applicable because trace-element concentrations are several times higher in sludges than in effluents.

Data presented in Table 13-8 are for grasses at the land-application site in Melbourne, Australia [27]. The authors collected green forage samples from grasses on nonirrigated (control) areas and in areas irrigated for 60 yr with wastewater that had been treated using limited settling. All constituents in the plants, except for nickel, are below the suggested tolerance limits for grasses proposed by Melsted [40]. The authors concluded that because nickel is poorly absorbed from ordinary diets and is relatively nontoxic, it would not pose a health problem to livestock [27].

13-3 IRRIGATION SYSTEMS

Irrigation—the controlled discharge of effluent, by sprinkling or surface spreading, onto land to support plant growth—is generally capable of producing the best results of all the land-treatment systems in terms of reliability and treated water quality. The wastewater is accounted for in plant uptake and evaporation (evapotranspiration) and in percolation through the soil. The detailed discussion in this section deals with (1) design objectives, (2) site selection, (3) preapplication treatment, (4) climate and storage, (5) loading rate, (6) land requirements, (7) crop selection, (8) distribution methods, (9) application cycles, (10) underdrainage, (11) surface runoff control, and (12) a case study.

Design Objectives

The principal objective in irrigation systems is treatment of the applied wastewater. Additional objectives can include: (1) economic return from use of water and nutrients to produce marketable crops, (2) water conservation, and (3) preservation and enlargement of greenbelts and open space. Treated effluent is generally used to achieve the latter two objectives.

When the objective is treatment, the hydraulic loading is usually limited either by the infiltration capacity of the soil or the nitrogen loading. If the hydraulic capacity of the site is limited by a relatively impermeable subsurface layer or by a high groundwater table, underdrains can be installed to increase the allowable loading. Grasses are usually selected for the vegetation because of their high nitrogen uptake and water tolerance.

When crop yields and economic returns are emphasized, hydraulic loadings are limited by the crop needs for water and nitrogen. In the western United States, application rates are generally 2 to 8 cm/week (0.8 to 3.1 in/week), reflecting the water needs of crops. Off-season storage may be used to allow irrigation of the maximum area in the summer. In the eastern United States, crops such as corn are selected to maximize economic returns.

Water conservation is an objective that can be achieved mainly in arid areas. Parks, golf courses, or landscape areas irrigated with potable water can be irrigated with treated effluent. The potable water previously used for irrigation can be conserved or used for other purposes.

When the irrigation objective is the preservation and enlargement of greenbelt and open space, the site is usually chosen on the basis of land-use planning considerations. Secondary effluent is generally used, and loading rates must be compatible with existing or planned vegetation.

Site Selection

The major factors and general criteria for site selection are listed in Table 13-9. Soil drainability is perhaps the primary factor because, coupled with the type of crop or vegetation selected, it directly affects the liquid loading rate. A moderately permeable soil capable of infiltrating approximately 5 cm/d (2 in/d) or more on an intermittent basis is preferable. In general, soils ranging from clays to sandy loams are suitable for irrigation. For cropland, agricultural extension service advisers or adjacent farmers should be consulted. For forest or landscape irrigation, university specialists should be consulted.

For crop irrigation, slopes should be limited to about 20 percent or less, depending on the type of farm equipment to be used. Forested hillsides up to 40 percent in slope have been irrigated successfully with sprinklers [54]. A suitable site for wastewater irrigation would preferably be located in an area where contact between the public and the irrigation water and land is controlled. In landscape irrigation, however, it is often difficult to meet this condition.

Table 13-9 Site-selection factors and criteria for effluent irrigation

Factor	Criterion
Soil	
Type	Loamy soils are preferred, but most soils from sands to clays are acceptable.
Drainability	Well-drained soil is preferred; consult experienced agricultural advisors.
Depth	Uniformly 1.5 to 1.8 m (5 to 6 ft) or more throughout sites is preferred.
Groundwater	
Depth to groundwater	A minimum of 1.5 m (5 ft) is preferred. Drainage to obtain this minimum may be required.
Groundwater control	Control may be necessary to ensure renovation if the water table is less than 3.1 m (10 ft) from the surface.
Groundwater movement	Velocity and direction of movement must be determined.
Slopes	Up to 20 percent are acceptable with or without terracing.
Underground formations	Formations should be mapped and analyzed with respect to interference with groundwater or percolating water movement.
Isolation	Moderate isolation from public is preferred; the degree of isolation depends on wastewater characteristics, method of application, and crop.
Distance from source of wastewater	An appropriate distance is a matter of economics.

The total amount of land required for a land-application system—though highly variable—primarily depends on application rates. Consideration must also be given to requirements for buffer zones, storage, access roads, dikes, flood-control structures, terraces, fencing, and space for an administration building. Because application rates vary significantly for each land application method, only typical ranges will be discussed here.

The land required for treatment in a 3800 m³/d (1 Mgal/d) irrigation system can vary from 22 to 226 ha (55 to 560 acres); the typical range is 40 to 80 ha (100 to 200 acres). Buffer zones (unirrigated strips of land around sites) are frequently required for sprinkling systems and can increase the total land requirement significantly, particularly in cases where the land requirement is otherwise small.

Preapplication Treatment

The degree of treatment required prior to land application depends on a number of factors, including public health regulations, the loading rate with respect to critical wastewater characteristics, and the desired effectiveness and dependability of the physical equipment. Biological stabilization may be needed to prevent odors from developing in storage ponds when long-term winter storage is required. Costs for increased treatment should be weighed against designing the storage ponds as stabilization ponds.

The required degree of preapplication treatment for crop irrigation will normally be based on a consideration of the state public health regulations or guidelines. Factors that should be considered in assessing the need for preapplication treatment include the type of crop grown, the intended use of the crop, the degree of contact by the public with the effluent, and the method of application. State regulations for preapplication treatment differ considerably. For example, the irrigation of certain crops to be eaten raw by humans may require either secondary or advanced wastewater treatment with disinfection, or it may be prohibited altogether [42].

Climate and Storage

For irrigation of annual crops, wastewater application is restricted to the growing season, and storage may be required for a period ranging from 1 to 3 months in moderate climates and 4 to 7 months in cold northern states. Irrigation of perennial grasses or double cropping annual crops can extend the period of application. Periods of snow cover and subfreezing conditions may limit the application to perennial grasses and forest land. This depends on water quality criteria for the percolating water because the treatment efficiency of soil systems decreases under these conditions.

With regard to temperature, it has been shown that irrigation systems can usually operate successfully below 0°C (32°F). A forest irrigaton system at Dover, Vt., operates at temperatures down to -12.2°C (10°F) [14].

The maximum precipitation allowed before application is stopped depends primarily on the maximum infiltration rates at the site and on storm water–runoff considerations. At sites with limited infiltration rates and where storm water runoff is of concern, as little as 0.5 cm/d (0.2 in/d) of precipitation may require suspension of application. In other cases, full-time operation may be assumed during days with 2.54 cm (1.0 in) or more of precipitation.

Irrigation of perennial grasses and forest lands can be conducted during snow cover provided that renovation is not critical under these conditions. When the built-up ice and snow melt, the runoff must be contained.

Loading Rate

To determine which characteristics of the wastewater will be limiting, balances should be made for water, nitrogen, phosphorus, organic matter, and other constituents of abnormally high concentration. From these balances, a loading rate can be established for each parameter. Each loading rate should then be used in calculating the required land area, and the critical loading rate is the one requiring the largest field area.

Hydraulic loading The elements to be considered in determining hydraulic loading are the quantity of effluent to be applied, precipitation, evapotranspiration, percolation, and runoff. For irrigation systems, the amount of effluent applied plus precipitation should equal the evapotranspiration plus a limited amount of percolation. In most cases, surface runoff from fields irrigated with municipal effluent is not allowed or must be controlled. The water balance will be

$$\frac{\text{Design}}{\text{precipitation}} + \frac{\text{wastewater}}{\text{applied}} = \text{evapotranspiration} + \text{percolation} \quad (13\text{-}2)$$

Seasonal variations in each of these values should be taken into account by evaluating the water balance for each month as well as the annual balance, as illustrated in Example 13-1.

The value for design precipitation should be determined from a frequency analysis of wetter-than-normal years. The wettest year in 10 is suggested as reasonable in most cases, but it is prudent to check the water balance using the range of precipitation amounts that may be encountered. For purposes of evaluating monthly water balances, the design annual precipitation can often be distributed over the year by means of the average distribution, which is the average percentage of the total annual precipitation that occurs in each month. Again, the range of monthly values that may be encountered should be analyzed, especially for the months when the storage reservoir is full.

Evapotranspiration will also vary from month to month, but the total for the year should be relatively constant. The amount of water lost to evapotranspiration each month should be entered in Eq. 13-2.

Percolation includes that portion of the water which, after infiltration into the soil, flows through the root zone and eventually becomes part of the groundwater. The percolation rate used in the design should be determined on the basis of a number of factors, including soil chacteristics, underlying geologic conditions, groundwater conditions, and the length of drying period required for satisfactory crop growth and wastewater renovation. The principal factor affecting the design percolation rate is the permeability or hydraulic conductivity of the most slowly permeable layer in the soil profile.

When irrigating in arid climates, it is necessary to remove the salts that accumulate in the root zone as a result of evaporation. Some amount of percolation is necessary to accomplish this leaching. Ayers [2] has calculated the leaching requirements for various crops, depending on crop tolerances and total dissolved solids in the effluent.

Example 13.1 Determination of the water balance for an irrigation system Determine the water balance for an irrigation system using the following assumptions:
1. The design precipitation is for the wettest year in 10, with average monthly distribution (see Table 13-10, column 5).
2. Use average monthly evapotranspiration rates (see Table 13-10, column 1).
3. The site is mostly flat and level.

4. The soil is a deep, sandy loam with a percolation rate of 25 cm/month (10 in/month) (see Table 13-10, column 3).
5. The crop is coastal Bermuda grass.
6. Storage will be provided for a portion of the flow during the winter.
7. Runoff, if any, will be collected and stored for reapplication.

SOLUTION

1   Compute the water losses. Add evapotranspiration (column 2) and percolation (column 3), and place the sum in column 4.
2   Compute the wastewater applied. Using Eq. 13-2, the design precipitation is subtracted from the total water losses to determine the amount of wastewater to be applied (column 6).

Comments

1. The maximum application of wastewater will be less than 10 cm/week (4 in/week) and will occur in July.
2. If the wastewater available equals the wastewater applied on a yearly basis, then 386 cm/yr (152 in/yr) divided by 12 months/yr equals 32.2 cm (12.7 in) of wastewater available each month.
3. Storage would be required for a portion of the flow for each month in which the wastewater available exceeded the wastewater applied (in this case, from approximately mid-November to mid-April).
4. The annual liquid loading of 386 cm (152 in) would place this land-treatment system within the normal loading range for irrigation of 61 to 600 cm/yr (24 to 240 in/yr).
5. The results obtained from this process would be used to determine land requirements and storage requirements.

Table 13-10 Water balance for Example 13-1

	Water losses, cm			Water applied, cm		
Month (1)	Evapo-transpiration (2)	Perco-lation (3)	Total (2) + (3) = (4)	Precipi-tation (5)	Wastewater applied (4) − (5) = (6)	Total (5) + (6) = (7)
Jan.	1.8	25.4	27.2	5.8	21.4	27.2
Feb.	3.8	25.4	29.2	5.8	23.4	29.2
Mar.	7.9	25.4	33.3	5.3	28.0	33.3
Apr.	9.9	25.4	35.3	4.1	31.2	35.3
May	13.2	25.4	38.6	1.0	37.6	38.6
June	16.5	25.4	41.9	0.5	41.4	41.9
July	17.8	25.4	43.2	0.3	42.9	43.2
Aug.	16.5	25.4	41.9	Trace	41.9	41.9
Sep.	11.2	25.4	36.6	0.5	36.1	36.6
Oct.	9.9	25.4	35.3	1.5	33.8	35.3
Nov.	3.8	25.4	29.2	2.6	26.6	29.3
Dec.	2.0	25.4	27.4	5.6	21.8	27.4
Total annual	114.3	304.8	419.1	33.0	386.1	419.1

Note: cm × 0.3937 = in

Nitrogen loading A total nitrogen balance is almost as important as a water balance, because nitrate ions are mobile in the soil and can affect the quality of the receiving water. On an annual basis, the applied nitrogen must be accounted for in crop uptake, denitrification, volatilization, addition to groundwater or surface water, or storage in the soil. The total nitrogen load is necessary because all forms—organic, ammonia, nitrate, and nitrite—interact in the soil. The total nitrogen loading will be

$$N = 0.1CL \qquad \text{(SI units)} \tag{13-3}$$

$$N = 2.7CL \qquad \text{(U.S. customary units)} \tag{13-3a}$$

where N = annual nitrogen loading, kg/ha · yr (lb/acre · yr)
C = total nitrogen concentration, mg/L
L = annual liquid loading, cm/yr (ft/yr)

The nitrogen uptake of most crops has been determined using fresh water for irrigation, and typical uptake values are given in Table 13-6. Nitrogen-uptake values may be higher when wastewater is applied instead of fresh water only because more nitrogen is available. For land treatment systems, few nitrogen-uptake values for crops currently exist. It is expected that definitive values will be established in the near future. Nitrogen uptakes for plants not listed in Table 13-6 can generally be obtained from agricultural extension service agents. When more than one crop per year is grown on one field, the total nitrogen uptake for the entire year should be determined. Nitrogen removal by crop uptake is a function of crop yield and requires the harvesting and physical removal of the crop to be effective.

The extent of denitrification and volatilization depends on the loading rate and characteristics of the wastewater to be applied, and the microbiological conditions in the active zones of the soil. Even in aerobic soils, denitrification may account for 15 to 25 percent of the applied nitrogen. Volatilization of ammonia will not be significant for effluents with a pH less than 7 or for nitrified effluents.

The soil mantle cannot hold nitrogen indefinitely, although organic nitrogen can be stored in the soil to a certain extent. The ammonium and organic nitrogen is ultimately converted to nitrate nitrogen, which can leach out of the soil. Unless nitrogen is taken up by crops and physically removed by harvesting, or the nitrates are converted to nitrogen gas by denitrification, the nitrogen will appear eventually in the runoff or percolate.

Phosphorus mass balance Phosphorus is removed from percolating wastewater by fixation and chemical precipitation. For irrigation, the phosphorus loading will usually be well below the capacity of the soil to fix and precipitate the phosphorus. Typically, less than 20 percent of the phosphorus applied is used by the crop, and the remainder stays in the topsoil. Soil-column tests or phosphorus-adsorption tests can be conducted to determine the fixation capacity of the soil, but the test results should be used with caution because long-term behavior and the effects of time cannot be duplicated in a short-term test.

Organic loading The average daily organic loading rate should be calculated from the liquid loading rate and the BOD_5 concentration of the applied effluent. Thomas has estimated that between 11.2 and 28.0 kg/ha · d (10 and 25 lb/acre · d) are needed to maintain a static organic matter content in the soil [66]. Additions of organic matter at these rates help to condition the soil and replenish the carbon oxidized by microorganisms, and would not be expected to pose problems of soil clogging. Higher loading rates can be managed, depending on the type of system and the resting period.

Using the range of 11.2 to 28.0 kg/ha · d of BOD_5 as a reference, the addition of 2.2 kg/ha · d (2 lb/acre · d) or less from a typical secondary effluent applied for irrigation will certainly not pose a problem of organic buildup in the soil. When primary effluent is used, organic loading rates may exceed 22.4 kg/ha · d (20 lb/acre · d) without causing problems [72].

Resting periods, standard with most irrigation techniques, give soil bacteria time to break down organic matter and allow the water to drain from the top few inches. Aerobic conditions are thus restored as air penetrates into the soil. Resting periods for sprinkler irrigation may range from less than a day to 14 d; periods of 5 to 10 d are common. The resting period for surface irrigation can be as long as 6 weeks but is usually between 6 and 14 d [51]. The resting period depends on the crop, the number of individual plots in the rotation cycle, and management considerations.

Land Requirements

The total land area required includes allowances for treatment, buffer zones and storage (if necessary), sites for buildings, roads and ditches, and land for emergencies or future expansion. If any on-site preapplication treatment is required, such as screening, sedimentation, biological or chemical treatment, or disinfection, an allowance must be made for the land needed.

The field area is that portion of the site in which the treatment process actually takes place. It is determined by comparing the areas, and it is calculated on the basis of acceptable loading rates for each different loading parameter (liquid, nitrogen, phosphorus, organic, or others) and then selecting the largest area. The loading parameter that corresponds to the largest field area requirements is the critical loading parameter. The field-area requirement based on the liquid loading rate is calculated by:

$$\text{Field area (ha)} = \frac{3.65Q}{L} \qquad \text{(SI units)} \qquad (13\text{-}4)$$

$$\text{or} \qquad \text{Field area (acres)} = \frac{1118Q}{L} \qquad \text{(U.S. customary units)} \qquad (13\text{-}4a)$$

where Q = flowrate, m³/d (Mgal/d)
L = annual liquid loading, cm/yr (ft/yr)

(Both 3.65 and 1118 are conversion factors.) For loadings of constituents such as nitrogen, the field-area requirement is calculated by:

$$\text{Field area (ha)} = \frac{0.365CQ}{L_c} \qquad \text{(SI units)} \qquad \qquad (13\text{-}5)$$

$$\text{or} \qquad \text{Field area (acres)} = \frac{3040CQ}{L_c} \qquad \text{(U.S. customary units)} \qquad (13\text{-}5a)$$

where C = concentration of constituent, mg/L
Q = flowrate, m³/d (Mgal/d)
L_c = loading rate of constituent, kg/ha · yr (lb/acre · yr)

Crop Selection

The important aspects of crops for irrigation systems are (1) nitrogen-removal capability, (2) water needs and tolerances, (3) sensitivity to wastewater constituents, (4) public health regulations, and (5) crop-management considerations. The three classes of crops to be discussed are (1) forage and field crops, (2) landscape vegetation, and (3) woodlands.

Forage and field crops Successful forages used in wastewater irrigation include Reed canary grass, brome grass, tall fescue, perennial rye grass, and coastal Bermuda grass. These grasses have high nitrogen uptakes (see Table 13-6), are water tolerant, and are relatively tolerant of high total dissolved solids and boron in wastewater.

Field crops that are most popular include barley, sorghum, corn, and milo. Water tolerances of these crops may be somewhat less than for forages [59]. Sensitivity of both forage and field crops to electrical conductivity (EC) is presented in Table 13-11.

For forage and field crops, primary treatment prior to application is often sufficient. Some state regulations require secondary treatment [42]. In California, irrigation of vegetable crops to be eaten raw requires a filtered and disinfected secondary effluent.

Forage crops are often perennial and require relatively little management. The grass can be grazed or cut and sold as hay. Field crops are annuals and therefore require more management (planting, cultivating, harvesting, and field preparation). Combinations of crops in sequence, such as corn (in the summer) followed by barley, oats, wheat, or rye grass (in the winter), can increase productivity and nutrient removal. Management techniques such as "no till" corn (not cultivated) reduce the labor involved and also reduce the potential for soil erosion.

Table 13-11 Yield decrement to be expected for forage and field crops resulting from high electrical conductivity in irrigation waters [3]

	EC_e values, in mmho/cm (saturated paste extract) for a reduction in crop yield of:		
	0%	25%	100%
Forage crops			
Alfalfa	2.0	5.4	15.5
Bermuda grass	6.9	10.8	22.5
Clover	1.5	3.6	10
Corn (forage)	1.8	5.2	15.5
Orchard grass	1.5	5.5	17.5
Perennial rye grass	5.6	8.9	19
Tall fescue	3.9	8.6	23
Vetch	3.0	5.3	12
Tall wheat grass	7.5	13.3	31.5
Field crops			
Barley	8.0[a]	13	28
Corn	1.7	3.8	10
Cotton	7.7	13	27
Potato	1.7	3.8	10
Soybeans	5.0	6.2	10
Sugarbeets	7.0	11.0	24
Wheat	6.0	9.5	20

[a] Barley and wheat are less tolerant during germination and seedling stage. EC_e should not exceed 4 or 5 mmho/cm.

Landscape Vegetation Application of effluent on landscape areas, such as highway median and border strips, airport strips, golf courses, parks and recreational areas, and wildlife areas, has several advantages. The areas irrigated are already owned, saving acquisition cost, and problems associated with crops for consumption are avoided. Additionally, the maintenance of landscape projects generally requires less water than other vegetation (since watering in these cases is based on vegetative maintenance rather than crop production); hence, the wastewater can be spread over a greater area. Irrigation of golf courses and parks has been practiced extensively [64, 70].

Woodlands irrigation Silviculture, the growing of trees, is being conducted with wastewater on at least 11 existing sites in Oregon, Michigan, Maryland, and Florida [64]. Forests offer several advantages as potential sites for land treatment:

1. Large forested areas exist near many sources of wastewater.
2. The infiltration properties of forest soils are often better than those of agricultural soils.
3. Site-acquisition costs for forest land are usually lower than site acquisition costs for agricultural land because of lower land values for forest lands.
4. During cold weather, soil temperatures are often higher in forest lands than in comparable agricultural lands.

The principal limitations on the use of wastewater for silviculture are that:

1. Water tolerances of the existing trees may be low.
2. Nitrogen removals are relatively low.
3. Fixed sprinklers, which are expensive, must generally be used.

Existing forests have adapted to the water supply from natural precipitation. Unless soils are well drained, the increase in hydraulic loading from wastewater application will drown existing trees. Approximate water use (evapotranspiration) and nutrient uptake of various tree species are summarized in Table 13-12. An irrigation system in woodlands is shown in Fig. 13-4.

Distribution Techniques

More than 20 distribution techniques for water are available for engineered wastewater-effluent applications. Distribution techniques for irrigation can be classified into two main groups: sprinkling and surface-application systems. Three techniques are illustrated in Fig. 13-5. Detailed criteria on these systems are available in Pair [44]. Many of the techniques developed in the irrigation

Table 13-12 Evapotranspiration and nitrogen uptake in woodlands [48, 49]

Species	Evapotranspiration, cm/yr		Nitrogen uptake, kg/ha · yr
	Range	Typical	
Pines	13–86	38	30–70
Mixed coniferous and deciduous	46–86	64	40–80
Deciduous	22–86	43	50–100

Note: cm/yr × 0.3937 = in/yr
kg/ha · yr × 0.8922 = lb/acre · yr

Figure 13-4 Woodlands irrigation at Seabrook Farms, N.J. [*courtesy of Belford L. Seabrook*]

industry have not been applied to wastewater. Drip irrigation, for example, would require a filtered effluent from crop irrigation and is therefore not economical for wastewater. Many of the sprinkling techniques described by Pair may be applicable to wastewater application and should be investigated with regard to economics, efficiency, operation and maintenance, and reliability.

Sprinkling Sprinkling systems are of two types, fixed and moving. Fixed sprinkling systems, often called solid-set systems, may be either on the ground surface or buried. Both types usually consist of impact sprinklers mounted on risers that are spaced along lateral pipelines, which are in turn connected to main pipelines. These systems are adaptable to a wide variety of terrains and may be used for irrigation of either cultivated land or woodlands. Portable aluminum pipe is normally used for above-ground systems. It has the advantage of a relatively low capital cost, but it is easily damaged, has a short expected life because of corrosion, and must be removed during cultivation and harvesting operations.

Plastic or asbestos-cement pipe is most often used for buried systems. Laterals may be buried as deep as 0.46 m (1.5 ft) below the surface; main pipelines, 0.76 to 0.91 m (2.5 to 3 ft) below the surface. Buried systems generally have the greatest capital cost of any of the irrigation systems. On the other hand, they are probably the most dependable and are well suited to automatic control. Typical design information for fixed sprinkling systems is presented in Table 13-13.

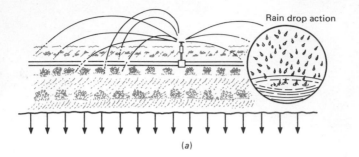

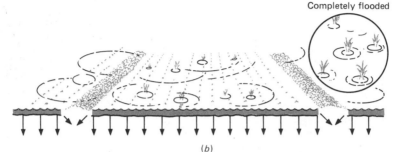

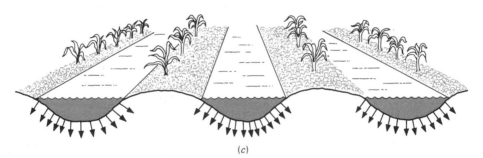

Figure 13-5 Irrigation techniques [45]. (*a*) Sprinkler. (*b*) Flooding. (*c*) Ridge and furrow.

There are a number of different moving sprinkling systems, including center-pivot, side-roll wheel-move, rotating-boom, and winch-propelled systems. The center-pivot system, which consists of a lateral suspended by wheel supports and rotating about a point, is the most widely used for wastewater irrigation (see Fig. 13-6). Typical design information is presented in Table 13-14.

Surface application The two main types of surface-application systems are ridge-and-furrow irrigation and border-strip flooding. Ridge-and-furrow irrigation is accomplished by gravity flow of effluent through furrows from which it seeps into the ground. Typically, water is applied to the furrows using gated aluminum pipe, as shown in Fig. 13-7. Border-strip irrigation consists of low parallel soil ridges constructed in the direction of slope. Typical design information for these systems is presented in Table 13-15.

Table 13-13 Typical design information for fixed sprinkling systems

Item	Value or description	
	Range	Typical
Sprinkler spacing, m (rectangular, square, or triangular)	12.2 by 18.3 to 30.5 by 30.5	18.3 by 24.4 to 24.4 by 30.5
Application rate, cm/h[a]	0.25–2.5 or more	0.4–0.6
Nozzles		
Size of openings, cm	0.64–2.54	
Discharge per nozzle, L/s	0.25–6.3	0.5–1.6
Discharge pressure, kN/m²[b]	200–700	350–400
Control systems[c]	Automatic, semiautomatic, manual	Automatic
Risers[d]		
Type	Galvanized pipe or polyvinyl chloride	
Height, m	. . . .	0.915–1.22

[a] The application rate is calculated using the following equation:

$$\text{Application rate, cm/h} = \frac{360 \text{ L/s per sprinkler}}{\text{Area (m}^2 \text{ covered)}}$$

[b] Single-nozzle sprinklers are preferred because of lesser clogging tendencies and larger spray diameters.

[c] Automatic valves may be operated either hydraulically or electrically.

[d] Height should be sufficient to clear the crop; the typical height given is for grass. The riser should be adequately staked because impact sprinklers cause vibrations that must be damped.

Note: m × 3.2808 = ft
          cm × 0.3937 = in
       L/s × 15.8508 = gal/min
    kN/m² × 0.1450 = lb$_f$/in²

Underdrainage

Underdrain refers to any type of buried conduit with open joints or perforations which collect and convey renovated water that has percolated through the soil during treatment. Underdrains should draw down the water table within a few days after effluent application or a major rainfall event.

Underdrains may be required in poorly drained soils or when groundwater levels affect wastewater renovation or crop growth. The topography of the land to be drained and the position, level, and annual fluctuation of the water table are factors to be considered in the preliminary design and layout of a drainage system for a given site. Detailed field investigations will be required before final design because subsoil and groundwater conditions are not evident from visual inspection of the site.

Figure 13-6 Center-pivot irrigation machine.

Table 13-14 Typical design information for center-pivot moving sprinkling system

| Item | Value or description |
	Range
Size	
Lateral length, m	180–425
Irrigation area per unit, ha	14–55
Propulsion	
Type of drive	Hydraulic or electric
Duration of 1 rotation	8 h–1 wk
Pressure[a]	
At the nozzle, kN/m^2	350–450
At the pivot, kN/m^2	550–650
Topography	Systems can be adapted to rolling terrain up to 15–20%

[a] Standard sprinkler nozzles or spray heads directed downward can be used.

Note: $kN/m^2 \times 0.1450 = lb_f/in^2$
$m \times 3.2808 = ft$
$ha \times 2.4711 = acre$

Figure 13-7 Gated aluminum pipe for wastewater distribution [*from Lockwood Corporation*].

Underdrain systems normally consist of a network of drainage pipes buried 1.2 to 3 m (4 to 10 ft) below the surface and intercepted at one end of the field by a cutoff ditch. The pipes normally range in diameter from 10 to 20 cm (4 to 8 in).

Underdrain spacing will be controlled by soil permeabilities and depth of the water table. Where high loadings occur, as in rapid infiltration, or where permeabilities are low, as for clay soils, underdrain spacing may be as close as 15 to 30 m (50 to 100 ft). For irrigation systems with moderate to rapid surface soil permeabilities, underdrains may be spaced much farther apart, up to 150 m (500 ft) or more [72].

Surface-Runoff Control

Requirements for control of surface runoff resulting from both applied effluent and storm water depend mainly on the expected quality of the runoff—for which few data exist. In surface-runoff control for irrigation systems, consideration must be given to tail water return, storm runoff, and system protection.

Table 13-15 Typical design information for surface-application systems

Item	Value or description	
	Range	Typical
Ridge-and-furrow system		
Topography[a]	Relatively flat to moderately sloped	
Dimensions		
Furrow length, m	183–427	
Furrow spacing, cm[b]	51–102	
Application[c]		
Pipe type	. . . .	Gated aluminum
Pipe length, m	24.4–30.5	
Rest periods	Up to 6 wk	7–14 d
Border-strip flooding		
Strip dimensions[d]		
Border widths, m	6.1–30.5	12.2–18.3
Slopes, %	0.2–0.4	
Strip lengths, m	183–427	
Method of distribution[e]	Concrete-lined ditch, underground pipe, or gated aluminum pipe	
Application rest periods	Up to 6 wk	7–14 d
Application rate per meter width of strip[f]		
Clay, L/m · s	2–4	
Sand, L/m · s	10–15	

[a] Ridge-and-furrow irrigation can be used on relatively flat land (less than 1 percent) with furrows running down the slope, or on moderately sloped land with furrows running along the contour.

[b] Furrow spacing depends on the crop.

[c] Short runs of pipe are preferred to minimize pipe diameter and head loss and to provide maximum flexibility. Surface standpipes are used to provide the head of 0.9 to 1.2 m (3 to 4 ft) necessary for even distribution. Application amounts of 7.6 to 10.2 cm (3 to 4 in) generally result in a matter of hours with both ridge-and-furrow systems and border-strip flooding.

[d] Strip dimensions vary with type of crop, type of soil, and slope. Relatively permeable soils require the steeper slopes.

[e] Distribution is generally by means of concrete-lined ditch with slide gates at the head of each strip, underground pipe with risers and alfalfa valves, or gated aluminum pipe.

[f] Application rates at the head of each strip will vary primarily with soil type. The period of application for each strip will vary with strip length and slope.

Note: m × 3.2808 = ft

Tail water return Surface runoff of applied effluent is usually designed into surface-application systems, such as ridge-and-furrow irrigation and border-strip flooding, because it is difficult to maintain even distribution across the field with these methods, and some excess water may accumulate at the ends of furrows or strips. Generally, this tail water is collected and returned by means of a series of collection ditches, a small reservoir, a float-actuated pumping station, and a force main to the main storage reservoir or distribution system.

The amount of tail water will vary between 10 and 40 percent of applied flows (depending on the management provided, the type of soil, and the rate of application). In humid climates, the tail water system design may be controlled by storm water–runoff flows.

Storm runoff For high-intensity rainfall, some form of storm-runoff control may be required for irrigation systems, except those with well-drained soils, relatively flat sites, or where the quality of the runoff is acceptable for discharge. Where runoff control is deemed necessary, it generally consists of the collection and treatment or return of the runoff from a storm of specified intensity, with a provision for the overflow of a portion of all larger flows.

The amount of runoff to be expected as a result of precipitation will depend on the infiltration capability of the soil, the antecedent moisture condition of the soil, the slope, the type of vegetation, and the temperature of both air and soil. The relationships between runoff and these factors are common to many other hydrologic problems and are adequately covered by Viessman [74].

Runoff quality during storms is essentially unknown for most parameters. To give some perspective to the magnitude of nitrogen and phosphorus concentrations measured in runoff from various rural areas, and until quantitative data from effluent-applied sites are available, the average values given in Table 13-16 may be useful.

It is important to note that the research work reported in Table 13-16 was aimed primarily at fertilizing practice and cultivation versus noncultivation as related to nutrient losses. Other factors in sediment and nutrient loss include contour planting versus straight-row planting and incorporation of plant residues to increase organic matter in the soil. Many additional factors that affect erosion losses were presented by Loehr [36].

System protection An additional requirement for runoff control often results from the need for protection from runoff caused by system failures. System failures may include ruptured sprinkler lines, inadvertent overapplication, or soil sealing as a result of wastewater constituents or frost. The requirements for this objective would probably be satisfied as part of the storm runoff–control system.

Case Study (Muskegon, Michigan)

The need for an alternative wastewater management program for the Muskegon County area became apparent in the late 1960s because of deterioration in the water quality of local surface waters. Fourteen municipalities and five major

Table 13-16 Average values of nitrogen and phosphorus measured in rural storm water runoff [36]^a

Source	Concentration, mg/L				
	COD	BOD	NO$_3$–N	Total N	Total P
Precipitation	9–16	12–13	0.14–1.1	1.2–0.04	0.02–0.04
Forested land			0.1–1.3	0.3–1.8	0.01–0.11
Agricultural cropland	80	7	0.4	9	0.02–1.7
Irrigation tile drainage, western United States					
Surface flow			0.4–1.5	0.6–2.2	0.2–0.4
Subsurface drainage			1.8–19	2.1–19	0.1–0.3
Cropland tile drainage				10–25	0.02–0.7
Seepage from stacked manure	25,900–31,500	10,300–13,800		1,800–2,350	190–280
Feedlot runoff	3,100–41,000	1,000–11,000	10–23	920–2,100	290–360

^{a} Data do not reflect the extreme ranges caused by improper waste management or extreme storm conditions.

Note: mg/L = g/m^3

industries were required to achieve an 80 percent phosphorus removal and to produce effluent that would not result in the degradation of water quality in Lake Michigan.

Areawide solutions were explored, and it was determined that the most cost-effective solution was to divert all the wastewater discharges from surface waters and to use undeveloped land as a major component of an areawide treatment system. The decision to undertake such a plan was based in part on economics and in part on a commitment to recycle nutrients as resources rather than discharge them to the environment in a nonbeneficial manner. Construction of the facilities was started in 1972, and operation began in stages starting in May 1974. The first full year of operation was 1975.

The Muskegon County Wastewater Project consists of two independent systems: the Muskegon Project and the smaller Whitehall Project. In the Muskegon Project, wastewater is collected from 13 municipalities and 5 industries and is transported to a treatment site approximately 24 km (15 mi) inland from Lake Michigan. The treatment system basically consists of biological treatment in aerated lagoons followed by sprinkler irrigation of land on which corn is presently grown. Although there are many interesting features of the system, its uniqueness lies primarily in the size of the facility. With a design capacity of 1.8 m^3/s (42 Mgal/d) and over 2020 ha (5000 acres) of land under irrigation, it is the largest operating facility in the United States for specifically designed land treatment of wastewater.

Description of facilities and design data A plan of the facilities is shown in Fig. 13-8. Incoming wastewater first enters three identical aerated lagoons, which may be operated in parallel or series. From the aerated lagoons, wastewater enters the two large storage lagoons. A separate settling pond, which can serve as a bypass to the storage lagoons, is also provided. During the irrigation season (April through November), water for irrigation is drawn from either the storage lagoons or the settling pond into a 5.7-ha (14-acre) outlet lagoon. The treated effluent released from the outlet lagoon is chlorinated in a mixing chamber prior to delivery via open channels to the two main distribution pumping stations. These stations pump the wastewater through a series of buried pipes to the irrigation equipment.

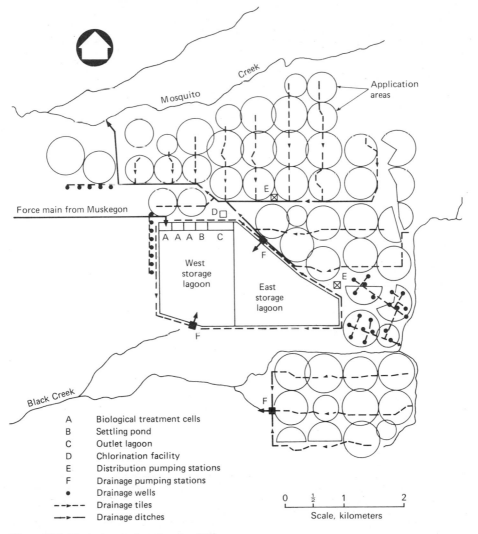

A	Biological treatment cells
B	Settling pond
C	Outlet lagoon
D	Chlorination facility
E	Distribution pumping stations
F	Drainage pumping stations
•	Drainage wells
--▶--	Drainage tiles
—▶—	Drainage ditches

Figure 13-8 Muskegon project site plan [49].

Figure 13-9 Corn crop being sprinkled by a center-pivot system at Muskegon, Mich. [*courtesy of Robert K. Bastian*].

Table 13-17 Distribution system data for Muskegon land-treatment system [17]

Item	Value
Pumping	
No. of vertical turbine pumps	17
Peak capacity, m^3/s	4.0
Piping size range, cm	20–90
Center-pivot irrigation rigs	
No. of rigs	54
Radius, m	215–430
Coverage range, ha	14–56
Operating pressure, kN/m^2	240–580
Nozzle pressure, kN/m^2	20–70
Application rate (continuous operation)	
cm/h	0.06
cm/wk	10

Note: $L/s \times 0.0228 = Mgal/d$
$cm \times 0.3937 = in$
$m \times 3.2808 = ft$
$ha \times 2.4711 = acre$
$k\dot{N}/m^2 \times 0.1450 = lb_f/in^2$

Wastewater is applied to the land by 54 center-pivot irrigation rigs rolling on pneumatic tires (see Fig. 13-9). Vertical turbine pumps at two main pumping stations discharge to an asbestos-cement-pipeline distribution network. Major design data for the distribution system are presented in Table 13-17.

Soils are mostly sands and sandy loams underlain by clay. The infiltration rates for sands range from 12.7 to more than 25.4 cm/h (5 to 10 in/h). The maximum design application rate is 10 cm/week (4 in/week), which includes an allowance for 1.9 cm/week (0.74 in/week) of precipitation. Application design data are summarized in Table 13-18.

A combination of drainage tiles, drainage wells, and natural drainage collects the subsurface water and discharges it to adjacent receiving surface waters. Most of the site is underlain with drainage tiles at approximately 152-m (500-ft) intervals at depths from 1.5 to 2.4 m (5 to 8 ft). The laterals, constructed of perforated polyethylene filtered by Fiberglas (fibrous glass) socks, conduct the water to main concrete drainage pipes. Concrete pipes are used to transport the water to open ditches, which in turn discharge to two receiving streams.

Operating characteristics and performance Irrigation with wastewater at Muskegon began in May 1974, and numerous temporary startup problems were encountered. Most of the problems, such as dike damage, breaks in irrigation pressure pipes, and electrical cable failure, have been resolved.

One operational problem has been the plugging of the irrigation nozzles with sand and weeds from the ditch upstream of the pumping stations. To alleviate this problem permanently, settling basins and screening systems were installed.

Corn planted in 1975 yielded an average of 148 bu/ha (60 bu/acre), only slightly less than the 160 bu/ha (65 bu/acre) average corn grain yield on operating farmland in Muskegon County. This is quite remarkable in light of the fact that most soils at the site are very poor and that wastewater renovation is the primary purpose of the system. Representative yields of corn grain from

Table 13-18 Application design data for Muskegon land-treatment system

Item	Value
Total project area, ha	4380
Field area, ha	2160
Application season, months	8
Weekly wastewater application, cm	
Average	7.5
Maximum	8.2
Weekly precipitation allowance, cm	1.8
Total weekly application rate, cm	10

Note: ha × 2.4711 = acre
cm × 0.3937 = in

Table 13-19 Representative yields of corn grain, Muskegon land-treatment site, 1975 [77]

Field soil type	Wastewater application, cm/yr	Supplemental nitrogen fertilizer, kg/ha · yr	Corn grain yield, bu/ha · yr
Roscommon sand	145	58	222
Rubicon sand	265	56	205
AuGres sand	150	62	175
Roscommon sand	175	36	170
Granby loamy sand	35	24	151
Rubicon sand	236	39	131
AuGres sand	35	9	89
Roscommon sand	35	0	77
Project average	135	39	148

Note: cm/yr × 0.3937 = in/yr
kg/ha · yr × 0.8922 = lb/acre · yr
bu/ha · yr × 0.4047 = bu/acre · yr

various fields at the Muskegon land treatment site for the 1975 season are presented in Table 13-19.

The wastewater provides an adequate amount of phosphorus and potassium for the corn crop [15], but the low levels of nitrogen would not be adequate without supplemental additions. From 0 to 100 kg/ha · yr (0 to 89 lb/acre · yr) of nitrogen fertilizer was added to the different irrigated fields, depending on the amount of wastewater applied and crop requirements. Nitrogen additions increased corn growth, which in turn stimulated increased removal of phosphorus, potassium, and other wastewater nutrients.

The average treatment results for 1975 are presented in Table 13-20. The discharge permit requirements are for 4 mg/L BOD, 10 mg/L suspended solids, 0.5 mg/L total phosphorus, and 200 fecal coliforms per 100 mL [77].

13-4 RAPID-INFILTRATION SYSTEMS

In rapid-infiltration land treatment, sometimes referred to as infiltration percolation, most of the applied wastewater percolates through the soil, and the treated effluent eventually reaches the groundwater. The wastewater is applied to rapidly permeable soils, such as sands and loamy sands, by spreading in basins or by sprinkling, and is treated as it travels through the soil matrix. Vegetation is not usually used, but there are some exceptions.

The typical hydraulic pathway for rapid infiltration is shown in the schematic view in Fig. 13-10a. A much greater portion of the applied wastewater percolates to the groundwater than with irrigation. There is little or no consumptive use by plants, and there is less evaporation in proportion to a reduced surface area. In

Table 13-20 Summary of 1975 average treatment performance at the Muskegon land-treatment site [49][a]

Parameter	Influent	Average storage lagoon effluent	Drain tiles	Mosquito Creek
BOD	205	13	1.2	3.3
pH, units	7.3	7.8	......	7.5
Specific conductance, μmhos	1049	825	599	574
Total solids	1093	691	......	466
Suspended solids	249	20	......	7
COD	545	118	......	33
TOC	107	38	11.6	15
Ammonia-N	6.1	2.4	0.29	0.6
Total Kjeldahl nitrogen	8.2	4.5	......	
Nitrate-N	Trace	1.1	2.2	1.9
Total P	2.4	1.4	......	0.05
Chloride	182	154	60	78
Sodium	166	144	42	66
Calcium	73	58	72	61
Magnesium	14	16	23	18
Potassium	11	9	2.6	4
Iron	0.8	1.0	7.7	1
Zinc	0.6	0.11	0.01	0.07
Manganese	0.28	0.16	0.20	0.11
Total coliforms, colonies/100 mL	......	$100–1.25 \times 10^8$	$<1–170$	$1–9.6 \times 10^4$
Fecal coliforms, colonies/100 mL	......	$4–1.2 \times 10^6$	$<1–32$	$<1–4.8 \times 10^3$
Fecal streptococci, colonies/100 mL	......	$2–38000$	$<1–47$	

[a] All units are expressed in mg/L (g/m^3), except as noted.

many cases, recovery of renovated water is an integral part of the system, and can be accomplished using underdrains or wells, as shown in Fig. 13-10b and c.

Design Objectives

The principal objective of rapid infiltration is wastewater treatment. The treated water can be used for (1) groundwater recharge, (2) recovery of renovated water by wells or underdrains with subsequent reuse or discharge, (3) recharge of surface streams by natural subsurface flow, or (4) temporary storage of renovated water in the aquifer.

Site Selection

Soils with infiltration rates of 10 to 61 cm/d (4 in to 2 ft/d) or more are necessary for successful rapid infiltration. Acceptable soil types include sand, sandy loams, loamy sands, and gravels. Very coarse sand and gravel are not ideal because they

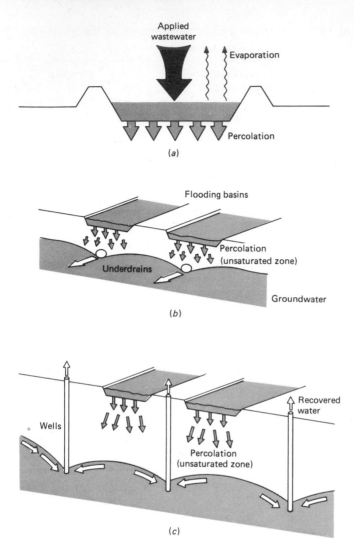

Figure 13-10 Rapid infiltration [49]. (*a*) Hydraulic pathway. (*b*) Recovery of renovated water by underdrains. (*c*) Recovery of renovated water by wells.

allow wastewater to pass too rapidly through the first few feet where the major biological and chemical action takes place.

Other important factors in site selection include percolation rates, depth, movement and quality of groundwater, topography, and underlying geologic formations. To control the wastewater after it infiltrates the surface and percolates through the soil matrix, the subsoil and aquifer characteristics must be known. Recharge should not be attempted without specific knowledge of the movement of the water in the soil system and the groundwater aquifer.

Preapplication Treatment

Reduction of suspended solids is the most important preapplication treatment criterion for rapid-infiltration systems, so that soil clogging and nuisance conditions from odors are minimized. Biological treatment is often provided for this purpose. Disinfection is generally not necessary, except possibly for sprinkling systems; it has been found in a number of studies that rapid infiltration quite effectively reduces pathogenic bacteria [34, 37].

Climate and Storage

Because rapid infiltration does not rely on vegetation, it is the land treatment process most adaptable to cold climates. Also, surface application by flooding basins is less susceptible to freezing than other distribution techniques. At Lake George, N.Y., and at Fort Devens, Mass., the systems are operated throughout the winter. When ice forms on the surface of the flooding basins, it is not removed but merely floated by the next application of wastewater. The ice serves to insulate the soil surface from further lowering of the temperature. Rapid-infiltration basins have also been operated successfully in the intermountain area of the northwest United States, where air temperatures can be as low as $-37°C$ ($-35°F$). No decline in renovation efficiency, as determined from monitoring wells, has been evident during periods of prolonged cold weather [75].

Loading Rate

The hydraulic loading rate and the loading rates for nitrogen, organics, and other constituents are important in rapid-infiltration systems.

Hydraulic loading Liquid loading rates generally range from 10 to 150 cm/week (4 to 60 in/week) for moderate-rate systems and 1.5 to 2.1 m/week (5 to 7 ft/week) for high-rate systems. The design rate must be based on the saturated hydraulic conductivity of the least permeable layer in the soil profile. It should be established for the poorest climatic conditions that can be reasonably expected.

Intermittent operation is required to maintain design loading rates and the renovative capacity of the soil. The resting period, which may vary from 5 to 20 d, is essential to allow atmospheric oxygen to penetrate the soil and reestablish aerobic conditions. As the surface dries, aerobic bacteria become active in organic-matter decomposition and nitrification. Organic-matter decomposition helps break up the clogging layer, and the microbial nitrification will free ammonium adsorption sites on clay and humus materials. When inundation begins again, the converted nitrate will be leached with the applied water until anaerobic conditions occur and denitrification begins.

Typical application rates and soil types are presented in Table 13-21. Annual application rates range from 6 to 120 m/yr (20 to 400 ft/yr).

Table 13-21 Typical application rates and soil types for rapid-infiltration systems [13, 49]

Location	Hydraulic loading rate, m/yr	Soil type	Type of effluent
Whittier Narrows (Los Angeles), Calif.	127	Sand	Secondary
Flushing Meadows (Phoenix), Ariz.	110	Sand	Secondary
Santee (San Diego), Calif.	81	Gravel	Secondary
Lake George, N.Y.	43	Sand	Secondary
Calumet, Mich.	34	Sand	Untreated
Ft. Devens, Mass.	29	Sand and gravel	Primary
Hemet, Calif.	33	Sand	Secondary
Westby, Wis.	11	Silt loam	Secondary

Note: m × 3.2808 = ft

Nitrogen loading The primary mechanism for nitrogen removal in rapid infiltration systems is denitrification. In high-rate systems, denitrification is the only significant mechanism of nitrogen removal. By managing the hydraulic loading cycle to create alternately anaerobic and aerobic conditions, Bouwer [7] obtained up to 80 percent nitrogen removal as a combined result of ammonia adsorption and denitrification during most of the period of inundation. Over a 4-yr period, the calculated removal was 30 percent at a loading rate of 23,450 kg/ha · yr (21,000 lb/acre · yr). By cutting the application rate in half, the removal was increased to 80 percent.

Organic loading Organic loading is important because it is related to the development of anaerobic conditions. To meet the oxygen demand created by the decomposing organic and nitrogenous material, an intermittent loading schedule is required. This allows air to penetrate the soil and supplies oxygen to the bacteria that oxidize the organic matter and ammonium. Existing organic loading rates are presented in Table 13-22.

Other constituents Because of the high liquid loadings involved, the mass loadings of constituents in even low concentrations can be considerable. Soils used for rapid infiltration usually have little capacity to retain soluble salts but retain large portions of the heavy metals and phosphorus. The concentrations of constituents such as sodium, chloride, or sulfate that are allowable in the renovated water may affect the design by requiring special controls on the use of the renovated water.

Adsorption and chemical precipitation are the primary phosphorus removal mechanisms in rapid-infiltration systems. Although all soil systems have a finite

Table 13-22 Organic loading for rapid-infiltration systems [13]

Location	BOD$_5$ loading rate, kg/ha · d	Ratio of drying time to application time
Food-processing wastewater		
Leicester, N.Y.	560	5:1
Delhi, N.Y.	270	3:1
Sumter, S.C.	120	2:1
Municipal wastewater		
Santee, Calif.	64	
Flushing Meadows, Ariz.	50	1:1
Whittier Narrows, Calif.	22	1.6:1
Lake George, N.Y.	21	13:1
Westby, Wis.	10	1:1

Note: kf/ha · d × 0.8922 = lb/acre · d

capacity to remove phosphorus, the capacity of many rapid-infiltration sites is quite large. After 88 years of rapid infiltration of untreated municipal wastewater at Calumet, Mich., concentrations of phosphorus in groundwater are still low (0.1 to 0.4 mg/L) [71]. However, long-term application has caused soil-soluble phosphorus to increase substantially in the top 30 cm (12 in), indicating that this layer is becoming saturated with phosphorus. At Lake George, N.Y., after 38 years of operation, phosphorus concentrations measured in seeps over 600 m (2000 ft) from the infiltration basins were 0.01 mg/L or less [1].

Distribution Techniques

Sprinkling and spreading basins are the two distribution techniques most suitable for rapid-infiltration systems. Factors that should be considered in the selection of the application technique include soil conditions, topography, climate, and economics.

Sprinkling Application of effluent at high rates using sprinkling has been accomplished. Systems with loading rates that exceed 10.2 cm/week (4 in/week) are included in this category. Normally, vegetation is necessary to protect the surface of the soil and to preclude runoff. Hydrophytic or water-tolerant grasses are usually chosen. Sprinkling of forest land may also be considered for rapid infiltration.

Spreading basins The surface of a rapid-infiltration basin should be designed to disperse the clogging solids [37]. This has been accomplished by growing vegetation or by adding a layer of graded sand or gravel to the surface. At Flushing Meadows, Ariz. (see Fig. 13-11), the vegetated basins were successful

Figure 13-11 Infiltration basins at Flushing Meadows (Phoenix, Ariz.).

[8]. At Whittier Narrows, Calif., adding a layer of pea gravel is reported to have increased the infiltration capacity [38]. In general, a bare or vegetated surface is preferable. The bare soil surface should be scarified or raked when solids accumulate. For vegetated surfaces, careful operation of the loading cycle is necessary in the spring until the vegetation is well established. The surface may be harrowed annually to break up any solids buildup.

Application Cycles

The hydraulic loading cycle, or alternation of flooding and drying periods, is essential to the restoration of aerobic conditions within the soil matrix. The drying periods are used to allow decomposition of the clogging materials and reaeration of the soil. Existing application cycles are presented in Table 13-23. The length of the drying period is a function of the hydraulic loading, the solids concentration in the applied effluent, and the climate.

Control of Underground Flow

Control of subsurface flow and recovery of renovated water are also essential for proper design of a rapid-infiltration system. If flow to groundwater is not desirable or not possible, recovery methods, such as the use of underdrains, pumped withdrawal, or natural drainage to surface waters, can be used.

Table 13-23 Existing application cycles for rapid-infiltration systems [49]

Location	Loading objective	Application period	Resting period	Bed surface
Calumet, Mich.	Maximize infiltration rates	1–2 d	7–14 d	Sand (not cleaned)
Flushing Meadows, Ariz.				
Maximum infiltration	Increase ammonium adsorption capacity	2 d	5 d	Sand (cleaned) and grass cover[a]
Summer	Maximize nitrogen removal	2 wk	10 d	Sand (cleaned) and grass cover[a]
Winter	Maximize nitrogen removal	2 wk	20 d	Sand (cleaned) and grass cover[a]
Fort Devens, Mass.	Maximize infiltration rates	2 d	14 d	Grass (not cleaned)
	Maximize nitrogen removal	7 d	14 d	Grass (not cleaned)
Lake George, N.Y.				
Summer	Maximize infiltration rates	9 h	4–5 d	Sand (cleaned)[a]
Winter	Maximize infiltration rates	9 h	5–10 d	Sand (cleaned)[a]
Tel Aviv, Israel	Maximize renovation	5–6 d	10–12 d	Sand[b]
Vineland, N.J.	Maximize infiltration rates	1–2 d	7–10 d	Sand (disked), solids turned into soil[c]
Westby, Wis.	Maximize infiltration rates	2 wk	2 wk	Grassed
Whittier Narrows, Calif.	Maximize infiltration rates	9 h	15 h	Pea gravel

[a] Cleaning usually involved physical removal of surface solids.
[b] Maintenance of sand cover is unknown.
[c] Solids are incorporated into surface sand.

If natural drainage to surface waters is planned, the water table must be controlled to prevent upward movement into the zone of filtration. Therefore, the aquifer should be able to transmit the renovated water readily from the infiltration sites. Bouwer [6] suggests the following equation for required elevation difference:

$$WI = KDH/L \qquad (13\text{-}6)$$

where W = width of infiltration area, m (ft)
I = hydraulic loading rate, m/d (ft/d)
K = hydraulic conductivity of aquifer, m/d (ft/d)
D = average thickness of zone below water table perpendicular to flow direction, m (ft)
H = elevation difference between water level in stream or lake and maximum allowable water-table level below infiltration area, m (ft)
L = distance of lateral flow, m (ft)

The product WI represents the amount of the applied water per m of axial extent for a given section and thereby controls the size of the infiltration basin (see Fig. 13-12). Thus, if the amount of applied water is restricted by the groundwater, relatively high hydraulic loading rates (I) may be used by designing basins of relatively narrow width (W).

Basin sizing includes consideration of the amount of usable land available, the hydraulic loading rate, topography, management flexibility, and groundwater conditions. To operate a system continuously, at least two basins are required, one for flooding and one for drying, unless sufficient storage is available elsewhere in the system. Multiple basins are desirable to provide flexibility in the management of the system.

The underdrainage system must provide sufficient soil detention time and underground travel distance to achieve the desired quality of renovated water. The quality of applied water, the application rate, soil renovation potential and permeability, aquifer conditions, and the use of a cover crop will determine the necessary detention time and travel distance.

Optimum depth and spacing of underdrains to recover renovated water from rapid-infiltration systems is mainly a matter of opinion. Lance et al. determined that water-table depths of more than 1.5 m (5 ft) would not greatly increase the depth of the aerobic zone during drying of infiltration basins [35].

Proper placement of underdrains to recover renovated water from rapid-infiltration systems is more critical than for irrigation systems. Bouwer [6] has developed an equation to determine the distance underdrains should be placed away from the infiltration area (see Fig. 13-13). The height H_c of the water table below the outer edge of the infiltration area can be calculated as follows:

$$H_c^2 = H_d^2 + IW(W + 2L)K \qquad (13\text{-}7)$$

where H_d = drain height above impermeable layer, m (ft)
$\quad I$ = infiltration rate, cm/d (in/d)
$\quad W$ = width of infiltration basin, m (ft)
$\quad L$ = distance to underdrain, m (ft)
$\quad K$ = hydraulic conductivity of the soil, cm/d (in/d)

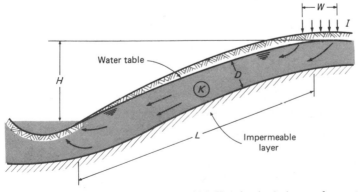

Figure 13-12 Natural drainage from rapid-infiltration basin into surface water [6].

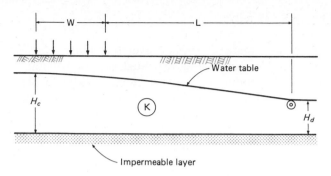

Figure 13-13 Collection of renovated water by drains [6].

The location of the drain is selected and H_c is calculated with Eq. 13-7. By adjusting variables L, W, and I, a satisfactory value of H_c is obtained. An L value less than the most desirable distance of underground travel may have to be accepted to obtain a workable system.

Plastic, concrete, and clay tile lines are used for underdrains. The choice usually depends on price and availability of materials. Most tile drains are laid in a machine-dug trench. Depending on soil conditions, covering the concrete or clay drains with coarse sand may be necessary to keep fine sands and silts out of the tile lines. Plastic drain lines are normally equipped with Fiberglas filter socks.

In organic soils and in loam and clay loam soils, a filter is not needed. The value of using a filter also depends on the cost of cleaning a plugged tile line versus the cost of the filter materials.

Recovery wells The use of wells to recover percolated wastewater is applicable only to rapid-infiltration systems. The percolation rates for other methods of land treatment are generally not high enough to make this process feasible. Recovery may be desired for reuse of the renovated water or to control the water table in order to increase the renovation distance and treatment effectiveness. The potential for percolate recovery at a site depends on several factors, such as the depth of the aquifer and the permeability and continuity of an aquiclude. The primary limitations to recovering percolated wastewater are the ability to maintain adequate depth to the groundwater-recharge mound and the ability to contain the percolate within a designated area.

Planning and design considerations for recovery well systems include spacing, depth, type of packing, and flowrate. These variables depend on the geology, soil, and groundwater conditions of the site, application rates, and the desired percentage of the renovated water to be recovered. The shape and configuration of the cone of depression after pumping must also be anticipated. To select the proper well spacing, the expected drawdown should be determined by installing test wells and making pumping tests.

13-5 OVERLAND-FLOW SYSTEMS

In overland-flow land treatment, wastewater is applied over the upper reaches of sloped terraces and allowed to flow across the vegetated surface to runoff collection ditches. The wastewater is renovated by physical, chemical, and biological means as it flows in a thin film down the relatively impermeable slope. Relatively little percolation is involved either because of an impermeable soil or a subsurface barrier (natural or constructed) to percolation.

Overland flow is a relatively new treatment process for domestic wastewater in the United States. The first overland-flow system was constructed in 1975 at Pauls Valley, Okla. In Melbourne, Australia, overland flow has been used to treat settled wastewater for several decades [30, 53]. The Campbell Soup Company treatment plant at Paris, Tex., is perhaps the best known of approximately 10 industrial systems in the country. Besides these full-scale examples, pilot-scale municipal studies are being conducted by the U.S. Environmental Protection Agency at Ada, Okla., and by the Corps of Engineers at Utica, Miss.

Design Objectives

The objectives of overland flow are wastewater treatment and, to a minor extent, crop production. Treatment objectives may be either (1) to achieve secondary or better effluent quality for screened wastewater that has received primary or lagoon treatment, or (2) to achieve high levels of nitrogen and BOD removals comparable to those achieved by conventional advanced wastewater treatment for wastewater that has received secondary treatment. Treated water is collected at the toe of the overland-flow slopes and can be either reused or discharged to surface water. Overland flow can also be used for production of forage grasses and the preservation of greenbelts and open space.

Site Selection

Soils with limited drainability, such as clays and clay loams, are suited to overland flow. The land should have a slope between 2 and 8 percent and a very smooth surface so that the wastewater will flow in a sheet over the ground surface. Slopes greater than 8 percent can be used successfully but may introduce problems, such as erosion, longer slopes to obtain adequate treatment, and difficulties in using farm machinery. Grass is planted to provide a habitat for the bacteria that provide the renovation. Because runoff is expected, a suitable means of final disposal should be provided.

Groundwater will not likely be affected by overland flow, so it is of minor concern in site selection. The groundwater table should be deeper than about 0.6 m (2 ft), however, so that the root zone is not waterlogged.

Preapplication Treatment

When overland flow is used as a secondary treatment process, the minimum pre-application treatment is screening and possibly grit and grease removal to avoid clogging the distribution system. No food crops are grown, and sprinkling systems can be designed to minimize the generation of mists by using rotating-boom sprays or by sprinkling at low pressures. In the pilot system at Ada, Okla., untreated comminuted wastewater, which was settled for 10 min for grease and grit removal, was applied successfully using rotating-boom sprays 9.1 m (30 ft) in diameter (see Fig. 13-14) discharging slightly downward at 10 N/cm^2 ($15 \text{ lb}_f/\text{in}^2$) [68].

Lagoon treatment or conventional secondary preapplication treatment may be advantageous when overland flow is used as an advanced wastewater-treatment process. Disinfection prior to application may avoid the need for postdisinfection of collected runoff and may allow sprinkling at higher pressures.

Climate and Storage

Because overland-flow treatment depends on microbiological activity at or near the surface of the soil, it is adversely affected by cold weather. As a result, storage is often necessary in the winter. In addition, reduction of application rates and longer drying periods may be necessary for periods of high precipitation.

The overland-flow system at Paris, Tex., has been studied extensively, and it

Figure 13-14 Overload-flow application at Ada, Okla. [*courtesy of Richard E. Thomas*].

has been confirmed that microorganisms normally found in the soil were responsible for oxidation or organic material. The removal efficiency of this system did not decrease during the winter months. Decreased metabolic activity was compensated for by an increase in microbial population [16], indicating that treatment can continue up to the point of freezing.

The storage design factors for overland flow are similar to those for irrigation of perennial crops, with one significant difference related to precipitation. The applied water is collected after treatment, and if runoff is to be strictly controlled or disinfected after collection, maximum daily application rates allowable during precipitation may be relatively small. On the other hand, if the wastewater is sufficiently treated and disinfected prior to application, stormwater runoff might not be of special concern, and storage requirements for periods of precipitation might be reduced.

Application Rates

Hydraulic and organic loadings are usually critical for overland flow. Removals of nitrogen and phosphorus are presented in Table 13-4.

Hydraulic loading Typical loading rates for overland-flow land-treatment systems range from 0.8 to 1.8 cm/d (0.3 to 0.7 in/d) for untreated wastewater or primary effluent, and from 2.1 to 5.7 cm/d (0.8 to 2.2 in/d) for secondary effluent. The water balance should be made mainly to determine the amount of runoff to be expected. The water-balance equation for overland flow is:

$$\begin{array}{c}\text{Design}\\\text{precipitation}\end{array} + \text{wastewater applied} = \text{evapotranspiration} + \text{percolation} + \text{runoff}$$

$$(13\text{-}8)$$

Design precipitation and evapotranspiration values are determined in the same manner as for irrigation systems. Losses to percolation will generally be on the order of 0.25 cm/d (0.1 in/d) or less. Percolation rates should be estimated under saturated or nearly saturated conditions. The runoff rate can be determined as the known values are entered into Eq. 13-8. A typical range of runoff values is from 40 percent (of the applied effluent plus precipitation) in the summer to 80 percent in the winter [16, 67, 69].

Nitrogen loading Two important mechanisms responsible for nitrogen removals in overland-flow systems are biological nitrification-denitrification and crop uptake. The overlying water film and organic matter and the underlying saturated soil form an aerobic-anaerobic double layer necessary for the completion of denitrification. These conditions are similar to those found in rice fields or marshes. The overland-flow process functions as a combination of nitrification and denitrification.

Organic loading The limits of organic loading for the overland-flow method are at present undefined. High-strength organic wastes have been treated at BOD_5 loadings of 44.8 to 112 kg/ha · d (40 to 100 lb/acre · d) [45]. Kirby [30] reports that the grass-filtration system at Melbourne, Australia, is loaded at 76.2 kg/ha · d (68 lb/acre · d) of BOD_5 with a 96 percent removal efficiency. Thomas [67] reports 92 to 95 percent removal of BOD_5 at loadings of 15.7 to 20.2 kg/ha · d (14 to 18 lb/acre · d). He observed the higher removals at the higher organic and liquid loading rates. Organic loading rates higher than 20 kg/ha · d can probably be used.

Because the organic matter is filtered out by the grass, litter, and topsoil, and is reduced by biological oxidation, the organic content of the soil is not affected substantially. However, high organic loadings may limit treatment efficiency as a result of the combination of effects of BOD_5 and liquid loading on the creation of anaerobic conditions. Because overland flow functions in a manner similar to a trickling filter, intermittent dosing with 6 to 8 h on and 16 to 18 h off has been used successfully [47]. In Australia, continuous dosing has been used for up to 6 months followed by 6 months of resting [30]. Provisions should be made to vary the resting period, depending on climatic conditions, harvesting requirements, and insect-control considerations.

Other elements Of the three major land-treatment processes, overland-flow systems are the least effective for phosphorus removal. Since there is very limited percolation of wastewater in these systems, the soil contact is limited to the soil surface area. The wastewater flowing over the soil surface does not have extensive contact with the iron and aluminum components of the soil that normally fix large amounts of phosphorus. In addition, the residence time on the soil is usually less than 24 h. However, some phosphorus appears to be removed by the organic layer on the surface of overland-flow slopes. This organic layer can be plowed under periodically to take advantage of the fixing capacity of the heavy-textured clay soil.

Trace-element removal by overland flow is very good. Hunt and Lee [25] report that rates of removal are over 90 percent for all metals and over 98 percent for some heavy metals. It is believed that most of the heavy metals are removed in the surface organic mat. As with phosphorus, the surface concentration of trace elements could be reduced periodically by plowing that layer under.

Land Requirements

Land requirements for a 0.044 m³/s (1 Mgal/d) overland-flow system typically range from 10 to 45 ha (25 to 110 acres). As with sprinkler irrigation, the area may be increased if buffer zones are required.

Terrace Characteristics

Uniform slopes between 2 and 6 percent with no depressions or gullies are preferred for overland-flow systems. The surface must be quite smooth to promote a thin sheet flow. A slope length of 53.3 m (175 ft) has been found to provide sufficient detention time to achieve effective treatment for the degradable food-processing wastewater at Paris, Tex. [20]. Typical lengths for domestic wastewater are 36 to 46 m (120 to 150 ft) [49].

Crop Selection

The cover crop is essential to the design because it serves as a habitat for the biota that are responsible for the oxidation of organic matter. The crop also serves to prevent erosion and to take up significant quantities of nutrients from the wastewater. Effective cover crops include Reed canary, tall fescue, trefoil, and Italian rye grasses. Italian rye grass is the dominant species in the grass-filtration system at Melbourne, Australia [30]. Other factors important in crop selection were discussed in Sec. 13-3.

Distribution Techniques

Sprinkling is the principal distribution technique presently practiced in the United States, but surface flooding is practicable for municipal wastewater. Factors that should be considered in the selection of the application technique include: topography, suspended solids in the wastewater, agricultural practices, and economics.

Sprinkling Wastewater is sprinkled on the upper reaches of a slope and allowed to flow downhill. Either fixed sprinklers or rotating-boom sprinklers may be used. Moving or portable systems are not used because a smooth surface must be maintained. Sprinklers are spaced from 18.3 to 24.4 m (60 to 80 ft) apart on the laterals.

Surface application Application by surface methods using a bubbling orifice at the top of the slope is shown in Fig. 13-15. Surface flooding has also been practiced successfully in Melbourne, Australia. If high concentrations of suspended solids are present, settling in the upper reaches may cause an odor problem.

Runoff Collection

More extensive runoff-control features are normally required for overland flow than for irrigation systems, because overland-flow systems are designed principally for runoff of applied effluent rather than percolation. Typically, 40 to 80 percent of the applied effluent runs off. The remainder is lost to percolation and evapotranspiration. In most cases, the runoff is collected in ditches at the toe of each terrace and then conveyed by open channel or gravity pipe to a discharge point, where it is monitored and, in some cases, disinfected. Discharge may be to surface waters, to reuse facilities, or sometimes to additional treatment facilities such as rapid infiltration.

Figure 13-15 Bubbling-orifice application to overland-flow slopes at Pauls Valley, Okla.

Storm runoff presents some special problems. Under conditions of light precipitation (about 0.13 to 0.25 cm/h (0.05 to 0.10 in/h), most overland-flow systems can be operated as usual. Although there is little documentation of the fact, the quality of runoff from greater amounts of precipitation should improve as a result of dilution, provided erosion is not caused. For the overland-flow system at Paris, Tex., rainfall events of 0.25 to 5.08 cm (0.1 to 2.0 in) increasingly reduced, from normal levels, the effluent conductivity in the runoff [16]. (Effluent conductivity was the only constituent recorded under these conditions.) Pathogenic organisms in the runoff should not generally cause serious problems if the effluent is disinfected before application.

13-6 OTHER SYSTEMS

The three principal land-treatment systems represent planned and engineered changes to the existing environment. Recently, the concept of using natural ecosystems, such as wetlands, for wastewater treatment has received considerable attention. Applications of wastewater to artificial and natural wetlands and the use of wastewater in aquaculture are described in this section.

Wetlands Application

Wetlands, which constitute 3 percent of the land area of the continental United States [79], are intermediate areas in a hydrological sense: they have too many plants and too little water to be called lakes, yet they have enough water to prevent most agricultural or silvicultural uses. The term wetlands is used here to encompass areas also known as marshes, bogs, wet meadows, peat lands, and swamps. The ability of wetlands to influence water quality is the reason for much current research on their use for wastewater management.

Three categories of wetlands are now used for municipal wastewater treatment: artificial wetlands, existing wetlands, and peat lands. Peat lands are discussed separately because these highly organic soils can be drained and managed in a manner similar to that used in irrigation.

Artificial wetlands Two artificial wetlands-treatment systems have been developed at the Brookhaven National Laboratory on Long Island, N.Y. [57]. Both are wetlands-pond systems. In the first system, the wetlands consist of wet meadows merging into a marsh followed by a pond (meadow marsh). In the second system, the wet meadows are eliminated.

These wetlands were formed in sandy soil by installing an impervious plastic liner under the soil. They were placed in operation in June 1973. Operating modes have evolved from the original recycling to the once-through approach with increasing loading rates. In April 1976 the rate of about 63 cm/week (25 in/week) was established. Wastewater is aerated for short periods prior to application. Typical averaged results for July through September 1975 for operation with a one-to-one recycling of pond effluent are presented in Table 13-24.

The wetlands area occupies 0.08 ha (0.2 acre) and is flooded to a depth of about 0.15 m (0.5 ft). Small recommends a 0.3-m (1-ft) depth or more to prevent volunteer weed growth and to prevent washout during storms [56]. Cattails were

Table 13-24 Treatment performance for two artificial wetland systems on Long Island, N.Y. [56][a]

Constituent	Influent	Meadow-marsh effluent	Wetlands effluent
BOD	520	15	16
Suspended solids	860	43[b]	57[b]
Total nitrogen	36	3	4
Fecal coliforms, count/100 mL	3000	17[c]	21[c]

[a] Units are expressed in mg/L, except as noted.
[b] Principally algae.
[c] Geometric mean.

Note: mg/L = g/m^3

Figure 13-16 Meadow-marsh system being harvested [*from Brookhaven National Laboratory*].

planted, and duckweed (*Lemna minor*) is prevalent. Regular harvest of cattails is not practiced, but weeds, grasses, and cattails were thinned out in March 1976 (see Fig. 13-16).

Existing wetlands The application of secondary effluent to existing freshwater and saltwater wetlands is being studied in Mississippi, as well as in California, Michigan, Louisiana, Florida, and Wisconsin. In Mississippi, Wolverton has studied the use of water hyacinths in secondary wastewater lagoons to effect removals of BOD, suspended solids, and nutrients [80]. A surface area of 0.28 ha (0.7 acre) was used, and detention times ranged from 14 to 21 d. The treatment performance of this system is compared to that of a control lagoon free of water hyacinths in Table 13-25.

Table 13-25 Treatment of oxidation-pond effluent by water hyacinths, mg/L [80]
(September 1975)

Constituent	Hyacinth pond		Control pond	
	Influent	Effluent	Influent	Effluent
BOD	22	7	27	30
Suspended solids	43	6	42	46
Total Kjeldahl				
nitrogen	4.4	1.1	4.5	4.5
Total phosphorus	5.0	3.8	4.8	4.6

Hyacinths must be harvested for effective nutrient removal. Wolverton suggests harvesting every 5 weeks during the warm growing season. The harvested plants may be processed into high-protein feed products, organic fertilizer and soil conditioner, or methane gas [81].

The use of existing wetlands appears to hold promise as an emerging technology for wastewater management. Management techniques for nutrient removal, loading rates, climatic constraints, and suitable site characteristics need further study.

Peat lands The use of peat lands or organic soils for land application has been studied by Kadlec in Michigan [28]. A system has been designed for the North Star Campground in Minnesota using sprinkler application [63]. This system was designed for 33.8 cm/week (13.3 in/week) and is underdrained at a depth of about 1 m (3 ft). Treatment efficiency for 1975 is summarized in Table 13-26. Secondary effluent was applied. Because of the high loading rate, the nitrogen uptake of the grass planted on the peat surface was exceeded. Although the peat pH was 4, the effluent pH was consistently between 6.5 and 7.5.

Aquaculture

Aquaculture is the term applied to the culture of aquatic organisms as a source of protein and food. Within the past 5 yr the potential of using aquaculture as a means of achieving both wastewater treatment and the production of usable plant and animal protein has received considerable attention. Based on what is now known, much remains to be done before it will be possible to meet both objectives in a single operation. Perhaps the most serious question is, "What are the health risks to animals and humans associated with the use of aquatic organisms grown in wastewater?"

Table 13-26 Treatment of wastewater on peat land in Minnesota [63][a]

Constituent	Influent	Effluent
BOD	...	5
Suspended solids	...	5
Total nitrogen	20–40	1–10
Total phosphorus	10	0.1
Fecal coliforms, count/100 mL	10^3–10^5	0–4

[a] Units are expressed in mg/L, except as noted.

Note: mg/L = g/m^3

Because it has been demonstrated that aquatic organisms (plants and animals) can be used to treat wastewater, it is important to note the distinction between the use of aquatic organisms for wastewater treatment and wastewater aquaculture. The objective of the former is treatment, whereas the objective of the latter is the production of usable fiber and food. In the future, it is anticipated that both the use of aquatic organisms for treatment and wastewater aquaculture will find wider application. The experiments at Woods Hole, Massachusetts, with shellfish and the experiments with catfish and shiners at Oklahoma City, Oklahoma, are examples of the types of research that must be done to define the application of such systems.

At the Woods Hole Oceanographic Institution, secondary effluent and seawater were blended and used to culture algae. The algae-laden water was then fed to oysters. Oysters require single-cell food, so the algae-laden water was passed through a homogenizer to break up any clumps of algal cells [22]. Other marine organisms that could be considered for aquaculture are listed in Table 13-27.

In the Quail Creek experiments at Oklahoma City, untreated wastewater was introduced into a serial lagoon system containing six cells. Catfish were added to the third and fourth cells, and golden shiners were added to the fifth and sixth cells. Although black bullhead, green sunfish, and mosquito fish (previously present) upset this initial balance and resulted in a more diverse polyculture than expected, the effects of the fish on water quality were impressive, as shown in Table 13-28. The detention time in the ponds was a total of 70 d.

Table 13-27 Some marine organisms that have commercial value and may be cultured in waste-recycling–aquaculture systems [22]

Food source	Species grown	Commercial value

Dissolved nutrients
   Domestic wastes
   Food-processing wastes
   Agricultural wastes
   Fertilizers
	Phytoplankton, benthic algae, seaweeds	Agar, food stabilizers, fertilizers, human food

Phytoplankton	Filter-feeding mollusks (oysters, clams, mussels, scallops); brine shrimp	Human food Tropical fish food
Seaweeds	Abalone, lobsters	Human food
Benthic algae	Omnivorous fish (grey mullet), shrimp	Human food
Brine shrimp	Fin fish (rainbow trout, flounders, puffers, juvenile lobsters); tropical fish	Human food Recreation
Detritus, feces, pseudofeces	Small crustaceans (polychaete worms)	Bait in sport fishing
Small crustaceans, worms, detritus	Carnivore fish (flounder), lobster, shrimp	Human food

Table 13-28 Wastewater-treatment performance of an aquaculture system at Oklahoma City [9]^a

Constituent	Untreated wastewater	Final cell effluent with fish	Final cell effluent without fish
BOD	184.0	6.0	13.0
Suspended solids	197.0	12.0	39.0
Total nitrogen	18.9	2.7	8.2
Total phosphorus	9.0	2.1	7.6
Fecal coliform, MPN/100 mL	3×10^6	20.0	<200.0
Turbidity	55.0	9.0	

^{a} All units are expressed in mg/L, except as noted.
  Note: mg/L = g/m^3

Fish were intensively sampled for bacterial contamination, but no pathogens were found. Additional studies were conducted without the fish; the results are shown in Table 13-28.

13-7 LAND APPLICATION OF SLUDGE

The discussion in this section deals with design objectives, sludge characteristics, site selection, market considerations, operational methods, and application rates. Sludge disposal in landfills is discussed in Chap. 11.

Design Objectives

Land application of sludge can have two objectives: (1) the disposal of waste products or residues, or (2) the use of nutrients and organic matter to fertilize crops and reclaim soil. In the first case, sanitary landfills or land spreading would be used. In the second case, sludge is considered a resource that can be used as a fertilizer, a soil conditioner, or (for liquid sludge) a source of irrigation.

Sludge Characteristics

Sludge quantities and characteristics that are important in the design of treatment facilities were discussed in Chap. 11. Sludge properties vary somewhat with the type of wastewater and type of treatment process. In the design of land-application systems, the most important sludge constituents are nutrients, such as nitrogen, phosphorus, and potassium, potentially toxic heavy metals and trace elements, and pathogens and parasites. Although minimum levels of metals, such as zinc, are required for plant growth, excessive quantities may be toxic to plants or detrimental to groundwaters and surface waters.

Municipal sludge usually contains most plant nutrients, but its fertilizer value varies according to the source of the waste and the wastewater treatment and sludge processes used. Common ranges and typical values of the principal nutrients in terms of percentages of the dry solids are shown in Table 13-29. The actual composition of the nutrients in the sludge as applied to the land is a function of the solids content and application rates.

Trace elements are essential to plant growth in minute quantities but toxic in higher concentrations. Their presence in wastewater sludge depends on the characteristics of the wastewater and the treatment processes used. Composition of wastewater is a function of the types of industrial activities in the service area. The wide range of composition of trace elements in various sludges is shown in Table 13-30. The median values are more useful than mean values because of the larger influence on mean values of a few high concentrations.

Site Selection

One of the most critical steps in land application of sludge is finding a suitable site. The characteristics of the site will determine the actual design and will influence the overall effectiveness of the land-application concept. The site-selection process should include an initial screening on the basis of the factors and criteria described in the following discussion. After the number of potential sites is narrowed, each site should be evaluated in detail, taking into account operational techniques and potential environmental impacts.

Proximity to critical areas Critical areas are those lands where application of wastewater sludge is precluded by major legal and institutional constraints. Isolation is an important characteristic of a proposed site. Actual distances from inhabited areas and bodies of water (surface and ground) must be known. Minimum requirements vary with climate, soil and geological characteristics, application techniques, and government regulations.

Accessibility The site should not be so isolated that it lacks access. Lack of nearby transport arteries, such as railroads, highways, or navigable waterways, may require construction of costly access roads or a pipeline.

Table 13-29 Common ranges and typical values for nutrients in digested sludge [5]

Nutrient	Digested sludge	
	Range	Typical
Volatile solids, %	30–60	40
Nitrogen (N), % total solids	1.6–6.0	3.0
Phosphorus (P_2O_5), % total solids	1.5–4.0	2.5
Potash (K_2O), % total solids	0.0–3.0	0.5

Table 13-30 Typical values for metals in various sludges [5]

Metal	Value, ppm Range	Mean	Median
Silver	nd[a]–960	225	90
Arsenic	10–50	9	8
Boron	200–1,430	430	350
Barium	nd–3,000	1,460	1,300
Beryllium	nd	nd	nd
Cadmium	nd–1,100	87	20
Cobalt	nd–800	350	100
Chromium	22–30,000	1,800	600
Copper	45–16,030	1,250	700
Mercury	0.1–89	7	4
Manganese	100–8,800	1,190	400
Nickel	nd–2,800	410	100
Lead	80–26,000	1,940	600
Strontium	nd–2,230	440	150
Selenium	10–180	26	20
Vanadium	nd–2,100	510	400
Zinc	51–28,360	3,483	1,800

[a] nd = not detected.

Slope Maximum ground slopes of 5 to 8 percent have been suggested [39, 52]. Steeper slopes could result in erosion and cause problems for operation of equipment.

Soil In terms of land reclamation or adding fertilizer, almost any soil—particularly agricultural soil—can be used for the application of sludge. The degree to which the soil can use the sludge depends on various physical and chemical properties of the soil. Soils should have the capability to filter, buffer, and absorb the sludge as well as to support crop growth. In general, desirable soils (1) have moderate permeabilities (1.5 to 15 cm/h or 0.6 to 6.0 in/h), (2) are well drained to moderately well drained, (3) are alkaline or neutral (pH > 6.5) to control heavy-metal solubility, and (4) are deep and relatively fine textured for high moisture and nutrient-holding capacity. Soils not having these properties can be used with modifications.

Geology A geologic investigation is particularly important in defining the nature of groundwater resources. The presence of faults, solution channels, or other similar connections between soil and groundwater will lessen the desirability of the site. The effect of land application on the quality and use of the groundwater must also be considered. Seismic hazards are another important factor in site selection.

Market Considerations

The primary markets for sludge are usually nearby farms. A cost-benefit analysis should be conducted to compare the value of the nutrients with the cost of delivery and application for various forms of sludge (e.g., liquid, de-watered). The actual price of the sludge should be no greater than that of commercial fertilizers, in terms of unit costs of available nutrients. Information on dried-sludge marketing is presented in Table 13-31.

Improvements in crop yield from sludge application can also be part of the marketing scheme. Many studies have shown moderate to dramatic increases in yields of corn, alfalfa, and forage crops when sludge is used as a fertilizer [26, 29, 76, 82]. In terms of productivity, an application rate of 10 to 22 Mg/ha (4.5 to 10 tons/acre) on a dry weight basis seems to be comparable to commercial fertilizer. The degree of increased productivity depends on soil types, sludge composition, climate, and overall management techniques, including actual application rates.

Operational Methods

Liquid sludges may be applied to land by methods similar to those used for wastewater application, i.e., sprinklers or ridge-and-furrow systems, or they may be directly spread by tank truck. Application of sludge in the liquid state is attractive because of its simplicity. Dewatering processes are not required, and inexpensive liquid-transfer systems can be used. Application of dewatered

Table 13-31 Marketing of dried sludge in the United States [5, 19]

Location	Treatment	Distribution	Product name
Amarillo, Tex.	Air-dried, pulverized, unbagged	User hauls	
Beltsville, Md.	Composted with wood chips	User hauls	
Boise, Id.		Retail outlets	BI Organic
Chicago, Ill.	Not composted; digested, dried and stockpiled	City hauls large loads	Nu Earth
Houston, Tex.	Heat-dried	Competitive bid; bagged for resale by contractor	Hou-Actinite
Los Angeles, Calif.	Windrow composted	Bagged for resale by Kellogg Co.	Nitrohumus
Milwaukee, Wis.	Heat-dried	Distributors	Milorganite
Philadelphia, Pa.		Given away	Philorganite
Winston-Salem, N.C.	Digested, air-dried, flash-dried	Distributors	Organiform-SS

sludge to the land is similar to an application of semisolid animal manure. This is an important advantage of dewatered sludge because private farmers can handle application on their lands with their own equipment.

Tank-truck spreading A common method of liquid sludge application is direct spreading by tank trucks with capacities ranging from 3.8 to 7.6 m³ (1000 to 2000 gal). Sludge is spread from a manifold on the rear of the truck as it is driven across the field. Application rates can be controlled either by valving on the manifold or by varying the speed of the truck. One modification of the basic process is to mount a spray apparatus on the truck so that a wider application area can be covered by each pass. The spray system can be refined so that larger tank trucks can operate from a network of roads at the application site.

As with haul, the principal advantages of a tank-truck system are low capital investment and ease of operation. The system is also flexible in that a variety of application sites can be served, such as pastures, golf courses, farmland, and athletic fields. Disadvantages include wet-weather problems and the high operating cost of the sludge haul. Tank trucks are not able to enter sites when the ground is soft; consequently, storage or wet-weather alternatives must be available. Special flotation tires used on the trucks can partially control this problem [78]. Repeated tank truck traffic may also reduce crop yields. This is primarily the result of damage to soil structure (higher bulk density, reduced infiltration) from the trucks rather than the sludge itself [29].

Sprinkling Wastewater sludge can be applied to the land by either fixed or portable sprinkler systems that have been designed to handle solids without clogging. The advantages of spraying include reduced operating labor, less land preparation, and use on a wide variety of plants. Operator attention is required to set portable sprinkler systems, but fixed units can be highly automated. Sprinklers can operate satisfactorily on land too rough or wet for tank trucks or injection equipment. The method can be used throughout the growing season.

Disadvantages include power costs of high-pressure pumps, contact of sludge with all parts of the crop, possible foliage damage to sensitive crops, and the potential for aerosol pollution from entrained pathogens. The problems of sludge contact with crops will limit the types of crops that can be grown. The aerosol problem can be controlled by buffer zones, low-pressure sprinkling, and operational control to avoid sprinkling on windy days.

Incorporation The principle of incorporation is to cut a furrow, deliver sludge into the furrow, and cover the sludge, all in one operation. A modification is an injection system in which the sludge is injected beneath the surface without turning over the soil. Sludge can also be trenched or plowed into the soil. The advantage of incorporation is the immediate mixture of sludge and soil; odor and vector problems that can arise from ponding sludge are eliminated. Surface runoff is also controlled.

The principal disadvantages of incorporation are its seasonal limitations and handling procedures. Application can be made only prior to the growing season or on noncultivated land; therefore, it would be difficult to sequence sites throughout the year. Handling procedure requires that a tank truck or trailer be part of the application system. Wet-weather operation is limited by the need to have equipment in the field.

Ridge-and-furrow methods Ridge-and-furrow sludge application is basically the same operation as ridge and furrow crop irrigation. Sludge flows in furrows between row crops, irrigating and fertilizing the soil. Advantages include the simplicity of the equipment involved and flexibility of use at existing sites.

The disadvantages are the settling of solids at the heads of the furrows and the need for well-prepared sites with proper gradients. Ponding of sludge in the furrows, which can result in odor problems, is also likely.

Application Rates

The rate at which sludge can be applied to agricultural land depends on sludge characteristics, soil types, climate, and crop to be grown. Nutrient content of the sludge, particularly nitrogen, and trace-element content are usually limiting factors in land application. The physical properties of the soil influence infiltration and water retention. Precipitation also affects the hydraulic characteristics of the soil. Performance of application equipment in wet or cold weather also affects the application rate.

Nitrogen The principal nutrients in sludge are the various forms of nitrogen and phosphorus. These elements are essential for plant growth but may also be serious water pollutants. The flow of nitrogen through the soil is a portion of the nitrogen cycle depicted in Fig. 3-18. In general, nitrogen in the soil is subject to plant uptake, adsorption or ion exchange, and volatilization or denitrification.

The three principal forms of nitrogen found in wastewater and sludge are organic, ammonia, and nitrate. In secondary digested sludge, about one-third of the total nitrogen is ammonia. The remaining fraction is in the organic form; there is very little nitrate. Organic nitrogen is generally in suspended form in wastewater and sludge. Consequently, it is filtered by the soil and decomposes to ammonia nitrogen, as shown in Fig. 13-3. Ammonia nitrogen is readily absorbed by soil and is thus resistant to leaching. Some ammonia nitrogen is lost to the atmosphere as volatilized ammonia gas. Some may also be nitrified to nitrates. Nitrates are available for plant uptake, but are not readily adsorbed by the soil. Nitrates not used by plants and not biologically denitrified can be leached to the groundwater.

The following equation can be used to determine nitrogen loading:

$$N = 0.1\ CL \tag{13-9}$$

where N = nitrogen loading, kg/ha · yr (lb/acre · yr)
  C = nitrogen content of sludge, mg/L
  L = application rate, cm/yr (in/yr)
  0.1 = conversion factor (0.225)

Total nitrogen must be used because all forms of nitrogen interact in the soil. The computed load should be compared with crop-uptake rates given in Table 13-6. Actual loading rates should be in excess of the values in Table 13-6, as not all nitrogen is available to plants. Some of the nitrogen is in organic or ammonia form and will be lost through volatilization and slowly mineralized into the nitrate form.

The rate at which organic nitrogen is converted to ammonia and nitrate nitrogen has been estimated to be 20 to 35 percent of the organic nitrogen in the first year, 5 to 15 percent of the residual organic nitrogen in the second year, and 1 to 5 percent in the third and subsequent years [26]. This "decay rate" is for a single application. To determine the total available nitrogen, applications in subsequent years must also be considered.

The nitrogen loading rate for the first year's application can be computed by the following equation:

$$N_1 = \frac{U}{r_1 D} - N_0 \qquad (13\text{-}10)$$

where N_1 = sludge in loading rate for the first year, dry kg/ha
  U = nitrogen uptake of the crop, kg/ha
  r_1 = fraction of applied nitrogen which is mineralized and available to plants in the first year
  D = percent (by weight) of nitrogen in dry sludge solids as applied to the soil, expressed as a fraction
  N_0 = initial available nitrogen content of the soil, kg/ha

For preliminary design, N_0 is usually assumed to be zero. Furthermore, in using Eq. 13-10, the residual nitrogen $N_1 D(1 - r_1)$ is assumed to be entirely the result of applied sludge N_1. In reality some is from crop residues, since not all the crop nitrogen is removed at harvest. This amount can be neglected for preliminary design.

The nitrogen content of the applied sludge D is the available nitrogen in that sludge. Losses from sprinkled or surface-applied sludge (not from incorporation) due to denitrification and volatilization could be significant [23] and should be evaluated.

Computation of nitrogen loading rates is illustrated in Example 13-2.

Example 13-2 Computation of nitrogen loading rate for sludge application Determine the dry-sludge application rate for a crop whose nitrogen uptake is 225 kg/ha · yr. Assume that 35 percent of the applied nitrogen is available to plants in the first year; 10 percent of the residual becomes available the second year; and 5 percent of the annual residual decays in the third and subsequent years. Also assume that initial nitrogen content of the soil is zero, that is, $N_0 = 0$. The sludge contains 2.5 percent nitrogen by weight, as applied to the soil.

SOLUTION
1 Determine the application rate for the first year (year 1).
    a   Solve Eq. 13-10 for N_1

$$N_1 = \frac{225 \text{ kg/ha}}{(0.35)(0.025)} - 0 \text{ kg/ha}$$

$$= 25,700 \text{ kg/ha}$$

$$= 25.7 \text{ Mg/ha (11.5 tons/acre)}$$

    This application rate is shown in column 1 of Table 13-32.
    b   Compute the actual amount of nitrogen applied.

$$(25,700 \text{ kg/ha})(0.025) = 643 \text{ kg/ha}$$

    This amount is shown in column 2 of Table 13-32.
2 Determine the residual nitrogen for year 1. The year 1 residual is 643 kg/ha $(1 - 0.35) = 418$ kg/ha and is shown in column 6 of Table 13-32.
3 Determine the residual nitrogen for year 2. In year 2, the residual 225 kg/ha from the year 1 application is transferred to column 3, as shown in the table, and decays at a rate of 0.10. Thus, $(418 \text{ kg/ha})(0.10) = 42$ kg/ha is available for crop uptake (column 5); $418 \text{ kg/ha} - 42 \text{ kg/ha} = 376$ kg/ha is the residual (column 6).
4 Determine the year 2 application and the residual for year 3. Since the total crop uptake has been given as 225 kg/ha and only 42 kg/ha is available from the first year's application, an application of sludge equivalent to $225 - 42 = 183$ kg/ha must be made in year 2. The 183 kg/ha decays at a rate of 0.35, so the nitrogen applied is $183/0.35 = 523$ kg/ha (column 2). At 2.5 percent nitrogen, the total dry-sludge application rate is $(523 \text{ kg/ha})/0.025 = 20,900$ kg/ha or 20.9 Mg/ha. The residual from this application is $(523 \text{ kg/ha})(0.65) = 340$ kg/ha (column 6).
5 Determine the application for year 3. In year 3, there are two residuals, each having a different decay rate. As with year 2, the residuals are first transferred to column 3. Crop uptakes are then computed in column 5, and the required new application determined as before.
6 Determine the application for year 4. The computation for year 4 and subsequent years proceeds exactly as above. Only year 4 is shown in this example.

Trace elements Trace elements, particularly heavy metals, are important not only because of their phytotoxicity but also because of their potential to accumulate in the soil and to enter the human food chain through crop uptake. Some elements, especially boron, are mobile and may contaminate groundwater. All trace elements can be phytotoxic in sufficient concentrations, but their toxicity to plants depends on several factors [11]:

1. Concentrations of such metals already in the soil
2. Soil characteristics such as pH and cation exchange capacity
3. Content of other substances, such as phosphates and organic matter
4. Length of time the elements have been in the soil

Furthermore, each crop has unique sensitivities to each element. In most crops, accumulation of most elements is greatest in leaves and edible roots and least in seed, fruit, and tubers. The effects on crops may also be evident in reduced

Table 13-32 Nitrogen balance for Example 13-2

| | Applied sludge Mg/ha[a,b] (1) | Nitrogen, kg/ha | | | | Nitrogen, kg/ha | |
		Applied in sludge[a,b] (2)	Residual from Previous year's application[a] (3)	Decay rates[c] (4)	Crop uptake, kg/ha (5)	Residual,[a] kg/ha (6)
Year 1	25.7	643	...	0.35	225[c]	418
Year 2						
Residual from year 1 application			418	0.10	42[a]	376
New application	20.9	523	...	0.35	183[a]	340
					225[c]	
Year 3						
Residual from year 1 application			376	0.05	19[a]	357
Residual from year 2 application			340	0.10	34[a]	306
New application	19.7	492	...	0.35	172[a]	320
					225[c]	
Year 4						
Residual from year 1 application			357	0.05	18[a]	339
Residual from year 2 application			306	0.05	15[a]	291
Residual from year 3 application			320	0.10	32[a]	288
New application	18.3	457	...	0.35	160[a]	297
					225[c]	

Note: Col. 2 = col. 1 × 0.025 × 1000
Col. 5 = col. 4 × col. 3 or col. 2
Col. 6 = col. 2 or col. 3 − col. 5
Col. 6 = col. 3 × (1 − col. 4)
kg/ha × 0.8922 = lb/acre
Mg/ha × 0.4461 = ton/acre

[a] Computed values.
[b] Assuming 2.5 nitrogen in dry-sludge solids.
[c] Given values.

yields. Data on trace elements in plants treated with wastewater sludge are summarized by Kirkham [32].

The complexity of soil–plant–trace metal interactions and the lack of data have resulted in a general lack of evaluative criteria for judging consequences of sludge application to land. The toxic effects of the trace elements themselves are understood, but there is no general agreement on how to relate trace-element content of sludge and soil to human health hazards [12]. Relating metal accumulation to cation exchange capacity, for example, is only approximate because sorption of metals is not strictly a function of ion exchange [49].

In some areas, plant tissues from crops grown in sludge-enriched soil have shown excessive concentrations of some trace metals [43]. However, data from the Chicago area indicate that accumulation of trace elements in plant tissues did not occur after 4 yr of irrigation with wastewater sludge [82]. Kirkham reported that after 35 yr of sludge application with 800 ppm cadmium, the corn had cadmium levels within the normal range [31].

DISCUSSION TOPICS AND PROBLEMS

13-1 An irrigation system for a flowrate of 30 L/s is designed for an application rate of 6 cm/week. For a year round operation, what is the required field area? If the system is designed for 36 week/yr of application, what is the required field area?

13-2 A sprinkler system is selected for application of the wastewater for irrigation. The sprinklers are spaced in a rectangular grid pattern of 12 m by 20 m, and each sprinkler nozzle discharges 2 L/s. What is the application rate in cm/h? How many hours must the system be operated in a single area each week to satisfy the application rate of 6 cm/week?

13-3 A rapid-infiltration system is designed for an application rate of 20 m/yr. The system is operated throughout the year on a cycle of 1 d of application followed by 7 d of drying. If the wastewater has a BOD of 60 mg/L, what is the average annual BOD loading rate in kg/ha? Over the 8-d cycle, what is the average BOD loading rate in kg/ha · d? For the first day of application, what is the loading rate in kg/ha · d?

13-4 Develop a water balance for an overland-flow system. Use the evapotranspiration and precipitation data from Table 13-10. Use a wastewater application rate of 80 cm/month, and assume a percolation rate of 10% of the application rate.

13-5 Determine the dry-sludge application rate for Reed canary grass on the basis of satisfying crop nitrogen uptake. Assume that a sludge containing 3 percent nitrogen by weight is applied to a soil that has an initial nitrogen content of zero. Use a decay rate of 30 percent for the first year, 15 percent for the second year, and 5 percent for the third and subsequent years.

13-6 A sludge containing 50 ppm of cadmium on a dry basis is to be applied to the land. If the limiting mass loading to the soil is set at 9 kg/ha, what would be the safe loading rate for 50 yr of application?

13-7 Using the data in Table 13-10, prepare a nitrogen balance for an irrigation system. The crop takes up 300 kg/ha (267 lb/acre) of nitrogen over the year. Distribute the uptake over 12 months in proportion to the percentage of total evapotranspiration occurring in that month. Assume that the applied wastewater contains 20 mg/L of total nitrogen and that loss to denitrification accounts for 25 percent of the applied nitrogen. In which month does the maximum load in the percolate occur? What is the concentration of total nitrogen in the percolate for this month?

13-8 A rapid infiltration system is designed to treat 60 L/s (1.37 Mgal/d) of primary effluent at a rate of 30 m/yr (98 ft/yr). What is the required field area? On the basis of a review of current literature, what should the soil permeability be to ensure a successful operation hydraulically? What would be the expected nitrogen removal?

13-9 An overload flow system is loaded at 20 cm/week (8 in/week) with a total nitrogen concentration in the applied wastewater of 25 mg/L. If the expected removal of nitrogen on a mass basis is 90 percent, estimate the amount of nitrogen that is removed each year. If the grass is coastal Bermuda grass, what percentage of this nitrogen removal can be accounted for in crop uptake?

13-10 Review at least four current articles on wetlands application and wastewater treatment with aquatic systems. What are the advantages and disadvantages of such treatment systems? What are the advantages and disadvantages of using water hyacinths for the partial treatment of wastewater? Cite the references reviewed.

REFERENCES

1. Aulenbach, D. B.: Thirty-five Years of Use of a Natural Sand Bed for Polishing a Secondary-Treated Effluent, in W. J. Jewell and R. Swan (eds.), *Water Pollution Control in Low Density Areas*, University Press of New England, Hanover, N.H., pp. 227–239, 1975.
2. Ayers, R. S., and R. L. Branson: *Guidelines for Interpretation of Water Quality for Agriculture*, University of California Cooperative Extension, 1975.
3. Ayers, R. S., and D. W. Westcot: *Water Quality for Agriculture*, Food and Agriculture Organization of the United Nations, Irrigation Drainage Paper No. 29, Rome, 1976.
4. Bagnall, L. O., et al.: Feed and Fiber from Effluent-grown Water Hyacinth, in *Wastewater Use in the Production of Food and Fiber*, U.S. Environmental Protection Agency, EPA 660/2-74-041, pp. 116–141, June 1974.
5. Bastian, R. K.: Municipal Sludge Management, in R. C. Loehr (ed.), *Land as a Waste Management Alternative*, Ann Arbor Science, Ann Arbor, pp. 673–689, 1977.
6. Bouwer, H.: Infiltration-Percolation Systems, in *Land Application of Wastewater*, Proceedings of a Research Symposium sponsored by the U.S. Environmental Protection Agency, Region III, Newark, Delaware, pp. 85–92, November 1974.
7. Bouwer, H., J. C. Lance, and M. S. Riggs: High-Rate Land Treatment II: Water Quality and Economic Aspects of the Flushing Meadows Project, *J. WPCF*, vol. 46, pp. 844–859, May 1974.
8. Bouwer, H., R. C. Rice, E. D. Escarcega, and M. S. Riggs: *Renovating Secondary Sewage by Ground Water Recharge with Infiltration Basins*, U.S. Environmental Protection Agency, Office of Research and Monitoring, Project No. 16060 DRV, March 1972.
9. Carpenter, R. L., M. S. Coleman, and R. Jarman: Aquaculture as an Alternative Wastewater Treatment System, in J. Tourbier and R. W. Pierson, Jr. (eds.), *Biological Control of Water Pollution*, University of Pennsylvania Press, Philadelphia, 1976.
10. Carroll, T. E., et al.: *Review of Landspreading of Liquid Municipal Sludge*, U.S. Environmental Protection Agency, Environmental Protection Technology Series, EPA-670/2-75-049, Cincinnati, June 1975.
11. Chaney, R. L.: Crop and Food Chain Effects of Toxic Elements in Sludges and Effluents, in *Proceedings of the Joint Conference on Recycling Municipal Sludges and Effluents on Land*, University of Illinois, Champaign, pp. 129–141, July 1973.
12. Chaney, R. L., S. B. Hornick, and P. W. Simon: Heavy Metal Relationships during Land Utilization of Sewage Sludge in the Northeast, in R. C. Loehr (ed.), *Land as a Waste Management Alternative*, Ann Arbor Science, Ann Arbor, pp. 283–314, 1977.
13. Crites, R. W.: Land Treatment of Wastewater by Infiltration-Percolation, in R. L. Sanks and T. Asano (eds.), *Land Treatment and Disposal of Municipal and Industrial Wastewater*, Ann Arbor Science, Ann Arbor, pp. 193–212, 1976.

14. Crites, R. W., and C. E. Pound: Present and Potential Land Treatment Practice in New England, paper presented at the annual meeting of the New England Water Pollution Control Association, Harwichport, Mass., October 1976.

15. Culp, G. L., and D. J. Hinrichs: *A Review of the Operation and Maintenance of the Muskegon County Wastewater Management System*, Muskegon County Board of Works, Muskegon, Mich., June 1976.

16. C. W. Thornthwaite Associates: An Evaluation of Cannery Waste Disposal of Overland Flow Spray Irrigation, *Publications in Climatology*, vol. 22, no. 2, September 1969.

17. Demirjian, Y. A.: *Land Treatment of Municipal Wastewater Effluents, Muskegon County Wastewater System*, U.S. Environmental Protection Agency, Office of Technology Transfer, Washington, D.C., 1975.

18. Ellis, B. G.: The Soil as a Chemical Filter, in W. E. Sopper and L. T. Kardos (eds.), *Recycling Treated Municipal Wastewater and Sludge through Forest and Cropland*, Pennsylvania State University Press, University Park, 1973.

19. Ettlich, W. F., and A. K. Lewis: Is There a "Sludge Market"? *Water Wastes Eng.*, vol. 13, no. 12, 1976.

20. Gilde, L. C., et al.: A Spray Irrigation System for Treatment of Cannery Wastes, *J. WPCF*, vol. 43, no. 10, October 1971.

21. Gilley, J. R.: Municipal Wastes as a Fertilizer Source, *Technology for a Changing World*, 1976 Technical Conference Proceedings, Sprinkler Irrigation Association, Silver Spring, Md., 1976.

22. Goldman, J. C., and J. H. Ryther: Waste Reclamation in an Integrated Food Chain System, in J. Tourbier and R. W. Pierson, Jr. (eds.), *Biological Control of Water Pollution*, University of Pennsylvania Press, Philadelphia, 1976.

23. Hartman, W. J., Jr.: *An Evaluation of Land Treatment of Municipal Wastewater and Physical Siting of Facility Installations*, U.S. Department of the Army, May 1975.

24. Heath, M. E., D. S. Metcalfe, and R. F. Barnes: *Forages*, 3d ed., Iowa State University Press, Ames, 1973.

25. Hunt, P. G., and C. R. Lee: Overland Flow Treatment of Wastewater—A Feasible Approach, in *Land Application of Wastewater*, Proceedings of a Research Symposium sponsored by the U.S. Environmental Protection Agency, Region III, Newark, Del., November 1974.

26. Hyde, H. C.: Utilization of Wastewater Sludge for Agricultural Soil Enrichment, *J. WPCF*, vol. 48, no. 1, January 1976.

27. Johnson, R. D., et al.: *Selected Chemical Characteristics of Soils, Forages, and Drainage Water from the Sewage Farm Serving Melbourne, Australia*, U.S. Army Corps of Engineers, January 1974.

28. Kadlec, J. A.: Dissolved Nutrients in a Peatland near Houghton Lake, Michigan, in *Proceedings of the National Symposium on Freshwater Wetlands and Sewage Effluent Disposal*, University of Michigan, Ann Arbor, pp. 25–50, May 1976.

29. Kelling, K. A., L. M. Walsh, and A. E. Peterson: Crop Response to Tank Truck Application of Liquid Sludge, *J. WPCF*, vol. 48, no. 9, September 1976.

30. Kirby, C. F.: *Sewage Treatment Farms*, Department of Civil Engineering, University of Melbourne, 1971.

31. Kirkham, M. B.: Trace Elements in Corn Grown on Long-Term Sludge Disposal Site, *Environ. Sci. Technol.*, vol. 9, pp. 765–768, August 1975.

32. Kirkham, M. B.: Trace Elements in Sludge on Land: Effect on Plants, Soils, and Ground Water, in R. C. Loehr (ed.), *Land as a Waste Management Alternative*, Ann Arbor Science, Ann Arbor, pp. 209–247, 1977.

33. Kirkwood, J. P.: *The Pollution of River Waters*, Seventh Annual Report of the Massachusetts State Board of Health, Wright & Potter, Boston, 1876 (reprint edition, Arno Press, Inc., 1970).

34. Krone, R. B.: The Movement of Disease-producing Organisms through Soils, in *Proceedings of the Symposium on Municipal Sewage Effluent for Irrigation*, Louisiana Polytechnic Institution, pp. 75–105, July 1968.

35. Lance, J. C., F. D. Whisler, and H. Bouwer: Oxygen Utilization in Soils Flooded with Sewage Water, *J. Environ. Qual.*, vol. 2, no. 3, 1973.

36. Loehr, R. C.: *Agricultural Waste Management—Problems, Processes and Approaches*, New York, Academic, 1974.

37. McGauhey, P. H., and R. B. Krone: *Soil Mantle as a Wastewater Treatment System*, Sanitary Engineering Research Laboratory Report No. 67-11, University of California, December 1976.

38. McMichael, F. C., and J. E. McKee: *Wastewater Reclamation at Whittier Narrows*, California State Water Quality Control Board Publication No. 33, 1966.

39. Manson, R. J., and C. A. Merrit: Land Application of Liquid Municipal Wastewater Sludges, *J. WPCF*, vol. 47, no. 6, 1975.

40. Melsted, S. W.: Soil-Plant Relationships (Some Practical Considerations in Waste Management), in *Proceedings of the Joint Conference on Recycling Municipal Sludges and Effluents on Land*, University of Illinois, Champaign, pp. 121–128, July 1973.

41. Metcalf & Eddy, Inc.: *Wastewater Engineering: Collection, Treatment, Disposal*, McGraw-Hill, New York, 1972.

42. Morris, C. E., and W. J. Jewell: Regulations and Guidelines for Land Application of Wastes, in R. C. Loehr (ed.), *Land as a Waste Management Alternative*, Ann Arbor Science, Ann Arbor, pp. 63–78, 1977.

43. Page, A. L.: *Fate and Effects of Trace Elements in Sewage Sludge When Applied to Agricultural Lands*, U.S. Environmental Protection Agency, Environmental Protection Series, EPA-670/2-74-005, Cincinnati, 1974.

44. Pair, C. H. (ed.): *Sprinkler Irrigation*, 4th ed., Sprinkler Irrigation Association, Silver Spring, Md., 1975.

45. Pound, C. E., and R. W. Crites: *Wastewater Treatment and Reuse by Land Application*, vols. I and II, U.S. Environmental Protection Agency, Office of Research and Development, August 1973.

46. Pound, C. E., R. W. Crites, and D. A. Griffes: *Costs of Wastewater Treatment by Land Application*, U.S. Environmental Protection Agency, Office of Water Program Operations, EPA 430/9-75-003, June 1975.

47. Pound, C. E., R. W. Crites, and D. A. Griffes: *Land Treatment of Municipal Wastewater Effluents, Design Factors—I*, U.S. Environmental Protection Agency, Technology Transfer Program, October 1975.

48. Powell, G. M.: Design Seminar for Land Treatment of Municipal Wastewater Effluents, U.S. Environmental Protection Agency Technology Transfer Program, presented at Technology Transfer Seminar, 1975.

49. *Process Design Manual for Land Treatment of Municipal Wastewater*, U.S. Environmental Protection Agency, Technology Transfer, October 1977.

50. Rafter, G. W.: *Sewage Irrigation, Part II*, U.S. Geological Survey Water Supply and Irrigation Paper No. 22, 1899.

51. Reed, S. C., et al.: *Wastewater Management by Disposal on the Land*, Corps of Engineers, U.S. Army Cold Regions Research and Engineering Laboratory, Hanover, N.H., May 1972.

52. Schmid, J., D. Pennington, and J. McCormick: Ecological Impacts of the Disposal of Municipal Sludge onto the Land, *Proceedings of the 1975 National Conference on Municipal Sludge Management and Disposal*, Information Transfer, Inc., Washington, D.C., 1975.

53. Seabrook, B. L.: *Land Application of Wastewater in Australia*, U.S. Environmental Protection Agency, Office of Water Programs, EPA-430/9-75-017, May 1975.

54. Sepp, E.: Disposal of Domestic Wastewater by Hillside Sprays, *J. Environ. Eng. Div., ASCE*, vol. 99, no. 2, pp. 109–121, April 1973.

55. Sepp, E.: *The Use of Sewage for Irrigation—A Literature Review*, Bureau of Sanitary Engineering, California State Department of Public Health, Berkeley, 1971.

56. Small, M. M.: Natural Sewage Recycling Systems, paper presented at the New York Water Pollution Control Association Winter Meeting, January 1977.

57. Small, M. M.: *Meadow/Marsh Systems as Sewage Treatment Plants*, Brookhaven National Laboratory, BNL20757, Upton, N.Y., November 1975.

58. Smith, J. L., and D. B. McWhorter: Continuous Subsurface Injection of Liquid Organic Wastes, in R. C. Loehr (ed.), *Land As a Waste Management Alternative*, Ann Arbor Science, Ann Arbor, pp. 643–656, 1977.

59. Soil-Plant-Water Relationship, Irrigation, chap. 1, *SCS National Engineering Handbook*, section 15, U.S. Department of Agriculture, Soil Conservation Service, March 1964.

60. Sorber, C. A.: Public Health Aspects of Land Application of Wastewater Effluents, in *Land Application of Wastewater*, Proceedings of a Research Symposium sponsored by the U.S. Environmental Protection Agency, Region III, Newark, Delaware, pp. 27–33, November 1974.

61. Sorber, C. A., et al.: A Study of Bacterial Aerosols at a Wastewater Irrigation Site, *J. WPCF*, vol. 48, no. 10, October 1976.

62. Spangler, F. L. W. E. Sloey, and C. W. Fetter, Jr.: *Wastewater Treatment by Natural and Artificial Marshes*, EPA-600/2-76-207, University of Wisconsin, Oshkosh, Wis., September 1976.

63. Stanlick, H. T.: Treatment of Secondary Effluent Using a Peat Bed, in *Proceedings of the National Symposium on Freshwater Wetlands and Sewage Effluent Disposal*, University of Michigan, Ann Arbor, pp. 257–268, May 1976.

64. Sullivan, R. H., et al.: *Survey of Facilities Using Land Application of Wastewater*, U.S. Environmental Protection Agency, Office of Water Program Operations, EPA-430/9-73-006, July 1973.

65. Thomas, R. E.: Land Disposal II: An Overview of Treatment Methods, *J. WPCF*, vol. 45, pp. 1476–1484, July 1973.

66. Thomas, R. E., and T. W. Bendixen: Degradation of Wastewater Organics in Soil, *J. WPCF*, vol. 41, pp. 808–813, 1969.

67. Thomas, R. E., B. Bledsoe, and K. Jackson: *Overload Flow Treatment of Raw Wastewater with Enhanced Phosphorus Removal*, U.S. Environmental Protection Agency, Office of Research and Development, EPA-600/2-76-131, June 1976.

68. Thomas, R. E., K. Jackson, and L. Penrod: *Feasibility of Overland Flow for Treatment of Raw Domestic Wastewater*, U.S. Environmental Protection Agency, Office of Research and Development, EPA-660/2-74-087, July 1974.

69. Thomas, R. E., J. P. Law, Jr., and C. C. Harlin, Jr.: Hydrology of Spray Runoff Wastewater Treatment, *J. Irrig. Drainage Div., ASCE*, vol. 96, no. 3, pp. 289–298, September 1970.

70. Uiga, A., I. K. Iskandar, and H. L. McKim: Wastewater Reuse at Livermore, California, in R. C. Loehr (ed.), *Land as a Waste Management Alternative*, Ann Arbor Science, Ann Arbor, pp. 511–532, 1977.

71. Uiga, A., and R. S. Sletten: An Overview of Land Treatment from Case Studies of Existing Systems, paper presented at the 49th Annual Water Pollution Control Federation Conference, Minneapolis, October 1976.

72. U.S. Environmental Protection Agency: *Evaluation of Land Application Systems*, Office of Water Program Operations, EPA-430/9-75-001, March 1975.

73. U.S. Environmental Protection Agency: *Proposed Environmental Protection Agency Regulations on Interim Primary Drinking Water Standards*, 40 CFR 141, Dec. 24, 1975.

74. Viessman, W., Jr., T. E. Harbaugh, and J. W. Knapp: *Introduction to Hydrology*, Intext Educational Publishers, New York, 1972.

75. Wallace, A. T.: Rapid Infiltration Systems—The Process of Site Selection, prepared for the U.S. Environmental Protection Agency Design Seminar for Land Treatment of Municipal Wastewater Effluents, 1975.

76. Walker, J. M.: Sewage Sludges—Management Aspects for Land Application, *Compost Science*, March-April, 1975.

77. Walker, J. M.: *Wastewater: Is Muskegon County's Solution Your Solution?* U.S. Environmental Protection Agency, Region V, Office of Research and Development, September 1976.

78. White, R. K.: Land Application of Sewage Sludge, Technology for a Changing World, *1976 Technical Conference Proceedings*, Sprinkler Irrigation Association, Silver Spring, Md., 1976.

79. Witter, J. A., and S. Croson: Insects and Wetlands, in *Proceedings of the National Symposium on Freshwater Wetlands and Sewage Effluent Disposal*, University of Michigan, Ann Arbor, pp. 269–295, May 1976.

80. Wolverton, B. C., and R. C. McDonald: *Water Hyacinths for Upgrading Sewage Lagoons to Meet Advanced Wastewater Treatment Standards*, Part 1, National Aeronautics and Space Administration Technical Memorandum TM-X-72729, Bay St. Louis, Miss., October 1975.

81. Wolverton, B. C., et al.: *Bio-conversion of Water Hyacinths into Methane Gas*, Part 1, National Aeronautics and Space Administration Technical Memorandum TM-X-72725, Bay St. Louis, Miss., July 1975.
82. Zenz, D. R., et al.: *U.S. Environmental Protection Agency Guideline on Sludge Utilization and Disposal—A Review of Its Impact upon Municipal Wastewater Treatment Agencies*, Metropolitan Sanitary District of Greater Chicago Report No. 75-20, October 1975.

FOURTEEN

EFFLUENT DISPOSAL AND REUSE

The sanitary engineer can design a treatment plant to accomplish as much removal of pollutants as may be required. Ultimate disposal of wastewater effluents will be by dilution in receiving waters; by discharge on land; or, in some cases in desert areas, by evaporation into the atmosphere as well as seepage into the ground. Disposal by dilution (after secondary treatment) in larger bodies of water, such as lakes, rivers, estuaries, or oceans, is by far the most common method.

The fundamental thesis governing the disposal of effluents and the regulation of pollution is to make the treatment plants do part of the work and to let nature complete it. In the past, however, the balance was often shifted and nature was called upon to do far more than its share of the work. As a consequence, the assimilative capacity of receiving waters was exceeded and conditions of pollution resulted. The amount of natural or self-purification that occurs in the receiving water depends on its flow or volume, its oxygen content, and its ability to reoxygenate itself. The proportion of the self-purification capacity, sometimes called the assimilative capacity, that can be safely utilized in rivers, lakes, or estuaries depends on the uses to which the water is subjected elsewhere, the desires of the people, and the assimilative capacity of the receiving-water system. An understanding of this subject is important to sanitary and public health engineers employed by government regulatory agencies who must set appropriate standards regulating waste discharges, and to sanitary engineers who must select the degree of treatment and type of plant required, based on the applicable receiving-water or effluent standards.

The purpose of this chapter is to introduce the reader to the general subject of wastewater disposal in the aquatic environment. The disposal of treated effluent on land was considered in detail in Chap. 13. This chapter is not meant to be an exhaustive treatment of the subject but is designed to expose the student to the important issues and some of the techniques and approaches used to assess the assimilative capacity of the aquatic environment. The topics discussed are (1) water pollution control, (2) disposal into lakes, (3) disposal into rivers, (4) disposal into estuaries, and (5) disposal into the ocean. The temporary disposal of treated wastewater by direct and indirect reuse is also considered briefly.

14-1 WATER-POLLUTION CONTROL

Water-pollution control is concerned with the protection of the aquatic environment and the maintenance of water quality in lakes, reservoirs, streams, rivers, estuaries, and the sea. The desired or required water quality that must be maintained depends on the uses to be made of the water. Therefore, water-quality criteria must be available for alternative beneficial uses if the adequacy of various pollution-control measures is to be assessed properly. Domestic water supply, industrial water supply, agricultural water supply, water for recreational use, and water for fish, other aquatic life, and wildlife are well-established beneficial uses.

Water-quality criteria and related information on these and other beneficial uses may be found in *Water Quality Criteria* by McKee and Wolf [10], *Water Quality Criteria* by the National Technical Advisory Committee [27], *Water Quality Criteria 1972* by the National Academy of Science [26], and *Quality Criteria For Water* [9]. Water-quality criteria for industrial use were suggested by a committee of the New England Water Works Association in 1940 and are still generally applicable [4]. When questions of water quality arise, these reference works should be consulted as a starting point (before the specialized literature).

Receiving-Water Standards

Once the criteria necessary for the protection of the various beneficial uses have been established, it is possible to set standards for surface waters with the stipulation that no discharge shall create conditions that violate them. These standards are known as receiving-water or stream standards. An example of such standards were those established by the Ohio River Sanitation Commission for surface waters to be used for various purposes in the Ohio River Valley. Some state and interstate commissions have taken the approach of classifying streams in several categories in accordance with the highest beneficial use to be made of the stream. This use is based, to a certain extent, on existing conditions. In the early 1970s, all states were ordered by the federal government to adopt standards, subject to federal approval, that would maintain or enhance the existing quality of receiving waters, and all have now complied.

Effluent Standards

A difficulty in enforcing receiving-water standards arises when the combined load of several dischargers exceeds the self-purification capacity of the receiving waters. In this case, it may be difficult for the regulatory agency to allocate the responsibility. It is also possible for the discharger farthest upstream in a river basin to preempt most of the assimilative capacity for his or her own use, leaving little or none for those located downstream, which is unfair. To avoid some of these problems, the U.S. Environmental Protection Agency has set effluent standards (see Table 4-1), which are relatively easy to enforce compared with receiving-water standards. In some instances, individual states have set even more restrictive effluent standards.

Ideally, effluent standards should be strict enough to protect the quality of the receiving waters, and should treat all dischargers fairly, but insofar as possible should be tailored to the character and volume of wastes at each point of discharge. A greater degree of treatment may logically be required of large waste flows than of small or insignificant waste flows, but this is not the current situation.

Setting of Standards

Although the setting of standards comes under the jurisdiction of state and local regulatory agencies, there are fundamental considerations applicable to the setting of standards. It should be obvious that oil, grease, and floating solids, especially those of wastewater origin, should be removed from wastes before discharge to receiving waters. Solids that may settle and form sludge banks should also be removed. These requirements should apply to all wastes and all receiving waters.

Degradable organic matter, through exertion of its BOD, uses the dissolved oxygen of the receiving waters to stabilize the wastes. The dissolved oxygen is replenished mainly by atmospheric reaeration, photosynthesis of algal life, and additional downstream dilution (see Sec. 14-3, "Oxygen Resources in Rivers"). The minimum dissolved oxygen necessary for fish life must be assured by the setting of proper standards. A common minimum requirement is 5.0 mg/L of dissolved oxygen. In canals used exclusively for drainage, where fish are not a factor, the dissolved oxygen must not be allowed to drop to zero or a nuisance due to foul odors may develop.

Where water supplies, bathing beaches, livestock, or water-contact sports may be affected, the bacterial content of the receiving waters, as measured by the number of coliform organisms, is of prime importance. Bacterial numbers are reduced by initial dilution and also by a relatively rapid death rate with the passage of time and distance from the point of discharge.

Conservative pollutants (those that do not decay), such as chemical constituents, are reduced in concentrations mainly by dilution. The dilution necessary for the discharge of toxic compounds may be assessed by the bioassay test (see Chap. 3).

The discharge of nutrients into small lakes and reservoirs and also into sluggish streams consisting of a series of pools or slack-water reaches, may promote excessive growths of algae and other forms of microscopic life. The result of this growth is undesirable turbidity and floating scum while the organisms are growing. As the organisms die, odors are produced, oxygen is consumed in the decay of cells, and the nutrients are returned to the watercourse. Where such conditions can develop, reduction of the nutrient content of wastewaters before discharge may be necessary. Fortunately, however, such conditions are not general.

Heated-water discharges, after initial mixing, should not increase the temperature of the main body of the receiving waters above 35°C (95°F), if fish life is to be preserved. An excessive increase in the temperature of the wastewater–diluting water mixture above that of the diluting water should not be permitted. Increases should be limited to 1 to 1.5°C (3 to 5°F), according to the report of the National Technical Advisory Committee [27]. At the present time, the permissible temperature increase is a matter of controversy. Detailed discussion of thermal pollution may be found in Krenkel and Parker [9, 16].

Effluent Disposal by Dilution

To describe the effects of a waste discharge on a body of water, the sanitary engineer must understand the physical phenomena that are occurring in that body of water. Any mathematical relationship, no matter how complicated, is simply an attempt to describe the physical phenomena. Therefore, a physical rather than purely mathematical understanding of a wastewater system is a prerequisite to intelligent analysis. The subjects discussed in the following sections include the special conditions applying to disposal by dilution in lakes, rivers, estuaries, and oceans, in that order, and the development of mathematical relationships that can be applied to the majority of cases arising in practice.

14-2 DISPOSAL INTO LAKES

In many inland locations where nearby streams are not available, it may be necessary to discharge treated wastewater into lakes or reservoirs. The analysis of the effects of such discharges is considered in this section.

Problem Analysis

Lakes and reservoirs are often subject to significant mixing due to wind-induced currents; therefore, it is often reasonable to assume that small lakes and reservoirs are completely mixed (i.e., as in a continuous-flow stirred-tank reactor). The theoretical model based on this assumption is shown in Fig. 14-1. To simplify the calculations, it is assumed that flowrates are constant and that the decay of

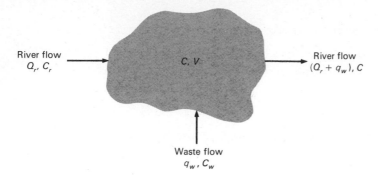

Figure 14-1 Schematic of complete-mix model for small lakes and reservoirs.

pollutants can be described with a first order reaction ($r_c = -K'C$). Writing a mass balance around the lake yields:

Accumulation = inflow − outflow + generation + utilization

$$V \frac{dC}{dt} = (Q_r C_r + q_w C_w) - (Q_r + q_w)C + 0 + (-K'C)V \qquad (14\text{-}1)$$

where V = lake volume
dC/dt = change in pollutant concentration in the lake
Q_r = river flowrate into lake
C_r = concentration of pollutant in river
q_w = wastewater flow rate into lake
C_w = concentration of pollutant in wastewater
K' = first-order decay constant (base e)

Now let

$$W = Q_r C_r + q_w C_w$$

and the detention time

$$t_0 = \frac{V}{Q}$$

where Q is the total flow into the lake $Q_r + q_w$. Substituting the foregoing values into Eq. 14-1 yields

$$\frac{dC}{dt} + C \left(\frac{1}{t_0} + K' \right) = \frac{W}{V} \qquad (14\text{-}2)$$

This is a first-order linear differential equation, which, upon integration (as outlined in Sec. 5-2), yields

$$C = \frac{W}{\beta V} (1 - e^{-\beta t}) + C_0 e^{-\beta t} \qquad (14\text{-}3)$$

where $\beta = 1/t_0 + K'$
$C_0 = $ concentration in the lake at time $t = 0$

The equilibrium concentration can be found by letting t equal ∞. Upon performing this operation, the equilibrium concentration C_e becomes

$$C_e = \frac{W}{\beta V} \qquad (14\text{-}4)$$

The determination of equilibrium waste concentration in a small lake is illustrated in Example 14-1.

Example 14-1 Radioactive-waste concentration in a lake A radioactive waste with a half-life of 1 d, a concentration of 10 mg/L, and a flow of 0.15 m³/s (5.3 ft³/s) is continually discharged to a small reservoir with an average length of 90 m (295 ft), an average width of 45 m (150 ft), and an average depth of 3 m (10 ft). Wind currents are such that the lake contents are completely mixed. Determine the equilibrium concentration of radioactive waste in the lake.

SOLUTION

1 Determine the decay constant for the waste.

$$T_{1/2} = \text{half-life} = 1 \text{ d}$$

$$L_t = L_0 e^{-K't} \qquad \text{(first-order decay rate)}$$

When $t = 1$ d,

$$L_t = \tfrac{1}{2}L_0$$
$$\tfrac{1}{2}L_0 = L_0 e^{-K'(1.0)}$$
$$\ln \tfrac{1}{2} = -K'(1.0)$$
$$K' = 0.693 \text{ d}^{-1}$$

2 Determine the lake volume.

$$V = (90 \text{ m})(45 \text{ m})(3 \text{ m}) = 12{,}150 \text{ m}^3 \ (429{,}065 \text{ ft}^3)$$

3 Determine the equilibrium concentration of the waste. Assume that the waste flow represents the total inflow to the lake.

$$t_0 = \frac{V}{Q} = \frac{12{,}150 \text{ m}^3}{(0.15 \text{ m}^3/\text{s})(86{,}400 \text{ s/d})} = 0.94 \text{ d}$$

$$W = q_w C_w = (0.15 \text{ m}^3/\text{s})(86{,}400 \text{ s/d})(10 \text{ g/m}^3)$$

$$= 129{,}600 \text{ g/d}$$

$$\beta = \frac{1}{t_0} + K' = \frac{1}{0.94 \text{ d}} + 0.693 \text{ d}^{-1} = 1.76 \text{ d}^{-1}$$

$$C_e = \frac{W}{\beta V} = \frac{129{,}600 \text{ g/d}}{(1.76/\text{d})(12{,}150 \text{ m}^3)}$$

$$C_e = 6.1 \text{ g/m}^3 = 6.1 \text{ mg/L}$$

Stratification in Large Lakes

The complete-mixing assumption cannot be applied to stagnant or extremely large lakes. In such cases, a different model should be used, based on the physical phenomena found to exist in the water system under study. (Disposal into large lakes, such as the Great Lakes, is more akin to ocean disposal and can be handled by a modification of the method of Example 14-4.)

Particularly significant is the vertical stratification common during certain seasons of the year. A complete-mix model would not be a good representation of a stratified lake because waste would not normally distribute itself over the entire lake volume. However, a soluble waste might be considered completely mixed in the upper layers (epilimnion) of a small, deep lake.

Stratification in lakes is the result of an increase in water density with depth caused by a decrease in temperature. The maximum density occurs at 4°C. During the spring, most lakes are of nearly uniform temperature and therefore are easily mixed by wind currents. As summer approaches, the upper waters are warmed, their density is thereby decreased, and a stable stratification is produced.

Three zones are normally present in a stratified lake: epilimnion, thermocline, and hypolimnion (see Fig. 14-2). The epilimnion may be 9 to 15 m (30 to 50 ft) deep and is fairly uniform in temperature because of mixing by wind action. The thermocline is a zone of significant temperature change and is extremely resistant to mixing. During the fall, temperatures drop, decreasing the amount of stratification until wind action may again completely mix the lake waters. This phenomenon is known as the fall turnover. In colder climates where the temperature drops below 4°C, a winter stratification may also occur because water density again decreases as the temperature falls below 4°C. It is also important to evaluate such parameters as wind, waves, and currents, because most mixing in a lake or reservoir body is the result of these forces.

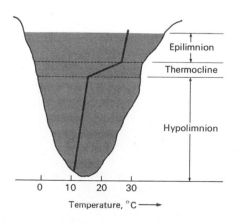

Figure 14-2 Schematic of summer stratification in a lake. Note: $1.8(°C) + 32 = °F$.

14-3 DISPOSAL INTO RIVERS

The late Earle B. Phelps, one of the early leaders in the field of stream sanitation, wrote: "A stream is something more than a geographic feature, a line on a map, a part of the fixed permanent terrain. It cannot be adequately portrayed in terms of topography and geology. A stream is a living thing, a thing of energy, of movement, of change" [18]. Streams or rivers are subject to much natural pollution because they serve as drainage channels for large areas of the countryside. In addition, rivers are capable of absorbing some pollution caused by humans because they possess the ability to purify themselves through the action of living organisms that consume organic matter and the sedimentation process that contributes to the river bottom.

Reoxygenation in Rivers

Aside from the oxygen contained in tributaries, surface drainage, and groundwater inflow, the sources of oxygen replenishment in a river are reaeration from the atmosphere and photosynthesis of aquatic plants and algae. The amount of reaeration is proportional to the dissolved-oxygen deficiency. The amount of oxygen supplied by photosynthesis is a function of the size of the algal population and the amount of sunlight reaching the algae. There is more incident radiation when the sun is high in the sky than when it is near the horizon; therefore, the rate of photosynthesis is assumed to be sinusoidal. Respiration, on the other hand, is assumed to be constant because it does not depend on light radiation. Where large populations of algae are present, a diurnal variation in dissolved-oxygen concentration occurs, as shown in Fig. 14-3.

The rate of reaeration r_R is defined as

$$r_R = K'_2(C_s - C) \tag{14-5}$$

where K'_2 = reaeration constant, d^{-1} (base e)
$\quad\quad C_s$ = dissolved oxygen saturation concentration, mg/L
$\quad\quad C$ = dissolved oxygen concentration, mg/L

The reaeration constant can be estimated by determining the characteristics of the stream and using one of the many empirical formulas that have been

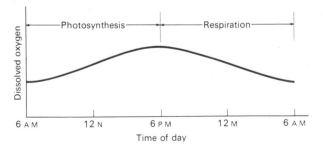

Figure 14-3 Diurnal variation of dissolved oxygen in water containing large algal populations.

proposed over the past 50 years. A generalized formula proposed by O'Conner and Dobbins for natural streams [15] is

$$K_2' = \frac{294(D_L U)^{1/2}}{H^{3/2}}$$ (14-6)

where D_L = molecular diffusion coefficient for oxygen, m²/d
  U = mean stream velocity, m/s
  H = average depth of flow, m

The variation of the coefficient of molecular diffusion with temperature can be approximated with the following expression

$$D_{L_T} = 1.760 \times 10^{-4} \text{ m}^2/\text{d} \times 1.037^{T-20}$$ (14-7)

where D_{L_T} = molecular diffusion coefficient for oxygen at temperature T, m²/d
  1.760×10^{-4} = molecular diffusion coefficient for oxygen at 20°C, m²/d
  T = temperature, °C

Values of K_2' have been estimated for various water bodies by the Engineering Board of Review for the Sanitary District of Chicago [18], as given in Table 14-1.

Table 14-1 Reacration constants [21]

Water body	Ranges of K_2' at 20°C (base e)[a]
Small ponds and backwaters	0.10–0.23
Sluggish streams and large lakes	0.23–0.35
Large streams of low velocity	0.35–0.46
Large streams of normal velocity	0.46–0.69
Swift streams	0.69–1.15
Rapids and waterfalls	> 1.15

[a] For other temperatures use $K_{2_T}' = K_{2_{20}}' 1.024^{T-20}$
  Note: 1.8(°C) + 32 = °F

Deoxygenation in Rivers

The oxygen in rivers and streams is depleted (1) by the bacterial oxidation of the suspended and dissolved organic matter discharged to them by both natural and man-made sources, and (2) by the oxygen demand of sludge and benthic deposits. Each is considered separately in the following discussion.

Deoxygenation due to organic matter The concept of biological oxidation of organic matter was introduced in Chap. 3. The amount of oxygen required to stabilize a waste is normally measured by the BOD_5 test; BOD_5 is therefore the primary source of oxygen depletion or utilization in a waterway. The rate of deoxygenation r_D is

$$r_D = -K'L \tag{14-8}$$

where K' = first order reaction rate constant, d^{-1}
L = ultimate BOD at point in question, mg/L

Substituting $L_0 e^{-K't}$ for L, Eq. 14-8 can be rewritten as

$$r_D = -K'L_0 e^{-K't} \tag{14-9}$$

where L_0 = ultimate BOD at the point of discharge, mg/L

Deoxygenation due to sludge deposits Matter that is too heavy to remain in suspension will settle out, forming a sludge deposit or benthic layer on the bottom of the stream. Sludge deposits in the bottoms of slow-moving rivers can exert a significant oxygen demand on the overlying water. Although most of the sludge will be undergoing anaerobic decomposition, which is a relatively slow process, aerobic decomposition can take place at the interface between the sludge and the flowing water.

Rates of deposition and scour vary with the velocity and turbulence of the river. At times, sedimentation may reduce the BOD load in the river, if settleable solids are being discharged or if coagulation of colloidal matter takes place. At other times, scour will increase the BOD load by returning these particles to the river. The benthic load must be assessed for each reach of river to determine its importance in the total oxygen balance. In many rivers, it is a factor that can be eliminated from consideration.

Where organic mud and sludge deposits are appreciable, their effect can be evaluated by means of an empirical equation developed by Fair, Moore, and Thomas [5]:

$$Y_m = 3.14(10^{-2}y_0)C_T w \frac{5 + 160w}{1 + 160w}\sqrt{t_a} \tag{14-10}$$

where Y_m = maximum daily benthal oxygen demand, g/m^2
y_0 = BOD_5 of benthal deposit, gm/kg of volatile matter at 20°C
w = daily rate of volatile-solids deposition, kg/m^2
t_a = time during which settling takes place, d
C_T = temperature correction factor
$= BOD_5 @ T/BOD_5 @ 20°C = (1 - e^{-5K_{T'}})/(1 - e^{-5K_{20'}})$

Development of Oxygen-Sag Model

In most river analyses, it is assumed that wastes that are discharged to the river are distributed evenly over the cross section of the river. This may be far from the truth in the immediate vicinity of the outlet, but the validity of the assumption improves in most cases as the wastes proceed downstream. It can also be assumed that no mixing occurs along the axis of the river, which is reasonable provided the river is not extremely turbulent.

If it is assumed that the river and waste are mixed completely at the point of discharge, then the concentration of a constituent in the river-waste mixture at $x = 0$ is

$$C_0 = \frac{Q_r C_r + q_w C_w}{Q_r + q_w} \tag{14-11}$$

where C_0 = initial concentration of constituent at point of discharge, mg/L
   Q_r = river flowrate, m³/s
   C_r = concentration of constituent in river before mixing, mg/L
   q_w = wastewater flowrate, m³/s
   C_w = concentration of constituent in wastewater, mg/L

The change in the oxygen resources of a river can be modeled by assuming that the river is essentially a plug-flow reactor (see Fig. 14-4). Over any incremental volume, the following mass balance can be written:

Accumulation = inflow − outflow + deoxygenation + reoxygenation

$$\frac{\partial \overline{C}}{\partial t} d\Psi = QC - Q\left(C + \frac{\partial C}{\partial x} dx\right) + r_D d\Psi + r_R d\Psi \tag{14-12}$$

Substituting for r_D and r_R yields

$$\frac{\partial \overline{C}}{\partial t} d\Psi = QC - Q\left(C + \frac{\partial C}{\partial x} dx\right) - K'L \, d\Psi + K_2'(C_s - \overline{C}) \, d\Psi \tag{14-13}$$

If steady-state conditions are assumed, $\partial C/\partial t = 0$ and the expression above can be simplified to

$$0 = -Q \frac{dC}{dx} dx - K'L \, d\Psi + K_2'(C_s - C) \, d\Psi \tag{14-14}$$

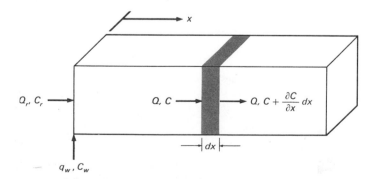

Figure 14-4 Definition sketch for plug-flow model used in river analysis.

Substituting $A\,dx$ for $d\Psi$ and dt for $A\,dx/Q$, Eq. 14-14 becomes

$$\frac{dC}{dt} = -K'L + K'_2(C_s - C) \qquad (14\text{-}15)$$

If the oxygen deficit D is defined as

$$D = (C_s - C) \qquad (14\text{-}16)$$

then the change in the deficit with time is

$$\frac{dD}{dt} = -\frac{dC}{dt} \qquad (14\text{-}17)$$

Using the relationships above, Eq. 14-15 written in terms of the oxygen deficit is

$$\frac{dD}{dt} = K'L - K'_2 D \qquad (14\text{-}18)$$

Substituting for L and rearranging yields

$$\frac{dD}{dt} + K'_2 D = K'L_0\, e^{-K't} \qquad (14\text{-}19)$$

Using the procedure detailed in Sec. 5-2 of Chap. 5, and noting that when $t = 0$, $D = D_0$ the integrated form of Eq. 14-19 is given by Eq. 14-20.

$$D_t = \frac{K'L_0}{K'_2 - K'}\,(e^{-K't} - e^{-K_2't}) + D_0\, e^{-K_2't} \qquad (14\text{-}20)$$

where D_t = oxygen deficit at time t, mg/L
$\quad\quad D_0$ = initial oxygen deficit at the point of waste discharge at time $t = 0$, mg/L

This is the classic Streeter-Phelps oxygen-sag equation, which is most commonly used in river analysis. It must be used with caution, however, because it applies to channels of uniform cross section where effects of algae and sludge deposits are negligible.

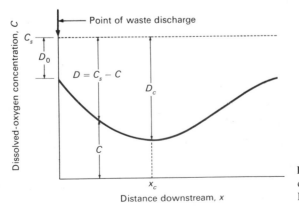

Figure 14-5 Characteristic oxygen-sag curve obtained using the Streeter-Phelps equation.

The graphical representation of the Streeter-Phelps oxygen sag equation is shown in Fig. 14-5. Active biological decomposition begins immediately after discharge. This decomposition utilizes oxygen. Because atmospheric reaeration is proportional to the dissolved-oxygen deficit, its rate will increase with increasing deficit. Finally a point is reached at which the rate of oxygen utilized for waste decomposition equals the rate of atmospheric reaeration. This is the point x_c in Fig. 14-5. Downstream from this point, the rate of reaeration is greater than the rate of utilization and the dissolved oxygen begins to increase. Eventually, the stream will show no effects due to the waste discharge. This is the phenomenon of natural stream purification.

The critical dissolved-oxygen deficit D_c at the point x_c is of engineering significance. This critical deficit can be determined by setting dD/dt in Eq. 14-19 to zero. When this is done the critical deficit is given by

$$D_c = \frac{K'}{K_2'} L_0 e^{-K't_c} \tag{14-21}$$

where $t_c =$ the time required to reach the critical point. The value of t_c can be determined by differentiating Eq. 14-20 with respect to t and setting dD/dt equal to zero.

$$t_c = \frac{1}{K_2' - K'} \ln \left[\frac{K_2'}{K'} \left(1 - \frac{D_0(K_2' - K')}{K'L_0} \right) \right] \tag{14-22}$$

The distance x_c is equal to

$$x_c = t_c v \tag{14-23}$$

where $v =$ velocity of flow in river.

The use of the Streeter-Phelps equation in determining the dissolved-oxygen sag in a river is illustrated in Example 14-2. Additional details on the oxygen-sag method of analysis may be found in Refs. 14, 18, 24, and 25.

The same approach can be used to determine the dissolved-oxygen distribution over the reach of a river downstream from a source of BOD loading when the dissolved-oxygen concentration is affected by the following factors

$$K'L = \text{BOD exertion}$$
$$K_2'(C_s - C) = \text{reaeration}$$
$$P = \text{photosynthesis}$$
$$R = \text{algal respiration}$$
$$S = \text{sludge deposits}$$

Performing a mass balance for the dissolved-oxygen concentration, the following steady-state differential equation is derived:

$$\frac{dC}{dt} = -K'L + K_2'(C_s - C) + P - R - S \tag{14-24}$$

Expressed in terms of the oxygen deficit, the integrated form of Eq. 14-24 is given by Eq. 14-25.

$$D = \frac{K'L_0}{K_2' - K'} (e^{-K't} - e^{-K_2't}) + D_0 e^{-K_2't} + \frac{S + R - P}{K_2'} (1 - e^{-K_2't}) \quad (14\text{-}25)$$

The use of this equation requires the evaluation of many parameters, in particular S, R, and P. The magnitude of the effects of algae and sludge deposits on the oxygen economy of a river can be determined only from detailed testing and analysis of the river in question.

Example 14-2 Dissolved-oxygen sag in a river A city discharges 115,000 $m^3/d = 1.33$ m^3/s (30.4 Mgal/d) of wastewater into a stream whose minimum rate of flow is 8.5 m^3/s (300 ft^3/s). The velocity of the stream is about 3.2 km/h (2 mi/h). The temperature of the wastewater is 20°C, and that of the stream is 15°C. The 20°C BOD_5 of the wastewater is 200 mg/L, and that of the stream is 1.0 mg/L. The wastewater contains no dissolved oxygen, but the stream is 90 percent saturated upstream of the discharge. At 20°C, K' is estimated to be 0.30 per day and K_2' is 0.7 per day. Determine the critical-oxygen deficit and its location. Also estimate the 20°C BOD_5 of a sample taken at the critical point. Use temperature coefficients of 1.135 for K' and 1.024 for K_2'. Also plot the dissolved-oxygen-sag curve.

SOLUTION
1   Determine the dissolved oxygen in the stream before discharge.
   a Saturation concentration, at 15°C = 10.2 mg/L (see Appendix C)
   b Dissolved oxygen in stream = 0.9(10.2 mg/L) = 9.2 mg/L
2   Determine the temperature, dissolved oxygen, BOD_5, and L_0 of the mixture.
   a Temperature of mixture =

$$\frac{(1.33 \text{ m}^3/\text{s})(20°\text{C}) + (8.5 \text{ m}^3/\text{s})(15°\text{C})}{1.33 \text{ m}^3/\text{s} + 8.5 \text{ m}^3/\text{s}} = 15.7°\text{C}$$

   b Dissolved oxygen of mixture =

$$\frac{(1.33 \text{ m}^3/\text{s})(0 \text{ mg/L}) + (8.5 \text{ m}^3/\text{s})(9.2 \text{ mg/L})}{1.33 \text{ m}^3/\text{s} + 8.5 \text{ m}^3/\text{s}} = 8.0 \text{ mg/L}$$

   c BOD_5 of mixture =

$$\frac{(1.33 \text{ m}^3/\text{s})(200 \text{ mg/L}) + (8.5 \text{ m}^3/\text{s})(1 \text{ mg/L})}{1.33 \text{ m}^3/\text{s} + 8.5 \text{ m}^3/\text{s}} = 27.9 \text{ mg/L}$$

   d L_0 of mixture =

$$\frac{27.9 \text{ mg/L}}{1 - e^{-0.3(5)}} = 35.9 \text{ mg/L}$$

3   Determine the initial dissolved oxygen deficit
   a Dissolved oxygen saturation concentration at 15.7°C = 10.1 mg/L (see Appendix C)
   b Initial deficit

$$D_0 = (10.1 - 8.0) \text{ mg/L} = 2.1 \text{ mg/L}$$

4   Correct the rate constants to 15.7°C.
   a $K' = 0.3(1.135)^{15.7-20} = 0.174/\text{d}$
   b $K_2' = 0.7(1.024)^{15.7-20} = 0.63/\text{d}$

5   Determine t_c and x_c.

a $t_c = \dfrac{1}{K_2' - K'} \ln \left[\dfrac{K_2'}{K'} \left(1 - \dfrac{D_0(K_2' - K')}{K'L_0} \right) \right]$

$t_c = \dfrac{1}{0.63 - 0.174} \ln \left[\dfrac{0.63}{0.174} - \left(\dfrac{(2.1 \text{ mg/L})(0.63 - 0.174)}{0.174(35.9 \text{ mg/L})} \right) \right]$

$t_c = 2.46$ d

b $x_c = vt_c = (3.2 \text{ km/h})(24 \text{ h/d})(2.46 \text{ d}) = 188.9 \text{ km } (117.4 \text{ mi})$

6   Determine D_c and the dissolved oxygen in the stream at x_c

a

$$D_c = \dfrac{K'}{K_2'} L_0 \, e^{-K'(x_c/v)}$$

$$D_c = \dfrac{0.174}{0.63} (35.9 \text{ mg/L})[e^{-0.174(2.46)}] = 6.5 \text{ mg/L}$$

b Dissolved oxygen in stream at x_c

$$DO_c = 10.1 - 6.5 = 3.6 \text{ mg/L}$$

7   Determine the BOD_5 of a sample taken at x_c.

$$L_t = (35.9 \text{ mg/L})[e^{-0.174(2.46)}] = 23.4 \text{ mg/L}$$

$$20°C \; BOD_5 = (23.4 \text{ mg/L})[1 - e^{-0.3(5)}] = 18.2 \text{ mg/L}$$

8   Compute some additional points on the curve and plot the oxygen sag curve (see Fig. 14-6).

Distance, km	0	60	120	188.9	240	300
DO, mg/L	8.0 (step 2b)	5.2	4.0	3.6 (step 6b)	3.8	4.1

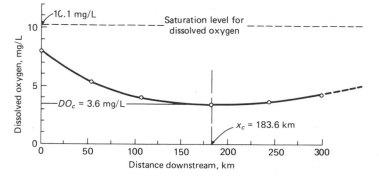

Figure 14-6 Oxygen-sag curve for Example 14-2.

Design of River Diffusers

In the previous discussion, a mathematical model was developed to describe the effects on the oxygen resources of domestic wastes discharged to rivers. Equally important to analysis is the design of the physical facilities used to introduce the wastes into the river. In the past, it was common practice to terminate the discharge pipeline from the treatment facilities at the bank of a river (see Fig. 14-7a). Under low-flow conditions, this usually led to the development of foam because the three essential elements necessary for foam formation were present: air, water, and foaming agents (almost always present in wastewater). The problem of foam can be overcome simply by submerging the pipe discharge below the low-water level. In some cases, this will be difficult to do, because the waste discharge may be larger than the flow in the stream.

Although the simple discharge arrangement shown in Fig. 14-7b will work, the preferred arrangement involves the use of some type of diffuser system. In its simplest form, a diffuser is a pipe with holes drilled in it at equal or varying spacings. Although the use of diffusers is desirable, there is a problem with their use, especially in streams classified as navigable by the Corps of Engineers where permission to place a pipeline across the bottom of the river or stream

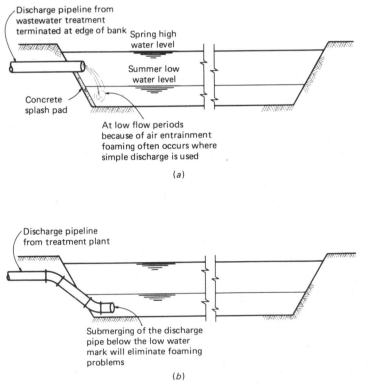

Figure 14-7 Simple river discharges for treated wastewater.

may not be granted. In such cases, it may be necessary to use a diffuser system that runs parallel to the bank of the river. By sizing the discharge ports properly, it is possible to distribute the wastewater across a larger section of the river. The subject of diffusers is considered further in Sec. 14-5, which deals with the discharge of wastewater into the ocean.

14-4 DISPOSAL INTO ESTUARIES

An estuary can roughly be defined as the zone in which a river meets the sea. There are between 850 and 900 estuary systems along the coastal periphery of the United States. They are becoming increasingly important to the sanitary engineer because nearly 50 million people in this country live near estuaries.

The analysis of estuaries is, in general, more complicated than the analysis of rivers or lakes. The ebb and flow of tides may cause significant lateral mixing in the reaches of rivers near the estuary. Indeed, the incoming tide often reverses the direction of flow in the section of river near the ocean. Usually, estuarine waters are vertically stratified. Seawater is heavier than fresh water; therefore, a layered system is often encountered, with fresh water riding above seawater.

Some estuaries, such as San Francisco Bay, are so large that complete-mix models are clearly inadequate. Where the physical processes are extremely complicated, it may be necessary to resort to physical rather than purely mathematical models of the basin. In any case, a physical understanding of the flow processes in an estuary is necessary before any rational analysis.

In many estuarine channels, tidal action merely increases the amount of mixing and dispersion of the waste along the length of the channel. The effect of this phenomenon is shown in Fig. 14-8. If a dye were released instantaneously in a river in which there was no mixing in the lengthwise direction and only advection (drift) occurred, the concentration of the waste at various points downstream would be represented by the rectangles. If mixing is significant, however, the plug of dye will diffuse outward and mix with surrounding waters as it moves downstream. Then the dye concentration will assume the shape of the bell-shaped areas of Fig. 14-8.

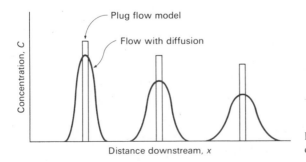

Figure 14-8 Effect of diffusion on flow characteristics.

Mathematical Analysis of Estuaries

The same type of analysis used for the case of flow with no mixing can be used in the case where dispersion is significant, provided a term is added that takes into account the effects of dispersion. The following equation is commonly used in describing these effects:

$$\frac{\partial M}{\partial t} = -EA\frac{\partial C}{\partial x} \qquad (14\text{-}26)$$

where $\partial M/\partial t$ = mass flow due to eddy diffusion
$\partial C/\partial x$ = concentration gradient
A = cross-sectional area
E = coefficient of eddy diffusion

Whenever a concentration gradient $\partial C/\partial x$ exists, a flow of mass $\partial M/\partial t$ occurs in such a way as to reduce the concentration gradient. It is assumed that the flowrate is proportional to the concentration gradient and the cross-sectional area over which this gradient acts. The proportionality constant is E and is commonly called the coefficient of eddy diffusion or turbulent mixing.

The plug-flow model shown in Fig. 14-9 will be used to illustrate the case of flow with diffusion. Assuming a first-order reaction $(r_c = -K'C)$ and performing a mass balance following the procedures delineated in Sec. 14-3 yields

$$\frac{\partial C}{\partial t} = E\frac{\partial^2 C}{\partial x^2} - v\frac{\partial C}{\partial x} - K'C \qquad (14\text{-}27)$$

where $v = Q/A$ = velocity of flow in the river.

Integrating for the steady-state case $(\partial C/\partial t = 0)$ results in the following equations:

1. For the instantaneous release of dye at $x = 0$ and $t = 0$, as shown in Fig. 14-10,

$$C = C_0\,e^{[-(x')^2/4Et]-Kt} \qquad (14\text{-}28)$$

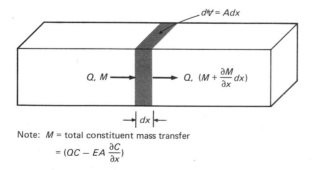

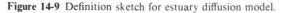

Note: M = total constituent mass transfer
$= (QC - EA\frac{\partial C}{\partial x})$

Figure 14-9 Definition sketch for estuary diffusion model.

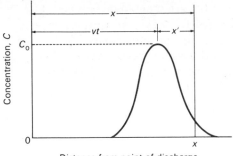

Concentration, C

Distance from point of discharge

Figure 14-10 Definition sketch for estuary diffusion model for instantaneous dye release.

where $C_0 = \dfrac{M}{2A \sqrt{\pi E t}}$

M = mass discharged at $t = 0$
$x' = x - vt$

2. For the continuous discharge of waste at a rate W,

$$C = C_0 \, e^{jx} \tag{14-29}$$

where $C_0 = \dfrac{W}{Q\sqrt{1 + 4KE/v^2}}$

$$j = \frac{v}{2E}\left(1 \pm \sqrt{1 + \frac{4KE}{v^2}}\right)$$

The positive root for j refers to the upstream $(-x)$ direction, and the negative root refers to the downstream $(+x)$ direction.

Determination of Coefficient of Eddy Diffusion

The foregoing analysis is most often applied to estuarine systems where significant mixing occurs as a result of tidal action. The coefficient of eddy diffusion is used as the measure of this mixing. Several approaches have been taken in the determination of the magnitude of the coefficient of eddy diffusion, including (1) mathematical formulations and (2) field measurements using Eq. 14-28 with measured values of conservative tracers, such as dyes or salt. A mathematical formulation by Harleman is given in Eq. 14-30 [6]:

$$E = C n v R^{5/6} \tag{14-30}$$

where E = coefficient of eddy diffusion, m^2/s (ft^2/s)
C = 63.2 (SI units); 77 (U.S. customary units)
n = Manning's roughness coefficient
v = velocity, m/s (ft/s)
R = hydraulic radius, m (ft)

Such an equation is useful in giving an order-of-magnitude answer. If E is significant, its magnitude should be determined from field measurements.

One method of measuring E in the field is the instantaneous-dye-release method. A conservative dye $(K = 0)$ is released from the point of waste discharge at slack tide, and the distribution of dye is measured at a later slack tide. Since t is known, E can be determined using Eq. 14-31.

$$\ln \frac{C}{C_0} = -\frac{x'^2}{4Et} \tag{14-31}$$

Values of $\ln C/C_0$ and x'^2 are plotted and the slope of the line is $-1/4Et$.

Another method of field measurement is the salinity intrusion method. The conservative substance is the salt concentration $(K = 0)$, and Eq. 14-29 is rewritten as

$$C = C_0\, e^{(v/E)x} \tag{14-32}$$

The plus root is used because distances are measured upstream from the sea. Values of C, x, and v can be measured and E can be determined as in the dye-release method using Eq. 14-33:

$$\ln \frac{C}{C_0} = \frac{v}{E} x \tag{14-33}$$

Measurements should be made preferably at slack tide but in any case at the same point in the tidal cycle. Additional details on the analysis of estuaries may be found in Refs. 8 and 24.

The use of the estuarine model with diffusion is illustrated in Example 14-3.

Example 14-3 Salinity intrusion in an estuary A city is located on an estuary, which it uses as a water supply. The city is 30 km (18.6 mi) upstream from the ocean. The estuary has a uniform cross-sectional area of 450 m² (4844 ft²), and E has been measured to be 25 km²/d (9.6 mi²/d). The chloride concentration in the ocean is 18,000 mg/L. At a freshwater outflow of 30 m³/s (1060 ft³/s), the chloride concentration at the city is 18.0 mg/L. A dam is to be built 30 km upstream of the city to prevent loss of fresh water to the ocean. If the resultant freshwater outflow is reduced to 3.0 m³/s (106 ft³/s), what will the chloride concentration be?

SOLUTION

1 Determine the net velocity of the freshwater outflow.

$$v = \frac{Q}{A} = \frac{3 \text{ m}^3/\text{s}}{450 \text{ m}^2} = 0.0067 \text{ m/s} = 0.576 \text{ km/d}$$

2 Determine j.

$$j = \frac{v}{E} = \frac{0.576 \text{ km/d}}{25 \text{ km}^2/\text{d}} = 0.023/\text{km}$$

3 Determine the chloride concentration.

$$C = C_0\, e^{jx} = (18{,}000 \text{ mg/L})[e^{0.023(-30)}] = 9018 \text{ mg/L}$$

This is decidedly too salty to drink and furthermore does not meet the U.S. Public Health Service standard of 250 mg/L; therefore, the city will have to build a pipeline to take its drinking water from above the proposed dam.

14-5 DISPOSAL INTO THE OCEAN

Ocean disposal is typically accomplished by submarine outfalls that consist of a long section of pipe to transport the wastewater some distance from shore and, in the best examples, a diffuser section to dilute the waste with seawater. A typical diffuser used for the discharge of wastewater into the ocean is shown in Fig. 14-11. Diffusers are one of the most efficient methods of providing initial dilution of a waste in any waterway; but, because most of the design parameters for diffusers originated from work done on ocean outfalls, their design and operation are discussed in this section.

At the end of the outfall, treated or untreated wastewater is released in a simple stream or jetted through a manifold or multiple-point diffuser (see Fig. 14-12). At this point, the wastewater mixes with surrounding seawater and rises to the surface where it drifts as a wastewater field in accordance with the prevailing ocean currents. This drift or movement with the currents is termed

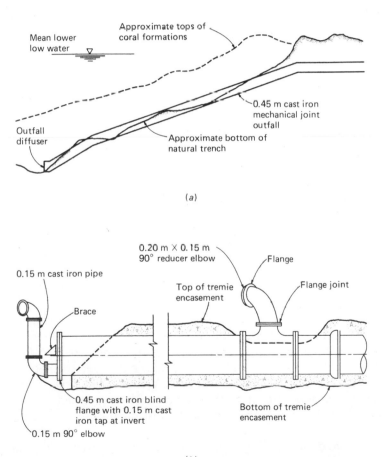

(a)

(b)

Figure 14-11 Small capacity ocean outfall. (a) General layout of outfall. (b) Typical diffuser section.

Figure 14-12 Underwater photograph of wastewater discharge from diffuser port.

advection. At the same time, the field is also diffusing outward into the surrounding water. Hence, the coefficient or eddy diffusion is important. If the ocean is sufficiently stratified at the point of discharge, it may be possible to maintain a submerged wastewater field.

The design of an outfall should meet applicable receiving-water standards. Because the initial dilution from an efficient diffuser is so large that the reduction in dissolved oxygen is usually of no significance, bacterial, floatable material, nutrient, and toxicity requirements will govern the design and location of most outfalls. Accurate estimation of the number of coliform bacteria requires taking into account their reduction due to die-off, flocculation, and settling.

In summary, where the wastewater field rises to the surface, the phenomena of importance in the design of ocean outfall systems are the initial dilution of the waste, the waste dispersion into surrounding waters, and the waste decay rate.

Estimation of Initial Dilution

When a waste is discharged from a single- or multiple-port diffuser, the velocity of the jet will cause turbulent mixing with the surrounding water. If the waste is of lower specific gravity than the dilution water, the mixing jet will bend upward and may eventually reach the surface. If, as previously noted, the ocean is vertically stratified with an upper layer of warm water riding over colder water, it is possible to dilute the waste sufficiently with cold water so that the specific gravity of the waste–cold water mixture is greater than that of the warm upper layer. In such a case, the waste plume will remain submerged under the upper layer. Such vertical stratification is most often observed during warm summer months.

Rawn, Bowerman, and Brooks [20] have developed curves from field data for determining the initial dilution D_1 in a waste jet issuing from a horizontal port. Abraham [1] extended these curves to show that the dilution is a function of the depth Y_0 of the discharge port, the diameter of the discharge orifice D, and the Froude number F, for a liquid-liquid system, as shown in Fig. 14-13.

The Froude number is defined as

$$F = \frac{V_j}{\sqrt{(\Delta S/S)gD}} \tag{14-34}$$

where V_j = jet velocity
ΔS = difference in specific gravity between the waste and the surrounding seawater
S = specific gravity of the waste
g = acceleration due to gravity
D = discharge-jet diameter

The specific gravity of seawater normally varies between 1.010 and 1.030, and that for wastewater ranges from 0.990 to 1.000. Along the California coast, the specific gravity of seawater is typically taken as 1.025 and that of wastewater is taken as 0.999; therefore, the value of $\Delta S/S$ typically used in California is 0.026.

When moderately strong currents are encountered, the initial dilution may also be estimated from the equation

$$D_1 = \frac{V_x bd}{Q} \tag{14-35}$$

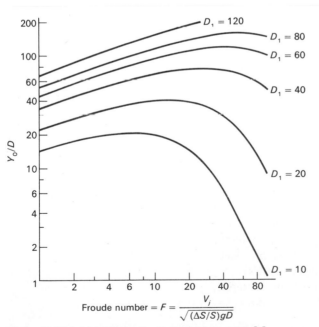

Figure 14-13 Initial dilution by turbulent jet mixing [1].

where V_x = current velocity, m/s or m/h (ft/s or ft/h)
 b = effective width of diffuser system, m (ft)
 d = average depth of the wastewater field, m (ft)
 Q = wastewater flow rate, m^3/s or m^3/h (ft^3/s or ft^3/h)

This equation is simply a continuity relation between the wastewater flowrate and the flowrate of fresh seawater over the outfall diffuser.

Dilution Due to Dispersion

After initial dilution, a rather uniform wastewater-seawater mixture is formed above the diffuser section. The wastewater field then begins to move in response to the prevailing current. As it moves, the outer edges of the field entrain seawater as a result of turbulent mixing. The wastewater field begins to diffuse outward and takes on the shape of a plume.

Float studies are frequently of value in determining both the strength and direction of prevailing currents as well as the extent of dispersion to be expected from a proposed outfall. For example, a description of floats and their use in studies made in the Great Lakes is given in Metcalf and Eddy [13].

Brooks has developed equations to describe the dispersion of a waste field by applying classical partial differential equations to the problem. The reader is referred to his excellent articles for a more detailed study [2, 20]. The result of his analysis for the case in which E is a function of the diffuser length raised to the four-thirds power is

$$D_2 = \frac{C_0}{C_t}$$

$$= \frac{1}{\text{erf} \sqrt{\dfrac{3/2}{[1 + \frac{2}{3}\beta(x/b)]^3 - 1}}} \tag{14-36}$$

where D_2 = dilution due to eddy diffusion after initial dilution
 C_t = maximum pollutant concentration at time t, mg/L
 C_0 = pollutant concentration after initial dilution, mg/L
 erf(x) = error function (x)
 $\beta = 12E/V_x b$, a dimensionless number
 $E = 4.53 \times 10^{-4}(b)^{4/3}$ m^2/s, b in meters
 $= 1.0 \times 10^{-3}(b)^{4/3}$ ft^2/s, b in feet
 V_x = current velocity, m/s (ft/s)
 x = distance along plume centerline, m (ft)
 b = effective diffuser system length, m (ft)

A nomograph for the solution of this equation is presented in Fig. 14-14.

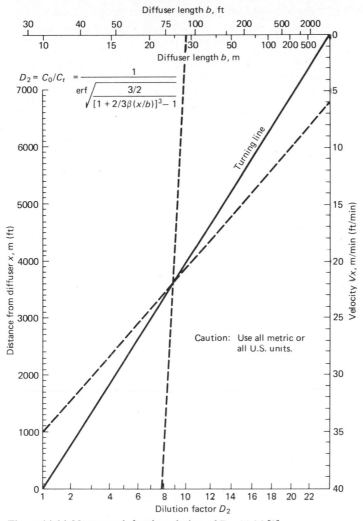

Figure 14-14 Nomograph for the solution of Eq. 14-36 [3].

Equation 14-36 gives the dilution at the centerline of the plume, assuming no longitudinal or vertical mixing. The former may be neglected, but the latter can be significant and should be allowed for where appropriate. The width of the plume is given by the equation

$$L = b\left(1 + \frac{2}{3}\frac{x}{b}\right)^{3/2}$$ (14-37)

assuming E varies with the length to the four-thirds power [2]. The concentration will be lower and the dilution greater at the edges of the plume. As the plume moves some distance away from the diffuser, the concentration curve may be assumed to have a bell-like shape represented by the normal probability equation

$$\frac{C}{C_t} = e^{-1/2(y/\sigma)^2} \qquad (14\text{-}38)$$

where y is the distance from the centerline of the plume and σ is the standard deviation, which was taken by Brooks to equal $L/2\sqrt{3}$ [2]. Where strong currents predominate in a direction more or less parallel to the shoreline, the edge of the plume may arrive at critical areas some hours before the centerline of the plume. Since bacterial decay increases with time as the portions of the plume reaching the shore or critical area are becoming more concentrated, a study may be required to determine the width y of the plume producing the greatest bacterial concentration.

Dilution Due to Decay

The third significant factor in waste dilution in the ocean is the decay rate of the waste. In the case of bacterial decay, this includes mortality as well as flocculation and sedimentation. Bacterial decay in most commonly assumed to follow first-order kinetics $(r_c = -kC)$. Thus,

$$C_t = C_0 e^{-kt} \qquad (14\text{-}39)$$

where C_t = bacterial concentration at time t
  C_0 = bacterial concentration after initial dilution
  k = bacterial decay constant
  t = time

Much research has been conducted to determine the decay constant k. It has been shown that a 90 percent reduction in bacterial numbers can usually be obtained in 2 to 6 h [7]. The variation in time is caused by differences in such characteristics of the seawater as temperature, salinity, and pH. The time in hours to achieve a 90 percent reduction in bacterial numbers is called the T_{90} time. If the decay constant k is changed to the equivalent T_{90} form, the dilution due to waste decay can be formulated as

$$D_3 = \frac{C_0}{C_t} = e^{2.3x/T_{90}60(V_x)} = e^{2.3t/T_{90}} \qquad (14\text{-}40)$$

In this equation V_x is the velocity in m/min (ft/min). A nomograph for this equation is presented in Fig. 14-15. Use of the nomograph is illustrated in Example 14-4. The practical aspects considered in design include the outfall pipe, the diffuser section, and the overall hydraulics.

Outfall Design

The outfall is used to convey the waste to the diffuser section. Its size is determined by the velocity, head loss, structural considerations, and economics of the situation. Velocities of 0.6 to 0.9 m/s (2 to 3 ft/s) at average flow are normally

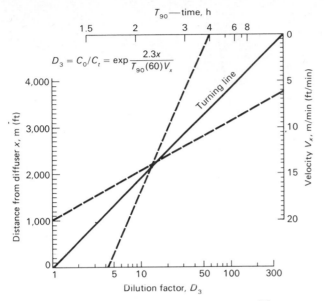

$$D_3 = C_0/C_t = \exp\frac{2.3x}{T_{90}(60)V_x}$$

Fig. 14-15 Nomograph for the solution of Eq. 14-40 [3].

recommended in pipeline design to avoid excessive head loss. Lower velocities will not be a problem provided the waste has received preliminary treatment to reduce the amount of settleable solids. On the other hand, velocities higher than 2.4 to 3.0 m/s (8 to 10 ft/s) should be avoided because of excessive head loss.

The diffuser section should be oriented perpendicular to the prevailing ocean current. If, as in most cases, the currents are not predominant in any one direction, a Y- or V-shaped diffuser is commonly used. Several possible diffuser arrangements are given in Burchett, Tchobanoglous, and Burdoin [3].

The size of diffuser ports varies from 7.6 to 23 cm (3 to 9 in). Port diameters of less than 7.6 cm (3 in) are being used on the fourth White's Point outfall of the Los Angeles County Sanitation Districts.

Diffuser ports are normally aligned horizontally, with ports alternating from side to side in order to avoid interference between adjacent jet plumes as they rise toward the surface. Experimental data have indicated that the diameter of the plume is approximately equal to $L/3$, where L is the length of the plume trajectory. Values of L can be determined from Fig. 14-16. With ports discharging alternately the individual ports can be as close as $L/6$. A rule of thumb, however, is that ports should be spaced 2.5 to 4.5 m (8 to 15 ft) on centers. Diffuser ports of many different shapes, including specially designed mixing nozzles, have been used. A simple but effective diffuser port is illustrated in Fig. 14-17.

The hydraulics of a diffuser section are somewhat complex and beyond the scope of this presentation. The problem centers around equalizing the flow from each diffuser port. This may require a number of different sizes of ports along the diffuser section. The port diameters chosen during the preliminary design

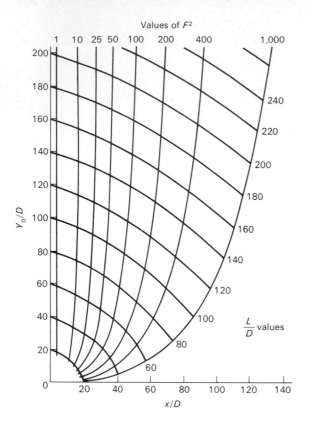

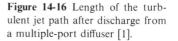

Figure 14-16 Length of the turbulent jet path after discharge from a multiple-port diffuser [1].

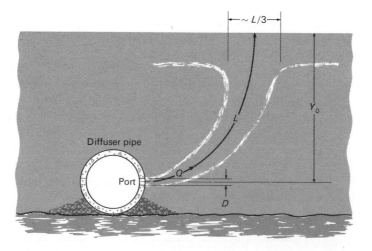

Figure 14-17 Definition sketch for turbulent jet mixing analysis with a diffuser [17].

represent an average value. The final hydraulic analysis will determine the range of port sizes required to discharge the waste flow uniformly along the length of the diffuser section. An analysis of such hydraulics is presented in the article by Rawn, Bowerman, and Brooks [20]. The design of an ocean outfall is illustrated in Example 14-4, which is adapted from Ref. 3.

Example 14-4 Ocean outfall design Design an ocean outfall for a wastewater flow of 40,000 m^3/d for the following conditions:

1. Peak flow = 80,000 m^3/d.
2. Bottom slope = 12.5 $m/10^3$ m.
3. Coliform concentration in untreated wastewater = $10^6/mL$.
4. Critical onshore current = 6 m/min.
5. Diffuser length = 1.0 m per 10^3 m^3/d = 40 m.
6. Diffuser spacing = 3 m.
7. Port discharge velocity at peak flow = 5 m/s.
8. Diffuser configuration is V-shaped with each side being 20 m in length. The diffusers are offset at an angle of 45° with respect to the shoreline.
9. T_{90} = 4 h.
10. Velocity in pipeline at average flow = 0.75 m/s.

SOLUTION

1 Determine the number of diffusers.

$$\text{Number of diffusers} = \frac{40 \text{ m}}{3 \text{ m}} = 13.3 \qquad \text{(use 14 ports)}$$

2 Determine the port diameter.

$$\text{Total area of ports} = \frac{80,000 \text{ m}^3/\text{d}}{(5 \text{ m/s})(86,400 \text{ s/d})} = 0.185 \text{ m}^2$$

$$\text{Area per port} = \frac{0.185 \text{ m}^2}{14} = 0.0132 \text{ m}^2/\text{port}$$

$$\text{Port diameter} = \left(\frac{0.0132 \text{ m}^2 \times 4}{3.14}\right)^{1/2} = 0.1298 \text{ m}$$

Use 14 ports with a 130-mm diameter.

3 Determine the outfall pipe size.

$$A = \frac{40,000 \text{ m}^3/\text{d}}{(0.75 \text{ m/s})(86,400 \text{ s/d})} = 0.62 \text{ m}^2$$

$$\text{Pipe diameter} = \left(\frac{0.62 \text{ m}^2 \times 4}{3.14}\right)^{1/2} = 0.887 \text{ m}$$

Use a 900-mm-diameter pipe. The head loss at peak flow ($f = 0.025$) is 3.00 $m/10^3$ m.

4 Determine the total dilution factor for various lengths of outfall. At 10^3 m from shore the depth is 12.5 m.

  a Compute the initial dilution.

$$\frac{Y_0}{D} = \frac{12.5 \text{ m}}{0.13 \text{ m}} = 96$$

$$F = \frac{5 \text{ m/s}}{\sqrt{0.026(9.81 \text{ m/s}^2)(0.13 \text{ m})}} = 27.5$$

From Fig. 14-13, $D_1 = 48$. (*Note:* The initial dilution should also be computed by Eq. 14-35, and the smaller of the two values should be used.)

  b Determine the dilution due to dispersion. The effective length of the diffuser is 28 m.

$$L_c = 2(20 \text{ m})(\cos 45°) = 28 \text{ m}$$

From Fig. 14-14, $D_2 = 7.9$

  c Determine the dilution due to decay. From Fig. 14-15, $D_3 = 4.9$.

  d Determine the total dilution.

$$D_t = D_1 D_2 D_3 = 48(7.9)(4.9) = 1860$$

5 Repeat the calculations in step 4 for several lengths of outfall. The dilutions obtained are shown in Table 14-2.

6 Using typical data for the percent removal of coliforms for various processes, determine the total dilution factor and required length of outfall required for several types of treatment to meet a standard of 10 coliforms per mL (see Table 14-3).

7 Using pertinent cost data for outfalls, treatment plants, and chlorine, combined with a suitable interest rate and estimated useful life, the most economical alternative can be selected.

Table 14-2 Dilution factors versus length of outfall for Example 14-4

Distance from shore, m	Dilution factor			
	D_1	D_2	D_3	D_T
300	17	2.1	1.6	57
600	30	4.4	2.6	340
1000	48	7.9	4.9	1,860
1500	70	13.0	11.0	10,000
2000	1000[a]		24.0	24,000
2500	1000		54.0	54,000
3000	1000		120.0	120,000

[a] Experience with existing outfall systems indicates that the maximum physical dilution achieved (initial dilution + transport dispersion) is in the neighborhood of 1000.

Note: m × 3.2808 = ft

Table 14-3 Required length of outfall for alternative types of treatment in Example 14-4

Type of treatment	Coliform reduction, %	Required dilution to meet standard[a]	Length of outfall required, m
Preliminary	25	75,000	2700
Preliminary + Cl_2	25 + 98	1500	950
Activated sludge	90	10,000	1500
Activated sludge + Cl_2	90 + 99	100	400

[a] Assumed to be 10 coliforms per mL.
Note: m × 3.2808 = ft

The methods described in this section are applicable to outfalls extending into large lakes, such as the Great Lakes, and other large bodies of water that cannot be considered completely mixed. The following differences should be noted and allowed for:

1. In Eq. 14-23, ΔS is mainly due to temperature differences in fresh water.
2. A significant level of general or background pollution may exist.
3. A sustained flow, varying by season, may exist from inlet to outlet.
4. The lake may serve as a source of public water supply, as the Great Lakes do, and wastewater outfalls must terminate at locations that will not cause contamination of the water supply.

14-6 DIRECT AND INDIRECT REUSE OF WASTEWATER

It is generally impossible to reuse a wastewater completely or indefinitely. The reuse of treated effluent by direct or indirect means is a method of disposal that complements the other disposal methods. The amount of effluent that can be reused is affected by the availability and cost of fresh water, transportation and treatment costs, water-quality standards, and the reclamation potential of the wastewater. A complete discussion and analysis of wastewater reuse is beyond the scope of this text, but some of the possible reuses for treated water are briefly reviewed in the following discussion. Additional details may be found in Chap. 13 and Ref. 22.

Water reuse may be classified according to use as (1) municipal, (2) industrial, (3) agricultural, (4) recreational, and (5) groundwater recharge. Direct and indirect reuse applications for these uses are shown in Table 14-4.

Municipal Reuse

Direct reuse of treated wastewater as drinking water, after dilution in natural waters to the maximum possible extent and after coagulation, filtration, and heavy chlorination for disinfection, is practicable on an emergency basis. This

Table 14-4 Potential uses of renovated water[a]

Use	Direct	Indirect
Municipal	Park or golf course watering; lawn watering with separate distribution system; potential source for municipal water supply	Groundwater recharge to reduce aquifer overdrafts
Industrial	Cooling tower water; boiler feed water; process water	Replenish groundwater supply for industrial use
Agricultural	Irrigation of certain agricultural lands, crops, orchards, pastures, and forests; leaching of soils	Replenish groundwater supply for agricultural overdrafts
Recreational	Forming artificial lakes for boating, swimming, etc.; swimming pools	Develop fish and waterfowl areas
Other	Groundwater recharge to control saltwater intrusion; salt balance control in groundwater; wetting agent–solid waste compaction	Groundwater recharge to control land subsidence problems; oil-well repressurizing; soil compaction

[a] Adapted in part from Ref. 23.

practice varies only in degree from the situation existing on many rivers that are used for both water supply and waste disposal. One example is the Merrimack River in Massachusetts. Advanced methods of wastewater and water treatment, such as demineralization and desalination, are capable of almost complete removal of impurities, and water treated by such methods, after chlorination, is safe to drink. These methods are very expensive and, where they are found to be necessary due to inadequate water supplies, may be economically feasible only if a dual supply system is adopted. In such cases, adequately treated and disinfected wastewater effluents could be reused for flushing toilets, yard watering, and other direct applications.

Industrial Reuse

Industry is probably the single greatest user of water in the world, and the largest of the industrial water demands is for process cooling water. Waters with a high mineral content and those that do not meet bacterial drinking water standards have been used by industry in some cases. Public health dangers and aesthetic concerns are generally eliminated because of the use of closed-cycle processes.

Agricultural Reuse

The types of crops that can be irrigated with reclaimed wastewater depend on the quality of the effluent, the amount of effluent used, and the health regulations concerning the use of treated and untreated wastewater on crops. Health considerations in this country have dictated against the use of untreated wastewater.

Furthermore, field crops that are normally consumed in a raw state cannot be irrigated with wastewater of any kind. Preliminary treated or undisinfected waste-water effluent is usually allowed for field crops, such as cotton, sugar beets, and vegetables for seed production. Additional data on agricultural reuse and waste-water disposal by irrigation may be found in Chap. 13 and in Ref. 13.

Recreational Reuse

Golf-course and park watering, establishment of ponds for boating and recreation, and maintenance of fish or wildlife ponds are methods for the recreational reuse of water. Today's technology allows the production of an excellent effluent that is well suited for the purposes described. The use of treated effluent for park watering has been practiced for many years in this country. The Santee Project outside San Diego is an example of recreational reuse of wastewater in forming a series of lakes suitable for boating, fishing, and other recreational purposes [12].

Groundwater Recharge

Groundwater recharge is one of the most common methods for combining water reuse and effluent disposal. Recharge has been used to replenish groundwater supplies in many areas. The effluent from the Whittier Narrows Plant, operated by the Los Angeles County Sanitation Districts, is used for replenishment of the groundwater in the Rio Hondo River Basin [11]. In New York, California, and other coastal areas, rapid development of industry and increase in population have caused a lowering of the groundwater table, resulting in saltwater intrusion into the freshwater aquifers. Treated effluent has been used to replenish the groundwater and to stop this intrusion. Another possible effluent use is in the recharging of oil-bearing strata. Oil companies have conducted much research on flooding techniques to increase the yield of oil-bearing strata.

DISCUSSION TOPICS AND PROBLEMS

14-1 Treated wastewater containing 25 mg/L of BOD_5 is discharged continuously at a rate of 3800 m^3/d into a small lake. The lake has a surface area of 20 ha, a drainage area of 25 km^2, and an average depth of 3 m; its contents can be considered to be completely mixed. Aerobic conditions prevail throughout its depth. The runoff from the drainage area, containing 1 mg/L of BOD_5, varies from 10^3 $m^3/km^2 \cdot d$ in the spring to 10^2 $m^3/km^2 \cdot d$ in the summer. The temperature of the lake contents is 5°C in the spring and 25°C in the summer. Determine the BOD_5 of the outlet stream in the spring and the summer. $K' = 0.3$/d, $\theta_{spring} = 1.135$, $\theta_{summer} = 1.056$.

14-2 Determine the reaeration coefficient K'_2 using the O'Connor-Dobbins formula (Eq. 14-6). The mean velocity of flow for a river is 0.06 m/s, the depth is 3 m, and the temperature of the river is 15°C.

14-3 Review the literature and find two additional expressions that can be used to estimate the reaeration constant K'_2. Using the expressions you have found and the O'Connor-Dobbins formula, estimate the value of K'_2 using the data from Prob. 14-2 for temperatures of 10, 20, and 30°C. Plot the results obtained (values of K'_2 versus temperature) using the three different equations.

14-4 A wastewater containing 130 mg/L of BOD_5 after preliminary treatment is discharged to a river at a rate of 75,000 m^3/d. The river has a minimum flowrate of 6 m^3/s, a BOD_5 of 2 mg/L, and a velocity of 2.4 km/h. After the wastewater is mixed with the river contents, the temperature is 20°C and the dissolved oxygen is 75 percent of saturation. Determine the oxygen sag at the critical point and at distances of $x_c/2$ above and below the critical point, and plot the curve. $K' = 0.25/d$; $K'_2 = 0.40/d$.

14-5 The waste from a small industry is discharged continuously into a nearby river. Using the following data, find (a) the DO deficit at a point 60 km downstream, (b) the location of the critical point on the oxygen-sag curve and the minimum DO in the river at that point, and (c) the 5-d BOD at a point 20 km downstream. Assume that the 5-d BOD in the river upstream of the point of waste discharge is equal to zero.
River characteristics just beyond the point of waste discharge

DO	= 6.0 mg/L
Velocity	= 0.3 m/s
Depth	= 2.0 m
Width	= 10.0 m
Chloride concentration	= 5000 mg/L
Temperature	= 25°C

Waste characteristics

BOD_5 = 5500 kg/d
K' = 0.25/d at 20°C

14-6 A stream with $K'_2 = 0.58$ d^{-1}, temperature = 15°C, and minimum flow = 10 m/s receives 0.4 m/s of wastewater from a city. The river water upstream of the point of waste discharge is 95 percent saturated with oxygen. What is the maximum permissible BOD_5 of the wastewater if the dissolved oxygen content of the stream is never to go below 4.0 mg/L? Assume K' of the river-waste mixture equals 0.35 d^{-1} at 20°C. What would the minimum DO in the stream be if the wastewater received secondary treatment as specified by EPA (BOD_5 in effluent = 30 mg/L)?

14-7 The oxygen resources for a small stream have been investigated and the following coefficients for oxygen production and consumption have been determined:

Organic degradation	$K' = 0.25$ d^{-1}
Reaeration	$K' = 0.45$ d^{-1}
Nitrification	$N = -3.0$ mg/L · d
Photosynthesis	$P_{max} = 5$ mg/L · d
Respiration	$R = -1$ mg/L · d

At some point, X, along the stream, the concentration of ultimate BOD present is 10 mg/L and the dissolved oxygen concentration is 5 mg/L. If the saturation value for dissolved oxygen is 10 mg/L, determine the following:
(a) The rate of dissolved oxygen change in mg/L · d at point X at midday when maximum photosynthesis occurs.
(b) The rate of dissolved oxygen change at the same point, but during the night when $P = 0$.
Assuming that the rate of dissolved oxygen change remains constant between point X and another point, Y, situated 1 h of stream flow time downstream from point X, determine the following:
(c) The dissolved oxygen concentration at point Y near midday.
(d) The dissolved oxygen concentration at point Y during the night.

(Courtesy P. L. McCarty)

14-8 The freshwater outflow in an estuary is 3 m³/s and the average cross-sectional area is 100 m². Assuming that seawater has a chloride concentration of 18,000 mg/L, determine E from the following data:

x, km	2	4	6	8	10
C, mg/L	16,000	11,500	8,350	6,000	4,350

14-9 A study of horizontal diffusion in a body of water consisted of determining the distribution of particles on the second and third days after their release from an initial point ($t = 0$). The particles assumed a gaussian or normal distribution, centered as they diffused outward, about the initial point. Determine, from the data reproduced in Fig. 14-18, the diffusion coefficient using the following formula:

$$D = \frac{\frac{1}{2} d(\sigma^2)}{dt}$$

where D = coefficient of diffusion
σ = standard deviation of distribution curve

Express your answer in units of centimeters and seconds. (*Courtesy G. T. Orlob*)

14-10 The freshwater runoff to an estuary has a chloride concentration of 30 mg/L and amounts to 30 m³/s. Assume that 3 m³/s of wastewater with an average chloride concentration of 50 mg/L is discharged to the estuary and that the chloride concentration at that point is 9000 mg/L. Determine the dilution available and the chloride concentration after mixing.

14-11 Derive Eq. 14-40 from Eq. 14-39.

14-12 Design an ocean outfall for an average wastewater flow of 120,000 m³/d and a peak flow of 180,000 m³/d. The bottom slope is 20 m/1000 m along the route of the outfall. The diffusers are to be located in 25 m of water. The prevailing current is 3.6 m/min parallel to the shore. Determine the dilution and coliform content at distances of 800 and 1600 m from the diffuser, assuming (*a*) the wastewater has had preliminary treatment and (*b*) the wastewater has had preliminary treatment plus chlorination. Use 3 m of diffuser length per 3800 m³/d. $T_{90} = 3$ h.

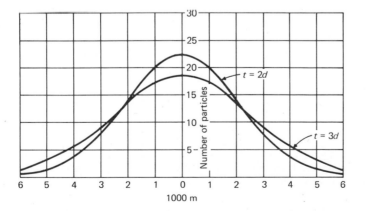

Figure 14-18 Particle distributions for Problem 14-9.

REFERENCES

1. Abraham, G.: *Jet Diffusion in Stagnant, Ambient Fluid*, Delft Hydraulics Laboratory Publication 29, 1963.
2. Brooks, N. H.: Diffusion of Sewage Effluent in an Ocean Current, *Proceedings First International Conference on Waste Disposal in the Marine Environment*, University of California, Berkeley, Pergamon, New York, 1960.
3. Burchett, M. E., G. Tchobanoglous, and A. J. Burdoin: A Practical Approach to Submarine Outfall Calculations, *Public Works*, vol. 98, no. 5, 1967.
4. Committee on Quality Tolerances of Water for Industrial Uses: Progress Report, *J. N. Engl. Water Works Assoc.*, vol. 54, p. 261, 1940.
5. Fair, G. M., E. W. Moore, and H. A. Thomas, Jr.: The Natural Purification of River Muds and Pollutional Sediments, *Sewage Works J.*, vol. 13, pp. 270, 756, 1941.
6. Harleman, D. R. F.: The Significance of Longitudinal Dispersion in the Analysis of Pollution in Estuaries, *Proceedings 2d International Conference on Water Pollution Research*, Tokyo, Pergamon, New York, 1964.
7. Hyperion Engineers: *Ocean Outfall Design*, Final Report to the City of Los Angeles, 1957.
8. Ippen, A. T. (ed.): *Estuary and Coastline Hydrodynamics*, McGraw-Hill, 1966.
9. Krenkel, P. A., and F. L. Parker (eds.): *Biological Aspects of Thermal Pollution*, Vanderbilt University Press, Nashville, Tenn., 1969.
10. McKee, J. E., and H. W. Wolf: *Water Quality Criteria*, 2d ed., California State Water Quality Control Board Publication 3A, 1963.
11. McMichael, F. C., and J. E. McKee: *Wastewater Reclamation at Whittier Narrows*, California State Water Quality Control Board Publication 33, 1965.
12. Merrell, J. E., et al.: *The Santee Recreation Project, Santee, California*, final report, U.S. Department of the Interior, Federal Water Pollution Control Administration, Cincinnati, 1967.
13. Metcalf, L., and H. P. Eddy: *American Sewerage Practice*, vol. III, 3d ed., McGraw-Hill, New York, 1935.
14. Nemerow, N. L.: *Scientific Stream Pollution Analysis*, McGraw-Hill, New York, 1974.
15. O'Conner, D., and W. Dobbins: The Mechanism of Reaeration in Natural Streams, *J. Sa. Eng. Div.*, ASCE, SA 6, 1956.
16. Parker, F. L., and P. A. Krenkel (eds.): *Engineering Aspects of Thermal Pollution*, Vanderbilt University Press, Nashville, Tenn., 1969.
17. Pearson, E. A.: *An Investigation of the Efficiency of Submarine Outfall Disposal of Sewage and Sludge*, California Water Pollution Control Board Publication 14, 1956.
18. Phelps, E. B.: *Stream Sanitation*, Wiley, New York, 1944.
19. *Quality Criteria For Water*, U.S. Environmental Protection Agency, Washington, D.C., 1976.
20. Rawn, A. M., F. R. Bowerman, and N. H. Brooks: Diffusers for Disposal of Sewage in Sea Water, *Proc. ASCE*, vol. 86, no. SA2, 1960.
21. *Report of the Engineering Board of Review*, Sanitary District of Chicago, part III, appendix I, 1925.
22. Shuval, H. I. (ed.): *Water Renovation and Reuse*, Academic Press, New York, 1977.
23 Tchobanoglous, G., and R. Eliassen: The Indirect Cycle of Water Reuse, *Water Wastes Eng.*, vol. 6, no. 2, 1969.
24. Thomann, R. V.: *Systems Analysis and Water Quality*, Environmental Science Service, New York, 1972.
25. Velz, C. J.: *Applied Stream Sanitation*, Wiley-Interscience, New York, 1970.
26. *Water Quality Criteria 1972*, National Academy of Science, Ecological Research Series, U.S. Environmental Protection Agency Report No. R3-73-033, Washington, D.C., 1973.
27. *Water Quality Criteria*, National Technical Advisory Committee, Federal Water Pollution Control Administration, Washington, D.C., 1968.

APPENDIXES

A CONVERSION FACTORS
Table A-1 Metric conversion factors (SI units to U.S. customary units)

Name	Multiply the SI unit Symbol	by	To obtain the U.S. customary unit Symbol	Name
Acceleration				
meters per second squared	m/s^2	3.2808	ft/s^2	feet per second squared
meters per second squared	m/s^2	39.3701	in/s^2	inches per second squared
Area				
hectare (10,000 m^2)	ha	2.4711	acre	acre
square centimeter	cm^2	0.1550	in^2	square inch
square kilometer	km^2	0.3861	mi^2	square mile
square kilometer	km^2	247.1054	acre	acre
square meter	m^2	10.7639	ft^2	square foot
square meter	m^2	1.1960	yd^2	square yard
Energy				
joule	kJ	0.9478	Btu	British thermal unit
joule	J	2.7778×10^{-7}	$kW \cdot h$	kilowatt-hour
joule	J	0.7376	$ft \cdot lb_f$	foot-pound (force)
joule	J	1.0000	$W \cdot s$	watt-second
joule	J	0.2388	cal	calorie
kilojoule	kJ	2.7778×10^{-4}	$kW \cdot h$	kilowatt-hour
kilojoule	kJ	0.2778	$W \cdot h$	watt-hour
megajoule	MJ	0.3725	$hp \cdot h$	horsepower-hour
Force				
newton	N	0.2248	lb_f	pound force
Flowrate				
cubic meters per day	m^3/d	264.1720	gal/d	gallons per day
cubic meters per day	m^3/d	2.6417×10^{-4}	Mgal/d	million gallons per day
cubic meters per second	m^3/s	35.3147	ft^3/s	cubic feet per second
cubic meters per second	m^3/s	22.8245	Mgal/d	million gallons per day
cubic meters per second	m^3/s	15,850.3	gal/min	gallons per minute
liters per second	L/s	22,824.5	gal/d	gallons per day
liters per second	L/s	0.0228	Mgal/d	million gallons per day
liters per second	L/s	15.8508	gal/min	gallons per minute

Length				
centimeter	cm	0.3937	in	inch
kilometer	km	0.6214	mi	mile
meter	m	39.3701	in	inch
meter	m	3.2808	ft	foot
meter	m	1.0936	yd	yard
millimeter	mm	0.03937	in	inch
Mass				
gram	g	0.0353	oz	ounce
gram	g	0.0022	lb	pound
kilogram	kg	2.2046	lb	pound
megagram (10^3 kg)	Mg	1.1023	ton	ton (short: 2000 lb)
megagram (10^3 kg)	Mg	0.9842	ton	ton (long: 2240 lb)
Power				
kilowatt	kW	0.9478	Btu/s	British thermal units per second
kilowatt	kW	1.3410	hp	horsepower
watt	W	0.7376	ft. lb_f/s	foot-pounds (force) per second
Pressure (force/area)				
pascal (newtons per square meter)	Pa (N/m^2)	1.4504×10^{-4}	lb_f/in^2	pounds (force) per square inch
pascal (newtons per square meter)	Pa (N/m^2)	2.0885×10^{-2}	lb_f/ft^2	pounds (force) per square foot
pascal (newtons per square meter)	Pa (N/m^2)	2.9613×10^{-4}	in Hg	inches of mercury (60°F)
pascal (newtons per square meter)	Pa (N/m^2)	4.0187×10^{-3}	in H$_2$O	inches of water (60°F)
kilopascal (kilonewtons per square meter)	kPa (kN/m^2)	0.1450	lb_f/in^2	pounds (force) per square inch
kilopascal (kilonewtons per square meter)	kPa (kN/m^2)	0.0099	atm	atmosphere (standard)
Temperature				
degree Celsius (centigrade)	°C	$1.8(°C) + 32$	°F	degree Fahrenheit
degree kelvin	°K	$1.8(°K) - 459.67$	°F	degree Fahrenheit
Velocity				
kilometers per second	km/s	2.2369	mi/h	miles per hour
meters per second	m/s	3.2808	ft/s	feet per second
Volume				
cubic centimeter	cm^3	0.0610	in^3	cubic inch
cubic meter	m^3	35.3147	ft^3	cubic foot
cubic meter	m^3	1.3079	yd^3	cubic yard
cubic meter	m^3	264.1720	gal	gallon
cubic meter	m^3	8.1071×10^{-4}	acre · ft	acre · foot
liter	L	0.2642	gal	gallon
liter	L	0.0353	ft^3	cubic foot
liter	L	33.8150	oz	ounce (U.S. fluid)

Table A-2 Metric conversion factors (U.S. customary units to SI units)

Multiply the U.S. customary unit			To obtain the SI unit	
Name	Symbol	by	Symbol	Name
Acceleration				
feet per second squared	ft/s^2	0.3048^a	m/s^2	meters per second squared
inches per second squared	in/s^2	0.0254^a	m/s^2	meters per second squared
Area				
acre	acre	0.4047	ha	hectare
acre	acre	4.0469×10^{-3}	km^2	square kilometer
square foot	ft^2	9.2903×10^{-2}	m^2	square meter
square inch	in^2	6.4516^a	cm^2	square centimeter
square mile	mi^2	2.5900	km^2	square kilometer
square yard	yd^2	0.8361	m^2	square meter
Energy				
British thermal unit	Btu	1.0551	kJ	kilojoule
foot-pound (force)	$ft \cdot lb_f$	1.3558	J	joule
horsepower-hour	$hp \cdot h$	2.6845	MJ	megajoule
kilowatt-hour	$kW \cdot h$	3600^a	kJ	kilojoule
kilowatt-hour	$kW \cdot h$	3.600×10^{6a}	J	joule
watt-hour	$W \cdot h$	3.600^a	kJ	kilojoule
watt-second	$W \cdot s$	1.000^a	J	joule
Force				
pound force	lb_f	4.4482	N	newton

Flow rate			
cubic feet per second	ft³/s	2.8317×10^{-2}	cubic meters per second
gallons per day	gal/d	4.3813×10^{-2}	liters per second
gallons per day	gal/d	3.7854×10^{-3}	cubic meters per day
gallons per minute	gal/min	6.3090×10^{-5}	cubic meters per second
gallons per minute	gal/min	6.3090×10^{-2}	liters per second
million gallons per day	Mgal/d	43.8126	liters per second
million gallons per day	Mgal/d	3.7854×10^{3}	cubic meters per day
million gallons per day	Mgal/d	4.3813×10^{-2}	cubic meters per second
Length			
foot	ft	0.3048^{a}	meter
inch	in	2.54^{a}	centimeter
inch	in	0.0254^{a}	meter
inch	in	25.4^{a}	millimeter
mile	mi	1.6093	kilometer
yard	yd	0.9144^{a}	meter
Mass			
ounce	oz	28.3495	gram
pound	lb	4.5359×10^{2}	gram
pound	lb	0.4536	kilogram
ton (short: 2000 lb)	ton	0.9072	megagram (10^{3} kilogram)
ton (long: 2240 lb)	ton	1.0160	megagram (10^{3} kilogram)
Power			
British thermal units per second	Btu/s	1.0551	kilowatt
foot-pounds (force) per second	ft · lb$_r$/s	1.3558	watt
horsepower	hp	0.7457	kilowatt

(Continued)

Table A-2 (Continued)

	Multiply the U.S. customary unit		To obtain the SI unit	
Name	Symbol	by	Symbol	Name
Pressure (force/area)				
atmosphere (standard)	atm	1.0133×10^2	kPa (kN/m²)	kilopascal (kilonewtons per square meter)
inches of mercury (60°F)	in Hg (60°F)	3.3768×10^3	Pa (N/m²)	pascal (newtons per square meter)
inches of water (60°F)	in H$_2$O (60°F)	2.4884×10^2	Pa (N/m²)	pascal (newtons per square meter)
pounds (force) per square foot	lb$_f$/ft²	47.8803	Pa (N/m²)	pascal (newtons per square meter)
pounds (force) per square inch	lb$_f$/in²	6.8948×10^3	Pa (N/m²)	pascal (newtons per square meter)
pounds (force) per square inch	lb$_f$/in²	6.8948	kPa (kN/m²)	kilopascal (kilonewtons per square meter)
Temperature				
degrees Fahrenheit	°F	$0.555(°F - 32)$	°C	degrees Celsius (centigrade)
degrees Fahrenheit	°F	$0.555(°F + 459.67)$	°K	degrees kelvin
Velocity				
feet per second	ft/s	0.3048^a	m/s	meters per second
miles per hour	mi/h	4.4704×10^{-1a}	m/s	kilometers per second
Volume				
acre-foot	acre-ft	1.2335×10^3	m³	cubic meter
cubic foot	ft³	28.3168	L	liter
cubic foot	ft³	2.8317×10^{-2}	m³	cubic meter
cubic inch	in³	16.3871	cm³	cubic centimeter
cubic yard	yd³	0.7646	m³	cubic meter
gallon	gal	3.7854×10^{-3}	m³	cubic meter
gallon	gal	3.7854	L	liter
ounce (U.S. fluid)	oz (U.S. fluid)	2.9573×10^{-2}	L	liter

[a] Indicates exact conversion.

Table A-3 Conversion factors for commonly used wastewater treatment plant design parameters

Parameter (in SI units)	SI units	To convert, multiply in direction shown by arrows		U.S. units
		→	←	
Screening				
m³ screenings/10³ m³ wastewater	m³/10³ m³	133.6806	7.4805×10^{-3}	ft³/Mgal
Grit removal				
Air supply				
m³ air/m of tank length·min	m³/m·min	10.7639	0.0929	ft³/ft·min
Grit removal				
g grit/m³ wastewater	g/m³	8.3454	0.1198	lb/Mgal
kg grit/m³ wastewater	kg/m³	8345.4	1.1983×10^{-4}	lb/Mgal
Surface overflow rate				
m³ flow/m² surface area·h	m³/m²·h	589.0173	0.0017	gal/ft²·d
m³ flow/m² surface area·d	m³/m²·d	24.545	0.0407	gal/ft²·d
Volume				
m³ grit/10³ m³ wastewater	m³/10³ m³	133.6806	7.4805×10^{-3}	ft³/Mgal
Flow equalization				
Air supply				
m³ air/m³ tank volume·min	m³/m³·min	133.6806	7.4805×10^{-3}	ft³/10³ gal·min
Mixing horsepower				
kW/m³ tank volume	kW/m³	5.0763	0.1970	hp/10³ gal
Sedimentation				
Particle settling rate				
m/h	m/h	3.2808	0.3048	ft/h
m/h	m/h	0.4090	2.4448	gal/ft²·min
Sludge scraper speed				
m/h	m/h	0.0547	18.2880	ft/min

(Continued)

Table A-3 (*Continued*)

Parameter (in SI units)	SI units	To convert, multiply in direction shown by arrows →	←	U.S. units
Solids loading				
kg solids/m² surface area · d	kg/m²·d	0.2048	4.8824	lb/ft²·d
Surface overflow rate				
m³ wastewater/m² surface area · d	m³/m²·d	24.5424	0.0407	gal/ft²·d
m³ wastewater/m² surface area · h	m³/m²·h	589.0173	0.0017	gal/ft²·d
Volume of sludge				
m³ sludge/10³ m³ wastewater	m³/10³ m³	133.6806	7.481×10^{-3}	ft³/Mgal
Weight of dry sludge solids				
g dry solids/m³ wastewater	g/m³	8.3454	0.1198	lb/Mgal
Weir overflow rate				
m³ wastewater/m weir length · d	m³/m·d	80.5196	0.0124	gal/ft·d
Activated sludge				
Aeration device mixing intensity, diffused aeration				
m³ air/m³ tank volume · min	m³/m³·min	1000.0	0.001	ft³/10³ ft³·min
Aeration device mixing intensity, mechanical aeration				
kW/10³ m³ tank volume	kW/10³ m³	0.0380	26.3342	hp/10³ ft³
Air flowrate				
m³ air/h	m³/h	0.5886	1.6990	ft³/min
Air requirements, organic removal				
m³ air/kg BOD₅ removed	m³/kg	16.0185	0.0624	ft³/lb
Air requirements, volume of wastewater				
m³ air/m³ wastewater	m³/m³	0.1337	7.4805	ft³/gal
Organic load				
kg BOD₅ applied/m³ aeration-tank volume · d	kg/m³·d	62.4280	0.0160	lb/10³ ft³·d
Oxygen requirements				
kg O₂/kg BOD₅ applied · d	kg/kg·d	1.0	1.0	lb/lb·d
Oxygen-transfer rate				
kg O₂ transferred/kW · h	kg/kW·h	1.6440	0.6083	lb/hp·h
kg O₂ transferred/m³ wastewater · h	kg/m³·h	0.0624	16.0185	lb/ft³·h

Trickling filters and rotating biological contactors

Hydraulic load

m³ wastewater/m² bulk surface area · d	24.5424	0.0407	gal/ft² · d
m³ wastewater/m² bulk surface area · h	589.0173	0.0017	gal/ft² · d
m³ wastewater/m² bulk surface area · d	1.0691	0.9354	Mgal/acre · d
L wastewater/m² bulk surface area · min	35.3420	0.0283	gal/ft² · d

Organic load

kg BOD_5/m³ filter-medium volume · d	62.4280	0.0160	lb/10^3 ft³ · d

Specific surface loading, hydraulic

m³ wastewater/m² filter medium surface area · d	24.5424	0.0407	gal/ft² · d
m³ wastewater/m² filter medium surface area · d	0.0170	58.6740	gal/ft² · min
m³ wastewater/m² filter medium surface area · h	589.0173	0.0017	gal/ft² · d

Specific surface loading, organic

kg BOD_5/m² filter medium surface area · d	0.2048	4.8824	lb/ft² · d

Tank volume

L/m² medium surface area (rotating biological reactor)	2.4542×10^{-2}	40.7458	gal/ft²

Stabilization ponds and lagoons

Organic loads

kg BOD_5/ha surface area · d	0.8922	1.1209	lb/acre · d

Volumetric load

kg BOD_5/m³ basin volume · d	62.4280	0.0160	lb/10^3 ft³ · d

Chlorination

Feed rate

kg chlorine/d	2.2046	0.4536	lb/d

Sludge thickening

Sludge loading

kg dry solids fed/m² surface area · d	0.2048	4.8824	lb/ft² · d

Surface overflow rate

m³ wastewater/m³ surface area · d	24.5424	0.0407	gal/ft² · d
m³, wastewater/m² surface area · d	0.0170	58.6740	gal/ft² · min

Sludge digestion

Gas production

m³ gas/kg volatile solids fed	16.0185	0.0624	ft³/lb
m³ gas/capita	35.3147	0.0283	ft³/capita

(Continued)

Table A-3 (*Continued*)

Parameter (in SI units)	SI units	To convert, multiply in direction shown by arrows →	←	U.S. units
Loading rate				
kg BOD_5/m³ digester volume · d	kg/m³ · d	62.4280	0.0160	lb/10³ ft³ · d
Sludge heating				
W/m² surface area · °C	W/m² · °C	0.1763	5.6735	Btu/ft² · °F · h
Volatile-solids loading				
kg volatile solids/m³ digester volume · d	kg/m³ · d	62.4280	0.0160	lb/10³ ft³ · d
Sludge drying beds				
Dry-solids loading				
kg dry solids/m² area · yr	kg/m² · yr	0.2048	4.8824	lb/ft² · yr
m² area/capita · yr	m²/capita · yr	10.7639	0.0929	ft²/capita · yr
Vacuum filtration				
Dry solids				
kg dry solids/m² surface area · h	kg/m² · h	0.2048	4.8824	lb/ft² · h
Pressure applied				
kPa (kN/m²) pressure	kPa	0.1450	6.8948	lb_f/in² (gage)
Sludge feed				
m³ wet sludge/m² surface area · h	m³/m² · h	3.2808	0.3048	ft³/ft² · h
Vacuum applied				
kPa (kN/m²) vacuum	kPa (kN/m²)	0.2961	3.3768	in Hg (60°F)
Heat drying				
kJ heat energy required/kg water evaporated (sludge cake)	kJ/kg	0.4303	2.3241	Btu/lb
kg water evaporated/h	kg/h	2.2046	0.4536	lb/h
kg wet sludge/m² heating surface · h	kg/m² · h	0.2048	4.8824	lb/ft² · h
Incineration				
kJ heat energy/kg moisture evaporated	kJ/kg	0.4303	2.3241	Btu/lb
kg sludge/m² heating surface area	kg/m²	0.2048	4.8824	lb/ft² · h
kg sludge/m³ combustion chamber volume · h	kg/m³ · h	0.0624	16.0185	lb/ft³ · h
Land disposal				
kg mass/ha field area	kg/ha	0.8922	1.1208	lb/acre
bu yield/ha field area · yr	bu/ha · yr	0.4047	2.4711	bu/acre · yr
Mg loading/ha field area	Mg/ha	0.4461	2.2417	tons/acre
m³ wastewater/ha field area · d	m³/ha · d	106.9064	0.0094	gal/acre · d
Surface or in-depth filters				
L wastewater (backwash)/m² surface area · min	L/m² · min	0.0245	40.7458	gal/ft² · min

Table A-4 Values of useful constants

Acceleration due to gravity, g = 9.807 m/s^2 (32.174 ft/s^2)
Standard atmosphere = 101.325 kN/m^2 (14.696 lb$_f$/in^2)
 = 101.325 kPA (1.013 bar)
1 bar = 10^5 N/m^2 (14.504 lb$_f$/in^2)
Standard atmosphere = 10.333 m (33.899 ft) of water
1 metre head of water (20°C) = 9,790 M/m^2 (1.420 lb$_f$/in^2)
 = 0.00979 N/mm^2 (1.420 lb$_f$/in^2)
 = 9.790 kN/m^2 (1.420 lb$_f$/in^2)

B PHYSICAL PROPERTIES OF WATER

The principal physical properties of water are summarized in Table B-1 in SI units and in Table B-2 in U.S. customary units. They are described briefly below [1].

B-1 Specific weight

The specific weight w of a fluid is its weight per unit volume. In the SI system, it is expressed in kilonewtons per cubic meter. The relationship between w, ρ, and the acceleration due to gravity g is w = ρg. At normal temperatures w is 9.81 kN/m^3 or 62.4 lb$_f$/ft^3.

Table B-1 Physical properties of water (SI units)^a

Temperature, °C	Specific weight, γ, kN/m^3	Density, ρ, kg/m^3	Modulus of elasticity,^{b} E/10^6, kN/m^2	Dynamic viscosity $\mu \times 10^3$, N · s/m^2	Kinematic viscosity $v \times 10^6$, m^2/s	Surface tension,^{c} σ, N/m	Vapor pressure, p_v, kN/m^2
0	9.805	999.8	1.98	1.781	1.785	0.0765	0.61
5	9.807	1000.0	2.05	1.518	1.519	0.0749	0.87
10	9.804	999.7	2.10	1.307	1.306	0.0742	1.23
15	9.798	999.1	2.15	1.139	1.139	0.0735	1.70
20	9.789	998.2	2.17	1.002	1.003	0.0728	2.34
25	9.777	997.0	2.22	0.890	0.893	0.0720	3.17
30	9.764	995.7	2.25	0.798	0.800	0.0712	4.24
40	9.730	992.2	2.28	0.653	0.658	0.0696	7.38
50	9.689	988.0	2.29	0.547	0.553	0.0679	12.33
60	9.642	983.2	2.28	0.466	0.474	0.0662	19.92
70	9.589	977.8	2.25	0.404	0.413	0.0644	31.16
80	9.530	971.8	2.20	0.354	0.364	0.0626	47.34
90	9.466	965.3	2.14	0.315	0.326	0.0608	70.10
100	9.399	958.4	2.07	0.282	0.294	0.0589	101.33

^{a} Adapted from Ref. 2.
^{b} At atmospheric pressure.
^{c} In contact with air.

Table B-2 Physical properties of water (U.S. customary units)[a]

Temperature, °F	Specific weight, γ, lb/ft^3	Density,[b] ρ, slug/ft^3	Modulus of elasticity,[b] $E/10^3$, lb$_f$/in^2	Dynamic viscosity, $\mu \times 10^5$, lb · s/ft^2	Kinematic viscosity, $v \times 10^5$, ft^2/s	Surface tension,[c] σ, lb/ft	Vapor pressure, p_v, lb$_f$/in^2
32	62.42	1.940	287	3.746	1.931	0.00518	0.09
40	62.43	1.940	296	3.229	1.664	0.00614	0.12
50	62.41	1.940	305	2.735	1.410	0.00509	0.18
60	62.37	1.938	313	2.359	1.217	0.00504	0.26
70	62.30	1.936	319	2.050	1.059	0.00498	0.36
80	62.22	1.934	324	1.799	0.930	0.00492	0.51
90	62.11	1.931	328	1.595	0.826	0.00486	0.70
100	62.00	1.927	331	1.424	0.739	0.00480	0.95
110	61.86	1.923	332	1.284	0.667	0.00473	1.27
120	61.71	1.918	332	1.168	0.609	0.00467	1.69
130	61.55	1.913	331	1.069	0.558	0.00460	2.22
140	61.38	1.908	330	0.981	0.514	0.00454	2.89
150	61.20	1.902	328	0.905	0.476	0.00447	3.72
160	61.00	1.896	326	0.838	0.442	0.00441	4.74
170	60.80	1.890	322	0.780	0.413	0.00434	5.99
180	60.58	1.883	318	0.726	0.385	0.00427	7.51
190	60.36	1.876	313	0.678	0.362	0.00420	9.34
200	60.12	1.868	308	0.637	0.341	0.00413	11.52
212	59.83	1.860	300	0.593	0.319	0.00404	14.70

[a] Adapted from Ref. 2.
[b] At atmospheric pressure.
[c] In contact with the air.

B-2 Density

The density ρ of a fluid is its mass per unit volume. In the SI system, it is expressed in kilograms per cubic meter. For water, ρ is 1000 kg/m^3 at 4°C. There is a slight decrease in density with increasing temperature.

B-3 Modulus of elasticity

For most practical purposes, liquids may be regarded as incompressible. The bulk modulus of elasticity K is given by

$$K = \frac{\Delta p}{\Delta V / V}$$

where Δp is the increase in pressure, which when applied to a volume V, results in a decrease in volume ΔV. For water, K is approximately 2.150 kN/m² at normal temperatures and pressures.

B-4 Dynamic Viscosity

The viscosity of a fluid μ is a measure of its resistance to tangential or shear stress. Viscosity is expressed in newton seconds per square meter in the SI system.

B-5 Kinematic Viscosity

In many problems concerning fluid motion, the viscosity appears with the density in the form μ/ρ, and it is convenient to use a single term v, known as the kinematic viscosity and expressed in square meters per second or stokes in the SI system. The kinematic viscosity of a liquid diminishes with increasing temperature.

B-6 Surface Tension

Surface tension is the physical property that enables a drop of water to be held in suspension at a tap, a glass to be filled with liquid slightly above the brim and yet not spill, or a needle to float on the surface of a liquid. The surface-tension force across any imaginary line at a free surface is proportional to the length of the line and acts in a direction perpendicular to it. The surface tension per unit length σ is expressed in newtons per meter. There is a slight decrease in surface tension with increasing temperature.

B-8 Vapor Pressure

Liquid molecules that possess sufficient kinetic energy are projected out of the main body of a liquid at its free surface and pass into the vapor. The pressure exerted by this vapor is known as the vapor pressure p_v. The vapor pressure of water at 15°C is 1.72 kN/m².

B-9 References

1. Webber, N. B.: *Fluid Mechanics for Civil Engineers*, SI ed., Chapman and Hall, London, 1971.
2. Vennard, J. K., and R. L. Street: *Elementary Fluid Mechanics*, 5th ed., Wiley, New York, 1975.

C DISSOLVED-OXYGEN SOLUBILITY DATA

Table C-1 Dissolved oxygen,[a] mg/L

Temperature, °C	Chloride Concentration, mg/L				
	0	5,000	10,000	15,000	20,000
0	14.62	13.79	12.97	12.14	11.32
1	14.23	13.41	12.61	11.82	11.03
2	13.84	13.05	12.28	11.52	10.76
3	13.48	12.72	11.98	11.24	10.50
4	13.13	12.41	11.69	10.97	10.25
5	12.80	12.09	11.39	10.70	10.01
6	12.48	11.79	11.12	10.45	9.78
7	12.17	11.51	10.85	10.21	9.57
8	11.87	11.24	10.61	9.98	9.36
9	11.59	10.97	10.36	9.76	9.17
10	11.33	10.73	10.13	9.55	8.98
11	11.08	10.49	9.92	9.35	8.80
12	10.83	10.28	9.72	9.17	8.62
13	10.60	10.05	9.52	8.98	8.46
14	10.37	9.85	9.32	8.80	8.30
15	10.15	9.65	9.14	8.63	8.14
16	9.95	9.46	8.96	8.47	7.99
17	9.74	9.26	8.78	8.30	7.84
18	9.54	9.07	8.62	8.15	7.70
19	9.35	8.89	8.45	8.00	7.56
20	9.17	8.73	8.30	7.86	7.42
21	8.99	8.57	8.14	7.71	7.28
22	8.83	8.42	7.99	7.57	7.14
23	8.68	8.27	7.85	7.43	7.00
24	8.53	8.12	7.71	7.30	6.87
25	8.38	7.96	7.56	7.15	6.74
26	8.22	7.81	7.42	7.02	6.61
27	8.07	7.67	7.28	6.88	6.49
28	7.92	7.53	7.14	6.75	6.37
29	7.77	7.39	7.00	6.62	6.25
30	7.63	7.25	6.86	6.49	6.13

[a] Saturation values of dissolved oxygen in fresh water and sea water exposed to dry air containing 20.90 percent oxygen under a total pressure of 760 mm of mercury.

Source: G. C. Whipple and M. C. Whipple: Solubility of Oxygen in Sea Water, J. Am. Chem. Soc., vol. 33, p. 362, 1911. Calculated using data developed by C. J. J. Fox: On the Coefficients of Absorption of Nitrogen and Oxygen in Distilled Water and Sea Water and Atmospheric Carbonic Acid in Sea Water, Trans. Faraday Soc., vol. 5, p. 68, 1909.

D MOST PROBABLE NUMBER OF COLIFORMS PER 100 mL OF SAMPLE

Number of positive tubes				Number of positive tubes				Number of positive tubes				Number of positive tubes				Number of positive tubes			
10 mL	1 mL	0.1 mL	MPN	10 mL	1 mL	0.1 mL	MPN	10 mL	1 mL	0.1 mL	MPN	10 mL	1 mL	0.1 mL	MPN	10 mL	1 mL	0.1 mL	MPN
0	0	0		2	0	0	4.5	3	0	0	7.8	4	0	0	13	5	0	0	23
0	0	1	1.8	2	0	1	6.3	3	0	1	11	4	0	1	17	5	0	1	31
0	0	2	3.6	2	0	2	9.1	3	0	2	13	4	0	2	21	5	0	2	43
0	0	3	5.4	2	0	3	12	3	0	3	16	4	0	3	25	5	0	3	58
0	0	4	7.2	2	0	4	14	3	0	4	20	4	0	4	30	5	0	4	76
0	0	5	9.0	2	0	5	16	3	0	5	23	4	0	5	36	5	0	5	95
0	1	0	1.8	2	1	0	6.8	3	1	0	11	4	1	0	17	5	1	0	33
0	1	1	3.6	2	1	1	9.2	3	1	1	14	4	1	1	21	5	1	1	46
0	1	2	5.5	2	1	2	12	3	1	2	17	4	1	2	26	5	1	2	64
0	1	3	7.3	2	1	3	14	3	1	3	20	4	1	3	31	5	1	3	84
0	1	4	9.1	2	1	4	17	3	1	4	23	4	1	4	36	5	1	4	110
0	1	5	11	2	1	5	19	3	1	5	27	4	1	5	42	5	1	5	130
0	2	0	3.7	2	2	0	9.3	3	2	0	14	4	2	0	22	5	2	0	49
0	2	1	5.5	2	2	1	12	3	2	1	17	4	2	1	26	5	2	1	70
0	2	2	7.4	2	2	2	14	3	2	2	20	4	2	2	32	5	2	2	95
0	2	3	9.2	2	2	3	17	3	2	3	24	4	2	3	38	5	2	3	120
0	2	4	11	2	2	4	19	3	2	4	27	4	2	4	44	5	2	4	150
0	2	5	13	2	2	5	22	3	2	5	31	4	2	5	50	5	2	5	180

(Continued)

APPENDIX D (*Continued*)

Number of positive tubes				Number of positive tubes				Number of positive tubes				Number of positive tubes				Number of positive tubes				Number of positive tubes			
10 mL	1 mL	0.1 mL	MPN	10 mL	1 mL	0.1 mL	MPN	10 mL	1 mL	0.1 mL	MPN	10 mL	1 mL	0.1 mL	MPN	10 mL	1 mL	0.1 mL	MPN	10 mL	1 mL	0.1 mL	MPN
0	3	0	5.6	1	3	0	8.3	2	3	0	12	3	3	0	17	4	3	0	27	5	3	0	79
0	3	1	7.4	1	3	1	10	2	3	1	14	3	3	1	21	4	3	1	33	5	3	1	110
0	3	2	9.3	1	3	2	13	2	3	2	17	3	3	2	24	4	3	2	39	5	3	2	140
0	3	3	11	1	3	3	15	2	3	3	20	3	3	3	28	4	3	3	45	5	3	3	180
0	3	4	13	1	3	4	17	2	3	4	22	3	3	4	31	4	3	4	52	5	3	4	210
0	3	5	15	1	3	5	19	2	3	5	25	3	3	5	35	4	3	5	59	5	3	5	250
0	4	0	7.5	1	4	0	11	2	4	0	15	3	4	0	21	4	4	0	34	5	4	0	130
0	4	1	9.4	1	4	1	13	2	4	1	17	3	4	1	24	4	4	1	40	5	4	1	170
0	4	2	11	1	4	2	15	2	4	2	20	3	4	2	28	4	4	2	47	5	4	2	220
0	4	3	13	1	4	3	17	2	4	3	23	3	4	3	32	4	4	3	54	5	4	3	280
0	4	4	15	1	4	4	19	2	4	4	25	3	4	4	36	4	4	4	62	5	4	4	350
0	4	5	17	1	4	5	22	2	4	5	28	3	4	5	40	4	4	5	69	5	4	5	430
0	5	0	9.4	1	5	0	13	2	5	0	17	3	5	0	25	4	5	0	41	5	5	0	240
0	5	1	11	1	5	1	15	2	5	1	20	3	5	1	29	4	5	1	48	5	5	1	350
0	5	2	13	1	5	2	17	2	5	2	23	3	5	2	32	4	5	2	56	5	5	2	540
0	5	3	15	1	5	3	19	2	5	3	26	3	5	3	37	4	5	3	64	5	5	3	920
0	5	4	17	1	5	4	22	2	5	4	29	3	5	4	41	4	5	4	72	5	5	4	1600
0	5	5	19	1	5	5	24	2	5	5	32	3	5	5	45	4	5	5	81				

E EXAMPLE OF SUMMARY TABLE LISTING BASIC DESIGN DATA FOR WASTEWATER TREATMENT PLANT[a]

Item	Design value or description	Item	Design value or description
Year	1985	Type of treatment	Secondary (activated sludge)
Population served		Expected average removal efficiencies	
Sewered	29,000	Percent BOD_5 removal	
Unsewered	13,000	Primary	30
Per capita contributions		Overall	90
Sewered		Percent suspended solids removal	
Average daily flow, L/capita · d	450	Primary	60
5-d BOD, g/capita · d	100	overall	90
Suspended solids, g/capita · d	120		
Unsewered		*Plant components*	
5-d BOD, g/capita · d	18	Communition equipment	
Suspended solids, g/capita · d	45	Number of units	2
		Size, mm	900
Total flows, m³/d		Maximum unit capacity, m³/d	57,000
Average daily	15,000	Bypass rack	yes
Maximum daily	30,000	Main pumping station	
Minimum daily	4,000	Number of variable speed pumps	3
Maximum hourly	45,000	Unit capacity range, m³/d	0–24,000
Total loadings, kg/d		Type of variable-speed drive	Wound rotor
5-d BOD (average daily)	3,100	Method of pump control	Liquid rheostat
Suspended solids (average daily)	4,100		

APPENDIX E (Continued)

Item	Design value or description	Item	Design value or description
Flow metering equipment		**Aeration tanks**	
Type	Flow tube	BOD$_5$ applied, kg/d	
Size, mm	600	Average	2,170
Flow range, m^3/d	2,500–50,000	Peak (1.5 × average)	3,255
Aerated grit chamber		Number of tanks	2
Number of units	1	Length, m	25
Length, m	9	Width, m	25
Width, m	3.5	Average water depth, m	3.5
Average water depth, m	3	Number of tanks in operation	2
Detention time at maximum hour flow, min.	3.0	Detention time, h	
Air-supply range, m^3/m · h	10–35	At average flow	7.0
Air blowers		At maximum daily flow	3.5
Number of units	2	Return sludge range, m^3/d	0–15,000
Type	Centrifugal	MLSS concentration, mg/L	2,500
Unit capacity range, m^3/h	100–350	MLVSS concentration, mg/L	2,000
Method of grit removal	Clam shell bucket	F/M ratio, kg BOD$_5$/kg MLVSS	0.25
		BOD$_5$ to volume ratio, kg/m^3 · d	0.50
Primary settling tanks		**Aeration equipment**	
Number of tanks	2	Type	Mechanical
Diameter, m	18	Number of aerators	8
Sidewall water depth, m	3	Unit power, kW	15
Bottom slope, mm/m	150	Total power, kW	120
Detention period, h		Oxygen transfer capacity, kg9d	5,000
At average flow	2.4	BOD$_5$ applied, kg/kg	
At maximum hour flow	0.8	Average	2.3
Overflow rate, m^3/m^2 · d		Peak	1.5
At average flow	29.5		
At maximum hour flow	88.4		

Final settling tanks

Number of tanks	2
Diameter, m	25
Sidewall water depth, m	3.5
Bottom slope, mm/m	20
Detention period, h	
At average flow	5.5
At maximum hour flow	1.8
Overflow rate, $m^3/m^2 \cdot d$	
At average flow	15.3
At maximum hour flow	45.8

Chlorination system

Chlorinators	
Number of units	2
Unit capacity range, kg/d	450
Dosage, mg/L with largest unit out of service	
At average flow	30
At maximum hour flow	10

Chlorine contact tank

Number of units	2
Length, m	12
Width, m	6
Depth, m	6.5
Detention time at maximum hour flow, min	
Contact tank	11.5
Outfall	4.5
Total	16

Sludge dewatering units

Type	Vacuum filters
Number of units	2
Unit filtration area, m^2	35
Unit filtration capacity, kg/h	650
Total filtration capacity, kg/h	1,300
Estimated operating period, h/wk	24

[a] Data for the process flowsheet shown in Fig. 4-1 for the wastewater treatment plant at Westfield, Mass.

NAME INDEX

NAME INDEX

Abraham, G., 864
Adams, C. E., 465, 570
Adelberg, E. A., 10, 117, 467
Allen, J. B., 277, 310
American Society of Civil Engineers, 543
American Society for Testing and Materials, 80, 115
American Water Works Association, 15, 368, 391
Applebaum, S. B., 743, 758
Aqua Aerobics, 343
Ardern, E., 430, 465
A.R.F. Products, Inc., 95
Arrhenius, S. A., 89, 145, 290
Asano, T., 824
Atkinson, B., 445, 446, 465
Aulenbach, D. B., 824
Autotrol Corp., 453
Ayers , R. S., 776, 824

Babbitt, H. E., 594, 694
Bagnall, L. O., 824
Barnes, R. F., 825
Barnhart, E. L., 527, 570
Bartsch, E. H., 570
Bastian, R. K., 792, 824
Baumann, E. R., 118, 255, 467, 571, 759
Beck, L., 759
Behrman, B. W., 255, 391
Bendixen, T. W., 827
Benjes, H., 570
Bennett, G. W., 759
Berg, G., 116, 310
Bernoulli, D., 42, 46
Bewtra, J. K., 570
BIF Division of New York Brake Co., 369
Bird Machine Co., 499, 647
Bishop, S. L., 255, 391
Biskner, C. D.., 255
Bjerrum, N., 266
Black, H. H., 116
Bledsoe, B., 827

Borland, S., 54
Bouwer, H., 798, 801, 802, 824, 825
Bowerman, F. R., 850, 857, 864
Bowlus, F. D., 54
Bradney, L., 117
Brady, J., 54, 392
Branson, R. L., 824
Brater, E. F., 54
Bregmen, J. I., 759
British Royal Commission of Sewage Disposal, 96
Brock, T. D., 116
Brookhaven National Laboratory, 810, 811
Brooks, N. H., 850, 852, 854, 857, 864
Bruce, A. M., 465, 538, 570
Brunauer, S., 280
Burchett, M. E., 855, 864
Burd, R. S., 694
Burdoin, A. J., 116, 855, 864
Burkhead, C. E., 465
Burns, R. G., 466, 759
Bursztynsky, T., 694

Caldwell, D. H., 594, 694
Caldwell-Connell Engineers, 570
Camp, T. R., 255, 338, 391
Campbell Soup Company, 804, 805, 808, 809
Capri, M. J., 694
Carman, P. C., 641, 694
Carollo, John, Engineers, 550
Caropreso, F., 570
Carpenter, R. L., 824
Carver-Greenfield, 659
Chaney, R. L., 824
Chapman, R. F., 742, 758
Chasick, A. H., 572
Chen, C. W., 112, 116
Chicago Bridge & Iron Company, 335, 370, 495, 511, 612
Chicago Pump Co., 319, 320, 495
Chick, H., 288, 310

SUBJECT INDEX

SUBJECT INDEX